CAMBRIDGE STUDIES IN ADVANCED MATHEMATICS 217

Editorial Board
J . BERTOIN, B. BOLLOBÁS, W. FULTON, B. KRA, I . MOERDIJK,
C PRAEGER, P. SARNAK, B. SIMON, B. TOTARO

ESSAYS IN CLASSICAL NUMBER THEORY

Offering a comprehensive introduction to number theory, this is the ideal book both for those who want to learn the subject seriously and independently, and for those already working in number theory who want to deepen their expertise. Readers will be treated to a rich experience, developing the key theoretical ideas while explicitly solving arithmetic problems, with the historical background of analytic and algebraic number theory woven throughout. Topics include methods of solving binomial congruences, a clear account of the quantum factorization of integers, and methods of explicitly representing integers by quadratic forms over integers. In the later parts of the book, the author provides a thorough approach towards composition and genera of quadratic forms, as well as the essentials for detecting bounded gaps between prime numbers that occur infinitely often.

Yoichi Motohashi is a mathematician and foreign member of the Finnish Academy of Science and Letters. He received his D.Sc. from the University of Tokyo. He is the author of *Lectures on Sieve Methods and Prime Number Theory* (Tata IFR, 1983) and *Spectral Theory of the Riemann Zeta-Function* (Cambridge, 1997) and the editor of *Analytic Number Theory* (Cambridge, 1997).

CAMBRIDGE STUDIES IN ADVANCED MATHEMATICS

Editorial Board

J. Bertoin, B. Bollobás, W. Fulton, B. Kra, I. Moerdijk, C. Praeger, P. Sarnak, B. Simon, B. Totaro

All the titles listed below can be obtained from good booksellers or from Cambridge University Press. For a complete series listing, visit www.cambridge.org/mathematics.

Already Published

177 E. Peterson *Formal Geometry and Bordism Operators*
178 A. Ogus *Lectures on Logarithmic Algebraic Geometry*
179 N. Nikolski *Hardy Spaces*
180 D.-C. Cisinski *Higher Categories and Homotopical Algebra*
181 A. Agrachev, D. Barilari & U. Boscain *A Comprehensive Introduction to Sub-Riemannian Geometry*
182 N. Nikolski *Toeplitz Matrices and Operators*
183 A. Yekutieli *Derived Categories*
184 C. Demeter *Fourier Restriction, Decoupling and Applications*
185 D. Barnes & C. Roitzheim *Foundations of Stable Homotopy Theory*
186 V. Vasyunin & A. Volberg *The Bellman Function Technique in Harmonic Analysis*
187 M. Geck & G. Malle *The Character Theory of Finite Groups of Lie Type*
188 B. Richter *Category Theory for Homotopy Theory*
189 R. Willett & G. Yu *Higher Index Theory*
190 A. Bobrowski *Generators of Markov Chains*
191 D. Cao, S. Peng & S. Yan *Singularly Perturbed Methods for Nonlinear Elliptic Problems*
192 E. Kowalski *An Introduction to Probabilistic Number Theory*
193 V. Gorin *Lectures on Random Lozenge Tilings*
194 E. Riehl & D. Verity *Elements of -Category Theory*
195 H. Krause *Homological Theory of Representations*
196 F. Durand & D. Perrin *Dimension Groups and Dynamical Systems*
197 A. Sheffer *Polynomial Methods and Incidence Theory*
198 T. Dobson, A. Malnič & D. Marušič *Symmetry in Graphs*
199 K. S. Kedlaya *p-adic Differential Equations*
200 R. L. Frank, A. Laptev & T. Weidl Schrödinger *Operators:Eigenvalues and LiebThirring Inequalities*
201 J. van Neerven *Functional Analysis*
202 A. Schmeding *An Introduction to Infinite-Dimensional Differential Geometry*
203 F. Cabello Sánchez & J. M. F. Castillo *Homological Methods in Banach Space Theory*
204 G. P. Paternain, M. Salo & G. Uhlmann *Geometric Inverse Problems*
205 V. Platonov, A. Rapinchuk & I. Rapinchuk *Algebraic Groups and Number Theory, I (2nd Edition)*
206 D. Huybrechts *The Geometry of Cubic Hypersurfaces*
207 F. Maggi *Optimal Mass Transport on Euclidean Spaces*
208 R. P. Stanley *Enumerative Combinatorics, II (2nd Edition)*
209 M. Kawakita *Complex Algebraic Threefolds*
210 D. Anderson & W. Fulton *Equivariant Cohomology in Algebraic Geometry*
211 G. Pineda *Villavicencio Polytopes and Graphs*
212 R. Pemantle, M. C. Wilson & S. Melczer *Analytic Combinatorics in Several Variables (2nd Edition)*
213 A. Yadin *Harmonic Functions and Random Walks on Groups*
214 Y. Kawamata *Algebraic Varieties: Minimal Models and Finite Generation*
215 J. Gillespie *Abelian Model Category Theory*
216 L. Anderson *Oriented Matroids*

Essays in Classical Number Theory

YOICHI MOTOHASHI
Finnish Academy of Science and Letters

Shaftesbury Road, Cambridge CB2 8EA, United Kingdom

One Liberty Plaza, 20th Floor, New York, NY 10006, USA

477 Williamstown Road, Port Melbourne, VIC 3207, Australia

314–321, 3rd Floor, Plot 3, Splendor Forum, Jasola District Centre, New Delhi – 110025, India

103 Penang Road, #05–06/07, Visioncrest Commercial, Singapore 238467

Cambridge University Press is part of Cambridge University Press & Assessment, a department of the University of Cambridge.

We share the University's mission to contribute to society through the pursuit of education, learning and research at the highest international levels of excellence.

www.cambridge.org
Information on this title: www.cambridge.org/9781009504553

DOI: 10.1017/9781009504522

© Yoichi Motohashi 2025

This publication is in copyright. Subject to statutory exception and to the provisions of relevant collective licensing agreements, no reproduction of any part may take place without the written permission of Cambridge University Press & Assessment.

When citing this work, please include a reference to the DOI 10.1017/9781009504522

First published 2025

A catalogue record for this publication is available from the British Library

A Cataloging-in-Publication data record for this book is available from the Library of Congress

ISBN 978-1-009-50455-3 Hardback

Cambridge University Press & Assessment has no responsibility for the persistence or accuracy of URLs for external or third-party internet websites referred to in this publication and does not guarantee that any content on such websites is, or will remain, accurate or appropriate.

For EU product safety concerns, contact us at Calle de José Abascal, 56, 1°, 28003 Madrid, Spain, or email eugpsr@cambridge.org

To Kazuko and Haruko

Contents

Preface		*page* xi
For Readers		xii
Table of Theorems		xv
1	**Divisibility**	**1**
	§1 Euclid's spirit. Divisors. Multiples	1
	§2 Deviation from being divisible. Division algorithm	3
	§3 Euclid's algorithm. Greatest common divisor. Coprimality	5
	§4 Integral modules. Inclusion and divisibility	8
	§5 Modular group. Group structure over the integers	10
	§6 Coprimality and divisibility	13
	§7 Greatest common divisor of many integers	14
	§8 Integral linear indefinite equations. Chaotic behavior	17
	§9 Canonical form of integral matrices. Free Abelian groups	21
	§10 Prime numbers. Least common multiple	24
	§11 Unique prime power decomposition. Euler product. Zeta-function	29
	§12 Prime number theorem. Riemann's paradigm and his hypothesis	35
	§13 Sums over prime numbers. Summation by parts	44
	§14 Least common multiple of many integers	49
	§15 Arithmetic functions. Multiplicative functions. Dirichlet series	51
	§16 Multiplicative convolution. Divisor problem	56
	§17 Möbius inversion. Chebyshev's logarithmic amplification	64
	§18 Separating coprime integers	69
	§19 Basic sieve identities. Two kinds of sieves	76
	§20 Fractions. Rational approximation	84

viii   *Contents*

§21	Euler–Lagrange theory of continued fractions. Convergents	88
§22	Infinite continued fractions. Extended Euclid's algorithm	93
§23	Lagrange's theorem on best rational approximation	96
§24	Legendre's criterion for convergents	100
§25	Modular property of continued fractions	104
§26	Saga of the *Elements*. History of continued fractions	106

2   **Congruences**   116

§27	Moduli. Residues	116
§28	Reduced residues. Fermat's theorem. Euler's proofs	119
§29	Linear congruence equations	125
§30	Rings of residue classes. Structure of finite Abelian groups	129
§31	System of linear congruence equations	133
§32	Groups of reduced residue classes. Small set of generators	135
§33	Congruence equations. Severe premise on moduli	138
§34	Prime moduli. Fundamental theorem of Lagrange	141
§35	Wilson's theorem proved in various ways	144
§36	Contraposition of Fermat's theorem	146
§37	Integers resembling to prime numbers	148
§38	ρ - algorithm for factoring integers	151
§39	Order of reduced residue. Primitive roots	153
§40	Existence of primitive roots	155
§41	Primitive roots and primality test	159
§42	Reduced residue system modulo a prime power	161
§43	Criteria for power residues. A reciprocity issue	163
§44	Discrete logarithm	169
§45	Solving binomial congruences (part 1)	173
§46	Probabilistic primality test and factorization of integers	178
§47	Quantum factorization of integers (phase 1)	183
§48	Quantum factorization of integers (phase 2)	190
§49	Basic methods of integer factorization. Public key cryptosystems	193

3   **Characters**   199

§50	Additive characters	199
§51	Multiplicative characters. Characters of finite Abelian groups	202
§52	Dirichlet characters and L-functions	207
§53	Primitive characters. Generalized Riemann hypothesis	211
§54	Structure of primitive characters. Primitive real characters	214
§55	Fourier expansion of primitive characters. Gauss sums	217

§56	Fourier expansion of arbitrary characters	221
§57	Quadratic residues and non-residues. Legendre symbol	225
§58	The quadratic reciprocity law. Gauss' 3rd proof	227
§59	Reciprocity and automorphic structure. Gauss' 4th proof	233
§60	Poisson's sum formula	235
§61	Digression related to Gauss sums and Jacobi sums	239
§62	Jacobi symbol. Contents of the reciprocity law	247
§63	Solving quadratic congruence	252
§64	Irreducibility of cyclotomic polynomials	256
§65	Quadratic decomposition of cyclotomic polynomials	262
§66	Gauss' 6th proof. Extension of the notion of integers	267
§67	Gauss' 7th and 8th proofs. Algebraic extensions of finite fields	270
§68	Solving binomial congruences (part 2)	278
§69	Discovery of the reciprocity law. Legendre's efforts	281
§70	Prehistory of Gauss sums. Vandermonde resolvents	285

4   Quadratic Forms   296

§71	Integral binary quadratic forms. The representation problem	296
§72	Lagrange's principle. Matrix modules, orders, and ideals	304
§73	Kronecker symbol. Rôle of the reciprocity law	313
§74	Classification of quadratic forms modulo the modular group	319
§75	Ambiguous forms and classes	326
§76	Class number. Automorphims and seed representations	329
§77	Definite forms. Fundamental domain of the modular group	334
§78	Sums of two squares	340
§79	The case of discriminant -20. Dirichlet identity	348
§80	The case of discriminant -231. Glimpse of genus groups	353
§81	Directly solving the representation problem (part 1)	356
§82	Indefinite forms. Periodic continued fractions. Closed geodesics	362
§83	Cycles of orbits. Parity of periods	369
§84	Lagrange's theorem on Pell equations	375
§85	Automorphism groups. Selberg's zeta-function (part 1)	387
§86	Cakravâla algorithm	395
§87	Directly solving the representation problem (part 2)	400
§88	Intermezzo. Legendre's proof of the reciprocity law	409
§89	Diagonal ternary quadratic forms	416
§90	Gauss' composition of quadratic forms. Class group	423

§91	Genus theory. Genus characters. Gauss' 2nd proof of the reciprocity law	438
§92	Dirichlet's class number formula	456
§93	Analytic approach to the theory of quadratic forms	470

5 Distribution of Prime Numbers — 487

§94	Riemann's article. Selberg's zeta-function (part 2)	487
§95	The explicit formula of Riemann	505
§96	Bounding exponential sums	518
§97	Vinogradov's prime number theorem	524
§98	Hoheisel scheme. Basic L^2-inequalities	535
§99	Mean values of the Riemann zeta-function	548
§100	Huxley's prime number theorem	562
§101	Exceptional Dirichlet characters. Siegel's theorem	568
§102	Duality between Linnik's and Selberg's sieves	581
§103	Blending dual sieves (part 1)	591
§104	Sifting arithmetic progressions	602
§105	Bombieri's prime number theorem	609
§106	Blending dual sieves (part 2). Linnik's prime number theorem	622
§107	Detecting infinitely often bounded gaps between prime numbers	638

Bibliography	653
Index	676

Preface

This is in essence a candid report of those memorable moments which I encountered early in my journey towards number theory. I have compiled it to share with readers the delight of learning classical number theory, while pondering modern developments so that readers will become prepared for advanced study.

I started my serious study of number theory with the 1863 edition of P.G.L. Dirichlet's *Vorlesungen*. The boy was absorbed in the treatise, as he could almost touch each subject; he realized how deeply the author cherished number theory.

The overall plan of the present essays is based on that of the *Vorlesungen*. I have thus tried to weave, into a tale, the fascinating developments of multiplicative theory of numbers which have been unfolding since the ancient thoughts of Euclid, via the works by Fermat, Euler, Lagrange, Legendre, Gauss, Dirichlet, Riemann,..., Linnik, Selberg until recent advancements in the theory of the distribution of prime numbers. May readers discover here a tapestry of timeless ideas and theorems!

I am profoundly indebted to my friends A. Ivić[†] and M. Jutila, to my mentors T. Tatuzawa[†] and P. Turán[†], for their lasting encouragement. I extend my special gratitude to R. Astley and his colleagues at Cambridge University Press for their thoughtful assistance in preparing the present volume. I thank also my family for the peaceful days and the music that have been sustaining my research life.

Tokyo
January, 2025   Y. M.

For Readers

CATEGORIES OF NUMBER THEORY

The material laid out in the present volume is the absolute minimum for anyone who desires to learn number theory. According to the common view in mathematics, there exist (a) elementary · number theory, (b) analytic · number theory, and (c) algebraic number · theory, although there are no explicit boundaries between these three. The sections §§1–91 roughly belong to (a), and §§92–107 to (b). To (c) occasional excursions will be made, with indications of further readings. Thus, I shall keep the whole of my reasoning strictly within (a)∪(b).

PREREQUISITES

No substantial experience in number theory;
Rudiments of linear algebra, groups, rings, fields;
Basic residue calculus and analytic continuation.

HOW TO PROCEED

Readers are expected to study the present volume without skipping. The NOTES placed at the end of each section are to be read as essential ingredients; later discussion will often require them and further readings are provided there. The variety and depth of subjects treated in the present volume is inevitably limited due to the nature of its aim; those treatises and articles recommended should indicate deeper topics.

EXERCISES AND EXAMPLES

No exercise is explicitly provided. Instead, easier parts of reasonings are often left for readers to fill in. A variety of numerical examples are given in order to make readers' understanding firm.

ALGORITHMS AND ESTIMATIONS

My principal aim is to develop clear and explicit reasoning which will possibly lead to practical algorithms; articles and treatises specialized in number

For Readers xiii

theoretical computations are appropriately indicated. As for estimations general, no effort to squeeze out numerically sharp bounds is exerted.

DEFINITIONS, NOTIONS, NOTATIONS, AND SYMBOLS

These are kept to a minimum collection. They are often embedded in the text, intentionally so as to stimulate readers to repeatedly look back on the discussion. The location of the first occurrence of a particular notion is indicated each time when it is reused after a relatively long pause of its use. Most of the notations and symbols are for local purposes, i.e., effective within a few adjacent sections. However, some are for global use; for instance:

(1) $\mathbb{N}$, $\mathbb{Z}$, $\mathbb{Q}$, $\mathbb{R}$, $\mathbb{C}$ are reserved as usual for positive integers, integers, rational numbers, real numbers, and complex numbers, respectively. It should be understood that discussions are developed mainly in $\mathbb{Z}$; uses of other numbers will be made with explicit mention.

(2) With a matrix A, its transpose is indicated by ${}^{t}A$. This applies also to tuples, so $\{a, b, c, \ldots\}$ is a row, and ${}^{t}\{a, b, c, \ldots\}$ a column.

(3) $|S|$ denotes the cardinality of a set S. Also, S^n indicates, according to the relevant local context, either the set of column vectors or the set of row vectors composed of n elements of S. An abbreviation such as $\{s_j\}$ is often applied to indicate an element of S^n, that is, the dimension of $\{s_j\}$ is left implicit as far as no confusion is anticipated.

(4) The αth power of a function f is denoted as f^{α} so that $f^{\alpha}(a) = (f(a))^{\alpha}$. If α is not an integer, then one should take care to choose the branch. The function $(f'/f)(z)$ is the logarithmic derivative of $f(z)$. With R an operator, R^{ℓ} indicates its ℓ-fold application; if R admits inversion R^{-1}, then ℓ can become negative.

(5) The adjective integral is used to indicate objects composed of elements in $\mathbb{Z}$; there should never be any confusion with integrals in analysis. For instance, integral polynomials and matrices are, respectively, of coefficients and of elements from $\mathbb{Z}$. However, functions over $\mathbb{Z}$ are exceptional in this respect, as they are called arithmetic. Dealing with equations, we shall say not integral but integer solutions to them.

BIBLIOGRAPHY

(1) The format Name(s) (year(s), page(s), etc.) is used in three ways: indicating either the article/treatise itself or its author(s) or both.

(2) Euclid's *Elements* and Gauss' *Disquisitiones Arithmeticae* are given special tags [Σ] and [DA], respectively. Gauss' *Tagebuch* (English translation: Dunnington (2004, pp.469–496)) is not taken into account save for a few

xiv   *For Readers*

exceptional cases, as I am concerned with the earliest published materials on each subject.

(3) The year labels of Euler's works are those when the respective items were either written or presented, according to *Commentationes Arithmeticae Collectae* (1849). They are mostly different from the dates of actual publication. In his time, discoveries were usually circulated via private letters years earlier than their publication in print. Only those items of his works which are not contained in *Collectae* are provided with the respective original publication data. The same practice is applied to the works by Lagrange, Gauss, and others, whose articles are available in the respective well-known collections; nowadays, there exists no need to try to go to their original publications, which are often quite hard to reach.

(4) Although not indicated in the text, I owe older references to the great *History of Number Theory* by Dickson (1919/1923). All the relevant data were examined via internet archives; and I am literally amazed by Dickson's strenuous efforts made a century ago.

BIBLIOGRAPHICAL ACKNOWLEDGEMENTS

Through the courtesy of Asakura Publishing I could freely use materials from my former publications in Japanese:

Analytic Number Theory. Vols.I, II. Tokyo, 2009/2011;
Basic Lectures on Number Theory. Tokyo, 2018.

Y.M.

Table of Theorems

1(§2)	2(§3)	3(§4)	4(§5)	5(§6)	6(§8)
7(§9)	8(§10)	9(§10)	10(§11)	11(§12)	12(§12)
13(§15)	14(§16)	15(§17)	16(§18)	17(§18)	18(§19)
19(§20)	20(§21)	21(§22)	22(§23)	23(§23)	24(§24)
25(§25)	26(§25)	27(§28)	28(§28)	29(§29)	30(§29)
31(§30)	32(§31)	33(§32)	34(§34)	35(§34)	36(§40)
37(§42)	38(§42)	39(§43)	40(§44)	41(§46)	42(§46)
43(§47)	44(§50)	45(§51)	46(§52)	47(§53)	48(§53)
49(§54)	50(§54)	51(§55)	52(§55)	53(§56)	54(§58)
55(§59)	56(§60)	57(§61)	58(§62)	59(§62)	60(§62)
61(§64)	62(§64)	63(§64)	64(§65)	65(§69)	66(§69)
67(§71)	68(§72)	69(§73)	70(§73)	71(§73)	72(§73)
73(§75)	74(§76)	75(§76)	76(§77)	77(§77)	78(§77)
79(§78)	80(§78)	81(§79)	82(§81)	83(§82)	84(§83)
85(§84)	86(§84)	87(§85)	88(§85)	89(§87)	90(§88)
91(§89)	92(§90)	93(§90)	94(§91)	95(§91)	96(§92)
97(§92)	98(§92)	99(§93)	100(§94)	101(§94)	102(§94)
103(§94)	104(§94)	105(§95)	106(§95)	107(§95)	108(§95)
109(§95)	110(§96)	111(§97)	112(§97)	113(§97)	114(§97)
115(§97)	116(§97)	117(§98)	118(§98)	119(§98)	120(§98)
121(§98)	122(§98)	123(§98)	124(§98)	125(§98)	126(§98)
127(§99)	128(§99)	129(§99)	130(§99)	131(§100)	132(§100)
133(§101)	134(§101)	135(§101)	136(§102)	137(§102)	138(§103)
139(§103)	140(§103)	141(§104)	142(§104)	143(§105)	144(§105)
145(§105)	146(§105)	147(§105)	148(§105)	149(§106)	150(§106)

1

Divisibility

§1

From time immemorial people's minds have dwelt, with irritation or delight, on thoughts about divisibility. Euclid devoted three books of his *Elements* to a theory of numbers based solely on the concept of divisors and multiples. His aim was to compile a treatise on the multiplicative nature of integers. We shall proceed in exactly the same spirit.

Thus, an integer a is said to be divisible by an integer $b \neq 0$ if there exists an integer q such that

$$a = qb. \tag{1.1}$$

This relation is denoted by either $b|a$ or $a \in b\mathbb{Z}$, where $b\mathbb{Z} = \{bn : n \in \mathbb{Z}\}$. We call b either a divisor or a factor of a, and also a a multiple of b; note that negative divisors may appear, although in §15 a stricter convention will be introduced. The situation where (1.1) does not hold for any integer q is indicated by either $b \nmid a$ or $a \notin b\mathbb{Z}$. If the integers a, b, c are such that $b|a$ and $c|b$, then $c|a$; that is, divisors of a divisor of a are divisors of a. By the same token, multiples of a multiple of b are multiples of b. Also, the set of all multiples of b is closed with respect to addition, subtraction, and multiplication by any integer; namely, if these operations are applied to the elements of $b\mathbb{Z}$, then the results are always in $b\mathbb{Z}$. One should note that the same additive property does not hold with divisors in general.

The most basic implication of divisibility is that

$$|a| < |b| \text{ in } (1.1) \implies a = 0. \tag{1.2}$$

For then $|q| < 1$, and hence $q = 0$, as q is an integer. The logic here may appear primitive, but the procedure will turn out to be effective at various crucial steps in our discussion.

2 *Divisibility*

The notion of divisibility of integers is among the simplest in the whole of mathematics; yet almost all of the deepest problems in number theory have roots in it.

Notes

[1.1] Euclid's *ΣΤΟΙΧΕΙΑ*, or the *Elements*, is no longer extant. As usual today, we refer thus to the Heiberg edition $[\Sigma]$ based on a Vatican codex via its English translation by Heath; see §26[b]. We read $[\Sigma]$ as a mathematical work, taking into account the essential contents of propositions but not every detail of their proofs. We also ignore historically inevitable limitations like the fact that Euclid dealt with only positive numbers. Furthermore, questions about the real authorship of each of the thirteen books of $[\Sigma]$ do not interest us. In any event, Euclid's vision emanates from $[\Sigma]$.

[1.2] We shall employ modern concepts, symbols, and common mathematical practices when explaining achievements of mathematicians in past ages. This will never alter their mathematical contents, and indeed there exists no other possible way available as possibility. For the same reason, we shall use the term algebra, regardless of the stages of its development: rhetorical, syncopated, and symbolic (Nesselmann (1842, p.302)).

[1.3] The birth of algebraic procedure could be traced back deep into times past, when chants or compositions of fixed patterns were most probably used to memorize and convey mathematical wisdom; some practices such as these continue today. In the time of Euclid, geometrical procedure was the main tool. For instance, a diagonal of a rectangle is used as a base line to continue construction with a straightedge without markings and a collapsing compass, which is indeed equivalent to solving quadratic equations and manipulating the roots algebraically. The fact that the words square and cube specify not only geometric figures but also numbers must be a remnant of ancient thoughts.

[1.4] The four basic arithmetic operations and their relations were already stated explicitly by Brahmagupta (628); negative numbers were treated in just the same way as today. The concept of zero as well can be traced back to ancient Indian mathematics. However, that divisors be non-zero did not become a postulate until relatively recent times; see Datta–Singh (1935: 1962, Chap.II, §15) for instance. As for the concept of negative numbers, two typical instances of stagnations caused by its dismissal will be indicated in Note [3.1] and §70(ii). Also, we should be aware that the ancient Maya concept of zero is said to be older than the Indian counterpart.

[1.5] The mathematical traditions of the ancient Greek, Indian, and their preceding civilizations were transmitted to the medieval Arab civilization;

see §26[a]–[c] for some details. Those people who traveled extensively and were thereby able to learn the Indian algorithm from the Arabs (al-Khwarizmi (ca 820b)) brought the greatly simplified calculation procedure based on the Indo–Arabic numeral system into daily transactions in Europe and elsewhere: *Nouem figure indorum he sunt 9,8,7,6,5,4,3,2,1,..., et cum hoc signo 0, quod arabice zephirum appellatur,...* (Fibonacci (1202: 1857, p.2)). Then a fundamental change took place in mathematics: the serene joy of arithmetic became shared by enthusiasts from various social milieux.

§2

In order to analyze the multiplicative nature of integers, we need also to deal with the deviation from being divisible. To this end, the following fundamental identity is employed:

Theorem 1 *Given two integers a and b ($\neq 0$), there exists a unique pair of integers q and r such that*

$$a = qb + r, \quad 0 \leq r < |b|. \tag{2.1}$$

Proof The set $\mathbb{R}$ is covered by the mutually disjoint intervals $[k|b|, (k+1)|b|)$, $k \in \mathbb{Z}$, and one of them must contain a; thus the existence of a required pair $\{q, r\}$ is evident. The uniqueness assertion is also obvious, but to make it certain, we apply (1.2): if $a = q'b + r'$ with $0 \leq r' < |b|$, then $|b(q - q')| = |r - r'| < |b|$, which implies that $r - r' = 0$, and $q - q' = 0$. This ends the proof.

The identity (2.1) is called the division algorithm; it is commonly expressed by stating that division of a by b yields the quotient q and the remainder or residue r. In terms of matrix algebra, (2.1) is the same as

$$(a, b) \begin{pmatrix} 1 & 0 \\ -q & 1 \end{pmatrix} = (r, b), \quad 0 \leq r < |b|, \tag{2.2}$$

or equivalently as

$$\begin{pmatrix} a \\ b \end{pmatrix} = \begin{pmatrix} q & 1 \\ 1 & 0 \end{pmatrix} \begin{pmatrix} b \\ r \end{pmatrix}, \quad 0 \leq r < |b|. \tag{2.3}$$

The division algorithm is thus expressed with an invertible integral matrix whose inverse is also an integral matrix. This is a viewpoint that will play a core rôle later on. A group-theoretic mechanism has been working through the history of number theory.

4 *Divisibility*

Notes

[2.1] It is a truly scientific strategy to attend equally to the deviation from being divisible when discussing divisibility. The origin of the division algorithm is unknown, however. In the three books [Σ.VII–IX], that is, Euclid's number theory, there are frequent applications of it, although they are in fact performed by means of repeated subtractions and comparisons; however, the uniqueness of the pair formed of quotient and remainder is not explicitly mentioned.

[2.2] The integer m such that $m \leq x < m+1$ is denoted by $[x]$ (Gauss (Werke II-1, p.5)); the same notation will be used for other purposes, but there should be no confusion. With this, (2.1) is expressed in the following form:

$$a = \begin{cases} [a/b]b + r & b > 0 \text{ or } b|a, \\ ([a/b] + 1)b + r & b < 0 \text{ and } b \nmid a. \end{cases}$$

[2.3] Given two integers a and b (≥ 2), the division algorithm can be applied repeatedly in the following fashion:

$$a = q_0 b + r_0$$
$$= (q_1 b + r_1)b + r_0$$
$$\cdots$$
$$= r_k b^k + \cdots + r_2 b^2 + r_1 b + r_0, \quad 0 \leq r_j < b, \ 0 \leq j \leq k,$$

where $b^k \leq a < b^{k+1}$, so $r_k \neq 0$. This is called the b-adic expansion of a, and denoted often by $a = (r_k r_{k-1} \ldots r_1 r_0)_b$; note that the arrangement of $\{r_j\}$ is opposite to that in the ascending expansion $a = r_0 + r_1 b + \cdots + r_k b^k$. The coefficients/digits r_j are uniquely determined. The 2-adic or the binary expansion is a useful tool in handling high powers of an integer; from §36 on, various examples will be given.

[2.4] Modern digital computers are designed on the basis of the binary system. They multiply numbers together using essentially the same method as that practiced in ancient Egypt (Ahmes (ca 1650 BCE): Chace (1927, p.3)). For example, if the multiplier is $(26)_{10} = (11010)_2$, then duplication is repeated four times; and the 1st, 3rd, and 4th results are added.

[2.5] The number of digits in the binary expansion of an integer $a > 0$ is equal to $[\log a / \log 2] + 1$; it can be employed for measuring the effort of writing down a. In the theory of computational complexity, the complexity of an algorithm, i.e., the number of steps it requires is estimated in terms of a function of the number of digits (bits) of input; see Note [47.3] for a clarification of what we mean here by the term steps. If that function is bounded above by a polynomial, then the algorithm is said to run in polynomial time or just called polynomial time. We shall use this expression to suggest intuitively that the algorithm under consideration is effective, thus, without explicitly estimating the number of steps. We shall in fact be more concerned whether an argument on a particular problem is constructive or existential. These two terms will also be used intuitively; however, where appropriate, explanations will be made.

[2.6] This might be counterintuitive: We do not know yet how to determine the number of divisors of an arbitrarily given integer or more basically how to find its divisors in polynomial time. In §§47–48, we shall describe an algorithm to run on quantum computers, the materialization of which should yield a practical solution to this seemingly innocent but actually extremely fundamental problem. The discussions in §§33 and 49 will indicate the reason behind the ongoing great efforts.

§3

A marvel is in (2.1)–(2.3): if $c|a$ and $c|b$, then $c|r$, that is, all common divisors of a,b are preserved in the set of divisors of the residue r. As is explained below, this fact can be exploited in order to effectively compute $\gcd(a,b)$ the greatest common divisor, that is, the largest among positive common divisors of a,b. Since $\gcd(a,b)$ is one of the most basic notions in number theory, we allocate a special notation to it: throughout the rest of the present essays,

$$\langle a,b\rangle \text{ denotes } \gcd(a,b). \tag{3.1}$$

We stress that the situation $\{a,b\} = \{0,0\}$ is naturally excluded and that $\langle a,b\rangle$ is always positive besides equal to $\langle |a|,|b|\rangle = \langle |b|,|a|\rangle$, although being the largest among common divisors implies that it is positive per se.

The relation between the division algorithm and the greatest common divisors is as follows: Since $\langle a,b\rangle$ is a common divisor of a,b, the identity (2.1) implies that it divides b,r, which means $\langle a,b\rangle \leq \langle b,r\rangle$. In the same token, (2.1) implies that $\langle b,r\rangle$ divides a,b, so $\langle b,r\rangle \leq \langle a,b\rangle$. Hence, the identity $\langle a,b\rangle = \langle b,r\rangle$ follows from (2.1). With this, if $r = 0$, then $\langle a,b\rangle = |b|$. If $r > 0$, then proceeding to the second application of the division algorithm we divide b by r, getting the quotient q' and the residue r', which gives $\langle a,b\rangle = \langle r,r'\rangle$. Also we have

$$(a,b)\begin{pmatrix} 1 & 0 \\ -q & 1 \end{pmatrix}\begin{pmatrix} 1 & -q' \\ 0 & 1 \end{pmatrix} = (r,r'), \quad 0 \leq r' < r < b. \tag{3.2}$$

6 *Divisibility*

By repeated applications of the division algorithm, residues decrease strictly and reach 0 eventually, while zero vector never results, as only invertible matrices are involved. In short, there exists an integral matrix U of unit determinant such that

$$(a,b)U = \text{either } (d,0) \text{ or } (0,d) \text{ with } d \in \mathbb{N}. \tag{3.3}$$

We find that $\langle a,b \rangle = d$.

This procedure to determine the greatest common divisor of any pair of integers $\{a,b\} \neq \{0,0\}$ is called Euclid's algorithm.

Although the above discussion suffices for our immediate purposes, it is expedient to look into (3.3) itself. That is, suppose we have got an integral matrix $U = \begin{pmatrix} u & v \\ w & z \end{pmatrix}$ of unit determinant with which (3.3) holds, regardless how it has been found. Then (3.3) gives either $au + bw = d$ or $av + bz = d$; in particular, $\langle a,b \rangle$ divides d, so $\langle a,b \rangle \leq d$. On the other hand, we have

$$(a,b) = \left\{ \text{either } (d,0) \text{ or } (0,d) \right\} \cdot \begin{pmatrix} z & -v \\ -w & u \end{pmatrix}, \tag{3.4}$$

so $d|a$, $d|b$. Hence, $d \leq \langle a,b \rangle$; and $d = \langle a,b \rangle$.

The use of the matrix notations (2.2) and (3.2) is convenient, for it allows one to work solely with matrices of unit determinant. However, for the sake of revealing the structure of Euclid's algorithm it is worth exploiting also the relation (2.3) which involves a matrix of determinant equal to -1; details will be provided in §21.

To specify a particularly important situation, we make the following definition:

$$\text{integers } a, b \text{ are coprime to each other} \iff \langle a,b \rangle = 1. \tag{3.5}$$

We shall use also mutually prime and relatively prime as the terms equivalent to coprime. Note that if one of a,b in this is equal to 0, then the other is equal to ± 1.

Theorem 2 *For any pair $\{a,b\} \neq \{0,0\}$ of integers, there exists a coprime pair $g,h \in \mathbb{Z}$ such that*

$$ag + bh = \langle a,b \rangle. \tag{3.6}$$

Thus

$$c|a, \ c|b \iff c|\langle a,b \rangle. \tag{3.7}$$

In other words, being a common divisor is equivalent to being a divisor of the greatest common divisor.

Proof The second assertion follows immediately from the first. The existence of g, h is implied by (3.3). That they are coprime follows from $\det U = 1$; also, it suffices to note the identity

$$(a/\langle a,b\rangle)g + (b/\langle a,b\rangle)h = 1. \tag{3.8}$$

This ends the proof.

There are, in fact, infinitely many pairs g, h that satisfy (3.6); see Note [6.2]. What Euclid's algorithm provides is a constructive process to find such a particular pair. Here, although trivial, it should be worth stressing that

$$\text{each procedure of the Euclid algorithm is unique.} \tag{3.9}$$

More precisely, it is fixed by the ordered pair of integers with which the algorithm starts.

Notes

[3.1] Euclid commenced his discussion of number theory with his algorithm or anthyphairesis. We hear echoes from remote antiquity: [Σ.VII, Prop.1]

Two unequal numbers being set out, and the less being continually subtracted in turn from the greater, if the number which is left never measures the one before it until an unit is left, then the original numbers will be prime to one another.

The famous phrase – the less being continually subtracted in turn from the greater (i.e., dividing; see Note [2.1]) has given rise to the term anthyphairesis, the substantive form of its Greek original (see Fowler (1979)). Antenaresis, which is also commonly used, indicates the same. [Σ.VII, Prop.2 and porism] correspond, respectively, to the second paragraph of the present section and to the equivalence statement (3.7). It seems, on the other hand, that the fundamental assertion (3.6) was grasped just before the time of Aryabhata I (499; Clark (1930, pp.42–50)) and was then shared among later authors in the classical era of Indian mathematics, or the classic Indian mathematics that flourished from the 5th to the 13th centuries CE; see Datta–Singh (1935: 1962, Chap.III, §13). It is, nevertheless, readily seen that Euclid's own argument can in fact yield (3.6); the reason that Euclid missed to assert it is in that there was no concept of negative numbers in his time. The origin of Euclid's algorithm has been discussed extensively, but no conclusion has been achieved yet. We choose to follow the tradition of attributing the algorithm to Euclid, even though either anthyphairesis or antenaresis might be considered a more historically justifiable term for it. The more that researches in number theory have been accumulated, the more impressive the simplicity and the penetrating

8 *Divisibility*

power of Euclid's algorithm have become. It is not merely an arithmetical device but a deep wisdom that we owe to ancient sages.

[3.2]   Lamé (1844). When applied to integers $a > b > 0$, Euclid's algorithm requires 5κ division steps at most, where κ is the number of digits in the decimal expansion of b. A proof will be given in Note [24.5]. Since the four basic arithmetic operations are all polynomial time as is evident via the inspection of the bit operations involved there, Euclid's algorithm is polynomial time. According to the criterion stated in Note [2.5], this ancient algorithm can be regarded as exceptionally effective. Moreover, Lamé's bound is sharp, for the computation of $\langle 144, 89 \rangle$ requires exactly 10 division steps: the relevant sequence of quotients is

$$\left\{ q, q', q'', \ldots, q^{(9)} \right\} = \left\{ 1, 1, 1, 1, 1, 1, 1, 1, 1, 2 \right\}$$

(see Note [24.2]).

[3.3]   For $a = 3041543$, $b = 1426253$, we have

$$\left\{ q, q', q'', \ldots, q^{(8)} \right\} = \left\{ 2, 7, 1, 1, 5, 13, 6, 4, 2 \right\}.$$

By the matrix argument (3.1)–(3.2),

$$(3041543, 1426253) \begin{pmatrix} 62011 & -27682 \\ -132241 & 59033 \end{pmatrix} = (0, 23), \quad \gcd = 23.$$

Hence,

$$3041543 \cdot (-27682) + 1426253 \cdot 59033 = 23,$$
$$3041543 = 23 \cdot 132241, \quad 1426253 = 23 \cdot 62011.$$

Also, with any $f \in \mathbb{Z}$,

$$3041543 \cdot (-27682 + 62011f) + 1426253 \cdot (59033 - 132241f) = 23,$$
$$\langle -27682 + 62011f, 59033 - 132241f \rangle = 1.$$

[3.4]   If $a, b \in \mathbb{N}$ are coprime, then $C(a,b) = (a+b-1)! \, / a! \, b! \in \mathbb{N}$: We have, for any $g, h \in \mathbb{Z}$,

$$(ag + bh)C(a,b) = g \cdot \binom{a+b-1}{a-1} + h \cdot \binom{a+b-1}{b-1},$$

which is an integer. It remains to take g, h so that $ag + bh = 1$.

§4

An alternative proof of the fundamental identity (3.6): Let $S = \{ax + by : x, y \in \mathbb{Z}\}$. Then S is an integral module, or a module in $\mathbb{Z}$. That is, $m, n \in S$

implies $m - n \in S$; see Note [4.2]. Let d_0 be the least positive element in S. Since there exist $x_0, y_0 \in \mathbb{Z}$ such that $d_0 = ax_0 + by_0$, we see that $\langle a, b \rangle | d_0$, so $\langle a, b \rangle \leq d_0$. On the other hand, an application of (2.1) to the pair $\{ax + by, d_0\}$, $x, y \in \mathbb{Z}$, gives the residue $r = a(x - qx_0) + b(y - qy_0)$, which is non-negative, less than d_0, and contained in S; hence, $r = 0$. Any element of S is divisible by d_0, or $S \subseteq d_0 \mathbb{Z}$; thus, $S = d_0 \mathbb{Z}$, as the opposite inclusion is obvious. Since $a, b \in S$, we see that d_0 is a common divisor of them, so $d_0 \leq \langle a, b \rangle$. Namely, $d_0 = \langle a, b \rangle$; and x_0, y_0 are coprime in view of (3.8). Therefore, Theorem 2 is equivalent to

Theorem 3 *For any non-zero a, b, we have*

$$a\mathbb{Z} + b\mathbb{Z} = \langle a, b \rangle \mathbb{Z}, \qquad (4.1)$$

in which the left side is the set $\{\alpha + \beta : \alpha \in a\mathbb{Z}, \ \beta \in b\mathbb{Z}\}$. In particular,

$$\langle a, b \rangle = 1 \ \Leftrightarrow \ a\mathbb{Z} + b\mathbb{Z} = \mathbb{Z}. \qquad (4.2)$$

A relation between Euclid's algorithm and the reasoning employed here will be discussed in the later part of §23. Euclid's algorithm is marvelously powerful and constructive. On the other hand, the present method often turns out to be extremely efficient in the cases where existential assertions suffice. See for instance proofs of assertions in (6.2).

Notes

[4.1]   The above argument indicates no particular way to reach d_0; we are certain only that d_0 exists, for it is obviously less than or equal to $\min\{|a|, |b|\}$. In contrast, this sort of vagueness is foreign to Euclid's algorithm. Thus, it should be worthwhile to consider modifying Euclid's algorithm so as to increase its efficiency. With this in mind, we drop the specification $0 \leq r < b$ in (2.1): with $a, b > 0$, we let $\tau \in \mathbb{N}$, $\rho \in \mathbb{Z}$ be such that

$$a = \tau b + \rho, \quad |\rho| \leq b/2,$$

which is called the division algorithm with the least absolute remainder. It is meant that the quotient q in (2.1) is to be increased by 1 if necessary; in the next step, i.e., if $\rho \neq 0$, then τ is divided by $|\rho|$ in the same way. Again we have $\langle a, b \rangle = \langle |\rho|, b \rangle$. Within $[\log b / \log 2] + 1$ division steps, we reach the greatest common divisor. By the way, in the present context the matrix representation (2.2) of the division algorithm needs to be modified appropriately; however, (2.3) is in fact more practical for this purpose, as is to be indicated in Note [21.1]. In the light of the assertion of Lamé (Note [3.2]), it might be surmised that such a modification of Euclid's algorithm would become visibly effective only when b is huge. To see whether this is correct or not, we return to the example in Note [3.3]. We get the new sequence of quotients $\{2, 8^{(-)}, 2, 5, 13, 6, 4, 2\}$, where $8^{(-)}$ indicates that the relevant residue is negative, and find that the improvement is insignificant, although $b = 1426253$ is certainly not small. On the other hand, with the example given in Note [3.2], we have the new sequence of quotients $\{2^{(-)}, 3^{(-)}, 3^{(-)}, 3^{(-)}, 2, 2\}$; thus the number of steps is almost halved; the improvement is significant, although $b = 89$ is a small number. In general a negative residue ρ results when the quotient 1 occurs in the original procedure; thus, the second example is an extreme case. The effect of the present modification of Euclid's algorithm depends not only on the size of b but also on the arithmetical nature of the pair $\{a, b\}$. A related discussion is in Note [21.4].

[4.2]   Following the proof for Theorem 3, one may show that if a subset K of $\mathbb{Z}$ is a module then $K = k\mathbb{Z}$ with a unique integer $k \geq 0$. On noting that any infinite cyclic group is isomorphic to $\mathbb{Z}$, the present assertion on modules can be expressed as follows: all non-trivial subgroups of an infinite cyclic group are isomorphic to each other, that is, to $\mathbb{Z}$. We note further that if K, L are non-zero modules such that $K = k\mathbb{Z}$ and $L = \ell\mathbb{Z}$, then $L \subseteq K \Leftrightarrow k|\ell$. This observation inspired Dedekind (1876: 1877a, §1) to invent the notions of ideals and their divisibility in algebraic number fields which generally lack the analogue of (2.1), the core of Euclid's algorithm; a glimpse is in Note [72.8].

[4.3]   We shall not, however, touch on the theory of algebraic number fields proper, save for occasional remarks, since one of the main aims of the present essays is to indicate what prepared the ground for such a modern development of number theory. Thus, Weber (1895/1908) and Hecke (1923) are recommended instead; these treatises are fascinating amalgams of the three categories of number theory mentioned in the preamble, although lacking the core of analytic number theory, i.e., what corresponds to our Chapter 5, which was of course historically inevitable.

§5

In the discussion of §3, elements of the full modular group $\Gamma = \mathrm{SL}(2, \mathbb{Z})$, which is the set of all 2×2 integral matrices of unit determinant, were used. The aim there was to reach the pair $\{g, h\}$ of (3.6) with a procedure which is both transparent and easy to handle. We shall show that the same argument can be used to prove also that the structure of Γ is relatively simple. Namely, the matrix formulation in §3 of Euclid's algorithm yields the following:

$$\S5 \qquad\qquad 11$$

Theorem 4 *The modular group Γ is generated by the two elements*

$$T = \begin{pmatrix} 1 & 1 \\ 0 & 1 \end{pmatrix}, \quad W = \begin{pmatrix} 0 & -1 \\ 1 & 0 \end{pmatrix}. \tag{5.1}$$

Proof If $A = \begin{pmatrix} a & b \\ * & * \end{pmatrix} \in \Gamma$, then $\langle a, b \rangle = 1$. The algorithm given in the second paragraph of §3 implies that $(a, b)U = (1, 0)$ or $(0, 1)$ with U being a certain product of $T^\alpha = \begin{pmatrix} 1 & \alpha \\ 0 & 1 \end{pmatrix}$, $WT^{-\beta}W^{-1} = \begin{pmatrix} 1 & 0 \\ \beta & 1 \end{pmatrix}$, and $W^2 = -\begin{pmatrix} 1 & 0 \\ 0 & 1 \end{pmatrix}$, where $\alpha, \beta \in \mathbb{Z}$. We see that $AU = \begin{pmatrix} 1 & 0 \\ * & 1 \end{pmatrix}$ or $\begin{pmatrix} 0 & 1 \\ -1 & * \end{pmatrix}$. In the latter case, we have $AUW = \begin{pmatrix} 1 & 0 \\ * & 1 \end{pmatrix}$. This ends the proof. See Note [25.1] for a more explicit argument; also, Note [9.4] gives a generalization of this theorem.

It should be remarked here that each element $\begin{pmatrix} a & b \\ c & d \end{pmatrix} \in \Gamma$ can appear in two different contexts, either a matrix or a linear fractional transformation, i.e.,

$$z \mapsto \frac{az + b}{cz + d}, \quad z \in \mathbb{C}. \tag{5.2}$$

In what follows, the term automorphic context will be used to indicate that the discussion at a particular spot concerns certain functions which do not essentially change their shapes while admitting actions of Γ; they are called automorphic functions, and automorphic forms sometimes. This term is used also to describe similar effects of actions of analogous groups. Further, it is attached to functions induced by automorphic functions and forms. It should, however, be noted that these terms are used actually in a restrictive way: in principle, the variables under consideration are required to be continuous like in (5.2). Thus, the relation (3.3) implies that the greatest common divisor is an automorphic function but only in a widely extended sense.

In passing, we remark that from Theorems 4 and 7, or rather Note [9.3] emerge multiplicative structures over $\mathbb{Z}$, which are in fact groups composed of integral matrices involved in Euclid's algorithm and its suitable extension to higher dimensions. This viewpoint will be maintained throughout the present essays:

$$\text{deviation from being divisible} \\ \rightarrowtail \text{ Euclid's algorithm } \rightarrowtail \text{ group structure.} \tag{5.3}$$

One may assert that tuples of coprime integers provide a parametrization of elements in these groups. See for instance the later part of §29.

Notes

[5.1]   Basic convention on matrix notation: Various 2×2 matrices will appear in our discussion. They are used in one of the following modes:

12 *Divisibility*

(1) matrices themselves,
(2) linear fractional transformations,
(3) mixed way.

Working under (1), we shall proceed mostly without mention. Occasionally we shall stay under (2); then probably the symbols like $\Gamma/\{\pm 1\}$ will be used, that is, two matrices $\pm A$ are identified. The situation (3) will become rather common from §77. In addition, be aware that we shall not try to keep consistency in the use of elements of matrices; for instance, in (5.2) above and in Note [5.4] below the elements a, b, c, d appear in different circumstances. Confusions will be avoided contextually.

[5.2]   Theorems 2 and 3 mean also that there exists a transformation of variables $^t(x, y) = U \cdot {}^t(X, Y)$, $U \in \Gamma$, such that the integral linear form $ax + by$ is moved to the simplified linear form $\langle a, b \rangle X$, which may be called an arithmetic base change. One may take, with g, h as in (3.6),

$$U = \begin{pmatrix} g & -b/\langle a, b \rangle \\ h & a/\langle a, b \rangle \end{pmatrix}.$$

[5.3]   The set of all 2×2 integral matrices of determinant $n \neq 0$ is decomposed as

$$\bigsqcup \Gamma \cdot \begin{pmatrix} a & b \\ 0 & d \end{pmatrix}, \ ad = n, \ a > 0, \ 0 \le b < |d|,$$

where the symbol $\sqcup$ denotes a sum of mutually disjoint subsets; the number of summands equals the sum of positive divisors of n (i.e., $(16.4)_{\alpha=1}$). This decomposition is due to Lipschitz (1857), and is used in defining Hecke operators (Hecke (1937, §2)), which is a core notion in the automorphic context. To prove the decomposition, let $n = \begin{pmatrix} u & v \\ w & z \end{pmatrix}$ be an integral matrix such that $\det n = n$. We take $\gamma = \begin{pmatrix} f & g \\ h & l \end{pmatrix} \in \Gamma$, $h = -w/\langle u, w \rangle$, $l = u/\langle u, w \rangle$; since $\langle h, l \rangle = 1$, Euclid's algorithm gives a required pair f, g. Then $\gamma n = \begin{pmatrix} a & b_1 \\ 0 & d \end{pmatrix}$, $a = \langle u, w \rangle$, $ad = n$; and it remains only to multiply γn by an appropriate power of T from the left. The disjointness assertion is easy to verify.

[5.4]   We apply the above decomposition to general quadruple sums: we write formally

$$S = \sum_{a,b,c,d \in \mathbb{Z}} f(a, b, c, d) = \sum_n f(n), \quad n = \begin{pmatrix} a & b \\ c & d \end{pmatrix}.$$

Then,

$$S = \sum_{\det n = 0} f(n) + \sum_{\substack{n=-\infty \\ n \neq 0}}^{\infty} {\sum_{\det n = n}}' \sum_{\gamma \in \Gamma} f(\gamma n),$$

where $\sum'$ indicates that n runs over the representative matrices of determinant n appearing in the preceding Note. Hence, quadruple sums in general can be regarded as relevant to the Hecke theory. See Note [99.3]. It should be added that the sum over n with det n $= 0$ is closely related to the contents of Note [18.7].

§6

The fundamental importance of Theorems 2 and 3 will be gradually revealed in the sequel; indeed Euclid's algorithm is exploited throughout the present essays either explicitly or implicitly.

To begin with, we state the following corollary:

Theorem 5 *If a, b are coprime, then*

$$a|bc \implies a|c. \tag{6.1}$$

Proof By (3.6) there are $g, h \in \mathbb{Z}$ such that $acg + bch = c$. The left side is divisible by a, so (6.1) follows. Alternatively, one may use (4.2): $a\mathbb{Z} + b\mathbb{Z} = \mathbb{Z} \implies ac\mathbb{Z} + bc\mathbb{Z} = c\mathbb{Z}$; and if $bc\mathbb{Z} \subseteq a\mathbb{Z}$, then $c\mathbb{Z} \subseteq a\mathbb{Z}$. This ends the proof.

We have

$$
\begin{aligned}
&(0) \quad \mathbb{Z} \ni f \implies \langle a, b \rangle = \langle a + bf, b \rangle, \\
&(1) \quad \mathbb{Z} \ni m \neq 0 \implies \langle am, bm \rangle = \langle a, b \rangle |m|, \\
&(2) \quad l | \langle a, b \rangle \implies \langle a/l, b/l \rangle = \langle a, b \rangle / |l|, \\
&(3) \quad \langle a, c \rangle = 1, \ \langle b, c \rangle = 1 \implies \langle ab, c \rangle = 1, \\
&(4) \quad \langle a, b \rangle = 1, a \neq 0, \mathbb{Z} \ni c \implies \langle a, bc \rangle = \langle a, c \rangle.
\end{aligned}
\tag{6.2}
$$

First, (0) is implied by Euclid's algorithm. As for (1), we have on one hand $\langle a, b \rangle |m| \leq \langle am, bm \rangle$ since $\langle a, b \rangle |m|$ is a common divisor of am and bm. On the other hand, by (3.6) we may choose g, h so that $\langle a, b \rangle |m| = amg + bmh$. Hence $\langle a, b \rangle |m|$ is divisible by $\langle am, bm \rangle$, and we have $\langle am, bm \rangle \leq \langle a, b \rangle |m|$, which gives (1). A quicker proof is to multiply both sides of (4.1) by m. The assertion (2) is a corollary of (1). As for (3), we note that if $\langle ab, c \rangle = d$, then $d | c$; thus $\langle d, a \rangle = 1$. This and (6.1) give $d | b$ for $d | ab$. Hence, d is a common divisor of b and c; that is, $d = 1$. If we employ (4.2), then $a\mathbb{Z} + c\mathbb{Z} = \mathbb{Z}$, and $ab\mathbb{Z} + bc\mathbb{Z} = b\mathbb{Z} \implies ab\mathbb{Z} + bc\mathbb{Z} + c\mathbb{Z} = b\mathbb{Z} + c\mathbb{Z} = \mathbb{Z}$; hence $ab\mathbb{Z} + c\mathbb{Z} = \mathbb{Z}$, which is equivalent to (3). To show (4), we let $\langle a, bc \rangle = d$; so $d \geq \langle a, c \rangle$. As $d | bc$, we get $d | c$ since $d | a$ induces $\langle b, d \rangle = 1$. That is, d is a common divisor of

14 *Divisibility*

a and c, which implies $d \le \langle a,c \rangle$, and thus (4). Or, by (4.2), $ac\mathbb{Z} + bc\mathbb{Z} = c\mathbb{Z}$; adding $a\mathbb{Z}$ to both sides, $a\mathbb{Z} + bc\mathbb{Z} = a\mathbb{Z} + c\mathbb{Z}$, which is equivalent to (4).

We have, more generally than (6.1),

$$a|bc \quad \Rightarrow \quad (a/\langle a,b \rangle)\big|c. \tag{6.3}$$

This is because $\big(a/\langle a,b \rangle\big)\big|\big(b/\langle a,b \rangle\big)c$ and $\langle a/\langle a,b \rangle, b/\langle a,b \rangle \rangle = 1$. Or just simply, $c(a\mathbb{Z} + b\mathbb{Z}) \subseteq a\mathbb{Z}$. Also, applying (3) repeatedly, we see that if $\langle a,b \rangle = 1$, then $\langle a^m, b^n \rangle = 1$ for any non-negative integers m, n.

Notes

[6.1]   Theorem 5 is equivalent to [Σ.VII, Prop.20] which is frequently applied in Euclid's number theory. The assertions in (6.2) are contained in [Σ.VII, Props.23–28].

Remark In spite of what we stated in Note [1.1], it should be appropriate to note here that the proof of [Σ.VII, Prop.20] raises a serious discussion concerning its logical structure. See Pengelley and Richman (2006) for details. See Note [11.4] as well.

[6.2]   Pairs $\{g, h\}$ which satisfy (3.6) exist infinitely many. With $a_1 = a/\langle a,b \rangle$, $b_1 = b/\langle a,b \rangle$, a particular pair g, h may be replaced by $g + b_1 t, h - a_1 t, t \in \mathbb{Z}$; note that $\langle g + b_1 t, h - a_1 t \rangle = 1$, as $a_1(g + b_1 t) + b_1(h - a_1 t) = 1$. That these exhaust all possibilities can be shown with (6.1): any pair $\{g_1, h_1\}$ gives $a(g - g_1) = b(h_1 - h)$; dividing both sides by $\langle a,b \rangle$, we get the confirmation.

§7

The notion of the greatest common divisor of two given integers is generalized, in an obvious way, to the greatest common divisor of an arbitrary number of given integers $a_k \in \mathbb{Z}$, $1 \le k \le K$, not all being equal to 0. It is denoted by $\langle a_1, a_2, \ldots, a_K \rangle$; note that this is also positive by definition. The proof of Theorem 3 is readily extended: there exist $\{h_k\} \in \mathbb{Z}^K$ such that

$$\sum_{k=1}^{K} a_k h_k = \langle a_1, a_2, \ldots, a_K \rangle. \tag{7.1}$$

A method to compute these h_j is indicated in Note [7.1]. The matrix argument given in §3 can also be extended, but the details are omitted since it is further extended in the proof of Theorem 7, §9. Corresponding to (3.7), we have

$$c|a_k, \ 1 \le k \le K \quad \Leftrightarrow \quad c|\langle a_1, a_2, \ldots, a_K \rangle. \tag{7.2}$$

As an application, we divide the tuple $\{a_1, a_2, \ldots, a_K\} \subset \mathbb{N}$ into L disjoint ones and denote by $\{d_1, d_2, \ldots, d_L\}$ the greatest common divisors of integers contained in respective sub-tuples. Then

$$\langle a_1, a_2, \ldots, a_K \rangle = \langle d_1, d_2, \ldots, d_L \rangle. \tag{7.3}$$

That each side divides the other follows immediately either from (7.1) or from (7.2). One may use also the relation

$$a_1 \mathbb{Z} + a_2 \mathbb{Z} + \cdots + a_K \mathbb{Z} = \langle a_1, a_2, \ldots, a_K \rangle \mathbb{Z}. \tag{7.4}$$

Generalizing the definition (3.5), we introduce the following:

$$\text{the tuple } \{a_1, a_2, \ldots a_K\} \text{ is coprime} \iff \langle a_1, a_2, \ldots, a_K \rangle = 1; \tag{7.5}$$

so $\{h_1, h_2, \ldots, h_K\}$ in (7.1) is coprime. Note that its maximally refined situation

$$\langle a_j, a_k \rangle = 1 \text{ for all pairs } \{j, k\}, j \neq k, \tag{7.6}$$

will always be specified.

Notes

[7.1] [Σ.X, Prop.4, Porism] corresponds to (7.2). Also, [Σ.VII, Prop.3] asserts that $\langle a, b, c \rangle = \langle \langle a, b \rangle, c \rangle$, which corresponds to (7.3). In particular, the coefficients $\{h_j\}$ of (7.1) can be obtained by multiple applications of (3.6), which means that the matrix argument in §3 can also be used with an obvious expansion of 2×2 matrices to $K \times K$ matrices.

[7.2] Mertens (1897a): Let $\{a_1, a_2, \ldots, a_K, m\} \in \mathbb{Z}^{K+1}$ with $K \geq 2$ be coprime in the sense of (7.5). Then one may choose $\{x_1, x_2, \ldots, x_K\} \in \mathbb{Z}^K$ such that

$$\langle a_1 + mx_1, a_2 + mx_2, \ldots, a_K + mx_K \rangle = 1.$$

To prove this, we first apply (7.3). There are two integers s, t such that $sb + tm = 1$, where $b = \langle a_1, a_2, \ldots, a_K \rangle$. We take $\{\lambda_k\}$ and $\{\mu_k\}$ so that

$$\sum_{k=1}^{K} a_k \lambda_k = b, \quad \sum_{k=1}^{K} a_k \mu_k = 0, \quad \begin{array}{l} \langle \lambda_1, \lambda_2, \ldots, \lambda_K \rangle = 1, \\ \langle \mu_1, \mu_2, \ldots, \mu_K \rangle = 1. \end{array}$$

The first sum is the same as (7.1). As for the second, if for instance $a_1 a_2 \neq 0$, then we put $\mu_1 = a_2/\langle a_1, a_2 \rangle$, $\mu_2 = -a_1/\langle a_1, a_2 \rangle$, and $\mu_k = 0$ for $k > 2$; if only one of a_k's is not equal to 0, then the choice of μ_k's is trivial. We now put $\gamma_k = \mu_k + s\lambda_k$. Then $\langle \gamma_1, \gamma_2, \ldots, \gamma_K \rangle = 1$. For on noting $\sum_{k=1}^{K}(a_k/b)\gamma_k = s$ we see that if $u | \gamma_k$, $1 \leq k \leq K$, then $u | s$; thus $u | \mu_k$, $1 \leq k \leq K$, which implies $u = 1$. Let ξ_k be such that $\sum_{k=1}^{K} \gamma_k \xi_k = 1$; and let $x_k = t\xi_k$. Then

$\sum_{k=1}^{K}(a_k + mx_k)\gamma_k = sb + tm = 1$, which ends the proof. An example: according to Note [3.3], we have $\langle 3041543, 1426253, 94 \rangle = 1$. In this case, $b = 23$, $m = 94$; $\lambda_1 = -27682$, $\lambda_2 = 59033$; $\mu_1 = 1426253/23 = 62011$, $\mu_2 = -3041543/23 = -132241$. Euclid's algorithm gives

$$(23, 94)\begin{pmatrix} 45 & -94 \\ -11 & 23 \end{pmatrix} = (1, 0);$$

that is, $s = 45$, $t = -11$; $\gamma_1 = -1183679$, $\gamma_2 = 2524244$. Again by Euclid's algorithm,

$$(1183679, 2524244)\begin{pmatrix} -59033 & -2524244 \\ 27682 & 1183679 \end{pmatrix} = (1, 0).$$

Hence, $\xi_1 = 59033$, $\xi_2 = 27682$; $x_1 = -649363$, $x_2 = -304502$; $a_1 + mx_1 = -57998579$, $a_2 + mx_2 = -27196935$. By yet another application of Euclid's algorithm, we find that

$$(57998579, 27196935)\begin{pmatrix} 1183679 & -27196935 \\ -2524244 & 57998579 \end{pmatrix} = (1, 0).$$

[7.3]   Gauss [DA, art.234]: With two integral matrices

$$A = \begin{pmatrix} a_1, & a_2, & \dots, & a_n \\ a'_1, & a'_2, & \dots, & a'_n \end{pmatrix}, \quad B = \begin{pmatrix} b_1, & b_2, & \dots, & b_n \\ b'_1, & b'_2, & \dots, & b'_n \end{pmatrix}, \quad n \geq 3,$$

assume that the set composed of all minors of order 2 of A is coprime in the sense of (7.5), and that every minor of order 2 of B is equal to the $f(\neq 0)$ times of the corresponding minor of A. Then there exists an integral square matrix F of order 2 such that

$$B = FA, \quad \det F = f.$$

To show this, we apply (7.1) so that

$$\sum_{i,j} \alpha_{ij} \begin{vmatrix} a_i & a_j \\ a'_i & a'_j \end{vmatrix} = 1, \quad \exists \alpha_{ij} \in \mathbb{Z}.$$

Then we put

$$u = \sum_{i,j} \alpha_{ij} \begin{vmatrix} b_i & b_j \\ a'_i & a'_j \end{vmatrix}, \quad v = \sum_{i,j} \alpha_{ij} \begin{vmatrix} a_i & a_j \\ b_i & b_j \end{vmatrix},$$

$$w = \sum_{i,j} \alpha_{ij} \begin{vmatrix} b'_i & b'_j \\ a'_i & a'_j \end{vmatrix}, \quad z = \sum_{i,j} \alpha_{ij} \begin{vmatrix} a_i & a_j \\ b'_i & b'_j \end{vmatrix}.$$

We have $ua_g + va'_g = b_g$ since

$$a_g \begin{vmatrix} b_i & b_j \\ a'_i & a'_j \end{vmatrix} + a'_g \begin{vmatrix} a_i & a_j \\ b_i & b_j \end{vmatrix} = b_i \begin{vmatrix} a_g & a_j \\ a'_g & a'_j \end{vmatrix} - b_j \begin{vmatrix} a_g & a_i \\ a'_g & a'_i \end{vmatrix}$$

$$= \frac{1}{f} \left(b_i \begin{vmatrix} b_g & b_j \\ b'_g & b'_j \end{vmatrix} - b_j \begin{vmatrix} b_g & b_i \\ b'_g & b'_i \end{vmatrix} \right) = \frac{b_g}{f} \begin{vmatrix} b_i & b_j \\ b'_i & b'_j \end{vmatrix} = b_g \begin{vmatrix} a_i & a_j \\ a'_i & a'_j \end{vmatrix}.$$

Analogously, we have $wa_g + za'_g = b'_g$. Hence, we may take $F = \left(\begin{smallmatrix} u & v \\ w & z \end{smallmatrix} \right)$. It is immediate that $\det F = f$.

[7.4] The difference between (7.5) and (7.6) should be noted. In [Σ.VII, Def.12], this separation is not made. In modern literature, the situation (7.6) is expressed often as mutually prime, relatively prime in pairs, etc. In Dirichlet (1871/1894, §6), (7.5) is described as *Zahlen ohne gemeinschaftlichen Theiler*, and (7.6) as *je zwei von ihnen relative Primzahlen sind* (abbreviation: *relative Primzahlen*). Thus the separation is precise. In Euler (1733b, p.20), (7.6) is *numeros inter se primos*. In [DA, art.19] the same is employed. In order to avoid any confusion, one should make the situation clear whenever not (7.5) but (7.6) is meant. For instance, if $a_1, a_2, a_3 > 1$ are mutually coprime, then $a_1 a_2, a_2 a_3, a_3 a_1$ are not, yet $\langle a_1 a_2, a_2 a_3, a_3 a_1 \rangle = 1$.

§8

An integral linear indefinite equation is defined to be

$$\sum_{l=1}^{n} c_l x_l = c_0, \tag{8.1}$$

where $n \geq 2$, and c_l, $l \geq 1$, are non-trivial, that is, not all of them are equal to zero, while c_0 is arbitrary. Our problem is to find integral tuples $\{x_l\} \in \mathbb{Z}^n$ with which this identity holds. The discussion in the preceding section implies immediately

Theorem 6 *The integral indefinite equation* (8.1) *is soluble in integers if and only if* $\langle c_1, c_2, \ldots, c_n \rangle$ *divides* c_0.

If this condition is satisfied, then a solution can be obtained via the matrix computation associated with an extension of Euclid's algorithm which will be developed in the next section, but the treatment of the case $n = 2$ serves already as a fine reference for how to deal with general cases. Thus, let U be as in Note [5.2]. Then, with ${}^t(x, y) = U \cdot {}^t(X, Y)$,

18   *Divisibility*

$$ax + by = c \;\Leftrightarrow\; (a,b) \cdot U \cdot \begin{pmatrix} X \\ Y \end{pmatrix} = c \;\Leftrightarrow\; (\langle a,b\rangle, 0) \cdot \begin{pmatrix} X \\ Y \end{pmatrix} = c. \qquad (8.2)$$

Since $U \in \Gamma$, integral ${}^t(X,Y)$ are mapped bijectively to integral ${}^t(x,y)$. This implies that X is equal to $c/\langle a,b\rangle$ and Y to any integer, provided $\langle a,b\rangle | c$. Hence all the solutions of equation $ax + by = c$, $\langle a,b\rangle | c$, are given by

$$\begin{pmatrix} x \\ y \end{pmatrix} = \frac{c}{\langle a,b\rangle} \begin{pmatrix} g \\ h \end{pmatrix} + \frac{k}{\langle a,b\rangle} \begin{pmatrix} -b \\ a \end{pmatrix}, \quad k \in \mathbb{Z}, \qquad (8.3)$$

which are lattice points forming an arithmetic progression on a straight line determined by a,b,c.

It should be worth stressing that the set of solutions to (8.1) is highly susceptible to the changes among the coefficients if subtle; an example of this phenomenon is given in Note [8.1].

Notes

[8.1]   Five sailors and a monkey were shipwrecked on a deserted island covered with a forest of coconut-palms. They collected coconuts for food in a heap and went to sleep. But they did not trust each other. At midnight, a sailor woke up; he found that he could divide the nuts into five equal heaps, save for one nut which he gave to the monkey. Then he hid his portion, gathered the remaining nuts in a heap, and went to sleep. In their turn, the other four sailors could do the same. In the morning, the sailors woke up; they could divide the nuts into five equal heaps, save for one nut which they gave to the monkey. Question: find the smallest possible number of nuts in the original heap. This is known as the quiz: *Coconuts, sailors, and a monkey*. We generalize it slightly, assuming there were m sailors. Let x be the number of nuts in question, and let $k_1, k_2, \ldots, k_m$ be the numbers of nuts which the sailors hid in turn; and k_{m+1} the number of nuts each sailor got in the next morning. Then, $x = mk_1 + 1$, $x - (k_1 + k_2 + \cdots + k_j + j) = mk_{j+1} + 1$. Thus, $k_{j+1} + 1 = (1 - 1/m)(k_j + 1)$. Denoting k_{m+1} by y, we have

$$(*) \qquad (m-1)^m x - m^{m+1} y = m^{m+1} - (m-1)^{m+1}.$$

When $m = 5$, this becomes $1024x - 15625y = 11529$. Euclid's algorithm yields

$$(1024, 15625) \begin{pmatrix} 15625 & -4776 \\ -1024 & 313 \end{pmatrix} = (0, 1).$$

According to Note [6.2], we have $x = 15625t - 4776 \cdot 11529$, $y = 1024t - 313 \cdot 11529$, $t \in \mathbb{Z}$. Hence, the positive minimum of x is given by $t = 3525$. That is, $15621 = 4147 + 3522 + 3022 + 2622 + 2302 + 6$, where the right side indicates

$$\S 8 \qquad\qquad 19$$

how the nuts are distributed among the sailors and the monkey. However, this result looks unnatural, for $x = 15621$ is just too big. Thus, suppose instead that in the morning the monkey got none; then

$$(**) \qquad (m-1)^m x - m^{m+1} y = (m^m - (m-1)^m)(m-1).$$

With $m = 5$, we have $3121 = 828 + 703 + 603 + 523 + 459 + 5$. Or rather, if the number of sailors is four, i.e., $m = 4$ in $(*)$, then we find $1021 = 335 + 271 + 223 + 187 + 5$. These results show clearly that seemingly trifle changes among coefficients of a linear indefinite equation can induce sudden and considerable changes in solutions. The reason for this phenomenon, which may be called chaotic, is in that the divisibility mechanism is working behind.

[8.2] We shall indicate how Euler (1733b; 1771/1822, Part II, Chapter I) would have solved $(*)_{m=5}$ in integers: First, note that $15625 = 15 \cdot 1024 + 265$, $11529 = 11 \cdot 1024 + 265$. That is, $x = 15y + 11 + 265(y+1)/1024$. The third term on this right side is an integer; thus there exists a $t \in \mathbb{Z}$ such that $265y = 1024t - 265$. In the same way, via $1024 = 4 \cdot 265 - 36$, we have $y = 4t - 1 - 36t/265$; and $t = 265u/36$, $u \in \mathbb{Z}$. Repeating the procedure, we have: $265 = 7 \cdot 36 + 13 \Rightarrow t = 7u + 13u/36 \Rightarrow 36 = 3 \cdot 13 - 3 \Rightarrow u = 36v/13 = 3v - 3v/13$, $v \in \mathbb{Z}$. Further, $13 = 4 \cdot 3 + 1 \Rightarrow v = 4w + w/3 \Rightarrow w = 3z$, $z \in \mathbb{Z}$. Hence, reversing the procedure, we find that $y = 1024z - 1$, $x = 15625z - 4$. The positive minimum of x is attained at $z = 1$, which yields naturally the same conclusion as in the preceding Note. This argument of Euler is equivalent to Euclid's algorithm.

[8.3] On noting that $(*)$ is equivalent to

$$(m-1)^m (x + m - 1) = m^{m+1}(y+1)$$

and $\langle (m-1)^m, m^{m+1} \rangle = 1$ (see the end of the text of §6), the general solution is obtained immediately:

$$x = m^{m+1} u - m + 1, \quad y = (m-1)^m u - 1,$$

with $u \in \mathbb{Z}$. Hence the positive minimum of x is $m^{m+1} - m + 1$. The number of coconuts in question increases explosively with the number of sailors, a fact that could be counterintuitive. By the way, the general solution to $(**)$ is given by

$$x = m^{m+1} v - (-1)^m m^m - m + 1,$$
$$y = (m-1)^m v + \big((-1)^{m+1}(m-1)^m - m + 1\big)/m,$$

with $v \in \mathbb{Z}$; observe that the second term on the right of the lower line is equal to $\sum_{j=1}^{m-1}(1-m)^j$, so it is of course an integer. The positive minimum of x is equal to $m^{m+1} - m^m - m + 1$ if $2 \mid m$, and to $m^m - m + 1$ if $2 \nmid m$.

20 *Divisibility*

[8.4] If each of $n, c_0, c_1, \ldots, c_n$ is large and the range of $\{x_l\}$ is restricted in an intricate manner, then it can be hard to solve (8.1) completely and express the solutions and their number in terms of an explicit function of $\{n, c_0, c_1, \ldots, c_n\}$ and the range. This applies to the case $n = 2$ already, even under quite a mild restriction. For instance, let the integers a, b, c be all positive, $\langle a, b \rangle = 1$, and let $A(c; a, b)$ be the number of solutions of the equation $ax + by = c$ under the fairly weak restriction $x, y \in \mathbb{N}$. Then, it is by no means straightforward to conclude that

$$|A(c; a, b) - \tau| \le 1, \quad \tau = [c/ab].$$

This bound is best possible, that is, it cannot be improved in general; in other words, $A(c; a, b)$ is equal to either $\tau - 1$ or τ or $\tau + 1$, and each takes place actually. For the confirmation, first let x_0, y_0 be a solution without any restriction. Then the general solution is, by Note [6.2], $x = x_0 + bt$, $y = y_0 - at$, $t \in \mathbb{Z}$. To this we impose the restriction: $x, y \in \mathbb{N}$, which is equivalent to $-x_0/b < t < y_0/a$. The length of this open interval is equal to c/ab. Thus, if $c/ab \le 1$, then $A = A(c; a, b)$ is equal to 0 or 1; and the bound in question is trivial. Next, if $c/ab > 1$, then $A \ge 1$. With this, (1) if both the end points of the interval are integers, that is, if $ab|c$, then $A = \tau - 1$. (2) if only one of the end points is an integer, that is, either if $a|c, b \nmid c$ or if $a \nmid c, b|c$, then $A = \tau$. (3) if neither end point is an integer, that is, if $a \nmid c, b \nmid c$, then let η_1 be the least distance between the lower end-point and integers contained in the interval; and let η_2 correspond analogously to the upper end-point. We have $A - 1 + \eta_1 + \eta_2 = c/ab \notin \mathbb{N}$. Hence, if $\eta_1 + \eta_2 < 1$, then $A - 1 = [A - 1 + \eta_1 + \eta_2] = \tau$, so $A = \tau + 1$; otherwise, $1 < \eta_1 + \eta_2 < 2$, and $A = [A + (\eta_1 + \eta_2 - 1)] = \tau$. For instance, $A(1001; 7, 11) = 12$ and $A(1008; 7, 11) = 13$ are the cases (1) and (2), respectively. Also, $A(1019; 7, 11) = 14$ corresponds to (3) with $\eta_1 + \eta_2 < 1$, as $\eta_1 = \frac{1}{11}, \eta_2 = \frac{1}{7}$. This ends the confirmation. Alternatively, one may argue as follows: Let $au - bv = 1$, with $u, v > 0$. Then, $x = cu - bt$, $y = at - cv$, $t \in \mathbb{Z}$. Hence, $A(c; a, b) = [cu/b] - [cv/a]$, on the understanding that if $b|c$, then this right side is to be decreased by one. We omit the deduction of the above inequality from this identity. A related discussion is given in Note [28.10].

[8.5] Paucity of soluble problems. Problems in number theory that are soluble with effective and explicit algorithms are very scarce. Integral linear indefinite equations belong to such exceptional cases, provided that the coefficients and the number of unknowns are not huge, and that ranges of solutions are not restricted in complicated ways. Another example is to express given intergers by a particular integral binary quadratic form in integral

$\S 9$ 21

variables. There exists a practical method to solve this highly interesting problem; see $\S\S 81$ and 87.

§9

Next we shall consider a system of integral linear indefinite equations:

$$\sum_{l=1}^{n} c_{kl} x_l = u_k, \quad 1 \le k \le m, \tag{9.1}$$

which is to be solved with integers $\{x_l\}$. The argument that we apply is a generalization of the one leading to (3.2). It yields the Smith canonical form of the coefficient matrix $C = (c_{kl})$:

Theorem 7 *There exist integral square matrices A and B of order m and n, respectively, such that* $\det A = \pm 1$, $\det B = \pm 1$, *and*

$$ACB = (g_k \delta_{k,l}). \tag{9.2}$$

Here $\delta_{k,l}$ is the Kronecker delta, and $g_k \ge 0$ depends only on C while

$$\begin{aligned} & g_k | g_{k+1}, \ 1 \le k \le r - 1, \\ & g_k = 0, r < k; \ r = \text{rank } C. \end{aligned} \tag{9.3}$$

Proof We may naturally assume that C is not a zero matrix. Let c_{uv} be any of the elements whose absolute values are equal to the minimum among non-zero $|c_{kl}|$. By (2.1), we take s_k, t_k, s_l', t_l' so that $c_{kv} = s_k c_{uv} + t_k$ and $c_{ul} = s_l' c_{uv} + t_l'$ with $0 \le t_k, t_l' < |c_{uv}|$. Then, for all $k \ne u$ and all $l \ne v$, we subtract s_k times the uth row from the kth row, and subsequently subtract s_l' times the vth column from the lth column. These row and column operations on C are realized by multiplying C on the left and the right by simple matrices which are of the same type as A and B, respectively. If any among t_k, t_l' is not equal to 0, then the resulting matrix has a non-zero element whose absolute value is less than $|c_{uv}|$; we proceed to a repetition of the procedure. Thus, we may assume, without loss of generality, that t_k, t_l' are all equal to 0; then, in the resulting matrix we exchange rows and columns to move c_{uv} to the upper-left corner, and further replace it by $|c_{uv}|$ if necessary; this procedure can also be realized by multiplying matrices as above. Hence, by induction, it suffices to deal with the situation in which C has been thus transformed into $(f_k \delta_{k,l})$, where $f_k > 0$ for $1 \le k \le r = \text{rank } C$ and $f_k = 0$ for $r < k$. Then, we have to show how to further transform the sequence $\{f_k\}$ into $\{g_k\}$ specified as in (9.3). To this end, we apply (3.6): There exist integers α, β such that $\alpha f_1 + \beta f_2 = \langle f_1, f_2 \rangle$. We note the identity

Divisibility

$$\begin{pmatrix} \alpha & 1 \\ -1+\alpha f_1' & f_1' \end{pmatrix} \begin{pmatrix} f_1 & 0 \\ 0 & f_2 \end{pmatrix} \begin{pmatrix} 1 & -f_2' \\ \beta & 1-\beta f_2' \end{pmatrix}$$
$$= \begin{pmatrix} \langle f_1, f_2 \rangle & 0 \\ 0 & f_1' f_2' \langle f_1, f_2 \rangle \end{pmatrix}, \tag{9.4}$$

where $f_1/\langle f_1, f_2 \rangle = f_1'$ and $f_2/\langle f_1, f_2 \rangle = f_2'$. We expand these four matrices in an obvious manner; note that after the expansion the first and the third matrices on the left side become matrices of the same type as A and B, respectively. Then we get a new $(f_k \delta_{k,l})$, where $f_1 | f_2$. We repeat the same with the pair f_1, f_3, getting a newer $(f_k \delta_{k,l})$, where $f_1 | f_2, f_1 | f_3$. In this way we are led to the situation $(f_k \delta_{k,l})$ with $f_1 | f_k$ for $2 \le k \le r$. Then, we apply the same transformation to the sequence $\{f_2, f_3, \ldots, f_r\}$, and so on. Eventually we reach (9.2)–(9.3), yet it remains to prove the uniqueness of the tuple $\{g_k\}$. For this sake, let $\{d_k^{(v)}\}$ be the set of all minors of order k of the original matrix C, and let d_k be their greatest common divisor. Then we have

$$d_k = g_1 g_2 \cdots g_k, \ 1 \le k \le r \, ; \text{ that is, } g_k = d_k/d_{k-1} \ (d_0 = 1). \tag{9.5}$$

To confirm this, let A, B be as in (9.2), and observe that each row of AC is an integral linear combination of the rows of C. Thus each minor of order k of AC is an integral linear combination of $\{d_k^{(v)}\}$. It is certainly a multiple of d_k. Next, each minor of order k of ACB is an integral linear combination of those of AC; and hence it is a multiple of d_k. Now, the greatest common divisor of the minors of order k of ACB is obviously equal to $g_1 g_2 \cdots g_k$; thus, by (7.2), we have $d_k | (g_1 g_2 \cdots g_k)$. Applying the same argument to $A^{-1}(ACB)B^{-1}$, we get $(g_1 g_2 \cdots g_k) | d_k$. This ends the proof.

The invariance of $\{d_k\}$ thus confirmed is a generalization of the automorphic property of the greatest common divisor of two integers that we mentioned in the ending lines of the second last paragraph of §5.

Returning to (9.1), we put

$$B^{-1} \cdot {}^t\{x_1, .., x_n\} = {}^t\{y_1, .., y_n\}, \quad A \cdot {}^t\{u_1, .., u_m\} = {}^t\{v_1, .., v_m\}. \tag{9.6}$$

Then we get the system of integral linear indefinite equations

$$g_k y_k = v_k, \ 1 \le k \le r \, ; \quad 0 = v_k, \ r < k, \tag{9.7}$$

which is equivalent to the original. The integer solution $\{y_k\}$ exists if and only if $g_k | v_k$ for $1 \le k \le r$ besides $v_k = 0$ for $r < k$. This is an extension of Theorem 6 and (8.2).

Notes

[9.1]   Theorem 7 is due to Smith (1861). The tuple $\{g_k\}$ is nowadays called the invariant factors of the integral matrix C; Smith himself did not give name to $\{g_k\}$. The usage of this term is confused in literature; see Note [30.7]. A detailed discussion of systems of integral linear indefinite equations over $\mathbb{Z}$ is developed in Bachmann (1898, Zweiter Abschnitt, Drittes Capitel).

[9.2]   If C is replaced by a matrix whose elements are polynomials over $\mathbb{C}$, then an analogue of the invariant factors is obtained by the polynomial division algorithm (see §67). An application of it is the transformation of square matrices over $\mathbb{C}$ into Jordanian canonical forms. That is, the basic theory of linear algebra is also related with Euclid's algorithm. See Gantmacher (1959, Chapter VI).

[9.3]   If the set $\{a_1, a_2, \ldots, a_K\}$ is coprime in the sense of (7.5), then there exists an integral square matrix Q with $\det Q = 1$, the first row of which is $\{a_1, a_2, \ldots, a_K\}$. To show this, we note that by the above theorem there exists a matrix B such that $(a_1, a_2, \ldots, a_K)B = (g_1, 0, \ldots, 0)$. Since g_1 is the greatest common divisor of $\{a_k\}$, we have $g_1 = 1$. Hence, if $\det B = 1$, then $Q = B^{-1}$. If $\det B = -1$, then change the sign of the second row of B^{-1}.

[9.4]   Let C be an arbitrary integral matrix. Transposing the argument of the preceding Note, we see that solely by row operation C can be transformed into the one whose first column is ${}^t\{v_1, 0, \ldots, 0\}$ with v_1 being the greatest common divisor of the elements of the original column; here $v_1 = 0$ is allowed. So, by induction, C can be transformed into an integral matrix which is upper-triangular. In particular, if C is a square matrix of determinant ± 1, then using row operations only, C can be transformed into an upper-triangular matrix whose diagonal entries are all equal to 1. Hence, further row operations yield the assertion that C is transformed into a unit matrix. Namely, any integral square matrix of determinant ± 1 is multiplicatively generated by those matrices, each of which corresponds to a row operation. Obviously the same can be said with respect to column operations. Whichever assertion is an extension of Theorem 4.

[9.5]   The set of all integral column vectors $\mathbf{x} = {}^t\{x_1, x_2, \ldots, x_n\}$, that is, $\mathbb{Z}^n$ is a typical Abelian group in terms of the vector addition and subtraction with integral coefficients. A set $\{\mathbf{x}_j : 1 \leq j \leq n\}$ is called a basis of $\mathbb{Z}^n$ if any $\mathbf{x} \in \mathbb{Z}^n$ can be uniquely expressed as $\mathbf{x} = \sum_j a_j \mathbf{x}_j$ with $a_j \in \mathbb{Z}$. For instance, the n columns of an integral square matrix of order n form a basis if and only if its determinant is equal to ± 1. To verify this, it suffices to see whether each of the columns ${}^t\{\delta_{1,j}, \delta_{2,j}, \ldots, \delta_{n,j}\}$, $1 \leq j \leq n$, can be expressed as an integral linear

24 *Divisibility*

combination of those of the given matrix. In other words, any integral matrix that transforms a basis into another is of determinant ± 1, and vice versa. By the way, the term Abelian group originated in Weber (1882, p.304).

[9.6] We shall give a few basic facts on Abelian groups here and in Notes [30.4]–[30.6], [51.2]–[51.6]: A free Abelian group G of rank n is a group isomorphic to $\mathbb{Z}^n$. Basis of G can be identified with that defined in the last Note. We assert that any subgroup H of G is also a free Abelian group. To confirm this, it is of course enough to treat $G = \mathbb{Z}^n$. First, Note [4.2] implies that $\{x_1 : \mathbf{x} \in H\} = h\mathbb{Z}$ with a certain integer $h \geq 0$. Let $\mathbf{h}$ be an element of H that yields h. Then $H = \mathbb{Z}\mathbf{h} \oplus H'$ (a direct sum/product), where $H' = \{\mathbf{x} \in H : x_1 = 0\}$. If $h = 0$, then $H = H'$. By induction, we reach $H \cong \mathbb{Z}^m$, $m \leq n$, which ends the confirmation. With this, let $\{\mathbf{c}_k \in G : 1 \leq k \leq m\}$ be a basis of H. We apply Theorem 7 to the $n \times m$ matrix $\{\mathbf{c}_1, \ldots, \mathbf{c}_m\}$ (note: here m, n are interchanged). We find that there exists a basis $\{\mathbf{y}_k\}$ of G such that $H = g_1\mathbb{Z}\mathbf{y}_1 \oplus \cdots \oplus g_m\mathbb{Z}\mathbf{y}_m$, $g_k|g_{k+1}$; note that $r = m$ here. The tuple $\{g_k\}$ is called the set of the invariant factors of the subgroup H, which is uniquely determined by H.

§10

We are now ready to introduce one of the most fundamental concepts in mathematics:

$$\text{an integer } p \geq 2 \text{ is a prime number } \Leftrightarrow \{1 < d < p : d|p\} = \varnothing. \quad (10.1)$$

Thus, if $p = ab$ with $a, b \in \mathbb{N}$, then one of a and b is equal to p, and the other to 1. We shall use the shortened terms a prime and primes in place of a prime number and prime numbers, respectively.

It is expedient to look into the traditional definition (10.1) from the viewpoint of coprimality (3.5). Thus, we note first the trivial fact that

$$\text{the integers which are coprime to all integers are } \pm 1. \quad (10.2)$$

On the other hand, any integer $q \geq 2$ is obviously not coprime to any element of $q\mathbb{Z}$. Hence, the following critical situation should be of special interest:

$$\text{an integer } \varpi \geq 2 \text{ is coprime to any integer not in } \varpi\mathbb{Z}. \quad (10.3)$$

These ϖ can be regarded as the most prime, non-trivial integers, which is a naïve expression though. We note that a positive divisor of ϖ is equal to either 1 or ϖ, for if $0 < d < \varpi$ is a divisor of ϖ, then $d \notin \varpi\mathbb{Z}$, which implies that $1 = \langle \varpi, d \rangle = d$. Conversely, if an integer $\varpi_1 \geq 2$ does not have any positive

$$\S10 \qquad\qquad 25$$

divisor which is greater than 1 and less than ϖ_1, then it satisfies (10.3). For if $n \notin \varpi_1\mathbb{Z}$, then $\langle n, \varpi_1 \rangle = 1$ since $\langle n, \varpi_1 \rangle$ is a divisor of ϖ_1 not equal to ϖ_1 as $\varpi_1 \nmid n$. In other words, one may define prime numbers by means of (10.3) instead of (10.1). That is, $\{\varpi\}$ is exactly the set of all primes.

Besides, primes are the most independent members of $\mathbb{N}$. To explain this claim, we observe that, with two positive integers which are coprime to each other,

$$\text{the divisibility by one of them} \atop \text{is independent of that by the other.} \qquad (10.4)$$

To clarify what we actually intend to assert here, we introduce a new notion: the least common multiple of two non-zero integers a and b is the positive integer that is the smallest among their common multiples. We denote it by $[a,b]$; any confusion with closed intervals should be avoided. Then, we have

$$a\mathbb{Z} \cap b\mathbb{Z} = [a,b]\mathbb{Z}. \qquad (10.5)$$

The left side coincides with the set of all common multiples of a, b; and if we divide any of these common multiples by $[a,b]$ in the sense of (2.1), then the residue is less than $[a,b]$ and still a common multiple; hence it is equal to 0 by (1.2). We get the equality in (10.5). Somewhat alternatively, one may argue as follows: according to Note [4.2] the left side is $t\mathbb{Z}$, and this t satisfies the definition of $[a,b]$, since $t = \min\{m \in (a\mathbb{Z} \cap b\mathbb{Z}) \cap \mathbb{N}\}$.

Further, it holds that

Theorem 8 *Given two non-zero integers a, b, we have*

$$\langle a,b \rangle [a,b] = |ab|. \qquad (10.6)$$

In particular, $\langle a,b \rangle = 1$ is equivalent to $[a,b] = |ab|$.

Proof Obviously we may suppose $a, b > 0$. Let $d = \langle a,b \rangle$ and $c = [a,b]$. By (6.2) (1) (2) we have $c = \langle ac/d, bc/d \rangle$. So $(ab/d)|c$ since ab/d is a common divisor of $ac/d, bc/d$; thus, $ab/d \leq c$. On the other hand ab/d is a common multiple of a, b, so $ab/d \geq c$. We end the proof. An alternative proof is given in [28.8].

Incidentally, we note that by the assertion (10.6) one may rewrite (9.4) as follows: there exist $A, B \in \Gamma$ such that

$$A \begin{pmatrix} f_1 & 0 \\ 0 & f_2 \end{pmatrix} B = \begin{pmatrix} \langle f_1, f_2 \rangle & 0 \\ 0 & [f_1, f_2] \end{pmatrix}. \qquad (10.7)$$

Also, observe that (10.6) provides a procedure to compute $[a,b]$ by means of Euclid's algorithm.

26 *Divisibility*

Returning to an explanation of (10.4), we note that in view of (10.6) we have

$$\langle a,b \rangle = 1, \ a,b > 0 \ \Leftrightarrow \ a\mathbb{N} \cap b\mathbb{N} = ab\mathbb{N}, \tag{10.8}$$

for $a\mathbb{N} \cap b\mathbb{N} = [a,b]\mathbb{N}$, a consequence of (10.5). Now, as an event taking place in $\mathbb{N}$, the probability of $a\mathbb{N}$ is equal to $1/a$ in an asymptotical sense since the number of integers divisible by a and less than a large $x \in \mathbb{N}$ is asymptotically equal to x/a. With this understanding, (10.8) states that the probability of the joint event of $a\mathbb{N}$ and $b\mathbb{N}$ is equal to the product of the respective probabilities, i.e., $1/ab = (1/a) \cdot (1/b)$. Hence, these events can be said to be independent of each other, adopting intuitively the independence notion in the general theory of probability.

Now, among the family of the events $\{n\mathbb{N} : n \geq 2\}$ inside $\mathbb{N}$, we have, for each prime p,

$$n\mathbb{N} \not\subset p\mathbb{N} \ \Rightarrow \ p\mathbb{N} \cap n\mathbb{N} = pn\mathbb{N}; \tag{10.9}$$

and $p\mathbb{N}$ may be regarded as a highly independent event, again adopting the independence notion in probability theory: The left side of the arrow means that $n \notin p\mathbb{N}$, so $\langle p,n \rangle = 1$; and (10.8) gives the right side. This closes our discussion on the fundamental definition (10.1).

Remark 1 It should, however, be stressed that the above independence property of prime numbers concerns strictly the divisibility inside $\mathbb{Z}$ or rather $\mathbb{N}$. Actually, there exist strong arithmetic interdependencies among prime numbers themselves; such an instance, an extremely important one, will be discussed in §57 and later on.

Remark 2 Moreover, prime numbers can cease to be prime, if looked in the extended algebraic environment. See Note [90.6].

Hereafter, we shall use the convention that

$$p, \text{ with or without subscripts, denotes a prime,} \tag{10.10}$$

unless otherwise stated.

Divisors which are primes are called prime factors, and a positive integer which either has two distinct prime factors or is divisible by the square of a prime is said to be a composite integer or just a composite. We have the classification

$$\mathbb{N} = \{1\} \sqcup \{\text{primes}\} \sqcup \{\text{composites}\}. \tag{10.11}$$

For if an integer $a > 1$ is not a prime, then by the contraposition of (10.1) there exists a b such that $b|a$ with $1 < b < a$; that is, $a = bc$ with $1 < b,c < a$,

$$\S10 \qquad\qquad 27$$

and by induction each of b, c is either a prime or a composite; hence a is a composite.

The wealth of number theory is based, to a great extent, on the following theorem of Euclid [Σ.IX, Prop.20]:

Theorem 9 *Let $\pi(x)$ denote the number of primes which do not exceed x a positive real number. Then we have*

$$\lim_{x\to\infty} \pi(x) = \infty. \qquad (10.12)$$

Remark 3 The famous notation $\pi(x)$ originated with Landau (1909, Vorwort).

Proof For an arbitrary $y \geq 2$, let

$$p(y) = \prod_{p\leq y} p + 1. \qquad (10.13)$$

According to (10.11), there exists a prime p' such that $p'|p(y)$. We have $p' > y$ since if $p' \leq y$, then dividing $p(y)$ by p' we get an immediate contradiction. Hence, there exists a prime number beyond any limit. This ends the proof.

If $\pi(x) < \infty$, then the multiplicative theory of integers would become just a collection of trivia.

Notes

[10.1]   In the Sumerian civilization which flourished for more than two thousand years well prior to the ancient Greek civilization, the division calculation was familiar among high scribes working at royal palaces and temples. For them, relatively small primes must have been the existence which they did not feel any necessity to specifically mention. Also, arithmetic was a major subject taught at schools in Sumer; see Nemet-Nejat (2002, Chapter 4). Experiences and observations on divisibility in times past were purified and crystalized by ancient Greek philosophers, and eventually the definition (10.1) was attained.

[10.2]   The reason that the integer 1 is not counted as a prime is contained in the discussion following (10.3), but more explicitly it is stated just after the proof for Theorem 10 in the next section.

[10.3]   Euclid placed the definitions of primes and coprimality in tandem, i.e., [Σ.VII, Defs.11, 12]; see also Note [11.1]. The integer 1 is not treated as a prime, although not stated explicitly. In the definition of divisors [Σ.VII, Def.3], he was concerned with the divisibility by a positive integer less than the integer in question. His divisor is an aliquot part in classical terms, that is, a proper divisor of today. Hence, his definition – a prime number is that which is measured by a unit alone – coincides with (10.1); observe that this and (10.3) match perfectly, for $\varpi\mathbb{N}$ obviously does not contain any integer k such that $1 < k < \varpi$. However, his definition of primes was not carried through centuries without intricacies. The following two beliefs, both of which are fallacies in modern mathematics, appeared time and again:

(a) 1 is a prime   (b) 2 is not a prime.

Concerning (a), there exists an extensive search by Caldwell et al (2012). We spotted also that the famous Goldbach conjecture, which is today understood to be the surmise that every even integer greater than 5 be a sum of two odd primes, was originally formulated with (a) (1742: Fuss (1843, Tome I, p.127)); Euler employed (a) in his unfinished manuscript (1849a, p.505) written around 1750, although later (1771, p.17) he adopted (10.1); Legendre (1798) employed (10.1) in p.6 but (a) in his table of primes (p.20); further, the epochal table of primes by Lehmer, D.N. (1914) begins with (a). Nonetheless, those misunderstandings never caused any serious flaw in the discussions of respective authors. As for (b), the text book (ca 100 CE) by Nicomachus (Gerasa) contains a relevant statement (see Note [19.3] as well). This book was translated into Latin by Boethius (ca 500) and circulated in Europe for more than 1,300 years without any correction being attempted on the fallacy – primes are odd integers. See its most recent edition (1867, p.30) as well as D'Ooge (1926, p.202). Far from such historical indifference to the true contents of (10.1), Euclid's view was ever standing out, for he treated 2 as a prime: a clear evidence for this is his reasoning in [Σ.IX, Prop.36], details of which are in our Notes [16.5]–[16.6].

[10.4] Nevertheless, the integer 2 is indeed a peculiar prime. Situations complicated by the fact that 2 is a prime are plenty. Such instances in the present essays are in Theorems 37, 39 as well as in §§72, 91.

[10.5] Although Euclid did not develop any specific discussion concerning his definition (10.1) itself, all facts that are needed to explain (10.3)–(10.9) are stated in [Σ.VII]. Thus, $p \nmid a \Rightarrow \langle p, a \rangle = 1$ (Prop.29); the relation between the greatest common divisor and the least common multiple of two integers (Prop.34); the relation between the least common multiple and any common multiple of two integers (Prop.35).

[10.6] The determination of whether a given integer $n \geq 2$ is a prime or a composite can be done by dividing n by integers less than or equal to $\sqrt{n}$ since $n = ab$ with $1 \leq a \leq b$ implies $a \leq \sqrt{n}$. Gauss put it *vulgo notum est* [DA, art.330]. See also Legendre (1785, p.7). However, this is never a practical method, for it demands a great amount of computation if n is large. In principle, any criterion should yield its conclusion without too much efforts. Here arises the issue to find a method to test any given integer for primality in polynomial time or rather with computation as little as possible. This is called the problem of primality test. As a matter of fact, a theoretical solution has already been achieved. See Notes [46.2]–[46.3].

[10.7]   As Dedekind (Dirichlet (1871, p.440)) observed, we are able to redefine the coprimality of a and b by the validity of the equality on the right side of (10.8). This new definition quickly yields the same conclusion as Theorem 5: If $a|bc$, then $bc \in a\mathbb{N}$; and $bc \in a\mathbb{N} \cap b\mathbb{N}$, which means $bc \in ab\mathbb{N}$; hence $ab|bc$, or $a|c$.

[10.8]   We identify the proof of Theorem 9 with that of [Σ.IX, Prop.20], and attribute the theorem to Euclid; however, note that he never applied this theorem in his number theory. He treated the case of the product of three primes: an illuminating argument is set out without frills. By the way, one of the earliest instances of the use of an expression close to (10.13) is in Prestet (1689, Premier vol., p.162), although a blemish related to (a) above is found on p.141.

[10.9]   Those p(p) which are primes are called primorial primes, a term analogous to factorial. An example is p(31) = 200560490131.

[10.10]   The extension of Euclid's use of (10.13) to arithmetic progressions is possible but only on a strongly restrictive condition: Murty (1988) proved that a Euclidean proof exists for the arithmetic progression $\{qn + h : n \in \mathbb{N}\}$, if and only if $q|(h^2 - 1)$.

§11

From what we have discussed so far, it transpires that prime numbers must be placed in the core of the theory on divisibility in $\mathbb{Z}$. Indeed we have the following assertion, or the fundamental theorem of arithmetic on which the multiplicative number theory stands:

Theorem 10 *Any integer $n \geq 2$ can be expressed as a product of powers of different prime numbers:*

$$n = p_1^{\alpha_1} p_2^{\alpha_2} \cdots p_J^{\alpha_J}, \quad \alpha_j \in \mathbb{N}. \tag{11.1}$$

The representation is unique up to the order of the factors.

30   *Divisibility*

Remark 1 This is the same as what Gauss stated in [DA, art.16]. We shall use the abbreviation a prime power to indicate a power of a prime. So this assertion can be called the unique prime power decomposition theorem as well.

Proof We quote first [Σ.VII, Prop.30]:

$$\text{if } p|ab, \text{ then either } p|a \text{ or } p|b. \tag{11.2}$$

This is a consequence of (6.1) since $p \nmid a$ implies that $\langle p, a \rangle = 1$. To show the possibility of a prime power decomposition of n, we may assume of course that n is not a prime but a composite. Thus, $n = kl$ with certain k, l such that $1 < k \leq l < n$. By induction, a decomposition of n follows from those of k, l (see [Σ.VII, Prop.31]). As for the uniqueness of the decomposition, we note that if $n = p_1^{\alpha_1} p_2^{\alpha_2} \cdots p_J^{\alpha_J} = p_1'^{\beta_1} p_2'^{\beta_2} \cdots p_K'^{\beta_K}$, then the prime p_1 must be equal to one of the primes $p_1', \ldots, p_K'$; the reason for this is in (11.2). Since $n/p_1 < n$, we may proceed by induction: from the uniqueness of the decomposition of n/p_1, the same assertion on n follows. We end the proof.

Here one should note that the premise $p \geq 2$ in (10.1), that is, $p \neq 1$ is paramount in order to get the above argument to go through.

In this way, we have now come to the point where arithmetic meets analysis, that is, Euler's proof of Theorem 9: It holds that for any $y \geq 2$

$$\prod_{p \leq y} \left(1 - \frac{1}{p}\right)^{-1} > \sum_{n \leq y} \frac{1}{n}. \tag{11.3}$$

To see this, we note that since

$$\left(1 - \frac{1}{p}\right)^{-1} = 1 + \sum_{v=1}^{\infty} \frac{1}{p^v}, \tag{11.4}$$

the left side of (11.3) is expanded into

$$1 + \sum_{j=1}^{\pi(y)} \sum_{p_1 < p_2 < \cdots < p_j \leq y} \sum_{v_1, v_2, \ldots, v_j = 1}^{\infty} \frac{1}{p_1^{v_1} p_2^{v_2} \cdots p_j^{v_j}}, \tag{11.5}$$

which is greater than the right side because (11.1) implies that every term $1/n$, $n \leq y$, appears at least once in (11.5). Since the harmonic series diverges, the left side of (11.3) cannot remain finite as y tends to infinity. This proves (10.12).

It should be observed that the uniqueness assertion on (11.1) is not needed in order to have (11.3); the possibility of the decomposition suffices, which is the same as with Euclid's proof of (10.12).

$$\S11 \qquad\qquad 31$$

The original version of the above argument is in Euler (1737b, Theorema 7). This thesis marked a truly seminal development in the entire history of mathematics because of the introduction of the zeta-function

$$\zeta(s) = \sum_{n=1}^{\infty} \frac{1}{n^s}, \quad \mathrm{Re}\, s > 1, \tag{11.6}$$

and because of the discovery of the Euler product expansion (Theorema 8)

$$\zeta(s) = \prod_{p} \left(1 - \frac{1}{p^s}\right)^{-1}, \quad \mathrm{Re}\, s > 1, \tag{11.7}$$

where p runs over all primes, although Euler considered only the case with $s \geq 1$ [sic] being an integer. If this product is restricted to those p not greater than y and expanded, then all the terms, which are s^{th} powers of those in (11.5), appear and the series converges to the expression (11.6) as y tends to infinity, provided $\mathrm{Re}\, s > 1$, because every term $1/n^s$ appears exactly once by virtue of the uniqueness assertion in the last theorem, and $\sum_{n=1}^{\infty} 1/n^{\sigma}$ is finite for each $\sigma > 1$. Therefore, the equality between (11.6) and (11.7) holds. Conversely, this equality implies the uniqueness of the decomposition (11.1); here we actually require the analytical uniqueness assertion (15.10) below, however. In passing, we remark that the absolute convergence of the infinite product (11.7) implies a very basic fact on the zeros of the zeta-function:

$$\zeta(s) \neq 0, \quad \mathrm{Re}\, s > 1, \tag{11.8}$$

since $\zeta(s) \prod_{p \leq y} (1 - p^{-s})$ is close to 1 if y is taken sufficiently large depending on s; see Note [12.6].

Remark 2 This use of the symbol s for a complex variable is made in accord with a tradition in analytic number theory. See Note [15.4]. Also the use of the notation $\zeta(s)$ to discuss the works prior to its introduction by Riemann (1860) is made of course for the sake of convenience.

Euler (1737b, Theorema 19) claimed that

$$\sum_{p \leq x} \frac{1}{p} \sim \log \log x, \quad \text{as } x \to \infty, \tag{11.9}$$

in which a minor modification has been applied to what he stated, i.e., (11.10); naturally the notation $f(x) \sim g(x)$ means $\lim f(x)/g(x) = 1$ (see Note [12.8]). To reach this, he rearranged terms of the logarithm of the right side of (11.7) with $s = 1$ [sic] in a way and indicated (11.9) symbolically (1737b, the ending statement):

$$\sum_p \frac{1}{p} \sim \log\left(\sum_n \frac{1}{n}\right). \tag{11.10}$$

We shall confirm (11.9) in §13, but in an alternative way. In Euler (1748b, Tomi primi, Caput XV) which is an expanded version of the thesis (1737b), the claims

$$\prod_p \left(1 - \frac{1}{p}\right)^{-1} = \infty, \quad \prod_p \left(1 - \frac{1}{p}\right) = 0. \tag{11.11}$$

are also made. For the latter see (95.18).

Even though Euler's argument for (11.7), (11.9)–(11.11) lacks both the uniqueness part of Theorem 10 and the rigorous notion of the convergence and the divergence of infinite sums and products, it is unquestionably deeper than Euclid's. From the product formula (11.7) a multitude of mathematical wealth is still flowing, with no indication of exhaustion. We greatly admire Euler's far-reaching insight.

Notes

[11.1] It should be stressed, particularly at this stage, that for an arbitrarily given n the identity (11.1) is no less easy to state but it can be extremely hard to execute the explicit decomposition. How to effectively achieve the decomposition (11.1) is a fundamental problem in number theory, and called the integer factorization problem. We shall develop very basic discussions on it in §§38, 46–49, and briefly elsewhere. Modern methods of integer factorization can be described as tours de force of the bests of number theory, algebra, and technology. However, those methods have all come out of truly elementary yet penetrating ideas, as is to be indicated in §49. Euclid seems to have put more weight in the concept of coprimality ([Σ.VII, Def.12]) than in that of prime numbers ([Σ.VII, Def.11]). The very fact that Euclid started his discussion (see Note [3.1]) with a relation between his algorithm and the notion of coprimality suggests that he was probably aware of the fundamental difficulty of integer factorization. Ancient Greeks inherited this difficulty from Babylonians and left it to us. In any event, it is a far-reaching view to place the two concepts primality and coprimality in tandem.

[11.2] The recognition of Theorem 10 can be traced back at least to al-Fārisī (ca 1300), according to Ağărgün–Özkan (2001). Continued in Note [16.6].

[11.3] The proof of (11.2) by Legendre (1798, pp.4–5): By a combination of (2.1) and (10.1), i.e., without recourse to (3.6), he showed the contraposition of (11.2), that is, if $p \nmid a_1$ and $p \nmid a_2$, then $p \nmid a_1 a_2$. Thus, first a_1, a_2 are divided by p in the sense (2.1); the remainders a_3, a_4 are such that $0 < a_3, a_4 < p$.

$$\S 11 \qquad\qquad 33$$

If it is assumed that $p|a_1a_2$, then $p|a_3a_4$; in particular, neither a_3 nor a_4 is equal to 1 (because of (1.2)). Next, p is divided by a_3, again in the sense (2.1), so that $p = a_3q + r_3$ with $0 < r_3 < a_3$ because of (10.1). Multiplying both sides by a_4 gives $p|a_4r_3$, and similarly $p|r_3r_4$ with r_4 corresponding to a_4. Neither r_3 nor r_4 is equal to 1. Obviously this procedure cannot be repeated indefinitely. Therefore, the assumption $p|a_1a_2$ is refuted, which ends the proof. Then, Legendre proceeded to the prime power decomposition (11.1) without explicitly mentioning the uniqueness, but see the top lines of Legendre (1798, p.8).

[11.4]   It is commonly asserted nowadays that Gauss [DA, art.16] established Theorem 10 for the first time in history. In this context, the preceding note should raise special interest because Gauss [DA, arts.13–14] stated a proof of (11.2) that is virtually the same as Legendre's, and proceeded to Theorem 10. The difference between their contributions to the matter lies just in that Gauss pointed out the uniqueness of the decomposition while Legendre did not, although the latter clearly indicated it by means of its proper use in numerous examples. In any event, it transpires that neither Legendre nor Gauss was content with the proof of [Σ.VII, Prop.20], or the corresponding assertions in the editions available to them: Euclid made [Σ.VII, Prop.30] depend on [Σ.VII, Prop.20] just in the same way as we did in the above; however, we had needed to provide [Σ.VII, Prop.20] with an alternative proof (in §6). Legendre and Gauss used instead a direct reasoning to dispense with [Σ.VII, Prop.20], which has a rather interesting implication as indicated in the next Note. Gauss, throughout [DA], did not quite explicitly apply Euclid's algorithm; this point is continued in Note [11.6].

[11.5]   Putting these episodes aside, we observe that the whole discussion in [DA] is standing on arts.13–16. However, Gauss' procedure in [DA] was different from ours: after stating an assertion equivalent to our Theorem 10, he proceeded to the notions of greatest common divisors, least common multiples, and so on; see (13.4).

[11.6]   Gauss stated in [DA, art.14] that the discussion in arts.13–14 should make it easier to understand the argument that he would develop in order to deal with the problems which he described as far more difficult: Apparently, he regarded arts.13–14 as a logical preparation for *Caput octavum* which he had to give up to include into [DA] because of a non-academical reason ([DA], pp.7–8). Glimpses of the contents of that missing chapter are in [DA, arts.61 and 335]: in the latter section they are described as *varia congruentiarum genera*, that is, in the modern terminology, a theory of polynomials over finite fields; a relevant basic discussion in the present essays is to be developed in §§64–67. In Gauss' posthumous manuscript (*Werke* II-1, pp.212–242) of

34   *Divisibility*

Caput octavum, the basis of the theory is placed in an analogue of Theorem 10 above. However, to establish it, an extension to polynomials of Euclid's algorithm exactly in the formulation (3.6) is employed; see arts.334–336 there. Obviously, Gauss made a basic change to his plan divulged in [DA. art.14]. By the way, Dirichlet's *Vorlesungen* (1863, §§1–4) begins with the notion of divisibility, the division algorithm, and an account of Euclid's algorithm (3.6), although not very explicitly yet. That signifies anyway the moment in history when number theory finally returned, in printed media, to the spirit of Euclid after more than two millennia. In today's common view, solely Euclid's algorithm is capable to open a really wide perspective.

[11.7]   One may ask if there exists any proper subset of $\mathbb{N}$ which is obtained by certain algebraic twists and still contains many primes. This problem was set out by Euler (1772f). He observed that the values of the polynomial $41 - x + xx$ for positive integers x up to 40 are all primes. See Note [77.7], and also Dress and Olivier (1999). By the way, quite an opposite situation has been discovered by McCurley (1984): The least positive integer x with which $x^{12} + 488669$ attains a prime value is 616980; this prime is of 70 decimal digits. In any event, any polynomial $f(x) \in \mathbb{Z}[x]$ cannot take prime values for all sufficiently large $x \in \mathbb{N}$. To show this, it suffices to consider $f(nf(a) + a)$, $a, n \in \mathbb{N}$; see Euler (1760, p.357, Theorema).

[11.8]   Bouniakowsky (1857) conjectured that if $f(x) \in \mathbb{Z}[x]$ satisfies the following three conditions, then there should exist infinitely many primes in the sequence $\{f(n) : n \in \mathbb{N}\}$:

(1) the coefficient of the highest power of x in $f(x)$ is positive;
(2) $f(x)$ is irreducible over $\mathbb{Q}$;
(3) 1 is the unique common divisor of all $f(n)$.

For the notion of being irreducible see the statement of Theorem 62. When f is linear, this is a conjecture, i.e., our (69.1), of Legendre concerning arithmetic progressions; it has been settled affirmatively by Dirichlet as we shall show in §88[A], but with a method different from his. When the degree of f is greater than 1, the above still remains unsolved; it is regarded as one of the central problems concerning the distribution of primes. By the way, the necessity of (3) is explained by Bouniakowsky with the example $f(x) = x^9 - x^3 + 2520$, which is non-trivial. Continued in Note [67.4].

[11.9]   However, if integral polynomials of two or more variables are taken into account, then the situation changes drastically. For instance, Legendre (1830, II, pp.102–104) conjectured that any integral binary quadratic form in integral variables should capture infinitely many prime values under an obvious prescription. This has turned out to be correct and has become the Dirichlet–Weber prime number theorem that is our Theorem 99, §93.

[11.10]   It must be a surprise that there exists an integral polynomial in many variables over $\mathbb{N} \cup \{0\}$, the set of whose positive values coincides exactly with the set of all prime numbers. This is a consequence of the MRDP Theorem, a negative answer to Hilbert's tenth problem. According to Jones et al (1976), $(a + 2)(1 - M)$ is such a polynomial, where M stands for the following polynomial of 26 integral variables $a, b, \ldots, z \geq 0$ (temporarily the convention (10.10) is dropped):

$$
\begin{aligned}
M = {} & \left(wz + h + j - q\right)^2 + \left((ga + 2g + a + 1)(h + j) + h - z\right)^2 \\
& + \left(2n + p + q + z - e\right)^2 + \left(16(a + 1)^3(a + 2)(n + 1)^2 + 1 - f^2\right)^2 \\
& + \left(e^3(e + 2)(k + 1)^2 + 1 - o^2\right)^2 + \left((k^2 - 1)y^2 + 1 - x^2\right)^2 \\
& + \left(16r^2y^4(k^2 - 1) + 1 - u^2\right)^2 + \left(n + l + v - y\right)^2 \\
& + \left(((k + u^2(u^2 - k))^2 - 1)(n + 4dy)^2 + 1 - (x + cu^2)^2\right)^2 \\
& + \left((k^2 - 1)l^2 + 1 - m^2\right)^2 + \left(ki + a + 1 - l - i\right)^2 \\
& + \left(p + l(k - n - 1) + b(2kn + 2k - n^2 - 2n - 2) - m\right)^2 \\
& + \left(q + y(k - p - 1) + s(2kp + 2k - p^2 - 2p - 2) - x\right)^2 \\
& + \left(z + pl(k - p) + t(2kp - p^2 - 1) - pm\right)^2.
\end{aligned}
$$

§12

The argument proving the fundamental identity (11.7) implies well that the assertion is an analytic expression of the arithmetical Theorem 10; here analytic indicates certain involvement of limit processes and arithmetic of finite constructions, both in an intuitive sense. One of our principal aims in the present essays is to appreciate the unification of arithmetic and analysis that is encapsulated in (11.7). However, we ought to develop a predominately arithmetical discussion until the end of §91, which might obscure our intension. Therefore, we make here a détour to a cursory exposition of the preliminary part of what we shall develop later in the analysis oriented context, mainly in Chapter 5:

Thus, Euler is said to have fused number theory and analysis by means of (11.7), opening a great mathematical field. This is very correct; indeed we have stated the same in the ending part of the text of the preceding section.

36 Divisibility

However, he dealt in fact with $s \in \mathbb{Z}$ mostly (details in Note [12.1]). It is a sharp difference from the modern thoughts, though historically inevitable. Nowadays, the true nature of the zeta-function is thought to be revealed only if it is considered as a creature living over the whole complex plane, which was not in Euler's view. In this context, the genuine analytic approach to the zeta-function was initiated by Riemann (1860); its sheer ten pages founded a firm paradigm in the theory of the distribution of prime numbers. Prime numbers are gems tightly adorning the visible tip of a huge placid entity. The notation $\zeta(s)$ was minted by Riemann, while the term *Riemann'schen Zetafunction* originated with Landau (1909, Vorwort, p.VI).

The function $\zeta(s)$ over $\mathbb{C}$ is defined by (11.6) for $\operatorname{Re} s > 1$ at the outset. It is regular there and admits the representation

$$\zeta(s) = \frac{s}{s-1} - s \int_1^{\infty} (x - [x]) \frac{dx}{x^{s+1}}. \tag{12.1}$$

This can be confirmed by dividing the range of integration into unit intervals $[n, n+1)$, $n \in \mathbb{N}$, and by computing explicitly each resulting integral. The above integral itself converges absolutely for $\operatorname{Re} s > 0$, and hence $\zeta(s)$ continues analytically (meromorphically) to this half of the complex plane. Euler's claim (11.11) can be understood as a consequence of the fact that $\zeta(s)$ has a pole at $s = 1$, which is evident in (12.1). A little more precisely, we have, with the Euler constant $c_E = 0.57721\ldots$,

$$\zeta(s) = \frac{1}{s-1} + c_E + O(|s-1|), \tag{12.2}$$

for

$$\lim_{s \to 1} \left(\zeta(s) - \frac{1}{s-1} \right) = 1 - \lim_{M \to \infty} \int_1^M (x - [x]) \frac{dx}{x^2}$$
$$= \lim_{M \to \infty} \left(\sum_{m=1}^M \frac{1}{m} - \log M \right). \tag{12.3}$$

As for the usage of the O-symbol, see Note [12.8].

Riemann started his discussion not with (12.1) but with the representation by Chebyshev (1851, §2):

$$\zeta(s) = \frac{1}{\Gamma(s)} \int_0^{\infty} \frac{x^{s-1}}{e^x - 1} dx, \quad s > 1, \tag{12.4}$$

where $\Gamma(s)$ is the Gamma-function whose basic properties are given in our §94; an inversion of the order of summation and integration involved here is justified by absolute convergence. Riemann transformed (12.4) into

$$\zeta(s) = \frac{-1}{\Gamma(s)(e^{\pi is} - e^{-\pi is})} \int_C \frac{(-x)^{s-1}}{e^x - 1} dx, \tag{12.5}$$

where the contour C starts at $+\infty$ on the real axis, goes down straightly to the point $\frac{1}{1859}$, encircles the origin once counterclockwise, and returns to $+\infty$ along the real axis, while $\arg(-x)$ changes continuously from $-\pi$ to π. By the way, C is called the Hankel path; see (94.31). The new integral converges absolutely for all $s \in \mathbb{C}$, and defines an entire function of s. Namely, (12.5) yields instantaneously an analytic continuation of $\zeta(s)$ to the whole of complex plane; here we ought to quote (94.28). Then, imposing the restriction $\operatorname{Re} s < 0$, Riemann expanded the circular part of C to infinity; by the theorem of residues he got

$$\int_C \frac{(-x)^{s-1}}{e^x - 1} dx = -2i(2\pi)^s \sin\left(\tfrac{1}{2}\pi s\right) \sum_{n=1}^{\infty} n^{s-1}, \quad \operatorname{Re} s < 0, \tag{12.6}$$

whereby he stated the functional equation (12.7) below; also, via the properties (94.25) and (94.30) of $\Gamma(s)$, he noticed that it can be expressed as (12.8):

Theorem 11 *The zeta-function $\zeta(s)$ defined by (11.6) continues analytically to the whole of $\mathbb{C}$ save for a simple pole with residue 1 at $s = 1$, and satisfies the functional equation*

$$\zeta(1 - s) = 2(2\pi)^{-s} \cos\left(\tfrac{1}{2}s\pi\right) \Gamma(s)\zeta(s) \tag{12.7}$$

or equivalently,

$$\pi^{-(1-s)/2} \Gamma\left(\tfrac{1}{2}(1 - s)\right) \zeta(1 - s) = \pi^{-s/2} \Gamma\left(\tfrac{1}{2}s\right) \zeta(s). \tag{12.8}$$

Remark It is customary to assert at this point some facts on the values of the zeta-function at negative integers. However, we skip it. See Note [60.4] as well as (94.7).

Riemann depicted then a sublime interplay, i.e., (94.11) in today's expression, between the prime numbers and the zeros of $\zeta(s)$, most spontaneously as if he had needed no effort to reach his groundbreaking viewpoint; see also its very brief description given in the later part of §17.

Based on the principle that Riemann thus rendered, Hadamard (1896) and de la Vallée Poussin (1896) independently established, as to be precisely explained in our §§94–95, the prime number theorem:

38   *Divisibility*

Theorem 12 *Let $\pi(x)$ be as in* (10.12). *Then we have*

$$\lim_{x \to \infty} \frac{\pi(x)}{x/\log x} = 1. \tag{12.9}$$

With this, some historical details are in order:

(1) The theorem (12.9) was conjectured, explicitly for the first time in history, by Legendre (1798, pp.18–19, footnote; 2nde éd., 1808, pp.394–398; 3e, éd., 1830, t. II, pp.65–70). More precisely, his conjectural statement (1808/1830) is

$$\pi(x) = \frac{x}{\log x - 1.08366} \quad \text{asymptotically,} \tag{12.10}$$

in modern notation. When published, this became a sensation. For instance, N. Abel wrote in a letter of his: ... *qui est certes le plus merveilleux de toutes les mathématiques* (Holst et al (1902, p.5)). By the way, the letter is dated *Année* $\sqrt[3]{6.064.321.219}$, i.e., August 4, 1823.

(2) Gauss conjectured (12.9) in his late teens: Werke II-1, p.444 and X-1, p.11. His view on the nature of $\pi(x)$ must have been known among people in close academical contact with him; see (4) below and Note [95.4].

(3) According to the footnote on p.372 of Dirichlet's Werke I, which was inserted by the editor Kronecker, Dirichlet had left a hand-written note on a page of the relevant complimentary offprint (1838) for Gauss: it states that the expression

$$\sum_{2 \le n \le x} \frac{1}{\log n} \tag{12.11}$$

should give a better approximation to $\pi(x)$. This is essentially equivalent to the use of the logarithmic integral

$$\mathrm{li}(x) = \int_2^x \frac{du}{\log u}, \quad x \ge 2, \tag{12.12}$$

as the summation by parts implies; see Note [13.8] below.

(4) Gauss' letter (Werke II-1, pp.444–447) of Christmas Eve, 1849, contains the claim that since the number of primes appears to decrease in the logarithmic rate, one should take $\int dn/\log n$ as an approximation to $\pi(x)$. He made a comparison between Legendre's and his own conjectures based on a numerical data. In any event, neither Legendre nor Gauss seems to have done more than empirical examination of the distribution of prime numbers.

(5) Chebyshev started his memorable article (1851; presented in 1848) praising Legendre and set out a decisive conclusion:

provided the existence of the limit on the left side of (12.9),
the equality should hold. $\tag{12.13}$

$$\S12 \qquad\qquad 39$$

For the proof, Chebyshev employed $\zeta(s)$, $\log \zeta(s)$ and their higher derivatives with $s > 1$, a fact that is of great significance because it pointed out, for the first time in history, a feasible way to a rigorous proof of the prime number theorem via the theory of the zeta-function, although as a matter of fact he never used any notation like $\zeta(s)$; instead he relied solely on the equality between the right sides of (11.6)–(11.7). It should be stressed the fact that Chebyshev tried his best to indicate that the logarithmic integral $\mathrm{li}(x)$ be chosen as the true approximation to $\pi(x)$. Moreover, in the work (1854; presented in 1850), he rigorously established besides other notable things, by means of a simple sieve procedure combined with careful numerical estimation, that there exist absolute constants c_1, $c_2 > 0$ such that for large enough x

$$c_1 \frac{x}{\log x} < \pi(x) < c_2 \frac{x}{\log x}. \tag{12.14}$$

He estimated c_1 to be close to 0.921292 and c_2 to $6c_1/5$. This was the unprecedented quantitative assertion on $\pi(x)$ that was established after more than 2100 years since Euclid's crude statement (10.12); see the ending lines of Note [19.6] below. While these are already impressive contributions, his article contained a fundamental mathematical legacy, which we shall be able to make precise in the last paragraph of §17. We note in addition that he began his article (1851) with (12.4); although he did not say anything about this, it should be worth mentioning that (12.4) is obviously equivalent to the integral representation

$$\zeta(s) = \frac{1}{\Gamma(s)} \int_0^1 \frac{(\log 1/x)^{s-1}}{1-x} dx, \quad s > 1, \tag{12.15}$$

due to Dirichlet (1837a, §2). Thus, Dirichlet considered the zeta-function solely for $s \in (1, \infty)$, and Chebyshev did the same; in the meantime the theory of analytic functions was started by Cauchy (1825). By the way, discussion in the next section contains alternative proofs of (12.13) and (12.14), although the latter with less precise constants.

(6) The article by Riemann (1860) was a true revolution in mathematics. Riemann was apparently inspired by Chebyshev (1851). The use of (12.4) in place of (12.15) might look as a meager matter. In fact, it led Riemann to taking x to $\mathbb{C}$, which is decisive in much the same sense as his taking s to $\mathbb{C}$. The reason for this view of ours is in that the representation (12.5) or rather its minor modification

$$\zeta(s) = \sum_{n \leq N} \frac{1}{n^s} + \frac{-1}{\Gamma(s)(e^{\pi i s} - e^{-\pi i s})} \int_C \frac{(-x)^{s-1} e^{-Nx}}{e^x - 1} dx, \tag{12.16}$$

40 *Divisibility*

with arbitrary $s \in \mathbb{C}$, $N \in \mathbb{N}$, made it possible for Riemann, in his *Nachlass* (Siegel (1932)), to employ the method of steepest descent, or the saddle point method which resulted in an amazingly sharp evaluation of $\zeta(\frac{1}{2}+it)$ for $t \in \mathbb{R}$; the saddle point is close to $(t/N)i$, and the optimal choice is $N = [(|t|/2\pi)^{1/2}]$. Indeed, having numerical evidences thus obtained in background, Riemann asserted that all zeros of $\zeta(s)$ in

$$\text{the critical strip } 0 \leq \text{Re}\, s \leq 1 \tag{12.17}$$

should probably be on the straight line, or the critical line $\text{Re}\, s = \frac{1}{2}$; here the word critical is used since for $\text{Re}\, s > 1$ and thus for $\text{Re}\, s < 0$ because of (12.7) the nature of $\zeta(s)$ is relatively clear, that is to say, the mystery is with s in (12.17). This is the second conjecture stated in Riemann's article, and called the Riemann Hypothesis today:

$$\text{RH}: \quad \text{If } \zeta(\rho) = 0 \text{ with } 0 \leq \text{Re}\, \rho \leq 1, \text{ then } \text{Re}\, \rho = \tfrac{1}{2}. \tag{12.18}$$

Its proof should lead us to an extraordinary conclusion on the distribution of primes: for $x \geq 2$

$$\text{RH} \iff \pi(x) = \text{li}(x) + O(x^{1/2} \log x), \tag{12.19}$$

the depth of which can be comprehended only after studying Chapter 5 and beyond. Note that Riemann himself did not state any consequence of RH. Despite of modern numerical evidences of gigantic magnitudes, RH still appears as if hidden deep in a terra incognita, although whether or not such a description of the hypothesis is appropriate will become apparent only when its proof is actually achieved. See also Notes [95.4]–[95.5]. Other conjectures stated by Riemann will be treated in §§94–95. In passing, we mention, for the sake of a later purpose, a triviality:

$$\zeta(s) \neq 0, \quad 0 < s < 1, \tag{12.20}$$

since (12.1) implies $\zeta(s) < 0$ on this interval. Alternatively, one may use the relation

$$(1 - 2^{1-s})\zeta(s) = \sum_{n=1}^{\infty} \frac{(-1)^{n-1}}{n^s} > 0, \quad s > 0, \tag{12.21}$$

where the sum converges conditionally.

Thus, in the first place, we have two formulas (11.6) and (11.7) for the zeta-function. There exists, however, a sharp difference between them. In the former we have a sum over all positive integers, and in the latter an infinite product over all primes. On the one hand, we see that $\zeta(s)$ is a highly smooth function throughout the complex plane, save for the unique pole at $s = 1$.

$§12$ 41

This smoothness, that is, the analyticity throughout $\mathbb{C} \setminus \{1\}$, comes from the fact that each member of $\mathbb{N}$ appears evenly in (11.6), without which it should obviously be impossible to connect $\zeta(s)$ to (12.1)–(12.8). It is known (§94) that the most essential analytic properties of $\zeta(s)$ are confined in the nature of (12.7)–(12.8) inside the critical strip (12.17). On the other hand, (11.7) is far less transparent than (11.6) since the structure of the sequence of all prime numbers is almost totally unknown if compared with that of all natural numbers, and moreover the region where one can be certain of the validity of (11.7) without any specific reasoning is only the half plane $\mathrm{Re}\,s > 1$, which is disjoint from the critical strip. Nevertheless, if it were not for (11.7), it would be impossible to connect $\zeta(s)$ to the set of all prime numbers. To investigate the distribution of prime numbers, one has anyway to combine these two sharply different appearances of the zeta-function, its Euler product expansion for $\mathrm{Re}\,s > 1$ and its analytic expression for $\mathrm{Re}\,s < 1$.

Here we repeat what is said in Note [11.1]: We often develop discussions in number theory taking the decomposition (11.1) for granted; however, in reality it can be extremely hard and laborious, if possible, to achieve this factorization for huge integers. That is to say, those facts, including (11.7), which are equivalent to the generic formula (11.1), should rather be regarded as abstract or theoretical. Therefore, if use of (11.7) leads us to a result that reveals a fact on the distribution of prime numbers so plain to grasp that it does not suggest any trace of the use, then one may assert that a successful derivation has been done. The theorem (12.9) is one of the most impressive achievements of this sort. For it is a result derived from (11.7) via analytic facts stemming from (11.6), yet it does not immediately indicate any use of (11.7). Besides, one should observe that this analytic reasoning is strictly one-way, that is, it proceeds from (11.6) to (11.7) and beyond but not readily in the reverse direction; see Note [94.3] for a clarification of what we actually intend to mean here. It is anyway needed to combine analytical facts thus obtained with the expression (11.7). Indeed, only ingenious combinations of analytical and arithmetical devices applicable via (11.6) and (11.7), respectively, have so far been able to successfully produce in-depth knowledge on the distribution of prime numbers, as the discussion in Chapter 5 will reveal. The principal aim of the analytic theory of the distribution of prime numbers is to squeeze plainly discernible facts out of the information that $\zeta(s)$ in the critical strip provides.

In the above we have placed ourselves strictly within Riemann's paradigm, i.e., looking into the set of prime numbers via his zeta-function. However, we are well aware that such a restriction has nothing to do with the true spirit of analytic number theory: Whichever methods are employed, deeper discoveries should be achieved. Yet, the beauty of the configuration that

42 *Divisibility*

Riemann discovered is irresistible. A further discussion will be made in Note [95.8].

Notes

[12.1]   The famous functional equation (12.7) had been conjectured by Euler (1749b, p.94); although it is stated only for a special set of real s, Euler's efforts were apparently directed towards (12.7). In any event, his reasoning is quite interesting: In order to evaluate, in his sense, $\sum_{n=1}^{\infty}(-1)^{n-1}n^{s-1}$ which diverges for $s \geq 1$ (but converges to $(1 - 2^s)\zeta(1 - s)$ for $s < 1$ because of (12.21)), Euler considered the power series

$$\sum_{n=1}^{\infty}(-1)^{n-1}n^{s-1}x^n.$$

He had an evident intention of introducing a new notion of sums of infinite series. Also with a good reason he attached the factor $(-1)^{n-1}$, namely, to attain a finite value at $x = 1$; indeed, when $s \in \mathbb{N}$, this is obviously a rational function of x, i.e., $(x \cdot d/dx)^{s-1}(x/(x + 1))$, and its value at $x = 1$, i.e., the Abelian sum of the original series is finite. To understand the nature of the value thus obtained, Euler transformed this power series by means of the summation formula invented by himself and C. Maclaurin (see Note [13.8]), and discovered that the value in question is undoubtedly connected with $(1 - 2^{1-s})\zeta(s)$. In this way he was led to the functional equation of $\zeta(s)$, although only the case $s \in \mathbb{Z} \cup (\mathbb{Z} + \frac{1}{2})$ was asserted in detail. This transformation argument is, in fact, essentially the same as multiple applications of integration by parts to the expression (12.1); that is one of the ways known today to attain analytic continuation of $\zeta(s)$ to the half plane $\mathrm{Re}\,(s) \leq 1$. Thus, it is quite natural that Euler reached (12.7), albeit from the direction opposite to that taken in the modern argument; see Titchmarsh (1951, §2.5). It is not known whether or not Riemann was aware of this fascinating historical fact. Riemann's discovery that (12.7)–(12.8) is valid throughout $\mathbb{C}$ and gives rise to a mysterious relation between prime numbers and zeros of $\zeta(s)$ must be one of the most impressive instances of applications of analytic continuation.

[12.2]   Viewing the matter through published articles and treatises, the very basic theory of the distribution of prime numbers was developed undoubtedly by Euler, Legendre, Dirichlet, Chebyshev, and above all, by Riemann. The monumental article of Riemann (1860) was communicated by Kummer at the assembly of the Royal Prussian Academy on November 3rd, 1859; it must have been a summary of an ongoing work that Riemann never completed in his life.

Interestingly, the title of this article is exactly the same as that of Chebyshev (1851), but in a different language; see also the title of de la Vallée Poussin (1900).

[12.3]   The application of the method of steepest descent to the zeta-function is explained in detail in Siegel (1932), Edwards (1974), and Motohashi (1987). Another instance is in the later part of the proof of Theorem 129, §99, below.

[12.4]   It should be noted that the genuine logarithmic integral is defined as

$$\mathrm{Li}(x) = \int_0^x \frac{du}{\log u},$$

in which the Cauchy principal value is taken into account concerning the singularity at $u = 1$. For the purpose of the present essays, $\mathrm{li}(x)$ defined as (12.12) suffices. However, the notation for logarithmic integral appears to be confused in literature; compare Riemann (1860, p.678), von Mangoldt (1895, p.300), Whittaker–Watson (1969, pp.341–342), Montgomery–Vaughan (2006, pp.189–192), and Gradshtein–Ryzhik (2007, p.887).

[12.5]   The theory of prime numbers and the theory of functions over $\mathbb{C}$ meet in Riemann's work which is an encounter of entirely different disciplines in mathematics. According to Hadamard (1954, p.123): *It has been written that the shortest and best way between two truths of the real domain often passes through the imaginary one.* We see that this is very true plainly because of multitude of successes achieved in science and technology by means of complex analysis; yet one might describe those trips to the whole $\mathbb{C}$ as bizarre, especially when encountered in dealing with issues related to the distribution of prime numbers.

[12.6]   In the light of RH the following fact should be of considerable interest: In the half plane $\mathrm{Re}\, s > 1$ the partial sum $\sum_{n=1}^N n^{-s}$ of $\zeta(s)$ has no zero if $1 \le N \le 18$ and $N = 20, 21, 28$, but has infinitely many for all other $N \in \mathbb{N}$. See Platt and Trudgian (2016).

Basic conventions on constants and bounds

Throughout the rest of the present essays the following two Notes will be observed:

[12.7]   The usage of the positive constants c and ε: The symbol ε denotes a sufficiently small positive constant, which in general takes different values at each occurrence. Actually, a quantum $\varepsilon_0 > 0$ could be chosen initially so that any local value of ε within the present essays would be an integral multiple of ε_0. On the other hand, the letter $c > 0$ with or without suffixes stands for an unspecified positive constant which does not need to be small but whose value

44   *Divisibility*

may also be different at each occurrence; at a particular instance the value may or may not be explicitly computed, an aspect which we shall make precise when appropriate to do so. We note also that c stands mostly for an absolute/universal constant but occasionally it depends on other constants like ε.

[12.8] Landau's O and o, and Vinogradov's $\ll$ (or $\gg$): Let A and B be possibly varying quantities. If there exists a constant $c > 0$ such that $|A| \leq c|B|$, then we write $A = O(B)$. Here the term constant or more precisely implied constant is used to indicate a quantity, the value of which does not change in the local environment under consideration. For instance, the Lamé bound mentioned in Note [3.2] may be expressed as $O(\log |b|)$. In this case the implied constant is absolute. Also, if a function $f(x, y)$ defined over a certain domain satisfies always the inequality $|f(x, y)| \leq g(x)h(y)$, then fixing the value of y we have $f(x, y) = O(g(x))$ with the implied constant being $h(y)$. The contents conveyed by a particular O-symbol depend much on the context of discussion; its use should be made cautiously so that any misunderstanding is avoided. This notation is usually called Landau's capital oh since its general use began with Landau (1909, Erster Bd., p.59). In fact, it had been introduced by Bachmann (1894, p.401); there exists a relevant historical note by Landau (1909, Zweiter Bd., p.883). By the way, the o-symbol (Landau's little oh) is introduced in Erster Bd., p.61; $A = o(B)$ indicates that A/B is infinitesimally small in a limit process. Consequentially $A = (1 + o(1))B$ implies that $A \sim B$ or A, B are asymptotically equal. Especially when U, V have involved expressions, we shall use the notation $U \ll V$ or $V \gg U$ instead of $U = O(V)$. This was introduced by I.M. Vinogradov (1937, p.5). We shall occasionally write $A \approx B$ to indicate that $A \ll B$ and $A \gg B$ hold simultaneously. Note that constants implied by Landau's and Vinogradov's symbols may depend on other constants like ε; such a dependency will be mentioned if there exists any possibility of confusion.

§13

Returning to the arithmetical aspect of Theorem 10, we see that there exists, for each $n \in \mathbb{N}$, a unique sequence $\{n(p)\}$ such that

$$n = \prod_p p^{n(p)}, \quad p^{n(p)} \| n. \tag{13.1}$$

Here p runs over all primes, and $p^\alpha \| n$ means that $p^\alpha | n$ but $p^{\alpha+1} \nmid n$, so

$$n(p) = \sum_{\nu=1}^{\infty} \iota(n/p^\nu), \tag{13.2}$$

where ι is the characteristic function for the set $\mathbb{N}$. These product and sum are both finite, of course: only a finite number of elements of $\{n(p)\}$ are non-zero.

Then, for $m, n \in \mathbb{N}$,

$$m|n \quad \Leftrightarrow \quad 0 \le m(p) \le n(p), \forall p. \tag{13.3}$$

For $n = p^{n(p)} n'$ means $\langle p, n' \rangle = 1$ and $\langle p^l, n' \rangle = 1$, $l \ge 0$; thus, $p^{m(p)} | p^{n(p)} n'$ and (6.1) give $p^{m(p)} | p^{n(p)}$, that is, (13.3). Provided prime power decomposition, divisibility can be determined by checking each prime power. This implies that for arbitrary $a, b > 0$

$$\langle a, b \rangle = \prod_p p^{\min\{a(p), b(p)\}}, \quad [a, b] = \prod_p p^{\max\{a(p), b(p)\}}. \tag{13.4}$$

In particular, (10.6) is equivalent to the trivial identity $\min\{\alpha, \beta\} + \max\{\alpha, \beta\} = \alpha + \beta$. Also,

$$\langle a, b \rangle = 1 \Leftrightarrow a(p)b(p) = 0, \forall p;$$
$$a \text{ is an } \ell^{\text{th}} \text{ power of an integer} \Leftrightarrow \ell | a(p), \forall p. \tag{13.5}$$

Here we may also introduce the notion of square-free:

$$a \text{ is sqf} \Leftrightarrow a(p) \le 1, \forall p. \tag{13.6}$$

That is, a is not divisible by any square greater than 1. We stress that in general it is never an easy matter to determine whether a given integer is sqf or not. See Note [17.3].

It should be interesting that the application of (13.2) to the factorial $n!$ yields

$$\sum_{p \le x} \frac{\log p}{p} = \log x + O(1), \quad \text{as } x \to \infty. \tag{13.7}$$

Thus, by summation by parts (see Note [13.8]), we have

$$\log N! = N \log N - \int_1^N \frac{[y]}{y} dy$$
$$= N \log N - N + 1 + \int_1^N \frac{y - [y]}{y} dy \tag{13.8}$$
$$= N \log N - N + O(\log N).$$

Then, on noting that the integer $\binom{2N}{N}$ is divisible by p such that $N < p < 2N$, we see that

$$\sum_{N < p < 2N} \log p \le \log \binom{2N}{N} = \log((2N)!) - 2 \log N! = O(N). \tag{13.9}$$

46   *Divisibility*

We let $N = 2^j$ in this, and sum over $j \leq \log x / \log 2$ for an arbitrary $x \geq 2$. We get

$$\sum_{p \leq x} \log p = O(x), \tag{13.10}$$

where the implied constant is absolute. On the other hand, by Legendre's prime power decomposition of $n!$, i.e., Note [13.5] below, we have

$$\log N! = \sum_{\substack{p, v \\ p^v \leq N}} \left[\frac{N}{p^v} \right] \log p = \sum_{p \leq N} \left[\frac{N}{p} \right] \log p + O(N)$$

$$= N \sum_{p \leq N} \frac{\log p}{p} - \sum_{p \leq N} \left(\frac{N}{p} - \left[\frac{N}{p} \right] \right) \log p + O(N). \tag{13.11}$$

Collecting these results, we obtain (13.7).

Next, we turn to Euler's (11.9). Again by summation by parts, we have

$$\sum_{p \leq x} \frac{1}{p} = \frac{S(x)}{\log x} + \int_2^x \frac{S(y)}{y(\log y)^2} dy, \quad S(y) = \sum_{p \leq y} \frac{\log p}{p}. \tag{13.12}$$

This integral is equal to

$$\int_2^x \left(\frac{1}{y \log y} + O\left(\frac{1}{y(\log y)^2} \right) \right) dy = \log \log x + O(1), \tag{13.13}$$

which proves (11.9).

Further, we shall prove Chebyshev's inequality (12.14) without explicitly estimating c_1, c_2: We note that for a sufficiently large $k > 0$

$$\sum_{N < p \leq kN} \frac{\log p}{p} = \log k + O(1). \tag{13.14}$$

The implied constant is absolute. Hence,

$$\tfrac{1}{2} \log k \cdot \frac{N}{\log N} < \pi(kN) - \pi(N) < 2 \log k \cdot \frac{kN}{\log kN}. \tag{13.15}$$

In this, we let $N = k^v$ and sum over $v \leq [\log x / \log k]$. We may omit the rest of proof. Also, one may prove Chebyshev's claim (12.13) by means of the identity

$$\sum_{p \leq x} \frac{\log p}{p} = \pi(x) \frac{\log x}{x} + \int_2^x \pi(u)((\log u - 1)/u^2) du, \tag{13.16}$$

which is of course the result of another application of summation by parts. If it is supposed that the value of the left side of (12.9) is equal to ϑ, then the

$$\S13 \qquad\qquad 47$$

right side of (13.16) is equal to $\vartheta \log x + O(1)$. That is, we should have $\vartheta = 1$ because of (13.7).

Notes

[13.1] If an integer $d > 0$ is not a square, then $\sqrt{d}$ is an irrational number. For if $\sqrt{d} = a/b$, $\langle a, b \rangle = 1$, then $b^2 d = a^2$, and in the prime power decomposition of d the exponent of each prime power is even; hence, d is a square, which is a contradiction. One may argue without the prime power decomposition of d: since $b | a^2$ and $\langle a^2, b \rangle = 1$, we get $b = 1$, a contradiction. More elementarily, if $\sqrt{d} = a/b \notin \mathbb{N}$, then $a/b = (db - [\sqrt{d}]a)/(a - [\sqrt{d}]b)$. This is also a contradiction because it means that a/b would have a denominator less than b.

[13.2] For arbitrary $a, b, c, d \in \mathbb{N}$, we have $\langle ab, cd, ac + bd \rangle | \langle ac, bd \rangle$. To prove this, let $a(p) = \alpha, b(p) = \beta, c(p) = \gamma, d(p) = \delta$ in the notation (13.1). Let $p^\lambda \| \langle ab, cd, ac + bd \rangle$. If $\alpha + \gamma < \beta + \delta$, then $p^{\alpha + \gamma} \| (ac + bd)$; thus $\lambda \le \alpha + \gamma$. If $\alpha + \gamma = \beta + \delta$, then $\min\{\alpha + \beta, \gamma + \delta\} \le (\alpha + \beta + \gamma + \delta)/2 = \alpha + \gamma$; thus, $\lambda \le \alpha + \gamma$ again. In these cases $p^{\alpha + \gamma} \| \langle ac, bd \rangle$, so $p^\lambda | \langle ac, bd \rangle$. By symmetry, we end the proof.

[13.3] For arbitrary $a, b, c, d \in \mathbb{N}$, we have

$$\langle ab, cd \rangle = \langle a, c \rangle \langle b, d \rangle \langle a/\langle a, c \rangle, d/\langle b, d \rangle \rangle \langle b/\langle b, d \rangle, c/\langle a, c \rangle \rangle.$$

To show this, we write $a = \langle a, c \rangle a'$, $b = \langle b, d \rangle b'$, $c = \langle a, c \rangle c'$, $d = \langle b, d \rangle d'$. We see that it suffices to show that $\langle ab, cd \rangle = \langle a, d \rangle \langle b, c \rangle$ provided $\langle a, c \rangle = 1$, $\langle b, d \rangle = 1$. Thus, let $\alpha, \beta, \gamma, \delta$ be as in the preceding Note. Then it suffices to show that $\min\{\alpha + \beta, \gamma + \delta\} = \min\{\alpha, \delta\} + \min\{\beta, \gamma\}$, provided $\alpha \gamma = 0$ and $\beta \delta = 0$; but this is immediate. Or rather apply (6.2) (4) to the second factor of the right side of the identity $\langle ab, cd \rangle = \langle b, c \rangle \langle ab/\langle b, c \rangle, dc/\langle b, c \rangle \rangle$.

[13.4] If $N \ge 2$, then the harmonic number $H_N = \sum_{n=1}^{N} 1/n$ cannot be an integer. For a confirmation, take the positive integer k such that $2^k \le N < 2^{k+1}$; then 2^k divides $n \le N$ only if $n = 2^k$. With this, let $2^\ell \| N!$. Then in the sum $\sum_{n=1}^{N} N!/n$ we have $2^{\ell - k} \| (N!/2^k)$ besides $2^{\ell - k + 1} | (N!/n)$ whenever $n \ne 2^k$; that is, $2^{\ell - k} \| N! H_N$. This ends the confirmation, for if H_N is an integer, then $2^\ell | N! H_N$.

[13.5] Legendre (1808, p.10). Let $(c_k c_{k-1} \ldots c_1 c_0)_p$ be the p-adic expansion of n. Then

$$n! = \prod_p p^{\alpha(n,p)}, \quad \alpha(n,p) = (n - c_k - \cdots - c_1 - c_0)/(p - 1),$$

48   *Divisibility*

since we have, by (13.2),

$$\alpha(n,p) = \sum_{m=1}^{n}\sum_{v=1}^{\infty} \iota(m/p^v) = \sum_{v=1}^{\infty}\sum_{m=1}^{n} \iota(m/p^v) = \sum_{v=1}^{\infty}[n/p^v]$$

$$= \sum_{v=1}^{k}\sum_{\mu=v}^{k} c_\mu p^{\mu-v} = \sum_{\mu=1}^{k} c_\mu \sum_{j=0}^{\mu-1} p^j = \sum_{\mu=0}^{k}(c_\mu p^\mu - c_\mu)/(p-1).$$

For instance, as $12345 = (343340)_5$, the exponent of the highest power of 5 dividing 12345! is equal to $(12345 - 17)/4 = 3082$; hence, the 3082 lower decimal-digits of 12345! are all equal to 0.

[13.6]   For arbitrary $a, b \in \mathbb{N}$, we have

$$\frac{(2a)!\,(2b)!}{a!\,b!\,(a+b)!} \in \mathbb{N}.$$

It suffices to show that $[2a/p^v] + [2b/p^v] \geq [a/p^v] + [b/p^v] + [(a+b)/p^v]$. In the sense of (2.1), let $a = up^v + r$ and $b = vp^v + s$. Then the left side of this inequality is $2(u+v)+[2r/p^v]+[2s/p^v]$; the right side is $2(u+v)+[(r+s)/p^v]$. Since we have $[2\xi] + [2\eta] \geq [\xi + \eta]$ for any $0 \leq \eta, \xi < 1$, we end the confirmation. Also, with any $a, b \in \mathbb{N}$, the following holds:

$$\frac{(4a)!\,(4b)!}{a!\,b!\,(2a+b)!\,(a+2b)!} \in \mathbb{N}.$$

See Bachmann (1902, pp.50–66).

[13.7]   [Σ.X, Prop.29, Lemma 1]: The relation $a^2 + b^2 = c^2$ with $\langle a, b, c \rangle = 1$ holds if and only if $c = u^2 + v^2$ and one of a, b is equal to $2uv$ and the other to $u^2 - v^2$, with $\langle u, v \rangle = 1$ and $2 \nmid (u - v)$. To prove this, we may suppose that $2|a$ and $2 \nmid b$ since $\langle a, b, c \rangle = 1$ implies that $2|a$, $2|b$ is impossible, and further if a, b were both odd, then $2\|c^2$ which is also impossible. Thus, in the identity $(a/2)^2 = ((c+b)/2)((c-b)/2)$, we have $\langle (c+b)/2, (c-b)/2 \rangle = 1$. Hence, by the lower line of (13.5) we have $(c + b)/2 = u^2$, $(c - b)/2 = v^2$, from which follow expressions for a, b, c in terms of u, v. The rest of the proof is omitted. By the way, these $\{a, b, c\}$ were called Pythagorean triples in the past. However, this attribution is nowadays regarded as inappropriate since the clay tablet *Plimpton 322* (ca 1800 BCE), excavated at Tell Es-Senkereh (Larsa), has been identified as a table of 15 sets of such triples. For example, the fourth entry on it is $13500^2 + 12709^2 = 18541^2$ (in fact, in the sexagesimal system: $(3\,31\,49)_{60} = 12709$, etc.), which is the case $u = 125, v = 54$ in the above. *Scripta manent!*

$$\S14 \qquad\qquad 49$$

[13.8] For a finite sequence $\{c_n\}$ and a continuously differentiable function f in the interval $[u, v]$ with $u, v \notin \mathbb{Z}$, we have the summation by parts:

$$\sum_{u<n<v} c_n f(n) = C(x)f(x)\Big|_u^v - \int_u^v C(x)f'(x)dx, \quad C(x) = \sum_{n\leq x} c_n.$$

To show this, we consider first the case $n - 1 < u < n < v < n + 1$ with an arbitrary $n \in \mathbb{Z}$. The integration interval is divided into two: $[u, n-0)\cup(n+0, v]$. We compute the respective integrals and get the identity in question for this special case. Then, summing the identity over n, we obtain the general case. If either of u, v is an integer, then one should take an appropriate limit. In particular, if $c_n \equiv 1$, then the above is the first step of the Euler–Maclaurin sum formula (Euler (1732b)), a typical instance of which is visible in (12.1). The full formula is the result of multiple applications of integration by parts. However, for our purpose the plain summation by parts and its consequence the sum formula of Poisson, i.e., our (60.1), suffice. See also Note [60.2] and Montgomery–Vaughan (2006, Appendix B).

[13.9] Also, in terms of Stieltjes integral which was introduced by him (1894, p.71) while dealing with a number theoretical problem, the above sum is expressed as

$$\int_u^v f(x)dC(x).$$

Then, summation by parts is the same as integration by parts. For the details, see Montgomery–Vaughan (2006, Appendix A).

§14

The least common multiple $[a_1, a_2, \ldots, a_K]$ of integers $a_k \neq 0$, $1 \leq k \leq K$, is the smallest among all positive common multiples of them. We have

$$a_k | l, \ 1 \leq k \leq K \ \Leftrightarrow \ [a_1, a_2, \ldots, a_K] | l, \qquad (14.1)$$

which can be regarded as dual to (7.2). To confirm this equivalence, it suffices to divide l by the least common multiple in the sense of (2.1). Analogously to (7.3), we divide the sequence $\{a_1, a_2, \ldots, a_K\} \subset \mathbb{N}$ into L disjoint subsequences, and let the least common multiples of elements in respective subsequences be $\{l_1, l_2, \ldots, l_L\}$. Then

$$[a_1, a_2, \ldots, a_K] = [l_1, l_2, \ldots, l_L], \qquad (14.2)$$

50   *Divisibility*

as each side divides the other. The same follows from the identity

$$a_1\mathbb{Z} \cap a_2\mathbb{Z} \cap \cdots \cap a_K\mathbb{Z} = [a_1, a_2, \dots, a_K]\mathbb{Z}, \tag{14.3}$$

which is an extension of (10.5). In particular, we see that

$$\langle a_j, a_k \rangle = 1, j \neq k \quad \Leftrightarrow \quad [a_1, a_2, \dots, a_K] = a_1 a_2 \cdots a_K, \tag{14.4}$$

an extension of the second assertion of Theorem 8. We skip the proof.

The expression (13.4) for $[a, b]$ generalizes readily to the present situation of many arguments. Prime-power decomposition simplifies discussions involving greatest common divisors and least common multiples. However, this applies either when prime power decompositions of integers under consideration are given ab initio or when theoretical prime power decompositions suffice to proceed discussion. In comparison, the explicit and efficient nature of Euclid's algorithm is quite significant.

Notes

[14.1]   In [Σ.VII, Prop.36] the identity $[a, b, c] = [[a, b], c]$ is shown, which corresponds to (14.2). A repeated use of the relation (10.6) allows one to compute $[a_1, a_2, \dots, a_K]$ by means of Euclid's algorithm for two arguments.

[14.2]   For arbitrary $a_k \in \mathbb{N}$, $1 \leq k \leq K$, we have the following decompositions of the greatest common divisor and the least common multiple:

$$\langle a_1, a_2, \dots, a_K \rangle = a_1^{(-)} a_2^{(-)} \cdots a_K^{(-)},$$
$$a_k^{(-)} | a_k, \ 1 \leq k \leq K, \quad \langle a_j^{(-)}, a_k^{(-)} \rangle = 1, j \neq k;$$
$$[a_1, a_2, \dots, a_K] = a_1^{(+)} a_2^{(+)} \cdots a_K^{(+)},$$
$$a_k^{(+)} | a_k, \ 1 \leq k \leq K, \quad \langle a_j^{(+)}, a_k^{(+)} \rangle = 1, j \neq k.$$

The first assertion can be shown by selecting, for each p, a single power $p^{a_v(p)}$ such that $a_v(p) = \min\{a_1(p), a_2(p), \dots, a_K(p)\}$; that is, if $k \neq v$, then $p^{a_k(p)}$ does not contribute to the prime power decomposition of the greatest common divisor. As for the second assertion, we consider $\max\{a_1(p), a_2(p), \dots, a_K(p)\}$ instead.

[14.3]   How to obtain the above decompositions without recourse to prime power decompositions: We deal with the case of two integers $a, b > 0$; the general case is analogous, though the actual treatment is involved. Thus, first compute $\delta = \langle a, b \rangle$ by Euclid's algorithm. Then put

$$\alpha_1 = \langle a, \delta^\infty \rangle, \ \alpha_0 = a/\alpha_1, \ \alpha_2 = \langle \alpha_1, (\alpha_1/\delta)^\infty \rangle, \ \alpha_3 = \alpha_1/\alpha_2,$$
$$\beta_1 = \langle b, \delta^\infty \rangle, \ \beta_0 = b/\beta_1, \ \beta_2 = \langle \beta_1, (\beta_1/\delta)^\infty \rangle, \ \beta_3 = \beta_1/\beta_2; \ \gamma = \langle \alpha_3, \beta_3 \rangle.$$

Here, $\langle m, n^\infty \rangle$ is $\langle m, n^\kappa \rangle$ with the least $\kappa = \kappa_0$ such that $\langle m, n^\kappa \rangle = \langle m, n^{\kappa+1} \rangle$. With this, we have

$$a^{(-)} = \alpha_3, \ b^{(-)} = \beta_3/\gamma; \quad a^{(+)} = \alpha_0\alpha_2, \ b^{(+)} = \beta_0\beta_2\gamma.$$

The proof is omitted. For $a = 26629837, b = 44074693$, we find that $\delta = 66079$, and that $\langle a, \delta^\infty \rangle = 859027 = \alpha_1 \ (\kappa_0 = 2); \ \alpha_0 = 31; \ \alpha_1/\delta = 13;$ $\langle \alpha_1, 13^\infty \rangle = 2197 = \alpha_2 \ (\kappa_0 = 3); \ \alpha_1/\alpha_2 = \alpha_3 = 391.$ Also, $\langle b, \delta^\infty \rangle = 1519817 = \beta_1 \ (\kappa_0 = 2); \ \beta_0 = 29; \ \beta_1/\delta = 23; \ \langle \beta_1, 23^\infty \rangle = 529 = \beta_2 \ (\kappa_0 = 2);$ $\beta_1/\beta_2 = \beta_3 = 2873; \ \gamma = \langle \alpha_3, \beta_3 \rangle = 17.$ Therefore, $a^{(-)} = 391, b^{(-)} = 169,$ $a^{(+)} = 68107, b^{(+)} = 260797.$ Indeed, $a^{(-)} \cdot b^{(-)} = 66079 = \langle a, b \rangle;$ $a^{(+)} \cdot b^{(+)} = 17762101279 = 26629837 \cdot 44074693/66079 = [a, b];$ further, $391|a, \ 169|b, \ \langle 391, 169 \rangle = 1; \ 68107|a, \ 260797|b, \ \langle 68107, 260797 \rangle = 1.$

[14.4] For any $\{q_1, \ldots, q_K\} \subset \mathbb{N}$, we have

$$\left[\langle q_1, q_K \rangle, \langle q_2, q_K \rangle, \ldots \langle q_{K-1}, q_K \rangle \right] = \langle [q_1, q_2, \ldots, q_{K-1}], q_K \rangle.$$

The case $K = 2$ is trivial. Thus, we proceed by induction. The left side with $K + 1$ in place of K is, by (14.2),

$$\left[\left[\langle q_1, q_{K+1} \rangle, \langle q_2, q_{K+1} \rangle, \ldots \langle q_{K-1}, q_{K+1} \rangle \right], \langle q_K, q_{K+1} \rangle \right]$$
$$= \left[\langle [q_1, q_2, \ldots, q_{K-1}], q_{K+1} \rangle, \langle q_K, q_{K+1} \rangle \right]$$
$$= \langle \left[[q_1, q_2, \ldots, q_{K-1}], q_K \right], q_{K+1} \rangle,$$

which ends the proof. In terms of prime power decomposition, the present assertion is equivalent to the identity

$$\max \left\{ \min\{\alpha_1, \alpha_K\}, \min\{\alpha_2, \alpha_K\}, \ldots, \min\{\alpha_{K-1}, \alpha_K\} \right\}$$
$$= \min \left\{ \max\{\alpha_1, \alpha_2, \ldots, \alpha_{K-1}\}, \alpha_K \right\}.$$

§15

The function

$$f : \mathbb{Z} \to \mathbb{C} \tag{15.1}$$

is called arithmetic. Often the set $\mathbb{N}$, instead of $\mathbb{Z}$, is the domain of definition. A typical example is the divisor function:

$$d(n) = \sum_{t|n} 1, \quad n, t \in \mathbb{N}. \tag{15.2}$$

This function is one of the most fundamental notions in number theory and has been attracting considerable interest since antiquity.

52   *Divisibility*

Hereafter we shall deal with various issues which are related to divisors. In order to avoid unnecessarily involved expressions, we introduce the convention

$$\text{divisors are generally assumed to be positive.} \qquad (15.3)$$

Thus, regardless of the sign of a, we let $t|a$ mean $t \geq 1$, unless otherwise stated.

We have already remarked at (13.3) that in the prime power decomposition of the divisor t in (15.2) there exists no restriction other than $t(p) \leq n(p)$, $\forall p$. Hence,

$$d(n) = \prod_p \big(n(p) + 1 \big). \qquad (15.4)$$

Thus, $d(ab) = \prod_p \big(a(p) + b(p) + 1 \big)$; and in particular,

$$\langle a, b \rangle = 1 \;\Rightarrow\; d(ab) = d(a)d(b), \qquad (15.5)$$

as the condition implies that $a(p)b(p) = 0$, so $a(p) + b(p) + 1 = (a(p) + 1)(b(p) + 1)$.

The identity (15.5) is a property that is shared by a great number of arithmetic functions; some typical among them will be investigated in the present essays. With this purpose in mind, we shall first make here a somewhat trivial but basic observation on the set of all divisors of a product of two coprime integers. We shall show in particular that (15.5) can be proved without recourse to the notion of prime power decomposition, a fact which will be seen to generalize to a considerable extent in the course of our discussion. In short, the prime power decomposition can often be dispensed with in discussing multiplicative problems in number theory.

Theorem 13 *Let* $\langle a, b \rangle = 1$. *Then the mapping*

$$t|a, \; u|b \;\mapsto\; tu \qquad (15.6)$$

is a bijection to the set of all divisors of ab. The injectivity of this mapping implies that a, b are coprime.

Proof This mapping is injective. For if $tu = t'u'$, $t'|a, u'|b$, then (6.1) implies $t|t'$, $t'|t$, and $t = t'$; thus, $u = u'$ as well. Also, it is surjective. For, taking any $v|ab$ we write $v = \langle v, a \rangle v'$, $a = \langle v, a \rangle a'$; then by (6.2) (1), $v = \langle v, ab \rangle = \langle v, a \rangle \langle v', a'b \rangle$, and a repeated use of (6.2)(4) implies $\langle v', a'b \rangle = \langle v', b \rangle = \langle v, b \rangle$; hence, $v = \langle v, a \rangle \langle v, b \rangle$. As for the second assertion, let c be a common divisor of a, b; then c is the image of $t = c, u = 1$ as well as of $t = 1, u = c$; hence, $c = 1$, if the mapping is injective. This ends the proof.

$$\S15 \qquad\qquad 53$$

Therefore, we have the decomposition mapping

$$n = ab, \ \langle a,b \rangle = 1: \quad v|n \ \mapsto \ \{\langle v,a \rangle, \langle v,b \rangle\}. \tag{15.7}$$

This configuration indicates well the existence of a kind of orthogonality in $\mathbb{N}$ which is equivalent to the notion of coprimality. One realizes, perhaps, that the integer world $\mathbb{N}$ resembles a structure of infinitely many orthogonal parts, an impression which will be definitely enhanced in the subsequent chapters. Indeed, the modern development of analytic number theory may be described as a marvelous collection of efforts of exploiting such quasi-orthogonalities.

In general, i.e., without the condition $\langle a,b \rangle = 1$, one can say only that the map (15.6) covers the set of all divisors of ab; that is,

$$\{v : v|ab\} \subseteq \{tu : t|a, u|b\}, \tag{15.8}$$

where the upper set may contain repeated elements. This is implied by the trivial identity $v = \langle a, v \rangle \cdot v / \langle a, v \rangle$ and (6.3) with an obvious replacement.

It is appropriate to introduce here the notion of the Dirichlet series $\lfloor f \rfloor (s)$ associated with an arithmetic function f:

$$\lfloor f \rfloor (s) = \sum_{n=1}^{\infty} \frac{f(n)}{n^s}, \quad s \in \mathbb{C}. \tag{15.9}$$

It is assumed that the sum converges absolutely, if $\mathrm{Re}\, s$ is greater than a certain lower bound, that is, if it is positive and sufficiently large. In the present essays we shall deal with a variety of Dirichlet series, among which is of course $\lfloor 1 \rfloor (s) = \zeta(s)$; as for their convergence, we shall follow the present practice, which should not cause any inconvenience as far as we discuss generalities. In a somewhat symbolical description, $\lfloor f \rfloor (s)$ is an extension of the sequence $\{f(n) : n \in \mathbb{N}\}$ to an existence over a half complex plane and possibly over the entire $\mathbb{C}$ via analytic continuation.

The most basic fact on Dirichlet series is the following analytic uniqueness assertion:

$$f(n) = g(n), \ \forall n \in \mathbb{N} \ \Leftrightarrow \ \lfloor f \rfloor (s) = \lfloor g \rfloor (s); \tag{15.10}$$

that is, $f \mapsto \lfloor f \rfloor$ is injective. To show the sufficiency, we let s tend to $+\infty$, and get $f(1) = g(1)$. Then, with the assumption $f(m) = g(m)$ for $m \leq n$, we do the same in the identity

$$(n+1)^s \left(\lfloor f \rfloor (s) - \sum_{m=1}^{n} f(m)m^{-s} \right)$$
$$= (n+1)^s \left(\lfloor g \rfloor (s) - \sum_{m=1}^{n} g(m)m^{-s} \right),$$

(15.11)

which gives $f(n+1) = g(n+1)$. This ends the confirmation of (15.10).

Let f be as in (15.1). It is said to be multiplicative, provided that

$$f(1) = 1 \text{ and } \langle a,b \rangle = 1 \Rightarrow f(ab) = f(a)f(b),$$

(15.12)

which is analogous to the property (15.5) of the divisor function; we shall use the term multiplicative functions instead of multiplicative arithmetic functions for the sake of brevity. If $f(a) \neq 0$, then $f(a) = f(a)f(1)$ implies $f(1) = 1$; so the first part of the condition is not actually a restriction. Also, $f(a) = f(-1)f(|a|)$ with $a < 0$; note that $f(-1) = \pm 1$, as $f(1) = f((-1)(-1)) = f(-1)^2$. This implies that since the decomposition (11.1) gives

$$f(n) = \prod_p f\left(p^{n(p)}\right), \quad n \in \mathbb{N},$$

(15.13)

f is fixed if its values at -1 and all prime powers are pre-defined.

Further, f is said to be completely multiplicative, provided that

$$f(1) = 1 \text{ and } f(ab) = f(a)f(b) \text{ without restriction.}$$

(15.14)

Thus,

$$f(n) = \prod_p \left(f(p)\right)^{n(p)}, \quad n \in \mathbb{N},$$

(15.15)

which means that f is fixed if its values at -1 and all primes are pre-defined.

It should be noticed that the definitions (15.12) and (15.14) are incomplete since the rôle of 0 is left obscure. For the time being, we shall restrict domains of multiplicative functions to $\mathbb{N}$, mostly; throughout the present essays these domains are not always explicitly prescribed but will be made clear in context.

Also,

$$f\text{: multiplicative} \Leftrightarrow \lfloor f \rfloor (s) = \prod_p \left(\sum_{j=0}^{\infty} \frac{f(p^j)}{p^{js}} \right);$$

(15.16)

$$f\text{: completely multiplicative} \Leftrightarrow \lfloor f \rfloor (s) = \prod_p \left(1 - \frac{f(p)}{p^s} \right)^{-1}.$$

(15.17)

Both identities hold provided Re $s > 0$ is sufficiently large. Following (11.7), these right sides are called also Euler products.

It should be added that if f is multiplicative and never vanishes over $\mathbb{N}$, then for any $a > 0$

$$f(an)/f(a) \text{ is a multiplicative function of } n \in \mathbb{N}. \qquad (15.18)$$

In view of (15.16), this is equivalent to the trivial relation: with $p^\alpha \| a$

$$\sum_{n=1}^{\infty} \frac{f(an)}{f(a)n^s} = \prod_p \left(\sum_{j=0}^{\infty} \frac{f(p^{\alpha+j})}{f(p^\alpha)p^{js}} \right) = \prod_p \left(\sum_{j=0}^{\infty} \frac{f(ap^j)}{f(a)p^{js}} \right). \qquad (15.19)$$

Notes

[15.1] There are a variety of deep problems concerning divisors and the divisor function, which remain still unsolved. For instance, this can be a surprise: If n varies in an arithmetic progression and a high level of uniformity with respect to the difference is required, then the asymptotic nature of the sum of $d(n)$ is extremely hard to analyze; see Note [104.6]. Also, counting integral matrices $\left(\begin{smallmatrix} u & v \\ w & z \end{smallmatrix} \right)$ is a typical example. On the condition $u, v, w, z \in \mathbb{N}$ with $vw = n$, the number of matrices with a positive determinant m is equal to $d(n)d(n + m)$, and the additive divisor problem of studying the asymptotic nature, as $N \to \infty$, of the sum

$$\sum_{n \leq N} d(n)d(n + m)$$

becomes an issue. In view of Note [5.3], in the background of this ostensibly accessible sum is the modular group Γ; and it is tightly connected with advanced subjects in the automorphic context. See Motohashi (1994) for instance, which is an exploitation of core innovations made by Kuznetsov (1977) and Bruggeman (1978). Moreover, if the divisor function $d(n)$ is replaced by a slightly generalized function $d_k(n)$, $k \geq 3$ (see (16.11) below), then a totally new problem emerges, the settlement of which appears to be extremely difficult in the present state of number theory. The divisor function which may appear to be a crude notion is, in fact, a gate to a deep realm of multiplicative number theory.

[15.2] Entanglement between addition and multiplication: Various problems in number theory are rooted in subtle relations between these two arithmetic operations. Concerning the additive divisor problem mentioned above, if $m = 0$, then it becomes a discussion on the multiplicative function $d^2(n) = (d(n))^2$; and with the aid of Note [18.7] below, it is possible to achieve

56   *Divisibility*

a relatively decent success. However, if $m > 0$, then at first glance it should be hard even to speculate on which means to employ. As a matter of fact, we are eventually led to a sequence of an infinite number of Dirichlet series (automorphic L-functions), all of which share a characteristic similar to that of the square of $\zeta(s)$. A seemingly harmless perturbation or like in an arithmetical problem may induce a genuinely new mathematical issue.

[15.3]   The *abc* conjecture: For each $\varepsilon > 0$, there exists a constant ℓ_ε such that if $a + b = c$, with $a, b > 0$, $\langle a, b \rangle = 1$, then it holds that

$$c < \ell_\varepsilon \Big(\prod_{p \mid abc} p \Big)^{1+\varepsilon}.$$

This was presented by D. Masser and J. Oesterlé in the middle of 1980s. If correct, then it should yield resolutions of a variety of extremely deep problems in number theory, in spite of its simplicity. We naturally do not dwell on the details but note just that the sheer difficulty of the conjecture indicates well a mysterious relation between addition and multiplication.

[15.4]   The term Dirichlet series commemorates the introduction of L-functions by Dirichlet (1837a); see (52.6) and Note [52.2]. The continuous real variable s appeared there for the first time. Riemann followed Dirichlet, and took s to $\mathbb{C}$, whence a firm tradition in analytic number theory was formed. As we have indicated in §§11–12, there are opinions attributing this origin instead to Euler (1737b, 1749b), for his discovery of the product expansion (11.7) and the functional equation (12.7)–(12.8) is of tremendous importance in analytic number theory. However, Euler considered these objects with discrete s solely, as remarked in §12 already. In view of later developments, the real fusion of analysis and arithmetic should better be located in investigations of number theoretical problems by means of continuous variables and functions. Then, the commencement is indisputably marked in Dirichlet (1837a, 1838, 1839). Specifically in this context, the present author appreciates Kummer's eulogy for Dirichlet; see Note [92.1].

§16

Any arithmetic function f can be used as a seed to generate a new arithmetic function over $\mathbb{N}$: Following the definition (15.2) of the divisor function, we let

$$F(n) = \sum_{t \mid n} f(t), \quad n \in \mathbb{N}. \tag{16.1}$$

$$\text{\S16} \qquad\qquad 57$$

If f is multiplicative, then F is so since with the aid of (15.6)–(15.7) we have, for $\langle a, b \rangle = 1$,

$$F(ab) = \sum_{t|a,\,u|b} f(tu) = \sum_{t|a,\,u|b} f(t)f(u) = F(a)F(b). \qquad (16.2)$$

In particular, we have, via (13.3),

$$\sum_{t|n} f(t) = \prod_{p} \left(\sum_{v=0}^{n(p)} f(p^v) \right); \qquad (16.3)$$

of course the same follows from (15.13).

For instance, with an arbitrary $\alpha \in \mathbb{C}$, the function $n \mapsto n^\alpha$ is multiplicative over $\mathbb{N}$; thus, the sum of powers of divisors

$$\sigma_\alpha(n) = \sum_{d|n} d^\alpha, \quad n \in \mathbb{N}, \qquad (16.4)$$

is also multiplicative; and

$$\sigma_\alpha(n) = \prod_{p} \frac{p^{(n(p)+1)\alpha} - 1}{p^\alpha - 1}, \quad \alpha \neq 0. \qquad (16.5)$$

As α tends to 0, the limit is (15.4).

Turning to Dirichlet series, the definition (16.1) is equivalent to the identity

$$\lfloor F \rfloor(s) = \zeta(s)\lfloor f \rfloor(s). \qquad (16.6)$$

In fact, in the region of absolute convergence, we insert (16.1) into each term of the series on the left side, exchange the order of summation, and obtain the right side. For example, we have

$$\lfloor \sigma_\alpha \rfloor(s) = \zeta(s)\zeta(s - \alpha), \quad \operatorname{Re} s > \max\{1, 1 + \operatorname{Re}\alpha\}. \qquad (16.7)$$

Further, let f_1, f_2 be arbitrary arithmetic functions. Then, generalizing (16.1), the multiplicative convolution

$$(f_1 * f_2)(n) = \sum_{t|n} f_1(t)f_2(n/t), \quad n \in \mathbb{N}, \qquad (16.8)$$

58 *Divisibility*

defines also a new arithmetic function over $\mathbb{N}$. If f_1, f_2 are multiplicative, then $f_1 * f_2$ is so. To see this, we again apply (15.6)–(15.7), and have, for $\langle a, b \rangle = 1$,

$$\begin{aligned}
(f_1 * f_2)(ab) &= \sum_{t|a} \sum_{u|b} f_1(tu) f_2((a/t)(b/u)) \\
&= \sum_{t|a} f_1(t) f_2(a/t) \sum_{u|b} f_1(u) f_2(b/u) \qquad (16.9) \\
&= (f_1 * f_2)(a)(f_1 * f_2)(b).
\end{aligned}$$

The operation $*$ is obviously commutative $f_1 * f_2 = f_2 * f_1$ and associative $(f_1 * f_2) * f_3 = f_1 * (f_1 * f_3)$; actually, the value at n of both sides of the latter equality is identical to $\sum_{abc=n} f_1(a) f_2(b) f_3(c)$. Also, the delta-function placed at the point 1:

$$\delta_1(n) = \begin{cases} 1 & n = 1, \\ 0 & n \neq 1, \end{cases} \qquad (16.10)$$

is the unit with respect to this operation. Namely, for any arithmetic function f, we have $f = \delta_1 * f = f * \delta_1$. The unit is obviously unique.

The divisor function d equals $\iota * \iota$ with ι introduced at (13.2); and the definition (16.1) can be expressed as $F = \iota * f$. Further, the k-fold divisor function

$$\begin{aligned}
d_k &= \overbrace{\iota * \cdots * \iota}^{k \text{ times}} ; \quad d_2 = d, \\
d_k(n) &= \left| \left\{ \{u_1, u_2, \ldots, u_k\} : u_1 u_2 \cdots u_k = n, \, u_j \in \mathbb{N}, 1 \leq j \leq k \right\} \right|,
\end{aligned} \qquad (16.11)$$

is also multiplicative; obviously $d_k(n) = \sum_{u|n} d_{k-1}(u)$. With the decomposition (13.1), we have

$$d_k(n) = \prod_p \binom{n(p) + k - 1}{k - 1}. \qquad (16.12)$$

To confirm this by induction, we note that $d_{k+1} = \iota * d_k$; and thus, for $k \geq 2$,

$$\begin{aligned}
d_{k+1}(p^{n(p)}) &= \sum_{v=0}^{n(p)} \binom{v + k - 1}{k - 1} \\
&= \sum_{v=0}^{n(p)} \left\{ \binom{v + k}{k} - \binom{v + k - 1}{k} \right\} = \binom{n(p) + k}{k}.
\end{aligned} \qquad (16.13)$$

Multiplicative convolution corresponds to ordinary multiplication among Dirichlet series: in the region of absolute convergence, it holds that

$$\lfloor f_1 * f_2 \rfloor(s) = \lfloor f_1 \rfloor(s) \cdot \lfloor f_2 \rfloor(s). \qquad (16.14)$$

A repeated application implies that (16.11) is equivalent to

$$\zeta^k(s) = \sum_{n=1}^{\infty} \frac{d_k(n)}{n^s}, \quad \mathrm{Re}\, s > 1. \tag{16.15}$$

Inserting the Euler product (11.7) into the left side of (15.16), we see that

$$\prod_p \left(1 - \frac{1}{p^s}\right)^{-k} = \prod_p \left\{ \sum_{j=0}^{\infty} \binom{-k}{j} \frac{(-1)^j}{p^{js}} \right\}. \tag{16.16}$$

This right side is equivalent to (16.12). Continued to Note [16.12].

Occasionally we need to have average/mean sizes of divisor functions, so we give here the following assertion: with the notation ε and $\ll$ being as in Notes [12.7]–[12.8],

Theorem 14 *For fixed integers k, $\lambda \geq 1$, we have*

$$\sum_{n \leq x} d_k^{\lambda}(n) \ll x(\log x)^{k^{\lambda}-1}, \tag{16.17}$$

$$\sum_{n \leq x} \frac{d_k^{\lambda}(n)}{n} \ll (\log x)^{k^{\lambda}}. \tag{16.18}$$

In particular, we have

$$d_k(n) \ll n^{\varepsilon}. \tag{16.19}$$

Proof The bound (16.19) follows from (16.17), as $d_k(m) \ll \left(m(\log m)^{k^{\lambda}-1}\right)^{1/\lambda}$ with $x = m$ there. The left side of (16.18) is

$$\sum_{n_1 n_2 \cdots n_k \leq x} \frac{d_k^{\lambda-1}(n_1 n_2 \cdots n_k)}{n_1 n_2 \cdots n_k} \tag{16.20}$$

$$\leq \left(\sum_{n \leq x} \frac{d_k^{\lambda-1}(n)}{n}\right)^k \leq \left(\sum_{n \leq x} \frac{1}{n}\right)^{k^{\lambda}} \ll (\log x)^{k^{\lambda}},$$

where we have used the inequality $d_k(mn) \leq d_k(m)d_k(n)$; to see this, note that $d_k(mn) = \sum_{u|mn} d_{k-1}(u) \leq \sum_{v|m,\, w|n} d_{k-1}(vw)$ because of (15.8); then proceed by induction. As for $(16.17)_{\lambda=1}$, we have thus

$$\sum_{n \leq x} d_k(n) \leq x \sum_{u \leq x} \frac{d_{k-1}(u)}{u} \ll x(\log x)^{k-1}. \tag{16.21}$$

With this, we consider the function

$$q(s) = \sum_{n=1}^{\infty} \frac{d_k^{\lambda}(n)}{n^s}, \tag{16.22}$$

Divisibility

which converges absolutely for $\operatorname{Re} s > 1$ because of (16.19). We have

$$q(s) = \prod_p \left(1 + \frac{k^\lambda}{p^s} + \sum_{j=2}^\infty \frac{d_k^\lambda(p^j)}{p^{js}} \right) = \zeta^{k^\lambda}(s) q_1(s). \tag{16.23}$$

It is readily seen that $q_1(s)$ converges absolutely for $\operatorname{Re} s > \frac{1}{2}$. Writing

$$q_1(s) = \sum_{u=1}^\infty \frac{r(u)}{u^s}, \tag{16.24}$$

we have, by (16.21),

$$\begin{aligned}
\sum_{n \leq x} d_k^\lambda(n) &= \sum_{n \leq x} \sum_{u \mid n} r(u) d_{k^\lambda}(n/u) \\
&\ll x \sum_{u \leq x} \frac{|r(u)|}{u} (\log x/u)^{k^\lambda - 1} \ll x (\log x)^{k^\lambda - 1},
\end{aligned} \tag{16.25}$$

since $\sum_{u=1}^\infty |r(u)|/u$ is finite. This ends the proof.

The above reasoning is a typical instance of exploiting the effect of having Euler products; see Titchmarsh (1951, Chapter I) for basic instances.

In passing, we show the asymptotic formula

$$\sum_{n \leq x} d(n) = x \left(\log x + 2c_E - 1 \right) + O\left(x^{1/2} \right), \tag{16.26}$$

where c_E is the Euler constant (see (12.2)). To this end we apply a splitting argument (Dirichlet's splitting method) to the left side:

$$\begin{aligned}
\sum_{kl \leq x} 1 &= \sum_{k \leq x^{1/2}} \sum_{l \leq x/k} 1 + \sum_{l \leq x^{1/2}} \sum_{k \leq x/l} 1 - \sum_{k \leq x^{1/2}} \sum_{l \leq x^{1/2}} 1 \\
&= 2 \sum_{k \leq x^{1/2}} \left[\frac{x}{k} \right] - \left[x^{1/2} \right]^2 \\
&= 2x \sum_{k \leq x^{1/2}} \frac{1}{k} - x + O\left(x^{1/2} \right),
\end{aligned} \tag{16.27}$$

and we get the right side of (16.26) since $\sum_{n \leq y} 1/n = \log y + c_E + O(1/y)$.

Notes

[16.1] For arbitrary $m, n \in \mathbb{N}$ and $\alpha \in \mathbb{C}$,

$$\sigma_\alpha(m)\sigma_\alpha(n) = \sum_{t \mid \langle m, n \rangle} t^\alpha \sigma_\alpha(mn/t^2),$$

$\S 16$ 61

which is a prototype of the famed identity of Hecke (1937, §2) for his operators. In view of (16.5), this is a consequence of the following identity: for $k, l \geq 0$

$$\frac{(x^{k+1} - 1)(x^{l+1} - 1)}{(x - 1)^2} = \sum_{j=0}^{\min\{k,l\}} x^j \cdot \frac{x^{k+l+1-2j} - 1}{x - 1}.$$

An alternative proof is given in Note [18.7].

[16.2] For any $n \in \mathbb{N}$ we have

$$d_3^2(n) = \sum_{u|n} d^3(u).$$

This is a consequence of the identity $\big((k + 2)(k + 1)/2\big)^2 = \sum_{\nu=0}^{k}(\nu + 1)^3$.

[16.3] In this and the next four Notes, we shall briefly discuss the perfect numbers. These rare integers $n > 0$ are to satisfy $\sigma_1(n) = 2n$ ([Σ.VII, Def.22]). According to Euler's famous theorem (1849a, p.514; 1849b, p.630),

n is even and perfect $\Leftrightarrow$ $n = 2^{\ell-1}(2^\ell - 1)$ with $2^\ell - 1$ being a prime.

The sufficiency part is easy but discussed in the next Note. As to the necessity, let an even integer $n = 2^{\ell-1}n'$, $2 \nmid n'$, $\ell \geq 2$, be perfect; then the multiplicativity of σ_1 implies $(2^\ell - 1)\sigma_1(n') = 2^\ell n'$ or $\sigma_1(n') = n' + n'/(2^\ell - 1)$. Hence, $(2^\ell - 1)|n'$; and $n'/(2^\ell - 1)$ is a divisor of n'; that is, n' has only two divisors; therefore, n' is a prime and equal to $2^\ell - 1$. This illuminating argument is due to Dickson (1911). On the other hand, Euler used the fact $\langle 2^\ell, 2^\ell - 1 \rangle = 1$. Thus, $n' = (2^\ell - 1)a$ and $\sigma_1(n') = 2^\ell a$ with a certain $a \geq 1$. If $a = 1$, then $\sigma_1(n') = n' + 1$, and n' is a prime. If $a > 1$, then $\sigma_1(n') \geq n' + a + 1 = 2^\ell a + 1$, which is a contradiction.

[16.4] The characterization of even perfect numbers can be traced back to Nicomachus (ca 100: D'Ooge (1926, p.210)). Fibonacci (1202: 1857, p.283) listed the cases $\ell = 2, 3, 5$, and continued: *poteris in infinitum perfectos numeros reperire*, which sounds like a conjecture. Concerning the existence of odd perfect numbers, extensive investigations have been conducted since Euler (1849a, pp.514–515); however, the problem still remains unsettled.

[16.5] The final assertion in Euclid's number theory is [Σ.IX, Prop.36]:
If as many numbers as we please beginning from an [sic] *unit be set out continuously in double proportion, until the sum of all becomes prime, and if the sum multiplied into the last make some number, the product will be perfect.*
Since the notation for powers was not known yet, Euclid defined the sequence $\{2^j : j \geq 0\}$ by repeated doubling; the number ℓ of repetitions is such that

62 *Divisibility*

$P = 2^\ell - 1$ is a prime; so $\ell \geq 2$. He discussed meticulously the aliquot divisors of $2^{\ell-1}P$ and demonstrated that they are exhausted by

$$\{1, 2, 4, \ldots, 2^{\ell-1}\} \sqcup \{P, 2P, 4P, \ldots, 2^{\ell-2}P\},$$

with which he ended his proof of the proposition. That is the same as we do nowadays. Namely, a logical procedure of finding all divisors of a composite integer, which is not a power of a single prime number, was explicitly described in [Σ.IX, Prop.36], most probably for the first time in history. Note that (16.5) is what Gauss [DA, art.17] showed immediately after his assertion of the uniqueness of prime power decomposition. He organized his discussion in view of the fact that in order to exhaust divisors of a given integer the uniqueness assertion of Theorem 10 is indispensable. In exactly the same sense, Euclid was perfectly aware that he had to clarify the way to exhaust the divisors of $2^{\ell-1}P$. It might be too haste to connect this historical fact with the notion of the uniqueness of prime power decomposition, as the integer $2^{\ell-1}P$ is of a very special shape. Nevertheless, it can be asserted that in the sense of the uniqueness of the set of divisors of each integer, Euclid knew well Theorem 10, at least empirically; more will be said in the next Note on this critical point. If it had not been for common awareness of the contents of Theorem 10, prime numbers would have been nothing but exotic objects which could scarcely gain such a distinctive treatment as in the *Elements*.

[16.6] In [Σ.IX, Prop.13] it is shown that the aliquot divisors of p^k are exhausted by p^s, for $s < k$, with which it is asserted in Prop.36 that the aliquot divisors of $2^{\ell-1}$ are 2^r, $r < \ell - 1$. That is, Euclid certainly treated 2 as a prime number (see Note [10.3]). Here we have a far more important inference: If Euclid had combined his Prop.13 with our Theorem 13 which was well within his reach, he would have been able to connect the prime power decomposition of an arbitrary integer and the set of its divisors. Since the latter is obviously unique for each integer, one may assert at least that Euclid was close to Theorem 10. If Euclid joined us and were shown Theorem 10, then he would say that the assertion is too trivial to be given the status of a theorem.

[16.7] To have a prime $2^\ell - 1$ with $\ell \geq 3$, the exponent ℓ needs to be a prime, for the polynomial $X^{ab} - 1$ is divisible by $X^a - 1$. Thus, numbers $M_p = 2^p - 1$ come into attention. Mersenne (1644, Præfatio generalis, XIX) claimed in effect that M_p are primes for $p = 2, 3, 5, 7, 13, 17, 19, 31, 67, 127, 257$; because of this, each M_p is called nowadays a Mersenne number. Later M_{67}, M_{257} were found to be composite; on the other hand, among $M_p, p < 257$, Mersenne missed primes M_{61}, M_{89}, M_{107}. It is conjectured that there should be infinity many Mersenne primes M_p; that is, there should be infinitely many even perfect

$\S16$ 63

numbers. About the primality test on Mersenne numbers, see Notes [41.1], [58.8], [84.10] below.

[16.8]   As for $2^\ell + 1$ whose construction is similar to M_p, it is composite if $\ell = ab$, $2 \nmid a$, $a \geq 3$, for the polynomial $(X^b)^a + 1$ is divisible by $X^b + 1$. Hence, whether or not $F_r = 2^{2^r} + 1$ is a prime becomes an issue. These are called Fermat numbers, and it is conjectured that only a finite number of them should be primes. Up to now, beyond the prime numbers $F_0 = 3$, $F_1 = 5$, $F_2 = 17$, $F_3 = 257$, $F_4 = 65537$, which were originally pointed out by Fermat (1640: 1679, p.115), none else has been confirmed to be a prime yet. The number F_5 is to be treated in Note [41.1]. By the way, the identity $F_0 \cdot F_1 \cdots F_r = F_{r+1} - 2$ implies that $\{F_r : r \geq 0\}$ are mutually coprime (i.e., (7.6)), which yields an alternative proof of Theorem 9. Thus, via prime power decompositions of F_r, we shall obtain a sequence of prime numbers which expands indefinitely with n. However, it is tremendously difficult to execute the factorization of F_r with r large. It should be added that Fermat primes F_r are closely connected with the problem of drawing regular convex polygons. See Notes [65.2]–[65.3] below.

[16.9]   On the attribution of (16.8): Dirichlet (1849) introduced the splitting method (16.27). It extends readily to the discussion of the generic sum

$$\sum_{kl \leq x} f(k)g(l) = \sum_{n \leq x} (f * g)(n).$$

Because of this identity, the attribution became a common practice, although Dirichlet's interest was in the left side, and the right side is in fact a later expression. For a typical application of his splitting method, see Motohashi (1987) and the proof of Theorem 148.

[16.10]   Here is another deep problem induced by the notion of divisors: The error term in (16.26) is commonly denoted by $\Delta(x)$. The bound $\Delta(x) \ll \sqrt{x}$ is due to Dirichlet (1849) as we have seen above. The divisor problem is to determine the best possible upper bound for $|\Delta(x)|$. Accordingly there arises one of the hardest problems in number theory: it is to validate

$$\text{conjecture:} \quad \Delta(x) \ll x^{1/4+\varepsilon}.$$

On this a multitude of investigations have been conducted until today; however, we shall not give details, as they are all highly technical, and moreover not much related to the principal purposes of the present essays. Nevertheless, it should be noted that there exist statistical assertions supporting this conjecture. Among them, the best result to date is

64 Divisibility

$$\int_1^x \{\Delta(y)\}^2 dy = \frac{\zeta^4(3/2)}{6\pi^2\zeta(3)} x^{3/2} + O\left(x(\log x)^3 \log\log x\right),$$

due to Lau and Tsang (2009). Their argument depends on the Fourier–Bessel expansion of the sum

$$\sum_{n=a}^b d(n)f(n)$$

by Voronoï (1904, p.529) and the inequality of Montgomery–Vaughan (1974) that is stated as (3) in our Note [103.2]; here f is a sufficiently smooth function in the closed interval $[a,b]$. By the way, the Voronoï expansion is equivalent to the functional equation for $\zeta^2(s)$ in much the same sense as the Poisson sum formula corresponds to (12.7); see Note [98.3].

[16.11]   To extend (16.17) to short intervals and arithmetic progressions is by no means trivial. See Theorem 142 below, in which the exponent λ is allowed to be any positive number.

[16.12]   It should be noted that although the ordinary product of multiplicative functions, i.e., $(f_1 \cdot f_2)(n) = f_1(n)f_2(n)$, is of course multiplicative, the Dirichlet series $\lfloor f_1 \cdot f_2 \rfloor(s)$ is usually not easy to handle. Nevertheless, if f_1, f_2 come from Fourier expansions of automorphic functions, then there exists a possibility to be able to deal with $\lfloor f_1 \cdot f_2 \rfloor(s)$ by means of the theory called the Rankin convolution. See for instance Motohashi (1997, pp.107–111). See also Note [18.7] below.

§17

The operation (16.1) is a linear transformation in the linear space spanned by all arithmetic functions over $\mathbb{C}$. It is invertible; and the inverse is called the Möbius inversion, a key notion in number theory:

In order to invert (16.1), as we shall see below, it suffices to fix an arithmetic function μ on $\mathbb{N}$ such that

$$\mu * \iota = \iota * \mu = \delta_1, \tag{17.1}$$

with δ_1 as in (16.10). Namely, we need to have

$$\sum_{d\mid n} \mu(d) = \begin{cases} 1 & n = 1, \\ 0 & n > 1. \end{cases} \tag{17.2}$$

§17   65

In view of (16.6), this is equivalent to the relation $1 \equiv \lfloor \delta_1 \rfloor (s) = \zeta(s) \lfloor \mu \rfloor (s)$; hence,

$$\sum_{n=1}^{\infty} \frac{\mu(n)}{n^s} = \frac{1}{\zeta(s)} = \prod_p \left(1 - \frac{1}{p^s}\right), \qquad \mathrm{Re}\, s > 1, \qquad (17.3)$$

where we have used (11.7). Expanding this infinite product and recalling (15.10), we are led to the definition of the Möbius function:

$$\mu(n) = \begin{cases} 1 & n = 1, \\ (-1)^J & n = p_1 p_2 \cdots p_J \ (\text{sqf}), \\ 0 & \text{otherwise}, \end{cases} \qquad (17.4)$$

which is of course multiplicative.

We now have the Möbius inversion formula:

Theorem 15 *Let $F(n) = \sum_{d|n} f(d)$ with an arbitrary arithmetic function f over $\mathbb{N}$. Then we have*

$$f(n) = \sum_{d|n} \mu(d) F(n/d). \qquad (17.5)$$

Proof For

$$\sum_{d|n} \mu(d) \sum_{t|n/d} f(t) = \sum_{t|n} f(t) \sum_{d|n/t} \mu(d) = f(n). \qquad (17.6)$$

The second equality is implied by (17.2). One may instead apply μ to $\iota * f = F$ and use (17.1), or equivalently, via a combination of (16.6), (16.14), and (17.3),

$$\lfloor f \rfloor (s) = \frac{\lfloor F \rfloor (s)}{\zeta(s)} = \lfloor \mu * F \rfloor (s) \;\Rightarrow\; f = \mu * F. \qquad (17.7)$$

We end the proof.

Note that the function f need not be multiplicative; neither $F(n)$ need be defined for all n: for a particular n such that $F(d)$ is defined for all divisors d of n, the inversion formula (17.5) holds.

The notion of the Möbius function can be generalized variously. For instance, if an arithmetic function f over $\mathbb{N}$ satisfies $f(1) \neq 0$, then there exists an arithmetic function μ_f such that $f * \mu_f = \delta_1$: define it inductively by

$$\mu_f(1) = 1/f(1), \quad \mu_f(n) = -\mu_f(1) \sum_{d|n, d<n} f(n/d) \mu_f(d). \qquad (17.8)$$

We have, of course, $\lfloor \mu_f \rfloor (s) = \left\{ \lfloor f \rfloor (s) \right\}^{-1}$, provided $\mathrm{Re}\, s$ is sufficiently large; see Note [17.7]. Hence, the set of all arithmetic functions f over $\mathbb{N}$ such that

66   *Divisibility*

$f(1) \neq 0$ is an Abelian group with respect to the operation $*$; in particular, $\mu_{f*g} = \mu_f * \mu_g$ since $(f * g) * (\mu_f * \mu_g) = (f * \mu_f) * (g * \mu_g) = \delta_1$. Then the set of all multiplicative functions over $\mathbb{N}$ is its subgroup. To confirm this, let f be multiplicative and let $a, b > 1$, $\langle a, b \rangle = 1$. We shall argue inductively; namely, we suppose that $\mu_f(tu) = \mu_f(t)\mu_f(u)$ for $t|a$, $u|b$, $tu < ab$: Then, by (15.6)–(15.7) and (17.8),

$$
\begin{aligned}
\mu_f(ab) &= - \sum_{\substack{t|a, u|b \\ tu < ab}} f(ab/tu)\mu_f(tu) \\
&= - \sum_{\substack{t|a, u|b \\ tu < ab}} f(a/t)f(b/u)\mu_f(t)\mu_f(u) \\
&= - \sum_{t|a} f(a/t)\mu_f(t) \sum_{u|b} f(b/u)\mu_f(u) + \mu_f(a)\mu_f(b) \\
&= -(f * \mu_f)(a)(f * \mu_f)(b) + \mu_f(a)\mu_f(b) = \mu_f(a)\mu_f(b).
\end{aligned}
\tag{17.9}
$$

Alternatively, use the Euler product (15.16).

For instance, with $f = d$ the divisor function (17.8) implies

$$
\mu_d(p^\nu) = \begin{cases} 1 & \nu = 0, 2, \\ -2 & \nu = 1, \\ 0 & \nu \geq 3, \end{cases} \quad \Leftrightarrow \quad \lfloor \mu_d \rfloor(s) = (\zeta(s))^{-2}.
\tag{17.10}
$$

Or one may use $\mu_d = \mu_{1*1} = \mu * \mu$.

Another way to extend the discussion prior to (17.5) is to apply a variation to (17.3): Regarding the latter as equivalent to the trivial relation $1 = \zeta(s)/\zeta(s)$, we consider instead $\zeta(s - \xi)/\zeta(s)$. That is, we exploit the relation

$$
\sum_{d|n} \mu(d)(n/d)^\xi = \prod_{p|n} \left(p^{n(p)\xi} - p^{(n(p)-1)\xi} \right).
\tag{17.11}
$$

Then, the Taylor expansion at $\xi = 0$ yields the von Mangoldt function Λ: from the left side of (17.11)

$$
\Lambda = \mu * \log : \quad \Lambda(n) = \sum_{d|n} \mu(d) \log(n/d), \quad n \in \mathbb{N},
\tag{17.12}
$$

and from the right side

$$
\Lambda(n) = \begin{cases} \log p & n \text{ is a power of a } p, \\ 0 & \text{otherwise.} \end{cases}
\tag{17.13}
$$

$$\S 17 \qquad 67$$

This is a fundamental arithmetic function in the theory of the distribution of primes: Its Dirichlet series expansion is, for $\mathrm{Re}\, s > 1$,

$$
\lfloor \Lambda \rfloor(s) = \sum_{n=1}^{\infty} \frac{\Lambda(n)}{n^s} = -\frac{\zeta'}{\zeta}(s)
$$

$$
= -\frac{d}{ds} \log \zeta(s) = \sum_{p} \sum_{j=1}^{\infty} \frac{\log p}{p^{js}} .
$$
(17.14)

The first line is of course a special case of (17.7), and the second line is equivalent to the Euler product (11.7). The main part in the rightmost side is the sum with $j = 1$; and (12.1) gives

$$
\sum_{p} \frac{\log p}{p^s} = \frac{1}{s-1} + O(1) \quad \text{as } s \to 1+0,
$$
(17.15)

which implies Theorem 9, of course. Here we actually need to argue a little more carefully, for (17.14) involves $\log \zeta(s)$, which can be multi-valued, although $(\zeta'/\zeta)(s)$ itself does not have such an ambiguity. One way to avoid the logarithm of $\zeta(s)$ is to use the representation $\zeta(s) = \prod_{p<x}(1 - p^{-s})^{-1} W(s,x)$ with $\mathrm{Re}\, s > 1$, so

$$
\frac{\zeta'}{\zeta}(s) = -\sum_{p<x} \frac{\log p}{p^s - 1} + \frac{W'}{W}(s,x).
$$
(17.16)

Then, observe that $\lim_{x\to\infty}(W'/W)(s,x) = 0$. A discussion concerning the definition of the logarithm of $\zeta(s)$ will be given in (94.48)–(94.52) below.

On the other hand, via the functional equation (12.7)–(12.8), the function $(\zeta'/\zeta)(s)$ continues meromorphically to the entire complex plane. The poles are located at the zeros of $\zeta(s)$ and at its sole pole $s = 1$. Therefore, we see immediately that the distribution of primes and that of zeros of $\zeta(s)$ should be closely related to each other. This is the grand perspective that Riemann (1860) opened, which we already indicated in §12. As a matter of fact, he dealt with not $(\zeta'/\zeta)(s)$ but $\log \zeta(s)$. The latter is harder than the former to handle, for the nature of its singularities is subtler; the details can be given only in the paragraph of §94 referred to above. The meromorphy of $(\zeta'/\zeta)(s)$ means that we should have more analytic tools applicable to it than to $\log \zeta(s)$. This is the reason that the whole discussion of Chapter 5 is concerned with the von Mangoldt function Λ.

Chebyshev (1854) was the first who recognized the fundamental fact that primes should better be counted with logarithmic amplification: $\sum_{p\leq x} \log p$, and still better with the weight Λ, i.e., via his psi-function

Divisibility

$$\psi(x) = \sum_{n \leq x} \Lambda(n) = \sum_{p^m \leq x} \log p = \sum_{p \leq x} (\log p)[\log x / \log p], \qquad (17.17)$$

although he used a notation other than Λ; here we suppose that $x \notin \mathbb{N}$, without any essential loss of generality. Well prior to the introduction of the notation Λ (see Note [17.4]), Chebyshev's $\psi(x)$ had become a standard function in the theory of the distribution of primes; this is the legacy mentioned in §12(5). The Euler product forced Chebyshev and forces us to ponder on the behavior of $\psi(x)$ instead of directly dealing with $\pi(x)$.

Notes

[17.1]   Euler (1748b, Tomi primi caput XV, §269) observed essentially the same as (17.3).

[17.2]   Möbius (1832) introduced the notion of his function μ; however, this familiar notation is due to Mertens (1874; 1897b). Möbius discussed actually the inversion of the power series $A(x) = \sum_{n=1}^{\infty} \alpha(n) x^n$ in the following sense: ignoring the convergence issue,

$$x = \sum_{n=1}^{\infty} \beta(n) A(x^n);$$

obviously $\alpha(1) \neq 0$ is necessary. This is equivalent to $\delta_1 = \alpha * \beta$; hence, $\beta = \mu_\alpha$ in the notation introduced above. What Möbius considered is, thus, more general than (17.1). Dedekind (1857a, p.21: the argument to deduce the same identity as our (67.5)) seems to be among the firsts who noticed (17.5), although he did not use the function μ but merely the right side of (17.4).

[17.3]   Since the definition of the Möbius function bases essentially on prime power decomposition, it might appear to be a concept deeper than the divisor function. However, repeating the last lines of Note [2.6], we stress that it can also be an extremely hard task in general to find a divisor of a given integer. Hence, neither of these arithmetic functions is a self-evident object at all. In the same token, the relation

$$|\mu(n)| = \sum_{d^2 | n} \mu(d) = \begin{cases} 1 & n : \text{sqf}, \\ 0 & \text{otherwise}, \end{cases}$$

never makes the notion of being sqf clearer. Indeed, there exists yet no known algorithm to test in polynomial time whether a given integer is sqf or not.

[17.4]   von Mangoldt (1895, pp.277–278) introduced (17.12), although he used the symbol L, not Λ. The latter originated with Landau (1909, p.125).

[17.5] For any real number $x \geq 1$, we have

$$K(x) = \sum_{m \leq x} k(x/m) \Rightarrow k(x) = \sum_{n \leq x} \mu(n)K(x/n),$$

as the latter sum is

$$\sum_{n \leq x} \mu(n) \sum_{m \leq x/n} k(x/mn) = \sum_{r \leq x} k(x/r) \sum_{mn=r} \mu(n) = k(x).$$

This reasoning is essentially due to Mertens (1897b); but see Riemann (1860, p.679). If $x \in \mathbb{N}$ and the function k is supported on $\mathbb{N}$, then we get (17.5). In the same article Mertens stated, based on the numerical data up to $x = 9998$, the following conjecture: with $M(x) = \sum_{n \leq x} \mu(n)$,

$$|M(x)| < \sqrt{x}, \quad \forall x > 1.$$

This would imply RH in particular, as remarked by himself (p.780); it is enough to note the representation, via Stieltjes integral,

$$\frac{1}{\zeta(s)} = \int_{1-0}^{\infty} x^{-s} dM(x) = s \int_{1}^{\infty} M(x) x^{-s-1} dx.$$

However, the conjecture is disproved by Odlyzko and te Riele (1984); yet any specific value of x such that $|M(x)| > \sqrt{x}$ has not been discovered. According to the historical details given by these authors, the conjecture should have been attributed to T.J. Stieltjes.

[17.6] On an obvious restriction,

$$H(n) = \prod_{d|n} h(d) \Rightarrow h(n) = \prod_{d|n} \left(H(n/d)\right)^{\mu(d)}.$$

For a proof, replace $H(n/d)$ by its definition and use (17.1).

[17.7] As for the convergence of the Dirichlet series $\lfloor \mu_f \rfloor(s)$ see Montgomery–Vaughan (2006, pp.162–163).

§18

One of the principal applications of the Möbius function is in the following formulation of the notion of coprimality:

$$(\mu * \iota)(\langle a, b \rangle) = \begin{cases} 1 & a, b \text{ are coprime,} \\ 0 & \text{otherwise.} \end{cases} \tag{18.1}$$

70 *Divisibility*

Rewriting the left side by (3.7),

$$\sum_{d|a,d|b} \mu(d) = \begin{cases} 1 & a,b \text{ are coprime,} \\ 0 & \text{otherwise.} \end{cases} \tag{18.2}$$

Or more generally, we have, for any arithmetic function f,

$$\sum_{d|a,d|b} (\mu * f)(d) = f(\langle a,b \rangle); \tag{18.3}$$

and (18.2) is the case with $f = \mu * \iota = \delta_1$. The merit of (18.3) is in its effect of separating the variables a, b trapped in $f(\langle a,b \rangle)$. Although this may appear trivial, it often turns out to be quite powerful in applications, especially to the issues related to sieves, which we shall briefly discuss in the next section.

A very basic consequence of (18.2) is the following explicit formula of the Euler phi-function φ. The value of $\varphi(n)$, $n \in \mathbb{N}$, is defined as the number of positive integers $h \leq n$ such that $\langle h,n \rangle = 1$.

Theorem 16 *We have, for any $n \in \mathbb{N}$,*

$$\varphi(n) = n \prod_{p|n} \left(1 - \frac{1}{p}\right), \tag{18.4}$$

where the product is with respect to all different prime factors of n.

Proof Suppose that $n \geq 2$. Then (18.2) gives

$$\varphi(n) = \sum_{h \leq n} \sum_{d|h,d|n} \mu(d) = \sum_{d|n} \mu(d) \sum_{\substack{h \leq n \\ d|h}} 1 = n \sum_{d|n} \frac{\mu(d)}{d}. \tag{18.5}$$

According to (16.2), the last sum is multiplicative, and by (16.3) we obtain (18.4). Also, $\varphi(1) = 1$ by definition; the product in (18.4)$_{n=1}$ is empty, and equal to 1; hence, (18.4) holds for all $n \in \mathbb{N}$. This ends the proof.

In particular, we find that

$$\langle m,n \rangle = 1 \implies \varphi(mn) = \varphi(m)\varphi(n). \tag{18.6}$$

That is, φ is multiplicative. Later, in §32, we shall show an alternative proof of this fact in a wider perspective.

The identity (18.5) is the same as

$$\varphi = \mu * I, \quad I(k) = k, \ \forall k \in \mathbb{N}. \tag{18.7}$$

Hence, $\iota * \varphi = I$; that is,

$$\sum_{d|n} \varphi(d) = n. \tag{18.8}$$

$$\S 18 \qquad\qquad 71$$

Alternatively, since (6.2) (2) implies that $|\{1 \le m \le n : \langle m,n \rangle = n/d\}| = \varphi(d)$ for any $d|n$, we get (18.8). Also, (18.8) is a consequence of writing each of the n fractions k/n, $1 \le k \le n$, in terms of irreducible fractions. Further, by (16.3) and (18.6), the sum (18.8) equals $\prod_{p^\alpha \| n} \sum_{j=0}^{\alpha} \varphi(p^j) = n$, as $\varphi(p^j) = p^j - p^{j-1}$. We note finally that (18.8) characterizes φ since one can return to (18.7) from (18.8), that is, $I = \iota * \varphi$ implies that $\mu * I = (\mu * \iota) * \varphi = \varphi$.

Another application of (18.2) concerns the Ramanujan sum which is defined as

$$c_q(n) = \sum_{\substack{h=1 \\ \langle h,q \rangle = 1}}^{q} e\big(nh/q\big), \quad e(x) = \exp(2\pi i x). \tag{18.9}$$

We shall show that

$$\langle q_1, q_2 \rangle = 1 \;\Rightarrow\; c_{q_1 q_2}(n) = c_{q_1}(n) c_{q_2}(n), \tag{18.10}$$

and

$$c_q(n) = c_q(\langle q,n \rangle). \tag{18.11}$$

These are immediate consequences of the following explicit formula:

Theorem 17 *For any $n,q \in \mathbb{N}$, we have*

$$c_q(n) = \frac{\mu(q/\langle q,n \rangle)}{\varphi(q/\langle q,n \rangle)} \varphi(q). \tag{18.12}$$

Proof We note that

$$\sum_{r|d} c_r(n) = \sum_{u=1}^{d} e(un/d) = \begin{cases} d & d|n, \\ 0 & d \nmid n. \end{cases} \tag{18.13}$$

Then the Möbius inversion (17.5) gives

$$c_q(n) = \sum_{d|\langle q,n \rangle} \mu(q/d)d. \tag{18.14}$$

Alternatively, we use (18.2), and have

$$c_q(n) = \sum_{h=1}^{q} e(nh/q) \sum_{d|q,\, d|h} \mu(d) = \sum_{d|q} \mu(d) \sum_{h=1}^{q/d} e(nh/(q/d))$$

$$= q \sum_{d|q,\, (q/d)|n} \frac{\mu(d)}{d} = \sum_{d|\langle q,n \rangle} \mu(q/d)d. \tag{18.15}$$

The assertions (18.10) and (18.11) are now obvious. Hence,

$$c_q(n) = \prod_{p} c_{p^{q(p)}}(n). \tag{18.16}$$

72 *Divisibility*

Here

$$c_{p^\alpha}(n) = \begin{cases} p^\alpha - p^{\alpha-1} & \langle p^\alpha, n \rangle = p^\alpha \\ -p^{\alpha-1} & \langle p^\alpha, n \rangle = p^{\alpha-1} \\ 0 & \text{otherwise} \end{cases} = \frac{\mu(p^\alpha/\langle p^\alpha, n \rangle)}{\varphi(p^\alpha/\langle p^\alpha, n \rangle)}\varphi(p^\alpha). \quad (18.17)$$

This ends the proof.

In particular, we have a trigonometrical or Fourier expansion of the Möbius function: Taking $n = 1$ in (18.12),

$$\mu(q) = \sum_{\substack{h=1 \\ \langle h, q \rangle = 1}}^{q} e(h/q). \quad (18.18)$$

Also it is worth remarking that (18.13) gives

$$\sigma_\alpha(n) = \zeta(1-\alpha)\sum_{q=1}^{\infty} \frac{c_q(n)}{q^{1-\alpha}}, \quad \operatorname{Re}\alpha < 0. \quad (18.19)$$

See Notes [50.4]–[50.5] for a basic rôle played by $c_q(n)$ in the divisibility issue general.

Notes

[18.1] The fact (18.6) was discovered by Euler (1758a, Theorema 5); the formula (18.4) was stated as a corollary. His proof is similar to that of our Theorem 33, §32. Gauss [DA, art.38] adopted that proof by Euler. The present proof is essentially due to Legendre (see Note [19.2]) and similar to Euler's second proof (1775). Euler used the symbol π, and Gauss ϕ. The notation φ originated with Dirichlet (1863, §11). Occasionally φ is called the totient function; it is a coinage by Sylvester (1879, p.361), although he used the symbol τ. The fact (18.8) was first stated by Gauss [DA, art.39]. It may be worth remarking that Euler (1775) pondered upon the distribution of irreducible fractions while dealing with φ, i.e., his function π.

[18.2] For arbitrary $m, n > 0$, we have $\varphi(mn) \geq \varphi(m)\varphi(n)$. The equality holds if and only if $\langle m, n \rangle = 1$. More precisely, we have

$$\varphi(mn) = \varphi(m)\varphi(n)\frac{\langle m, n \rangle}{\varphi(\langle m, n \rangle)}.$$

[18.3] The Euler function $\varphi(n)$ may appear simple, at least outwardly. In fact, its nature has not been fully understood yet. A reason for this is in that the prime power decomposition of the value $\varphi(n)$ can be far more involved than that of n; see §46 below. Consequently, for instance, the Carmichael conjecture (1907/1922)

$$|\{n : \varphi(n) = \varphi(m)\}| \geq 2, \quad \forall m \geq 1$$

remains open. Also, the totient problem of Lehmer, D.H. (1932) is in the same situation: it concerns whether or not there exists any composite n such that

$$\varphi(n)|(n - 1);$$

see Note [37.2]. We add that the formula (18.4) is not practical but theoretical, for it requires the prime power decomposition of the argument n which is most always extremely hard to achieve, if n is huge. This fact is exploited in modern cryptography; see Notes [49.2]–[49.3].

[18.4] If $a, b \in \mathbb{N}$ are such that $a\varphi(a) = b\varphi(b)$, then $a = b$. To prove this, let $a_1 = a/a_2$ with $a_2 = \langle a, \langle a, b \rangle^{\infty} \rangle$; $b_1 = b/b_2$ with $b_2 = \langle b, \langle a, b \rangle^{\infty} \rangle$. The assumption gives

$$(a_1 a_2)^2 \prod_{p|a_1}(1 - 1/p) = (b_1 b_2)^2 \prod_{p|b_1}(1 - 1/p).$$

Let $a_3 = a_1^2/(\prod_{p|a_1} p)$, $b_3 = b_1^2/(\prod_{p|b_1} p)$. Then

$$a_2^2 a_3 \prod_{p|a_1}(p - 1) = b_2^2 b_3 \prod_{p|b_1}(p - 1).$$

Since $\langle a_3, b_2 b_3 \rangle = 1$, we have $a_3 | \prod_{p|b_1}(p - 1)$, and analogously $b_3 | \prod_{p|a_1}(p - 1)$. Hence, if $a_3 \neq 1$, then neither a_1 nor b_1 is equal to 1; and

$$a_3 < \prod_{p|b_1} p \leq b_1 \leq b_3 < \prod_{p|a_1} p \leq a_1 \leq a_3,$$

which is a contradiction. Therefore, we should have $a_3 = 1$, i.e., $a_1 = 1$; and $b_3 = 1$, i.e., $b_1 = 1$. The rest of the argument can be omitted. See Niven et al (1991, p.74, Problem 41).

[18.5] When an arbitrary pair $a, b \in \mathbb{N}$ is chosen, the probability that $\langle a, b \rangle = 1$ is $6/\pi^2 > 3/5$. To explain this somewhat obscure claim, we note that (18.2) gives

$$\sum_{\substack{\langle a,b \rangle=1 \\ a,b \leq N}} 1 = \sum_{a,b \leq N} \sum_{d|a, d|b} \mu(d)$$

$$= \sum_{d \leq N} \mu(d)[N/d]^2 = N^2 \sum_{d=1}^{\infty} \frac{\mu(d)}{d^2} + R_N,$$

74 *Divisibility*

where

$$R_N = \sum_{d \le N} \mu(d)\left([N/d]^2 - (N/d)^2\right) - N^2 \sum_{d > N} \frac{\mu(d)}{d^2}.$$

On noting the inequality $|[\theta]^2 - \theta^2| \le 2\theta + 1$, we see that for any sufficiently large N

$$|R_N| \ll N \sum_{d \le N} \frac{1}{d} + N^2 \sum_{d > N} \frac{1}{d^2} \ll N \log N,$$

with $\ll$ as in Note [12.8] above. That is,

$$\lim_{N \to \infty} \frac{1}{N^2} \sum_{\substack{\langle a,b \rangle = 1 \\ a,b \le N}} 1 = \sum_{d=1}^{\infty} \frac{\mu(d)}{d^2} = \prod_p \left(1 - \frac{1}{p^2}\right).$$

By (17.3) and by Euler (1735, §18), the last product equals $1/\zeta(2) = 6/\pi^2$; see Euler (1737b, Th.8, Cor.1) as well as Note [60.1] below.

[18.6] For every $n \in \mathbb{N}$

$$\varphi(n) \gg \frac{n}{\log \log 3n}.$$

To confirm this, we note that

$$\log(n/\varphi(n)) = -\sum_{p|n} \log\left(1 - \frac{1}{p}\right) = \sum_{p|n} \frac{1}{p} + O(1).$$

The sum on the right side is

$$\le \sum_{p \le \log n} + \sum_{\substack{p|n \\ p > \log n}} < \log \log \log n + O(1) + \frac{v(n)}{\log n}, \quad v(n) = \sum_{p|n} 1,$$

where we have applied the asymptotic formula for $\sum_{p \le x} 1/p$ (Note [13.8]), assuming that n is sufficiently large. Also, since $2^{v(n)} \le n$, we have $v(n) \le \log n / \log 2$. This ends the proof.

[18.7] Ramanujan (1916). If $\operatorname{Re} s > 1 + \max\{0, \operatorname{Re} \alpha, \operatorname{Re} \beta, \operatorname{Re} (\alpha + \beta)\}$, then

$$(\dagger) \qquad \lfloor \sigma_\alpha \sigma_\beta \rfloor(s) = \frac{\zeta(s)\zeta(s - \alpha)\zeta(s - \beta)\zeta(s - \alpha - \beta)}{\zeta(2s - \alpha - \beta)}.$$

The present proof of this famous identity is an application of (18.3); it is different from the known one that relies on the Euler product (11.7) (see Titchmarsh (1951, pp.8–9)): Thus, by $(14.1)_{K=2}$,

$$\sigma_\alpha(n)\sigma_\beta(n) = \sum_{u|n,\, v|n} u^\alpha v^\beta = \sum_{[u,v]|n} u^\alpha v^\beta;$$

so

$$\lfloor\sigma_\alpha\sigma_\beta\rfloor(s) = \zeta(s)\sum_{u,v}\frac{u^\alpha v^\beta}{[u,v]^s} = \zeta(s)\sum_{u,v}\frac{\langle u,v\rangle^s}{u^{s-\alpha}v^{s-\beta}},$$

in which (10.6) is applied. Next, we let $f(n) = n^s$ in (18.3); then

$$\lfloor\sigma_\alpha\sigma_\beta\rfloor(s) = \zeta(s)\sum_{u,v}\frac{1}{u^{s-\alpha}v^{s-\beta}}\sum_{d|u,d|v}(\mu*f)(d)$$

$$= \zeta(s)\zeta(s-\alpha)\zeta(s-\beta)\sum_d\frac{(\mu*f)(d)}{d^{2s-\alpha-\beta}}.$$

This ends the proof, for (17.7) gives $\lfloor\mu*f\rfloor(w) = \zeta(w-s)/\zeta(w)$, provided $\operatorname{Re} w > 1 + \operatorname{Re} s$. In particular, with $\alpha = \beta = 0$, we obtain a typical instance of Ramanujan's beautiful formulas:

$$\sum_{n=1}^{\infty}\frac{d^2(n)}{n^s} = \frac{\zeta^4(s)}{\zeta(2s)}, \quad \operatorname{Re} s > 1.$$

Since the function σ_α appears in the Fourier expansion of Eisenstein series (see Note [61.8] below), these results can be understood via the mechanism of the Rankin convolution mentioned in Note [16.12]. By the way, the assertion of Note [16.1] can be proved alternatively as follows: We have, in the region of absolute convergence,

$$\sum_{m=1}^{\infty}\sum_{n=1}^{\infty}\frac{\sigma_\alpha(m)\sigma_\alpha(n)}{m^{s_1}n^{s_2}} = \zeta(s_1)\zeta(s_2)\zeta(s_1-\alpha)\zeta(s_2-\alpha)$$

$$= \zeta(s_1+s_2-\alpha)\sum_{l=1}^{\infty}\frac{\sigma_\alpha(l)\sigma_{s_1-s_2}(l)}{l^{s_1}}$$

because of (†); and this is equal to

$$\zeta(s_1+s_2-\alpha)\sum_{m=1}^{\infty}\sum_{n=1}^{\infty}\frac{\sigma_\alpha(mn)}{m^{s_1}n^{s_2}}$$

$$= \sum_{t=1}^{\infty}\sum_{m=1}^{\infty}\sum_{n=1}^{\infty}\frac{t^\alpha\sigma_\alpha(mn)}{(tm)^{s_1}(tn)^{s_2}} = \sum_{m=1}^{\infty}\sum_{n=1}^{\infty}\frac{1}{m^{s_1}n^{s_2}}\sum_{t|\langle m,n\rangle}t^\alpha\sigma_\alpha(mn/t^2),$$

which ends the proof. Here we have used Dirichlet series of two variables; to them the analytic uniqueness assertion (15.10) extends in an obvious way.

[18.8] Smith (1876) discovered the identity

$$\det\left(f(\langle m,n\rangle)\right) = \prod_{a=1}^{N}(\mu*f)(a), \quad 1 \le m,n \le N,$$

76   *Divisibility*

where the function f and the order N of the matrix are arbitrary. For the proof, we temporarily extend the definition of the function μ so that $\mu(x) = 0$ if $x \notin \mathbb{N}$. The matrix $(\mu(n/m))$, $1 \le m, n \le N$, is upper triangular; the diagonal entries are all equal to 1. Then, the product $(f(\langle m,n \rangle)) \cdot (\mu(n/m))$ is lower triangular; the diagonal entries are $(\mu * f)(a)$, $a \le N$. For the $\{k,l\}$-element of this matrix product is

$$\sum_{r=1}^{N} f(\langle k, r \rangle) \mu(l/r) = \sum_{s|l} f(\langle k, l/s \rangle) \mu(s)$$

$$= \sum_{s|l} \mu(s) \sum_{t|\langle k, l/s \rangle} (\mu * f)(t) = \sum_{t|\langle k, l \rangle} (\mu * f)(t) \sum_{s|l/t} \mu(s),$$

where (18.3) is used. Only when $l | \langle k, l \rangle$, i.e., $l | k$, the last sum is equal to $(\mu * f)(l)$; otherwise it vanishes. This ends the proof.

[18.9]   The sum $c_q(n)$ was introduced by Ramanujan (1918). The explicit formula (18.12) is due to Hölder (1936). The formula (18.19) is called the Ramanujan expansion of σ_α; see his assertion (1918, (6.1)). This formula might look somewhat bizarre; in fact, it is a fundamental means in the theory of automorphic forms, which will be made explicit in Note [61.8] as has been suggested already in Note [18.7]. For a typical exploitation of (18.19), see Motohashi (1993; 1997, Chapter 4). See also Note [50.5].

§19

This section provides very basic facts on sieves in general; knowledge of these will become helpful only in §102, however. More precisely, our purpose here is to show purely logical identities (19.9) and ($\natural$) of Note [19.8], which are in fact elaborations of (18.1) and on which we shall indicate the modern sieve theory stands.

Thus, let Ω be a map of the set of all primes into the family of all subsets of $\mathbb{Z}$. We shall then consider, for an arbitrary sequence $\mathcal{A}$ of integers,

$$\mathcal{S}(\mathcal{A}, \Omega, z) = \{a \in \mathcal{A} : a \notin \Omega(p), \forall p < z\};$$
$$\text{abbreviated to } \mathcal{S}(\mathcal{A}, z), \tag{19.1}$$

where z is a parameter supposed to tend to infinity. This set is called the resultant of sifting $\mathcal{A}$ by Ω up to z. For instance, let $\mathcal{A} = \mathbb{N}$ and $\Omega(p) = p\mathbb{Z} \cup \{2 + p\mathbb{Z}\}$ for all p. Then, according to Note [10.6], the set $\mathcal{S}(\mathcal{A}, z)$ contains twin primes, i.e., pairs of primes with differences equal to 2, which are in the interval $[z, z^2)$, if ever exist.

$$\S19 \qquad\qquad 77$$

We extend the domain of Ω to all square-free integers $u > 0$ by defining $\Omega(u)$ as $\bigcap_{p|u} \Omega(p)$, while $\Omega(1) = \mathbb{Z}$. Then we have the following identity originating in Legendre (1785):

Theorem 18 *Let* $\mathcal{A}_u = \{a \in \mathcal{A} : a \in \Omega(u)\}$. *Then, for any finite* $\mathcal{A}$,

$$|S(\mathcal{A}, z)| = \sum_{u|P(z)} \mu(u)|\mathcal{A}_u|, \quad P(z) = \prod_{p<z} p. \tag{19.2}$$

Proof The sum is equal to

$$\sum_{\substack{a \in \mathcal{A} \\ u|P(z) \\ a \in \Omega(u)}} \sum \mu(u) = \sum_{a \in \mathcal{A}} \sum_{u|U(a)} \mu(u), \tag{19.3}$$

where $U(a)$ is the least common multiple of all $u|P(z)$ such that $a \in \Omega(u)$, since $a \in \Omega(u_1) \cap \Omega(u_2)$ is the same as $a \in \Omega([u_1, u_2])$ because of (10.6). The fact that $U(a) = 1$ is identical to that $a \notin \Omega(p), \forall p < z$, which is to be compared with (18.1). This gives (19.2). Alternatively one may argue as follows: We have

$$S(\mathcal{A}, z) = \mathcal{A} - \bigcup_{p<z} \mathcal{A}_p \quad \text{(set-minus)}. \tag{19.4}$$

On noting $\mathcal{A}_u = \bigcap_{p|u} \mathcal{A}_p$, we find that the definition (17.4) corresponds precisely to that of the plus-minus signs in the inclusion–exclusion principle (Sylvester's identity). We end the proof.

We shall elaborate (19.2). Thus, we first classify integers $a \in \mathcal{A}$ with respect to the least p such that $a \in \Omega(p)$; then, we have, in place of (19.4),

$$S(\mathcal{A}, z) = \mathcal{A} - \bigsqcup_{p<z} S(\mathcal{A}_p, p);$$

$$|S(\mathcal{A}, z)| = |\mathcal{A}| - \sum_{p<z} |S(\mathcal{A}_p, p)|. \tag{19.5}$$

We modify the lower line: with any function η which is arbitrary save for $\eta(1) = 1$,

$$|S(\mathcal{A}, z)| = |\mathcal{A}| - \sum_{p<z} \eta(p)|S(\mathcal{A}_p, p)| - \sum_{p<z} (1 - \eta(p))|S(\mathcal{A}_p, p)|. \tag{19.6}$$

78 *Divisibility*

This is the case $\ell = 1$ of the identity

$$\left|S(\mathcal{A},z)\right| = \sum_{\substack{u|P(z) \\ \nu(u)<\ell}} \mu(u)\rho(u)|\mathcal{A}_u| + \sum_{\substack{u|P(z) \\ \nu(u)\leq\ell}} \mu(u)\tau(u)\left|S(\mathcal{A}_u,p(u))\right|$$
$$+ (-1)^{\ell} \sum_{\substack{u|P(z) \\ \nu(u)=\ell}} \rho(u)\left|S(\mathcal{A}_u,p(u))\right|, \quad \ell \geq 1, \tag{19.7}$$

where $\nu(u)$ is as in Note [18.6], and with $u = p_1p_2\cdots p_l$ with $p_1 > p_2 > \cdots > p_l$, $p(u) = p_l$, and

$$\rho(u) = \eta(p_1)\eta(p_1p_2)\cdots\eta(p_1p_2\cdots p_l), \quad \rho(1) = 1;$$
$$\tau(u) = \rho(u/p(u)) - \rho(u), \quad \tau(1) = 0. \tag{19.8}$$

To confirm (19.7), we apply, to (19.6), the replacement $\mathcal{A} \mapsto \mathcal{A}_u, z \mapsto p(u)$, $\eta(p) \mapsto \eta(up)$, and insert the result into the third sum on the right side of (19.7); then ℓ is increased by one; and we see by induction that (19.7) holds for any $\ell \geq 1$. Taking $\ell > \pi(z)$, we obtain the following elaboration or rather refinement of (19.2):

$$|S(\mathcal{A},z)| = \sum_{u|P(z)} \mu(u)\rho(u)|\mathcal{A}_u| + \sum_{u|P(z)} \mu(u)\tau(u)\left|S(\mathcal{A}_u,p(u))\right|; \tag{19.9}$$

see Note [19.5].

Obviously,

$$|S(\mathcal{A},z)| - \sum_{u|P(z)} \mu(u)\rho(u)|\mathcal{A}_u| \begin{cases} \leq 0 & \text{if } \mu(u)\tau(u) \leq 0, \\ \geq 0 & \text{if } \mu(u)\tau(u) \geq 0, \end{cases} \quad \forall u|P(z). \tag{19.10}$$

The first case is called an upper-bound sieve, and the second a lower-bound one. All known sieves are the results of specifications of ρ so that the sieve main term, i.e., the last sum over u, becomes non-trivial.

In the combinatorial sieve method that was initiated by Brun in his seminal works (1915; 1919; 1920; 1925) and greatly deepened by Iwaniec (1971; 1980a, 1980b, 1980c; 1981), the discussion starts with (19.9) in which ρ is the characteristic function of a set of divisors of $P(z)$. A variety of results achieved by means of the combinatorial sieve belong to the heart of the modern analytic number theory. However, we shall not dwell on the combinatorial sieve in the present essays, save for an indication, in Notes below, of a few of its rudimental points. The reason for this is as follows: Although we shall in fact employ sieve arguments in Chapter 5, they will be developed in the framework that originated in the sieve ideas of Linnik (1941) and Selberg (1947), which are drastically different from being combinatorial but amalgamate excellently with analytic properties of the zeta and allied functions which admit Euler product expansions. Outstanding achievements in number theory, including those resulted via combinatorial sieve, are often consequences of applications of the theory of these functions, and one of the main aims of the present essays is to provide truly basic material to understand their exquisite characteristics. For this purpose, the Linnik–Selberg sieve offers us a well structured and readily discernible approach, as we shall indicate in §§98–107. Specifically, in the final section it will be shown that the recent major development on bounded gaps between prime numbers is completely within this non-combinatorial sieve framework.

Notes

[19.1] There exist a number of monographs and expository articles dealing with sieves. Among them, Halberstam–Richert (1974), Motohashi (1983), Greaves (2001), and Friedlander–Iwanec (2010) contain accounts of both combinatorial and other sieves.

[19.2] The first explicit application of (19.2) was made by Legendre (1785, pp.471–472); naturally he did not use the notion of the Möbius function. Later, in (1798, p.14), he supplied more details and proved (18.4). Thus, he considered the case where $\mathcal{A}$ is the set of all positive integers not greater than n and $\Omega(p) = p\mathbb{Z}$ for $p|n$ and $= \varnothing$ otherwise. See also his proof of Theorem 36 below together with Note [40.4].

[19.3] The Eratosthenes sieve: Suppose that $x > 0$ be sufficiently large and not in $\mathbb{N}$. Let $\Omega(p) = p\mathbb{Z}$ for all p. With $\mathcal{A} = \{n < x : n \in \mathbb{N}\}$ we have $\mathcal{S}(\mathcal{A}, \sqrt{x}) = \{1\} \sqcup \{\sqrt{x} < \text{primes} < x\}$ according to Note [10.6]; namely, if we sift out all integers that are divisible by any prime $< \sqrt{x}$, then those left in the interval $[1, x]$ are 1 and the primes in the sub-interval $(\sqrt{x}, x)$. The identity (19.2) gives

$$\pi(x) - \pi(\sqrt{x}) + 1 = \sum_{u|P(\sqrt{x})} \mu(u) \left[\frac{x}{u}\right];$$

see Legendre (1808, p.5; Quatrième partie, §XI). One might notice here a method of searching for prime numbers. Nicomachus (ca 100 CE: D'Ooge (1926, p.204)) attributed this famous sifting procedure to the ancient Greek all-arounder Eratosthenes of Cyrene (b.ca 275 BCE, the third chief librarian of Alexandria Mouseion); hence, (19.2) is often called the Eratosthenes–Legendre sieve. As a matter of fact, Nicomachus applied the method to the set of odd integers ≥ 3, which seems to have caused the confusion (b) mentioned in Note [10.3]. Also, in Euler (1849a (posth.), Caput I), this sieve is explained,

80   *Divisibility*

though without any attribution. Near the end of that chapter Euler stated that prime numbers become sparse along with searching, and *nullum plane ordinem apparere*. It is a surprise that he missed surmising the prime number theorem and even Chebyshev's bound (12.14). In any event, it is practically impossible to determine $\pi(x)$ for an arbitrary x by means of the above explicit formula. For, as is often mentioned in literature, the computation of the sum on the right side becomes extremely laborious and useless as x grows indefinitely since the error caused by replacing $[x/u]$ by x/u increases uncontrollably.

[19.4]   The fundamental identities (19.5) are due to Buchstab (1937, p.1241), which considerably simplified Brun's arguments.

[19.5]   The identity (19.9) holds in fact with any ρ such that $\rho(1) = 1$ and with τ being defined by the second line of (19.8): The right side of (19.9) is equal to

$$\begin{aligned}
|\mathcal{A}| + &\sum_{1<u|P(z)} \mu(u)\rho(u)\left(|\mathcal{A}_u| - |\mathcal{S}(\mathcal{A}_u,p(u))|\right) \\
+ &\sum_{1<u|P(z)} \mu(u)\rho(u/p(u))|\mathcal{S}(\mathcal{A}_u,p(u))| \\
= |\mathcal{A}| + &\sum_{1<u|P(z)} \mu(u)\rho(u) \sum_{p<p(u)} |\mathcal{S}(\mathcal{A}_{pu},p)| \\
- &\sum_{1<v|P(z)} \mu(v)\rho(v) \sum_{p<p(v)} |\mathcal{S}(\mathcal{A}_{pv},p)| - \sum_{p<z} |\mathcal{S}(\mathcal{A}_p,p)|,
\end{aligned}$$

since in the second line on the left side we have $u = pv$, $p = p(u)$. This confirms the claim; note that the reversed reasoning yields an alternative proof of (19.9). For a consequence of this fact, see Note [102.5]. It should, however, be added that the specification of ρ by means of (19.8) often turns out to be expedient.

[19.6]   Efforts to overcome the wastefulness, noted above, of the Eratosthenes–Legendre sieve resulted in Brun's contributions. He seems to have come to the view that (19.2) is often too exact to produce anything meaningful. In short, (19.2) contains noise which is to be filtered out; hence, one ought to introduce truncation procedures into (19.3) so that (19.2) is replaced by effective inequalities, which is of course the same as taking ρ in (19.9) to be characteristic functions of certain well-chosen subsets of the set of all divisors of $P(z)$. Most probably in this way, he (1919) was initially led to the following truncation: With any $L \in \mathbb{N}$, we set either

$$\rho(u) = \begin{cases} 1 & v(u) \le 2L - 1, \\ 0 & v(u) \ge 2L, \end{cases} \quad \forall u | P(z),$$

or

$$\rho(u) = \begin{cases} 1 & v(u) \le 2L, \\ 0 & v(u) \ge 2L + 1, \end{cases} \quad \forall u | P(z).$$

In the former $\tau(u) = 1$ only if $v(u) = 2L$, and in the latter $\tau(u) = 1$ only if $v(u) = 2L + 1$; otherwise $\tau(u) = 0$. Then (19.10) yields

$$\sum_{\substack{u|P(z) \\ v(u) \le 2L-1}} \mu(u)|\mathcal{A}_u| \le |\mathcal{S}(\mathcal{A}, z)| \le \sum_{\substack{u|P(z) \\ v(u) \le 2L}} \mu(u)|\mathcal{A}_u|$$

for any $L \in \mathbb{N}$. This is called the pure sieve of Brun; its merit is in that the size of u is less than z^{2L}, so both of the last sums over u are manageable, provided z^{2L} is small compared with $|\mathcal{A}|$. Such a restriction of the size of u is not available in (19.2), which causes the wastefulness mentioned above. Applying the upper bound part of this inequality, Brun discovered that

$$\sum \frac{1}{p} < \infty, \quad \text{(sum over twin primes)}.$$

That is, in view of (11.9) (or Note [13.8]), twin primes, if ever exist, should appear definitely less frequently than primes themselves (an alternative proof is given in Note [102.8]). In his article (1920), Brun refined his pure sieve by means of a more elaborate truncation procedure, and demonstrated among other things that any sufficiently large even integer $2N$ can be expressed as $2N = n_1 + n_2$ with $v(n_1), v(n_2) \le 9$ and that there exist infinitely many n such that $v(n), v(n + 2) \le 9$; see p.32 there. These are the first real advancements towards the Goldbach conjecture and the twin prime conjecture, respectively. It was the dawn of the modern theory of sieve methods, and was precisely *cutting the Gordian knot*, as until then people had been wondering solely how to deal with the exact formula (19.2). However, one may see as well Chebyshev's argument for the inequality (12.14) to be the earliest stroke. Incidentally, here one may be reminded of *sparse-modeling*.

Remark We shall not give any discussion on the famous result of Chen (1973): $2N = n_1 + n_2$, with $v(n_1) = 1, v(n_2) \le 2$, for all sufficiently large N. It requires a preparation which is excessive for our main purpose, although all basic means are contained in the present essays. See instead Halberstam–Richert (1974, Chapter 11).

82   *Divisibility*

[19.7] The use of (19.10) requires appropriate information about the size of $\mathcal{A}_u$. It is formulated typically as

$$|\mathcal{A}_u| = \frac{\omega(u)}{u}X + E_u,$$

where X is a large parameter, ω a multiplicative function, and E_u an error term. It should be stressed, specifically at this moment, that what is really needed in applications is to have not the bounds for individual E_u but rather the statistical bound for them, i.e., the estimation of

(a) $$\sum_{u|P(z)} \mu(u)\rho(u)E_u.$$

In other words, sieve problems are intrinsically of statistical nature, which may sound unexpected though. Indeed, thanks to this very nature of sieve mechanisms, real possibilities emerge to avoid grand conjectures like GRH (see (53.9)) in investigating classical problems such as Goldbach's. Thereby analytic means, like those to be developed in Chapter 5, come to play an essential rôle in pursuing sieve arguments. As for ω, it is usually supposed to satisfy

$$\sum_{p<z} \frac{\omega(p)}{p} \log p = \big(1 + o(1)\big)\kappa \log z,$$

as $z \to \infty$; this is to be compared with (13.7). The constant κ is called the dimension of sieve problem $|\mathcal{S}(\mathcal{A},z)|$. The linear sieve means $\kappa = 1$; it includes a variety of problems in analytic number theory. In the example given after (19.1) we have $\kappa = 2$, so one may classify the twin prime conjecture as a two-dimensional sieve problem; however, there exists a way to treat it as a linear sieve problem by virtue of a great discovery, i.e., Theorem 147 below, made by analytic means, on the statistical distribution of primes in arithmetic progressions with variable differences.

[19.8] Here is a way to elaborate (19.9): We divide the interval $[2,z)$ into subintervals I so that

$$[2,z) = \bigsqcup I, \quad I = [s,t), \quad (I) = t,$$

where $2 \le s < t$ are integers; z is now supposed to be in $\mathbb{N}$. We consider set-theoretic direct products of mutually disjoint I's:

$$K = I_1 I_2 \cdots I_l, \quad (I_1) > (I_2) > \cdots > (I_l), \quad \nu(K) = l;$$

$\S 19$ 83

and we make the definitions:

$$I < K \;\Leftrightarrow\; (I) < (I_l);$$
$$u \in K \;\Leftrightarrow\; u = p_1 p_2 \cdots p_l, \; p_j \in I_j;$$
$$1 \in K \;\Leftrightarrow\; K = \varnothing, \; v(\varnothing) = 0.$$

We write the second line of (19.5) as

$$|\mathcal{S}(\mathcal{A},z)| = |\mathcal{A}| - \sum_I \sum_{p \in I} |\mathcal{S}(\mathcal{A}_p, p)|.$$

We let $\varkappa$ be a function on the set of all K's. Analogously to (19.6), we have

$$|\mathcal{S}(\mathcal{A},z)| = |\mathcal{A}| - \sum_I \varkappa(I) \sum_{p \in I} |\mathcal{S}(\mathcal{A}_p, p)| - \sum_I (1 - \varkappa(I)) \sum_{p \in I} |\mathcal{S}(\mathcal{A}_p, p)|.$$

Imitating (19.7) we perform iteration and get the following elaboration of (19.9):

$$
\begin{aligned}
(\natural) \quad |\mathcal{S}(\mathcal{A},z)| &= \sum_K (-1)^{v(K)} \Theta(K) \sum_{u \in K} |\mathcal{A}_u| \\
&\quad + \sum_K \sum_{I < K} (-1)^{v(K)} \Theta(KI) \sum_{\substack{p' < p \\ p, p' \in I}} \sum_{u \in K} |\mathcal{S}(\mathcal{A}_{upp'}, p')| \\
&\quad + \sum_K (-1)^{v(K)} \Psi(K) \sum_{u \in K} |\mathcal{S}(\mathcal{A}_u, p(u))|,
\end{aligned}
$$

with

$$\Theta(K) = \varkappa(I_1) \varkappa(I_1 I_2) \cdots \varkappa(I_1 I_2 \cdots I_l); \quad \varkappa(\varnothing) = 1,$$
$$\Psi(K) = \Theta(I_1 I_2 \cdots I_{l-1}) - \Theta(I_1 I_2 \cdots I_l); \quad \Psi(\varnothing) = 0.$$

This logical identity is devised by Motohashi (1983, Chapters 2–3, but z_1 there is taken to be 1 here) in order to supply an accessible proof to a marvel in number theory due to Iwaniec (1980c): the combinatorial linear sieve whose error-term is expressed as a flexible bilinear form. The core of Iwaniec's discovery is roughly rendered as follows: The optimal weight ρ in the combinatorial linear sieve has an infinitely divisible structure that allows one to deal with, instead of (a) above, the double sum

$$\text{(b)} \qquad \sum_{u < y_1, v < y_2} \alpha(u) \beta(v) E_{uv},$$

where y_j, α, β are arbitrary as far as $y_1 y_2 = z$ and $|\alpha(u)|, |\beta(v)| \leq 1$. Hence, one is able to employ the double sum strategy (Note [97.3] below) so that the limitation inherent in the estimation of the plain $\sum_{u < z} |E_u|$ may be overcome.

84   *Divisibility*

The identity (♮) makes readily discernible such a structure in error terms of sieves in general; now (a) is replaced by

$$\sum_K (-1)^{\nu(K)} \Theta(K) \sum_{u \in K} E_u.$$

Then appropriate divisions of K yield (b), in the case of combinatorial sieve, i.e., with Θ a characteristic function. It should be noted that prior to the above developments in combinatorial sieve a double sum structure of the error term in Selberg sieve had already been discovered by Motohashi (1974). See Notes [100.6] and [104.3].

[19.9] Finding integers composed of relatively large prime factors is the principal aim of applying sieves. In a contrasting context, integers composed of relatively small prime factors as well occupy a significant place in number theory; so the elements of the set $\{n \in \mathbb{N} : p|n \Rightarrow p \le B\}$ are specifically termed B-smooth. This notion has indeed a variety of fascinating applications as is displayed in Granville (2008). Thus, we make an addition to his collection: there is a good reason to suppose that smooth integers should play a basic rôle in sieve arguments in general. For instance, dealing with the sifting procedure that is applied in the study of the distribution of pairs of primes with exceptionally small gaps, such as twin primes, smooth integers have been found to play a remarkable rôle. See Note [107.3](b).

§20

We now return to the division algorithm $(2.1)_{b>0}$. We shall reconsider it via the modified form $a/b = q + r/b$; that is, we are about to discuss fractions in general. We use the term fraction, though rational number or just rational will sometimes be better. To recognize integers, intuitive perception may suffice, whereas fractions require extra care because of their opacity, or dense existence. Note also that fractions could be viewed as objects not only in one-dimensional $\mathbb{Q}$ but also in two-dimensional $\mathbb{Z}^2$ equipped with an obvious equivalence via the notion of ratio; see Note [20.1].

Since it suffices to make observation on fractions in any interval of length 1, we first collect all irreducible fractions, i.e., those a/b with $\langle a,b \rangle = 1$, in the interval $[0, 1]$ whose denominators $b > 0$ are less than or equal to $N \in \mathbb{N}$. The resulting set is called the Farey sequence $\mathcal{F}_N$ of order N:

$$\mathcal{F}_N = \left\{ m/n : 1 \le n \le N; \ 0 \le m \le n, \langle m,n \rangle = 1 \right\}. \tag{20.1}$$

$$\S20 \qquad\qquad 85$$

Theorem 19 *For three consecutive terms $a/b, m/n, c/d$ in $\mathcal{F}_N$, we have*

$$\text{(1)} \quad b+n \geq N+1, \ n+d \geq N+1,$$
$$\text{(2)} \quad bm - an = 1, \ cn - dm = 1, \qquad\qquad (20.2)$$
$$\text{(3)} \quad m/n = (a+c)/(b+d).$$

Proof We assume the validity of (2); then, by an elimination argument, we have

$$m(bc - ad) = a + c, \ n(bc - ad) = b + d, \qquad\qquad (20.3)$$

which gives (3). To prove (1), (2) for the pair $\{m/n, c/d\}$, we consider the indefinite equation $nx - my = 1$. According to Theorem 2, an integer solution $\{x_0, y_0\}$ exists; and the general solution is given by $x = x_0 + mu$, $y = y_0 + nu$, $u \in \mathbb{Z}$, (Note [6.2]). In particular, there exists a solution $\{x_1, y_1\}$ such that $N - n < y_1 \leq N$. If $x_1/y_1 = c/d$, then by (6.1) we get $x_1 = c$, $y_1 = d$, and $cn - dm = 1$, $d + n > N$, which means (1), (2) hold. Otherwise, $x_1/y_1 \neq c/d$, and we note that $y_1 \leq N$ implies $x_1/y_1 \in \mathcal{F}_N$. Since m/n, c/d are adjacent in $\mathcal{F}_N$ and $m/n < x_1/y_1$, we should have $c/d < x_1/y_1$. Hence,

$$\frac{x_1}{y_1} - \frac{c}{d} = \frac{dx_1 - cy_1}{dy_1} \geq \frac{1}{dy_1}. \qquad\qquad (20.4)$$

This implies a contradiction:

$$\frac{1}{ny_1} = \frac{x_1}{y_1} - \frac{m}{n} = \frac{x_1}{y_1} - \frac{c}{d} + \frac{c}{d} - \frac{m}{n}$$
$$\geq \frac{1}{dy_1} + \frac{1}{dn} = \frac{n + y_1}{dny_1} > \frac{N}{dny_1} \geq \frac{1}{ny_1}. \qquad\qquad (20.5)$$

We end the proof.

The assertion (20.2) yields the following procedure: For $N \geq 2$,

find all adjacent terms a/b, c/d in $\mathcal{F}_{N-1}$ such that $b + d = N$
and augment $\mathcal{F}_{N-1}$ by adding these mediants $(a+c)/(b+d)$. $\qquad (20.6)$

Then one gets $\mathcal{F}_N$. To confirm this, note that terms in $\mathcal{F}_N - \mathcal{F}_{N-1}$ are of the form h/N, $\langle h, N \rangle = 1$, and by (20.2) (2) any two of them do not fall simultaneously in an interval whose end points are adjacent terms of $\mathcal{F}_{N-1}$. Therefore, by (3), these h/N must be generated from $\mathcal{F}_{N-1}$ by the procedure (20.6).

As an application of the Farey sequence, we consider rational approximations to real numbers, that is, approximations by irreducible fractions. Thus, for any $\xi \in \mathbb{R} \setminus \mathbb{Q}$ there exist infinitely many irreducible fractions a/b with which the inequality

86 Divisibility

$$\left| \xi - \frac{a}{b} \right| < \frac{1}{b^2} \qquad (20.7)$$

holds. For a proof, we may restrict ourselves to the case $0 < \xi < 1$. It suffices to show that for any given $N \geq 2$ there exists an irreducible fraction h/q such that $q \leq N$ and

$$\left| \xi - \frac{h}{q} \right| \leq \frac{1}{q(N+1)} . \qquad (20.8)$$

To this end, we divide $[0, 1]$ into subintervals by the terms of $\mathcal{F}_N$ and their mediants. There must exist a subinterval which contains ξ; let its end point in $\mathcal{F}_N$ be h/q. Then, by the last theorem, we have (20.8). Note that the intermediate assertion (20.8) holds for rational ξ as well.

Notes

[20.1] Positive fractions are the same as rational ratios (ratios between two positive integers); the latter is a concept originating in remote antiquity. Comparisons either among fractions or among rational ratios gave rise to the notions of common divisors and multiples. [Σ.V] deals with rational (commensurable) ratios and continues, via Euclid's algorithm [Σ.VII], to the impressive [Σ.X] devoted to irrational (incommensurable) ratios, which contains 115 propositions. However, this is our view. One should be aware that it is a matter of a lasting debate among historians how Euclid saw the relation between his number theory and his theory of irrational numbers. Continued in Note [23.4].

[20.2] The property (20.2) of $\mathcal{F}_N$ was first discovered and reported by Haros (1802). Later Cauchy (1816) gave a proof ascribing (20.2) to Farey (1816); apparently Cauchy did not notice the fact that Farey's finding was a rediscovery of Haros'. The wrong attribution thus caused has remained uncorrected to this day. The above proof is adopted from Landau (1927, Dritter Teil, Kapitel 1). See [DA, art.190] as well.

[20.3] The formula (18.18) implies that the sequence $\mathcal{F}_N$ is related to the distribution of values of the Möbius function; consequently, to the zeta-function as (17.3) indicates. Indeed one is led to an alternative formulation of RH in terms of an asymptotic nature of $\mathcal{F}_N$; see Franel (1925) and Landau (1925). Also, it should be remarked that there exists a tangible relation between sieves and Farey sequences; this is of course not unexpected in view of (18.2), (18.18), and (19.2). See Note [50.4]. The distribution of fractions or rational numbers is certainly a deep subject.

[20.4] Egyptian fractions: The oldest among mathematical books and articles whose authors are known is said to be the text book by Ahmes (ca 1650

BCE: Rhind papyrus; in fact, a copy of an older book). It contains a table of fractions $2/n$, $2 \nmid n$, $3 \le n \le 101$. Each of these is represented as a sum of distinct unit-fractions; such a sum is called an Egyptian fraction today. For instance

$$\frac{2}{83} = \frac{1}{60} + \frac{1}{332} + \frac{1}{415} + \frac{1}{498}$$

is given (Ahmes; Chace (1927, p.22)). Any positive rational number can be represented as an Egyptian fraction; see Botts (1967). By the way, the Erdős–Straus conjecture (Erdős (1950, p.210)) states that for any integer $n \ge 2$ the indefinite equation

$$\frac{4}{n} = \frac{1}{x} + \frac{1}{y} + \frac{1}{z}, \quad x, y, z \in \mathbb{N},$$

should have a solution. This remains unsolved. Naturally it is enough to consider the case where n is a prime; then, see Elsholtz–Tao (2013). It should be interesting to know that their argument depends on Theorem 147, §105, one of the very best results in the theory of the distribution of prime numbers.

[20.5] To prove (20.7) alternatively, Dirichlet (1863, §141) applied the pigeon box principle: divide the interval $[0, 1]$ into N subintervals evenly. For each $b \in \{0, 1, \ldots, N\}$, choose an integer a so that $0 \le b\xi - a < 1$. Then, since the number of distinct elements in the set $\{b\xi - a\}$ equals $N + 1$, there should be at least one subinterval in which two different points $b_1\xi - a_1$, $b_2\xi - a_2$ fall. Namely, we have $|(b_1 - b_2)\xi - (a_1 - a_2)| < 1/N$, which gives (20.8).

[20.6] A natural question concerning (20.7) is whether it is best possible or not; or more precisely, one may wonder if it is possible to improve the inequality for every ξ in a sufficiently wide class of real numbers. This problem originated in the discovery of Liouville (1844) that there exists a clear limitation in rational approximations to any real algebraic number $\xi \notin \mathbb{Q}$ (an irrational root of an algebraic equation with integer coefficients): if we replace the right side of (20.7) by $1/b^\lambda$ with λ greater than a certain lower bound, then there exist only a finite number of a/b satisfying the inequality. To make the situation more precise, we define that a real number ξ has an irrationality measure $\tau > 2$ if $|\xi - a/b| > b^{-\tau}$ holds for any integers $\{a, b\}$ save for a finite number of exceptions. Then, our present problem is equivalent to finding $\tau(\xi) = \liminf \tau$ for each ξ. Here we skip an interesting history but quote only the definitive result of Roth (1955): $\tau(\xi) = 2$ for any algebraic irrational ξ. Therefore, one may say that (20.7) is essentially best possible for such ξ's.

88 *Divisibility*

§21

We shall now take a practical point of view: We are concerned with rational approximation processes to individual real numbers. It is true that the division of the unit interval by means of the sequences $\mathcal{F}_N$ yields rational approximations like (20.7); however, it is not specified how to locate those fractions a/b when a particular ξ is given. The argument there indicates only the existence of the relevant fractions; it cannot be said constructive. In the present and the subsequent four sections we shall develop a method that starts with ξ itself and enables one to construct those fractions which are rational approximations to ξ as good as (20.7). This is the theory of continued fractions. In §48 and in Chapter 4 we shall have important applications of them.

It will become apparent later on that the notion of continued fractions can be regarded as Euclid's algorithm generalized to arbitrary real numbers. This might sound inconsistent with what developed in §3, for Euclid's algorithm concerns originally pairs of integers while here we have individual numbers, not pairs. Because of this, we shall first show a modified approach to Euclid's algorithm: We rewrite $(2.1)_{b>0}$ in terms of fractions as we did already in the last section, and if $r > 0$, then $a/b = q + 1/(b/r)$; note that here a/b is not necessarily in lowest terms, i.e., possibly $\langle a, b \rangle \neq 1$. Further, if $r' > 0$ in (3.1), then $a/b = q + 1/(q' + 1/(r/r'))$. Euclid's algorithm yields the greatest common divisor $\langle a, b \rangle$ after a certain number of repetitions of this procedure, and it stops. If the end step is the $(k+1)^{\text{st}}$, $k \geq 0$, then we get the regular continued fraction expansion of a/b:

$$\frac{a}{b} = a_0 + \cfrac{1}{a_1 + \cfrac{1}{a_2 + \cfrac{\ddots}{+\cfrac{1}{a_k}}}} \qquad (21.1)$$

$$= a_0 + \frac{1}{a_1} + \frac{1}{a_2} + \cdots + \frac{1}{a_k}, \qquad (21.2)$$

where $a_0 = q$, $a_1 = q', \dots$, and the second line is in accord with common practice; note that a_0 is $[a/b]$ which is possibly not positive, while $a_j \in \mathbb{N}$ for $j \geq 1$. The term regular indicates that unit fractions $1/a_j$ appear consecutively; of course they do not if $b|a$, that is, if $k = 0$; but see Note [21.2]. Throughout the present essays, continued fractions are mostly regular ones; in some exceptional cases we shall deal also with continued fractions which are not regular.

If continued fractions are given initially, then it is often too laborious to transform them into ordinary fractions by chain of fraction calculation.

$$\S21 \qquad\qquad 89$$

To ease this intricacy, we shall employ the matrix version (2.3) of the division algorithm. Namely, we shall identify continued fractions as products of linear transformations of a special type. With this in mind, we make the following definitions and observations:

$$\begin{pmatrix} A_j & A_{j-1} \\ B_j & B_{j-1} \end{pmatrix} = \begin{pmatrix} a_0 & 1 \\ 1 & 0 \end{pmatrix} \begin{pmatrix} a_1 & 1 \\ 1 & 0 \end{pmatrix} \cdots \begin{pmatrix} a_j & 1 \\ 1 & 0 \end{pmatrix}, \quad -1 \leq j \leq k; \quad (21.3)$$

$$A_{-2} = 0,\ B_{-2} = 1;\quad A_{-1} = 1,\ B_{-1} = 0;\quad A_0 = a_0,\ B_0 = 1; \qquad (21.4)$$

$$A_j = a_j A_{j-1} + A_{j-2},\ B_j = a_j B_{j-1} + B_{j-2},\quad 0 \leq j \leq k; \qquad (21.5)$$

$$A_j B_{j-1} - A_{j-1} B_j = (-1)^{j-1},\quad -1 \leq j \leq k. \qquad (21.6)$$

The first line can be called either the matrix representation of continued fraction expansions or just the matrix form of continued fractions; the product for $j = -1$ is empty and equal to the identity matrix of order 2; (21.4) contains relevant conventions.

Theorem 20 *With (21.3)–(21.4), we have $a/b = A_k/B_k$ in (21.2) or more generally*

$$\frac{A_j}{B_j} = a_0 + \frac{1}{a_1 +} \frac{1}{a_2 +} \cdots + \frac{1}{a_j}, \quad 0 \leq j \leq k. \qquad (21.7)$$

The fractions A_j/B_j, $j \geq 0$, are all irreducible; hence A_k/B_k is the same as the result of simplifying a/b to lowest terms.

Remark 1 With this, we introduce the basic notion:

$$A_j/B_j \text{ is called the } j^{\text{th}} \text{ convergent of } a/b. \qquad (21.8)$$

Proof The second assertion follows from (21.6). As for (21.7), we may suppose $j \geq 1$. By induction, the fraction on the right side is equal to the result of replacing a_{j-1} by $a_{j-1} + 1/a_j$ in A_{j-1}/B_{j-1}, that is,

$$\frac{(a_{j-1} + 1/a_j)A_{j-2} + A_{j-3}}{(a_{j-1} + 1/a_j)B_{j-2} + B_{j-3}} = \frac{A_j}{B_j}. \qquad (21.9)$$

We end the proof. See Remark 2 below.

Now, let $\{r_j\}$ be the remainders at respective division steps in the expansion (21.1)–(21.2) so that $r_{j-1} = a_j r_j + r_{j+1}, r_j > r_{j+1}, 0 \leq j \leq k$. By the discussion in §3 we have

$$\langle a, b \rangle = \langle r_0, r_1 \rangle = \cdots = \langle r_k, r_{k+1} \rangle = r_k,$$
$$r_{-1} = a,\ r_0 = b,\ r_{k+1} = 0. \qquad (21.10)$$

90 *Divisibility*

Here one may suppose that $a, b \in \mathbb{N}$ without loss of generality. As in (2.3) we have $\binom{r_{j-1}}{r_j} = \binom{a_j\ 1}{1\ 0}\binom{r_j}{r_{j+1}}$. Hence, according to the definition (21.3),

$$\binom{a}{b} = \begin{pmatrix} A_j & A_{j-1} \\ B_j & B_{j-1} \end{pmatrix} \binom{r_j}{r_{j+1}}, \quad -1 \le j \le k. \tag{21.11}$$

Then, on noting (21.6),

$$\binom{r_j}{r_{j+1}} = (-1)^{j-1} \begin{pmatrix} B_{j-1} & -A_{j-1} \\ -B_j & A_j \end{pmatrix} \binom{a}{b}, \quad -1 \le j \le k. \tag{21.12}$$

Thus

$$|aB_{j-1} - bA_{j-1}| = r_j > r_{j+1} - |aB_j - bA_j|, \quad 0 \le j \le k; \tag{21.13}$$

and

$$\langle a, b \rangle = (-1)^{k-1}\left(aB_{k-1} - bA_{k-1}\right). \tag{21.14}$$

Note that in (21.13) we have $j \ge 0$ in general, though if $a > b$, then $j \ge -1$.

If $b|a$, then $k = 0$. If a/b is not an integer, then $k \ge 1$, and $a_k \ge 2$ since $r_{k-1} = a_k r_k + r_{k+1}$ with $r_{k-1} > r_k > r_{k+1} = 0$. This procedure can be reversed. Namely, the chain of division algorithm $r_{j-1} = a_j r_j + r_{j+1}$ with $r_j > r_{j+1}$ can proceed backwards from $j = k \ge 1$ to $j = 0$, only if $a_k \ge 2$; note that if $a_k = 1$, then the premise $r_{k-1} > r_k$ would be violated. Namely,

$$\begin{aligned} & \text{the expansion (21.2) of } a/b \text{ is unique,} \\ & \text{provided } a_k \ge 2 \text{ if } a/b \text{ is not an integer,} \end{aligned} \tag{21.15}$$

because then the expansion is the same as the Euclid algorithm applied to $\{a, b\}$; see (3.9) as well as (22.18). In particular, the bound for the length $k + 1$ of the expansion is the same as the one asserted in Note [3.2]. A subtler relation between Euclid's algorithm and continued fraction expansion is discussed in the second half of §23.

Remark 2 According to (21.15), the right side of (21.7) is not generally the same as the expansion of A_j/B_j via Euclid's algorithm applied to $\{A_j, B_j\}$ since a_j is possibly equal to 1.

The combination of (21.3) and (21.14) provides a way to solve the linear indefinite equation $ax + by = c$, in view of the explanation given in §8 (here $n = 2$): If $\langle a, b \rangle | c$, then (8.3) can be replaced by

$$\binom{x}{y} = (-1)^{k-1} \frac{c}{\langle a, b \rangle} \begin{pmatrix} B_{k-1} \\ -A_{k-1} \end{pmatrix} + t \begin{pmatrix} B_k \\ -A_k \end{pmatrix}, \quad t \in \mathbb{Z}, \tag{21.16}$$

which often turns out to be quite practical.

Notes

[21.1] The generic continued fraction

$$a_0 + \cfrac{\tau_1}{a_1 + \cfrac{\tau_2}{a_2 + \cfrac{\ddots}{\quad + \cfrac{\tau_k}{a_k}}}} = a_0 + \frac{\tau_1}{a_1} + \frac{\tau_2}{a_2} + \cdots + \frac{\tau_k}{a_k},$$

where a_j, τ_j are arbitrary unless division by zero takes place, corresponds to the product

$$\begin{pmatrix} a_0 & 1 \\ 1 & 0 \end{pmatrix} \begin{pmatrix} a_1 & 1 \\ \tau_1 & 0 \end{pmatrix} \cdots \begin{pmatrix} a_k & 1 \\ \tau_k & 0 \end{pmatrix}.$$

This can be proved by a minor modification of the proof of Theorem 20. The case in which $\tau_j = \pm 1$, $\mathbb{N} \ni a_j \geq 2$ for $j \geq 1$, is called half-regular and of special interest; see Lagrange (1798, pp.14–16) as well as Euler (1771/1822, pp.470–472). Two examples are in Note [21.4]. In §86, we shall encounter fascinating instances of such expansions while discussing an amazing achievement of classic Indian mathematics. By the way, Euler dictated the draft of the famous textbook (1771) to his assistant; he had already lost his eyesight.

[21.2] If a/b is an integer, then (21.1)–(21.2) terminates with the first term a_0, i.e., $k = 0$. We may alter this situation into $a/b = (a_0 - 1) + 1/a_1$ with $a_1 = 1$ so that the number of terms of the expansion becomes two. If a/b is not an integer, then Euclid's algorithm yields $a_k \geq 2$, as noted already. Hence, we may replace the last term a_k by $(a_k - 1) + 1/a_{k+1}$ with $a_{k+1} = 1$. That is, we are able to adjust, as we wish, the parity of the number of terms of the regular continued fraction expansion of any rational number. As for the relation between this modification and Euclid's algorithm see Note [23.3]. In §§24–25 we shall use it as a device.

[21.3] By the computation made in Note [3.3], we have

$$\frac{3041543}{1426253} = 2 + \frac{1}{7} + \frac{1}{1} + \frac{1}{1} + \frac{1}{5} + \frac{1}{13} + \frac{1}{6} + \frac{1}{4} + \frac{1}{2}.$$

In terms of ordinary fractions, the expansion proceeds as follows:

$$\frac{2}{1}, \frac{15}{7}, \frac{17}{8}, \frac{32}{15}, \frac{177}{83}, \frac{2333}{1094}, \frac{14175}{6647}, \frac{59033}{27682}, \frac{132241}{62011}.$$

These fractions are obtained by computing the products (21.3) with the current specification. The last one is the reduced form of the original fraction. Thus, as an example of (21.14), we find that

$$-(3041543 \cdot 27682 - 1426253 \cdot 59033) = 23,$$
$$\langle 3041543, 1426253 \rangle = 23.$$

Also, we have

$$\begin{pmatrix} 3041543 \\ 1426253 \end{pmatrix} = \begin{pmatrix} 2 & 1 \\ 1 & 0 \end{pmatrix}\begin{pmatrix} 7 & 1 \\ 1 & 0 \end{pmatrix}\begin{pmatrix} 1 & 1 \\ 1 & 0 \end{pmatrix}^2\begin{pmatrix} 5 & 1 \\ 1 & 0 \end{pmatrix}$$
$$\times \begin{pmatrix} 13 & 1 \\ 1 & 0 \end{pmatrix}\begin{pmatrix} 6 & 1 \\ 1 & 0 \end{pmatrix}\begin{pmatrix} 4 & 1 \\ 1 & 0 \end{pmatrix}\begin{pmatrix} 2 & 1 \\ 1 & 0 \end{pmatrix}\begin{pmatrix} 23 \\ 0 \end{pmatrix}.$$

[21.4]   If we express, with a non-regular continued fraction, what is stated in Note [4.1], then we get the expansion

$$\frac{3041543}{1426253} = 2 + \frac{1}{8} + \frac{-1}{2} + \frac{1}{5} + \frac{1}{13} + \frac{1}{6} + \frac{1}{4} + \frac{1}{2}.$$

The convergent $\frac{15}{7}$ is skipped. This acceleration of approximation takes place at the term $\frac{1}{1}$ in the regular expansion; see Note [24.4]. In terms of matrices, one may move reversibly between these two types of expansions by means of the relations

$$\begin{pmatrix} a & 1 \\ \pm 1 & 0 \end{pmatrix}\begin{pmatrix} 1 & 1 \\ 1 & 0 \end{pmatrix} = \begin{pmatrix} a+1 & 1 \\ \pm 1 & 0 \end{pmatrix}\begin{pmatrix} 1 & 1 \\ 0 & -1 \end{pmatrix},$$

$$\begin{pmatrix} 1 & 1 \\ 0 & -1 \end{pmatrix}\begin{pmatrix} b & 1 \\ 1 & 0 \end{pmatrix} = \begin{pmatrix} b+1 & 1 \\ -1 & 0 \end{pmatrix},$$

where double-signs correspond. For instance, by this we readily get the second line from the first in the following expression and vice versa:

$$\frac{33142721}{15541427} = 2 + \frac{1}{7} + \frac{1}{1} + \frac{1}{1} + \frac{1}{5} + \frac{1}{13} + \frac{1}{1} + \frac{1}{14} + \frac{1}{1} + \frac{1}{1} + \frac{1}{1} + \frac{1}{15} + \frac{1}{1} + \frac{1}{16}$$
$$= 2 + \frac{1}{8} + \frac{-1}{2} + \frac{1}{5} + \frac{1}{14} + \frac{-1}{16} + \frac{1}{3} + \frac{-1}{17} + \frac{-1}{17}.$$

[21.5]   Any prime number which leaves 1 when divided by 4 is a sum of two squares (Euler's theorem: see (78.3)). There are a variety of proofs for this famous fact; see Note [78.2]. Here we shall show a somewhat exotic proof due to Smith (1855) but with a modification by means of the matrix form of continued fractions. Thus, we note first that if an irreducible fraction $a/b > 2$ with $b > 1$ has a symmetric continued fraction expansion, then there are two possibilities (without the adjustment mentioned in Note [21.2]):

(1) $\quad \dfrac{a}{b} = a_0 + \dfrac{1}{a_1} + \dfrac{1}{a_2} + \cdots + \dfrac{1}{a_h} + \dfrac{1}{a_h} + \dfrac{1}{a_{h-1}} + \cdots + \dfrac{1}{a_1} + \dfrac{1}{a_0},$

(2) $\quad \dfrac{a}{b} = a_0 + \dfrac{1}{a_1} + \dfrac{1}{a_2} + \cdots + \dfrac{1}{a_h} + \dfrac{1}{f} + \dfrac{1}{a_h} + \dfrac{1}{a_{h-1}} + \cdots + \dfrac{1}{a_1} + \dfrac{1}{a_0}.$

By the definition (21.3),

$$\begin{pmatrix} a_0 & 1 \\ 1 & 0 \end{pmatrix}\begin{pmatrix} a_1 & 1 \\ 1 & 0 \end{pmatrix}\cdots\begin{pmatrix} a_h & 1 \\ 1 & 0 \end{pmatrix} = \begin{pmatrix} A_h & A_{h-1} \\ B_h & B_{h-1} \end{pmatrix},$$

$$\begin{pmatrix} a_h & 1 \\ 1 & 0 \end{pmatrix}\cdots\begin{pmatrix} a_1 & 1 \\ 1 & 0 \end{pmatrix}\begin{pmatrix} a_0 & 1 \\ 1 & 0 \end{pmatrix} = \begin{pmatrix} A_h & B_h \\ A_{h-1} & B_{h-1} \end{pmatrix},$$

in which the second expression is the transpose of the first. Hence, (1) corresponds to

$$\begin{pmatrix} A_h & A_{h-1} \\ B_h & B_{h-1} \end{pmatrix}\begin{pmatrix} A_h & B_h \\ A_{h-1} & B_{h-1} \end{pmatrix} = \begin{pmatrix} A_h^2 + A_{h-1}^2 & * \\ A_h B_h + A_{h-1} B_{h-1} & * \end{pmatrix}.$$

That is, $a/b = (A_h^2 + A_{h-1}^2)/(A_h B_h + A_{h-1} B_{h-1})$, and by the irreducibility of both sides

$$(1) \Rightarrow a = A_h^2 + A_{h-1}^2.$$

Also,

$$(2) \Rightarrow a = A_h(fA_h + 2A_{h-1}).$$

With this, we consider the set $P = \{p/2, p/3, \ldots, p/2\ell\}$ for a prime $p = 4\ell + 1$. We may obviously suppose that $\ell \geq 3$. We expand each $p/w \in P$ into a continued fraction without applying the adjustment of Note [21.2], and let its matrix representation be

$$\begin{pmatrix} u_0 & 1 \\ 1 & 0 \end{pmatrix}\begin{pmatrix} u_1 & 1 \\ 1 & 0 \end{pmatrix}\cdots\begin{pmatrix} u_t & 1 \\ 1 & 0 \end{pmatrix} = \begin{pmatrix} U_t & U_{t-1} \\ V_t & V_{t-1} \end{pmatrix}, \quad t \geq 1; \quad \frac{U_t}{V_t} = \frac{p}{w},$$

where $u_0, u_t \geq 2$. We transpose this product of matrices, and let p/w' be the value of the corresponding continued fraction; note that the numerator is $U_t = p$. We have $w' = U_{t-1} \geq U_{t-1}/V_{t-1} \geq u_0$. Also, since $p/w' > u_t$, we have $w' < (4\ell + 1)/u_t \leq 2\ell + \frac{1}{2}$. Thus, $2 \leq w' \leq 2\ell$, i.e., $p/w' \in P$. Namely, this transpose operation is an involution. Since $|P|$ is odd, there must be a fixed point in P. The continued fraction expansion of this particular rational number must be of the type (1), for otherwise it would be of type (2), and then $A_h|p$ which yields a contradiction. We end the proof. See Notes [78.3]–[78.4].

§22

In the present and the next sections, we shall give salient points of the theory of continued fractions that Lagrange developed in his renowned treatise (1798;

94 *Divisibility*

Euler (1771/1822, Additions)). The difference between his and our accounts is solely in our use of matrices.

To begin with, we define the transformation R of an arbitrary $\eta \in \mathbb{R}$:

$$R : \eta \mapsto (\eta - [\eta])^{-1} \Leftrightarrow \eta = \frac{[\eta]R(\eta) + 1}{R(\eta)} = \begin{pmatrix} [\eta] & 1 \\ 1 & 0 \end{pmatrix} (R(\eta)), \quad (22.1)$$

where the matrix on the right side stands for a linear fractional transformation, i.e., Note [5.1](2); this should be compared with (2.3). Note here that if η is an integer, then $R(\eta)$ is not defined. It is understood that one should avoid such cases. However, this causes minor inconveniences when dealing with rational numbers, which will be encountered, for instance, in the proof of Theorem 23 in the next section. Thus, for the sake of simplicity, we shall suppose that

$$\text{within the present section, } \eta \in \mathbb{R} \text{ is irrational.} \quad (22.2)$$

Then, iterating (22.1), we get, with $R^0(\eta) = \eta$ and $a_j = [R^j(\eta)]$,

$$\eta = \begin{pmatrix} a_0 & 1 \\ 1 & 0 \end{pmatrix} \begin{pmatrix} a_1 & 1 \\ 1 & 0 \end{pmatrix} \cdots \begin{pmatrix} a_j & 1 \\ 1 & 0 \end{pmatrix} (R^{j+1}(\eta)), \quad (22.3)$$

$$\eta = a_0 + \frac{1}{a_1 +} \frac{1}{a_2 +} \cdots + \frac{1}{a_j +} \frac{1}{R^{j+1}(\eta)}, \quad (22.4)$$

$$\eta = \begin{pmatrix} A_j & A_{j-1} \\ B_j & B_{j-1} \end{pmatrix} (R^{j+1}(\eta)) = \frac{A_j R^{j+1}(\eta) + A_{j-1}}{B_j R^{j+1}(\eta) + B_{j-1}}, \quad (22.5)$$

where $j \geq -1$ under an obvious convention. Naturally, $a_0 \in \mathbb{Z}$, $a_j \in \mathbb{N}, j \geq 1$; and A_j, B_j are as in (21.3)–(21.7). With (22.2), j is unbounded since $R^j(\eta)$ never becomes an integer. One should note that (21.5) implies that

$$\eta > [\eta] + \tfrac{1}{2} \;\Rightarrow\; B_1 = B_0 = 1, \; B_{j+1} \geq B_j + 1, \, j \geq 1;$$
$$\eta < [\eta] + \tfrac{1}{2} \;\Rightarrow\; B_{j+1} \geq B_j + 1, \, j \geq 0; \quad (22.6)$$

see Note [22.3]. It is also appropriate to extend the notion (21.9) to irrational numbers: with (22.5),

$$A_j/B_j, j \geq 0, \text{ is called the } j^{\text{th}} \text{ convergent of } \eta. \quad (22.7)$$

By construction, the jth-convergent, which is an irreducible fraction, becomes less and greater than η depending on j being even and odd, respectively. More precisely, we have, by (21.6) and (22.5),

$$\eta - \frac{A_j}{B_j} = \frac{(-1)^j}{B_j(B_j R^{j+1}(\eta) + B_{j-1})}, \quad j \geq 0. \quad (22.8)$$

$$\S 22 \qquad\qquad 95$$

Since

$$B_j R^{j+1}(\eta) + B_{j-1} < B_j(a_{j+1} + 1) + B_{j-1} = B_{j+1} + B_j,$$
$$B_j R^{j+1}(\eta) + B_{j-1} > B_j a_{j+1} + B_{j-1} = B_{j+1},$$
$$j \geq 0, \qquad (22.9)$$

we see that

$$\frac{1}{B_j(B_{j+1} + B_j)} < \left| \eta - \frac{A_j}{B_j} \right| < \frac{1}{B_j B_{j+1}}, \quad j \geq 0. \qquad (22.10)$$

We have, by (22.6), $B_j \to \infty$ as $j \to \infty$, and

$$\eta = \lim_{j\to\infty} A_j/B_j : \quad \begin{array}{l}\text{irrational numbers are expanded} \\ \text{into infinite continued fractions.}\end{array} \qquad (22.11)$$

An infinite continued fraction is denoted by

$$a_0 + \frac{1}{a_1 +} \frac{1}{a_2 +} \cdots + \frac{1}{a_j +} \cdots, \quad \mathbb{N} \ni a_j, \ j \geq 1. \qquad (22.12)$$

Or more precisely, we let it mean, independently of (22.11), the infinite sequence

$$\frac{A_j}{B_j} = a_0 + \frac{1}{a_1 +} \frac{1}{a_2 +} \cdots + \frac{1}{a_j}, \quad j \geq 0. \qquad (22.13)$$

Theorem 21 *Between irrational numbers and infinite continued fractions there exists a one-to-one correspondence.*

Proof We have (22.11) already; thus it suffices to prove that

$$\begin{array}{c}\text{with (22.13), } A_j/B_j \text{ converges to an irrational number,} \\ \text{whose } j^{\text{th}}\text{-convergent is } A_j/B_j.\end{array} \qquad (22.14)$$

Given (22.13), we have, by (21.6),

$$\frac{A_j}{B_j} - \frac{A_{j-1}}{B_{j-1}} = \frac{(-1)^{j-1}}{B_j B_{j-1}}, \quad j \geq 1. \qquad (22.15)$$

So

$$\frac{A_j}{B_j} = a_0 + \frac{1}{B_1 B_0} - \frac{1}{B_2 B_1} + \cdots + \frac{(-1)^{j-1}}{B_j B_{j-1}}, \quad j \geq 1. \qquad (22.16)$$

As B_j tends to infinity, the convergence of this right side is immediate. Let ω be the limit. Then, since $a_0 + 1/(a_1 + 1) < A_j/B_j < a_0 + 1/(a_1 + 1/(a_2 + 1))$ for $j \geq 3$, we have $[\omega] = a_0$. Thus $\omega = a_0 + 1/R(\omega)$, where $R(\omega)$ is well-defined, as ω is not an integer because of this inequality. On the other hand, let

$$\omega' = \lim_{j\to\infty} \left\{ a_1 + \frac{1}{a_2 +} \frac{1}{a_2 +} \cdots + \frac{1}{a_j} \right\}. \qquad (22.17)$$

96 Divisibility

Then we get $[\omega'] = a_1$ in the same way; obviously, $\omega' > 1$. Also, by the continuity of $a_0 + 1/x$ for $x > 0$, we have $\omega = a_0 + 1/\omega'$. Hence, $\omega' = R(\omega)$, and $\omega = a_0 + \frac{1}{a_1} + \frac{1}{R^2(\omega)}$. By induction, we see that the jth-convergent of ω is A_j/B_j. Then, although ω has not been shown yet to be irrational, (22.10) holds for ω in place of η; and $0 < |B_j\omega - A_j| < 1/B_{j+1}$ for all sufficiently large j. Hence, if ω is ever equal to a rational number a/b, then $1 \le |aB_j - bA_j| < |b|/B_{j+1}$, which yields a contradiction. We end the proof.

Here is a trivial remark: Regardless whether η is irrational or not,

$$\eta = a_0 + \cfrac{1}{a_1 +} \cfrac{1}{a_2 +} \cdots \cfrac{1}{a_m +} \cfrac{1}{\lambda}, \qquad \Rightarrow \quad R^{m+1}(\eta) = \lambda. \qquad (22.18)$$
$$a_0 \in \mathbb{Z}, \{a_1, a_2, \ldots, a_m\} \subset \mathbb{N}, \ 1 < \lambda$$

Notes

[22.1] [Σ.X, Prop.2] essentially coincides with Theorem 21, only if Euclid's argument is expressed in terms of continued fraction expansions. As for the assumption (22.2), one should be aware that to determine whether a given number is rational or irrational can be an immensely difficult task. Thus, the above discussion may appear just theoretical; but, see Note [23.4].

[22.2] Euler (1737a, p.112) called convergents *fractiones principales*. The theory of continued fractions *theoria fractionum continuarum* originated in this article of his.

[22.3] The situation $B_1 = B_0 = 1$ noticed in (22.6) is a particular case, i.e., $j = 0$, of the more general $B_{j+1} = B_j + B_{j-1}$. This equality takes place if and only if $a_{j+1} = [R^{j+1}(\eta)] = 1$, which is equivalent to $R^j(\eta) > [R^j(\eta)] + \frac{1}{2}$, that is, the nearest integer to $R^j(\eta)$ is not a_j but $a_j + 1$. The observation of this subtlety in the nature of continued fraction expansions will play an important rôle in §86. See also Note [24.4].

§23

We shall show a fundamental relation between continued fraction expansions and rational approximations, which is due to Huygens (ca 1685 (1728, posth.)) and Lagrange (1798, pp.45–57; Euler (1771/1822, pp.495–504)).

To begin with, let (22.4) be the continued fraction expansion of an irrational number $\eta \in \mathbb{R}$. Then, it holds analogously to (21.13) that

$$|B_{j-1}\eta - A_{j-1}| > |B_j\eta - A_j|, \quad j \ge 0, \qquad (23.1)$$

since inverting (22.5) we get

$$R^{j+1}(\eta) = -\frac{B_{j-1}\eta - A_{j-1}}{B_j\eta - A_j}, \quad j \geq -1, \tag{23.2}$$

and $R^{j+1}(\eta) > 1$, $j \geq 0$. Hence, one may infer that the distribution of values of the linear form $x\eta - y$ with $x, y \in \mathbb{Z}$ should be closely related to the characterization of convergents of η. The following assertion of Lagrange substantiates this view:

Theorem 22 *Let $\eta \in \mathbb{R}$ be an arbitrary irrational number, and let*

$$L(V) = \min\{|t\eta - s| : s, t \in \mathbb{Z}, 1 \leq t \leq V\}. \tag{23.3}$$

Then, with the convergents A_j/B_j of η, we have, for each $v \geq 1$ such that $B_v \geq 2$,

$$B_{v-1} \leq V < B_v \implies L(V) = |B_{v-1}\eta - A_{v-1}|. \tag{23.4}$$

Further,

$$s \neq A_v \implies |B_v\eta - s| > |B_{v-1}\eta - A_{v-1}|. \tag{23.5}$$

Remark 1 In (23.4), $V \geq 1$ is assumed to be strictly less than B_v; hence, $B_v \geq 2$ ought to be supposed. This causes a minor complication: By (22.6), if $\eta < [\eta] + \frac{1}{2}$, then $B_1 \geq 2$, so $v \geq 1$ and in particular $L(1) = |\eta - [\eta]| = |B_0\eta - A_0|$. On the other hand, if $\eta > [\eta] + \frac{1}{2}$, then $B_1 = 1$, $B_2 \geq 2$, so $v \geq 2$ and $L(1) = |\eta - [\eta] - 1| = |B_1\eta - A_1| < |B_0\eta - A_0|$; see Note [22.3]. Anyway, the function $L(V)$ is a decreasing step-function. The assertion (23.1) and the present theorem imply that at each $B_v \geq 2$ it is discontinuous from the left but continuous from the right with $L(B_v) = |B_v\eta - A_v|$. This characterizes convergents A_v/B_v provided $B_v \geq 2$; see Note [23.3].

Proof To prove (23.4) it suffices to show that if $|t\eta - s| < |B_{v-1}\eta - A_{v-1}|$ with $B_v \geq 2$, then $t \geq B_v$; note that $B_v \geq 2$ implies $v \geq 1$ obviously. We apply the following change of variables to $\{s, t\}$: with any $j \geq -1$

$$s = MA_j - NA_{j-1}, \quad t = MB_j - NB_{j-1}, \tag{23.6}$$

which is inverted into

$$M = (-1)^{j-1}(sB_{j-1} - tA_{j-1}), \quad N = (-1)^{j-1}(sB_j - tA_j); \tag{23.7}$$

in particular, $M, N \in \mathbb{Z}$. For $j \geq 0$,

$$\begin{aligned} t\eta - s &= M(B_j\eta - A_j) - N(B_{j-1}\eta - A_{j-1}) \\ &= (-1)^j(M|B_j\eta - A_j| + N|B_{j-1}\eta - A_{j-1}|); \end{aligned} \tag{23.8}$$

98 *Divisibility*

the lower line depends on (22.8). With this, let $j = v \geq 1$. If $MN > 0$, then $|t\eta - s| > |B_{v-1}\eta - A_{v-1}|$, which contradicts the current assumption. Also, if $M = 0$, then $-NB_{v-1} = t \neq 0$ by (23.6); thus $N \neq 0$, and $|t\eta - s| \geq |B_{v-1}\eta - A_{v-1}|$, again a contradiction. Hence, we may suppose that $MN \leq 0$, $M \neq 0$. That is, $M > 0$, $N \leq 0$ since (23.6) implies $M > 0$. Then, $t \geq B_v$. This proves (23.4). As for (23.5), we apply (22.10) to see that for $s \neq A_v$

$$
\begin{aligned}
|B_v\eta - s| &> |s - A_v| - |B_v\eta - A_v| \\
&> 1 - 1/B_{v+1} > 1/B_v > |B_{v-1}\eta - A_{v-1}|,
\end{aligned}
\tag{23.9}
$$

in which we have used $B_{v+1} > B_v \geq 2$. We end the proof.

Now, adopting the above reasoning, we shall show the following refinement of Theorem 3:

Theorem 23 *Let $\{a,b\}$ be an arbitrary pair of positive integers, and let*

$$
E(V) = \min\{|ta - sb| : s, t \in \mathbb{Z}, 1 \leq t \leq V\}.
\tag{23.10}
$$

Then, with A_j/B_j and r_j as in (21.7) and (21.10), respectively, we have, for each $v \in [1, k+1]$ such that $B_v \geq 2$,

$$
B_{v-1} \leq V < B_v \implies E(V) = r_v.
\tag{23.11}
$$

Further,

$$
s \neq A_v \implies |B_v a - sb| > r_v.
\tag{23.12}
$$

Remark 2

(1) It is understood, exclusively here, that $R^{k+1}(a/b) = \infty$ and $B_{k+1} = \infty$.
(2) The device given in Note [21.2] should not be applied; thus $a_k \geq 2$ if $k \geq 1$.
(3) The theorem holds for these v :

$$
\begin{aligned}
&k = 0, \text{ i.e., } b|a: && v = 1, \\
&k = 1, \text{ then } B_1 = a_1 \geq 2: && v = 1, \\
&r \leq b/2 \text{ in (2.1), then } B_1 \geq 2: && v \geq 1, \\
&r > b/2 \text{ in (2.1), then } B_1 = 1: && v \geq 2.
\end{aligned}
\tag{23.13}
$$

See Note [23.3].

Proof If $k = 0$, then $v = 1$, $B_0 = 1$, and $B_1 = \infty$. With this, (23.11)–(23.12) holds since $r_1 = 0$. Thus we may assume that $k \geq 1$. By (21.12) and (23.6), we have, for $-1 \leq j \leq k$,

$$ta - sb = (-1)^j \left(Mr_{j+1} + Nr_j \right). \tag{23.14}$$

Since for $v = k + 1$ the assertions of the theorem are trivial, we assume that $|ta - sb| < r_v$ with a $v \in [1,k]$. Then $(23.14)_{j=v}$ leads us to the situation $MN \leq 0$, $M \neq 0$; hence $t \geq B_v$ by (23.6), which proves (23.11). As for (23.12), we note that

$$\frac{a}{b} - \frac{A_{j-1}}{B_{j-1}} = \frac{(-1)^{j-1}}{B_{j-1}(B_{j-1}R^j(a/b) + B_{j-2})}, \quad 1 \leq j \leq k+1, \tag{23.15}$$

which corresponds to (22.8). Thus,

$$\left| \frac{a}{b} - \frac{A_{j-1}}{B_{j-1}} \right| \leq \frac{1}{B_{j-1}B_j}, \quad 1 \leq j \leq k+1. \tag{23.16}$$

In view of (21.13), we may express this as

$$r_j \leq \frac{b}{B_j}, \quad 1 \leq j \leq k+1. \tag{23.17}$$

Also, for $1 \leq v \leq k$,

$$s \neq A_v \implies |B_v a - sb| = |B_v a - A_v b + (A_v - s)b| \geq b - r_{v+1}. \tag{23.18}$$

Then, by (23.17),

$$b - r_{v+1} \geq b(1 - 1/B_{v+1}) > b/B_v \geq r_v, \tag{23.19}$$

since $B_{v+1} > B_v \geq 2$. This confirms (23.12). We end the proof.

We see that strata structures exist both in $\{t\eta - s \; : \; s,t \in \mathbb{Z}\}$ and in $\{ta - sb : s,t \in \mathbb{Z}\}$. In particular, (23.11)–(23.12) indicates how the path of the application of Euclid's algorithm penetrates the strata and reaches the greatest common divisor $\langle a,b \rangle$. Theorem 23 will be applied in §81 as an essential means.

Notes

[23.1] Following Lagrange, we call $L(V)$ defined by (23.3) the best approximation within the limit V to $\eta \in \mathbb{R}\setminus\mathbb{Q}$. The change of variables (23.6) is due to Lagrange (1798, p.46; Euler (1771/1822, p.496)). See also Legendre (1798, p.27).

[23.2] From Theorem 22, together with Remark 1, it follows that we have, for any $\eta \in \mathbb{R}\setminus\mathbb{Q}$ and $v \geq 1$,

$$s/t \neq A_v/B_v \text{ and } 1 \leq t \leq B_v \implies |\eta - A_v/B_v| < |\eta - s/t|.$$

Namely, it may be appropriate to call each convergent A_v/B_v, $v \geq 1$, the best approximation fraction within the limit B_v to η. There exists a difference

100 *Divisibility*

between this notion and the one introduced in the preceding Note, since, for instance, the 1st convergent of π equals $22/7$, which has been known from antiquity to be a good approximation to π, but if denominators are restricted to be less than or equal to 6, that is, if $\min\{|\pi - s/t| : s,t \in \mathbb{Z}, 1 \leq t \leq 6\}$ is considered, then the best approximation fraction equals $19/6$ which is not a convergent of π; on the other hand, the best approximation within 6 takes place at $t = 1, s = 3$, and 3 is the 0th convergent of π. In order to describe fractions such as $19/6$ for π, one ought to use the notion of semiconvergents. However, we shall not dwell on this as it has no relevance to what is to be discussed in the rest of the present essays.

[23.3]   The $L(V)$ in (23.4) is taken by the unique pair $\{s,t\} = \{B_{\nu-1}, A_{\nu-1}\}$. As for $E(V)$ in (23.11) with $k \geq 2$, the situation is the same for $\nu \leq k - 1$; however, if $\nu = k$, then the minimum is attained by two pairs

$$\{s,t\} = \{B_{k-1}, A_{k-1}\}, \{B_k - B_{k-1}, A_k - A_{k-1}\}.$$

[23.4]   In [Σ.X, Prop.2] the following statement is made, apparently with the intention of extending anthyphairesis/antenaresis to incommensurable ratio between two numbers:
If, when the less of two unequal magnitudes is continually subtracted in turn from the greater, that which is left never measures the one before it, then the magnitudes will be incommensurable.
This can be regarded as the infinite Euclid algorithm applied to $\{\alpha, \beta\} \in \mathbb{R}^2$ such that $\alpha/\beta \in \mathbb{R} \setminus \mathbb{Q}$. Here a comparison may be made between the Euclid–Lagrange approach to the linear continuum and that of Dedekind (1892) via the notion *Schnitt*. The latter should appear more abstract than the former. However, the process via regular continued fraction expansion as well contains a non-constructible aspect. Namely, the determination of the value of $[R^j(\eta)]$ is by no means a trivial task because there exists no known general algorithm other than computing it via numerical values of η, which might sound like a tautology. Hence, one may say that the theory of continued fraction expansions has a limited scope of applicability. Nevertheless, in the context of quadratic irrationals its exploitation by classic Indian mathematicians, Euler, Lagrange, and Legendre resulted in a variety of impressive practical successes as we shall show in Chapter 4.

§24

Now, in view of the last two theorems, it should be highly worthwhile to have a criterion with which one can assert, without recourse to continued fraction

§24   101

expansions, that a particular irreducible fraction is a convergent of a given real number.

The following assertion, which is an answer to this issue, is due to Legendre (1798, p.29). It will have important applications in §48 and §84.

Theorem 24 *If $\eta \in \mathbb{R}$ and an irreducible fraction A/B with $B > 0$ are such that*

$$\left| \eta - \frac{A}{B} \right| < \frac{1}{2B^2}, \tag{24.1}$$

then A/B is a convergent of η.

Remark Here η is not required to be irrational.

Proof If $B = 1$, then we have two cases: either $[\eta] = A$ or $[\eta] = A - 1$. With the first case, $\eta = A + 1/R(\eta)$, $R(\eta) > 2$, and $A = A/B$ is the 0th convergent, including the case $\eta = A$, i.e., $R(\eta) = \infty$. As for the second case, we have $\eta - [\eta] > \frac{1}{2}$; thus, $1 < R(\eta) < 2$, and $\eta = A - 1 + 1/(1 + 1/R^2(\eta))$, $1 < R^2(\eta) < \infty$, which means that $A/B = A$ is the 1st convergent. Hence, we may assume that $B \geq 2$. Then, in the expansion, via Euclid's algorithm,

$$\frac{A}{B} = s_0 + \frac{1}{s_1 +} \frac{1}{s_2 + \cdots +} \frac{1}{s_g},$$
$$\begin{pmatrix} S_g & S_{g-1} \\ T_g & T_{g-1} \end{pmatrix} = \begin{pmatrix} s_0 & 1 \\ 1 & 0 \end{pmatrix} \begin{pmatrix} s_1 & 1 \\ 1 & 0 \end{pmatrix} \cdots \begin{pmatrix} s_g & 1 \\ 1 & 0 \end{pmatrix}, \tag{24.2}$$

we have $g \geq 1$, $s_g \geq 2$ (see Note [21.2]); and $A = S_g$, $B = T_g$ since $B > 0$. If $\eta - A/B = 0$, then A/B is a convergent of η. On the other hand, if $\mathrm{sgn}\,(\eta - A/B) = (-1)^g$, then the condition (24.1) can be written as

$$\eta - \frac{S_g}{T_g} = (-1)^g \frac{\theta}{T_g^2}, \quad 0 < \theta < \tfrac{1}{2}. \tag{24.3}$$

With

$$\tau = \frac{1}{\theta} - \frac{T_{g-1}}{T_g} > 1, \tag{24.4}$$

we have

$$s_0 + \frac{1}{s_1 +} \frac{1}{s_2 + \cdots +} \frac{1}{s_g +} \frac{1}{\tau} = \frac{S_g \tau + S_{g-1}}{T_g \tau + T_{g-1}} = \eta. \tag{24.5}$$

Therefore, $\tau = R^{g+1}(\eta)$ because of (22.18), and A/B is the gth convergent S_g/T_g of η. On the other hand, if $\mathrm{sgn}\,(\eta - A/B) = (-1)^{g+1}$, then we modify (24.2) by $s_g \mapsto s_g - 1 + 1/s_{g+1}$, $s_{g+1} = 1$ (Note [21.2]):

102 Divisibility

$$\frac{A}{B} = s_0 + \frac{1}{s_1} + \frac{1}{s_2} + \cdots + \frac{1}{s_g - 1} + \frac{1}{s_{g+1}},$$

$$\begin{pmatrix} S_{g+1}^* & S_g^* \\ T_{g+1}^* & T_g^* \end{pmatrix} = \begin{pmatrix} s_0 & 1 \\ 1 & 0 \end{pmatrix}\begin{pmatrix} s_1 & 1 \\ 1 & 0 \end{pmatrix}\cdots\begin{pmatrix} s_g - 1 & 1 \\ 1 & 0 \end{pmatrix}\begin{pmatrix} s_{g+1} & 1 \\ 1 & 0 \end{pmatrix}. \tag{24.6}$$

Here we have $S_{g+1}^* = S_g = A$, $T_{g+1}^* = T_g = B$. In place of (24.3) we have

$$\eta - \frac{S_{g+1}^*}{T_{g+1}^*} = (-1)^{g+1}\frac{\theta^*}{(T_{g+1}^*)^2}, \quad 0 < \theta^* < \tfrac{1}{2}; \tag{24.7}$$

Corresponding to (24.4)–(24.5), with

$$\tau^* = \frac{1}{\theta^*} - \frac{T_g^*}{T_{g+1}^*} > 1, \tag{24.8}$$

we have

$$s_0 + \frac{1}{s_1} + \frac{1}{s_2} + \cdots + \frac{1}{s_{g-1}} + \frac{1}{s_g} - 1 + \frac{1}{s_{g+1}} + \frac{1}{\tau^*} = \frac{S_{g+1}^*\tau^* + S_g^*}{T_{g+1}^*\tau^* + T_g^*} = \eta. \tag{24.9}$$

Hence A/B is the $(g+1)^{\text{st}}$ convergent of η. We end the proof.

Notes

[24.1]   We have the expansion

$$\sqrt{163} = 12 + \frac{1}{1} + \frac{1}{3} + \frac{1}{3} + \frac{1}{2} + \frac{1}{1} + \frac{1}{1} + \frac{1}{7} + \frac{1}{1} + \frac{1}{11} + \frac{1}{1}$$
$$+ \frac{1}{7} + \frac{1}{1} + \frac{1}{1} + \frac{1}{2} + \frac{1}{3} + \frac{1}{3} + \frac{1}{1} + \frac{1}{12 + \sqrt{163}}.$$

The 15th, 16th, 17th convergents are, respectively,

$$\frac{14921333}{1168729}, \frac{49158693}{3850406}, \frac{64080026}{5019135}.$$

Thus, by (22.9),

$$\left|\sqrt{163} - \frac{49158693}{3850406}\right| < \frac{1}{3850406 \cdot 5019135}.$$

The left side is close to $4.974 \cdot 10^{-14}$, and the right side to $5.174 \cdot 10^{-14}$; the approximation is indeed excellent. This does not satisfy (24.1), though. Nevertheless, the 15th and the 17th convergents both fulfill (24.1), a fact which can easily be generalized by observing

$$|\eta - A_{v+1}/B_{v+1}| + |\eta - A_v/B_v| = 1/(B_{v+1}B_v) < \tfrac{1}{2}B_{v+1}^{-2} + \tfrac{1}{2}B_v^{-2},$$

provided $B_v < B_{v+1}$. By the way, the continued fraction expansion of $12 + \sqrt{163}$ is cyclic; moreover, the period is palindromic, that is, reversing the order of its terms does not alter it. Furthermore, the 17th convergent of $\sqrt{163}$ yields the relation

$$6408026^2 - 163 \cdot 5019135^2 = 1.$$

These remarkable phenomena will be discussed in §84 in a general setting.

[24.2]   The simplest periodic continued fraction is

$$\phi = 1 + \frac{1}{1+} \frac{1}{1+} \frac{1}{1+} \cdots + \frac{1}{1+} \cdots.$$

Since $\phi = 1 + 1/\phi$, this equals the golden ratio $(1 + \sqrt{5})/2$; that is, $a : b = (a+b) : a$ (the extreme and mean ratio ([Σ.VI, Def.3])). As $\phi = 2\cos(\pi/5)$, it has a close relation with the construction of the pentagon ([Σ.IV, Prop.11]). By $(22.3)_{\eta=\phi}$, the jth convergent of ϕ has the expression $f_{j+2}/f_{j+1}, j \geq 0$, where f_j are defined by

$$\begin{pmatrix} 1 & 1 \\ 1 & 0 \end{pmatrix}^j = \begin{pmatrix} f_{j+1} & f_j \\ f_j & f_{j-1} \end{pmatrix}, \quad j \geq 0.$$

We have $f_{j+1} = f_j + f_{j-1}$,

$$\{f_j : j \geq -1\} = \{1, 0, 1, 1, 2, 3, 5, 8, 13, 21, 34, 55, 89, 144, 233, \dots\}.$$

The sub-sequence $\{f_j : j \geq 1\}$ was discussed by Fibonacci (1202: 1857, pp.283–284), and is named accordingly. However, the generating method of it was known already in classic Indian mathematics (ca 700; see Singh (1985)). The number ϕ is hidden in ancient architectures and fine arts. Above all, the nature has long rendered the numbers f_j in a variety of plants such as *Broccolo Romanesco*.

[24.3]   We have

$$\langle f_a, f_b \rangle = f_{\langle a, b \rangle}.$$

To show this, in the above matrix identity we set $j = l + m + 1, l \geq 1$, $m \geq 0$, and divide the power of matrix on the left side into two parts according to the exponents $m + 1$ and l, which gives $f_{m+l} = f_{m+1}f_l + f_m f_{l-1}$. Letting $m = 0, l, 2l, \dots$, we find that $l|m \Rightarrow f_l|f_m$. Further, with $a = qb + r, 0 \leq r < b$, we have

$$\langle f_a, f_b \rangle = \langle f_{qb+1}f_r + f_{qb}f_{r-1}, f_b \rangle = \langle f_{qb+1}f_r, f_b \rangle = \langle f_r, f_b \rangle,$$

since $f_b|f_{qb}$ and $\langle f_{qb+1}, f_{qb} \rangle = 1$, so $\langle f_{qb+1}, f_b \rangle = 1$. The rest of the argument is omitted.

104 *Divisibility*

[24.4] Intuitively, a weak approximation by continued fraction expansion ought to take place when the subtle situation noted in Note [22.3] happens, for then the relevant $R^j(\eta)$ is closer to $[R^j(\eta)] + 1$ than to $[R^j(\eta)]$. If we take the former in place of the latter or the ordinary, then we are led to continued fraction expansions that admit negative terms; see Note [26.3]. Therefore, the worst approximation via regular continued fractions occurs when $a_j = 1$ for all $j \geq 1$. In this sense ϕ is extremal.

[24.5] Proof of the bound of Lamé (Note [3.2]). We may suppose that $b \nmid a$. Using the notation of §21, we have $B_0 \geq f_1$ and $B_1 \geq f_2$; note that $k \geq 1$ presently. By the fact that $B_j = a_j B_{j-1} + B_{j-2} \geq B_{j-1} + B_{j-2}$ for $j \geq 1$ and by induction, we have $B_j \geq f_{j+1}, j \geq 0$. Also, $(21.11)_{j=k}$ gives $b = r_k B_k$. Hence, $b \geq f_{k+1}$. On the other hand, $f_j \geq \phi^{j-2}$ for $j \geq 2$. In fact, this is trivial for $j = 2, 3$, and by induction $f_j = f_{j-1} + f_{j-2} \geq \phi^{j-3} + \phi^{j-4} = \phi^{j-2}$. Namely, $b \geq \phi^{k-1}$. From $\log_{10} b < \kappa$ and $\log_{10} \phi = 0.20898764 \cdots$, we get $k - 1 < 5\kappa$, i.e., $k \leq 5\kappa$, which ends the proof.

§25

Closing our account of the fundamentals of the theory of continued fractions, we shall show a modular property of the expansion. To this end, we first give a useful assertion of Serret (1849a, pp.212–215): We suppose that irrational numbers η and ω are connected by the relation

$$\omega = \frac{\alpha\eta + \beta}{\gamma\eta + \delta}, \quad \alpha\delta - \beta\gamma = (-1)^v, \ \alpha, \beta, \gamma, \delta \in \mathbb{Z}. \tag{25.1}$$

Theorem 25 *If $\eta > 1$, $\gamma > 0$, $\delta > 0$, then α/γ is a convergent of ω.*

Proof We write

$$\frac{\alpha}{\gamma} = a_0 + \frac{1}{a_1 +} \frac{1}{a_2 +} \cdots + \frac{1}{a_k}, \quad (-1)^{k+1} = (-1)^v, \quad k \geq 0, \tag{25.2}$$

in which Note [21.2] may become relevant in order to adjust the parity of k. Using the notation (21.4), we have $\alpha = A_k$, $\gamma = B_k$. Hence,

$$\begin{pmatrix} A_k & A_{k-1} \\ B_k & B_{k-1} \end{pmatrix}^{-1} \begin{pmatrix} \alpha & \beta \\ \gamma & \delta \end{pmatrix} = \begin{pmatrix} 1 & \tau \\ 0 & 1 \end{pmatrix}. \tag{25.3}$$

In particular, $\tau B_k + B_{k-1} = \delta$; and by the condition $\delta > 0$, we have $\tau \geq 0$, for $B_k \geq B_{k-1} \geq 0$. Hence, $\eta + \tau > 1$, and

$$\omega = \frac{A_k(\eta + \tau) + A_{k-1}}{B_k(\eta + \tau) + B_{k-1}} = a_0 + \frac{1}{a_1 +} \frac{1}{a_2 +} \cdots + \frac{1}{a_k +} \frac{1}{\eta + \tau}, \tag{25.4}$$

so $\eta + \tau = R^{k+1}(\omega)$ because of (22.18). We end the proof.

Theorem 26 *If* (25.1) *holds, then there exists a pair of positive integers* $\{h, h'\}$ *such that* $2 \mid (h + \nu - h')$ *and* $\mathrm{R}^h(\omega) = \mathrm{R}^{h'}(\eta)$.

Proof We have, using (22.5),

$$\omega = \frac{(\alpha A_j + \beta B_j)\mathrm{R}^{j+1}(\eta) + \alpha A_{j-1} + \beta B_{j-1}}{(\gamma A_j + \delta B_j)\mathrm{R}^{j+1}(\eta) + \gamma A_{j-1} + \delta B_{j-1}}. \tag{25.5}$$

Here $\det\left(\begin{pmatrix} \alpha & \beta \\ \gamma & \delta \end{pmatrix} \begin{pmatrix} A_j & A_{j-1} \\ B_j & B_{j-1} \end{pmatrix}\right) = (-1)^{j+\nu+1}$ with ν as above. For a sufficiently large j, we have $\gamma A_j + \delta B_j \sim (\gamma \eta + \delta) B_j$ and $\gamma A_{j-1} + \delta B_{j-1} \sim (\gamma \eta + \delta) B_{j-1}$. We may assume that $\gamma \eta + \delta > 0$; otherwise, change the signs of $\alpha, \beta, \gamma, \delta$ simultaneously. Hence, the same situation as in the last theorem arises but with $\nu \mapsto j + \nu + 1$ and $\eta \mapsto \mathrm{R}^{j+1}(\eta)$. Thus, as a counterpart of (25.2), we write

$$\frac{\alpha A_j + \beta B_j}{\gamma A_j + \delta B_j} = t_0 + \cfrac{1}{t_1} + \cfrac{1}{t_2} + \cdots + \cfrac{1}{t_J}, \quad (-1)^{J+1} = (-1)^{j+\nu+1}. \tag{25.6}$$

Corresponding to (25.4), we have

$$\omega = t_0 + \cfrac{1}{t_1} + \cfrac{1}{t_2} + \cdots + \cfrac{1}{t_J} + \cfrac{1}{\mathrm{R}^{j+1}(\eta) + \sigma} \tag{25.7}$$

with an integer $\sigma \geq 0$. We have $\mathrm{R}^{J+1}(\omega) = \mathrm{R}^{j+1}(\eta) + \sigma$ by (22.18). That is, $\mathrm{R}^{J+2}(\omega) = \mathrm{R}^{j+2}(\eta)$, which ends the proof.

In passing, we remark a trivial fact: provided (22.5),

$$\text{if } \eta > 1, \text{ then the } j^{\text{th}} \text{ convergent of } 1/\eta \text{ is equal to } B_{j-1}/A_{j-1}. \tag{25.8}$$

For

$$\frac{1}{\eta} = 0 + \cfrac{1}{a_0} + \cfrac{1}{a_1} + \cdots + \cfrac{1}{a_j} + \cfrac{1}{\mathrm{R}^{j+1}(\eta)} \tag{25.9}$$

is the continued fraction expansion of $1/\eta$, as $a_0 \geq 1$, so $\mathrm{R}^{j+1}(\eta) = \mathrm{R}^{j+2}(1/\eta)$; that is,

$$
\begin{aligned}
\frac{1}{\eta} &= \begin{pmatrix} 0 & 1 \\ 1 & 0 \end{pmatrix} \begin{pmatrix} a_0 & 1 \\ 1 & 0 \end{pmatrix} \begin{pmatrix} a_1 & 1 \\ 1 & 0 \end{pmatrix} \cdots \begin{pmatrix} a_j & 1 \\ 1 & 0 \end{pmatrix} (\mathrm{R}^{j+1}(\eta)) \\
&= \begin{pmatrix} B_j & B_{j-1} \\ A_j & A_{j-1} \end{pmatrix} (\mathrm{R}^{j+2}(1/\eta)).
\end{aligned}
\tag{25.10}
$$

Note

[25.1] An explicit version of Theorem 4: Let $\begin{pmatrix} a & b \\ c & d \end{pmatrix} \in \Gamma$, under Note [5.1](1). If $c = 0$, then this is equal to either T^b or $W^2 T^{-b}$. Thus we may assume that $c \neq 0$, and have

$$\frac{a}{c} = a_0 + \cfrac{1}{a_1} + \cfrac{1}{a_2} + \cdots + \cfrac{1}{a_k}, \quad 2 \nmid k,$$

106　　　　　　　　　　　　　　*Divisibility*

where the parity of k is adjusted by means of Note [21.1]. Then there exists a unique integer τ such that with A_j, B_j as in (25.3)

$$\begin{pmatrix} a & b \\ c & d \end{pmatrix} = \begin{pmatrix} A_k & A_{k-1} \\ B_k & B_{k-1} \end{pmatrix} \cdot \begin{cases} T^\tau & \text{if } c > 0, \\ W^2 T^{-\tau} & \text{if } c < 0. \end{cases}$$

On noting that

$$\begin{pmatrix} \alpha & 1 \\ 1 & 0 \end{pmatrix} \begin{pmatrix} \beta & 1 \\ 1 & 0 \end{pmatrix} = T^\alpha W T^{-\beta} W^{-1}, \quad \alpha, \beta \in \mathbb{Z},$$

we find that

$$\begin{pmatrix} a & b \\ c & d \end{pmatrix} = \left(\prod_{j=0}^{(k-1)/2} T^{a_{2j}} W T^{-a_{2j+1}} W^{-1} \right) \cdot \{\text{either } T^\tau \text{ or } W^2 T^{-\tau}\}.$$

§26

As far as the description of the multiplicative nature of the integers is concerned, we may dare to dispense with any mention of historical matters after Euclid and before Fermat, save for the contributions of Brahmagupta and Diophantus. However, it should be as well to try to view our very basic subjects from their historical roots, if brief.

[a] *Everything is number*: Pythagoras (b. ca 570 BCE, Samos) and his followers initiated mathematical debates on the numbers and the universe. According to a biography (Iamblichus (ca 300 CE)), Pythagoras went to Egypt to learn astronomy and geometry from sages at temples on the recommendation of his mentor Thales (b. ca 620 BCE, Miletus). After an extensive pilgrimage for 22 years he was captured during the Egyptian campaign of Cambyses II (525 BCE) and brought to Babylon. There he involved himself for 12 years in arithmetic, music, and mysticism, until repatriated in his mid-50s. Then he moved to Croton and founded a renowned school embracing his philosophy of mathematical harmony, or his view that the whole universe is tightly associated with natural numbers. The school soon became a presence ruling the city with Pythagoras' authority. Then, an influential but tyrannical man wished to join the school. Pythagoras rejected him. The incident caused eventually a fierce rebellion, and the tragic death of the most venerable sage. Nonetheless, his teachings were disseminated for generations and developed into a core of the ancient Greek philosophy. This is a tale where deep respect and romantic imagination merge. In spite of being inconsistent with modern scholarly

accounts of Pythagoras' life, it at least tells greatly how the Egyptian and Babylonian mathematical traditions for some two thousand years flowed into succeeding civilizations.

[b]   In his monumental *Elements*, Euclid rendered the thoughts of his mathematical antecedents who were active during the two centuries after Pythagoras and before himself. One should note that the *Elements* carries no trace of mysticism like the numerology of the Pythagorean school; it was compiled instead with deductive logic to the extent that Euclid would need only minor changes in the way of expressing his reasoning if he were ever here to discuss mathematics with us. It is beyond doubt that mathematics in the *Elements* will remain intact forever. Hence, the historical details about individual assertions or books of the *Elements* do not essentially interest us, as has been indicated already in Note [1.1]. However, it would be a different matter to know how the *Elements* has endured physically through time. Thus, although neither the year nor the place of Euclid's birth nor those of his death have been clarified yet, it is evident that his extraordinary project was completed in Alexandria around 300 BCE under the reign of Ptolemaios I. The original text of the *Elements* was apparently lost long ago; but the edition $[\Sigma_\theta]$ by Theon (ca 380 CE) circulated widely, and a copy of it at the imperial Byzantium library was brought, on a diplomatic agreement, to Bayt al-Hikma at Baghdad to be translated into Arabic under a decree of Calif Harun al-Rashid (ca 800 CE).
  Besides, there exists a Vatican library codex (*vat.gr.190*, ca 900 CE), which count de Péluse (G. Monge) conveyed from Rome to Paris and F. Peyrard in 1808 identified as a pristine edition, that is, free of Theon's editing: It turned out that Theon had applied to the *Elements* changes of text and rearrangements of propositions. That was apparently a result of a mathematician's critical reading of the original. Anyway, there are considerable differences between the edition $[\Sigma]$ by J.L. Heiberg, which is mainly based on *vat.gr.190* and currently regarded as definitive, and various translations published prior to the 19th century. For instance, the *Elements* to which Gauss referred appears to be one of such translations; $[\Sigma.\text{VII}, \text{Prop.30}]$ is given the confusing label 32 in [DA, art.14]. A similar discrepancy is in Euler (1755, the proof of Theorema 1). In the present essays, we have been referring to $[\Sigma]$ via its English translation by Heath (1956), as mentioned beforehand in Note [1.1]. However, one ought to be aware that investigations on some extant Arabic manuscripts of the *Elements* are going on; and the possibility cannot be denied that texts closer to the genuine original might be reconstructed in the future. The success of this endeavor will be of great importance; however, it is true also that such a reconstruction will never bring about any change in Euclid's

108   *Divisibility*

mathematics itself. By the way, the first printed Latin and English translations of the *Elements* are due to Campano da Novara (1482) and Billingsley (1570; Lord Mayor of London, 1596–1597), respectively; the former is via an Arabic edition, and the latter is from a copy, of $[\Sigma_\theta]$. Both printings are gorgeous indeed; however, their title pages contain the same blemish, that is, Euclid (Megara), a revered philosopher (b. ca 435 BCE) but not our great geometer.

[c]   The transmission of mathematical traditions that is touched on in Note [1.5] is a fascinating subject to discuss. Here we shall, however, indicate just the most salient aspect: Translation into the then linga-franca, Arabic, of a great number of older and contemporary Greek, Indian, as well as Persian mathematical works was conducted at the Bayt al-Hikma during the Abbasid dynasty, which secured ancient wisdoms against imminent erosion following the fall of preceding empires. An eternal outcome achieved during this great cultural movement is the development of algebra due to al-Khwarizmi (ca 820a) with which he liberated arithmetic from the rigid geometrical argument of ancient Greeks, while advocating the Indian numeral and calculation system. All generations after al-Khwarizmi owed fundamentally to his legacy; see §70(ii) as well. Besides, Byzantium monasteries maintained invaluable volumes during those centuries of turmoils; one of the most significant among them was Diophantus' Arithmetica (ca 250 CE). A later copy of it was eventually acquired by Bachet and resulted in his Latin translation (1621); see Heath (1910, Introduction, Chapter II) for relevant historical details. That yielded a tremendous impact on number theory: Fermat's earnest study (1679) of problems and assertions left by Diophantus opened classical multiplicative number theory which has gone a long way towards the modern number theory.

[d]   The theory of continued fractions up to Lagrange: Since the continued fraction expansion of a fraction is a result of repeated applications of the division algorithm, and because of the existence of the identity (21.14), the anthyphairesis or Euclid's algorithm can be regarded as its origin. In fact, we are disposed to assert more explicitly that the origin is in the tenth book of the *Elements*. Thus, repeating what is stated in Note [23.4], the impressive $[\Sigma.X]$ is a firm evidence that Euclid tried hard to describe the incommensurability (irrational numbers) by employing anthyphairesis and by so doing suggested infinite continued fraction expansion; one may trace such endless applications of anthyphairesis further back to the treatment of $\sqrt{2}$ by the Pythagorean school or rather by much more older traditions (see §84). Also, the ancient Indian method to solve linear indefinite equations as well lies in between Euclid's algorithm and the continued fraction expansion (Note [3.1]). These arithmetical expertises were sophisticated by the medieval Arab

mathematicians and brought to Europe in the late Renaissance. Thus, Bombelli (1579, pp.35–38) devised a method to compute $\sqrt{n}$ for $n \in \mathbb{N}$, which is said to be a beginning of the theoretical study of continued fractions; see Note [26.1]. Then, the first practical notation for continued fractions was invented by Cataldi (1613, p.70):

$$4.\&\frac{272}{1121} = 4.\&\frac{2}{8} \cdot \&\frac{2}{8} \cdot \&\frac{2}{8} \cdot \&\frac{2}{8}, \tag{26.1}$$

which is of course the same as $4 + \frac{2}{8} + \frac{2}{8} + \frac{2}{8} + \frac{2}{8}$, and is a very good approximation to $3\sqrt{2}$; see Note [26.1] again. Also, the first instance of the other representation indicated in Note [21.1] is visible at the same place, although nowadays it is commonly attributed to W. Brouncker who gave a beautiful expansion of $4/\pi$ (Wallis (1656, p.192)). Then, Huygens (ca 1685 (1728, posth.)) grasped the fundamental relation between convergents and best approximation fractions (Note [23.2]; see also Lagrange (1769, p.424; 1798, pp.43–44)). Based on these developments, the modern theory of continued fractions was started by Euler (1737a). He showed (22.15)–(22.16), which are called Euler identities today; see Euler (1748b, Caput XVIII) as well. Moreover, Euler (1759) applied continued fractions to quadratic indefinite equations and opened a way to their general solutions; see Note [82.3]. That evolved into superb number of theoretical investigations of Lagrange (1768/1798), as is to be shown in our Chapter 4. To this we must add the truly amazing fact that Brahmagupta (628) anticipated Euler and Lagrange by some 1,100 years in dealing with the quadratic indefinite equation $x^2 - dy^2 = 1$ ($d,x,y \in \mathbb{N}$; d is fixed) by a marvelously practical method which is in fact a procedure based on none other than the half-regular continued fractions mentioned in Note [21.1] if expressed with modern terminology; our account of it will be given in §86; see also Datta–Singh (1935: 1962: 1962, Chap.III, §§16–17). The investigation on the propagation of this and other advanced methods of classic Indian mathematic is an active area of current historical research.

[e]   Brouncker's continued fraction mentioned above is

$$\frac{4}{\pi} = 1 + \frac{1}{2} + \frac{3^2}{2} + \frac{5^2}{2} + \frac{7^2}{2} + \frac{9^2}{2} + \cdots, \tag{26.2}$$

a proof of which is given in Note [26.2]. Like this, continued fractions in general attract our attention in a rather mysterious way. Those quantities which look chaotic in decimal expansions could often appear highly structural once expanded into continued fractions, as has been seen in Note [24.1] already. Moreover, one may obtain surprisingly good rational approximations from the expansions. The same applies equally to rational numbers; fractions with

110   *Divisibility*

huge denominators can often be represented with handy sequences of integers (Note [21.3]). Because of these characteristics, continued fractions have a variety of applications. Thus, in nature, there are a great number of phenomena which change continuously with periods. Periodic repetitions are counted in terms of positive integers; and relations between them can be represented by ratios or fractions. However, in general, periods do not punctuate movements exactly; and these fractions become approximations. Hence, in the balance of precision and convenience, efficient fractions are to be selected; being efficient means that those selected are easy to handle and provide good approximation simultaneously. Here emerges the importance of Theorems 22–23 and Note [23.2]. Since ancient times, various fractions have been supplied for practical use; and many of them have turned out to be convergents of some real numbers. We shall show three typical examples:

(1) To simulate the motion of heavenly bodies by means of combinations of gears must have been dreamed of since great antiquity. Sumerians were, perhaps, observing the celestial rotations as if watching a mysteriously smooth mechanism. Although it was constructed a few thousand years later in ancient Greek, the *Antikythera mechanism* is an evidence of such a dream of ancients (Nature, **444** (2006), iss.7119, pp.587–591). However, one of the earliest examples of similar mechanisms, which are well documented, is the design of a planetarium by Huygens (1728). As we have already suggested above, his design describes clearly the gear mechanism on a sound speculation that convergents should yield best approximation fractions. His task was to find cog-numbers, within the technological limit in his time, which provide excellent approximations to ratios of cycles of revolutions of planets around the Sun. According to the astronomical observation data (January 1682: Huygens (1728, p.172)), the ratio of cycles of Saturn and the Earth is approximately

$$\frac{77708431}{2640858} = 29 + \frac{1}{2+} \frac{1}{2+} \frac{1}{1+} \frac{1}{5+} \frac{1}{1+} \frac{1}{4+} \cdots \qquad (26.3)$$

(p.174); and taking the third convergent $206/7$ Huygens gave Saturn a gear of 206 cogs and the Earth a gear of 7 cogs (p.175). If the fourth convergent $1177/40$ had been chosen, then his design would have been dismissed as impossible to materialize.

(2) If a solar eclipse takes place somewhere, then at the 223rd new moon after it a similar eclipse will likely be observed elsewhere. This way of predicting solar eclipses is stated, for instance, in Plinius (ca 77 CE: Spira edition, p.5): *DEfectus* ccxxiii *mensibus redire* ...; E. Halley named it as the Saros period. It is certainly a knowledge from Babylonian astronomy. We give here a brief account of how this famous number 223 arises: A solar eclipse occurs only

§26   111

when the Sun, the Moon, and the Earth are aligned in syzygy. This is a configuration concerning three bodies with some extents. So, on the inner surface of the celestial sphere, a solar eclipse can take place when the Sun and the Moon are in a quite narrow neighborhood, that is, close to a node where the Moon crosses the ecliptic in either an ascending or a descending mode; note that nodes move in the opposite direction to that of the Sun (the lunar nodal precession). With these facts, the periods to be taken into account are: S = Draconic Year, in which the Sun proceeds from a node to the same node; D = Draconic Month, in which the Moon traverses once the celestial sphere from a node to the same node; and M = Synodic Month, in which the lunar phase completes a cycle of waxing and waning. Also, the depth of solar eclipses is of course closely related to the apparent diameter of the Moon; thus, we need to take into account P = Anomalistic Month in which the Moon returns to a perigee which is also moving. There are further aspects like the changes of the apparent diameter of the Sun, which make the theory of solar eclipses quite involved. To avoid excessive complications, here we use mean values extracted from the data J2000.0: measuring in the number of days,

$$S = 346.620076, \ D = 27.212221, \ M = 29.530589, \ P = 27.554550. \quad (26.4)$$

Now, suppose that after a solar eclipse having occurred close to a certain node the Sun meets the same node u times, and the Moon traverses the celestial sphere v times while becoming the new phase w times and attaining the same apparent diameter z times. With this, when $|uS - vD|$, $|uS - wM|$, $|uS - zP|$ become all smaller than one (a day), we may expect a similar eclipse. In this way, we are led to the expansions

$$\frac{S}{D} = 12 + \frac{1}{1} + \frac{1}{2} + \frac{1}{1} + \frac{1}{4} + \frac{1}{3} + \frac{1}{5} + \frac{1}{1} + \frac{1}{27} + \cdots,$$

$$\frac{S}{M} = 11 + \frac{1}{1} + \frac{1}{2} + \frac{1}{1} + \frac{1}{4} + \frac{1}{3} + \frac{1}{5} + \frac{1}{1} + \frac{1}{30} + \cdots, \quad (26.5)$$

$$\frac{S}{P} = 12 + \frac{1}{1} + \frac{1}{1} + \frac{1}{2} + \frac{1}{1} + \frac{1}{1} + \frac{1}{1} + \frac{1}{5} + \frac{1}{3} + \cdots.$$

Here $S/D - S/M = 1$ essentially, i.e., a beat period. Fortunately, both the fourth and the sixth convergents of S/D and S/P, respectively, have the same denominator 19:

$$\frac{242}{19} = 12 + \frac{1}{1} + \frac{1}{2} + \frac{1}{1} + \frac{1}{4}, \quad \frac{239}{19} = 12 + \frac{1}{1} + \frac{1}{1} + \frac{1}{2} + \frac{1}{1} + \frac{1}{1} + \frac{1}{1}. \quad (26.6)$$

Therefore, we take $u = 19$, $v = 242$, $w = v - 19$, $z = 239$, finding that

$$\begin{aligned} 19S &= 6585.781444, \quad 242D = 6585.357482, \\ 223M &= 6585.321347, \quad 239P = 6585.537450. \end{aligned} \quad (26.7)$$

112   *Divisibility*

We have detected the Saros period $223M$ which is about $18^y 11^d 8^h$ plus 0 or ± 1 days, depending on leaps. If one wants to predict similar solar eclipses at almost the same location, then 3 Saros = 1 Exeligmos is the right choice of period since the 8-hour remainder in one Saros means the $120°$ advance of longitude, ignoring the latitude aspect. It should be no surprise that the Antikythera mechanism is reported to contain a gear corresponding to the Saros period.

(3) Another typical instance of the problem of representing relations between continuity and periodic repetitions is a mode of dividing one-octave interval to accomplish a harmonious scale. Here we shall deal with a minimum of this delicate subject in musicology. Discussions in past times were developed in terms of the positions of a movable bridge on a monochord; we shall use instead the ratios of frequencies; with whichever means, the results are of course the same.

Pythagoras scale: Tones mix harmoniously if their pitches are integer multiples of a basis tone. Thus, dividing the interval $[1, 2]$ (one octave), the point $\frac{3}{2}$ plays a basic rôle, for its octave (twice) overtone corresponds to the triple of the basis. The tone corresponding to $\frac{3}{2}$ is called the perfect fifth with respect to the basis (it is the fifth from the left in the division (26.8) below). The plan attributed to the Pythagorean school, which is in fact of the Babylonia origin or older, was to find a meaningful division of the interval $[1, 2]$ by exploiting the mixture of powers of $\frac{3}{2}$ (repetition of shifts each in perfect fifth) and powers of 2 (repetition of shifts each in octave); it depends on the coprimality of 2 and 3. For instance, we repeat twice the ascending shift of perfect fifth starting at $\frac{3}{2}$, getting $\left\{\frac{9}{4}, \frac{27}{8}\right\}$; and then taking one octave descending shift of each, we get the division $\left\{1, \frac{9}{8}, \frac{3}{2}, \frac{27}{16}, 2\right\}$. Further, starting from $\frac{27}{8}$ we repeat again twice the ascending shift of perfect fifth and take two octave descending shift of each. Combining these together we get the division $\{1, \frac{9}{8}, \frac{81}{64}, \frac{3}{2}, \frac{27}{16}, \frac{243}{128}, 2\}$. Finally, using the ratio of the narrow interval $\{\frac{243}{128}, 2\}$, we divide the wide interval $\{\frac{81}{64}, \frac{3}{2}\}$, getting the point $\frac{4}{3}$. In this way, we obtain

the Pythagoras major scale :
$$\left\{1, \ \frac{9}{8}, \ \frac{81}{64}, \ \frac{4}{3}, \ \frac{3}{2}, \ \frac{27}{16}, \ \frac{243}{128}, \ 2\right\}$$
$$\Delta\Delta\delta\Delta\Delta\Delta\delta = 2, \ \Delta = \frac{3^2}{2^3}, \ \delta = \frac{2^8}{3^5} \, .$$
(26.8)

This is composed of the perfect fourth $\Delta\Delta\delta = \frac{4}{3}$ and the perfect fifth $\Delta\Delta\Delta\delta = \frac{3}{2}$, which are pure pitches realized by plain ratios. Repeating the perfect fifth is employed in tuning the four strings G(sol)D(ré)A(la)E(mi) of the violin: starting with the string A, i.e., $\frac{27}{16}$, provided C(do) corresponds to 1 in (26.8), one downward shift ($\frac{9}{8}$:the string D) and another ($\frac{3}{4}$:the string G); also one upward shift ($\frac{81}{32}$:the string E). As a consequence, for instance, if {C♯, D♭} is indicated on a sheet of music, then it is meant that ascending and descending shifts are applied, respectively, in the following manner

$$(26.8) \Rightarrow C < D♭ = \delta \cdot C < C♯ = \delta^{-1} \cdot D < D, \qquad (26.9)$$

with an obvious abbreviation. Thus, between C and D there appear different tones $D♭ = \frac{2^{11}}{3^7} \cdot D$ and $C♯ = \frac{3^7}{2^{11}} \cdot C$; and the ratio $\frac{C♯}{D♭} = \Delta/\delta^2 = \frac{3^{12}}{2^{19}}$, which equals $1.0136\ldots$, is called the Pythagoras comma today. This gap, i.e., $\delta^2 < \Delta$ is well discernible. However, in reality tonality and melody matter, according to playing in either solo or accompanied or ensemble, and violinists are led to subtle modulations of these shifts.

Equal temperament scale: Tuning to the piano, we see quite a different situation. On the keyboard there exits only one black key between C and D. This is due to the trivial fact that the keys on keyboard instruments need to be fixed at the outset. The division of the octave has to be made on this severe premise; hence, each ideal tone is to be tempered to achieve the aim. This is indeed an approximation problem concerning continuity and periodical division; and a variety of proposals were made both in the East and in the West. A well-known quasi-solution is the equal temperament where no gap similar to the Pythagoras comma appears. This is thus to divide the one octave by the ratio $2^{1/12}$ (see Mersenne (1636, Libri IV, pp.18–19) for instance):

$$\text{the equal temperament major scale}: \quad \begin{array}{c} \Delta'\Delta'\delta'\Delta'\Delta'\Delta'\delta' = 2, \\ \Delta' = \delta'^2, \delta' = 2^{1/12}. \end{array} \qquad (26.10)$$

In particular

$$(26.10) \Rightarrow C < D♭ = \delta' \cdot C = C♯ = \delta'^{-1} \cdot D < D, \qquad (26.11)$$

which is different from the construction (26.9); note also that the D there is higher than D here. The three equalities in (26.11) are the meaning of the black key between C and D. That is, the piano is basically tuned on the equal temperament. The new perfect fourth is $\Delta'\Delta'\delta' = 2^{5/12} \doteqdot 1.335$, and the new perfect fifth is $\Delta'\Delta'\Delta'\delta' = 2^{7/12} \doteqdot 1.498$. These are not pure pitches and are less harmonious than those corresponding on the violin; this time, only the two terminal tones in one octave integrate well. Nevertheless, under the condition to make the failure relatively unnoticeable, the equal temperament offers probably the best division, as we have the expansion

$$\frac{\log(3/2)}{\log 2} = 0 + \frac{1}{1} + \frac{1}{1} + \frac{1}{2} + \frac{1}{2} + \frac{1}{3} + \frac{1}{1} + \frac{1}{5} + \cdots. \qquad (26.12)$$

The convergents are $A_4/B_4 = 7/12$, $A_5/B_5 = 24/41$, $A_6/B_6 = 31/53$, etc. Hence, the fact that the new perfect fifth corresponds to $2^{7/12}$ is the same as

114 *Divisibility*

that it is assigned to this fourth convergent. The choice of the fifth convergent would yield a keyboard whose octave is composed of 41 keys, which would be impossible for human hands to play; see Mersenne (1636, Libri IV, p.129) for a compromising yet complicated design of a keyboard with 31 keys in an octave. By the way, when the present author was a junior pupil at a primary school, he thought that any music could be written without using ♭ since on the piano he discovered that D♭ = C♯, etc. The enigma had persisted in his mind until his daughter started serious scale studies on the violin. However, he is now well aware that accomplished pianists are able to cope with this mechanical limitation of their instruments.

The comparative study of tones, through the ages, by dividing the taut string on a monochord with a movable bridge must have made a basis of Euclid's number theory. On the other hand, sound as a fundamental physical object gave rise to Fourier analysis, a basis of modern mathematics. This aspect in number theory will be revealed in our later chapters.

Notes

[26.1] Bombelli (1579, pp.35–38) described how to compute fractions approximating $\sqrt{13}$. His argument is quite general, for it readily implies the following: Suppose $n \in \mathbb{N}$ is non-square. Let a^2 be the square closest to n from below. Let $b = n - a^2$. Then

$$\sqrt{n} = a + \frac{b}{a + \sqrt{n}} = a + \frac{b}{2a} + \frac{b}{2a} + \frac{b}{2a} + \cdots$$

The approximation $a + b/2a$ for $\sqrt{n}$ has been known since antiquity.

[26.2] A proof of (26.2). With variables $\{x_j\}$ such that $x_j \neq x_{j+1}$, $x_j \neq 0$, $j \geq 1$, we have

$$\sum_{j=1}^{n} \frac{(-1)^{j-1}}{x_j} = 0 + \frac{1}{x_1} + \frac{x_1^2}{(x_2 - x_1)} + \frac{x_2^2}{(x_3 - x_2)} + \cdots + \frac{x_{n-1}^2}{(x_n - x_{n-1})}.$$

To confirm this, apply the replacement $x_n \mapsto (1/x_n - 1/x_{n+1})^{-1}$. We then set $x_j = 2j - 1$, and let $n \to \infty$:

$$1 - \frac{1}{3} + \frac{1}{5} - \frac{1}{7} + \frac{1}{9} - \cdots = 0 + \frac{1}{1} + \frac{1}{2} + \frac{3^2}{2} + \frac{5^2}{2} + \frac{7^2}{2} + \frac{9^2}{2} + \cdots.$$

See Euler (1748b, Tomus primus, p.372). As for the left side, see Note [78.7] below.

[26.3] Here we add the notion of negative continued fraction expansion: In the definition (22.1) we replace $R(\eta)$ by $R_*(\eta) = ([\eta]+1-\eta)^{-1}$. Then we have

$$\eta = a_0^* + \cfrac{-1}{a_1^* + \cfrac{-1}{a_2^* + \cdots + \cfrac{-1}{a_j^* + \cfrac{-1}{R_*^{j+1}(\eta)}}}}, \quad a_j^* = \left[R_*^j(\eta)\right] + 1;$$

the right side can be computed by means of Note [21.1]. This expansion does not terminate as R_* is applicable to integers unlike R; thus we stop the expansion as soon as $R_*^j(\eta)$ becomes an integer. The relation between the Euclid algorithm and the negative continued fraction expansion is as follows: Let $a, b \in \mathbb{N}$. We define the sequence $\{s_j : j \geq -1\}$ by

$$\begin{pmatrix} s_{j-1} \\ -s_j \end{pmatrix} = \begin{pmatrix} c_j & 1 \\ -1 & 0 \end{pmatrix} \begin{pmatrix} s_j \\ -s_{j+1} \end{pmatrix}, \quad j \geq 0; \quad s_{-1} = a, s_0 = b,$$

with $c_j = [s_{j-1}/s_j] + 1$. Then we have $\langle a, b \rangle = \langle s_j, s_{j+1} \rangle, j \geq -1$; the analogue of (21.14) can be deduced in much the same way. In the case of Note [21.3], we have

$$\begin{pmatrix} 3041543 \\ -1426253 \end{pmatrix} = \begin{pmatrix} 3 & 1 \\ -1 & 0 \end{pmatrix} \begin{pmatrix} 2 & 1 \\ -1 & 0 \end{pmatrix}^6 \begin{pmatrix} 3 & 1 \\ -1 & 0 \end{pmatrix} \begin{pmatrix} 7 & 1 \\ -1 & 0 \end{pmatrix}$$
$$\times \begin{pmatrix} 2 & 1 \\ -1 & 0 \end{pmatrix}^{12} \begin{pmatrix} 8 & 1 \\ -1 & 0 \end{pmatrix} \begin{pmatrix} 2 & 1 \\ -1 & 0 \end{pmatrix}^3 \begin{pmatrix} 3 & 1 \\ -1 & 0 \end{pmatrix} \begin{pmatrix} 23 \\ 0 \end{pmatrix}.$$

This time we need 26 steps while formerly only 9 steps: $s_{25} = 23 = \langle a, b \rangle$. Thus the negative continued fraction expansion can be a wasteful procedure. Nevertheless, mixed uses of positive and negative terms, like in the examples of Note [21.4], can result in impressive improvements of convergence speed, as has been already suggested in Note [21.1].

2

Congruences

§27

There exist multiplicative periodicities among deviations from being divisible. This is the fundamental structure in the set of all integers that Fermat in the 17th century and Euler in the 18th century discovered while immersing themselves in analysis of pairs of coprime integers. In this chapter we shall give an account of their findings, relying on the viewpoints and means that were thought up later by their successors. We shall then proceed to some important consequences, which constitute a substantial part of the very basis of today's number theory. Further, as an application of these discussions, we shall show a set of foundational approaches to the problem of integer factorization.

Thus, we first introduce the notion of congruence due to Gauss [DA, arts.1–2]: Two integers a and b are said to be congruent relative to a modulus q a positive integer, provided $q|(a-b)$ or rather $(a-b) \in q\mathbb{Z}$. We denote this fact by

$$a \equiv b \mod q. \tag{27.1}$$

If a number of consecutive congruences are all with respect to a particular modulus, then we shall indicate the modulus appropriately, not strictly keeping the format (27.1). Also, the situation contrary to (27.1) is denoted by

$$a \not\equiv b \mod q, \tag{27.2}$$

which is equivalent to $q\nmid(a-b)$ or $(a-b) \notin q\mathbb{Z}$; we state that a,b are incongruent modulo q.

Remark 1 We shall suppose that $q \geq 2$ unless otherwise stated. Nevertheless, it is logically possible to take 1 as a modulus; see Note [27.3]. Gauss did not impose any restriction on moduli, except for that they should be positive ([DA, art.1: footnote]).

§27   117

It is readily seen that the congruence satisfies the classification axiom:

$$a \equiv a; \quad a \equiv b \Rightarrow b \equiv a; \quad a \equiv b, b \equiv c \Rightarrow a \equiv c \quad \mod q. \qquad (27.3)$$

Hence (27.1) is equivalent to classifying elements of $\mathbb{Z}$ as follows:

$$\mathbb{Z} = \bigsqcup_{j=0}^{q-1} C_j, \quad C_j = j + q\mathbb{Z}. \qquad (27.4)$$

The subset C_j is the arithmetic progression with the initial term j and the difference q. So $C_j \cap C_{j'} = \varnothing$ for $j \neq j'$ in (27.4). Since dividing integers by q we get residues $0, 1, \ldots, q - 1$, the equality in (27.4) holds. More generally, if $r_j \equiv j \bmod q$, then $C_j = C_{r_j}$. The set $\{r_0, r_1, \ldots, r_{q-1}\}$ is called a residue system mod q and each C_r is a residue class, where r stands for one of r_j; these r_j should not be confused with r_j of (21.10). The term residues might suggest integers whose absolute values are less than the relevant modulus. In fact, the size of residues is mostly immaterial in discussing congruences. What is essential is that each residue class is an arithmetic progression in $\mathbb{Z}$.

Remark 2 The term complete residue system is usually used instead of our term residue system. We shall often say just residues instead of residue classes modulo an integer in order to avoid verbose expressions, provided no misunderstanding is anticipated.

Thus, we can write instead

$$\mathbb{Z} = \bigsqcup_r C_r, \quad \{r\}: \text{a residue system mod } q. \qquad (27.5)$$

Also, we shall use the expression

$$\mathbb{Z}/q\mathbb{Z} = \Big\{ \text{all residues mod } q \Big\}. \qquad (27.6)$$

The left side means that $a, b \in \mathbb{Z}$ are identified if $(a - b) \in q\mathbb{Z}$, that is, $a \equiv b \bmod q$; the right side means that $\mathbb{Z}$ splits into q residue classes which are mutually disjoint. This is of course the coset decomposition of the Abelian group $\mathbb{Z}$ with respect to its subgroup $q\mathbb{Z}$; the notion of coset decomposition itself can in fact be traced back to the procedure employed by Euler (1755), as is explained in Notes [27.4] and [28.3].

Since periodic phenomena and events are plenty both in nature and in human activities, congruences between integers became a major interest among mathematicians long before the publication of [DA]; however, none of them grasped this notion so penetratingly as Gauss did by introducing the notation (27.1). In what follows, we shall employ (27.1) in discussing the works of

118 *Congruences*

mathematicians who did not use it but devised some alternative ways to express their thoughts in the congruence context; naturally, this will not cause any change in the mathematical contents of their works.

Notes

[27.1] It is often said that (27.1) is a typical instance demonstrating the importance of correct choices of notations in science. Indeed, if it were not for, then the whole of today's works in number theory would look an accumulation of horrible confusion. Although Gauss's original purpose of (27.1) was to ease complexities of number theoretical discussions, the effect of his invention turned out to be far-reaching. It should, however, be noted that Gauss was not the first who gave a notation to congruences: As is remarked in [DA, art.2: footnote], Legendre (1785, p.466) came quite close to (27.1); Gauss admitted that he introduced his notation in order to avoid the ambiguity which might be caused by Legendre's, that is, the equality symbol but with a specific usage. Also an unpublished treatise of Euler contains similar efforts. See Note [27.4].

[27.2] Congruences among integral polynomials can also be discussed: we define the congruence of $f(x), g(x) \in \mathbb{Z}[x]$ by

$$f(x) \equiv g(x) \bmod q \quad \text{(as polynomials)}$$
$$\Leftrightarrow \text{ there exists an } h(x) \in \mathbb{Z}[x] \text{ such that } f(x) = g(x) + qh(x).$$

One may write instead $f(x) \equiv g(x) \bmod q$ in $\mathbb{Z}[x]$. Note that this includes the situation $\deg f(x) \neq \deg g(x)$. We shall be more explicit in §33.

[27.3] Further, there is no necessity to restrict the notion of congruence to integral moduli. For instance, within $\mathbb{R}$, one may define $\alpha \equiv \beta \bmod \gamma$ to be equivalent to $(\alpha - \beta) \in \gamma\mathbb{Z}$, i.e., $\mathbb{R}/\gamma\mathbb{Z}$. Thus, mod 1 is often used to indicate that the difference of two given numbers is an integer: $\mathbb{R}/\mathbb{Z}$ is essentially the same as the one-dimensional torus. By the same token, the polynomial congruence of the preceding Note is the same as $(f(x) - g(x)) \in q\mathbb{Z}[x]$, i.e., $\mathbb{Z}[x]/q\mathbb{Z}[x]$. A more essential extension will be discussed in §67.

[27.4] In the context of the present chapter, Euler's posthumous article *Tractatus* (1849a) draws a special attention. Euler in the prime of his life got a plan to compile a treatise on number theory, based on his own investigations for some two decades. His notations and expressions are different from ours naturally, but the contents of his manuscript include essentially the whole of our sections §§27–43. What he missed is Theorem 34, §34, due to Lagrange (1770b) and the view point of Note [40.2]. If Euler had acquired it, then he could have obtained a complete proof of the extremely important Theorem 36 (see Note [40.3]). Theorem 34 was a pivotal assertion for a new development of number theory started by Legendre (1785); see Note [34.1]. Euler gave up his plan of a number theory treatise; instead he published parts of it. His two articles (1755, 1758a) are such instances; they are regarded as the beginning of the theory of groups. Details are to be given in Notes [28.2]–[28.3].

§28

Considering arithmetical issues by means of the reduction with respect to appropriately selected moduli is called modular arithmetic; we are often led to highly interesting situations.

Congruences react to summation, subtraction, and multiplication in just the same way as ordinary equalities do: it is easily seen that

$$a \equiv b, \ c \equiv d \ \Rightarrow \ a \pm c \equiv b \pm d, \ ac \equiv bd \ \text{mod} \ q; \tag{28.1}$$

the last congruence follows from $ac - bd = (a-b)c + b(c-d)$. Consequently, for any integral polynomial $f(x_1, x_2, \dots, x_N)$,

$$\begin{aligned} a_\nu &\equiv b_\nu, \quad 1 \le \nu \le N, \\ \Rightarrow f(a_1, a_2, \dots, a_N) &\equiv f(b_1, b_2, \dots, b_N) \ \text{mod} \ q. \end{aligned} \tag{28.2}$$

Also, we have, by (14.1),

$$a \equiv b \ \text{mod} \ q_k, \ 1 \le k \le K \ \Leftrightarrow \ a \equiv b \ \text{mod} \ [q_1, q_2, \dots, q_K], \tag{28.3}$$

for $a - b$ is a common multiple of q_k, $1 \le k \le K$.

On the other hand, congruences react to division in the following way:

Theorem 27 *We have the equivalence relation*

$$sa \equiv sb \ \text{mod} \ q \ \Leftrightarrow \ a \equiv b \ \text{mod} \ \frac{q}{\langle s, q \rangle}. \tag{28.4}$$

Proof This is a consequence of $q | s(a - b)$ and (6.3). We end the proof.

Thus,

$$sa \equiv sb \ \text{mod} \ q, \ \langle s, q \rangle = 1 \ \Rightarrow \ a \equiv b \ \text{mod} \ q. \tag{28.5}$$

Because of this fact, integers prime to a particular modulus attract special attention. According to (6.2) (0), for any s in a class C_r mod q the value of $\langle s, q \rangle$ is equal to $\langle r, q \rangle$, that is, determined by the class itself. Hence, we are led to the following notion: From a residue system mod q, we collect all residues which are coprime to q; and we call the resulting set a reduced residue system mod q.

Remark The term complete reduced residue system is customarily used instead of our reduced residue system. We shall often say just reduced residues instead of reduced residue classes mod an integer in order to avoid verbose expressions, provided no misunderstanding is anticipated.

With this, we introduce

$$(\mathbb{Z}/q\mathbb{Z})^* = \Big\{\text{all reduced residues mod } q\Big\}. \tag{28.6}$$

By the definition (§18) of the Euler φ-function, we have

$$\big|(\mathbb{Z}/q\mathbb{Z})^*\big| = \varphi(q). \tag{28.7}$$

Via (28.4)–(28.5), we see that $\langle a, q \rangle = 1$ if and only if either of the following assertions holds:

$$\begin{aligned} \{r\} \text{ is a residue system mod } q &\Rightarrow \{ar\} \text{ is the same,} \\ \{s\} \text{ is a reduced residue system mod } q &\Rightarrow \{as\} \text{ is the same.} \end{aligned} \tag{28.8}$$

The latter depends also on (6.2)(3). In particular, by (28.2),

$$\prod s \equiv \prod as \equiv a^{\varphi(q)} \prod s \mod q \, ; \tag{28.9}$$

and we obtain, again by a combination of (6.2)(3) and (28.5), a fundamental assertion of Euler in number theory that has been suggested at the beginning of the present chapter:

Theorem 28 *Provided $\langle a, q \rangle = 1$, we have*

$$a^{\varphi(q)} \equiv 1 \mod q. \tag{28.10}$$

Euler's original proof (1755, 1758a) of this is involved; nonetheless, it is superior than the one given above since it depicts the multiplicative nature of the assertion which is not well captured by (28.9); the details are given in Note [28.3].

As a corollary of (28.10), we get, for any prime p,

$$p \nmid a \;\Rightarrow\; a^{p-1} \equiv 1 \mod p. \tag{28.11}$$

This is a famous theorem commonly attributed to Fermat; see Note [28.1]. However, it should be stressed that (28.10) and (28.11) are in fact equivalent. To show this, let $p^\alpha \| q$; then we have

$$\begin{aligned} a^{\varphi(q)} &= \big((a^{\varphi(q/p^\alpha)})^{p-1}\big)^{p^{\alpha-1}} = (1 + p\gamma_p)^{p^{\alpha-1}} \quad (\gamma_p \in \mathbb{Z}) \\ &\equiv 1 \mod p^\alpha. \end{aligned} \tag{28.12}$$

The first line is due to (18.4) and (28.11); the second line is due to the fact that the binomial expansion gives $(1 + p^\nu u)^p = 1 + p^{\nu+1}v$, $u, v \in \mathbb{Z}$; the rest of the

$\S 28$ 121

proof is only to quote (28.3). By the way, dropping the condition $p \nmid a$, we have more generally than (28.11)

$$a^p \equiv a \bmod p, \quad a: \text{arbitrary.} \tag{28.13}$$

The converse of the congruence (28.11) is not valid; namely, the congruence $a^{q-1} \equiv 1 \bmod q$ does not necessarily imply that q is a prime, a fact which induces various ideas of integer factorization. Details will be given in §37.

By the way, Euler's (28.10) indicates that there exists a close resemblance between $(\mathbb{Z}/q\mathbb{Z})^*$ and the unit circle on $\mathbb{C}$. This view point was exploited extensively by the young Gauss and became the basis of his fundamental research [DA, Sectio VII] and his posthumous article (1863a). Our discussion in Chapter 3 will be developed in this perspective.

Notes

[28.1] The beautiful congruence (28.11) is asserted in Fermat's letter (1679, p.163: October 18, 1640; Œuvres II, p.209). From the way by which he stated this great discovery of his, it transpires that a group-theoretical reasoning was already in his unpublished proof: indeed, he quoted the example that if 3^n ($n = 1, 2, 3, \ldots$) are divided by 13, then the residue 1 repeats with the period 3 which divides $13 - 1 = 12$. Euler (1732a) conjectured not only (28.11) but also (28.10) for $q = p^m$ and $q = p_1 p_2 \cdots p_k$ (sqf). As to the latter, his observation was more precise: $a^{[p_1-1, p_2-1, \ldots, p_k-1]} \equiv 1 \bmod q$; see Note [42.2] below. For instance, according to (28.10), we have $7^{192} \equiv 1 \bmod 221$, but in fact $7^{48} \equiv 1 \bmod 221$. It is claimed sometimes that Euler discovered (28.11) independently of Fermat, perhaps because in this article he referred to Fermat but only in the context of the issue quoted in our Note [16.8].

[28.2] The first published proof of (28.13) is in Euler (1736) (this time, he credited the discovery of (28.11) to Fermat). He started with the trivial fact that $\binom{p}{j} \equiv 0 \bmod p$, $1 \le j \le p - 1$; so $(a + 1)^p - a^p - 1 \equiv 0 \bmod p$, which implies that $a^p - a \equiv 0 \Rightarrow (a+1)^p - (a+1) \equiv 0 \bmod p$. Then, by induction, the proof is completed. A similar argument is said to be in an obscure note of G.W. Leibniz: Specializing the following by $x_j = 1$, $1 \le j \le k$,

$$(x_1 + x_2 + \cdots + x_k)^p - (x_1^p + x_2^p + \cdots + x_k^p) \in p\mathbb{N}[x_1, x_2, \ldots, x_k]$$

we get (28.13). For the coefficients of this polynomial are

$$C(j_1, j_2, \ldots, j_k) = \frac{p!}{j_1! j_2! \cdots j_k!}, \quad j_1 + j_2 + \cdots + j_k = p,$$

which is of course in $\mathbb{N}$; and $p \mid j_1! j_2! \cdots j_a! \, C(j_1, j_2, \ldots, j_k)$ but $p \nmid j_1! j_2! \cdots j_k!$, for j_l are all less than p. Euler (1755, Theoremata 13, 14) gave an alternative

122   *Congruences*

proof of (28.11), which we shall indicate in the next Note. That argument of Euler is truly a multiplicative reasoning which is the same as the one that is employed in the modern theory of finite groups; and it must be what g anticipated a century earlier (Notes [28.1], [41.1]). Euler began his note (1755, p.269, 53. Scholion) stating *En ergo novam demonstrationem theorematis eximii, a Fermatio quondam prolati, ...*; his deep delight of discovery is apparent, for he continued that his new proof is *magis naturalis* compared with that depending on binomial expansion. Indeed, the multiplicative nature of reduced residue classes does not seem to be revealed by binomial expansion. Later Euler (1758a, Theorema 11) proved the general assertion (28.10) with the same argument as that employed in his article (1755). These two articles of Euler are outstanding in the history of number theory. Naturally, Gauss [DA, art.49] adopted them. By the way, the observation (28.8) is traced back to Euler (1758a, Theoremata 1, 2); the well-known reasoning (28.9) is due to Ivory (1806). Further, the same observation as (28.12) is in Dirichlet (1863, §20)). In Note [35.3] we shall show yet another proof of (28.11) that is due to Lagrange (1771b).

[28.3]   Euler (1758a) proved Theorem 28 in the following way (the argument of the article (1755) is essentially the same, as has been remarked above): Since the set $S_0 = \{a^j \bmod q : j = 0, 1, 2, \ldots\}$ is finite, there exist $0 \leq u < v$ such that $a^u \equiv a^v \equiv a^{u+(v-u)} \bmod q$. Thus, by (28.5), $a^{v-u} \equiv 1 \bmod q$. This means that there are $l > 0$ such that $a^l \equiv 1 \bmod q$; then let ℓ be the least among them. With this, $S_0 = \{a^j \bmod q : 0 \leq j < \ell\}$, and the ℓ elements of S_0 represent mutually different reduced classes mod q. If $\ell = \varphi(q)$, then we get (28.10). Otherwise, taking any reduced residue class $a_1 \bmod q$ outside S_0, we consider $S_1 = a_1 S_0 = \{a_1 a^j \bmod q : 0 \leq j < \ell\}$. Then $S_0 \cap S_1 = \varnothing$. If not, then there exist $0 \leq w, z < \ell$ such that $a^w \equiv a_1 a^z \bmod q$, so $a^{w+\ell} \equiv a_1 a^z \bmod q$; and by (28.5) we have $a_1 \equiv a^{w+\ell-z} \bmod q$, which is in S_0, a contradiction. Thus, if $S_0 \sqcup S_1$ is a reduced residue system mod q, then $2\ell = \varphi(q)$, and we get (28.10). Otherwise, taking any reduced residue $a_2 \bmod q$ outside $S_0 \sqcup S_1$, we consider $S_2 = a_2 S_0$. By definition, $S_0 \cap S_2 = \varnothing$. Also $S_1 \cap S_2 = \varnothing$ since if $a_1 a^{j_1} \equiv a_2 a^{j_2} \bmod q$, $0 \leq j_1, j_2 < \ell$, then $a_1 a^{j_1+\ell-j_2} \equiv a_2 \bmod q$, which is a contradiction. The same procedure cannot be repeated indefinitely; and a reduced residue system mod q will eventually be attained, which ends the proof. This argument of Euler was revived in the theory of finite groups a century later. By the way, we could make the discussion shorter using the convention (29.1) of the next section; but Euler's original way of reasoning ought to be treated as it is.

$$\S 28 \qquad\qquad 123$$

[28.4] For $p \nmid a$, the quantity $(a^{p-1} - 1)/p$ is called the Fermat quotient with respect to the base a and modulus p. The main interest is in the peculiar situation where this quotient is divisible by p; that is, $a^{p-1} \equiv 1 \bmod p^2$. The search for such pairs $\{a, p\}$ began with a problem set out by Abel (1828). The first two instances are $18^{36} \equiv 1 \bmod 37^2$ (Jacobi (1828)), $10^{486} \equiv 1 \bmod 487^2$ (Desmarest (1852)); and quite a few more pairs are known by now. A singular case is $68^{112} \equiv 1 \bmod 113^3$ (Hertzer (1908)). However, no general solution has been achieved yet. By the way, in the case $a = 2$, the relevant p's are called the primes of Wieferich (1909); only two have been identified so far: 1093 (Meissner (1913)) and 3511 (Beeger (1922)); see Granville (2008, p.313).

[28.5] An extension of (28.13) by Grandi (1883). For an arbitrary pair $a, q \in \mathbb{N}$,

$$\Delta_a(q) = \sum_{d|q} \mu(d) a^{q/d} \equiv 0 \bmod q,$$

where μ is the Möbius function. If q is a prime, this coincides with (28.13). To prove the general case, we note that for any $p|q$

$$\Delta_a(q) = \sum_{d|q,\, p \nmid d,} \mu(d) \big(a^{q/d} - a^{q/pd} \big).$$

Each summand is divisible by p^α, $\alpha = q(p)$ since if $p \nmid a$, then by (28.10) we have $a^{(q/d)(1-1/p)} \equiv 1 \bmod p^\alpha$; and if $p|a$, then it suffices to observe $p^{\alpha-1} \geq \alpha$. This ends the proof. We note that

$$\Delta_a(q) \geq a^{q/2} \big(a^{q/2} - q/2 \big),$$

for $|\{d|q : d \geq 2\}| = |\{f|q : f \leq q/2\}| \leq q/2$; hence, provided $a \geq 2$, we have $\Delta_a(q) > 0$, i.e., $\geq q$. See §67(7).

[28.6] Schönemann (1846a, §13). Let $\{s_k\}$ be the set of all elementary symmetric polynomials of variables $\{x_j\}$; and let $\{s_k^{(p)}\}$ be its counterpart for $\{x_j^p\}$, where p is an arbitrary prime. Then we have

$$s_k^{(p)} = (s_k)^p + p t_k,$$

with t_k an integral symmetric polynomial of $\{x_j\}$. This can be confirmed as in Note [28.2]. Here is an application: Let $f(x) \in \mathbb{Z}[x]$ be monic, i.e., the coefficient of the highest power of x is equal to 1; and let $f(x) = \prod(x - \xi_j)$ so that $\{\xi_j\}$ is the set of all roots in $\mathbb{C}$ of $f(x) = 0$. Also, let $F(x) = \prod(x - \xi_j^p)$. Then $F(x) \in \mathbb{Z}[x]$, and we have $F(x) \equiv f(x) \bmod p$ in the sense of Note [27.2] or more explicitly

$$F(x) - f(x) \in p\mathbb{Z}[x].$$

124 *Congruences*

This is because the coefficients $(-1)^k s_k$ and $(-1)^k s_k^{(p)}$ of $f(x)$ and $F(x)$, respectively, are integers by the theory of symmetric polynomials (see (64.5)), and $(-1)^k s_k^{(p)} \equiv ((-1)^k s_k)^p \equiv (-1)^k s_k \bmod p$ by the combination of the preceding Note and (28.13). In addition, we note, for any $g(x) \in \mathbb{Z}[x]$,

$$g(x)^p - g(x^p) \in p\mathbb{Z}[x].$$

[28.7] The additive group $\mathbb{Z}/q\mathbb{Z}$ is cyclic; any reduced residue $\bmod q$ is its generator, and vice versa, as the first line of (28.8) implies. Since any cyclic group of order q is isomorphic to $\mathbb{Z}/q\mathbb{Z}$, it has $\varphi(q)$ generators. This is a group-theoretical characterization of the Euler function.

[28.8] For any $a, q \in \mathbb{N}$, it holds that

$$a\mathbb{Z}/[a,q]\mathbb{Z} \cong \langle a, q \rangle \mathbb{Z}/q\mathbb{Z}.$$

The relevant map is $an \mapsto an \bmod q$, which is obviously a homomorphism from $a\mathbb{Z}$ to $\mathbb{Z}/q\mathbb{Z}$. Its kernel and image are $[a,q]\mathbb{Z}$ and $\langle a, q \rangle \mathbb{Z}/q\mathbb{Z}$, respectively. For, since $m = an \equiv 0 \bmod q$ is equivalent to $[a,q]|m$, the kernel is $[a,q]\mathbb{Z}$. Also, the image is obviously in $\langle a, q \rangle \mathbb{Z}/q\mathbb{Z}$. Conversely, let n be arbitrary. By (4.1) divided by $\langle a,b \rangle$, we have $n = (a/\langle a,q \rangle)g + (q/\langle a,q \rangle)h$ with $g,h \in \mathbb{Z}$, which gives $\langle a,q \rangle n \equiv ag \bmod q$. This ends the confirmation. It remains to apply the fundamental theorem on homomorphisms. In particular, we have got an alternative proof of (10.6): use (28.4) to find the cardinalities of the sets on the two sides of the isomorphism.

[28.9] It is of course possible to replace a, b in (27.1) by integral matrices of the same dimension with congruence being considered element-wise; then, (28.2) extends to matrix arguments in an obvious way. For instance, the congruence properties of the Fibonacci sequence can be extracted from the powers of matrices used in Note [24.2]: Since the number of matrices among $\begin{pmatrix} 1 & 1 \\ 1 & 0 \end{pmatrix}^k$, $k \in \mathbb{Z}$, which are different with respect to the modulus q, is obviously finite, there exists $k_1 \neq k_2$ such that $\begin{pmatrix} 1 & 1 \\ 1 & 0 \end{pmatrix}^{k_1} \equiv \begin{pmatrix} 1 & 1 \\ 1 & 0 \end{pmatrix}^{k_2} \bmod q$. Hence, there exist integers $\ell \neq 0$ such that $\begin{pmatrix} 1 & 1 \\ 1 & 0 \end{pmatrix}^{\ell} \equiv \begin{pmatrix} 1 & 0 \\ 0 & 1 \end{pmatrix} \bmod q$. The least among $|\ell|$ is called the qth Pisano period, with which the Fibonacci sequence $\bmod q$ repeats. In a wider context, Lagrange (1798, §§78–79) noted the existence of this period.

[28.10] Let $a, b \in \mathbb{N}$, $\langle a,b \rangle = 1$. Then $ax + by = ab - a - b + h$ is soluble in integers $x, y \geq 0$ whenever $h \in \mathbb{N}$ but not when $h = 0$: To prove this assertion, we note first that if either a or b is equal to 1, then it is trivial. Thus we may suppose that $2 \leq a < b$. We consider the segment

$$x = b - 1 + (h - b - by)/a, \quad 0 \leq y \leq a - 1 + (h - a)/b.$$

It is needed to check if x becomes an integer, that is, if $h - b - by \equiv 0 \bmod a$ for any integer y in the indicated range. We consider the case $h = 0$. The possible values of y are $\{0, 1, \ldots, a - 2\}$. Since $by \bmod a$ are distinct for these y because of (28.8), if we had $a|(-b - by)$ for such a y, then it would follow that $a \nmid (-b - b(a - 1))$, which is but a contradiction. We have settled the case $h = 0$. Next, if $1 \le h \le a - 1$. Then the possible values of y are the same as with the case $h = 0$. However, this time we have $a \nmid (h - b - b(a - 1))$, so there must be a required solution pair $\{x, y\}$. The case $h \ge a$ can be skipped. See Niven et al (1991, p.219: Problem 16).

§29

Continuing the discussion of the preceding section, we consider the problem to find an integer x that satisfies $ax \equiv 1 \bmod q$ on the condition $\langle a, q \rangle = 1$. Obviously the assertion (28.10) implies that $x = a^{\varphi(q)-1}$ is a solution. Also, by (3.6) there are g, h such that $ag + qh = 1$, and $x = g$ is a solution. As for the relation between these two solutions, we note that (28.5) implies that the residue class $\bmod q$ to which a solution belongs is unique since $ax \equiv 1 \equiv ax' \bmod q$ implies $x \equiv x' \bmod q$. With this, the residue class that is represented by either $a^{\varphi(q)-1}$ or g is called the inverse residue class of the reduced residue class $a \bmod q$, and is denoted by either $\bar{a} \bmod q$ or $a^{-1} \bmod q$. Hence, replacing $\bar{a}$, a^{-1} by an appropriate representative integer, the following congruences hold:

$$a \cdot \bar{a} \equiv 1, \ a \cdot a^{-1} \equiv 1 \bmod q. \tag{29.1}$$

These new notations should not be confused with complex conjugates or with fractions either. Hereafter

we shall use integers and residue classes in mixed ways. $\qquad$ (29.2)

For instance, the inverse residue class of $3 \bmod 17$ is represented by $6, -11$, 1757, etc.; and sometimes we shall use one of these integers and some other time we shall just indicate it by either $\bar{3}$ or 3^{-1}; thus $(5 \cdot 3^{-1})^5 \equiv (5 \cdot 6)^5 \equiv 13 \bmod 17$.

We introduce then the following basic definition: Regardless of the value of $\langle a, q \rangle$,

a solution of a linear congruence equation $ax \equiv b \bmod q$ is a residue class $c \bmod q$ such that $ac \equiv b \bmod q$. $\qquad$ (29.3)

126 *Congruences*

When $\langle a,q \rangle = 1$, the solution is unique mod q: that is, $x \equiv a^{-1}b$ mod q. In general we have:

Theorem 29

$$ax \equiv b \bmod q \text{ has a solution } \Leftrightarrow \langle a,q \rangle | b ;$$
$$\text{the number of solutions mod } q \text{ is } \langle a,q \rangle. \tag{29.4}$$

Proof This congruence equation is equivalent to the linear indefinite equation $ax + qy = b$; by Theorem 6 a solution exists if and only if $\langle a,q \rangle | b$. Thus, dividing both sides by $\langle a,q \rangle$, we get the situation that can be viewed through (28.5). The solution x is unique with respect to the modulus $q/\langle a,q \rangle$. Namely, with a certain c we have $x = c + (q/\langle a,q \rangle)u, u \subset \mathbb{Z}$. In this we insert $u = 1,2,\ldots,\langle a,q \rangle$; and we obtain all different solutions mod q, for if $c + (q/\langle a,q \rangle)u \equiv c + (q/\langle a,q \rangle)u'$ mod q, then (28.4) gives $u \equiv u'$ mod $\langle a,q \rangle$. We end the proof.

An additional discussion: Generalizing (29.4), we consider the linear congruence equation of two unknowns

$$ax + by \equiv 0 \bmod q. \tag{29.5}$$

Although this is equivalent to the linear indefinite equation $ax + by + qz = 0$ of three unknowns to which the discussion in §§8–9 is closely related, we shall take a somewhat different procedure: Let the matrix U be the same as that in Note [5.2] so that $^{\mathrm{t}}\{x,y\} = U \cdot {}^{\mathrm{t}}\{X,Y\}$. If (29.5) holds, then we have $\langle a,b \rangle X = ax + by \equiv 0$ mod q, which is equivalent to $X \equiv 0$ mod $q/\langle a,b,q \rangle$, where we have used (28.4) as well as Note [7.1]. Here Y can be any integer. Hence, $^{\mathrm{t}}\{x,y\} = U \cdot {}^{\mathrm{t}}\{mq/\langle a,b,q \rangle,n\}, m,n \in \mathbb{Z}$. Namely, we have for any $\{a,b\} \neq \underline{0}$ and modulus q,

$$\begin{pmatrix} x \\ y \end{pmatrix} = m\mathbf{u} + n\mathbf{v}, \ m,n \in \mathbb{Z} ;$$
$$\mathbf{u} = \frac{q}{\langle a,b,q \rangle} \begin{pmatrix} g \\ h \end{pmatrix}, \ \mathbf{v} = \frac{1}{\langle a,b \rangle} \begin{pmatrix} -b \\ a \end{pmatrix}, \tag{29.6}$$

which is to be compared with (8.3). Obviously, any $\{x,y\}$ defined by this formula is a solution to (29.5). We have proved the following assertion:

Theorem 30 *The set of all solutions of* (29.5) *is a free Abelian group of rank 2 with the basis* $\{\mathbf{u}, \mathbf{v}\}$ *given in* (29.6); *i.e., the set of all solutions is*

$$\mathbb{Z}\mathbf{u} + \mathbb{Z}\mathbf{v}. \tag{29.7}$$

Remark Moving the pair $\{\mathbf{u}, \mathbf{v}\}$ to $\{\mathbf{u}', \mathbf{v}'\}$ with

$$\mathbf{u}' = \alpha\mathbf{u} + \beta\mathbf{v}, \ \mathbf{v}' = \gamma\mathbf{u} + \delta\mathbf{v}, \quad \begin{pmatrix} \alpha & \beta \\ \gamma & \delta \end{pmatrix} \in \Gamma, \tag{29.8}$$

we get a new basis. Among these $\{\mathbf{u}', \mathbf{v}'\}$, there exists a unique pair which can be regarded as fundamental; see Note [77.6]. Also, the set of solutions to (29.5) can be seen to form a lattice on plane. The area of its basic parallelogram equals $q/\langle a,b,q \rangle$.

In view of the explanation leading to the definition (29.1), it should be appropriate to make here an observation on the parameterization of elements of the group Γ, under Note [5.1](2): We have the following double coset decomposition of Γ by $\Gamma_\infty = \{T^n : n \in \mathbb{Z}\}$ with T as in (5.1):

$$\Gamma_\infty \backslash \Gamma / \Gamma_\infty = \{1\} \sqcup \{\gamma_{c,d} : c > 0, \ d \bmod c, \ \langle c,d \rangle = 1\}, \tag{29.9}$$

where

$$\gamma_{c,d} = \begin{pmatrix} \bar{d} & (d\bar{d} - 1)/c \\ c & d \end{pmatrix}, \ d\bar{d} \equiv 1 \bmod c; \tag{29.10}$$

the right side of (29.9) stands for the full set of representatives of the decomposition, with an obvious abuse of notation. The verification is immediate since $\det \begin{pmatrix} a & b \\ c & d \end{pmatrix} = 1$ implies that either $c = 0, a = d = 1$ or $ad \equiv 1 \bmod c$. One should know that (29.9) is an overall expression of the application of Euclid's algorithm to all coprime pairs of integers. It is a fundamental means in the spectral theory of automorphic functions, a glimpse of which we shall have in the later part of §94; see also Note [61.7].

Notes

[29.1] For $\langle a,q \rangle = 1$ and any $k \geq 1$, let $a_k = \left(1 - (1 - a \cdot a^{-1})^k\right)/a$, where $\ldots/a$ is an ordinary division. Then we have $aa_k \equiv 1 \bmod q^k$. Also, with

$$\prod_j \left(1 - a^{p_j - 1}\right)^{\alpha_j} = 1 - aA, \quad q = \prod_j p_j^{\alpha_j},$$

we have $aA \equiv 1 \bmod q$. Namely, the solution of $ax \equiv 1 \bmod q$ can be constructed from those of $ax \equiv 1 \bmod p$, $p|q$. This observation is due to Binet (1831). For example, in the case of $3x \equiv 1 \bmod 4459$ the modulus is decomposed into $7^3 \cdot 13$. Thus,

$$\left(1 - 3^6\right)^3 \cdot \left(1 - 3^{12}\right) = 205044619386880 = 1 + 3 \cdot 68348206462293$$

$$\Rightarrow \quad 3^{-1} \equiv -68348206462293 \equiv 2973 \bmod 4459.$$

Indeed, $3 \cdot 2973 = 1 + 2 \cdot 4459$. Equivalently, one may replace $a^{p_j - 1}$ by ay_j which satisfies $ay_j \equiv 1 \bmod p_j$. Thus, on noting $3 \cdot 5 \equiv 1 \bmod 7$ and $3 \cdot 9 \equiv 1 \bmod 13$, we have

$$\left(1 - 3 \cdot 5\right)^3 \cdot \left(1 - 3 \cdot 9\right) = 71344 = 1 + 3 \cdot 23781$$
$$\Rightarrow \quad 3^{-1} \equiv -23781 \equiv 2973 \bmod 4459.$$

[29.2] For $\langle a, b \rangle = 1$ and an arbitrary $c \in \mathbb{N}$, there are infinitely many $n \in \mathbb{Z}$ such that $\langle a + bn, c \rangle = 1$. To prove this, we put $c_1 = \langle c, b^\infty \rangle$, $c_2 = c/c_1$; here the definition of c_1 means that $c_1 = \prod_p p^\alpha$ with $p | b$ and $p^\alpha \| c$ (see (13.1)). Then for any x, we have $\langle a + bx, c \rangle = \langle a + bx, c_2 \rangle$. Since $\langle b, c_2 \rangle = 1$, there exists a class $n \bmod c_2$ such that $a + bn \bmod c_2$ is a reduced class. This ends the proof. For instance, for $a = 140$, $b = 297$, $c = 117$, we have $297 \equiv 17 \bmod 140$, so $\langle a, b \rangle = 1$. Also, $c_1 = 9$, $c_2 = 13$. We have $297 \equiv -2 \bmod 13$ and $\langle b, c_2 \rangle = 1$. The congruence $1 \equiv 140 + 297n \bmod 13$ is equivalent to $-2n \equiv 4 \bmod 13$. Multiplying both sides by 6, we find that $n \equiv 11 \bmod 13$. This gives $140 + (11 + 13k) \cdot 297 = 3407 + 117 \cdot 33 \cdot k$ which is coprime to 117 for all $k \in \mathbb{Z}$, as $\langle 3407, 117 \rangle = 1$.

[29.3] Thue (1902). For coprime $a, q > 0$, there exist integers s, t such that $0 < |s|, |t| \leq \sqrt{q}$ and $sa \equiv t \bmod q$. We have two proofs:

(A) We apply the pigeon box principle to the set $\{ua - v : 0 \leq u, v \leq [\sqrt{q}]\}$. Since $([\sqrt{q}] + 1)^2 > q$, there exist different pairs $\{u_1, v_1\}$, $\{u_2, v_2\}$ such that $u_1 a - v_1 \equiv u_2 a - v_2 \bmod q$. Then we may set $s = u_1 - u_2$, $t = v_1 - v_2$. For if $u_1 = u_2$, then $v_1 \equiv v_2 \bmod q$, which implies $v_1 = v_2$. This can be reversed, and we end the proof.

(B) We may assume that $a < q$. Applying (21.10) to the pair $\{a, q\}$, we choose $v \geq 0$ such that $r_{v+1} \leq \sqrt{q} < r_v$. This is possible, as $r_0 = q$ and $r_k = 1 < \sqrt{q}$. Then, by (21.11), $q = r_v B_v + r_{v+1} B_{v-1} \geq r_v B_v$. Hence, $B_v < \sqrt{q}$. Further, by (21.12), $r_{v+1} = (-1)^v a B_v + (-1)^{v+1} q A_v \Rightarrow r_{v+1} \equiv (-1)^v a B_v \bmod q$. Therefore, we may set $s = (-1)^v B_v$ and $t = r_{v+1}$.

Here is an application: If $a^2 \equiv -1 \bmod q$, and if q is not a square, then $s^2 + t^2 = q$. To prove this, we note that $s^2 + t^2 < 2q$ since $[\sqrt{q}]^2 < q$. Also, $s^2 + t^2 \equiv s^2(a^2 + 1) \equiv 0 \bmod q$. That is, $(s^2 + t^2)/q$ is a positive integer less than 2, which ends the proof. (A) reminds us of Note [20.5]. (B) is contained in Cornacchia (1908). In fact, the latter seems to have been induced by (B). Obviously (B) is superior to (A), as (B) gives an algorithm to compute s, t explicitly but (A) does not. See Note [58.6] and §81.

[29.4] Number theory in weaves: the three principal weaves are plain, twill, and satin. Lucas (1867) discussed the construction of satin weave by means of

$$\S30 \qquad\qquad 129$$

Theorem 30. That mysterious luster of satin is a work of Euclid's algorithm. See also [DA, (d), VI.2] by A.-M. Décaillot.

[29.5] A system of linear congruences of several unknowns, that is, (9.1) with the equalities being replaced by congruences with respect to a single modulus q, can be treated by (9.2) and by applying Theorem 29 to each of the resulting single-unknown linear congruence equations mod q. In particular, an extension of Theorem 30 should follow, so that if $q|u_k$, $1 \leq k \leq m$, then the solutions form a lattice of dimension n, or an Abelian group of rank n:

$$\sum_{j=1}^{n} \frac{q}{\langle g_j, q \rangle} \mathbb{Z}\mathbf{b}_j,$$

where the matrix $B = \{\mathbf{b}_1, \ldots, \mathbf{b}_n\}$ (column vectors) and g_j are as in Theorem 7.

§30

We shall now consider the algebraic structure of the set $\mathbb{Z}/q\mathbb{Z}$ under the convention (29.2). First, for arbitrary integers a, b let $C_a + C_b$ and $C_a \cdot C_b$ be equal to C_{a+b} and C_{ab}, respectively, in the notation of (27.4). By (28.1) the resulting residue classes are independent of the choice of the representatives of the operand classes; hence, these two operations between classes are well-defined. Of course they satisfy the associative law. In particular, the classes C_0 and C_1 are units with respect to addition $+$ and multiplication $\cdot$, respectively. Obviously the distributive law $C_c \cdot (C_a + C_b) = C_c \cdot C_a + C_c \cdot C_b$ holds. In other words, $\mathbb{Z}/q\mathbb{Z}$ is a commutative ring with multiplicative unit; however, as is readily seen, in general, this ring can have zero-divisors, i.e., $C_u \cdot C_v = C_0$ with neither C_u nor C_v being equal to C_0. With this understanding, we call $\mathbb{Z}/q\mathbb{Z}$ the residue class ring mod q. Its basic structure is given in (30.1) below; the symbol $\oplus$ stands for the ordinary direct product of Abelian groups and also for a direct product of commutative rings, as is indicated in the proof.

Theorem 31 *For $q = \prod_{k=1}^{K} q_k$, with $\{q_k\}$ being coprime in the sense of (7.6), we have a ring isomorphism*

$$\mathbb{Z}/q\mathbb{Z} \cong (\mathbb{Z}/q_1\mathbb{Z}) \oplus (\mathbb{Z}/q_2\mathbb{Z}) \oplus \cdots \oplus (\mathbb{Z}/q_K\mathbb{Z}). \qquad (30.1)$$

In particular, with the prime power decomposition $q = \prod_{j=1}^{J} p_j^{\alpha_j}$,

$$\mathbb{Z}/q\mathbb{Z} \cong (\mathbb{Z}/p_1^{\alpha_1}\mathbb{Z}) \oplus (\mathbb{Z}/p_2^{\alpha_2}\mathbb{Z}) \oplus \cdots \oplus (\mathbb{Z}/p_J^{\alpha_J}\mathbb{Z}). \qquad (30.2)$$

130 *Congruences*

Proof We let $t_k = q/q_k$; then $\langle t_1, t_2, \ldots, t_K \rangle = 1$. By (7.1) there exist integers u_k such that

$$\sum_{k=1}^{K} t_k u_k = 1. \tag{30.3}$$

Let $\{a_1, a_2, \ldots, a_K\}$, $a_k \bmod q_k$, be the coordinate of a generic element in the product set on the right side of (30.1), and let

$$a \equiv \sum_{k=1}^{K} a_k t_k u_k \bmod q. \tag{30.4}$$

This is a well-defined mapping from the right side of (30.1) to the left side: if $a_k \equiv a'_k \bmod q_k$, $1 \leq k \leq K$, then a' corresponding to $\{a'_1, a'_2, \ldots, a'_K\}$, $a'_k \bmod q_k$, satisfies $a \equiv a' \bmod q$ because of (14.4) and (28.3). Conversely, for any given $a \bmod q$, we let $a_k \equiv a \bmod q_k$; then multiplying a on both sides of (30.3) we get (30.4); moreover, if $a \equiv \sum_{k=1}^{K} a'_k t_k u_k \bmod q$, then $a_k \equiv a'_k \bmod q_k$, $1 \leq k \leq K$, since $q_k | t_{k'}$, $k' \neq k$, and $t_k u_k \equiv 1 \bmod q_k$. That is, (30.4) gives rise to a bijection between the two sides of (30.1). Also, since (30.4) is a linear combination of a_k, addition of residue classes is preserved coordinate-wise. As to multiplication, we have, for $b \equiv \sum_{k=1}^{K} b_k t_k u_k \bmod q$,

$$ab \equiv \sum_{k=1}^{K} a_k b_k (t_k u_k)^2 \bmod q, \tag{30.5}$$

since $t_k u_k t_{k'} u_{k'} \equiv 0 \bmod q$ for $k \neq k'$. Further, since (30.3) implies that $t_k u_k \equiv 1 \bmod q_k$, we have $(t_k u_k)^2 \equiv t_k u_k \bmod q$. Hence,

$$ab \equiv \sum_{k=1}^{K} a_k b_k t_k u_k \bmod q. \tag{30.6}$$

We may omit the rest of our argument. We end the proof.

The expression (30.4) means also that

$$\mathbb{Z}/q\mathbb{Z} = \bigoplus_{k=1}^{K} t_k u_k \mathbb{Z}/q\mathbb{Z}. \tag{30.7}$$

Namely the group $\mathbb{Z}/q\mathbb{Z}$ is decomposed into the direct sum of its subgroups $t_k u_k \mathbb{Z}/q\mathbb{Z}$ as $(t_k u_k \mathbb{Z}/q\mathbb{Z}) \cap (t_{k'} u_{k'} \mathbb{Z}/q\mathbb{Z}) = \{0 \bmod q\}$; and (30.1) is the same as $t_k u_k \mathbb{Z}/q\mathbb{Z} \cong \mathbb{Z}/q_k \mathbb{Z}$; see Note [30.3]. Under the convention introduced in Note [27.3], we have, also from (30.4),

$$a \bmod q \quad \Leftrightarrow \quad \frac{a}{q} \equiv \sum_{k=1}^{K} \frac{v_k}{q_k} \bmod 1, \quad v_k \equiv a_k u_k \bmod q_k. \tag{30.8}$$

Thus, if v_k runs over a residue system $\bmod \ q_k$ for each k, then a_k does the same as $\langle u_k, q_k \rangle = 1$ because of (30.3), and this sum represents, without overlapping, all fractions with denominator equal to q, if integral differences are ignored.

Notes

[30.1] The use of the relation (30.3) is the same as in Euler (1733b, §29)); see [DA, art.36]. In fact, this argument seems to have been known since ancient time in computations of calendar, distribution of goods, logistics, etc. However, for real-life use the theorem in the next section should be more practical.

[30.2] $355642261 = 629 \cdot 713 \cdot 793$. By Euclid's algorithm we see that $\langle 629, 713 \rangle = 1$, $\langle 629, 793 \rangle = 1$, $\langle 713, 793 \rangle = 1$. In particular,

$$\langle 565409, 498797, 448477 \rangle = \langle 713 \cdot 793, 629 \cdot 793, 629 \cdot 713 \rangle = 1$$

Thus the indefinite equation

$$565409y_1 + 498797y_2 + 448477y_3 = 1$$

has integer solutions. Following the argument applied in §9, we get, by Euclid's algorithm,

$$(565409, 498797, 448477) \begin{pmatrix} -1258 & -558 & 1887 \\ 3565 & 1466 & -4991 \\ -2379 & -927 & 3172 \end{pmatrix} = (0, 1, 0).$$

Hence, we choose $\{y_1, y_2, y_3\} = \{-558, 1466, -927\}$. Therefore, if $a \equiv b_1 \bmod 629$, $a \equiv b_2 \bmod 713$, $a \equiv b_3 \bmod 793$, then we have

$$a \equiv -558 \cdot 565409b_1 + 1466 \cdot 498797b_2 - 927 \cdot 448477b_3 \bmod 355642261.$$

Thus, for instance,

$$\frac{123456789}{355642261} = \frac{3}{629} + \frac{49}{713} + \frac{217}{793}.$$

[30.3] Any subgroup of $\mathbb{Z}/q\mathbb{Z}$ is isomorphic to $\mathbb{Z}/h\mathbb{Z}$ with an $h|q$. To prove this, let $H \neq \{0 \bmod q\}$ be a subgroup, and let $s = \min\{0 < r < q : r \bmod q \in H\}$. Then a combination of (1.2) and (2.1) gives that $n \bmod q \in H \Leftrightarrow s|n$. Hence, $s|q$ and $H = \{sm \bmod q\}$. The mapping $sm \bmod q \mapsto m \bmod q/s$ implies that $H \cong \{m \bmod h\}$, $h = q/s$ because of $(28.4)_{s=q/h}$, which ends the proof. We have proved also that for each divisor h of q, there exists a unique subgroup of order h. Naturally, this assertion translates into the characterization of subgroups of any cyclic group of finite order.

132 *Congruences*

[30.4] The structure of a finite Abelian group A: Let $N = |A|$. We consider the homomorphism

$$\omega: \quad \mathbb{Z}^N \rightarrow A = \{\mathbf{a}_1, \mathbf{a}_2, \ldots, \mathbf{a}_N\},$$

$$\omega(\mathbf{x}) = \begin{cases} \sum_j x_j \mathbf{a}_j & \text{(additive)}, \\ \prod_j \mathbf{a}_j^{x_j} & \text{(multiplicative)}, \end{cases} \quad \mathbf{x} = {}^{t}\{x_1, x_2, \cdots, x_N\}.$$

We have $A \cong \mathbb{Z}^N / \ker \omega$. We apply [9.6] with $H = \ker \omega$. We discard those $g_k = 1$ because of an obvious reason, and have, regardless of being additive or multiplicative,

the Structure Theorem v.1 for finite Abelian groups:
$$A \cong (\mathbb{Z}/f_1\mathbb{Z}) \oplus (\mathbb{Z}/f_2\mathbb{Z}) \oplus \cdots \oplus (\mathbb{Z}/f_L\mathbb{Z}),$$

where $1 < f_l | f_{l+1}$, $|A| = \prod_{l=1}^{L} f_l$. These integers $\{f_l\}$ are called the invariant factors of A; their uniqueness is shown in the next Note. Thus, there exists a basis $\{\mathbf{b}_1, \mathbf{b}_2, \ldots, \mathbf{b}_L\}$ in A such that $\mathfrak{b}_l$ is of order f_l, and every $\mathbf{a} \in A$ is uniquely represented in the form

$$\mathbf{a} = \begin{cases} t_1 \mathbf{a}_1 + t_2 \mathbf{b}_2 + \cdots + t_L \mathbf{b}_L & \text{(additive)}, \\ \mathbf{b}_1^{t_1} \mathbf{b}_2^{t_2} \cdots \mathbf{b}_L^{t_L} & \text{(multiplicative)}, \end{cases} \quad t_l \bmod f_l.$$

In particular

$$\mathbf{a} = \text{unit} \iff t_l \equiv 0 \bmod f_l, \ 1 \le l \le L.$$

[30.5] Although we have yet to show the uniqueness of the above decomposition of A, we may anyway apply (30.2) to each $\mathbb{Z}/f_l\mathbb{Z}$ and are led to the following assertion: regardless of being multiplicative or additive,

the Structure Theorem v.2 for finite Abelian groups:
$$A \cong \text{a direct sum of } \mathbb{Z}/p^\nu\mathbb{Z};$$

note that the same p^ν may appear multiple times. In other words, any finite Abelian group is decomposed into a direct sum/product of cyclic subgroups of prime power order. We need to confirm the uniqueness of this decomposition. Thus, let $\alpha_p^{(\nu)}$ be the number of all cyclic groups of order p^ν appearing in such a decomposition of A. In the multiplicative case, we have, for $\nu \ge 1$,

$$\alpha_p^{(\nu)} = \frac{1}{\log p} \log \left(\frac{|A^{p^{\nu-1}}| \, |A^{p^{\nu+1}}|}{|A^{p^\nu}|^2} \right),$$

with $A^{\ell} = \{\mathbf{a}^{\ell} : \mathbf{a} \in A\}$; the additive case is analogous. Here we have used the fact that if $\mathbf{a}^m = 1$ and $p \nmid m$, then via Euclid's algorithm applied to $\{p, m\}$ we see that there exists a μ such that $\mathbf{a}^{\mu p} = \mathbf{a}$; namely, the cyclic group of order m is not altered by raising its elements to the pth power. So the numbers $\alpha_p^{(\nu)}$ are determined solely by A, which obviously ends the required confirmation. The prime powers $\{p^\nu\}$ appearing in the above decomposition are called the elementary divisors of A. We are now able to show the uniqueness of $\{f_l\}$ of the preceding Note: Let $\{p_1, p_2, \ldots, p_J\}$ be the set of all distinct prime divisors of $|A|$. For each p_j, let the highest elementary divisor be $p_j^{\beta_j}$; if there are many such elementary divisors, then we take one of them. We see that there exists a cyclic subgroup B of order $\prod_j p_j^{\beta_j}$ since the product/sum of two cyclic groups of coprime orders is readily seen to be cyclic. We get the direct decomposition $A = B \cdot C$, where C corresponds to all remaining elementary divisors. By induction we finish the proof since the construction implies readily that $B \cong \mathbb{Z}/f_L\mathbb{Z}$.

[30.6]   The number α_p of elementary divisors divisible by p is called the p-rank of the group. In the multiplicative case, it equals

$$\alpha_p = \sum_{\nu=1}^{\infty} \alpha_p^{(\nu)} = \frac{1}{\log p} \log\left(\frac{|A|}{|A^p|}\right) : \quad p^{\alpha_p} = |A/A^p|.$$

In particular, A is cyclic if and only if $\alpha_p = 1$, that is, $x^p = 1$ has exactly p roots, for all p dividing $|A|$. This observation has an important implication; see Note [40.1].

[30.7]   There exists a confusion in the usage of the two terms: invariant factors and elementary divisors. Namely, in some literature the $\{g_k\}$ of Smith (Note [9.7]) are still called elementary divisors following an old practice: Frobenius (1879, p.148) called each g_k *die Elementartheiler*, which was followed by Bachmann (1898, Zweiter Abschnitt, Zweites Capitel); and an inconvenience started, as they left those prime power factors of g_k without name. In the present essays we adopt the relevant practice in linear algebra so that these two concepts are provided with decent names; see Gantmacher (1959, Chapter VI)), for instance. One should be aware that a subtle modification of the notion of invariant factors has been made when adopting it for finite Abelian groups: that is, $\{f_l\} \subseteq \{g_k\}$; however, the notion of elementary divisors is not affected.

§31

We have treated congruences with respect to a single modulus. We shall here consider a system of linear congruence equations each of which is with respect to an individual modulus:

$$c_k x \equiv s_k \bmod q_k, \quad 1 \le k \le K; \tag{31.1}$$

134 Congruences

note that this $\{q_k\}$ is different from $\{q_k\}$ of Theorem 31. We first apply Theorem 29 to each congruence equation, and then appeal to

Theorem 32 *For an arbitrary set of moduli $\{q_k\}$, the following equivalence holds:*

$$\text{the system } x \equiv a_k \bmod q_k, \ 1 \le k \le K, \text{ has a solution}$$
$$\Leftrightarrow \ a_j \equiv a_k \bmod \langle q_j, q_k \rangle, \ 1 \le j, k \le K. \tag{31.2}$$

If it exists, the solution is unique with respect to the modulus $[q_1, q_2, \ldots, q_K]$.

Proof The second assertion is the same as (28.3). Hence, we shall deal with the first assertion only. In view of the discussion in §8, the necessity part of the assertion is immediate since the indefinite equation $x = a_j + q_j y_j = a_k + q_k y_k$ should have a solution. To prove the sufficiency part, we argue by induction on K. Thus, we suppose that we have got $c \bmod [q_1, q_2, \ldots, q_K]$ such that $c \equiv a_k \bmod q_k, \ 1 \le k \le K$. Then, provided that

$$c \equiv a_{K+1} \bmod \langle [q_1, q_2, \ldots, q_K], q_{K+1} \rangle, \tag{31.3}$$

the system with K replaced by $K + 1$ has a solution. We need to confirm this congruence holds. According to Note [14.4] the modulus in (31.3) is equal to

$$\left[\langle q_1, q_{K+1} \rangle, \langle q_2, q_{K+1} \rangle, \ldots, \langle q_K, q_{K+1} \rangle \right]. \tag{31.4}$$

By the assumption (the lower line of (31.2)), we have $c \equiv a_k \equiv a_{K+1} \bmod \langle q_k, q_{K+1} \rangle$. Hence, via (28.3), we see that (31.3) holds. We end the proof.

As for the construction of a solution to (31.2), we begin with the first two congruence equations, and inductively proceed. In the example Note [31.3], we shall indicate how to handle this initial step.

Notes

[31.1] The system (31.1) was successfully treated in classic Indian mathematics already in the 9th century or earlier; see Datta–Singh (1935: 1962, Chap.III, §15)). The argument is essentially the same as above. In particular, the rôle of the least common multiple of moduli was precisely recognized.

[31.2] In [DA, arts.33–35] a procedure is shown to deal with (31.1) on the assumption that the prime power decomposition of each modulus is given. Then we are led to an expanded system of congruences in which each modulus is a prime power. Among those congruences whose moduli are powers of an identical prime, it may be readily checked if there exists any inconsistency; if it exists, then the solubility of the original system is denied. If all of those congruences survive, then we eliminate redundant equations. In the system resulting after this procedure, moduli are powers of different primes; and a

$$\begin{cases} 26x \equiv 65 \mod 3887 & (1), \\ 18x \equiv 321 \mod 4301 & (2). \end{cases}$$

solution of the original system is composed using (30.4). Nevertheless, if any of the original moduli is huge, then it can be extremely hard to factor it. Therefore, the above argument which relies solely on Euclid's algorithm is certainly more practical.

[31.3]   Consider the system:

$$\begin{cases} 26x \equiv 65 \mod 3887 & (1), \\ 18x \equiv 321 \mod 4301 & (2). \end{cases}$$

First, in (1) we have $\langle 26, 3887 \rangle = \langle 26, 13 \rangle = 13$ and $13|65$. By Theorem 29, we replace (1) by the equivalent congruence $2x \equiv 5 \mod 299$. Since $2^{-1} \equiv 150$, we get $x \equiv 150 \cdot 5 \equiv 152 \mod 299$. As for (2), Euclid's algorithm gives $18 \cdot 239 - 4301 = 1$; thus, multiplying both sides of (2) by 239, we get $x \equiv 321 \cdot 239 \equiv 3602 \mod 4301$. Hence, the original system is equivalent to

$$x \equiv 152 \mod 299, \quad x \equiv 3602 \mod 4301.$$

With this, we have $\langle 299, 4301 \rangle = 23$, and note that $23|(152 - 3602)$ which is required by Theorem 32; hence, the system (1)–(2) is solvable. Now, $x = 152 + 299y_1 = 3602 + 4301y_2$ with $y_1, y_2 \in \mathbb{Z}$, which is the same as $13y_1 - 187y_2 = 150$. Again by Euclid's algorithm or rather by (22.12), we have $13 \cdot 72 - 187 \cdot 5 = 1$. Multiplying both sides by 150, we get $y_1 \equiv 72 \cdot 150 \equiv 141 \mod 187$. Therefore,

$$x \equiv 152 + 299 \cdot 141 \equiv 42311 \mod 55913; \quad 55913 = 299 \cdot 187 = [299, 4301].$$

Indeed we have, with $u \in \mathbb{Z}$,

$$26 \cdot (42311 + 55913u) = 65 + 3887 \cdot (283 + 374u),$$
$$18 \cdot (42311 + 55913u) = 321 + 4301 \cdot (177 + 234u).$$

§32

We shall next consider $(\mathbb{Z}/q\mathbb{Z})^*$ the set of all reduced residue classes mod q. We note first that if $\langle a, q \rangle = 1$ and $\langle b, q \rangle = 1$, then by (6.2) (3) we have $\langle ab, q \rangle = 1$, whence $(\mathbb{Z}/q\mathbb{Z})^*$ is closed with respect to the multiplication introduced in §31. Moreover, on the convention (29.1), we have $C_a \cdot C_{\bar{a}} = C_a \cdot C_{a^{-1}} = C_1$ in the notation of (27.4), which means that every element in $(\mathbb{Z}/q\mathbb{Z})^*$ has its inverse in the same set; also the associative law holds of course. Namely, $(\mathbb{Z}/q\mathbb{Z})^*$ is a multiplicative Abelian group of order $\varphi(q)$. So the assertion (28.10) is an instance of the most basic fact in the general theory of finite groups: if the order of a group is k, then the kth power of any

136   *Congruences*

of its elements is equal to unit. In what follows, $(\mathbb{Z}/q\mathbb{Z})^*$ will be called the reduced residue class group mod q. In particular, with $\langle ab, q \rangle = 1$, we have $(ab)^{-1} \equiv a^{-1}b^{-1} \bmod q$, so $(a^\kappa)^{-1} \equiv (a^{-1})^\kappa \bmod q$ for each $\kappa \in \mathbb{N}$; the expression $a^{-\kappa} \bmod q$ is defined as $(a^\kappa)^{-1} \bmod q$, and then $a^\alpha \equiv a^\beta$ implies $a^{\alpha-\beta} \equiv 1 \bmod q$, for any $\alpha, \beta \in \mathbb{Z}$.

Corresponding to (30.1), the following assertion holds:

Theorem 33 *For $q = \prod_{k=1}^{K} q_k$, with $\{q_k\}$ as in Theorem 31, we have*

$$(\mathbb{Z}/q\mathbb{Z})^* \cong (\mathbb{Z}/q_1\mathbb{Z})^* \times (\mathbb{Z}/q_2\mathbb{Z})^* \times \cdots \times (\mathbb{Z}/q_K\mathbb{Z})^*, \tag{32.1}$$

with an obvious restriction of the isomorphism in (30.1). The right side of this isomorphism is a direct product of multiplicative groups. Corresponding to (30.2), we have

$$(\mathbb{Z}/q\mathbb{Z})^* \cong \left(\mathbb{Z}/p_1^{\alpha_1}\mathbb{Z}\right)^* \times \left(\mathbb{Z}/p_2^{\alpha_2}\mathbb{Z}\right)^* \times \cdots \times \left(\mathbb{Z}/p_J^{\alpha_J}\mathbb{Z}\right)^*. \tag{32.2}$$

Proof It suffices to show that if $\langle a_k, q_k \rangle = 1$, $1 \le k \le K$, on the right side of (30.4), then $\langle a, q \rangle = 1$, and the converse assertion holds. Thus, suppose that $p | \langle a, q \rangle$; then there exists a unique k such that $p | q_k$; hence, $p | a_k t_k u_k$, and $p | a_k$, so $p | \langle a_k, q_k \rangle$. On the other hand, if $\langle a, q \rangle = 1$ on the left side of (30.4), then $1 = \langle a, q_k \rangle = \langle a_k, q_k \rangle$, $1 \le k \le K$. This ends the proof.

In particular, from (28.7) and (32.1) we get $\varphi(q) = \prod_{k=1}^{K} \varphi(q_k)$, which means that the Euler φ-function is multiplicative; and $\varphi(p^\alpha) = p^\alpha - p^{\alpha-1}$ gives (18.4).

In addition, we have the following restricted version of (30.8):

$$\langle a, q \rangle = 1 \quad \Leftrightarrow \quad \frac{a}{q} \equiv \sum_{k=1}^{K} \frac{v_k}{q_k} \bmod 1, \quad \langle v_k, q_k \rangle = 1. \tag{32.3}$$

Notes

[32.1] The recognition of the structure (32.1) can be traced back to Euler (1758a, Theorema 5). Outwardly the isomorphisms (30.1) and (32.1) look similar to each other; however, in general, the actual structure of the right side of (32.1) is far more involved than that of (30.1), as will become apparent in §46. While (30.1) is a decomposition of a finite cyclic group into cyclic subgroups, (32.1) is not so in general. Nevertheless, we shall show later (Theorem 38) that (32.2) is a decomposition into cyclic sub-groups, provided $2^\lambda \| q, \lambda \le 2$.

[32.2] The case $q = p$ a prime is of special importance. The set $\mathbb{Z}/p\mathbb{Z} = (\mathbb{Z}/p\mathbb{Z})^* \sqcup \{0 \bmod p\}$ is a field of p elements. That is, $\mathbb{Z}/p\mathbb{Z}$ is a ring, and the subset of all elements save for the class $0 \bmod p$ is a multiplicative group. Throughout the rest of the present essays, we shall signify this fact by allocating the symbol $\mathbb{F}_p$ to $\mathbb{Z}/p\mathbb{Z}$.

[32.3] $1234567^{1234567} = \cdots 223$. Since $\varphi(10^3) = 400$, it suffices, in view of Theorem 28, to show that $567^{167} \equiv 223 \bmod 10^3$. First, we note that $567^{167} \equiv (-1)^{167} \equiv -1 \bmod 2^3$. Also, as $\varphi(5^3) = 100$, we see that $567^{167} \equiv 67^{67} \bmod 5^3$. To deal with the right side, we may use the binary expansion $67 = 1+2+2^6$ as is indicated in Note [2.3], but here we shall proceed with an alternative way. Thus, by the binomial expansion $67^{67} = (2 + 13 \cdot 5)^{67} \equiv 2^{67} + 67 \cdot 13 \cdot 5 \cdot 2^{66} + 33 \cdot 67 \cdot (13 \cdot 5)^2 \cdot 2^{65} \bmod 5^3$. In this, $67 \cdot 13 \equiv 17 \cdot 13 \equiv -4 \bmod 5^2$ and $33 \cdot 67 \cdot 13^2 \equiv -1 \bmod 5$. Hence, $67^{67} \equiv 2^{67} - 4 \cdot 5 \cdot 2^{66} - 5^2 \cdot 2^{65} \bmod 5^3$. Also, on noting $2^7 \equiv 3 \bmod 5^3$, we have $2^{67} \equiv 3^9 \cdot 2^4 \equiv 53 \bmod 5^3$. Further, applying Theorem 28 with the moduli 5^2 and 5, we get, respectively, $2^{68} \equiv 2^8 \equiv 6 \bmod 5^2$ and $2^{65} \equiv 2 \bmod 5$; so $67^{67} \equiv 53-5\cdot6-5^2\cdot2 \equiv -27 \bmod 5^3$. Therefore, we are led to the system of congruence equations: $x \equiv -1 \bmod 2^3$, $x \equiv -27 \bmod 5^3$. By Euclid's algorithm, we have $2^3 \cdot 47 - 5^3 \cdot 3 = 1$. Hence, $x \equiv -2^3\cdot47\cdot27+5^3\cdot3 = -9777 \equiv 223 \bmod 2^3 5^3$. This ends the confirmation.

[32.4]   An observation on the structure of the reduced residue system mod q: With an obvious abbreviation,

$$d|q \;\Rightarrow\; \text{reduced system mod } q = \bigsqcup \{\text{reduced system mod } d\}.$$

In other words, if we look at an arbitrary reduced residue system $\bmod\, q$ via the modulus d dividing q, then the system decomposes into $\varphi(q)/\varphi(d)$ reduced residue systems mod d: With an arbitrary l, $\langle d, l \rangle = 1$, the number of reduced residues mod q which are congruent to $l \bmod d$ is

$$\sum_{u \bmod q/d} \sum_{\langle l+du, q\rangle=1} 1 = \sum_{u \bmod q/d} \sum_{\langle l+du, q_1\rangle=1} 1,$$

where $q_1 = q/\langle q, d^\infty \rangle$; see Note [29.2]. Then, by an application of (18.2), the second dosuble sum is equal to

$$\sum_{u \bmod q/d} \sum_{\substack{s|q_1 \\ s|(l+du)}} \mu(s) = \sum_{s|q_1} \mu(s) \sum_{\substack{u \bmod q/d \\ l+du\equiv 0 \bmod s}} 1$$

$$= \frac{q}{d} \sum_{s|q_1} \mu(s)/s = \frac{q}{d} \prod_{\substack{p|q \\ p\nmid d}} (1 - 1/p) = \varphi(q)/\varphi(d),$$

which ends the proof.

[32.5]   Alternatively, one may consider the mapping

$$(\mathbb{Z}/q\mathbb{Z})^* \to (\mathbb{Z}/d\mathbb{Z})^* : \quad k \bmod q \mapsto k \bmod d.$$

138   *Congruences*

Since $d \mid q$, this is a homomorphism. Also, by Note [29.2], it is surjective. So the fundamental theorem on homomorphisms implies that the cardinality of the kernel is equal to $\varphi(q)/\varphi(d)$, which ends the proof since each of the cosets of the kernel corresponds to a unique element of $(\mathbb{Z}/d\mathbb{Z})^*$.

[32.6] The group $(\mathbb{Z}/q\mathbb{Z})^*$ is expected to have a surprisingly small generating set: Under the generalized Riemann hypothesis (GRH: (53.9)), Ankeny (1952) proved that there exists an absolute constant A such that reduced residues $a \bmod q$, $0 < a \le A(\log q)^2$, generate $(\mathbb{Z}/q\mathbb{Z})^*$. More precisely, he showed that with an arbitrarily given proper subgroup H of $(\mathbb{Z}/q\mathbb{Z})^*$

$$\min\{a > 0 : \langle a, q \rangle = 1,\, a \bmod q \notin H\} \le A(\log q)^2.$$

A proof will be given in Note [101.7]. Because of the basic nature of the group, this bound has connections with a variety of issues in number theory; a typical instance is treated in Note [101.8]. It should be added that Bach (1990) obtained $A \le 2$.

§33

So far we have considered linear congruence equations. We shall now move to quadratic, cubic, quartic, or higher degree situations. Thus, let $f(x) = \sum_{k=0}^{K} a_k x^k \in \mathbb{Z}[x]$. Then the congruence

$$f(x) \equiv 0 \bmod q \qquad (33.1)$$

which contains an unknown integer x is called a congruence equation mod q.

With this, if an integer a is such that $f(a) \equiv 0 \bmod q$, then $f(a') \equiv 0 \bmod q$ for any $a' \equiv a \bmod q$. Or rather $f(C_a) = C_{f(a)} = C_0$ in the notation of §30. In other words, with (33.1) we are actually considering the equation $f(x) = 0$ in the ring $\mathbb{Z}/q\mathbb{Z}$. Then, the notion $f(x) \bmod q$ that has been introduced in Note [27.2] needs to be redefined:

$$\text{to view } f(x) \in \mathbb{Z}[x] \text{ as an element of the ring } (\mathbb{Z}/q\mathbb{Z})[x]. \qquad (33.2)$$

One would say that the ring $\mathbb{Z}[x]$ is mapped homomorphically onto the ring $(\mathbb{Z}/q\mathbb{Z})[x]$ in a natural way. So we define the degree of $f(x) \bmod q$ by

$$\deg\left(\sum_{k=0}^{K} C_{a_k} x^k\right) \text{ in } (\mathbb{Z}/q\mathbb{Z})[x],$$
$$\text{that is, } \max\{k : a_k \not\equiv 0 \bmod q\}: \qquad (33.3)$$
$$\text{denoted by } \deg(f(x) \bmod q).$$

In particular, $\deg(f(x) \bmod q) = 0$ means that $a_k \equiv 0$, $k \ge 1$, but $a_0 \not\equiv 0 \bmod q$. Also, if $a_k \equiv 0 \bmod q$, $0 \le k \le K$, that is, $f(x) \in q\mathbb{Z}[x]$ or

$f(x) = 0$ in $(\mathbb{Z}/q\mathbb{Z})[x]$, then we have $\deg(f(x) \bmod q) = -\infty$, following a usual convention.

Returning to (33.1), one should note this: the situation that $f(x) \equiv 0 \bmod q$ for every integer x or that $f(x)$ is identically equal to zero over $\mathbb{Z}/q\mathbb{Z}$ is different from that $f(x) = 0$ in $(\mathbb{Z}/q\mathbb{Z})[x]$. Namely, equal to 0 in $(\mathbb{Z}/q\mathbb{Z})[x] \Rightarrow$ identically equal to zero over $\mathbb{Z}/q\mathbb{Z}$; but the converse is not always true. For instance, by (28.13), $x^p - x$ is identically equal to 0 over $\mathbb{F}_p = \mathbb{Z}/p\mathbb{Z}$, whereas $x^p - x \neq 0$ in $\mathbb{F}_p[x]$.

Here, although it might look abrupt, we introduce the following severe premise on a generic modulus q:

$$\text{the prime power decomposition } q = \textstyle\prod_{j=1}^{J} p_j^{\alpha_j} \text{ is given.} \qquad (33.4)$$

This is made because of the present state of number theory: Linear congruence equations, with respect to whatever moduli, can be treated solely by means of Euclid's algorithm, without recourse to (33.4). In a sharp contrast to this, the discussion on non-linear congruence equations cannot, in general, be conducted without having (33.4), at least presently. The cause of this huge difference between the linear and the non-linear situations will become apparent in the course of discussion; the first paragraph of §46 gives a summary of it.

Then, suppose that the congruence equation

$$f(x) \equiv 0 \bmod p_j^{\alpha_j}, \quad 1 \leq j \leq J, \qquad (33.5)$$

has a root $c_j \bmod p_j^{\alpha_j}$. Using the notation of (30.4) with $q_k = p_j^{\alpha_j}$ we let

$$c \equiv \sum_{j=1}^{J} c_j t_j u_j \bmod q. \qquad (33.6)$$

We have obviously $f(c) \equiv 0 \bmod q$. All the roots $c \bmod q$ of (33.1) are given in this way, and vice versa. To solve (33.1) is the same as to solve (33.5) for all j, provided (33.4). Thus, with the definition

$$\varkappa_f(q) \;:\; \begin{array}{c} \text{the number of mutually incongruent} \\ \text{roots } \bmod q \text{ of (33.1),} \end{array} \qquad (33.7)$$

we have, on (33.4),

$$\varkappa_f(q) = \prod_{j=1}^{J} \varkappa_f(p_j^{\alpha_j}), \quad \varkappa_f(1) = 1. \qquad (33.8)$$

In particular, the arithmetic function $\varkappa_f$ is multiplicative. See Note [33.1].

140   *Congruences*

Further, we impose the following restriction to the scope of our investigation on (33.1):

we shall stay within the context centering on Theorem 28.   (33.9)

In other words, we shall study (33.1) via various structures in $\mathbb{Z}$ which are connected either immediately or remotely but indispensably with the Fermat–Euler theorem (28.10)–(28.11). Expressing our intention more explicitly, we shall discuss (33.1) mostly via the knowledge on binomial congruence equations

$$x^\ell \equiv a \bmod q, \tag{33.10}$$

where $\ell \geq 2$, $q \in \mathbb{N}$, and $a \in \mathbb{Z}$ are arbitrary. The reason for this is closely related to the historical developments of the theory of algebraic equations over $\mathbb{Q}$, which can be summarized, though quite roughly, as an accumulation of strenuous efforts of solving those equations by means of nested radicals. We thus regard (33.1) as an analog of algebraic equations over $\mathbb{Q}$; and we are led to the discussion of its solubility in terms of analogues of ordinary radicals; and these analogs are precisely the roots of (33.10).

Therefore, we need, above all, to have a means to see whether (33.10) has a solution or not; this is the feature in which (33.10) differs from its counterpart over $\mathbb{C}$. Fortunately, to a limited extent though, if the modulus is a power of a prime, then there exists an effective criterion (Theorem 39, §43) which is basically due to Euler (1747); and this restriction of moduli to prime powers is the main reason for supposing (33.4). Hence, we shall develop our discussion towards Theorem 39. The findings thus achieved will play a basic rôle in the rest of the present essays.

Here, it should be stressed that we are concerned not only with the cases where (33.1) is soluble but also with the contrary cases. Since the latter raises the basic necessity of pondering the situation where (33.1) with $q = p$ does not have any root in the field $\mathbb{F}_p$, we are naturally led to the notion of algebraic extensions of $\mathbb{F}_p$ with generic p, in much the same spirit as algebraic extensions of $\mathbb{Q}$. Details will be given in §67, and further in Note [70.2].

Mainly because we are currently unable to replace (33.4) by anything less restrictive, the issue of achieving an effective method for prime power decompositions in general is regarded as one of the most fundamental problems in number theory, especially from the practical point of view. It should, therefore, be highly significant that by reversing, in a sense, the reasoning up to Theorem 39, we shall be led to an approach to the settlement of the problem of integer factorization in polynomial time, as has been indicated at the beginning of the present chapter. This will be discussed in §§46–48.

$$\S34 \qquad\qquad 141$$

It is thus understood, throughout the rest of the present essays, that (33.4) is supposed when dealing with congruences; we shall not mention this detail unless doing so is essential in order to make our relevant reasoning precise.

Notes

[33.1] Note that the definition (33.7) concerns mutually incongruent residue classes; multiplicities of roots are not taken into account there. Of course, it is natural to consider multiplicities especially in the case of $(33.1)_{q=p}$; see Note [34.4]. However, that is to be conducted algebraically, separating it from the issue of counting roots in the set $\mathbb{Z}/q\mathbb{Z}$. A reason for this is, for instance, in sieve theory: In the definition (19.1), the set $\Omega(p)$ is often defined to be $\{a : f(a) \equiv 0 \bmod p\}$ where f is fixed in $\mathbb{Z}[x]$. Then, multiple exclusion does not make sense.

[33.2] The origin of (33.10) is, of course, in the ordinary equation $x^{\ell} = a$ over $\mathbb{C}$, as is indicated above. We shall develop a discussion on this in the case $a = 1$, that is, cyclotomic equations, in §64 and later, from a number theoretical point of view; the term cyclotomy indicates divisions of the circle into a given number of equal segments. That is a wonderland discovered by Gauss [DA, Sectio VII]. As indicated in the last paragraph of our §28, he was fascinated first by the analogy between the roots of $x^{\ell} - 1 \equiv 0 \bmod q$ and those of the ordinary $x^{\ell} - 1 = 0$, especially by the multiplicative nature shared by them.

[33.3] Any practical algorithm to solve (33.10) is of great importance. In this respect, the existence of the algorithms of Tonelli (§45, §63) and Cipolla (§68) concerning $(33.10)_{q=p}$ is a precious fact in number theory, even though both are probabilistic in the sense to be clarified in due course. See also Note [68.6].

§34

The most fundamental assertion concerning the congruence equation (33.1) is the following theorem of Lagrange (1770b). It cannot be simpler; yet its effect is far-reaching.

Theorem 34 *For $f(x) = \sum_{k=0}^{K} a_k x^k \in \mathbb{Z}[x]$, it holds that*

$$p \nmid a_K \implies \varkappa_f(p) \le K = \deg(f(x) \bmod p), \qquad (34.1)$$

with the notation introduced in (33.3). Hence, if $\varkappa_f(p) > K$, then $f(x) \in p\mathbb{Z}[x]$.

Proof This depends on the definition $(33.7)_{q=p}$. Let $\{u_1, u_2, \ldots, u_K\}$ be a set of mutually incongruent roots $\bmod p$. The ordinary division algorithm for

142 *Congruences*

polynomials gives $f(x) = (x - u_1)f_1(x) + f(u_1)$, so $f(x) \equiv (x - u_1)f_1(x) \bmod p$ as polynomials, where the highest term of $f_1(x)$ is $a_K x^{K-1}$. We take $x = u_2$. Then, because $u_1 \not\equiv u_2 \bmod p$, we get $f_1(u_2) \equiv 0 \bmod p$. So $f_1(x) \equiv (x - u_2)f_2(x) \bmod p$ as polynomials; and the highest term of $f_2(x)$ is $a_K x^{K-2}$. In this way we are led to

$$f(x) \equiv (x - u_1)(x - u_2) \cdots (x - u_\kappa)f_\kappa(x) \bmod p, \tag{34.2}$$

The highest term of $f_\kappa(x)$ is $a_K x^{K-\kappa}$, which obviously yields (34.1). We end the proof.

Note that this argument is valid in general only for prime moduli (see Notes [34.2]–[34.3]); and then the situation becomes analogous to that of algebraic equations over $\mathbb{Q}$. Namely, in a somewhat formal expression, the assertion (34.1) is due to the fact that $\mathbb{F}_p$ is a field. One may discuss the analogue of ordinary irrational and complex zeros by algebraically extending the field $\mathbb{F}_p$, but it is postponed to §67, as has been suggested already in the preceding section. By the way, when $\deg(f(x) \bmod p) \geq p$, we apply the ordinary division algorithm in $\mathbb{Z}[x]$, and have $f(x) = (x^p - x)q(x) + r(x)$, $\deg r(x) < p$; then $f(x) \equiv 0 \bmod p$ is the same as $r(x) \equiv 0 \bmod p$. In particular, we have the following assertion:

Theorem 35 *If $f(x)$ is identically equal to zero over $\mathbb{Z}/p\mathbb{Z}$, then there exists a $q(x) \in \mathbb{Z}[x]$ such that*

$$f(x) - (x^p - x)q(x) \in p\mathbb{Z}[x]. \tag{34.3}$$

Namely, mapped into $\mathbb{F}_p[x]$, $f(x)$ is divisible by $x^p - x$.

The rest is a trivial remark: Despite what was stated in the preceding section, it is, of course, natural to consider the possibility of multiple roots $\bmod\, p$ or the decomposition

$$f(x) \equiv (x - a)^h g(x) \bmod p \tag{34.4}$$

with $a \in \mathbb{Z}$, $h \geq 2$, and $g(x) \in \mathbb{Z}[x]$. We have then

$$f'(x) - (x - a)^{h-1}(hg(x) - (x - a)g'(x)) \in p\mathbb{Z}[x], \tag{34.5}$$

so $f'(a) \equiv 0 \bmod p$. On the other hand, if $h = 1$ and $f'(a) \equiv 0 \bmod p$, then $g(a) \equiv 0 \bmod p$, that is $g(x) \equiv (x - a)g_1(x) \bmod p$ with $g_1(x) \in \mathbb{Z}[x]$; namely $f(x) \equiv (x - a)^2 g_1(x) \bmod p$. Hence, $f(x) \equiv 0 \bmod p$ has $a \bmod p$ as a multiple root if and only if $f(a) \equiv f'(a) \equiv 0 \bmod p$. Continued to Note [34.4].

Notes

[34.1]   Euler (1772c, p.519) stated (34.1) for $f(x) = x^K - 1$. Lagrange (1770b, pp.667–669: Corollaire V) gave a proof of Theorem 34 only in the case of $K = 3$, which reminds us of Euclid's style (Note [10.8])), but Lagrange explicitly remarked that his argument is general. His proof is essentially a repeated application of the differencing argument of Euler (1755, Theorema 19) (see Note [43.2]). Legendre (1785, pp.466–467) appears to be the first who perceived Theorem 34 to be a fundamental building block of number theory; see Note [40.2]. Indeed the above proof, so the modern proof, is due to Legendre. See also Gauss [DA, art. 44]. Nowadays Theorem 34 is included in the theory of fields which emerged, no doubt, from Lagrange's discovery of (34.1); see §67 for details.

[34.2]   For the polynomial $f(x) = x^3 - x + 7$, we have $\varkappa_f(43) = 0$. On the other hand, $\varkappa_f(53) = 3$. Roots are $27, 30, 49 \bmod 53$; thus $(x-27)(x-30)(x-49) = x^3 - 106x^2 + 3603x - 39690$, and $106 \equiv 0$, $3603 \equiv -1$, $39690 \equiv -7 \bmod 53$. If the modulus is composite, then as is easily inferred from (33.8) the number of roots of a congruence equation can be greater than its degree. This was already remarked by Lagrange (1770b, p.669). For example, $x^3 - x + 7 \equiv 0 \bmod 53 \cdot 71$ has 9 roots: 560, 610, 844, 1249, 1533, 1670, 2309, 2593, 3684 mod 3763.

[34.3]   The Hensel lifting (1901). Let f be an integral polynomial. If $f(x) \equiv 0 \bmod p^2$ has a solution $u \bmod p^2$, then writing $u = u^{(0)} + u^{(1)}p$ we have obviously $f(u^{(0)}) \equiv 0 \bmod p$. Also, by the Taylor expansion

$$f(u^{(0)} + u^{(1)}p) \equiv f(u^{(0)}) + f'(u^{(0)})u^{(1)}p \bmod p^2$$
$$\Rightarrow f'(u^{(0)})u^{(1)} \equiv -f(u^{(0)})/p \bmod p,$$

for $f^{(j)}(u^{(0)})/j! = \sum_{k=j}^{K} a_k \binom{k}{j}(u^{(0)})^{k-j} \in \mathbb{Z}$. On the assumption that $u^{(0)} \bmod p$ is simple, i.e., $f'(u^{(0)}) \not\equiv 0 \bmod p$, we see that $u^{(1)} \bmod p$ is uniquely determined by $u^{(0)} \bmod p$. Repeating the same, we are able to lift any simple root of $f(x) \equiv 0 \bmod p$ to a unique root of $f(x) \equiv 0 \bmod p^\ell$ for each $\ell = 2, 3, \ldots$ Hence, if roots of $f(x) \equiv 0 \bmod p$ are all simple, then $\varkappa_f(p^\ell) = \varkappa_f(p)$, $\forall \ell \geq 1$. In the case of the last Note, with the modulus 53, we have $f(30)/53 = 509 \equiv 32$ and $f'(30) \equiv 49$. Thus, applying Euclid's algorithm to $49u^{(1)} \equiv -32$, we find $u^{(1)} \equiv 8$. Hence, $30 + 8 \cdot 53 = 454$ should be a solution mod 53^2: indeed, $f(454) = 33313 \cdot 53^2$. By the way, the lifting of 27 mod 53 to mod 53^2 is 27, for $f(27) = 19663 = 7 \cdot 53^2$ and $f'(27) = 2186 \equiv 13 \bmod 53$. Also, the lifting of 49 mod 53 to mod 53^2 is 2328. In addition, the lifting of 27 mod 53 to mod 53^4 is $27 + 28 \cdot 53^2 + 9 \cdot 53^3 = 1418572$, and so on. In Lagrange (1769, pp.500–501) and Legendre (1798, pp.419–420) there are prototypes of this lifting argument. See Notes [45.5] and [68.1]. See also

144   *Congruences*

Gauss [DA, art.101]. Further, note that Smith (1859/1869, art.72) gave a deeper discussion. Nevertheless, lifting is attributed to Hensel. This is to commemorate his theory of p-adic numbers which is a natural way of justifying $f(u) \equiv 0 \bmod p^{\infty}$, $u = \sum_{v=0}^{\infty} u^{(v)} p^v$.

[34.4]   Although the preceding Note suffices for our application of lifting, we add that a detailed discussion on the case of multiple roots can be found in Niven et al (1991, pp.88–90, 487–489), where the issue is related to how the discriminant of f is divisible by the relevant prime powers.

[34.5]   An integral polynomial $f(x)$ is called intersective if (i) its coefficient of highest degree is equal to 1, (ii) it has no zero in $\mathbb{Q}$, (iii) it has zeros in $\mathbb{Z}/q\mathbb{Z}$ for every $q > 1$. See Hyde et al. (2014).

§35

We shall give a typical application of Theorem 34. Thus, the congruence equation

$$x^{p-1} - 1 - \prod_{j=1}^{p-1}(x-j) \equiv 0 \ \bmod p \tag{35.1}$$

has roots $1, 2, \ldots, p-1 \bmod p$ by (28.11); the number of mutually different roots $\bmod p$ is greater than the degree of the polynomial on the left side. Hence, by Theorem 34 this polynomial is in $p\mathbb{Z}[x]$. In particular, the constant term gives Wilson's theorem:

$$(p-1)! \equiv -1 \ \bmod p. \tag{35.2}$$

Here is an alternative proof of this famous congruence: We may assume naturally that $p \geq 5$. Then for $1 \leq r \leq p-1$ we choose $1 \leq r' \leq p-1$ such that $rr' \equiv 1 \bmod p$. We have

$$(p-1)! = \prod_{r \neq r'} r \prod_{r=r'} r \equiv \prod_{r^2 \equiv 1 \bmod p} r \ \bmod p. \tag{35.3}$$

The condition $r^2 \equiv 1 \bmod p$ is equivalent to $p|(r-1)(r+1)$, and $r = 1, p-1$. Hence, $(p-1)! \equiv 1 \cdot (p-1) \equiv -1 \bmod p$. Extensions are given in Notes [42.4] and [44.5].

Notes

[35.1]   It is known nowadays that the assertion (35.2) was stated by the great polymath al-Haytham (ca 1025: Rashed (1980)). The present name for this congruence has come from the fact that Lagrange (1771b) gave its first proof with the remark that Waring had reported that (35.2) was found by Wilson (Lagrange's success is mentioned in Waring's later work (1782, p.xxxii)). This somehow reminds us of the wrong attribution of the sequence $\mathcal{F}_N$ (Note [20.2]). The first of the above two proofs is due to Chebyshev (1848, §19), and the second to Gauss [DA, art.77]. Also Euler (1773c) gave a proof (see Note [44.5]), referring to Lagrange but not to Wilson.

[35.2]   An extension of the argument (35.3) to composite moduli is mentioned in [DA, art.78].

[35.3]   Lagrange's original proof (1771b) of (35.2) does not depend on Theorem 34 but on $\binom{p}{f} \equiv 0 \bmod p$ for $1 \le f \le p - 1$, which was applied in our Note [28.2]: It starts with the expansion

$$\prod_{j=1}^{p-1}(x+j) = \sum_{k=0}^{p-1} a_k x^{p-k-1}.$$

Replacing x by $x + 1$, one gets

$$\sum_{j=0}^{p-1} a_j(x+1)^{p-j} = (x+p)\sum_{k=0}^{p-1} a_k x^{p-k-1}.$$

This is the same as

$$\sum_{h=0}^{p} x^h \sum_{j=0}^{p-h} a_j \binom{p-j}{h} = \sum_{k=0}^{p}(a_k + pa_{k-1})x^{p-k}, \quad a_{-1}, a_p = 0.$$

Equating the coefficients of x^{p-v-1}, we find that

$$v a_v = \sum_{j=0}^{v-1} \binom{p-j}{p-v-1} a_j, \quad 1 \le v \le p - 1.$$

It follows that $a_0 = 1; a_1, \ldots, a_{p-2} \equiv 0, a_{p-1} \equiv -1 \bmod p$. Hence

$$\prod_{j=1}^{p-1}(x+j) \equiv x^{p-1} - 1 \bmod p. \quad \text{(as polynomials)}$$

Therefore, for any reduced residue system $\{\alpha_j\} \bmod p$, we have

$$\prod_{j=1}^{p-1}(x-\alpha_j) \equiv x^{p-1} - 1 \bmod p \ \Rightarrow \ \alpha_1\alpha_2\cdots\alpha_{p-1} \equiv -1 \bmod p,$$

which is equivalent to (35.2). With $x = \alpha_h$, we get also an alternative proof of (28.11). See Gauss [DA, art.76].

146 Congruences

[35.4] Lagrange (1771b, p.431). With a prime number $p \geq 3$, we write the reduced residue system in the form

$$1, 2, \ldots, \tfrac{1}{2}(p-1), p - \tfrac{1}{2}(p-1), \ldots, p-2, p-1.$$

Hence $\{((p-1)/2)!\}^2 \equiv (-1)^{(p+1)/2} \bmod p$. In particular, if $p \equiv 1 \bmod 4$, then the solutions of $x^2 \equiv -1 \bmod p$ are $x \equiv \pm((p-1)/2)! \bmod p$. Conversely, if there exists a solution $x_0 \bmod p$, then either $p \equiv 1 \bmod 4$ or $p = 2$. Otherwise, raising both sides of $x_0^2 \equiv -1 \bmod p$ to the power $(p-1)/2$, we get a contradiction. As for the case $p \equiv -1 \bmod 4$, we have $((p-1)/2)! \equiv \pm 1 \bmod p$. This right side can be made explicit; see Note [92.9].

[35.5] Lagrange (1771b, p.432). As is readily seen, the converse of (35.2) is correct. Namely, if $(m-1)! \equiv -1 \bmod m$, then m is a prime number.

[35.6] Prime numbers satisfying $(p-1)! \equiv -1 \bmod p^2$ are called Wilson primes. So far only three instances $5, 13, 563$ have been identified. $12! = 479001600 = -1 + 2834329 \cdot 13^2$ (Mathews (1892, p.318)). The case of 563 is a result of early electric-computer searches (Goldberg (1952)).

[35.7] J. Wolstenholme (1862). For any prime number $p \geq 5$, we have

$$\sum_{j=1}^{p-1} \frac{1}{j} = \frac{a}{b}, \quad \langle a, b \rangle = 1 \quad \Rightarrow \quad p^2 | a.$$

To prove this, we note that $a/b = a_{p-2}/(p-1)!$, with $\{a_k\}$ as in Note [35.3]. So it suffices to show $p^2 | a_{p-2}$ for $p \geq 5$. We then consider $\prod_{j=1}^{p-1}(-p+j)$. We have

$$a_0(-p)^{p-1} + a_1(-p)^{p-2} + \cdots + a_{p-3}(-p)^2 + a_{p-2}(-p) = 0,$$

and end the proof.

§36

In the present and the next sections we shall consider basic facts concerning the contraposition of Fermat's theorem (28.11):

$$\text{If there exists a reduced residue } a \bmod q \text{ such that}$$
$$a^{q-1} \not\equiv 1 \bmod q, \text{ then } q \text{ is composite.} \tag{36.1}$$

In Note [36.3] we shall show an example of this situation. The difficulty with the criterion (36.1) for compositeness is that we are not always able to attain a clear conclusion. Namely, if it turns out that $a^{q-1} \equiv 1 \bmod q$, then we are unable to say whether q is a prime or not; in this respect the example shown in Note [36.4] belongs to an extreme situation that will be considered in the

§36 147

next section. In any event, we shall see that (36.1) is a starting point of a highly practical primality test that will be discussed in the first half of §46; until then, we shall develop some preparations which are in fact various approaches to the set of powers in the group $(\mathbb{Z}/q\mathbb{Z})^*$.

In applying the criterion (36.1) to a large modulus, we have to deal with high powers of integers. To this end we may employ a device called the modular binary exponentiation, which will play an important rôle also in other places of the present essays. Thus, in order to compute $a^c \bmod q$, we first expand c into $(r_k r_{k-1} \ldots r_1 r_0)_2$. Then $a_j \equiv a^{2^j}$ is computed inductively by the relation $a_{j+1} \equiv a_j^2$, and

$$\text{modular binary exponentiation:} \quad a^c \equiv \prod_{\substack{j=0 \\ r_j=1}}^{k} a_j \bmod q, \tag{36.2}$$

which should be compared with the ancient Egyptian algorithm explained in Note [2.4]. In computing this product we use modular reduction appropriately. The reason that specifically the binary expansion is used is in the simplicity of the procedure. By the way, (36.2) implies that the criterion (36.1) is polynomial time with respect to q; see Crandall and Pomerance (2005, §2.1.2).

Notes

[36.1]   The first instance of the use of (36.1) seems to be due to Lambert (1770, §§50–54).

[36.2]   An application of (36.2) is in Euler (1755, 11. Scholion). He computed $7^{2^n} \bmod 641$ and obtained $7^{160} = 7^{2^5} \cdot 7^{2^7} \equiv -1 \bmod 641$. See also Legendre (1785, p.473) and (1798, p.229); in the latter, via the expansion $506 = (111111010)_2$, he got $601^{506} \equiv -1 \bmod 1013$; see Note [58.5] below.

[36.3]   As for the modulus 5293, we have $5292 = (1010010101100)_2$; and

$$2^2 = 4, \ 2^{2^2} = 16, \ 2^{2^3} = 16^2 = 256, \ 2^{2^4} = 256^2 \equiv 2020, \ 2^{2^5} \equiv 2020^2 \equiv 4790,$$

$$2^{2^6} \equiv 4790^2 \equiv 4238, \ 2^{2^7} \equiv 4238^2 \equiv 1495, \ 2^{2^8} \equiv 1495^2 \equiv 1379,$$

$$2^{2^9} \equiv 1379^2 \equiv 1454, 2^{2^{10}} \equiv 1454^2 \equiv 2209,$$

$$2^{2^{11}} \equiv 2209^2 \equiv 4828, \ 2^{2^{12}} \equiv 4828^2 \equiv 4505.$$

Thus,

$$2^{5292} \equiv 16 \cdot 256 \cdot 4790 \cdot 1495 \cdot 2209 \cdot 4505 \equiv 2890 \bmod 5293.$$

Hence, 5293 is composite. In fact, $5293 = 67 \cdot 79$.

148   *Congruences*

[36.4]   As for the modulus 8911, we have $8910 = (1000101\,1001110)_2$; and

$$2^2 = 4,\ 2^{2^2} = 16,\ 2^{2^3} = 16^2 = 256,\ 2^{2^4} = 256^2 \equiv 3159,\ 2^{2^5} \equiv 3159^2 \equiv -1039,$$

$$2^{2^6} \equiv 1039^2 \equiv 1290,\ 2^{2^7} \equiv 1290^2 \equiv -2257,\ 2^{2^8} \equiv 2257^2 \equiv -3043,$$

$$2^{2^9} \equiv 3043^2 \equiv 1320,\ 2^{2^{10}} \equiv 1320^2 \equiv 4755,\ 2^{2^{11}} \equiv 4755^2 \equiv 2818,$$

$$2^{2^{12}} \equiv 2818^2 \equiv 1423,\ 2^{2^{13}} \equiv 1423^2 \equiv 2132 \ \bmod 8911.$$

Thus,

$$2^{8910} \equiv 4 \cdot 16 \cdot 256 \cdot 1290 \cdot (-2257) \cdot 1320 \cdot 2132 \equiv 1 \bmod 8911.$$

This conforms to (28.11). However, $8911 = 7 \cdot 19 \cdot 67$.

§37

The last Note implies that the converse of Fermat's theorem (28.11) does not hold. We thus introduce the following notion: Let $q \geq 3$ be odd and let $a \geq 2$ be coprime to q; then,

$$q \text{ is a probable prime to base } a \ \Leftrightarrow\ a^{q-1} \equiv 1 \bmod q. \tag{37.1}$$

This situation is denoted by $q \in \mathrm{pp}(a)$. If $q \notin \mathrm{pp}(a)$, then q is composite; Note [36.3] states that $5293 \notin \mathrm{pp}(2)$. On the other hand, if $q \in \mathrm{pp}(a)$, then we have yet the possibility that q is composite. So we introduce the following notion:

$$\begin{aligned} &\text{an odd } q \text{ is a pseudoprime to base } a \\ &\Leftrightarrow\ \mathrm{pp}(a) \ni q \text{ is composite.} \end{aligned} \tag{37.2}$$

This situation is denoted by $q \in \mathrm{psp}(a)$; that q is indeed composite and has to be verified by an independent argument. The example in Note [36.3] means that $8911 \in \mathrm{psp}(2)$.

    It might be expected that if several bases are taken, then one would be able to make a precise conclusion about whether the relevant modulus is a prime or not. However, there exist pseudoprimes which do not react to the changes of bases. Thus, we introduce further the following notion:

$$\text{an odd } q \text{ is a Carmichael number } \Leftrightarrow\ q \in \bigcap_{\substack{2 \leq a < q \\ \langle a,q \rangle = 1}} \mathrm{psp}(a). \tag{37.3}$$

As a matter of fact, 8911 (Note [36.3]) is a Carmichael number: If $\langle a, 8911 \rangle = 1$, then $a^6 \equiv 1 \bmod 7$, $a^{18} \equiv 1 \bmod 19$, $a^{66} \equiv 1 \bmod 67$. So $a^{8910} = (a^6)^{1485} \equiv 1 \bmod 7$, $a^{8910} = (a^{18})^{495} \equiv 1 \bmod 19$, $a^{8910} = (a^{66})^{135} \equiv 1 \bmod 67$. Hence, by (28.3), we find that $a^{8910} \equiv 1 \bmod 8911$.

$$\S37 \qquad\qquad 149$$

Therefore, the converse of (28.11) cannot be used as a primality test in general. Nevertheless, the following trivial observation on (28.11) induces an effective primality test: With any prime $p \geq 3$ such that $2^h \| (p-1)$, we may put (28.11) into a more precise form:

$$a^{p-1} - 1 = \left(a^{(p-1)/2^h} - 1\right) \prod_{j=1}^{h} \left(a^{(p-1)/2^j} + 1\right) \equiv 0 \bmod p. \qquad (37.4)$$

Thus, at least one of the $h+1$ factors on the right side is divisible by p. That is, the definition (37.1) can be made stricter. With an odd $q \geq 3$ such that $2^h \| (q-1)$ we introduce the following notion:

an odd q is a strong probable prime to base a

$$\Leftrightarrow \begin{cases} \text{either } a^{(q-1)/2^h} \equiv 1 \bmod q \\ \text{or } a^{(q-1)/2^u} \equiv -1 \bmod q,\ 1 \leq \exists u \leq h. \end{cases} \qquad (37.5)$$

This situation is denoted by $q \in \mathrm{spp}(a)$. Strong probable primes have more similarity to primes than mere probable primes. Those $q \notin \mathrm{spp}(a)$ are of course composite; see Note [37.4] for decomposing such integers. Following the definition (37.2), we introduce

an odd q is a strong pseudoprime to base a

$$\Leftrightarrow \mathrm{spp}(a) \ni q \text{ is composite.} \qquad (37.6)$$

This situation is denoted by $q \in \mathrm{spsp}(a)$. Thus, strong pseudoprimes are integers which are proved to be composite by some means but have strong similarity to primes.

Then, it should be highly interesting that changing bases in (37.5) is indeed effective: We shall prove in §46 that

$$\text{for any odd } q > 0 \text{ it holds that } q \notin \bigcap_{\substack{0<a<q \\ \langle a,q\rangle=1}} \mathrm{spsp}(a). \qquad (37.7)$$

That is, if $q > 0$ is composite, then there should exist $a \geq 2$, $\langle a,q \rangle = 1$, such that $q \notin \mathrm{spp}(a)$. Hence (37.5) can play the rôle of a definitive primality test. We shall show, also in §46 (see specifically Note [46.1]), that the number of residue classes mod q to be utilized in the test is practically $O((\log q)^2)$ on GRH, which can be regarded as extremely small compared with q.

Notes

[37.1] Our discussion on primality tests is limited because of our view to be expressed in Note [46.3]. For most practical purposes, what is given in the first half of §46 suffices. A thorough discussion of modern primality tests is in Crandall–Pomerance (2005, Chapter 4).

150 *Congruences*

[37.2]   The naming (37.3) is known to be inconsistent with the historical fact: the first instances of integers, i.e.,

$$561, 1105, 1729, 2465, 2821, 6601, 8911,$$

that satisfy the definition (37.3) were discovered by Šimerka (1885, p.224); independently Carmichael (1910, p.238) reported $561, 1105, 2821, 15841$. According to Alford et al (1994), there exist infinitely many Carmichael numbers. Any example for the totient problem of Lehmer (Note [18.3]) is obviously a Carmichael number, although none of known Carmichael numbers satisfy Lehmer's condition. See Note [43.12] below.

[37.3]   Let P_J be the set of the first J primes, and let

$$q_0(J) \text{ be the least element in } \bigcap_{p \in P_J} \mathrm{spsp}(p).$$

Suppose that $m < q_0(J)$ is not divisible by any prime in P_J. If there exists a $p \in P_J$ such that $m \notin \mathrm{spp}(p)$, then m is composite. Otherwise, m is a prime, for then there should exist a $p \in P_J$ such that $m \in \mathrm{spp}(p) \backslash \mathrm{spsp}(p)$. Hence, in order to determine, by means of (37.5), whether $m < q_0(J)$ is a prime or not, one needs to test it against the prime bases $p \in P_J$ only: if $m \in \bigcap_{p \in P_J} \mathrm{spp}(p)$, then m must be a prime. Because of this practical consequence, considerable efforts have been made to find strong pseudoprimes with respect to small prime bases. For instance, $q_0(13)$ is of 25 decimal digits, so huge despite of such a small set of bases. See Sorenson–Webster (2017).

[37.4]   If $q \in \mathrm{pp}(a) \backslash \mathrm{spp}(a)$, then $q \in \mathrm{psp}(a)$, and a non-trivial decomposition of q should follow from the identity

$$q = \langle a^{(q-1)/2^h} - 1, q \rangle \prod_{u=1}^{h} \langle a^{(q-1)/2^u} + 1, q \rangle, \quad 2^h \| (q - 1);$$

note that the odd factors of $a^{(q-1)/2^h} - 1$ and $a^{(q-1)/2^u} + 1$, $1 \leq u \leq h$, are mutually prime. An example is $q = 745889$. We have $2^5 \| (q - 1)$ and

$$
\begin{array}{ll}
\langle 2^{(q-1)/2} + 1, q \rangle = 1, & \langle 2^{(q-1)/4} + 1, q \rangle = 1, \\
\langle 2^{(q-1)/8} + 1, q \rangle = 353, & \langle 2^{(q-1)/16} + 1, q \rangle = 2113, \\
\langle 2^{(q-1)/32} + 1, q \rangle = 1, & \langle 2^{(q-1)/32} - 1, q \rangle = 1.
\end{array}
$$

Hence, $q \in \mathrm{pp}(2) \backslash \mathrm{spp}(2)$ and $q = 353 \cdot 2113$; both factors are prime numbers.

§38

If an integer is found to be composite without having been factored, then it is, of course, highly worth trying to identify a factor of it. To this end, one may apply the ρ algorithm for factorization:

Let q be known to be composite, and let k_0, c be chosen appropriately. Then

$$\text{generate the sequence } k_\nu, \; \nu \geq 0,$$
$$\text{via the recurrence relation } k_{\nu+1} \equiv k_\nu^2 + c \mod q, \qquad (38.1)$$
$$\text{and observe the sequence } g_t = \langle k_{2t} - k_t, q \rangle.$$

This is to exploit the periodicity $k_\nu \equiv k_{\nu+\ell} \mod p$, where p is a prime factor of q. Naturally, we are unable to compute $k_\nu \mod p$, so $k_\nu \mod q$ is employed instead. Prime factors of q are acting covertly. The number of those residues $\mod p$ is, of course, not greater than q, and the cyclic repetition should take place. In particular, with respect to multiples u of this cycle we have $k_{2u} \equiv k_u \mod p$; and hence either p or its multiples are expected to appear in the sequence $\{g_t\}$. However, it is not always a good strategy to observe only $\{g_t\}$; see Notes [38.4], [38.5].

The ρ algorithm is a device based on trial and error. The aim of the use of quadratic congruence is to randomize the generating procedure of the sequence $\{k_\nu\}$ so that the probability of hitting a good g_t be increased. However, as is shown in Note [38.3], applications of the method are not always successive. In case of failure, alter the initial $\{k_0, c\}$ appropriately and resume the trial.

To this day, no deterministic factoring algorithm of polynomial time has been discovered. However, as has been remarked already in Note [2.6] it is known that on quantum computers, if they are materialized, factorization of integers in polynomial time is possible, although probabilistically. We shall give, in §§47–48, an account of this fascinating development.

Notes

[38.1] The ρ algorithm for factorization of integers is an invention of Pollard (1975). Its simplicity is highly appreciated; in applications it has a charm like treasure hunting. The name of the method came from the periodicity in the sequence $\{k_\nu \mod q\}$. The cycle starts in the middle of sequence; thus the orbit of procedure may look like the letter ρ, although the hand-writing of this Greek character proceeds oppositely. The idea to use $\{g_t\}$ came from an algorithm in computer science to detect cycles in sequences. For more details, see Niven et al (1991, pp.80–81); there an expected number of the recursion steps is given that is needed for the method to yield a proper divisor of q.

[38.2]   According to Note [36.3] the modulus $q = 5293$ is composite. We apply the ρ algorithm with $c = 1, k_0 = 1$. Then, $k_1 = 2, k_2 = 5, k_3 = 26$, $k_4 = 677, k_5 = 3132, k_6 = 1496, k_7 = 4371, k_8 = 3205, k_9 = 3606$, $k_{10} = 3629, k_{11} = 658, k_{12} = 4232, k_{13} = 3606$. Hence, $g_4 = 79$; and the decomposition $5293 = 67 \cdot 79$ follows. By the way, observing the sequence with respect to the prime factor 79, we see that $k_4 \equiv 45, k_5 \equiv 51$, $k_6 \equiv 74, k_7 \equiv 26, k_8 \equiv 45 \bmod 79$, and there appears a cycle composed of four terms.

[38.3]   One finds that $q = 266537 \notin \mathrm{pp}(2)$, in the same way as in Note [36.3]. Thus, we apply the ρ algorithm with $c = 1, k_0 = 1$. Then $k_1 = 2, k_2 = 5$, $k_3 = 26, k_4 = 677, k_5 = 191793, k_6 = 50017, k_7 = 250545, k_8 = 135082$, $k_9 = 23705, k_{10} = 67030, k_{11} = 6692, k_{12} = 4649, k_{13} = 23705$. This gives $g_6 = 5671$, and $q = 47 \cdot 5671$. Here, $5671 \notin \mathrm{pp}(2)$, but for this composite integer the choice $c = 1, k_0 = 1$ does not work well, as $k_1 = 2, k_2 = 5$, $k_3 = 26, k_4 = 677, k_5 = 4650, k_6 = 4649, k_7 = 1021, k_8 = 4649$, which does not yield a desired decomposition. Then, we replace c by 3. This time we get $k_0 = 1, k_1 = 4,, k_2 = 19, k_3 = 364, k_4 \equiv 2066, k_5 \equiv 3767, k_6 \equiv 1450$, $k_7 \equiv 4233, k_8 \equiv 3603$. Hence, $g_4 = 53$, which is a prime divisor of 5671. In this way, we obtain the complete factorization $266537 = 47 \cdot 53 \cdot 107$.

[38.4]   In detecting cycles, the use of the sequence $\{g_t\}$ is not always effective, as there is a possibility that the cycle starts at the very beginning. Thus, not only (38.1) but also $\langle k_v - k_0, q \rangle$ is better to be taken into account. For instance, in the case of $q = 1000009$, we first check $q \notin \mathrm{pp}(2)$ and start the ρ method with $c = 1, k_0 = 1$. Then, $k_1 = 2, k_2 = 5, k_3 = 26, k_4 = 677, k_5 = 458330, k_6 = 498325, k_7 = 570701, k_8 = 700138, k_9 = 807353, k_{10} = 293$, $k_{11} = 85850$. We find that $\langle k_{11} - k_0, q \rangle = 293$; namely, $k_j \equiv 1 \bmod 293$, $j \geq 11$. We get anyway the decomposition $1000009 = 293 \cdot 3413$, where both factors are primes. Incidentally, 1000009 has an interesting history: Euler (1749a, p.170) obtained the decomposition with a method presented in Note [78.11] below. However, later Euler (1774, p.91) erroneously included it in his table of prime numbers. A few years later he became aware of his oversight and published a correction (1778b). It is, thus, bizarre that Dickson (1919, p.361) made a remark on this article of Euler that is opposite to Euler's intension; an extremely rare error in the famous *History*.

[38.5]   A situation somewhat different from the above examples occurs with $q = 5428681$: We check $q \notin \mathrm{pp}(2)$ and start with $c = 1, k_0 = 1$ as usual. We have $k_{20} \equiv 5001013, k_{21} \equiv 2226654 \bmod q$; and $\langle k_{21} - k_{20}, q \rangle = 307$. The decomposition $k = 307 \cdot 17683$ follows. See Note [63.6] where 17683 is shown to be a prime.

§39

We return to the issue (33.10). We refine the condition on the residue a mod q; and on the premise (33.4) we shall consider

$$x^\ell \equiv a \bmod q, \quad \langle a, q \rangle = 1, \quad \ell \geq 2. \tag{39.1}$$

We start here an extensive discussion that will continue to the first half of the next chapter; then we shall begin at §57 a detailed study of the quadratic case, i.e., $\ell = 2$. Solutions to (39.1) are called ℓ^{th} roots of a mod q; and if roots exist, then a is called an ℓth power residue mod q.

Note that we are dealing with reduced residues. This is not really a restriction. For, with (33.10) lacking the condition $\langle a, q \rangle = 1$, we have $p^\nu \| \langle a, q \rangle \Rightarrow x = p^\eta y, p \nmid y, \ell \eta \geq \nu$. Thus, $p^{\ell \eta - \nu} y^\ell \equiv a p^{-\nu} \bmod q p^{-\nu}$. If $p | q p^{-\nu}$, then $\ell \eta - \nu = 0$; that is, $y^\ell \equiv a p^{-\nu} \bmod q p^{-\nu}$. On the other hand, if $p \nmid q p^{-\nu}$, then $y^\ell \equiv (\bar{p})^{\ell \eta - \nu} a p^{-\nu} \bmod q p^{-\nu}$, where $p \bar{p} \equiv 1 \bmod q p^{-\nu}$. Repeating this reduction procedure, we find that (39.1) is sufficiently general.

Next, if x_1, x_2 mod q satisfy (39.1), then $(x_1 x_2^{-1})^\ell \equiv 1 \bmod q$. This means that with a special solution x_0 mod q we may write all solutions in the form $x_0 \xi$ while ξ is to satisfy

$$x^\ell \equiv 1 \bmod q. \tag{39.2}$$

Hence, it becomes a basic issue to devise a method that yields a special solution. However, this is postponed to §45, and we shall consider (39.2) first. We then introduce the following notions (39.3) and (39.5): For an integer c, $\langle c, q \rangle = 1$,

$$\text{the order of } c \bmod q \text{ is } g \Leftrightarrow g = \min \{ u \in \mathbb{N} : c^u \equiv 1 \bmod q \};$$
$$\text{if } c^k \equiv 1 \bmod q, \text{ then } g | k; \text{ in particular, } g | \varphi(q). \tag{39.3}$$

According to Theorem 28, the order g exists always. In Note [32.4] we have used already the term order, but that was a use of a concept in group theory; as is well-known, the notion (39.3) came first in history and was followed by its generalization in group theory. As for the second line of (39.3), it suffices to divide k by g in the sense of (2.1). Also, we have

$$\text{the order of } c^f \bmod q \text{ is equal to } g / \langle f, g \rangle. \tag{39.4}$$

In fact, if $(c^f)^u \equiv 1 \bmod q$, then $g | fu$, and (6.3) gives this assertion.

Now, intuitively it is desirable that we are able to choose an element whose order is as large as possible, since then the structure of $(\mathbb{Z}/q\mathbb{Z})^*$ would become easier to describe. Hence, we make the following definition:

$$c \bmod q \text{ is a primitive root} \Leftrightarrow \text{the order of } c \bmod q \text{ equals } \varphi(q). \tag{39.5}$$

154   *Congruences*

In other words,

$$c \text{ is a primitive root mod } q$$
$$\Leftrightarrow \ \left\{ c^k \bmod q : k \bmod \varphi(q) \right\} = (\mathbb{Z}/q\mathbb{Z})^* . \tag{39.6}$$

Obviously, here $c^u \equiv c^v \bmod q \Leftrightarrow u \equiv v \bmod \varphi(q)$. That is, if a primitive root mod q exists, then $(\mathbb{Z}/q\mathbb{Z})^*$ is a cyclic group; and there are $\varphi(\varphi(q))$ primitive roots in total as is readily seen from (39.4) or rather Note [28.7].

However,

$$\text{not all the moduli have primitive roots.} \tag{39.7}$$

Details of this assertion will be given in Theorem 38, §42. Any modulus q with which $(\mathbb{Z}/q\mathbb{Z})^*$ is a cyclic group needs to belong to a special subset of $\mathbb{N}$. Reversing the way of reasoning, one may use this fact in order to specify the nature of the modulus, provided the existence of a primitive root is shown by some means. See the remark following (41.1); its applications are in Notes [41.3]–[41.6], which are also to augment the discussion in §§36–37.

Notes

[39.1]   In the case of a prime modulus, the concept of the order of a reduced residue class gradually became visible in Euler's articles, and in (1755, Theorema 5) it was clearly set forth, together with (39.3); he used the term *minima potestas*. Also, he (1772c, p.518) stated explicitly the concept of primitive roots with the coinage *radices primitivas*. That was a decisive step in the history of number theory.

[39.2]   If $a^{\alpha_j} \equiv 1 \bmod q$ for several exponents α_j, then by (7.1) we have $a^\beta \equiv 1 \bmod q$ with the greatest common divisor β of $\{\alpha_j\}$.

[39.3]   If the orders α, β of $a, b \bmod q$, respectively, are coprime, then the order of $ab \bmod q$ is equal to $\alpha\beta$. In fact, $(ab)^\gamma \equiv 1 \ \Rightarrow \ (ab)^{\alpha\gamma} \equiv 1 \ \Rightarrow \ b^{\alpha\gamma} \equiv 1 \bmod q$, and by the second line of (39.3) we have $\beta | \gamma$. In the same way, we have $\alpha | \gamma$. Hence, $\alpha\beta | \gamma$.

[39.4]   Let $\langle q_1, q_2 \rangle = 1$, $\langle a, q_1 q_2 \rangle = 1$, and γ_ν be the order of $a \bmod q_\nu$, $\nu = 1, 2$. Let γ be the order of $a \bmod q_1 q_2$. Then we have $\gamma = [\gamma_1, \gamma_2]$. For since $a^\gamma \equiv 1 \bmod q_\nu$, $\nu = 1, 2$, we see that γ is a common multiple of γ_1, γ_2; thus, $[\gamma_1, \gamma_2] | r$. On the other hand, $a^{[\gamma_1, \gamma_2]} \equiv 1 \bmod q_\nu$, $\nu = 1, 2$. By (28.3), $a^{[\gamma_1, \gamma_2]} \equiv 1 \bmod q$. Namely, $\gamma | [\gamma_1, \gamma_2]$.

§40

We now establish one of the most fundamental facts in number theory: For any prime modulus there exist primitive roots. In the proof below, we shall closely follow the argument of Legendre (1785, Théorème II).

Theorem 36 *For each $d|(p-1)$, there exist exactly $\varphi(d)$ residue classes mod p whose order is equal to d. In particular, there are $\varphi(p-1)$ primitive roots mod p.*

Proof First, we remark that Theorem 34 yields the following consequence:

$$
\text{the congruence equation } x^d \equiv 1 \bmod p
$$
$$
\text{has exactly } d \text{ mutually incongruent roots.}
$$
(40.1)

For we have the decomposition

$$
x^{p-1} - 1 = \left(x^d - 1\right)A(x), \quad A(x) = \sum_{v=0}^{(p-1)/d-1} x^{dv}. \tag{40.2}
$$

In this the congruence equation $x^{p-1} - 1 \equiv 0 \bmod p$ has exactly $p-1$ simple roots. On the other hand, by Theorem 34 the equation $x^d - 1 \equiv 0 \bmod p$ and $A(x) \equiv 0 \bmod p$ have at most d and $p-1-d$ roots, respectively. Hence, $x^d - 1 \equiv 0 \bmod p$ must have exactly d roots. More precisely,

$$
\text{any residue class whose } d^{\text{th}} \text{ power is equal to } 1 \bmod p
$$
$$
\text{is contained in these } d \text{ roots,}
$$
(40.3)

for otherwise $x^d - 1 \equiv 0 \bmod p$ would have more than d roots.

With this, let $p_1, p_2, \ldots, p_J$ be all the different prime factors of d. Also, for $t|d$ let $T(t)$ be the set of all roots of $x^t - 1 \equiv 0 \bmod p$. Then the set

$$
U(d) = T(d) - \bigcup_{j=1}^{J} T(d/p_j) \quad \text{(set-minus)} \tag{40.4}
$$

coincides with the set of all residue classes of order d. Indeed, if $U(d) \ni \omega \bmod p$ has the order $g < d$, then since (39.3) gives $g|d$ there exists p_k such that $p_k|(d/g)$ and $\omega^{d/p_k} = (\omega^g)^{d/gp_k} \equiv 1 \bmod p$. Namely, $T(d/p_k) \ni \omega \bmod p$, a contradiction. Conversely, any residue class of order d is contained in $U(d)$ since it is contained in $T(d)$ but not in any of $T(d/p_j)$.

Next, we have

$$
\bigcap_{h=1}^{H} T(d/p_{j_h}) = T(d/t), \quad t = p_{j_1}p_{j_2}\cdots p_{j_H}, \ |\mu(t)| = 1. \tag{40.5}
$$

156 *Congruences*

The right side is obviously contained in the left side. Conversely, if $\eta \bmod p$ is contained in the left side, then by Note [39.2] we have $\eta^f \equiv 1$ with $f = \langle d/p_{j_1}, d/p_{j_2}, \ldots, d/p_{j_H} \rangle = d/t$; hence, $\eta \in T(d/t)$. Thus, in much the same way as in the proof of Theorem 18, we obtain

$$|U(d)| = \sum_{t|d} \mu(t)|T(d/t)| = \sum_{t|d} \mu(t)\frac{d}{t} = \varphi(d). \qquad (40.6)$$

This ends the proof. An alternative proof is indicated in Note [40.1].

Notes

[40.1]   The group $(\mathbb{Z}/p\mathbb{Z})^*$ is cyclic. According to Note [30.6], this is equivalent to that $x^q \equiv 1 \bmod p$ has exactly q roots for any prime factor q of $p-1$. Hence, (40.1) immediately yields the existence of the primitive roots $\bmod p$. Then the combination of Notes [28.7] and [30.3] can be utilized to prove the first assertion of the last theorem. Compared with this, the above proof of the theorem is certainly more involved; however, it has its own merit, as will be seen later.

[40.2]   The assertion (40.1) is due to Lagrange (1773/1775, pp.777–778). The original statement is a little more general: If polynomials $A(x), B(x), C(x) \in \mathbb{Z}[x]$ satisfy the relation

$$x^{p-1} - 1 = A(x)B(x) + pC(x),$$
$$\deg(A) = m, \ \deg(B) = n; \ m + n = p - 1,$$

then the numbers of mutually incongruent roots of $A(x) \equiv 0, B(x) \equiv 0 \bmod p$ are m, n, respectively. Proof is the same as above. Lagrange himself made an application only to a problem concerning quadratic residues (see Note [43.4]). As has been already remarked in Note [34.1], Legendre (1785, pp.466–467) seems to be the first who explicitly recognized the fundamental importance of Theorem 34 in its generality together with the consequence (40.1) and its exploitation shown above.

[40.3]   The main part of Theorem 36 was stated by Euler (1772b, §53). In the subsequent article (1772c) he developed a discussion in the effort to prove his claim; but see the opinion of Gauss [DA, art.56].

[40.4]   In modern literature, Theorem 36 is attributed to Gauss because of his three proofs (see the succeeding two Notes). However, Legendre (1785) had already a proof which with a minor augmentation is shown above; he did not mention the notion of primitive roots, but it is of course immaterial here. Anyhow, those proofs are nothing else than easy consequences of Lagrange's observation (40.1). What is really essential is that these developments must have made young Gauss (1863a, (1) (posth.)) start his thorough investigation on (39.2); see the first paragraphs of our §64. Following Euler (1755, 1758a, 1772b, 1772c), he looked, from the outset, into the multiplicative structure among the solutions of $(39.2)_{q=p}$. A highly notable outcome of Gauss' investigation on this line is his theory of cyclotomic equations [DA, Sectio VII] (see our Note [33.2]); it depends indispensably on the existence of primitive roots modulo any prime, as will be shown in §65.

[40.5]   [DA, art.55] and Poinsot (1845, pp.65–67): For a prime number ϖ such that $\varpi^\nu \| d$, $d|(p-1)$, the number of solutions of $x^{\varpi^\nu} - 1 \equiv 0 \bmod p$ which does not satisfy $x^{\varpi^{\nu-1}} - 1 \equiv 0 \bmod p$ is $\varpi^\nu - \varpi^{\nu-1} = \varphi(\varpi^\nu)$, by (40.3). The residues thus counted are all of order ϖ^ν. The rest of the argument is an application of Note [39.3]. Still another proof is as follows ([DA, art.54]): If $\eta(h)$ denotes the number of residue classes of order h, then for any $d|(p-1)$ we have $\sum_{h|d} \eta(h) = d$. Hence, using the notation of §16, we have $\iota * \eta = \iota * \varphi$. Applying μ both sides, we get $\eta = \varphi$. See also Note [64.5].

[40.6]   In [DA, art.73] there is given an algorithm to find a primitive root, by an argument of trial and error: First, let the order of $a \bmod p$ be $\alpha > 1$. If $\alpha = p-1$, then $a \bmod p$ is a primitive root. Otherwise, let $b \bmod p$ be not contained in the set $\{1, a, \dots, a^{\alpha-1} \bmod p\}$. If its order is β, then by (40.3) we have $\beta \nmid \alpha$, which implies $[\alpha, \beta] > \alpha$. With this, construct $c \bmod p$ whose order is equal to $[\alpha, \beta]$: By either Note [14.2] or Note [14.3] we have the decomposition $[\alpha, \beta] = uv$ such that $\langle u, v \rangle = 1$, $u|\alpha$, $v|\beta$; then $c = a^{\alpha/u} b^{\beta/v}$ is the residue in question (Note [39.3]). The rest of the argument is omitted.

[40.7]   The discussion in the preceding Note indicates that the search for primitive roots $\bmod\ p$ is related to the prime power decomposition of $p-1$. To see this more closely, we consider here the modulus 983; in the next section we shall give further examples. Thus, we inspect $2^\nu \bmod 983$, but we do not need to compute all exponents ν since the order should be a divisor of $982 = 2 \cdot 491$ with 491 a prime. Namely, we need to compute $2^{491} \bmod 983$. To this end, we use (36.2). Thus $491 = (111101011)_2$; and

$$2^1, 2^2 = 4, 2^{2^2} = 16, 2^{2^3} = 16^2 = 256, 2^{2^4} = 256^2 \equiv -325, 2^{2^5} \equiv 325^2 \equiv 444,$$

$$2^{2^6} \equiv 444^2 \equiv -447, 2^{2^7} \equiv 447^2 \equiv 260, 2^{2^8} \equiv 260^2 \equiv -227 \bmod 983;$$

$$2^{491} \equiv 2 \cdot 4 \cdot 256 \cdot 444 \cdot (-447) \cdot 260 \cdot (-227) \equiv 1 \bmod 983.$$

Hence, the order of 2 mod 983 is 491. In the same way we find that 3 mod 983 also has the order 491. On the other hand,

$$5^1, 5^2 = 25, 5^{2^2} \equiv -358, 5^{2^3} \equiv 358^2 \equiv 374, 5^{2^4} = 374^2 \equiv 290,$$

$$5^{2^5} \equiv 290^2 \equiv 545, 5^{2^6} \equiv 545^2 \equiv 159, 5^{2^7} \equiv 159^2 \equiv -277, 5^{2^8} \equiv 277^2 \equiv 55;$$

$$5^{491} \equiv 5 \cdot 25 \cdot 374 \cdot 545 \cdot 159 \cdot (-277) \cdot 55 \equiv -1 \bmod 983.$$

That is, the order of 5 mod 983 is 982, and a primitive root has been found. Hence, 10 mod 983 is also a primitive root. In other words, starting from $v = 1$, the integer $10^v - 1$ becomes a multiple of 983 at $v = 982$ for the first time. As is explained in the next Note, this is the same as the fact that the decimal expansion of the fraction $\frac{1}{983}$ has the period which starts with the beginning of expansion and is of length 982.

[40.8] With $p \neq 2, 5$, the decimal expansion of $1/p$ is nothing else than to compute the quotients and residues of the division of 10^v, $v \geq 0$, by p. The residue cannot be 0 and their number is less than p, so $1/p$ is expanded into an infinite decimal number with periods; and if the order of 10 mod p is ℓ, the residues from the first to ℓ^{th} are mutually different, and only the first and the $(\ell + 1)^{\text{st}}$ are equal to 1. That is, the decimal expansion of $1/p$ has the period, which starts with the first digit, and is of length ℓ. More generally, for any $1 \leq a < p$ we have the periodic decimal expansion

$$\frac{a}{p} = \frac{at}{10^\ell - 1} = \sum_{v=1}^{\infty} \frac{at}{10^{v\ell}},$$

where $10^\ell - 1 = pt$, so $at < 10^\ell$. See Gauss [DA, arts.312–318]. For example, 10 mod 61 is a primitive root, and

$$\frac{1}{61} = 0.01639344262295081967213114754098360655737704918032786885245901 6...$$

Interestingly, in these 60 digits of the period each of $0, 1, 2, \ldots, 9$ appears 6 times. This is a typical instance of the following assertion (Anonymous (1864)):

> If 10 is a primitive root modulo a prime $p = 10s + 1$,
>
> then in the period of $1/p$ each digit appears s times.

To prove this, let r be the least residue among those immediately before the digit $\gamma \geq 1$ appears; note that $r \geq 2$, for $10r > \gamma p \geq 11$. Then $10r = \gamma p + u$, $1 \leq u \leq 9$, since if $10r = \gamma p + v$, $10 \leq v < p$, then $v = 10$ is naturally discarded, so $v \geq 11$, which would imply that $10(r - 1) = \gamma p + (v - 10)$, a contradiction to the choice of r. With this, let $r + w$ be a residue corresponding to the digit γ so that $10(r + w) < (\gamma + 1)p$. We have $10w < p - u$, and $w < s - (u - 1)/10$ or $w \leq s - 1$. Conversely, with $0 \leq w \leq s - 1$, we have

$$\gamma p + u \le 10(r+w) \le (\gamma+1)p - (11-u).$$ That is, $\left[10(r+w)/p\right] = \gamma$. We have shown that the set of residues immediately before getting the digit $\gamma \ge 1$ are exactly $\{r, r+1, \ldots, r+s-1\}$; indeed, all of these residues appear because $10 \bmod p$ is assumed to be a primitive root. It remains to consider the case $\gamma = 0$, but it now follows from the other nine cases. The rest of the argument is omitted. The cases $p \equiv 3, 7, 9 \bmod 10$ are discussed by Sardi (1869).

[40.9] The identification of primitive roots is one of *profundissima numerorum mysteria* if we borrow an expression of Euler (1772b, §48; see [DA, art.73] as well). It is widely conjectured that the size of the least primitive root $g_p \bmod p$ should be $O((\log p)^2)$, that is, extremely small against p. According to T. Oliveira e Silva (2015),

$$\frac{|\{g_p = 2 : p < 10^{16}\}|}{\pi(10^{16})} = \frac{104422801358700}{279238341033925} \doteqdot 0.37395581485.$$

[40.10] In 1927 E. Artin set out the conjecture that any integer which is equal to neither ± 1 nor a square should be a primitive root modulo infinitely many primes. This is called the Artin conjecture on primitive roots, and still remains open as one of the major problems in number theory. If $a \ge 2$ is square-free and $\not\equiv 1 \bmod 4$, then the conjecture states in fact that

$$\lim_{x\to\infty} |\{\text{prime} \le x \text{ which has } a \text{ as a primitive root}\}|/\pi(x)$$

$$= \prod_p \left(1 - \frac{1}{p(p-1)}\right) \doteqdot 0.37395581362.$$

This is to be compared with the above statistical data. Hooley (1967) proved the conjecture under an extension of RH to certain algebraic number fields.

§41

The following simple observation, which in this general expression is essentially due to Legendre (1785) and was already used in confirming (40.4), has important applications:

$$\begin{aligned} &\text{the order of } a \bmod q \text{ is } d \\ \Leftrightarrow\quad &a^d \equiv 1 \bmod q \text{ and } a^{d/\varpi} \not\equiv 1 \bmod q \text{ for any prime } \varpi\,|\,d. \end{aligned} \tag{41.1}$$

In particular, if this turns out to hold with $d = q-1$, then $\varphi(q) = q-1$, and q is concluded to be a prime; also $a \bmod q$ is a primitive root. Note that then we have $a^{(q-1)/2} \equiv -1 \bmod q$ since $(a^{(q-1)/2} - 1)(a^{(q-1)/2} + 1) \equiv 0 \bmod q$.

Notes

[41.1] Two famous applications of (41.1) prior to Legendre's above assertion:

$$M_{37} = 2^{37} - 1 = 223 \cdot 616318177 \quad \text{Fermat (1640; 1679, p.177)},$$
$$F_5 = 2^{2^5} + 1 = 641 \cdot 6700417 \quad \text{Euler (1732a)}.$$

First, let s be a prime divisor of M_{37}; then by (41.1) the order of 2 mod s is equal to 37 since 37 is a prime. Hence, $s \equiv 1$ mod 37; and more precisely $s \equiv 1$ mod 74, as s is odd. Now, M_{37} is not divisible by the prime 149, and the prime 223 is to be checked, which was Fermat's reasoning (Œuvres II, p.199). Next, let t be a prime factor of F_5; then $2^{64} \equiv 1$ mod t. Hence by (41.1) the order of 2 mod t is 64; that is, $t \equiv 1$ mod 64. Primes to be checked are $193, 257, 449, 577, 641, \ldots$; see (Euler, 1747, Th. 8 and p.60; 1760). In order to show that Euler's quotient 6700417 is also a prime, it suffices to check that it is not divisible any of the seven primes $641, 769, 1153, 1217, 1409, 1601, 2113 \leq \left[6700417^{1/2} \right]$ (Note [10.6]). Actually, we could impose the stricter condition $t \equiv 1$ mod 128, so only $257, 641, 1409$ are relevant; see Note [58.8]. As for $M_{37}/223$, see Note [41.6].

[41.2] The above test for the primality of the modulus q is sometimes called Lucas' test (1891, p.441) despite an apparent historical inconsistency. Although this test requires the prime power decomposition of $q - 1$, it often turns out to be effective. We shall give a few examples below. See also Note [63.7].

[41.3] We shall show that $q = 430883$ is a prime. By Note [10.6] one may confirm this by using 119 primes or rather 327 positive odd integers less than $[q^{1/2}] = 656$. Here we shall instead apply (41.1) and the modular exponentiation (36.2), although we skip details: We have the decomposition $q - 1 = 2 \cdot 17 \cdot 19 \cdot 23 \cdot 29$, and

$$2^{(q-1)/2} \equiv -1, \ 2^{(q-1)/17} \equiv -17592, \ 2^{(q-1)/19} \equiv 205285,$$
$$2^{(q-1)/23} \equiv -111925, \ 2^{(q-1)/29} \equiv -59674 \ \text{mod } q.$$

Hence, the order of 2 mod q is equal to $q - 1$, and q is a prime. In particular, 2 mod q is a primitive root. In the same way one may conclude that 5 mod q is also a primitive root. Since $2^{(q-1)/2}, 5^{(q-1)/2} \equiv -1$ mod q, we have $10^{(q-1)/2} \equiv 1$; and confirming that $10^{(q-1)/34} \not\equiv 1$, etc., we find the order of 10 mod q is equal to $(q-1)/2$. The decimal expansion of the fraction $1/430883$ has the period of 215441 digits. By the way, because of $31^{(q-1)/2} \equiv -1$, one might infer that 31 mod q were a primitive root; but this is wrong, as $31^{(q-1)/17} \equiv 1$. In fact, the order of 31 mod q is equal to $(q-1)/17$. See Note [43.9].

[41.4]   As for $q = 430897$, the order of 10 mod q is equal to $q - 1$; hence, q is a prime. Indeed, $q - 1 = 2^4 \cdot 3 \cdot 47 \cdot 191$, and

$$10^{(q-1)/2} \equiv -1, \ 10^{(q-1)/3} \equiv 30302,$$
$$10^{(q-1)/47} \equiv 196589, \ 10^{(q-1)/191} \equiv 72658 \bmod q.$$

[41.5]   As for $q = 122761$, we have $q - 1 = 2^3 \cdot 3^2 \cdot 5 \cdot 11 \cdot 31$, and find that by (41.1) any of a mod q, $1 < a \leq 12$, is not a primitive root since $a^{(q-1)/2} \equiv 1 \bmod q$. Then, 13 mod q turns out to be of order $q - 1$. Hence, 122761 is a prime.

[41.6]   That $q = 616318177 = M_{37}/223$ is a prime is due to the fact that the order of 5 mod q is equal to $q - 1$. This time we take 10^2-adic expansion. Appropriate combinations of $5^{10^2} \equiv 113866312$, $5^{10^4} \equiv 274589582$, $5^{10^6} \equiv 247016482$, $5^{10^8} \equiv 119353590 \bmod q$ and the decomposition $q - 1 = 2^5 \cdot 3 \cdot 37 \cdot 167 \cdot 1039$ yield

$$5^{(q-1)/2} \equiv -1, \ 5^{(q-1)/3} \equiv 46907798, \ 5^{(q-1)/37} \equiv 4194304,$$
$$5^{(q-1)/167} \equiv 373111825, \ 5^{(q-1)/1039} \equiv 136077913 \bmod q.$$

§42

For a prime power modulus, the assertion of Theorem 36 is replaced by the following. A singular nature of the prime 2 becomes apparent.

Theorem 37 *For any odd prime p,*

there exists a primitive root r mod p such that $r^{p-1} \not\equiv 1 \bmod p^2$.   (42.1)

With this, r mod p^α is a primitive root for any $\alpha \geq 1$; and a reduced residue system is given by $\{r^u \bmod p^\alpha : u \bmod \varphi(p^\alpha)\}$. On the other hand, although 1 mod 2 and -1 mod 4 are primitive roots, there exists no primitive root with respect to the modulus 2^α, $\alpha \geq 3$, and a reduced residue system is given by

$$(-1)^v 5^w \bmod 2^\alpha, \quad v \bmod 2, \ w \bmod 2^{\alpha-2}.$$ (42.2)

Proof   For an odd prime p, if $r^{p-1} \equiv 1 \bmod p^2$, $1 < r < p$, then one may replace r by $p - r$, for

$$(p - r)^{p-1} \equiv r^{p-1} - (p - 1)r^{p-2}p \equiv 1 + r^{p-2}p \not\equiv 1 \bmod p^2.$$ (42.3)

Then, under (42.1) we have $r^{p^\beta (p-1)} = 1 + t_\beta p^{\beta+1}$, $p \nmid t_\beta$. This is obvious when $\beta = 0$. Also, $r^{p^{\beta+1}(p-1)} = (1 + t_\beta p^{\beta+1})^p \equiv 1 + t_\beta p^{\beta+2} \bmod p^{\beta+3}$, for $\beta \geq 1$.

162 *Congruences*

Hence, by induction, the order of r mod p^α is equal to $p^{\alpha-1}(p-1) = \varphi(p^\alpha)$. This ends the proof of the first half of the assertion of the theorem. As for the second half, we need to consider the case $\alpha \geq 3$. We have $(1 + 2s)^{2^\beta} \equiv 1 \bmod 2^{\beta+2}$ for any $\beta \geq 1$ and $s \in \mathbb{Z}$, as this holds with $\beta = 1$, and by induction

$$\begin{aligned}(1 + 2s)^{2^{\beta+1}} &= (1 + 2^{\beta+2}s_1)^2 \\ &= 1 + 2^{\beta+3}s_1(1 + 2^{\beta+1}s_1) \equiv 1 \bmod 2^{\beta+3}.\end{aligned} \tag{42.4}$$

That is, $(1 + 2s)^{2^{\alpha-2}} \equiv 1 \bmod 2^\alpha$ for $\alpha \geq 3$. We see that with respect to the modulus 2^α, $\alpha \geq 3$, no odd integer can have the order $\varphi(2^\alpha) = 2^{\alpha-1}$; hence, primitive roots mod 2^α do not exist. We then take $s = 2$. By induction, we see readily that s_1 is odd. Hence, the order of 5 mod 2^α is equal to $\varphi(2^\alpha)/2$. On noting that any reduced residue system mod 2^α is composed of odd integers which are congruent to $\pm 1 \bmod 4$, we conclude that they are congruent to either 5^w or -5^w. This ends the proof.

With this, we are now able to make the assertion (39.7) precise: [DA, arts.82–92] states that

Theorem 38 *Primitive roots* mod q *exist if and only if* q *is equal to one of*

$$\{2, 4, p^\alpha, 2p^\alpha\}, \tag{42.5}$$

where $p \geq 3$, $\alpha \geq 1$ *are arbitrary. In other words,* $(\mathbb{Z}/q\mathbb{Z})^*$ *is cyclic with these* q *only.*

Proof The existence of primitive roots mod q is trivial for $q = 2, 4$, and has been proved for $q = p^\alpha$ in Theorem 37. Since $\varphi(2p^\alpha) = \varphi(p^\alpha)$, we see that $\{r^u : 1 \leq u \leq \varphi(p^\alpha)\}$ with a primitive root r mod p^α, $2 \nmid r$, is a reduced residue system mod $2p^\alpha$; if $2|r$, then replace r by $p^\alpha - r$. Hence, for moduli of these four types, there exist primitive roots. The case $q = 2^\beta$, $\beta \geq 3$, is excluded by Theorem 37. If $q > 1$ does not belong to any of these five types of positive integers, we have a decomposition $q = q_1 q_2$, $\langle q_1, q_2 \rangle = 1$, $q_1, q_2 \geq 3$. For any a coprime to q, we have $a^{[\varphi(q_1), \varphi(q_2)]} \equiv 1 \bmod q$ by (28.3) and (28.10); the order of a mod q is not greater than $[\varphi(q_1), \varphi(q_2)]$. According to (10.6), we have $[\varphi(q_1), \varphi(q_2)] = \varphi(q)/\langle \varphi(q_1), \varphi(q_2) \rangle < \varphi(q)$ since $\langle \varphi(q_1), \varphi(q_2) \rangle \geq 2$. This ends the proof.

Notes

[42.1] Although 18 mod 37 is a primitive root, we have $18^{36} \equiv 1 \bmod 37^2$ (Note [28.4]). Then, $(18 + 37)^{36} \equiv 1 + 2 \cdot 37 \bmod 37^2$. By the way, in the case of Note [40.7], $10^{982} \not\equiv 1 \bmod 983^2$. Hence, 10 mod 983^2 is a primitive root, and the period of the decimal expansion of $\frac{1}{983^2}$ is equal to $983 \cdot 982 = 965306$.

[42.2]   An alternative proof of Theorem 38: Let $q = p_1^{q(p_1)} p_2^{q(p_2)} \cdots p_J^{q(p_J)}$. For any a coprime to q, we have $a^{\varphi(p_j^{q(p_j)})} \equiv 1 \bmod p_j^{q(p_j)}$. Thus, with the least common multiple L of $\{\varphi(p_j^{q(p_j)}) : j \leq J\}$, we have $a^L \equiv 1 \bmod q$. Hence, if $a \bmod q$ is a primitive root, then we should have $L = \varphi(q)$ since $\varphi(q)|L$ and $L \leq \varphi(q)$. It follows, via (14.4), that $\langle \varphi(p_j^{q(p_j)}), \varphi(p_k^{q(p_k)}) \rangle = 1, j \neq k$, which implies that q cannot have two different odd prime factors; and $q = 2^a p^b$, $p \geq 3$. The rest of argument is omitted.

[42.3]   Still alternatively, one may use the theory of Abelian groups: Take $A = (\mathbb{Z}/q\mathbb{Z})^*$ in Note [30.6]. Then, $\alpha_2 \geq 2$ if q has two different odd prime factors, as is readily seen via (32.2). The rest of argument is omitted.

[42.4]   If q is as in (42.5), then we have

$$\prod_{\substack{1 \leq a < q \\ \langle a, q \rangle = 1}} a \equiv -1 \bmod q.$$

This is a part of an extension of (35.2) by Gauss [DA, art.78]. We shall give a proof different from his. We treat only the case $q = 2p^\alpha$, as other cases are analogous and simpler. Thus, let r be as in the proof of the last theorem. Then the product is congruent to r^Q, with $Q = \sum_{j=1}^{\varphi(q)} j = (\varphi(q)/2)(\varphi(q)+1)$. On the other hand, $\left(r^{\varphi(q)/2}+1\right)\left(r^{\varphi(q)/2}-1\right) \equiv 0 \bmod q$. Here $r^{\varphi(p^\alpha)/2} \not\equiv 1 \bmod p$. For otherwise the order of $r \bmod p$ would be a divisor of $(p-1)/2$, a contradiction. Hence, $r^{\varphi(q)/2} \equiv -1 \bmod p^\alpha$; and thus $r^{\varphi(q)/2} \equiv -1 \bmod q$, as r is odd. On noting that $\varphi(q) + 1$ is odd, we end the proof. We have got an alternative poof of (35.2) as well.

§43

Having established the existence of primitive roots with respect to the moduli specified in Theorem 38, we may now return to the basic problem (39.1).

Since the premise (33.4) is in effect, the following theorem asserts that one may decide quantitatively the existence of solutions solely from the parameters $\{a, \ell, q\}$; when dealing with (39.1) for composite moduli, one should apply (33.6). We stress that the issue to construct explicit solutions belongs to a different context; two general methods for that purpose will be given in §§45 and 68.

164 *Congruences*

Theorem 39 *Let $2 \leq \ell \in \mathbb{N}$, and let q be one of the following integers*

$$(1)\ 2,\ 4,\ p^{\alpha},\ 2p^{\alpha}\ \text{with}\ p \geq 3,\ \alpha \geq 1,$$
$$(2)\ 2^{\alpha}\ \text{with}\ \alpha \geq 3,\ \langle \ell, 2^{\alpha-2} \rangle = 2^{\gamma}. \qquad (43.1)$$

Then the congruence equation $x^{\ell} \equiv a \bmod q$, $\langle a, q \rangle = 1$, has

$$(1): \begin{cases} \text{(A)} \ \langle \ell, \varphi(q) \rangle \ \text{solutions} & \text{if } a^{\varphi(q)/\langle \ell, \varphi(q) \rangle} \equiv 1 \bmod q, \\ \text{(B)} \ \text{no solution} & \text{otherwise.} \end{cases} \qquad (43.2)$$

$$(2): \begin{cases} \text{(C)} \ \text{only one solution} & \text{if } \gamma = 0, \\ \text{(D)} \ 2^{\gamma+1} \ \text{solutions} & \text{if } a \equiv 1 \bmod 2^{\gamma+2},\ \gamma \geq 1, \qquad (43.3) \\ \text{(E)} \ \text{no solution} & \text{otherwise.} \end{cases}$$

Proof

(1) Let $r \bmod q$ be a primitive root. If we put $x \equiv r^{y}$, $a \equiv r^{h} \bmod q$, then by (39.6) the congruence equation is equivalent to $\ell y \equiv h \bmod \varphi(q)$. Thus by (29.4) we need to have $\langle \ell, \varphi(q) \rangle | h$. With this, the number of solutions $y \bmod \varphi(q)$ is $\langle \ell, \varphi(q) \rangle$, which means that the congruence equation in question has $\langle \ell, \varphi(q) \rangle$ solutions $x \bmod q$. Further, if $\langle \ell, \varphi(q) \rangle | h$, then $a^{\varphi(q)/\langle \ell, \varphi(q) \rangle} \equiv (r^{h/\langle \ell, \varphi(q) \rangle})^{\varphi(q)} \equiv 1 \bmod q$. Conversely, if $1 \equiv r^{h\varphi(q)/\langle \ell, \varphi(q) \rangle} \bmod q$, then $\varphi(q)|(h\varphi(q)/\langle \ell, \varphi(q) \rangle)$; that is, $\langle \ell, \varphi(q) \rangle | h$. Assertions (A), (B) have been proved.

(2) If we put $x \equiv (-1)^{s}5^{t}$, $a \equiv (-1)^{f}5^{g} \bmod 2^{\alpha}$, then $\ell s \equiv f \bmod 2$, $\ell t \equiv g \bmod 2^{\alpha-2}$.

(C) Since $\gamma = 0$ implies $2 \nmid \ell$, both $s \bmod 2$ and $t \bmod 2^{\alpha-2}$ are uniquely determined.

(D) We have $2|\ell$, $f = 0$, and $2^{\gamma}|g$, i.e., $\langle \ell, 2^{\alpha-2} \rangle | g$. So $s \equiv 0, 1 \bmod 2$ are both solutions; and there are 2^{γ} solutions $t \bmod 2^{\alpha-2}$. The congruence equation in question has $2^{\gamma+1}$ solutions. See Note [57.6].

(E) We have $\gamma \geq 1$, $a \not\equiv 1 \bmod 2^{\gamma+2}$. If $f = 1$, then $s \bmod 2$ does not exist since $2|\ell$. On the other hand, if $f = 0$, then $t \bmod 2^{\alpha-2}$ does not exist since $a \equiv 5^{g} \not\equiv 1 \bmod 2^{\gamma+2}$ implies that $2^{\gamma} \nmid g$, i.e., $\langle \ell, 2^{\alpha-2} \rangle \nmid g$.

This ends the proof.

Now, with $q = p \geq 3$, the essential content of the theorem becomes as follows: If $p - 1 = \ell w$, then

$$x^{w} \equiv 1 \bmod p \text{ has } w \text{ solutions: } \{s_1, s_2, \ldots, s_w\} \bmod p;$$
$$y^{\ell} \equiv s_j \text{ has } \ell \text{ solutions: } \{t_{j,1}, t_{j,2}, \ldots, t_{j,\ell}\} \bmod p, \text{ and}$$
$$\text{these } w \text{ sets of } \ell \text{ residues cover a reduced} \qquad (43.4)$$
$$\text{residue system} \bmod p \text{ without overlapping.}$$

$$\S43 \qquad\qquad 165$$

Namely, if we raise each of $1, 2, \ldots, p - 1$ to the ℓth power, then we get w incongruent ℓ^{th} power residues mod p. For instance, there are $(p - 1)/2$ quadratic ($\ell = 2$) residues; and if $p \equiv 1 \bmod 3$, then there are $(p - 1)/3$ cubic ($\ell = 3$) residues, so on. When a prime modulus $p \equiv 1 \bmod \ell$ is taken, it is a simple matter to see whether a given residue mod p is an ℓth residue or not. However, here a deep reciprocity issue arises: when an arbitrary pair $\{\ell, a\}$ of two positive integers is given,

> find an algorithm to fix the set of all primes p such that
>
> $a \bmod p$ is an ℓth power residue. $\qquad(43.5)$

For the case of quadratic residues, we shall give a complete solution to this by the discussion which is developed in §§57–69. That is one of the outstanding achievements in number theory, mainly due to Gauss (the contents of our Theorem 59) and Dirichlet (the statement (62.9)). On the other hand, to deal with any exponent $\ell \geq 3$ one has to step in an advanced theory of algebraic numbers, which is beyond the scope of the present essays; but see Notes [43.11], [65.6], [90.6], and [93.4].

Notes

[43.1] Theorem 39 is customarily attributed to Euler. However, he considered only the case of prime moduli. The necessity assertion is shown in Euler (1747, Theorema 13), which is of course a simple application of (28.11). In the same article (64. Scholion) the sufficiency is conjectured, and it is confirmed in Euler (1755, Theorema 19); note that he did not use primitive roots (see the next Note). As is stated clearly in Legendre (1785, Théorème I), an application of Lagrange's theorem (Note [40.2]) gives a simple proof, even including the number of solutions (see Note [43.4]). Theorem 39 itself is contained in [DA, Sectio III], but the above presentation is adopted from Arndt (1846).

[43.2] We shall show a slightly modified version of the argument of Euler (1755, Theorema 19); note that this works only for prime moduli. We assume that $p - 1 = \ell w$ and $a^w \equiv 1 \bmod p$. We are going to show that $x^\ell \equiv a \bmod p$ has a solution. We suppose that there exists no solution. Then, since $(y^\ell)^w - a^w \equiv 0 \bmod p$ holds for $y = 1, 2, \ldots, p - 1$, we find via the decomposition $(y^\ell)^w - a^w = (y^\ell - a)F_0(y)$ that $F_0(y) \equiv 0 \bmod p$ for any of these y. Hence, in the difference $F_0(y) - F_0(1) = (y - 1)F_1(y)$ we see that $F_1(y) \equiv 0 \bmod p$ for $y = 2, 3, \ldots, p - 1$. In the same way, in $F_1(y) - F_1(2) = (y - 2)F_2(y)$ we have $F_2(y) \equiv 0 \bmod p$ for $y = 3, 4, \ldots, p - 1$. Repeating the procedure, we reach a contradiction. Euler was proud of his differencing argument; see the closing remark of this remarkable article (1755).

166 *Congruences*

[43.3] The statement (43.4) is taken from Smith (1859, art.12). In the footnote there the articles Euler (1736, 1755, 1758a, 1772b, 1772c) are listed up. As already suggested in Notes [40.3]–[40.4], these form the core of Euler's theory of numbers; namely, an indisputable source of today's number theory. In passing, we remark that in terms of the theory of cyclic groups the assertion (43.4) is a triviality; however, that is not all of the story.

[43.4] Lagrange (1773/1775, pp.778–779) dealt with the case $\ell = 2, q = p$ of (43.2). It is easy to extend this argument of his to the general situation with $\ell | (p - 1)$. Indeed, provided $a^{(p-1)/\ell} \equiv 1 \bmod p$, we have $x^{p-1} - 1 \equiv (x^\ell)^{(p-1)/\ell} - a^{(p-1)/\ell} \bmod p$ and the right side is divisible by $x^\ell - a$; hence, by Note [40.2], $x^\ell = a \bmod p$ has ℓ solutions. The difference between this and the above argument of Euler might look small, but the clarity of the conclusion by Lagrange's argument is remarkable. Yet, his argument works only for prime moduli, which is the same as Euler's; to deal with prime-power moduli one needs Theorem 38. By the way, Lagrange's theory of numbers came into existence out of Euler's strenuous researches. In his articles and letters Lagrange often acknowledged Euler's contributions with deep admiration. Lagrange's discussions are transparent and penetrating; studying them is a delight. This will become just more apparent in the discussion on quadratic forms which will be developed in Chapter 4.

[43.5] For primes ℓ and p, the congruence equation $x^\ell \equiv a \bmod p$ with $p \nmid a$ becomes worth considering only when $p \equiv 1 \bmod \ell$ since if $p \not\equiv 1 \bmod \ell$, then every a, $p \nmid a$, is an ℓth power residue $\bmod p$. This is an immediate consequence of (43.2), but we may argue directly: If $\langle p - 1, \ell \rangle = 1$, then there exists a μ such that $\ell \mu \equiv 1 \bmod (p - 1)$; and with $x_0 \equiv a^\mu \bmod p$, we have $x_0^\ell \equiv a^{\ell\mu} \equiv a \bmod p$. More generally, if $\langle k, \varphi(q) \rangle = 1$, then taking $k\lambda \equiv 1 \bmod \varphi(q)$ we find that a solution of $x^k \equiv a \bmod q$ is $x \equiv a^\lambda \bmod q$. Here q is arbitrary, that is, not restricted as in Theorem 39; however, this is superficial, for the validity of $\langle k, \varphi(q) \rangle = 1$ cannot be checked without having the prime power decomposition of q, at least presently.

[43.6] Continuing the above, we consider $x^\ell \equiv a \bmod p^\alpha$. Provided $p \equiv 1 \bmod \ell$, this congruence is reduced to the case $\alpha = 1$. Namely, it is solvable if and only if $a^{(p-1)/\ell} \equiv 1 \bmod p$, and the number of solutions $\bmod p^\alpha$ is equal to ℓ for any $\alpha \geq 1$ because of the Hensel lifting (Note [34.3]).

[43.7] The congruence equation $x^7 \equiv 7 \bmod 983$: This belongs to the case (1) of Theorem 39 or rather Note [43.5]. Hence, it has only one solution. Recycling the computation in Note [40.7] concerning the primitive root $5 \bmod 983$, we find rather fortunately that $5^{100} \equiv 7 \bmod 983$, which is due to $100 = (1100100)_2$. We solve $7y \equiv 100 \bmod 982$ by Euclid's algorithm; since

$7 \cdot 421 - 982 \cdot 3 = 1$, we get $y \equiv 421 \cdot 100 \equiv 856 \bmod 982$. Then, on noting $856 = (1101011000)_2$, we find that $x \equiv 5^{856} \equiv 589 \bmod 983$. The confirmation is: $589^7 \equiv 589 \cdot 589^2 \cdot (589^2)^2 \equiv 589 \cdot (-78) \cdot 78^2 \equiv 259 \cdot 186 \equiv 7 \bmod 983$.

[43.8] The congruence equation $x^6 \equiv 89 \bmod 1072$ has 24 solutions. To see this, we note first the prime power decomposition $1072 = 2^4 \cdot 67$. We are led to (i) $x_1^6 \equiv 9 \bmod 2^4$ and (ii) $x_2^6 \equiv 22 \bmod 67$. As for (i), (43.3)(D) is applicable with $\gamma = 1$. Hence, the number of solutions $x_1 \bmod 2^4$ is 4. In fact, $x_1 \equiv \pm 5^u$ and $5^{6u} \equiv 5^2 \bmod 2^4$; equivalently, $6u \equiv 2 \bmod 2^2$. Thus $u = 1,3 \bmod 2^2$, and $x_1 \equiv \pm 5, \pm 5^3 \bmod 2^4$. As for (ii), since $22^{11} \equiv 1 \bmod 67$, we see, by (43.2)(A), the number of solutions is 6. In fact, 2 mod 67 is a primitive root; and on noting $2^{60} \equiv 22 \bmod 67$, we are led to $6y \equiv 60 \bmod 66$, where $x_2 = 2^y$. We find that $y \equiv 10, 21, 32, 43, 54, 65 \bmod 66$; and $x_2 \equiv 19, 52, 33, 48, 15, 34 \bmod 67$, respectively. We combine these assertions by means of (30.4). By Euclid's algorithm or (21.14), we have $21 \cdot 16 - 5 \cdot 67 = 1$; and $x \equiv 21 \cdot 16 \cdot x_2 - 5 \cdot 67 \cdot x_1 \bmod 1072$. Therefore, $x \equiv \pm 19, \pm 101, \pm 115, \pm 149, \pm 235, \pm 253, \pm 283, \pm 301, \pm 387, \pm 421, \pm 435, \pm 517 \bmod 1072$. For instance, $387^6 = 89 + 3133798041160 \cdot 1072$.

[43.9] The congruence equation $x^{17} \equiv 31 \bmod p$ with $p = 430883$. Using the computation of Note [41.3] again, we find that $17 \mid (p - 1)$ and $31^{(p-1)/\langle 17, p-1 \rangle} \equiv 1 \bmod p$. Hence, by (43.2)(A), there exists 17 solutions. However, this time the situation is different from those of the last two Notes: it is not easy to explicitly compute the solutions. The reason for this is in the difficulty to find $u \bmod p - 1$ such that $2^u \equiv 31 \bmod p$. Here we shall treat this problem by exploiting the peculiar situation provided by $p - 1 = 2 \cdot 17 \cdot 19 \cdot 23 \cdot 29$. Thus, we put $u \equiv u_2 \bmod 2, u \equiv u_{17} \bmod 17, u \equiv u_{19} \bmod 19, u \equiv u_{23} \bmod 23, u \equiv u_{29} \bmod 29$. Then,

$$2^{u_2(p-1)/2} \equiv 2^{u(p-1)/2} \equiv 31^{(p-1)/2} \equiv -1,$$

$$2^{u_{17}(p-1)/17} \equiv 2^{u(p-1)/17} \equiv 31^{(p-1)/17} \equiv 1,$$

$$2^{u_{19}(p-1)/19} \equiv 2^{u(p-1)/19} \equiv 31^{(p-1)/19} \equiv 261760,$$

$$2^{u_{23}(p-1)/23} \equiv 2^{u(p-1)/23} \equiv 31^{(p-1)/23} \equiv 379692,$$

$$2^{u_{29}(p-1)/29} \equiv 2^{u(p-1)/29} \equiv 31^{(p-1)/29} \equiv 139143 \bmod p.$$

Since $2^{(p-1)/2} \equiv -1 \bmod p$, we have $u_2 = 1$. Also, $u_{17} = 0$ is obvious. Next, from Note [41.3] we have $2^{(p-1)/19} \equiv 205285$. Computing $205285^v \bmod p$, we find that $205285^5 \equiv 261760 \bmod p$; that is, $u_{19} = 5$. In just the same way, we get $u_{23} = 12$, $u_{29} = 10$. From these we shall compute $u \bmod (p - 1)$ by (30.4). Euclid's algorithm is used to compute

$$\left\langle \tfrac{1}{2}(p-1), \tfrac{1}{17}(p-1), \tfrac{1}{19}(p-1), \tfrac{1}{23}(p-1), \tfrac{1}{29}(p-1) \right\rangle$$

so that

$$215441 - 25346 - 12 \cdot 22678 + 25 \cdot 18734 - 26 \cdot 14858 = 1.$$

Hence,

$$\begin{aligned}
u &\equiv 215441 \cdot u_2 - 25346 \cdot u_{17} - 12 \cdot 22678 \cdot u_{19} \\
&\quad + 25 \cdot 18734 \cdot u_{23} - 26 \cdot 14858 \cdot u_{29} \\
&\equiv 180999 \bmod (p-1) \quad \Rightarrow \quad 2^{180999} \equiv 31 \bmod p.
\end{aligned}$$

Returning to the original congruence equation, we put $x \equiv 2^\xi \bmod p$; then $17\xi \equiv 180999 \bmod (p-1)$. Solving this, we have $\xi \equiv 10647 + 25346j \bmod (p-1)$, $0 \leq j \leq 16$. In this way, we find the following 17 solutions:

$$\begin{aligned}
x \equiv \ &22837, 31639, 34899, 65067, 107548, 154358, 197467, 250066, \\
&265335, 369281, 370165, 386426, 389613, 413102, \\
&413177, 414868, 422982 \ \bmod 430883.
\end{aligned}$$

We shall show an alternative argument in Note [45.4], which is much simpler than the present one.

[43.10]   The algorithm of the last Note can be extended to the general situation $(43.1)_{q=p}$ where $p-1$ has the prime power decomposition $p_1^{e_1} p_2^{e_2} \cdots p_k^{e_k}$. Those coefficients corresponding to u_{17}, u_{29}, etc., are represented as the expansions $u \equiv \lambda_{j,0} + \lambda_{j,1} p_j + \cdots + \lambda_{j,e_j-1} p_j^{e_j-1} \bmod p_j^{e_j}$, $0 \leq \lambda_{j,v} < p_j$. Taking a primitive root $r \bmod p$, one may fix the coefficients $\lambda_{j,v}$ recursively via

$$a^{(p-1)/p_j^{v+1}} \equiv r^{(\lambda_{j,0} + \lambda_{j,1} p_j + \cdots + \lambda_{j,v} p_j^v)(p-1)/p_j^{v+1}} \bmod p, \quad 0 \leq v \leq e_j - 1.$$

Hence, provided $r \bmod p$ and the prime power decomposition of $p-1$ are known, the exponent u such that $r^u \equiv a \bmod p$ can be identified, at least theoretically.

[43.11]   Euler considered the problem (43.5) for $\ell \geq 3$, although with limited combinations of ℓ and a: In his posthumous article (1849a), it is apparently conjectured that

$$\begin{aligned}
x^3 &\equiv 2 \bmod p \ \Leftrightarrow \ p \equiv 1 \bmod 6, \ p = u^2 + 27v^2, \\
x^4 &\equiv 2 \bmod p \ \Leftrightarrow \ p \equiv 1 \bmod 8, \ p = w^2 + 64z^2,
\end{aligned}$$

with $u, v, w, z \in \mathbb{Z}$; see §§408, 458 there. The necessity of these algebraic constraints in terms of quadratic forms induced a deep development in algebraic number theory that is called the theory of general reciprocity laws.

Gauss (1828/1832) was the first who achieved a real contribution in this field. See Notes [65.6] and [93.4] below.

[43.12] Definition (37.3) means that

$$\text{an odd composite } q \text{ is a Carmichael number}$$
$$\Leftrightarrow \{a : a^{q-1} \equiv 1 \bmod q\} = (\mathbb{Z}/q\mathbb{Z})^*.$$

Let $q = p_1^{\alpha_1} p_2^{\alpha_2} \cdots p_J^{\alpha_J}$. Then by Theorem 39 (a) the number of solutions of the congruence equation $x^{q-1} \equiv 1 \bmod q$ is equal to $\prod_{j=1}^{J} \langle q - 1, \varphi(p_j^{\alpha_j}) \rangle$. This product is not divisible by any p_j, while it needs to be divisible by each $\varphi(p_j^{\alpha_j})$. Hence,

$$\text{an odd composite } q \text{ is a Carmichael number}$$
$$\Leftrightarrow q : \text{sqf and } (p_j - 1)|(q - 1), 1 \le j \le J.$$

In particular, $J \geq 3$. For if $q = p_1 p_2$, then since $(p_1 - 1)|(p_2(p_1 - 1) + (p_2 - 1))$ we have $(p_1 - 1)|(p_2 - 1)$, and similarly $(p_2 - 1)|(p_1 - 1)$, which gives a contradiction. This criterion is attributed to Korselt (1899).

§44

We take a primitive root r_p mod p for each $p \geq 3$, observing the prescription (42.1). Then, under the convention (29.1)–(29.2), we introduce the discrete logarithm Ind: For $p \nmid a$,

$$\text{Ind}_{r_p}(a) \equiv u \bmod \varphi(p^\alpha) \quad \Leftrightarrow \quad r_p^u \equiv a \bmod p^\alpha. \tag{44.1}$$

In this,

$$\text{the exponent } \alpha > 0 \text{ can be supposed to be arbitrary.} \tag{44.2}$$

In other words, the function $\text{Ind}_{r_p}(a)$ is defined for $a \bmod p^\lambda$ with a sufficiently large λ, and one gets $\text{Ind}_{r_p}(a) \equiv u \bmod \varphi(p^\alpha)$ locally. Obviously, we have $\text{Ind}_{r_p}(ab) \equiv \text{Ind}_{r_p}(a) + \text{Ind}_{r_p}(b) \bmod \varphi(p^\alpha)$. That is, Ind_{r_p} is an isomorphism from the cyclic group $(\mathbb{Z}/p^\alpha\mathbb{Z})^*$ to the additive group $\mathbb{Z}/\varphi(p^\alpha)\mathbb{Z}$. If r_p is replaced by a different primitive root r_p', then $\text{Ind}_{r_p'}(a) \equiv \text{Ind}_{r_p'}(r_p) \cdot \text{Ind}_{r_p}(a) \bmod \varphi(p^\alpha)$, $\langle \text{Ind}_{r_p'}(r_p), \varphi(p^\alpha) \rangle = 1$.

For the modulus 2^α, we define

$$\text{Ind}^{(2)}(a) = \begin{cases} 0 & a \equiv 1 \bmod 2, \alpha = 1, \\ v \bmod 2 & a \equiv (-1)^v \bmod 4, \alpha = 2, \\ \{v \bmod 2, w \bmod 2^{\alpha-2}\} & a \equiv (-1)^v 5^w \bmod 2^\alpha, \alpha \geq 3, \end{cases} \tag{44.3}$$

170   Congruences

where the lower line is a two-dimensional vector. Then Theorem 37 can be expressed as follows:

Theorem 40 *We denote by* $Z_{p^\alpha}^+$ *the additive group* $\mathbb{Z}/\varphi(p^\alpha)\mathbb{Z}$ *and* $(\mathbb{Z}/2\mathbb{Z}) \oplus (\mathbb{Z}/2^{\alpha-2}\mathbb{Z})$ *provided either* $p \geq 3, \alpha \geq 0$ *or* $p = 2, \alpha \leq 2$ *and provided* $p = 2, \alpha \geq 3,$ *respectively. Then we have the following group-isomorphism:*

$$(\mathbb{Z}/q\mathbb{Z})^* \cong Z_{2^{q(2)}}^+ \oplus \cdots \oplus Z_{p^{q(p)}}^+ \oplus \cdots, \qquad q = \prod_p p^{q(p)}. \qquad (44.4)$$

The right side is a direct sum as additive groups; and the terms $Z_1^+ = \{0\}$ *are ignored. Thus, for a reduced residue class a* mod *q, the map*

$$a \bmod q \;\mapsto\; \left\{ \mathrm{Ind}^{(2)}(a), \ldots, \mathrm{Ind}_{r_p}(a), \ldots \right\} \qquad (44.5)$$

gives rise to an isomorphism. Here an obvious convention on the range is applied to each element on the right side.

Remark One should be aware that the isomorphism (44.5) depends on the choice of primitive roots r_p. Although this fact will not cause any complexity in our theoretical discussion (especially in §51), in explicit numerical computations it certainly matters. Gauss seems to have paid attention to this point; see Note [44.2].

In passing, we stress, for the sake of later purposes, the following: Let $p \geq 3$, and let r_p be as above, and $p \nmid ab$. Then for any $\alpha \geq 1$,

$$a \equiv b \bmod p^\alpha \;\Leftrightarrow\; \mathrm{Ind}_{r_p}(a) \equiv \mathrm{Ind}_{r_p}(b) \bmod \varphi(p^\alpha), \qquad (44.6)$$

which is of course the same as the trivial fact that $c \equiv 1 \bmod p^\alpha \Leftrightarrow \mathrm{Ind}_{r_p}(c) \equiv 0 \bmod \varphi(p^\alpha)$. In particular,

$$\mathrm{Ind}_{r_p}(-1) \equiv \tfrac{1}{2}\varphi(p^\alpha) \bmod \varphi(p^\alpha), \qquad (44.7)$$

which has been applied in Note [42.4]. The analog of (44.6) for the modulus 2^α is that for any $2 \nmid ab$ and $\alpha \geq 3$

$$a \equiv b \bmod 2^\alpha \;\Leftrightarrow\; \begin{matrix} v(a) \equiv v(b) \bmod 2, \\[4pt] w(a) \equiv w(b) \bmod 2^{\alpha-2}, \end{matrix} \qquad (44.8)$$

with v, w as in (44.3). In addition, we emphasize that the last theorem is of highly theoretical nature, for it depends fully on the premise (33.4).

Now, when a primitive root r mod p and an exponent u mod $(p-1)$ are given, there is no particular difficulty to compute r^u mod p, especially with the aide of (36.2). However, when a primitive root r mod p and a, $p \nmid a$, are given, it can be quite laborious in general to compute $\mathrm{Ind}_r(a)$. Even in the case of small modulus p it is hard to find any regular pattern among the sequence $\{a, \text{Ind}_r(a)\}$, except for a few trivia. In Note [43.9] we could find $\text{Ind}_2(31)$ for $p = 430883$; in fact, it was due to the fortunate fact that $p - 1$ is composed of relatively small primes.

Because of this difficulty, a method which may cope with general situations has been devised; it is an analogue of the ρ algorithm concerning integer factorization (§38), and again probabilistic:

Let a primitive root $r \bmod p$, $2 \le a < p$, and $x_0 = 1$, $\alpha_0 = 0$, $\beta_0 = 0$ be given. Then we examine the tuples $\{x_\nu, \alpha_\nu, \beta_\nu\}$ in which $x_\nu \equiv a^{\alpha_\nu} r^{\beta_\nu} \bmod p$ and

$$
\begin{aligned}
0 < x_\nu < \tfrac{1}{3}p &\Rightarrow x_{\nu+1} = ax_\nu : \ \alpha_{\nu+1} = \alpha_\nu + 1, \ \beta_{\nu+1} = \beta_\nu, \\
\tfrac{1}{3}p < x_\nu < \tfrac{2}{3}p &\Rightarrow x_{\nu+1} = x_\nu^2 : \ \alpha_{\nu+1} = 2\alpha_\nu, \ \beta_{\nu+1} = 2\beta_\nu, \qquad (44.9) \\
\tfrac{2}{3}p < x_\nu < p &\Rightarrow x_{\nu+1} = rx_\nu : \ \alpha_{\nu+1} = \alpha_\nu, \ \beta_{\nu+1} = \beta_\nu + 1.
\end{aligned}
$$

If there is a pair such that $x_\eta \equiv x_\tau \bmod p$, then $a^{\alpha_\eta} r^{\beta_\eta} \equiv a^{\alpha_\tau} r^{\beta_\tau} \bmod p$, and thus $(\alpha_\eta - \alpha_\tau)\text{Ind}_r(a) \equiv \beta_\tau - \beta_\eta \bmod (p - 1)$, from which one may find $\text{Ind}_r(a)$. If this fails, then start with a new choice of $\{\alpha_0, \beta_0\}$. The same as the method for integer factorization, this is based on trial and error. The sequence of tuples $\{x_\nu, \alpha_\nu, \beta_\nu\}$ is generated with the aim to achieve randomness.

Presently no deterministic algorithm is known with which one may compute the discrete logarithm in polynomial time. This issue is called the discrete logarithm problem. It is known, nevertheless, that there exists a probabilistic algorithm of polynomial time for discrete logarithm which runs but on quantum computers. See Note [49.5].

Notes

[44.1] The arithmetic function Ind is introduced in [DA, art.57] where the analogy to ordinary logarithms is stressed. To solve binomial congruence equations $\bmod p$, the table of Ind values $\bmod (p - 1)$ is an efficient means. However, in order to perform the arithmetic based on the Ind calculation, one needs to compose these tables for each p, which is by no means a realistic approach to binomial congruences. Thus, as Gauss stated in [DA, art.61], the concept of Ind appears to have merits mainly in theoretical discussions of issues of multiplicative nature. He emphasized instead the importance of obtaining a constructive method of finding special solutions to binomial congruence equations. In the next section, we shall show an answer to this fundamental problem.

[44.2] [DA, arts.70–71]. This is a generalization of $(44.7)_{\alpha=1}$: If the order of $a \bmod p$ is equal to t, then regardless of the choice of a primitive root $r \bmod p$ we have

172 *Congruences*

$$\langle \mathrm{Ind}_r(a), p - 1 \rangle = (p - 1)/t,$$

since if $1 \equiv a^f \equiv r^{f\,\mathrm{Ind}_r(a)} \bmod p$, then $f \equiv 0 \bmod (p-1)/\langle \mathrm{Ind}_r(a), p-1 \rangle$, that is, $t = (p-1)/\langle \mathrm{Ind}_r(a), p-1 \rangle$, which is obviously independent of the choice of r. By the way,

if $\langle t, u \rangle = 1$, then with a certain primitive root $r \bmod p$
$$\mathrm{Ind}_r(a) \equiv ud \bmod (p-1), \quad d = (p-1)/t.$$

To show this, we note that $\mathrm{Ind}_s(a) \equiv vd \bmod (p-1)$, $\langle t, v \rangle = 1$, for an arbitrary primitive root $s \bmod p$, as $d = \langle \mathrm{Ind}_s(a), p-1 \rangle$ and $1 = \langle v, (p-1)/d \rangle$. Let $w \bmod (p-1)$ be such that $uw \equiv v \bmod z$ and $w \equiv 1 \bmod (p-1)/z$ with $z = \langle p-1, t^\infty \rangle$. Then $\langle w, p-1 \rangle = 1$ and $udw \equiv vd \bmod (p-1)$. We may now take $r \equiv s^w \bmod p$ which is obviously a primitive root. We have $a \equiv s^{vd} \equiv s^{udw} \equiv r^{ud} \bmod p$.

[44.3] [DA, art.80] Provided $p \neq 3$, the product of all incongruent primitive roots $\bmod\, p$ is congruent to $1 \bmod p$. To show this, we take an arbitrary primitive root $\bmod\, p$; then the problem is equivalent to considering the sum $\sum j$, where $1 \leq j < p-1$, $\langle j, p-1 \rangle = 1$. Looking at the pairs $\{j, p-1-j\}$, we see that this sum is a multiple of $p-1$ save for the case $\langle (p-1)/2, p-1 \rangle = 1$. The exception occurs only when $p = 3$, which ends the proof. Alternatively, invoking (18.2), one may consider the sum $\sum_{j=1}^{p-1} j \sum_{d|\langle j, p-1 \rangle} \mu(d)$. See Note [64.5].

[44.4] For any prime $p \geq 3$, the sum of all incongruent primitive roots $\bmod\, p$ is congruent to $\mu(p-1) \bmod p$. To prove this, we take a primitive root $r \bmod p$; then the sum in question is, by (18.2), congruent to

$$\sum_{\substack{1 \leq k < p-1 \\ \langle k,\, p-1 \rangle = 1}} r^k \equiv \sum_{\substack{1 \leq k < p-1 \\ d|k \\ d|(p-1)}} r^k \sum \mu(d) \equiv \sum_{\substack{d|(p-1) \\ d < p-1}} \mu(d) \sum_{j=1}^{(p-1)/d-1} r^{jd}$$

$$\equiv \sum_{\substack{d|(p-1) \\ d < p-1}} \mu(d)(r^{p-1} - r^d)\omega_d \equiv - \sum_{\substack{d|(p-1) \\ d < p-1}} \mu(d) \equiv \mu(p-1) \bmod p,$$

where $(r^d - 1)\omega_d \equiv 1 \bmod p$; note that $r^d \not\equiv 1 \bmod p$. We end the proof. This assertion should be compared with (18.18) which concerns the sum of qth primitive roots of 1. See Note [64.5]. Alternatively, following the argument of [DA, art.81], we use the prime power decomposition $p - 1 = p_1^{\alpha_1} \cdots p_J^{\alpha_J} \geq 2$. Then the sum in question is decomposed into $\prod_{j=1}^{J} S_j$. Here S_j is the sum of residues $\bmod\, p$ of order $p_j^{\alpha_j}$ because of Note [39.3]; that is, the difference between the sum of solutions of $x^{p_j^{\alpha_j}} - 1 \equiv 0 \bmod p$ and that of solutions of

$x^{p_j^{\alpha_j-1}} - 1 \equiv 0 \bmod p$. The former is $\equiv 0$, and the latter is $\equiv 1$ only when $\alpha_j = 1$ and $\equiv 0 \bmod p$ otherwise. Hence $\sum_j \equiv \mu(p_j^{\alpha_j}) \bmod p$, which ends Gauss' argument.

[44.5] [DA, art.75] Let α be the order of $a \bmod p$. Then

$$\prod_{j=0}^{\alpha-1} a^j \equiv (-1)^{\alpha-1} \bmod p.$$

It suffices to note that if $2|\alpha$, then $a^{\alpha/2} \equiv -1 \bmod p$. Alternatively, we may use Note [44.2] with $u = 1$. Thus, with an appropriate primitive root $r \bmod p$, we have $\mathrm{Ind}_r(a) \equiv (p-1)/\alpha \bmod (p-1)$. The product in question is congruent to $r^{(\alpha-1)(p-1)/2} \equiv (-1)^{\alpha-1} \bmod p$. By the way, if $a \bmod p$ is a primitive root, then the product is congruent to $(p-1)!$, which proves (35.2). This is the same as the argument of Euler (1773c, I), although he did not explicitly mention his use of a primitive root $\bmod\ p$.

[44.6] The ρ algorithm for discrete logarithm is due to Pollard (1978). We show here the details of the procedure (44.7) in the case $p = 43$, $a = 7$, $r = 5$; this example is never impressive numerically but appropriate enough to indicate how the method works: We have the sequence

$$\{1,0,0\}, \{7,1,0\}, \{6,2,0\}, \{42,3,0\}, \{38,3,1\}, \{18,3,2\}, \{23,6,4\},$$
$$\{13,12,8\}, \{5,13,8\}, \{35,14,8\}, \{3,14,9\}, \{21,15,9\}, \{11,30,18\},$$
$$\{34,31,18\}, \{41,31,19\}, \{33,31,20\}, \{36,31,21\}, \{8,31,22\}, \{13,32,22\}.$$

Hence, $x_7 = x_{18} = 13$. That is, $a^{12}r^8 \equiv a^{32}r^{22} \bmod 43$ or $20\,\mathrm{Ind}_5(7) \equiv -14 \bmod 42$. By Euclid's algorithm, $\mathrm{Ind}_5(7) \equiv 14$ or $35 \bmod 42$. A verification implies that $\mathrm{Ind}_5(7) \equiv 35 \bmod 42$; by the way, $5^{14} \equiv -7 \bmod 43$. Alternatively, one may use $x_8 = 5$. Thus, $r \equiv a^{13}r^8 \bmod 43$ or $13\,\mathrm{Ind}_5(7) \equiv -7 \bmod 42$, which gives $\mathrm{Ind}_5(7) \equiv 35 \bmod 42$.

§45

Now, assume that via an application of Theorem 39 we have been able to conclude that the binomial congruence equation

$$x^\ell \equiv a \bmod p, \quad p \nmid a, \quad \ell \geq 2, \tag{45.1}$$

has a solution. Suppose either that we have not found any primitive root $\bmod$ p or that we have got a primitive root $r \bmod p$ but have faced difficulties to

174 *Congruences*

compute $\text{Ind}_r(a)$. Then, one may ask how an explicit solution to (45.1) can be constructed.

We shall provide an answer to this truly basic problem under the premise

$$\text{the exponent } \ell \text{ is a prime.} \tag{45.2}$$

Obviously, (45.1) with (45.2) is the most basic in discussing binomial congruence equations in view of Notes [34.3] and [45.5], albeit in addition to (33.4) the prime power decomposition of general exponents needs to be presupposed.

In Note [43.5] we already finished treating the case $\langle \ell, p-1 \rangle = 1$; so we may assume that $p \equiv 1 \bmod \ell$. Thus, we consider

$$x^\ell \equiv a \bmod p, \quad p \nmid a; \quad p - 1 = \ell^g t, \quad \ell \nmid t, \ g \geq 1. \tag{45.3}$$

Because of either (28.11) or Theorem 39, we should have

$$a^{(p-1)/\ell} \equiv 1 \bmod p. \tag{45.4}$$

We put

$$X_0 \equiv a^{(ht+1)/\ell}, \quad Y_0 \equiv a^{ht} \bmod p, \quad ht \equiv -1 \bmod \ell. \tag{45.5}$$

Then

$$X_0^\ell \equiv aY_0 \bmod p, \text{ and the order of } Y_0 \bmod p \text{ is } \ell^{g_0}, 0 \leq g_0 < g, \tag{45.6}$$

for (45.4) gives $Y_0^{\ell^{g-1}} \equiv (a^{(p-1)/\ell})^h \equiv 1 \bmod p$. Hence,

$$\text{if } g_0 = 0, \text{ then } X_0 \bmod p \text{ is a solution of (45.1).} \tag{45.7}$$

Otherwise, we enter into a probabilistic mode:

$$\text{choose an } \ell^{\text{th}} \text{ power non-residue } Z \bmod p. \tag{45.8}$$

The reason that we regard this as probabilistic is given in Note [45.1]. The order of $Z^t \bmod p$ is obviously equal to ℓ^g. Hence, powers of $Z^t \bmod p$ exhaust the solutions of $x^{\ell^g} \equiv 1 \bmod p$. As $Y_0 \bmod p$ is such a solution, there exists $u \bmod \ell^g$ with which $Y_0 \equiv (Z^t)^u \bmod p$ (we have used (40.3) with obvious replacements). We have $((Z^t)^u)^{\ell^{g_0}} \equiv Y_0^{\ell^{g_0}} \equiv 1 \bmod p$; hence $u \equiv 0 \bmod \ell^{g-g_0}$. This means that there should exist a k such that

$$Y_0 \equiv (Z^t)^{k\ell^{g-g_0}} \bmod p, \quad 0 < k < \ell^{g_0}. \tag{45.9}$$

Therefore, we get a special solution of (45.1):

$$X_1 \equiv (Z^t)^{(\ell^{g_0}-k)\ell^{g-g_0-1}} a^{(ht+1)/\ell} \bmod p; \tag{45.10}$$

and all solutions are

$$\left\{ X_1 \cdot Z^{j(p-1)/\ell} \bmod p : j \bmod \ell \right\}, \tag{45.11}$$

since $\left\{ Z^{j(p-1)/\ell} \bmod p : j \bmod \ell \right\}$ exhaust the solutions of $x^\ell \equiv 1 \bmod p$.

$§45$ 175

Remark One may replace (45.10) by $Z^{-kt\ell^{g-g_0-1}} a^{(ht+1)/\ell} \mod p$, of course; however, the formulation (45.10) is easier to deal with in applications.

Note that (45.7) holds for the case $g = 1$. In particular, we see that

$$\begin{aligned} &\text{provided } \ell|(p-1) \text{ and } \ell^2 \nmid (p-1), \\ &a^{(ht+1)/\ell} \mod p \text{ is a solution of } x^\ell \equiv a \mod p. \end{aligned} \tag{45.12}$$

Thus, it is relatively easy to solve (39.1) with a prime ℓ, if the modulus q is composed solely of those primes p_j such that $\ell^2 \nmid (p_j - 1)$; besides, the use of Note [34.3] may also be necessary. See Note [45.6].

Notes

[45.1] This section is an adaptation of Tonelli (1891) with minor changes; see Note [45.7] for a group theoretical approach. Tonelli considered only the case $\ell = 2$, which makes no essential difference (see §63), however. The core of his idea is in (45.8), and thus his method is based on the assumption that an ℓth power non-residue $Z \mod p$ is available, which is, as it were, a weak point of the method. It is, nevertheless, probabilistically an easy task to find a $Z \mod p$ since according to (43.4) the density of such residues is equal to $1 - 1/\ell$, which is never small. Although the estimation of the least positive ℓ^{th} power non-residue is regarded as a very deep problem, in practice a Z can be found most always in the set $\{c \mod p : 1 < c \leq D_p\}$ with a relatively small D_p; see the next Note. On the other hand, the search of the exponent k in (45.9) is a kind of a discrete logarithm problem. Naturally, it is not difficult if ℓ, g are not too large; especially in the case $\ell = 2$, at the corresponding stage we are able to introduce a further algorithm, i.e., (63.10) below. By the way, in §68 we shall present an alternative method due to Cipolla (1903) to solve (45.1)–(45.2), which is again probabilistic.

[45.2] The ℓth power residues $\mod p$ is a proper subgroup of $(\mathbb{Z}/p\mathbb{Z})^*$, outside of which is filled with ℓ^{th} power non-residues. Hence, by Note [32.6] we have $D_p \leq 2(\log p)^2$ under GRH.

[45.3] Let $p = 8101$.
(1) Consider the congruence equation $x^3 \equiv 7 \mod p$. We use the modular binary exponentiation (36.2), although omitting the details. First, we have $p - 1 = 3^4 \cdot 100$; thus $g = 4, t = 100$. Also, $7^{(p-1)/3} \equiv 1 \mod p$, and $7 \mod p$ is a cubic residue (Theorem 39). We have $h = 2$. Hence, $X_0 \equiv 7^{67} \equiv 5038$ and $Y_0 \equiv 7^{200} \equiv 2013 \mod p$. Next, $Y_0^3 \equiv 5883$, $Y_0^9 \equiv 1 \mod p$, which gives $g_0 = 2$. Consequently, we need to choose a cubic non-residue. On noting $3^{(p-1)/3} \not\equiv 1 \mod p$, we may take $Z = 3$. Then we compute $(3^t)^{s \cdot 3^2} \mod p$

176 *Congruences*

for $s \bmod 3^2$, and find that at $s = 7$ we reach $Y_0 \bmod p$; that is, $k = 7$. By (45.9), we are led to $X_1 \equiv (3^t)^{(3^2-7)\cdot 3}X_0 \equiv 6901 \bmod p$. Therefore, computing $3^{j(p-1)/3}X_1 \bmod p$ for $j = 0, 1, 2$, we conclude that the solutions are 4472, 4829, 6901 mod p. Namely, if one divides n^3 by 8101 one by one, starting from $n = 1$, then finally at $n = 4472$ the residue 7 appears:

$$4472^3 = 7 + 11039941 \cdot 8101.$$

(2) Consider the congruence equation $x^{3^2} \equiv 7 \bmod p$. Since $7^{(p-1)/3^2} \equiv 1$, this has solutions. Taking into account the result in (1), we consider the congruence $x^3 \equiv 4472 \equiv A \bmod p$. With an obvious allocation of parameters, $X_0 \equiv A^{67} \equiv -525$ and $Y_0 \equiv A^{200} \equiv 454$. So $Y_0^9 \equiv -2218$, $Y_0^{27} \equiv 1$; that is, $g_0 = 3$. Computing $(3^t)^{s\cdot3}$, $0 < s < 3^3$, we find that $(3^t)^{16\cdot3} \equiv Y_0$. Hence $X_1 \equiv (3^t)^{3^4-16}A^{67} \equiv 7764 \equiv -337 \bmod p$ should be a solution. Indeed,

$$337^{3^2} + 7 = 6920388498448091384 \cdot 8101.$$

Other eight roots are $1722, 2103, 2174, 4276, 4326, 4617, 6264, 7259 \bmod p$.
(3) By the way, despite $3^3 | (p-1)$, the congruence $x^{3^3} \equiv 7 \bmod p$ has no solution since $7^{(p-1)/3^3} \equiv 2217$.

[45.4] Independently of Note [43.9], we shall treat $x^{17} \equiv 31 \bmod p$ with $p = 430883$. According to Note [41.3], this modulus is a prime. We have $p - 1 = 17 \cdot 25346$, $17^2 \nmid (p-1)$; thus $g = 1$, $t = 25346$. Also, $31^{(p-1)/17} \equiv 1 \bmod p$, and 31 mod p is a 17th power residue mod p (Theorem 39). Next, on noting $t \equiv -1 \bmod 17$, we take $h = 1$. We may appeal to (45.12), and get a solution $X_0 \equiv 31^{(t+1)/17} \equiv 31^{1491} \equiv 197467 \bmod p$. This is one of the solutions which we obtained before. The rest of the argument is to find all solutions of $x^{17} \equiv 1 \bmod p$. Following the discussion after (45.8), we see that we need a 17th power non-residue $Z \bmod p$. Then, by the fact that $2^{(p-1)/17} \equiv -17592 \bmod p$, we take $Z \equiv 2 \bmod p$ (Theorem 39). Therefore, all the solutions are $\{197467 \cdot (-17592)^j \bmod p : j = 0, 1, \ldots 16\}$. This conclusion naturally coincides with that of Note [43.9]. Certainly the present argument is much simpler than that of Note [43.9].

[45.5] Let p, ℓ, a be as in (45.1)–(45.2), and suppose that $x^\ell \equiv a \bmod p$ has a solution $\xi \bmod p$. In order to lift this to a solution of $x^\ell \equiv a \bmod p^\alpha$, one may appeal to Note [34.3] (especially, Legendre (1798, pp.419–420)). Alternatively, a minor generalization of the practical method of Tonelli (1891, p.345), which he indicated for the case $\ell = 2$, can be employed to show that

$$\xi \bmod p \mapsto \xi^{p^{\alpha-1}} a_*^{(p^{\alpha-1}-1)/\ell} \bmod p^\alpha, \ aa_* \equiv 1 \bmod p^\alpha.$$

$$\S 45 \qquad\qquad 177$$

For a confirmation, it suffices to note that we get $(\xi^\ell a_*)^{p^{\alpha-1}} \equiv 1 \bmod p^\alpha$ by an application to $\xi^\ell a_* \equiv 1 \bmod p$ of the discussion following (42.3). For example, in the case of the preceding Note we have $a_* \equiv 17967112228 \bmod p^2$ by Euclid's algorithm. Hence, $\xi \equiv 197467 \bmod p \mapsto \xi^p a_*^{(p-1)/17} \equiv 68321005947 \bmod p^2$.

[45.6]   As an application of the observation (45.12), we consider the congruence equation $x^7 \equiv 307 \bmod q$ with $q = 470119$: This modulus is easily decomposed into $13 \cdot 29^2 \cdot 43$. Ignoring Euler's criterion (Theorem 39), we proceed to solve

(1) $x_1^7 \equiv 307 \equiv 8 \bmod 13$,
(2) $x_2^7 \equiv (y_1 + 29y_2)^7 \equiv 307 \bmod 29^2$,
(3) $x_3^7 \equiv 307 \equiv 6 \bmod 43$,

where (2) is related to the Hensel lifting (Note [34.3]). As for (1), since $7 \cdot 7 \equiv 1 \bmod 12$, we have, by Note [43.5], $x_1 \equiv 8^7 \equiv 5 \bmod 13$. As for (2), since $29 - 1 = 7 \cdot 4$ and $5 \cdot 4 + 1 \equiv 0 \bmod 7$, we have, by (45.12), $y_1 \equiv 307^3 \equiv 12 \bmod 29$. Then, $7 \cdot 12^6 \cdot y_2 \equiv (307 - 12^7)/29 \equiv 5 \bmod 29$ or $y_2 \equiv -9 \bmod 29$, so $x_2 \equiv 12 - 9 \cdot 29 \equiv -249 \bmod 29^2$. Further, as for (3), $43 - 1 = 7 \cdot 6$ and $6 + 1 \equiv 0 \bmod 7$; so $x_3 \equiv 6 \bmod 43$ by (45.12). Then, on noting (30.3), we observe, by Euclid algorithm, that $29^2 \cdot 43 \cdot 4 \equiv 1 \bmod 13$; $13 \cdot 43 \cdot (11 + 17 \cdot 29) \equiv 1 \bmod 29^2$; $13 \cdot 29^2 \cdot 4 \equiv 1 \bmod 43$. Hence $x \equiv 29^2 \cdot 43 \cdot 4 \cdot x_1 + 13 \cdot 43 \cdot 504 \cdot x_2 + 13 \cdot 29^2 \cdot 4 \cdot x_3 \equiv 411000 \bmod q$. According to (43.2), there exist 48 more solutions mod q.

[45.7]   This partly group-theoretic argument should be compared with (45.1)–(45.11): Let a, t, ℓ, g, h, Z be as above, and consider, with an obvious abbreviation, the subgroups

$$U = \{x : x^t = 1\}, \quad V = \{x : x^{\ell^g} = 1\} = \{(Z^t)^k : k \bmod \ell^g\}$$

of $G = (\mathbb{Z}/p\mathbb{Z})^*$; note that $G = U \times V$, a direct product decomposition. The condition $a \in G^\ell$ is supposed. Let $\alpha, \beta \in \mathbb{Z}$ be such that $\alpha \ell^{g-1} + \beta t = 1$. Then $a = uv$ with $u = a^{\alpha \ell^{g-1}} \in U$ and $v = a^{\beta t} \in V^\ell$, for $u^t = a^{\alpha(p-1)/\ell} = 1$, and $v^{\ell^{g-1}} = a^{\beta(p-1)/\ell} = 1$ because of (45.4). In particular, if $g = 1$, then $a = u$, and $a^{(ht+1)/\ell}$ is a solution to (45.1), which is the same as (45.12). Hence, we may suppose $g \geq 2$. Then, u is an ℓth residue, and v ought to be so. An ℓth root of u is $a^{\alpha \ell^{g-2}}$. We need to find an ℓth root of v. For this sake, we note that $v = (Z^t)^k$ with $k = \gamma \ell^\tau$, $\tau \geq 1$, $\ell \nmid \gamma$ because $1 = v^{\ell^{g-1}} = (Z^t)^{k\ell^{g-1}}$, and $\ell^g | (k \ell^{g-1})$, i.e., $\ell | k$. One may find τ by determining the order of v which is equal to ℓ^c, $c \geq 1$, such that $(Z^t)^{k\ell^c} = 1$ and $(Z^t)^{k\ell^{c-1}} \neq 1$. Namely, $\ell^g \| k\ell^c$, so

178 *Congruences*

$\tau = g - c$. Then γ is obtained by raising $(Z^t)^{\ell^\tau}$ to powers until reaching v. In this way, we are led to the special solution $a^{\alpha \ell g - 2}(Z^t)^{\gamma \ell^{\tau-1}}$ to (45.1), provided $g \geq 2$. For instance, in the case of Note [45.3](1), we have $\alpha = -37$ and $\beta = 10$, so that $u \equiv 7^{-37 \cdot 3^3}$, $v \equiv 7^{1000} \equiv 315$, where $7^{-1} \equiv 3472$. A cubic root of u is $7^{-37 \cdot 3^2} \equiv 6677$. On the other hand, since $315^3 \not\equiv 1$, $315^9 \equiv 1$, we get $\tau = 4 - 2$, i.e., $315 \equiv (3^{100})^9{}^\gamma$, as $Z = 3$. Raising $(3^{100})^9$ to powers, we find that $\gamma = 8$; and a cubic root of 315 is $(3^{100})^{24} \equiv 4872$. Hence, $6677 \cdot 4872 \equiv 4829$ is one of the cubic roots of $7 \bmod 8101$.

§46

Our discussion on binomial congruence equations (39.1) has been conducted on the premise (33.4). The main reason for this is in that in the current state of number theory the relevant argument makes it necessary for us to exploit primitive roots either directly or indirectly and in that the primitive roots exist if and only if the moduli are in the set of the special integers which are given in Theorem 38. Hence, as far as the stance towards (33.1) that is given immediately after (33.10) is taken, it is virtually impossible for us to discuss congruence equations of higher degrees without assuming the decomposition of corresponding moduli into products of those special integers.

We then notice an interesting fact: discussions in §§37 and 41 on primality test are contained in the theory of power residues. That is, there appears to exist a possibility of being able to devise practical methods for primality test and perhaps even for integer factorization as well by reversing somehow the reasoning developed so far. We shall show that a kind of reversion is indeed possible, though in a certain restricted sense of possibility. The discussion will occupy the present and the succeeding two sections.

We shall first give a probabilistic method for primality test, which depends on the notion of strong probable primes (spp: (37.5)). This method is, more precisely, a consequence of the explicit formula (46.2) below.

Theorem 41 *Let $q \geq 3$ be an odd integer, and assume that*

$$q = p_1^{\alpha_1} p_2^{\alpha_2} \cdots p_J^{\alpha_J}, \; q - 1 = 2^\kappa t, \; 2 \nmid t, \tag{46.1}$$
$$p_j - 1 = 2^{\beta_j} u_j, \; 2 \nmid u_j; \; \beta_0 = \min\{\beta_j : j \leq J\}.$$

Then we have

$$\left| \{a \bmod q : q \in \mathrm{spp}(a)\} \right| = \left(1 + (2^{J\beta_0} - 1)/(2^J - 1)\right) \prod_{j=1}^{J} \langle t, u_j \rangle. \tag{46.2}$$

$§46$ 179

Proof To begin with, we note that $\beta_0 \leq \kappa$ since $q \equiv 1 \bmod 2^{\beta_0}$ follows from $p_j^{\alpha_j} \equiv 1 \bmod 2^{\beta_0}, j \leq J$. Also, the definition (37.5) implies that

$$\{a \bmod q: q \in \operatorname{spp}(a)\}$$

$$= \{a \bmod q: a^t \equiv 1 \bmod q\} \sqcup \bigsqcup_{\gamma=0}^{\kappa-1} \{a \bmod q: a^{2^\gamma t} \equiv -1 \bmod q\}. \qquad (46.3)$$

In this we have

$$\left|\{a \bmod q: a^t \equiv 1 \bmod q\}\right| = \prod_{j=1}^{J} \langle t, u_j \rangle; \qquad (46.4)$$

and

$$\left|\{a \bmod q: a^{2^\gamma t} \equiv -1 \bmod q\}\right| = \begin{cases} 0 & \gamma \geq \beta_0, \\ 2^{J\gamma} \prod_{j=1}^{J} \langle t, u_j \rangle & \gamma < \beta_0. \end{cases} \qquad (46.5)$$

The assertion (46.4) follows from the fact that according to Theorem 39 the number of solutions of $x^t \equiv 1 \bmod p_j^{\alpha_j}$ is equal to $\langle t, \varphi(p_j^{\alpha_j}) \rangle = \langle t, u_j \rangle$. On the other hand, according to (43.2), $x^{2^\gamma t} \equiv -1 \bmod q$ has a solution if and only if $\varphi(p_j^{\alpha_j})/\langle 2^\gamma t, \varphi(p_j^{\alpha_j}) \rangle \equiv 0 \bmod 2$ for all $j \leq J$. This is equivalent to $\gamma < \beta_j$ for all $j \leq J$, that is, to $\gamma < \beta_0$. Then, the number of solutions of $x^{2^\gamma t} \equiv -1 \bmod p_j^{\alpha_j}$ is equal to $\langle 2^\gamma t, \varphi(p_j^{\alpha_j}) \rangle = 2^\gamma \langle t, u_j \rangle$, which gives (46.5). Summing the results (46.4) and (46.5) for $\gamma < \beta_0$ and on noting (46.3), we end the proof.

An important consequence of (46.2) is:

for any odd composite $q \geq 15$,

$$\left|\{a \bmod q: q \in \operatorname{spp}(a)\}\right| \leq \tfrac{1}{4}\varphi(q). \qquad (46.6)$$

To show this, we note that since $1 + (2^{J\beta_0} - 1)/(2^J - 1) \leq 2^{J(\beta_0-1)+1}$ we have

$$\frac{\varphi(q)}{\left|\{a \bmod q: q \in \operatorname{spp}(a)\}\right|} \geq \tfrac{1}{2} \prod_{j=1}^{J} \frac{2^{\beta_j} u_j p_j^{\alpha_j-1}}{2^{\beta_0-1} \langle t, u_j \rangle}. \qquad (46.7)$$

Each factor of this product is even, and if $J \geq 3$, then $(46.7) \geq 2^{J-1} \geq 4$. If $J = 2$ and $p_k^2 | q$, with either $k = 1$ or 2, then the k-factor is not less than $2p_k^{\alpha_k-1}$, and $(46.7) \geq 2p_k > 4$. Further, if $q = p_1^{\alpha_1} \geq 15, \alpha_1 \geq 2$, then $(46.7) \geq p_1^{\alpha_1-1} > 4$. In the remaining case, we have $q = p_1 p_2, p_1 < p_2$. With this, if $\beta_0 < \beta_2$, then $2^{\beta_0-1} \langle t, u_2 \rangle \leq \tfrac{1}{4} 2^{\beta_2} u_2$, and $(46.7) \geq 4$. Thus, we assume that $\beta_0 = \beta_2$. Then, on noting that $q - 1 = p_1(p_2 - 1) + p_1 - 1$ and $\beta_0 \leq \kappa$, we have $2^{\beta_0-1} \langle t, u_2 \rangle = \tfrac{1}{2} \langle q - 1, p_2 - 1 \rangle = \tfrac{1}{2} \langle p_1 - 1, p_2 - 1 \rangle$.

Congruences

This implies that $2^{\beta_2} u_2 / (2^{\beta_0 - 1} \langle t, u_2 \rangle) = 2(p_2 - 1) / \langle p_1 - 1, p_2 - 1 \rangle \geq 4$ since $(p_2 - 1) / \langle p_1 - 1, p_2 - 1 \rangle > 1$. We end the proof of (46.6).

A probabilistic method for testing primality:

We develop a heuristic discussion based on (46.6). Thus, we try to see whether a large odd integer q is composite or not. We take $1 < a_1 < q$ randomly. If $\langle a_1, q \rangle > 1$, then our procedure ends. Otherwise, we check whether $q \in \mathrm{spp}(a_1)$ or not. If not, then q is not prime, and our procedure ends. If yes, then provided q is composite, the probability of hitting such an a_1 is not greater than $1/4$ because of (46.6). We then take a_2 randomly and repeat the same procedure. Either if $\langle a_2, q \rangle > 1$ or if $\langle a_2, q \rangle = 1$ and $q \notin \mathrm{spp}(a_2)$, then q is composite. Otherwise, $q \in \mathrm{spp}(a_1) \cap \mathrm{spp}(a_2)$. Provided q is composite, the probability of hitting such a_1, a_2 is not greater than $1/4^2$. We continue the procedure by taking $a_3, \ldots$ randomly. Suppose then that we have reached the situation $q \in \cap_{r=1}^{R} \mathrm{spp}(a_r)$. Provided q is composite, the probability of this event is not greater than $1/4^R$. If R is sufficiently large, then the plausibility of this situation is extremely low. Therefore, one may determine in polynomial time whether q is prime or composite with a high probability.

Now, if it is known solely that q is composite, then one may ask how an explicit factorization of q can be achieved. We shall show an approach to this fundamental problem:

By an obvious reason we exclude the case where q is equal to one of $[q^{1/\nu}]^\nu$, $\nu = 2, 3 \ldots \leq [\log q / \log 2]$. Thus, q is assumed not to be a power of any integer. Then we start our discussion with the assumption (46.1); thus,

$$q \text{ is odd with } J \geq 2. \tag{46.8}$$

With this, we observe the following:

if the order of $a \bmod q$ is even ($2s$, say), and $a^s \not\equiv -1 \bmod q$, then $q = \langle a^s - 1, q \rangle \langle a^s + 1, q \rangle$ is a non-trivial factorization. $\tag{46.9}$

For since $(a^s - 1)(a^s + 1) \equiv 0 \bmod q$ and $\langle a^s - 1, a^s + 1 \rangle$ is equal to either 2 or 1, we have $q = \langle q, (a^s - 1)(a^s + 1) \rangle = \langle q, a^s - 1 \rangle \langle q, a^s + 1 \rangle$, and neither of the factors on the right side can be equal to 1.

Hence, if there exists an $a \bmod q$ which satisfies the condition stated in (46.9), then we shall be able to achieve a non-trivial factorization of q. We shall thus show an analogue of (46.6) by computing the number of $a \bmod q$ that do not satisfy this condition; namely, we shall prove the probability that the procedure (46.9) fails is small. This time the necessary quantitative facts are deduced not from Theorem 39 but from a few assertions obtained prior to it, although the argument itself is analogous to that of the last proof.

Theorem 42 *Provided* (46.1) *and* (46.8), *we have*

$$\left|\left\{a \bmod q: \begin{array}{l} \text{either the order of } a \bmod q \text{ is odd} \\ \text{or even } 2s \text{ and } a^s \equiv -1 \bmod q \end{array}\right\}\right|$$

$$= \frac{\left(1 + (2^{J\beta_0} - 1)/(2^J - 1)\right)}{2^{\beta_1 + \beta_2 + \cdots + \beta_J}} \varphi(q) \le \varphi(q)/2^{J-1}. \tag{46.10}$$

Proof Because of the inequality mentioned immediately after (46.6), it suffices to prove this equality part. First, we take a primitive root $r_j \bmod p_j^{\alpha_j}$ (Theorem 37), and suppose that the order of $a \bmod p_j^{\alpha_j}$ is equal to $2^{\eta_j}\xi_j$, $2 \nmid \xi_j$. Naturally, we have $2^{\eta_j}\xi_j | \varphi(p_j^{\alpha_j})$; in particular

$$\eta_j \le \beta_j, j \le J. \tag{46.11}$$

More precisely, we have, for reduced classes $a \bmod q$

$$a \equiv (r_j^{\nu_j})^{w_j} \bmod p_j^{\alpha_j}, v_j = \varphi(p_j^{\alpha_j})/2^{\eta_j}\xi_j,$$
$$w_j \bmod 2^{\eta_j}\xi_j, \langle w_j, 2^{\eta_j}\xi_j \rangle = 1. \tag{46.12}$$

According to Note [39.4], the order of $a \bmod q$ is equal to $2^\eta \xi$ where $\eta = \max\{\eta_j\}$ and $\xi = [\xi_1, \xi_2, \ldots, \xi_J]$. In particular,

$$\text{the order of } a \bmod q \text{ is odd} \Leftrightarrow \eta_j = 0, \forall j \le J. \tag{46.13}$$

In this case, we have $\xi_j | u_j p_j^{\alpha_j - 1}$ for each j, and the lower line of (46.12) becomes $w_j \bmod \xi_j$, $\langle w_j, \xi_j \rangle = 1$. Hence, we have

$$\left|\left\{a \bmod q : \text{ of odd order}\right\}\right|$$

$$= \prod_{j=1}^J \sum_{\xi_j | u_j p_j^{\alpha_j - 1}} \varphi(\xi_j) = \prod_{j=1}^J \frac{\varphi(p_j^{\alpha_j})}{2^{\beta_j}} = \frac{\varphi(q)}{2^{\beta_1 + \beta_2 + \cdots + \beta_J}}. \tag{46.14}$$

Turning to the case $\eta \ge 1$, the order of $a \bmod q$ is equal to $2s, s = 2^{\eta-1}\xi$. Here, if there exists a j such that $\eta_j < \eta$, then $a^s \equiv r_j^{\tau_j} \bmod p_j^{\alpha_j}$ where $\tau_j = 2^{\eta - \eta_j - 1}(\xi/\xi_j)w_j \varphi(p_j^{\alpha_j}) \equiv 0 \bmod \varphi(p_j^{\alpha_j})$, and $a^s \equiv 1 \bmod p_j^{\alpha_j}$, that is, $a^s \not\equiv -1 \bmod q$; hence, this situation is to be discarded. If we have $\eta_j = \eta$ instead, then on noting (44.7) we have $a^s \equiv r_j^* \equiv -1 \bmod p_j^{\alpha_j}$ with $* = \varphi(p_j^{\alpha_j})w_j\xi/2\xi_j$. Therefore, $a^s \equiv -1 \bmod q$ is equivalent to $\eta_j = \eta$ for all $j \le J$. We then return to (46.12), and since the number of w_j is equal to $\varphi(2^\eta \xi_j)$ we see that the number of relevant $a \bmod p_j^{\alpha_j}$ is equal to $\sum \varphi(2^\eta \xi_j)$ where $\xi_j | u_j p_j^{\alpha_j - 1}$. This sum equals $2^{\eta-1} u_j p_j^{\alpha_j - 1} = 2^{\eta - \beta_j - 1}\varphi(p_j^{\alpha_j})$. Hence,

$$\left|\left\{a \bmod q: \begin{array}{l} \text{the order of } a \bmod q \text{ is } 2s \\ 2^{\eta-1} \| s, \text{ and } a^s \equiv -1 \bmod q \end{array}\right\}\right| = \frac{2^{(\eta-1)J}\varphi(q)}{2^{\beta_1 + \beta_2 + \cdots + \beta_J}}. \tag{46.15}$$

Observing the independence of events, we sum (46.14) and $(46.15)_{1 \leq \eta \leq \beta_0}$ since (46.11) implies that $\eta \leq \min_{j \leq J} \beta_j$. This ends the proof.

A probabilistic method for factoring integers:
We shall make an intuitive discussion based on (46.10). On the supposition (46.8) we take $1 < a_1 < q$ randomly. If $\langle a_1, q \rangle > 1$, then we are done. If $\langle a_1, q \rangle = 1$, then we determine the order of a_1 mod q; and if the condition stated in (46.9) is satisfied, then we are done. The probability of the contrary situation is not greater than $1/2$ by (46.10). Under this circumstance, we take a_2 randomly. If $\langle a_2, q \rangle > 1$, then our procedure ends. If $\langle a_2, q \rangle = 1$, then we determine the order of a_2 mod q; and if the condition stated in (46.9) is satisfied, then we are done. The probability of the contrary situation is not greater than $1/2^2$. The probability that we fail the factorization for R times consecutively is not greater than $1/2^R$. Therefore, we shall be able to achieve a factorization of q almost certainly.

However, this factorization argument works fine only if for an arbitrary given a, $\langle a, q \rangle = 1$, we are able

$$\text{to determine the order } \lambda \text{ of } a \text{ mod } q. \tag{46.16}$$

Continued to the next section.

Notes

[46.1] On GRH (53.9), the above primality test becomes a highly effective deterministic method. Namely, if q is not a power of an integer, then on GRH

$$q \in \mathrm{spp}(a), \ \forall a \leq 2(\log q)^2, \ \langle a, q \rangle = 1 \ \Rightarrow \ q \text{ is a prime.}$$

This is due to Bach (1990, Theorem 2). For a proof see Note [101.8].

[46.2] The important assertion (46.6) is due to Monier (1980, Proposition 2) and Rabin (1980, Theorem 1), independently. For the proof of (46.3) see also Monier (1980, Proposition 1) and Crandall–Pomerance (2005, §3.5). If one appeals to this probabilistic primality test, then the required number of trials is remarkably small; moreover, it can be carried out without any assumption or hypothesis. However, one should be aware that the primality thus inferred remains mathematically inconclusive.

[46.3] A deterministic primality test of polynomial time has already been discovered by Agrawal et al in their renowned work (2004). That was a remarkable progress in number theory. Nevertheless, we ought to point it out that if a gigantic integer has been proved to be a prime by a test and if its verification becomes necessary at an independent occasion, then one cannot do anything but repeating the same test or rather try other tests. In a strong contrast, any factorization of an integer can be verified by just multiplying the claimed factors; indeed, it does not matter whether the factorization is achieved either by a probabilistic method or by anything else. A factorization itself stands for an absolute proof.

[46.4] The crucial bound $\varphi(q)/2^{J-1}$ in (46.10) was stated by Shor (1997, p.1498) with a sketch of the proof; see also Crandall–Pomerance (2005, p.435).

§47

According to Shor (1994, 1997),

Theorem 43 *On quantum computers there exists an algorithm that probabilistically solves the problem* (46.16) *in polynomial time.*

Remark 1 We shall use the term Shor's algorithm to indicate the procedure that is described in the present and next sections. In a wider sense it includes the discussion following the proof of Theorem 42. A historical account of this fascinating development in number theory and technology is given in Shor (1994).

Remark 2 Shor's algorithm is not deterministic but probabilistic in two reasons: its dependency on the second half of the preceding section and on the intrinsic nature of quantum mechanics. However, any probabilistic integer factorization can deliver decisive results nonetheless, as we have stressed in Note [46.3]. Therefore, the present theorem implies that the factorization of a given integer in polynomial time should be achieved once quantum computers of appropriate capacity become available.

Remark 3 The strictly quantum-theoretical part of Shor's algorithm is developed in the present section; an analysis of resulting outputs will be carried out in the next section, which is an application of the basic theory of continued fractions. Our quantum computer is defined below in a conceptual way. We assert that detailing its physics-theoretical structure is not of absolute necessity, save for very basics given in Notes at the end of the present section, because our account is restricted to a purely mathematical aspect of the algorithm, that is, a manipulation of certain unitary operators on a finite dimensional Hilbert space. In particular, it is assumed that we are given an ideal quantum computer which is free of noise and errors. Also, in order to stay in mathematical comfort, we dispense with tempting terms such as superposition, parallel computing, etc., which are commonly used in literature on quantum computers; see Note [47.1].

184   *Congruences*

Proof

(1) To begin with, let **r** denote a generic binary quantum register; see Note [47.8]. Let $L \in \mathbb{N}$ be arbitrary, and let $\mathbf{r}^L$ be an array of L registers numbered from 0 (right) to $L - 1$ (left). We define our quantum computer as $\mathbf{r}^L$. The set of basic states of **r** is denoted by $\mathbf{b} = \{|0\rangle, |1\rangle\}$; the term basic refers to being physically observable in the classical sense, that is, **b** is a bit. Then the set of all basic states of $\mathbf{r}^L$ is

$$\mathbf{b}^L = \{|x_{L-1}\rangle \otimes \cdots \otimes |x_1\rangle \otimes |x_0\rangle : \{x_{L-1}, x_{L-1}, \ldots, x_0\} \in \{0, 1\}^L\}, \quad (47.1)$$

where indexing is the same as in $\mathbf{r}^L$. We shall use Dirac's ket-notation $|\cdot\rangle$ in a mixed way: A generic element in $\mathbf{b}^L$ is represented as either the tensor product given in (47.1) or $|x_{L-1}\rangle \cdots |x_1\rangle |x_0\rangle$ or $|x_{L-1} \cdots x_1 x_0\rangle$ or just $|x\rangle$ with $x = (x_{L-1}, x_{L-2}, \ldots, x_1, x_0)_2$ which is an integer in the interval $[0, 2^L)$; see Note [2.3]. Thus each integer in this interval can be identified with an element of $\mathbf{b}^L$. The array or bit size L will always be prescribed contextually.

(2) Let $\mathsf{q} = \mathbb{C}|0\rangle + \mathbb{C}|1\rangle$ be the two-dimensional Hilbert space spanned by the orthonormal basis $\mathbf{b} = \{|0\rangle, |1\rangle\}$; then q is the same as the set of all quantum states of a register **r**. We call q a qubit, as an analogue of a bit **b**. The tensor product q^L of L copies of q which are numbered as above is the 2^L-dimensional Hilbert space with the orthonormal basis $\mathbf{b}^L$. Quantum mechanics postulates that q^L is the same as the set of all quantum states of $\mathbf{r}^L$.

(3) With this, we aim to exploit the Hilbert space q^L as a resource of computation. The same as in the classical case, a quantum computation maps an element of $\mathbf{b}^L$, i.e., an integer in $[0, 2^L)$, to another. These two elements are of course physically observable. However, the core part of our map is to act actually on the space q^L spreading over $\mathbf{b}^L$ and the physical reality of the action is not observable because of the quantum observation axiom (47.4) below, which signifies a major difference between quantum and classical computers; see Note [47.2]. Thus, we have the quantum computation scheme:

$$\begin{aligned} \mathbf{b}^L \ni \text{ an input vector } &\Rightarrow \text{ a unitary mapping-process in } \mathsf{q}^L \\ &\Rightarrow \text{ observation of the image } \Rightarrow \text{ output} \in \mathbf{b}^L. \end{aligned} \quad (47.2)$$

This might remind one of an application of complex analysis to a computation of a real variable integration; see Note [12.5]. Thus, $\mathbf{b}^L$ and q^L correspond to $\mathbb{R}$ and $\mathbb{C}$, respectively.

(4) In the case of classical computers with L registers, each output is a unique integer $\xi \in [0, 2^L)$ with which the task ends; by the way, hereafter either ξ or $|\xi\rangle$ stands for a generic element of $\mathbf{b}^L$ in the same way as either x or $|x\rangle$ does. We have stated at (47.2) that in the case of the quantum computer $\mathbf{r}^L$ as well as a computation or a program results in an integer $\xi \in [0, 2^L)$. However, it is of probabilistic nature, i.e., not unique in general, in the sense to be made precise

in a moment. Hence, the program needs to be run repeatedly, so the output is actually an accumulation of certain ξ's which bears a probabilistic distribution in $[0, 2^L) \cap \mathbb{Z}$. The statistical data thus acquired is to be analyzed through the theoretical rendition of the distribution: If $\mathbf{r}^L$ is mathematically supposed to be in the image state $\sum_x c_x |x\rangle$ after the mapping-process part in (47.2), then the statistical data is to coincide well with the probabilistic distribution which is predicted by the measurement axiom

> the probability of observing the particular output ξ
> is equal to $|c_\xi|^2 / \left(\sum_x |c_x|^2 \right)$. $\qquad$ (47.3)

Further, it is understood that

> once an output ξ is observed,
> then the state of $\mathbf{r}^L$ becomes $|\xi\rangle \in \mathbf{b}^L$; $\qquad$ (47.4)

this conforms with common sense. We repeat: the statistical analysis of the set of out-put ξ's in the light of the theoretical (47.3) should give an assertion concerning the original computation issue.

(5) Now, our computation (47.2) ought to be a time evolution on a quantum system $\mathbf{r}^L$. Hence, by an axiom of quantum mechanics,

> the mapping-process part U in (47.2) is
> unitary on the space q^L. $\qquad$ (47.5)

In particular, it should be reversible, while computations on classical computers in general are irreversible; see Note [47.4] for more details on this important aspect. A little more generally, we suppose that our computer $\mathbf{r}^L$ is in the state $|\Psi\rangle = \sum_x \alpha_x |x\rangle$. We apply a unitary operator V to $\mathbf{r}^L$; then the linearity of V implies that the state of $\mathbf{r}^L$ becomes

$$V |\Psi\rangle = \sum_{x=0}^{2^L - 1} \alpha_x V |x\rangle. \qquad (47.6)$$

We specify V and $|\Psi\rangle$ to be the process U and the input vector $|\xi_{\text{in}}\rangle \in \mathbf{b}^L$, both of which are of course fixed. Then we let (47.5) run, observe $U |\xi_{\text{in}}\rangle$, and plot an output $|\xi_{\text{out}}\rangle \in \mathbf{b}^L$. The same is repeated many times. We obtain a probabilistic distribution of integers $\{\xi_{\text{out}}\}$. This physical result should well coincide with what mathematical computation, i.e., (47.3) yields via (47.6) with the current specification. As asserted in Note [47.2], here it is irrelevant to our purpose whether (47.6) is indeed realized physically or not. To perform a quantum computation is the same as to compose an appropriate unitary map and exploit the logical identity (47.6) in order to interpret the statistical data or the probabilistic distribution of the outputs.

186   Congruences

(6)   Now, any bounded function $f : \mathbb{N} \cup \{0\} \to \mathbb{N} \cup \{0\}$ can be regarded as a unitary map. To confirm this fundamental fact, suppose that both the domain and the range of f are within L binary digits. We then define the linear map U_f acting on the space q^{2L} by

$$U_f : \ |x\rangle \, |y\rangle \mapsto |x\rangle \, |f(x) \oplus y\rangle \,. \tag{47.7}$$

Here $|x\rangle$, $|y\rangle$ correspond to the upper and lower L digits of a basis in $\mathbf{b}^{2L}$; and the operation $\oplus$ is addition mod 2 performed digit-wise or addition without carrying, i.e., $1 + 1 = 0$. This ends the confirmation since U_f is obviously unitary. Note that the unitarity of U_f is gained by the pairing $\{x, f(x)\}$; that is, without erasing.

(7)   Needless to say, this construction of U_f should not be the result of having computed all the values of $f(x)$ for individual x. The function f ought to be such that U_f is effectively programable by means of universal quantum logic gates in much the same sense as computations on classical computers are restricted to those programable by means of universal classical logic gates. See Notes [47.3]–[47.7] for more details.

(8)   We are now ready to enter into the core of Shor's theory. To begin with, we let L satisfy

$$2^{L-1} \le q^2 < 2^L \tag{47.8}$$

for the modulus q in our original problem (46.15); the reason for this specification of L will become apparent at (48.2)–(48.4) in the next section. We prepare two arrays of L quantum registers and let $\{|x\rangle \, |y\rangle : |x\rangle, |y\rangle \in \mathbf{b}^L\}$ be the orthonormal basis of q^{2L}. Also, let $g(x)$ be positive and less than q, and that $g(x) \equiv a^x \bmod q$. We shall consider the map U_g in the space q^{2L}. For this sake, we initialize the computer $\mathbf{r}^{2L}$ so that its state becomes $|0\rangle \, |0\rangle$ with $|0\rangle \in \mathbf{b}^L$, that is, $|\xi_{\text{in}}\rangle = |0\rangle \, |0\rangle$. Then, starting the construction of the mapping-process part of (47.2), each of the upper L qubits is modified by the Hadamard operator: $q \mapsto q$

$$\mathrm{H}: \ |0\rangle \mapsto \frac{1}{\sqrt{2}}(|0\rangle + |1\rangle), \ |1\rangle \mapsto \frac{1}{\sqrt{2}}(|0\rangle - |1\rangle). \tag{47.9}$$

This is the same as sending $|x\rangle \in \mathbf{b}$ to $2^{-1/2} \sum_{y=0,1} e(-xy/2) \, |y\rangle$, with $e(\eta) = \exp(2\pi i\eta)$. Hence, the state of $\mathbf{r}^{2L}$ becomes $2^{-L/2} \sum_{x=0}^{2^L - 1} |x\rangle \, |0\rangle$. Then we apply the operator U_g. By (47.6) the state of $\mathbf{r}^{2L}$ is supposed to become

$$\frac{1}{2^{L/2}} \sum_x |x\rangle \, |g(x)\rangle = \frac{1}{2^{L/2}} \sum_{\rho=0}^{\lambda-1} \ \sum_{x \equiv \rho \bmod \lambda} |x\rangle \, |g(\rho)\rangle \,. \tag{47.10}$$

$$\S 47 \qquad\qquad 187$$

Here, λ is the order of $a \bmod q$ as in (46.15); we know only the existence of λ. The construction of U_g in terms of universal quantum logic gates is indicated in Note [47.6].

(9) Further, we apply the finite Fourier transform

$$F_L |x\rangle = \frac{1}{2^{L/2}} \sum_{\xi} e\left(-x\xi/2^L\right) |\xi\rangle \qquad (47.11)$$

to (47.10). This is obviously unitary. Its construction in terms of universal quantum logic gates is explained in Note [47.7]. The state of our $\mathbf{r}^{2L}$ becomes now

$$\frac{1}{2^L} \sum_{\xi} \sum_{\rho=0}^{\lambda-1} \left(\sum_{x \equiv \rho \bmod \lambda} e(-x\xi/2^L) \right) |\xi\rangle |g(\rho)\rangle . \qquad (47.12)$$

Then, we make an observation of this image vector. By the axiom $(47.3)_{L \mapsto 2L}$ the output $\{\xi, g(\rho)\}$ is predicted to arise with the probability

$$P(\xi, \rho) = \frac{1}{2^{2L}} \left| \sum_{x \equiv \rho \bmod \lambda} e(-x\xi/2^L) \right|^2 . \qquad (47.13)$$

This closes the application of a quantum computer to the problem (46.16). It may be worth remarking that if we do observation at the stage (47.10), then we get each output $a^x \bmod q$ evenly, and to find the period λ, we have to repeat, in general, the operation huge times, which is absurd. Therefore, we have amplified (47.10) by means of the Fourier transform F_L so that the probabilistic distribution of the outputs gains well-discernible shading. It will be confirmed at (48.8) in the next section that F_L induces indeed a remarkable amplification effect.

The proof of Shor's Theorem continues to the next section.

Notes

[47.1] Depicted above is a scene of a finite dimensional Hilbert space enchanted with quantum mechanics axioms. Shor's algorithm is marvelously simple, and thus it ought to be presented in a way as natural as possible. Hence, as has been mentioned at the end of Remark 3 above, we have avoided those technical terms by understanding them as follows:

superposition of states $\equiv$ linear combination of vectors,

quantum parallel computing $\equiv$ application of a unitary operator.

This is mathematically formalized in (47.6): because of its linearity, a unitary operator acts on the set $\mathbf{b}^L$, thus, on the space q^L as a single entity, i.e., as

188   *Congruences*

if in parallel. We do not inquire whether in reality an application of a huge unitary operator takes effect instantaneously or not. In mathematics general the concept of time is missing, that is, operations like applying a linear mapping in a vector space do not demand us to discuss how long they take. Neither we take into account the technical problems of quantum encoding of inputs, etc., which can actually be time consuming.

[47.2]   The guiding principle we followed in this section is this: it is impossible to see/check the states of a quantum computer in the middle of computation; also, repeated runs of the same quantum computation may yield different outputs. Nevertheless, it is possible to compute mathematically the probabilistic distribution of the set of all possible outputs; and its comparison with that of the physical outputs can be used to squeeze out the answer to the original computational problem. *All I'm concerned with is that the theory should predict the results of measurements* (S.W. Hawking: Penrose (2007, p.785)).

[47.3]   Circuits in classical computers can be composed solely of NOT's and AND's; because of this fact these logic gates are called universal. The number of steps of an algorithm can be identified as that of universal logic gates used in the corresponding circuit. If this number is $O(L^d)$ with d a constant and L the bit-size of input, then the algorithm is said to be polynomial time. In the same way, circuits in quantum computers can be composed of universal quantum logic gates operating on a few qubits. This is equivalent to decomposing a unitary operator into a tensor product of small unitary operators which are to be made precise in Note [47.6]. That a quantum computation is polynomial time is defined analogously to the classical counterpart.

[47.4]   The general design of quantum circuits stemmed from the theory of reversible computing on classical computers; the term reversible means that the relevant computation is time-reversible. One may wonder whether or not it is possible to replace any computation by a reversible one without too much complication. This problem originated in the principle of Landauer (1961): the heat dissipated by a circuit is caused by erasing information at gates. Since circuit chips become quite vulnerable to heat as they shrink in physical size, the theory of reversible computation gained fundamental importance; a reversible circuit erases no information and dissipates no heat, in principle; see the next Note. Then, Lecerf (1963) and Bennett (1973) established the existence of this desired replacement. It remained yet to find reversible universal logic gates. NOT is reversible, but AND is irreversible, as it yields one bit output from input of two bits, erasing one bit. Hence, AND should be replaced by a reversible gate. A solution was invented by Toffoli (1980):

$$\text{CCNOT}: |x\rangle\,|y\rangle\,|z\rangle \mapsto |x\rangle\,|y\rangle\,|xy \oplus z\rangle \text{ on } \mathbf{b}^3. \quad (\oplus : \text{addition mod } 2)$$

This is reversible since $\text{CCNOT}^2 = 1$; moreover, specifically

$$|1\rangle\,|1\rangle\,|z\rangle \mapsto |1\rangle\,|1\rangle\,\text{NOT}(|z\rangle), \quad |x\rangle\,|y\rangle\,|0\rangle \mapsto |x\rangle\,|y\rangle\,\text{AND}(|x\rangle\,|y\rangle).$$

Therefore, any computation on classical computer can be made time-reversible, as it can be programmed solely with CCNOT's.

[47.5] The loss of information on quantum computer is postponed to the last step of computation, i.e., observation in (47.2).

[47.6] As for quantum circuits, Barenco et al (1995) established that the universal quantum logic gates are

$$\text{all unitary operators on q; and}$$
$$\text{CNOT}: |x\rangle\,|y\rangle \mapsto |x\rangle\,|x \oplus y\rangle \text{ on } q^2.$$

In particular, CCNOT which is unitary can be composed by three CNOT and four simple one-qubit operators. Consequently, any computation of polynomial time on classical computers can be translated into a polynomial time procedure on quantum computers; so, U_g is polynomial time in view of what is noted after (36.2). However, one should know that the real supremacy of quantum computing over classical cannot always be expected; the success of Shor's algorithm is due to the very fact that the integer factorization via Theorem 42 is a cycle detecting procedure and hence can be enhanced with Fourier transform which by definition acts in Hilbert spaces where quantum mechanics exactly bases itself. Namely, if an algorithm admits similar enhancements by means of Fourier transform, then its quantum version is expected to output marvels.

[47.7] An explicit decomposition of the Fourier transform F_L is as follows (Coppersmith (1994)):

$$F_L = \text{P} \cdot \text{H}_L \cdot (\text{Y}_{L-1,L} \cdot \text{H}_{L-1}) \cdot (\text{Y}_{L-2,L} \cdot \text{Y}_{L-2,L-1} \cdot \text{H}_{L-2}) \cdots$$
$$\cdots (\text{Y}_{2,L} \cdot \text{Y}_{2,L-1} \cdots \text{Y}_{2,4} \cdot \text{Y}_{2,3} \cdot \text{H}_2) \cdot (\text{Y}_{1,L} \cdot \text{Y}_{1,L-1} \cdots \text{Y}_{1,3} \cdot \text{Y}_{1,2} \cdot \text{H}_1);$$

the operand vector in q^L is to be placed after H_1. Here, P is to reverse the order of qubits; H_j is to apply the Hadamard operator H to the qubit j^{th} from the left; if the qubit k^{th} from the left is $|0\rangle$, then $\text{Y}_{j,k}$ is to do nothing, and otherwise to apply $Z(-1/2^{k-j+1})$ with

$$Z(u): \quad |0\rangle \mapsto |0\rangle, \quad |1\rangle \mapsto e(u)\,|1\rangle, \quad e(u) = \exp(2\pi i u),$$

190 *Congruences*

to the qubit j^{th} from the left. For confirmation, use the following two expressions and proceed by induction:

$$F_L\,|x\rangle \;=\; \frac{1}{2^{L/2}}(|0\rangle + e(-x/2)\,|1\rangle)(|0\rangle + e(-x/2^2)\,|1\rangle)\cdots$$
$$\cdots(|0\rangle + e(-x/2^{L-1})\,|1\rangle)(|0\rangle + e(-x/2^L)\,|1\rangle)$$

and

$$(Y_{1,L}\cdot Y_{1,L-1}\cdots Y_{1,3}\cdot Y_{1,2}\cdot H_1)\,|x_{L-1}\rangle\,|x_{L-2}\rangle\cdots|x_0\rangle$$
$$= \frac{1}{\sqrt{2}}(|0\rangle + e(-x/2^L)\,|1\rangle)\,|x_{L-2}\rangle\cdots|x_0\rangle\,.$$

According to Barenco et al (1995, Corollary 5.3), $Y_{j,k}$ is composed of a few universal quantum logic gates. Also, P is decomposed into a product of $[L/2]$ swaps of qubits, and a swap into a product of three CNOT. Hence, F_L is polynomial time.

[47.8]   The rôle of a binary quantum register can be played by electron via its spin and by photon via its polarization, etc. Physical realization of universal quantum logic gates is discussed, for instance, in Tsirelson (1997).

[47.9]   In the above we have ignored an important aspect: programs, both classical reversible and quantum, require the resource, a sort of a notepad, called ancilla bits or qubits, in order to execute jobs like C in CNOT. To construct stable quantum systems of substantial size, including ancilla qubits, is a huge technological challenge. The operator U_g, i.e., the modular exponentiation is known to demand ample ancilla resource.

§48

Now, in the identity (47.13), the left side or rather the value ξ is the result of a time evolution of the quantum system $\mathbf{r}^{2L}$, and it is physically observable. On the other hand, the right side is the result of a mathematical deduction starting at (47.3). We shall analyze this right side and squeeze out the period λ from the observed data set $\{\xi\}$.

Mathematical Discussion We put, in (47.13), $x = \rho + w\lambda$, $0 \le w < T_\rho$ with $T_\rho = |\{0 \le x < 2^L : x \equiv \rho \bmod \lambda\}|$, and sum over w, getting

$$P(\xi, \rho) = \frac{1}{2^{2L}} \left(\frac{\sin\left(\pi T_\rho \lambda \xi / 2^L\right)}{\sin\left(\pi \lambda \xi / 2^L\right)} \right)^2. \tag{48.1}$$

This right side gives a sharp local maximum when $\lambda\xi/2^L$ is close to an integer. In such situations, λ, ξ are in a peculiar relation, from which one might reach λ. Namely, we are interested in ξ with which there exists an $h \in \mathbb{N}$ such that $\lambda\xi/2^L - h$ is exceptionally small. In other words, h/λ should be an exceptionally good approximative fraction to $\xi/2^L$; note that it is possibly not irreducible. Namely, the value of h/λ is expected to appear among convergents of the continued fraction expansion of $\xi/2^L$.

Indeed, each h/λ with h, $1 \le h \le \lambda - 1$, does appear as a convergent of $\xi/2^L$ with a $\xi < 2^L$. To show this fact, we put

$$2^L h = \lambda\xi_h + \lambda\omega_h, \quad 1 \le \xi_h < 2^L, \ |\omega_h| \le \tfrac{1}{2}, \tag{48.2}$$

(Note [4.1]); here it should be remarked that $\xi_h \le 2^L(1 - 1/\lambda) + \tfrac{1}{2} < 2^L$ by (47.8). Then, since $\lambda|\xi_h - \xi_k| \ge 2^L|h - k| - \lambda$,

$$h \ne k \ \Rightarrow \ \xi_h \ne \xi_k; \tag{48.3}$$

and

$$\left| \frac{\xi_h}{2^L} - \frac{h}{\lambda} \right| = \frac{|\omega_h|}{2^L} \le \frac{1}{2^{L+1}} < \frac{1}{2q^2} < \frac{1}{2\lambda^2}. \tag{48.4}$$

In particular, by Legendre's criterion (Theorem 24), h/λ is a convergent of $\xi_h/2^L$.

Further, the irreducible fraction equal to h/λ can be determined by means of the relation

$$\left| \frac{\xi_h}{2^L} - \frac{\alpha_h}{\beta_h} \right| \le \frac{1}{2^{L+1}}, \ \langle \alpha_h, \beta_h \rangle = 1, \ \beta_h < q \ \Rightarrow \ \frac{\alpha_h}{\beta_h} = \frac{h}{\lambda}, \tag{48.5}$$

as $|\alpha_h/\beta_h - h/\lambda| < 1/q^2$ and consequently $|\alpha_h \lambda - \beta_h h| < 1$.

In passing, we make a useful observation:

$$\langle h, k \rangle = 1 \ \Rightarrow \ [\beta_h, \beta_k] = \lambda. \tag{48.6}$$

This is due to the fact that from $\lambda = d_h \beta_h = d_k \beta_k$, $d_h | h$, $d_k | k$, we get $\beta_h = d_k \tau$, $\beta_k = d_h \tau$, and thus $[\beta_h, \beta_k] = [d_h, d_k]\tau = d_h d_k \tau = \lambda$.

Thus one may wonder if any of $\{\xi_h\}$ is observed as a result of quantum computation developed in the preceding section. We now need to compute the total probability of the appearances of the numbers $\{\xi_h\}$. For this sake, we note that since $2^L/\lambda - 1 < T_\rho < 2^L/\lambda + 1$, we have $T_\rho \lambda \omega_h/2^L = \omega_h + O(|\omega_h|\lambda/2^L)$, and $\sin\left(\pi T_\rho \lambda \omega_h/2^L\right) = \sin(\pi\omega_h)(1 + O(\lambda/2^L))$. We have, with an absolute constant $c > 0$,

$$
\begin{aligned}
P(\xi_h, \rho) &\geq \frac{1}{2^{2L}}\left(\frac{\sin\left(\pi\omega_h\right)}{\sin\left(\pi\lambda\omega_h/2^L\right)}\right)^2\left(1 - c\lambda/2^L\right) \\
&\geq \frac{4}{\pi^2\lambda^2}\left(1 - c\lambda/2^L\right),
\end{aligned}
\tag{48.7}
$$

where we have used the fact that $|\sin(\pi\omega_h)| \geq 2|\omega_h|$ and $|\sin\left(\pi\lambda\omega_h/2^L\right)| \leq \pi\lambda|\omega_h|/2^L$. Hence, as q is supposed to be sufficiently large, i.e., via (47.8), we have

$$
\sum_{h=1}^{\lambda-1}\sum_{\rho=0}^{\lambda-1} P(\xi_h, \rho) > \tfrac{2}{5}.
\tag{48.8}
$$

This inner sum is the predicted probability of the event that ξ_h appears as a result of observation. On noting (48.3), or the independence of these events, we conclude that the probability of one of $\{\xi_h\}$ being observed is theoretically expected to be greater than $\tfrac{2}{5}$.

Measurement implication Based on (48.8) we shall make an intuitive discussion concerning physically observed integers $\{\xi\}$. Thus, if the quantum computation (47.9)–(47.12) is repeated 12 times, then we get 12 outputs $\{\xi\}$. We check whether or not a particular ξ corresponds to a convergent κ/γ such that

$$
\left|\frac{\xi}{2^L} - \frac{\kappa}{\gamma}\right| \leq \frac{1}{2^{L+1}}, \quad \gamma < q.
\tag{48.9}
$$

This can be done in polynomial time on classical computers (Note [3.2]). Those ξ's which survive the screening (48.9) have a unique κ/γ, as can be seen in the same way as what follows (48.5). Denote these convergents by $\kappa_1/\gamma_1, \kappa_2/\gamma_2, ..., \kappa_s/\gamma_s, s \leq 12$. Then almost certainly $s \geq 4$, since by (48.8) the number of $\{\xi_h\}$ which are actually observed is almost certainly greater than 4, and these ξ_h satisfy (48.9) because of the first inequality in (48.4). Further, by Note [18.5], it is almost certain that at least two of these ξ_h satisfy the condition in (48.6). Namely, the period λ should be found in the set $\{[\gamma_j, \gamma_k] : j, k \leq s\}$ with a probability greater than a positive absolute constant.

Finally, via Notes [47.3]–[47.7], the operators U_g, F_L are polynomial time in terms of universal quantum logic gates, and we end the proof of Theorem 43.

The repetition of 12 times is for illustration; we should say actually that the quantum computation needs to be repeated sufficiently many times. Also, one may use Note [18.6] instead of Note [18.5]. Then the estimation (48.8) is replaced by

$$\sum_{\substack{h=1 \\ \langle h, \lambda \rangle = 1}}^{\lambda-1} \sum_{\rho=0}^{\lambda-1} P(\xi_h, \rho) > \frac{2}{5} \cdot \frac{\varphi(\lambda)}{\lambda} > \frac{c}{\log \log q} \tag{48.10}$$

with an absolute constant $c > 0$. That is, if our quantum computation is repeated more than $[c^{-1} \log \log q]$ times, then the period λ should be detected almost certainly.

We restate what has been indicated already in Note [47.6]: The merit of Shor's theory lies in its ability of detecting cycles. Thus, among the steps in the last section, the application of the Fourier transform F_L is the most decisive; it enhances biases hidden in (47.10) so that useless data are virtually suppressed. States of a quantum system are thought to span a Hilbert space, whence the application of F_L becomes possible.

Fourier analysis or harmonic analysis is one of the fundamentals throughout our life because of its typical applications like medical scanning and remote sensing. As a matter of fact, an origin of Fourier analysis is in the theory of algebraic equations and number theory, or more precisely, in the theory of Lagrange–Vandermonde resolvents, as is to be described in the next chapter.

§49

The list of gigantic integers conquered by modern factorization methods is impressive indeed. The practical algorithms thus devised to run on classical computers are naturally highly sophisticated both mathematically and technologically. However, the basic principles from which these methods stemmed are quite simple, and it appears that they can be traced back largely to the pioneering contributions by

$$\text{(i) Fermat (ii) Euler (iii) Legendre} \\ \text{(iv) Kraïtchik (v) Pollard,} \tag{49.1}$$

in historical order. In the present section we shall give brief accounts of (i) (iv) (v); note that Pollard's ρ-method has been presented in §38, so we shall treat below another method of his. Other two need knowledge available later in the present essays; thus, (ii) will be touched on in Notes [78.11] and [81.8], and (iii) in Note [84.11]. In addition, an idea of Šimerka is mentioned in Note [92.7]; see also Note [75.3].

(i) Fermat (ca 1643: Œuvres II, pp.256–258): For two odd integers $a > b > 2$, let $\alpha = (a+b)/2$, $\beta = (a-b)/2$; then $ab = \alpha^2 - \beta^2$. Thus, provided $|a-b|$ is small, the odd integer $n = ab$ becomes a square, if a small square is added. In other words, if we can find a relatively small integer $v > 0$ such that

$$([\sqrt{n}] + v)^2 - n \tag{49.2}$$

is a square, then we shall be able to obtain a factorization of n. Fermat's own example is $n = 2027651281$. We have $[\sqrt{n}] = 45029$, $([\sqrt{n}] + 12)^2 - n = 1040400 = 1020^2$. Hence $n = (45041 + 1020)(45041 - 1020)$; that is, the factorization $n = 46061 \cdot 44021$ results. If $|a-b|$ is supposed to be large, then one may try to choose an integer t such that $|ta - b|$ should become small; adding a small square to $tn = tab$ we get a square. That is, we check whether $([\sqrt{tn}] + v)^2 - tn$ becomes a square. For instance, with $n = 94734901$ we have $[\sqrt{17n}] = 40130$, and $([\sqrt{17n}] + 1)^2 - 17n = 3844 = 62^2$. Hence, $17n = (40131 + 62)(40131 - 62) = 40193 \cdot 2357 \cdot 17$, which gives $n = 40193 \cdot 2357$. If the multiplier 17 is not used, then $([\sqrt{n}] + 11542)^2 - n = 18918^2$. It is generally a hard task to find convenient multipliers.

(iv) Kraïtchik (1926, Chapitre XIV): Viewing Fermat's method from a different angle, we note that under the assumption $\alpha^2 \equiv \beta^2$, $\alpha \not\equiv \pm\beta \bmod n$, we see that $\langle \alpha + \beta, n \rangle$ is a non-trivial factor of n. Kraïtchik devised the following method to choose α, β. Take first integers g_j which are close to $\sqrt{n}$ (can be assumed naturally to be coprime to n); then collect relatively small quadratic residues $g_j^2 \equiv h_j \bmod n$; these h_j are required to be decomposable into products of small primes. Looking into their prime power decompositions, try to find $\eta_j = \pm 1$ such that

$$\alpha^2 \equiv \prod_j g_j^{2\eta_j} \equiv \prod_j h_j^{\eta_j} \equiv \beta^2 \bmod n; \tag{49.3}$$

note that Kraïtchik himself did not use the exponents η_j but an equivalent arrangement which is indicated by (49.4) below. Then, $\langle \alpha + \beta, n \rangle$ could be a factor of n. If this choice of $\{h_j, \eta_j\}$ does not work well, then other candidates of quadratic residues are to be tried. Kraïtchik gave a variety of examples. One of them is as follows: For $n = 453 \cdot 2^{30} + 1 = 486405046273$, he found (but we have made some corrections) that

$$692697^2 \equiv -2^4 \cdot 11^2 \cdot 19^2 \cdot 97^2,$$
$$-2^5 \cdot 3^4 \cdot 11^2 \cdot 19 \cdot 31 \equiv 697295^2,$$
$$697457^2 \equiv 2^5 \cdot 3^7 \cdot 19 \cdot 31,$$
$$2^4 \cdot 3 \cdot 23^2 \cdot 127^2 \equiv 697721^2; \tag{49.4}$$

$$\S49 \qquad\qquad 195$$

that is, with $\eta_1 = 1, \eta_2 = -1, \eta_3 = 1, \eta_4 = -1$, one gets

$$(692697 \cdot 697457 \cdot 23 \cdot 127)^2 \equiv (19 \cdot 97 \cdot 697295 \cdot 3 \cdot 697721)^2,$$
$$692697 \cdot 697457 \cdot 23 \cdot 127 + 19 \cdot 97 \cdot 697295 \cdot 3 \cdot 697721 \qquad (49.5)$$
$$= 4101166640634864;$$

and

$$\langle q, 4101166640634864 \rangle = 135433,$$
$$\Rightarrow 453 \cdot 2^{30} + 1 = 135433 \cdot 3591481. \qquad (49.6)$$

These two factors are both prime numbers. The device of applying a small multiplier as in (i) is effective. Continued to Note [84.12].

(v) Pollard (1974), or the $(p-1)$-method: We observe first that if $\langle a, n \rangle = 1$ and $p|n$, then $p|\langle a^{p-1} - 1, n \rangle$. We shall exploit this triviality as a means of factoring n. Thus, we choose an $a = a_0$, and assuming $\langle a_0, n \rangle = 1$ naturally, we examine the sequence

$$\langle a_j - 1, n \rangle, \quad a_{j+1} \equiv a_j^{j+1} \equiv a^{(j+1)!} \bmod n, \, j \geq 0. \qquad (49.7)$$

Behind the computation mod n, the prime modulus p is covertly working; it is the same as in the ρ algorithm (§38). Namely, as soon as $J! \equiv 0 \bmod (p-1)$, we shall have $\langle a_J - 1, n \rangle \neq 1$, and a possibility of extracting a factor of n arises. If this procedure is viewed through modulo p, then the sequence $\{a_j\}$ reaches the stagnant state $a_j \equiv 1, \forall j \geq J$. If the procedure fails, that is, if $\langle a_J - 1, n \rangle = n$, then the base a is to be changed. For instance, with $n = 143771437961$ and $a = 2$, we have $\langle a_j - 1, n \rangle = 1, j \leq 28$, and

$$a_{29} \equiv 124688060775, \langle a_{29} - 1, n \rangle = 430883 \Rightarrow n = 430883 \cdot 333667;$$
$$(49.8)$$

these two factors are primes. The reason that the procedure has worked well is in the special structures $430883 - 1 = 2 \cdot 17 \cdot 19 \cdot 23 \cdot 29$, which was exploited in Note [43.9], and $333667 - 1 = 2 \cdot 3^3 \cdot 37 \cdot 167$. In this case, since 2 is a primitive root with respect to both factors, we had to examine up to a_{29}. However, if we choose $a = 10$, then we get $\langle a_6 - 1, n \rangle = 333667$. This is due to the fact that the order of 10 mod 333667 is equal to 9, and $j! \equiv 0 \bmod 9$ is attained at $j = 6$. Thus, one may infer that the $(p-1)$-method is particularly effective when n has a prime factor of a peculiar structure like 430883.

Notes

[49.1]   With the aim and the scope of the present essays in mind, we have restricted ourselves to a brief account of the origins of factorization methods. For a thorough discussion of factorization algorithms both old and modern, see Crandall–Pomerance (2005, Chapters 5–7).

196 *Congruences*

[49.2] Once an integer factorization method of polynomial time, such as the large scale implementation of Shor's algorithm on quantum computers, which seems yet to have a long way to go though, is achieved, the public key cryptosystem RSA, which is ubiquitously applied in our daily life, will become insecure since it is based on the apparent difficulty of factoring huge integers by means of classical computing systems.

[49.3] The core of RSA is as follows:

(a) The recipient (= X) of a message takes two different large primes p_1, p_2 and computes $q = p_1 p_2$ as well as $\varphi(q)$. Also, X takes a positive $\alpha < \varphi(q)$, $\langle \alpha, \varphi(q) \rangle = 1$, and fix β such that $\alpha\beta \equiv 1 \bmod \varphi(q)$, $0 < \beta < \varphi(q)$. X keeps the set $\{p_1, p_2, \varphi(q), \alpha\}$ in secrecy. The integers α, β are called the secret and the public exponents, respectively.

(b) X lets only $\{\beta, q\}$ be available to the sender (= Y) of a message via a public, i.e., insecure, channel.

(c) Y transforms the message into an integer A according to ASCII code system; then Y computes $A^\beta \bmod q$, and sends the result to X via a public channel.

(d) X computes $(A^\beta)^\alpha \bmod q$; then Euler's theorem (28.10) yields $A \bmod q$. X gains the message from Y via ASCII table, provided $0 < A < q$, $\langle A, q \rangle = 1$.

This works only under certain premise like that the size of A should be smaller than $\min\{p_1, p_2\}$; thus, it is preferable to have $p_2/2 < p_1 < p_2$ and $A < (q/2)^{1/2}$, say. Also, as (i)–(v) above indicate, the selection of p_1, p_2 should be made with considerable care. Moreover, it is known that the size of the secrete exponent α does matter; see Note [49.6]. Despite these and further advanced precautions, there are still various ways to attack RSA. Yet, the difficulty to decompose q into the product $p_1 p_2$ will guard the system because the decomposition is necessary to acquire the value of $\varphi(q)$ without which anyone is unable to perform the decryption procedure (d). However, this is believed only within the present state of the art in conventional computing technology.

[49.4] As is well known nowadays, prior to its publication by Rivest et al (1978), RSA had already been invented secretly by J.H. Ellis and C.C. Cocks at an agency of the British government. Ellis (1987) gives a fascinating historical account of the birth of non-secret encryption: Ellis got his idea late in 1969. It came to him from *the discovery of a wartime, Bell-Telephone report* by an unknown author describing an ingenious idea for secure telephone speech. It proposed that the recipient should mask the sender's speech by adding noise to the line. He could subtract the noise afterward since he had added it and* *therefore knew what it was* (* Final report on project C43. Bell Lab., 1944). A mathematical implementation of Ellis' idea was achieved in 1973 by his colleague Cocks. Thus, in terms of the above protocol of RSA, (c) or A^β mod q corresponds to the masking by noise, and (d) to the subtraction of the noise. However, their invention documents were declassified only in 1997. In the meanwhile, RSA was invented independently, and since then it has been flourishing as a fundamental device supporting the modern digital life.

[49.5] RSA is actually an implementation of the theoretical protocol due to Diffie–Hellman (1976) for public key cryptosystem. Their article also contains a proposal of a public key distribution system which depends on the apparent difficulty of computing the discrete logarithm on classical computers. Again the same had been invented by M.J. Williamson in 1974 secretly at the agency indicated above, as is described in Ellis (1987). By the way, Shor (1994, 1997) presents a polynomial time computation of discrete logarithms on quantum computers; his idea is not much different from that on integer factorization. Thus, the fate of the Diffie–Hellman public key distribution system will be just the same as that of RSA.

[49.6] The attack by Wiener (1990) on RSA is interesting, as it relies on the theory of continued fractions. A simplified account of it is as follows: If $p_1 < p_2$ and $\alpha < (p_1/4)^{1/2}$ in the above formulation of RSA, then the secret exponent α can be detected in polynomial time via the public data $\{\beta, q\}$. For the confirmation, let $\alpha\beta = 1 + k\varphi(q)$ and observe that

$$\begin{aligned}
|\beta/q - k/\alpha| &= (p_1 + p_2)(k - (k+1)/(p_1 + p_2))/(\alpha q) \\
&< (p_1 + p_2)(k + 1/\varphi(q))/(\alpha q) = (p_1 + p_2)(\beta/\varphi(q))/q \\
&< (p_1 + p_2)/q < 2/p_1.
\end{aligned}$$

Hence,

$$\alpha < (p_1/4)^{1/2} \implies |\beta/q - k/\alpha| < 1/(2\alpha^2).$$

According to Legendre's criterion (Theorem 24), the irreducible fraction k/α is a convergent of β/q; that is, by the discussion in §21 it is possible to find α and k by Euclid's algorithm applied to the pair $\{\beta, q\}$, thus in polynomial time with respect to q. This implies further that $\varphi(q) = (\alpha\beta - 1)/k$ and consequently the decomposition $q = p_1 p_2$ can be obtained solely from $\{\beta, q\}$ since p_1, p_2 are to satisfy the quadratic equation $x^2 - (q + 1 - \varphi(q))x + q = 0$. Here is an example, although too small for a real-life warning: with

$$p_1 = 320009, p_2 = 430897, \alpha = 281, \beta = 51524795561,$$

we have $\alpha < (p_1/4)^{1/2}$ and

$$\frac{\beta}{q} = 0 + \frac{1}{2} + \frac{1}{1} + \frac{1}{2} + \frac{1}{11} + \frac{1}{3} + \frac{1}{5} + \cdots + \frac{1}{6}.$$

The fifth convergent is $105/281 = k/\alpha$; and $(\alpha\beta - 1)/k = 137890167168$, so $q + 1 - \varphi(q) = 750906$. Solving $x^2 - 750906x + 137890918073 = 0$, we recover the current p_1, p_2. Note that as the above proof indicates those $\alpha > (p_1/4)^{1/2}$, $\langle \alpha, \varphi(q) \rangle = 1$, do not automatically guarantee secure secret exponents. For instance, with $\alpha = 671 > (p_1/4)^{1/2}$ we have $\beta = 8630979167$, and the third convergent of β/q is $42/671 = k/\alpha$.

3

Characters

§50

We shall further investigate the multiplicative structure among deviations from being divisible. To this end, we shall exploit Fourier analysis on the additive group $\mathbb{Z}/q\mathbb{Z}$ and on the multiplicative group $(\mathbb{Z}/q\mathbb{Z})^*$. The discussion of the preceding and present chapters yields, among other things, a solution to the fundamental problem $(43.5)_{\ell=2}$.

A function f on $\mathbb{Z}/q\mathbb{Z}$ is the same as an arithmetic function of period q. It can be expanded into Fourier series. The relevant complete system of basic harmonic waves is $\Psi_q = \{\psi_{a/q} : a \bmod q\}$, $\psi_{a/q}(n) = e(an/q)$, where $e(x) = \exp(2\pi i x)$ as before. Namely, any such function f can be expressed uniquely as a linear combination of the elements of Ψ_q: We have

$$f(n) = \frac{1}{\sqrt{q}} \sum_{a \bmod q} \widehat{f}(a) \psi_{a/q}(n), \tag{50.1}$$

where the Fourier coefficients $\widehat{f}(a)$ are given by

$$\widehat{f}(a) = \frac{1}{\sqrt{q}} \sum_{k \bmod q} f(k) \overline{\psi}_{a/q}(k), \quad \overline{\psi}_{a/q}(k) = e(-ak/q). \tag{50.2}$$

One may recover the left side of (50.1) by inserting (50.2) into the right side and by observing the orthogonal relation

$$\frac{1}{q} \sum_{a \bmod q} \psi_{a/q}(u) \overline{\psi}_{a/q}(v)$$

$$= \frac{1}{q} \sum_{a \bmod q} e((u-v)a/q) = \begin{cases} 1 & u \equiv v \bmod q, \\ 0 & u \not\equiv v \bmod q. \end{cases} \tag{50.3}$$

As a consequence, the Parseval identity

$$\sum_{n \bmod q} |f(n)|^2 = \sum_{a \bmod q} |\widehat{f}(a)|^2 \tag{50.4}$$

holds: In the triple sum obtained by inserting (50.2) into this right side, move the summation over a innermost, and apply (50.3).

Changing the point of view, each function $\psi_{a/q}$ can be regarded as a homomorphism from an additive group to a multiplicative group:

$$\psi_{a/q} : \mathbb{Z}/q\mathbb{Z} \to \{z \in \mathbb{C} : |z| = 1\}. \tag{50.5}$$

That is, for any $m, m', n \in \mathbb{Z}$,

$$\begin{aligned} m \equiv m' \bmod q &\Rightarrow \psi_{a/q}(m) = \psi_{a/q}(m'), \\ |\psi_{a/q}(m)| = 1, \quad &\psi_{a/q}(m+n) = \psi_{a/q}(m) \cdot \psi_{a/q}(n). \end{aligned} \tag{50.6}$$

Conversely, if an arithmetic function ψ satisfies these conditions, then $\psi(0) = 1$ since $\psi(0) \neq 0$ and $(\psi(0))^2 = \psi(0+0) = \psi(0)$; thus $(\psi(1))^q = \psi(q) = \psi(0) = 1$, and there exists $a \bmod q$ such that $\psi(1) = e(a/q)$; hence, $\psi(n) = (\psi(1))^n = \psi_{a/q}(n)$ for any $n \in \mathbb{N}$. Also, for $n < 0$, we have $\psi(n)\psi(-n) = 1$, and $\psi(n) = \psi_{a/q}(n)$. Namely, (50.6) uniquely determines the system Ψ_q. With this construction, we call $\psi_{a/q}$ an additive character $\bmod q$ or a character of the additive group $\mathbb{Z}/q\mathbb{Z}$. The set Ψ_q is closed with respect to the multiplication:

$$\begin{aligned} (\psi_{a/q}\psi_{b/q})(n) &= \psi_{a/q}(n)\psi_{b/q}(n) = \psi_{(a+b)/q}(n), \\ &\Rightarrow \psi_{a/q}\psi_{b/q} = \psi_{h/q}, \ h \equiv a+b \bmod q \,; \end{aligned} \tag{50.7}$$

hence, it is an Abelian group and called the additive character group of $\mathbb{Z}/q\mathbb{Z}$; its unit element is ψ_0, i.e., $\psi_0(n) = 1$ for any integer n, and the inverse element of $\psi_{a/q}$ is $\overline{\psi}_{a/q}$. We also have the orthogonality in Ψ_q

$$\frac{1}{q} \sum_{n \bmod q} (\psi_{a/q}\overline{\psi}_{b/q})(n) = \begin{cases} 1 & a \equiv b \bmod q \ \Leftrightarrow \ \psi_{a/q} = \psi_{b/q}, \\ 0 & a \not\equiv b \bmod q \ \Leftrightarrow \ \psi_{a/q} \neq \psi_{b/q}. \end{cases} \tag{50.8}$$

The following assertion should now be obvious:

Theorem 44 *We have the isomorphism:*

$$\begin{aligned} \Psi_q &\cong \mathbb{Z}/q\mathbb{Z}, \\ \psi_{a/q} &\mapsto a \bmod q. \end{aligned} \tag{50.9}$$

This relation together with (50.3) and (50.8) is called as the duality between the additive group $\mathbb{Z}/q\mathbb{Z}$ and its additive character group Ψ_q, a concept which readily generalizes to finite Abelian groups.

Here is a trivial remark:

$$h \equiv 0 \bmod q \;\Leftrightarrow\; \psi(h) = 1 \text{ for all } \psi \in \Psi_q,$$
$$\Psi_q \ni \psi = 1 \;\Leftrightarrow\; \psi(h) = 1 \text{ for all } h \bmod q. \tag{50.10}$$

Notes

[50.1]   It is Dirichlet (1835) who connected number theory with the harmonic analysis of Fourier and Poisson, the two prominent disciples of Lagrange. In 1825, Dirichlet, then a student, made their acquaintance. See Elstrodt (2007), which is a rich and touching biography of the founder of analytic number theory.

[50.2]   The Ramanujan sum c_q is of period q, and

$$\widehat{c_q}(a) = \begin{cases} \sqrt{q} & \langle a, q \rangle = 1, \\ 0 & \langle a, q \rangle \neq 1. \end{cases}$$

Hence, (50.4) gives

$$\frac{1}{q} \sum_{n=1}^{q} |c_q(n)|^2 = \varphi(q).$$

Alternatively, according to the explicit formula (18.12), this left side equals

$$\frac{\varphi^2(q)}{q} \sum_{n=1}^{q} \frac{\mu^2(q/\langle n, q \rangle)}{\varphi^2(q/\langle n, q \rangle)} = \frac{\varphi^2(q)}{q} \sum_{d \mid q} \frac{\mu^2(q/d)}{\varphi^2(q/d)} \varphi(q/d)$$

$$= \frac{\varphi^2(q)}{q} \prod_{p \mid q} \left(1 + \frac{1}{\varphi(p)}\right) = \varphi(q).$$

[50.3]   Let $J, q \geq 1$, and let R_j, $1 \leq j \leq J$, be J residue systems $\bmod q$ which are chosen arbitrarily. Then, each residue class $\bmod q$ is represented q^{J-1} times by the q^J sums $s = \sum_j r^{(j)}$, $r^{(j)} \in R_j$. A confirmation can be made by setting $x = 1$ in the identity

$$\frac{1}{q} \sum_{h=0}^{q-1} \overline{\psi}_{h/q}(k) \prod_{j=1}^{J} \left(\sum_{r \in R_j} \psi_{h/q}(r) x^r\right) = \sum_{s \equiv k \bmod q} x^s.$$

Alternatively, observe that $R_j \bmod q$ is an Abelian group, which is of course isomorphic to $\mathbb{Z}/q\mathbb{Z}$. With this, consider the surjective map

$$(R_1 \bmod q) \oplus \cdots \oplus (R_J \bmod q) \;\to\; \mathbb{Z}/q\mathbb{Z},$$
$$\left(r^{(1)} \bmod q, \ldots, r^{(J)} \bmod q\right) \mapsto \sum_{j=1}^{J} r^{(j)} \bmod q.$$

Then, apply the fundamental theorem on homomorphisms.

202 — Characters

[50.4] The relation between divisibility and (50.3) has been already indicated at (18.13). We shall extend it as follows: The function $\iota(\cdot/d)$, $d \in \mathbb{N}$, with the ι as in (13.2) signals the divisibility by d. Hence, the system $\{\iota(\cdot/d) : d \leq D\}$ can be regarded as a basic means in discussing the divisibility with respect to those divisors not greater than D. On the other hand, $\iota(\cdot/d)$ is of period d, and its Fourier expansion is

$$\iota(n/d) = \frac{1}{d}\sum_{u=1}^{d} \psi_{u/d}(n) = \frac{1}{d}\sum_{q|d}\sum_{\substack{h=1 \\ \langle h,q \rangle = 1}}^{q} \psi_{h/q}(n) = \frac{1}{d}\sum_{q|d} c_q(n),$$

where $c_q(n)$ is the Ramanujan sum defined by (18.9). Then, invoking the definition (20.1), it can be claimed that divisibility, in general, is closely related to the system $\{\psi_\tau : \tau \in \mathcal{F}_D\}$ with variable D, where $\mathcal{F}_D$ is as in (20.1). In other words, issues involving divisibility may be translated into those concerning the distribution of irreducible fractions in the unit interval. The recognition of this fundamental structure was first made and successfully exploited in Linnik (1941). The large sieve, which originated in the title of this article of his, has since then evolved into one of the most fundamental methods, or more precisely, a major paradigm in analytic number theory. What Linnik actually exploited is, in modern terminology, the quasi-orthogonality of the system $\{\psi_\tau : \tau \in \mathcal{F}_D\}$. A precise treatment of this notion will be given in Chapter 5.

[50.5] Fourier expansion of an arbitrary arithmetic function f: We have

$$f(n) = \sum_{d|n}(\mu * f)(d) = \sum_{d}\iota(n/d)(\mu * f)(d)$$

$$= \sum_{d}\frac{(\mu * f)(d)}{d}\sum_{q|d}c_q(n) = \sum_{q}\widetilde{f}(q)c_q(n),$$

where $\widetilde{f}(q) = \sum_d(\mu * f)(d)/d$, $d \equiv 0 \bmod q$, which is assumed to be convergent. For instance, with $\operatorname{Re}\alpha < 0$, we have $\widetilde{\sigma}_\alpha(q) = \zeta(1-\alpha)q^{\alpha-1}$, as $(\mu * \sigma_\alpha)(d) = d^\alpha$; this proves (18.19) again. Since c_q is a sum of additive characters, one may exploit this expansion of f in order to separate the two integers trapped in $f(m+n)$, a fact relevant to the basic issue raised in Note [15.2].

§51

Next, we shall consider Fourier analysis on the group $(\mathbb{Z}/q\mathbb{Z})^*$. In view of the discussion of the preceding section, it suffices to determine a relevant complete system of basic harmonic waves.

$$\S51 \qquad\qquad 203$$

To this end, we may start with a multiplicative analogue of (50.5), but we shall adopt rather the constructive reasoning of Dirichlet (1837a, §8; 1863, §§131, 133), appreciating pioneer's contribution. We shall thus exploit the structure theorem (44.4) and the argument of the preceding section. Namely, on noting that each $Z^+_{p^{q(p)}}$ is a group similar to that treated there, we define the following function ξ on $(\mathbb{Z}/q\mathbb{Z})^*$ via the mapping (44.5): Providing $\langle n, q \rangle = 1$,

$$\xi(n) = \begin{cases} e\left(\sum_{p\geq 2} h_p \mathrm{Ind}_{r_p}(n)/\varphi(p^{q(p)})\right) & \text{if } q(2) \leq 2, \\ \xi^{(2)}(n)e\left(\sum_{p\geq 3} h_p \mathrm{Ind}_{r_p}(n)/\varphi(p^{q(p)})\right) & \text{if } q(2) \geq 3, \end{cases} \tag{51.1}$$

$$\xi^{(2)}(n) = e\left(k_1 v/2 + k_2 w/2^{q(2)-2}\right), \quad n \equiv (-1)^v 5^w \bmod 2^{q(2)}, \tag{51.2}$$

where $h_p \bmod \varphi(p^{q(p)})$, $k_1 \bmod 2$, $k_2 \bmod 2^{q(2)-2}$ are arbitrary. (51.3)

Those p such that $q(p) = 0$ do not participate in (51.1). The set Ξ_q of all ξ's does not depend on the selection of primitive roots $r_p \bmod p$, $p \geq 3$, as far as the condition (42.1) is fulfilled. That is, the use of other primitive roots $r'_p \bmod p$ causes only a permutation in Ξ_q because h_p is then replaced by $\mathrm{Ind}_{r'_p}(r_p)h_p$, which implies a permutation in $\{h_p \bmod \varphi(p^{q(p)})\}$ since $\langle \mathrm{Ind}_{r'_p}(r_p), \varphi(p^{q(p)}) \rangle = 1$.

On this construction, the mapping

$$\xi : (\mathbb{Z}/q\mathbb{Z})^* \rightarrow \{z \in \mathbb{C} : |z| = 1\}, \tag{51.4}$$

is a homomorphism, for $\xi(mn) = \xi(m)\xi(n)$, $\langle mn, q \rangle = 1$, as is readily seen from the definitions (44.1)–(44.3). Also, ξ's are different from each other as functions on $(\mathbb{Z}/q\mathbb{Z})^*$. To see this, suppose $2 \nmid q$. Let $\xi, \xi' \in \Xi_q$ be such that there exists a $p|q$ for which $h_p \not\equiv h'_p \bmod \varphi(p^{q(p)})$ with an obvious correspondence. We take an n such that $n \equiv r_p \bmod p^{q(p)}$, $n \equiv 1 \bmod q/p^{q(p)}$. We have then $\xi(n) = e\left(h_p/\varphi(p^{q(p)})\right) \neq e\left(h'_p/\varphi(p^{q(p)})\right) = \xi'(n)$, which proves the claim. The case $2|q$ is similar. In particular, we have

$$|\Xi_q| = \varphi(q). \tag{51.5}$$

Conversely, such a homomorphism η like (51.4) should belong to Ξ_q. To confirm this, let

$$\varrho_p \equiv r_p \bmod p^{q(p)}, \quad \varrho_p \equiv 1 \bmod q/p^{q(p)}$$
$$\varrho_p^{\varphi(p^{q(p)})} \equiv 1 \bmod q, \tag{51.6}$$

where r_p is as before. Provided $2 \nmid q$, we have, for any reduced residue $n \bmod q$,

$$n \equiv \prod_p \varrho_p^{\mathrm{Ind}_{r_p}(n)} \bmod q. \tag{51.7}$$

On noting that $(\eta(\varrho_p))^{\varphi(p^{q(p)})} = \eta(\varrho_p^{\varphi(p^{q(p)})}) = \eta(1) = 1$, we see that

$$\eta(n) = \prod_p \left(\eta(\varrho_p)\right)^{\operatorname{Ind}_{r_p}(n)} = \prod_p e\left(f_p \operatorname{Ind}_{r_p}(n)/\varphi(p^{q(p)})\right) \tag{51.8}$$

with certain f_p's. This means that $\eta \in \Xi_q$. As for the case $2|q$, if $q(2) \le 2$, then (51.6) can be supposed for $p = 2$ as well. If $q(2) \ge 3$, then we choose $\lambda_1, \lambda_2 \bmod q$ such that

$$\begin{aligned} \lambda_1 &\equiv -1, & \lambda_2 &\equiv 5 & &\bmod 2^{q(2)}, \\ \lambda_1 &\equiv 1, & \lambda_2 &\equiv 1 & &\bmod q/2^{q(2)}. \end{aligned} \tag{51.9}$$

By Theorem 40, we have, for any reduced residue class $n \bmod q$,

$$n \equiv \lambda_1^v \lambda_2^w \prod_{p>2} \varrho_p^{\operatorname{Ind}_{r_p}(n)} \bmod q, \quad v \bmod 2, \ w \bmod 2^{q(2)-2}. \tag{51.10}$$

Hence

$$\eta(n) = \eta^v(\lambda_1)\eta^w(\lambda_2) \prod_{p>2} \left(\eta(\varrho_p)\right)^{\operatorname{Ind}_{r_p}(n)}, \tag{51.11}$$

$$\left(\eta(\lambda_1)\right)^2 = 1, \quad \left(\eta(\lambda_2)\right)^{2^{q(2)-2}} = 1.$$

We have $\eta \in \Xi_q$ and end the confirmation.

We now call ξ a multiplicative character $\bmod q$ or just a character of the group $(\mathbb{Z}/q\mathbb{Z})^*$. The set Ξ_q is closed with respect to the multiplication

$$(\xi\xi')(n) = \xi(n)\xi'(n), \quad \langle n,q \rangle = 1, \tag{51.12}$$

and it becomes an Abelian group, called the character group of $(\mathbb{Z}/q\mathbb{Z})^*$. Its unit is the identity map: $n \mapsto 1$ for $\langle n,q \rangle = 1$; and the inverse of ξ is the complex conjugate $\overline{\xi}$. More precisely, the product (51.12) is the same as the coordinate-wise sum of h_p and k_1, k_2 in (51.1)–(51.2). Therefore, via (44.4), we obtain the duality assertion:

Theorem 45 *With the construction* (51.1), *we have the isomorphism*

$$\Xi_q \cong (\mathbb{Z}/q\mathbb{Z})^*. \tag{51.13}$$

Moreover, we have the orthogonal relations corresponding to (50.3) and (50.8):

$$\frac{1}{\varphi(q)} \sum_{\xi \in \Xi_q} \xi(m)\overline{\xi}(n) = \begin{cases} 1 & m \equiv n \bmod q, \\ 0 & m \not\equiv n \bmod q, \end{cases} \quad \langle mn, q \rangle = 1 \tag{51.14}$$

$$\frac{1}{\varphi(q)} \sum_{\substack{n=1 \\ \langle n,q \rangle=1}}^{q} \xi(n)\overline{\xi'}(n) = \begin{cases} 1 & \xi = \xi', \\ 0 & \xi \neq \xi'. \end{cases} \tag{51.15}$$

The sum in (51.14) is equivalent to the multiple sums over h_p and k_1, k_2; for each $p > 2$ we have $\sum_{h_p=1}^{\varphi(p^{q(p)})} e\big(h_p(\mathrm{Ind}_{r_p}(m) - \mathrm{Ind}_{r_p}(n))/\varphi(p^{q(p)})\big)$, which is equal to $\varphi(p^{q(p)})$ if $\mathrm{Ind}_{r_p}(m) \equiv \mathrm{Ind}_{r_p}(n) \bmod \varphi(p^{q(p)})$, that is, if $n \equiv m \bmod p^{q(p)}$; otherwise the sum vanishes. The treatment of the case $p = 2$ may be skipped. As for (51.15), the argument is similar.

Finally, the completeness of Ξ_q, that is, the fact that any function g on $(\mathbb{Z}/q\mathbb{Z})^*$ can be expressed uniquely as a linear combination of the elements of Ξ_q is confirmed via (51.14): for $\langle n,q \rangle = 1$,

$$g(n) = \frac{1}{\sqrt{\varphi(q)}} \sum_{\xi \in \Xi_q} \widehat{g}(\xi)\xi(n), \tag{51.16}$$

where

$$\widehat{g}(\xi) = \frac{1}{\sqrt{\varphi(q)}} \sum_{\substack{k=1 \\ \langle k,q \rangle=1}}^{q} g(k)\overline{\xi}(k). \tag{51.17}$$

We also have the Parseval identity

$$\sum_{\substack{n \bmod q \\ \langle n,q \rangle=1}} |g(n)|^2 = \sum_{\xi \in \Xi_q} |\widehat{g}(\xi)|^2. \tag{51.18}$$

Here is a trivial remark, an analogue of (50.10):

$$\begin{aligned} h \equiv 1 \bmod q &\Leftrightarrow \xi(h) = 1 \text{ for all } \xi \in \Xi_q, \\ \Xi_q \ni \xi = 1 &\Leftrightarrow \xi(h) = 1 \text{ for all reduced } h \bmod q. \end{aligned} \tag{51.19}$$

In passing, we stress that any $\xi \in \Xi_q$ is not an arithmetic function since it is not defined over $\mathbb{Z}$. In particular, it does not have its Dirichlet series.

Notes

[51.1]   The introduction of (51.1)–(51.3) by Dirichlet (1837a) was a pivotal event in the history of number theory. It should, however, be noted that the function $e(h\mathrm{Ind}_r(n)/(p-1)) \in \Xi_p$ had already appeared in [DA, art.360]. See §70(viii) below.

[51.2]   We shall extend the discussion §§50–51 to a generic finite Abelian group A; see Notes [30.4]–[30.6]: Let A be multiplicative. A character $\varkappa$ of A is a homomorphism

$$\varkappa : A \to \{z \in \mathbb{C} : |z| = 1\}.$$

The set $\widehat{A}$ of all characters of A is an Abelian group: $(\varkappa \cdot \varkappa')(\mathbf{a}) = \varkappa(\mathbf{a})\varkappa'(\mathbf{a})$. We call $\widehat{A}$ either the dual group or the character group of A. The unit of $\widehat{A}$ sends all elements of A to 1, and the inverse element of $\varkappa \in \widehat{A}$ is the complex conjugate $\overline{\varkappa}$. According to the structure theorem (Note [30.4]), we have

$$\varkappa(\mathbf{a}) = \varkappa(\mathbf{b}_1)^{t_1} \varkappa(\mathbf{b}_2)^{t_2} \cdots \varkappa(\mathbf{b}_L)^{t_L}.$$

Since $\varkappa(\mathbf{b}_l)^{f_l} = 1$, there exists a certain $s_l \bmod f_l$ such that $\varkappa(\mathbf{b}_l) = e(s_l/f_l)$. Hence, if we let $\varkappa_l \in \widehat{A}$ be such that $\varkappa_l(\mathbf{b}_l) = e(1/f_l)$, $\varkappa_l(\mathbf{b}_k) = 1$, $l \neq k$, then $\varkappa = \varkappa_1^{s_1} \varkappa_2^{s_2} \cdots \varkappa_L^{s_L}$. We see readily that the set $\{\varkappa_l\}$ is an invariant factor basis of the group $\widehat{A}$ such that $\varkappa_j^{f_j} = 1$. More precisely, the correspondence

$$\mathbf{b}_1^{t_1} \mathbf{b}_2^{t_2} \cdots \mathbf{b}_L^{t_L} \mapsto \varkappa_1^{t_1} \varkappa_2^{t_2} \cdots \varkappa_L^{t_L}$$

is an isomorphism; in particular, analogously to (51.19), we have

$$A \ni \mathbf{a} = 1 \iff \varkappa(\mathbf{a}) = 1 \text{ for all } \varkappa \in \widehat{A},$$
$$\widehat{A} \ni \varkappa = 1 \iff \varkappa(\mathbf{a}) = 1 \text{ for all } \mathbf{a} \in A.$$

So the duality

$$A \cong \widehat{A}$$

holds as a generalization of (50.9) and (51.13). In particular $\widehat{\widehat{A}} \cong A$. On the other hand, the mapping $\varkappa \mapsto [\mathbf{a}](\varkappa) = \varkappa(\mathbf{a})$ with a fixed $\mathbf{a} \in A$ defines a character of the group $\widehat{A}$; and $\mathbf{a} \mapsto [\mathbf{a}]$ is an isomorphism. Hence,

$$\widehat{\widehat{A}} \text{ can be identified as } A.$$

This is an instance of the Pontrjagin duality. By the way, the concept of characters of a finite Abelian group and the relevant duality theorem were described by Weber (1882) for the first time in history.

[51.3] Also, the orthogonal relations (51.14)–(51.15) generalize as follows:

$$\frac{1}{|A|} \sum_{\varkappa \in \widehat{A}} \varkappa(\mathbf{a})\overline{\varkappa}(\mathbf{a}') = \begin{cases} 1 & \mathbf{a} = \mathbf{a}', \\ 0 & \mathbf{a} \neq \mathbf{a}'; \end{cases}$$

$$\frac{1}{|A|} \sum_{\mathbf{a} \in A} \varkappa(\mathbf{a})\overline{\varkappa'}(\mathbf{a}) = \begin{cases} 1 & \varkappa = \varkappa', \\ 0 & \varkappa \neq \varkappa'. \end{cases}$$

[51.4] With the notation as above, let $\mathbf{a} \in A$ be of order g. Then the set $\{\varkappa(\mathbf{a}) : \varkappa \in \widehat{A}\}$ splits into $|A|/g$ copies of $\{e(u/g) : u = 1, 2, \dots g\}$. We shall show two proofs: First, we consider the sum

$$\sum_{\varkappa \in \widehat{A}} \sum_{r=1}^{g} e(-ur/g)\varkappa(\mathbf{a}^r).$$

The inner sum is equal to g if $\varkappa(\mathbf{a}) = \exp(u/g)$ and to 0 otherwise, for $(\varkappa(\mathbf{a}))^g = \varkappa(\mathbf{a}^g) = 1$. Exchanging the order of summation, the new inner sum is equal to $|\widehat{A}| = |A|$ if $\mathbf{a}^r = 1$, and to 0 otherwise, so the double sum equals $|A|$. Equating these results, we end the first proof. Next, we consider the homomorphism of $\widehat{A}$: $\varkappa \mapsto \varkappa(\mathbf{a})$. Let E be its kernel. Then

$$\widehat{A}/E \cong \text{the group generated by } \{\varkappa(\mathbf{a}) : \varkappa \in \widehat{A}\}.$$

The size of the right side is not greater than g, for $(\varkappa(\mathbf{a}))^g = 1$. Thus, with $h = |\widehat{A}/E|$, we have $h \leq g$. On the other hand, $\varkappa^h \in E$ for any $\varkappa$; so $\varkappa(\mathbf{a}^h) = \varkappa^h(\mathbf{a}) = 1$ for all $\varkappa \in \widehat{A}$. Namely, $\mathbf{a}^h = 1$, or $g|h$. Hence $g = h$. This ends the second proof.

[51.5]   Let B, C be subgroups of A such that $C \subset B$. Then, under the natural identification $\widehat{\widehat{A}} = A$, i.e., $[\mathbf{a}] = \mathbf{a}$ in the notation of Note [51.2], it holds that with $B^\perp = \{\varkappa \in \widehat{A} : \varkappa(\mathbf{b}) = 1, \forall \mathbf{b} \in B\}$

(1) $B^\perp \cong \widehat{A/B}$,   (2) $\widehat{A}/B^\perp \cong \widehat{B}$,   (3) $(B^\perp)^\perp = B$,   (4) $C^\perp/B^\perp \cong \widehat{B/C}$.

To prove (1), it suffices to consider the homomorphism

$$B^\perp \ni \varkappa \mapsto \tilde{\varkappa} \in \widehat{A/B} : \tilde{\varkappa}(\mathbf{a}B) = \varkappa(\mathbf{a}), \, \mathbf{a} \in A.$$

For (2), consider $\widehat{A} \ni \varkappa \mapsto \varkappa|_B \in \widehat{B}$, the restriction of $\varkappa$ to B, which induces an injection $\widehat{A}/B^\perp$ into $\widehat{B}$. This is surjective as well since $|\widehat{A}/B^\perp| = |\widehat{A}|/|B^\perp| = |A|/|B^\perp| = |B|$ because of (1). For (3), note that $B \subset (B^\perp)^\perp$ and that $|(B^\perp)^\perp| = |\widehat{A}/B^\perp| = |B|$ by (1) and (2). Note that (3) means that B defines $B^\perp$, and vice versa; this fact is applied in the next Note. The assertion (4) is a refinement of (1): consider

$$C^\perp \ni \varkappa \mapsto \tilde{\varkappa} \in \widehat{B/C} : \tilde{\varkappa}(\mathbf{b}C) = \varkappa(\mathbf{b}), \, \mathbf{b} \in B.$$

[51.6]   Let $A^p = \{\mathbf{a}^p : \mathbf{a} \in A\}$ with p dividing $|A|$. Then $(A^p)^\perp = \{\varkappa \in \widehat{A} : \varkappa^p = 1\}$. By Note [30.5], there exists an integer α such that $(A^p)^\perp$ is isomorphic to the direct product of α copies of $\mathbb{Z}/p\mathbb{Z}$. By (1) above but with $B = A^p$, we have $\alpha = \alpha_p$, the p-rank of A (Note [30.6]). This and (3) above imply that $\widehat{A}$ has a set $\{\theta_j : j \leq \alpha_p\}$ of independent characters that defines A^p. In other words, $\mathbf{a} \in A$ is a pth power if and only if $\theta_j(\mathbf{a}) = 1$ for all $j \leq \alpha_p$.

§52

Here is a milestone in the development of number theory: We now introduce a Dirichlet character χ with respect to a modulus q. This is a completely multiplicative function on $\mathbb{Z}$ that satisfies

$$\chi(n) = \begin{cases} \chi(m) & n \equiv m \ \text{mod} \ q, \\ 0 & \langle n, q \rangle \neq 1. \end{cases} \tag{52.1}$$

Euler's theorem (28.10) implies that $1 = \chi(n^{\varphi(q)}) = \{\chi(n)\}^{\varphi(q)}$ for $\langle n, q \rangle = 1$; thus, $|\chi(n)| \in \{0, 1\}$. If χ takes real values only, then it is called a real Dirichlet character; otherwise, a complex Dirichlet character. The definition (52.1) implies that for $q = 1$ we have $\chi(n) = 1$ throughout $\mathbb{Z}$; in particular, $\chi(0) = 1$; that is, this χ is the same as the characteristic function of $\mathbb{Z}$.

In connection with the remark at the end of the text of the preceding section, we stress that each Dirichlet character is an arithmetic function and has the associated Dirichlet series; see (52.6). This distinction between multiplicative characters mod q and Dirichlet characters mod q is subtle to recognize.

Remark Nevertheless, for the sake of brevity, we shall often say a character $\text{mod} \, q$ or even just a character instead of a Dirichlet character $\text{mod} \, q$. Thus, throughout the rest of our discussion, a character means a Dirichlet character modulo a certain integer, provided no misunderstanding is anticipated.

With this, let $\mathcal{X}_q$ be the set of all characters $\text{mod} \, q$. We employ the expression

$$\chi \in \mathcal{X}_q \tag{52.2}$$

for the sake of notational simplicity. This is also to avoid a possible confusion with elements of Ξ_q introduced in the preceding section. Obviously $\mathcal{X}_q$ is an Abelian group with respect to the multiplication $(\chi \chi')(n) = \chi(n)\chi'(n)$. Let J_q be the unit element of this group; it is called the principal character mod q; naturally, $J_q(n) = 1$, $\langle n, q \rangle = 1$, and $J_q(n) = 0$, $\langle n, q \rangle \neq 1$, that is, J_q is the same as the characteristic function of the set $\{n \in \mathbb{Z} : \langle n, q \rangle = 1\}$. The inverse element of χ is the complex conjugate $\overline{\chi}$. If χ is a complex character, then $\chi \neq \overline{\chi}$; thus, the number of complex characters in $\mathcal{X}_q$ is even. We note in addition that if $\chi_j \in \mathcal{X}_{q_j}, j = 1, 2$, then $\chi_1 \chi_2 \in \mathcal{X}_{q_1 q_2}$; that is, $(\chi_1 \chi_2)(n) = \chi_1(n)\chi_2(n)$ is completely multiplicative and satisfies (52.1) with the modulus $q_1 q_2$, or more precisely $\chi_1 \chi_2 \in \mathcal{X}_{[q_1, q_2]}$.

Theorem 46 *The set $\mathcal{X}_q$ is an Abelian group of order $\varphi(q)$. The following orthogonality relations hold:*

$$\frac{1}{\varphi(q)} \sum_{\chi \in \mathcal{X}_q} \chi(m)\overline{\chi}(n) = \begin{cases} J_q(m) & m \equiv n \ \text{mod} \ q, \\ 0 & m \not\equiv n \ \text{mod} \ q, \end{cases} \tag{52.3}$$

$$\frac{1}{\varphi(q)} \sum_{n \ \text{mod} \ q} \chi(n)\overline{\chi}'(n) = \begin{cases} 1 & \chi = \chi', \\ 0 & \chi \neq \chi', \end{cases} \quad \chi, \chi' \in \mathcal{X}_q. \tag{52.4}$$

$\S52$ 209

Proof Let $\xi \in \Xi_q$, and let $\chi(n) = \xi(n)$, $\langle n,q \rangle = 1$ while $\chi(n) = 0$, $\langle n,q \rangle > 1$. Then we have $\chi \in \mathcal{X}_q$. Conversely, restricting $\chi \in \mathcal{X}_q$ to $(\mathbb{Z}/q\mathbb{Z})^*$, we get an element of Ξ_q. This ends the proof.

Remark This is trivial but worth noting: It follows from (52.4) with $\chi' = J_q$ that the sum $\sum_{n \leq N} \chi(n)$, $\chi \neq J_q$, is uniformly bounded as N varies.

Because of the last proof, we redefine (52.1) as follows:

$$\chi = \iota_q \xi, \ \xi \in \Xi_q \ \Leftrightarrow \ \begin{cases} \chi(n) = \xi(n) & \text{if } \langle n,q \rangle = 1, \\ \chi(n) = \ \ 0 & \text{if } \langle n,q \rangle \neq 1. \end{cases} \tag{52.5}$$

Here $\iota_q \xi$ denotes not a product of characters but an extension of the domain of ξ: note that a character mod q is a function not on $(\mathbb{Z}/q\mathbb{Z})^*$ but on $\mathbb{Z}/q\mathbb{Z}$.

The Dirichlet series $\lfloor \chi \rfloor(s)$ associated with the arithmetic function $\chi \in \mathcal{X}_q$ is called the Dirichlet L-function or just an L-function if no ambiguity is possible, and it is denoted by $L(s, \chi)$. We have, for $\mathrm{Re}\, s > 1$,

$$L(s, \chi) = \sum_{n=1}^{\infty} \frac{\chi(n)}{n^s} = \prod_p \left(1 - \frac{\chi(p)}{p^s} \right)^{-1}. \tag{52.6}$$

This will turn out to be an entire function of s, provided $\chi \neq J_q$; see Theorem 52 together with (53.8) below. Just the same as (11.8), we have

$$L(s, \chi) \neq 0, \quad \mathrm{Re}\, s > 1. \tag{52.7}$$

Also, following (17.14), we see that

$$-\frac{L'}{L}(s, \chi) = \sum_{n=1}^{\infty} \frac{\chi(n) \Lambda(n)}{n^s} = \sum_p \sum_{j=1}^{\infty} \frac{\chi(p^j) \log p}{p^{js}}. \tag{52.8}$$

For each ℓ, $\langle \ell, q \rangle = 1$, we have, by (52.3),

$$-\frac{1}{\varphi(q)} \sum_{\chi \bmod q} \overline{\chi}(\ell) \frac{L'}{L}(s, \chi) = \sum_{\substack{p,j \\ p^j \equiv \ell \bmod q}} \frac{\log p}{p^{js}}. \tag{52.9}$$

The most significant part of the right side is the sum

$$\sum_{p \equiv \ell \bmod q} \frac{\log p}{p^s}, \tag{52.10}$$

for the part with $j \geq 2$ in (52.9) is obviously bounded if $\mathrm{Re}\, s > 1/2$. Hence, the distribution of prime numbers in an arithmetic progression is closely related to analytic properties of L-functions. Details will be developed in Chapter 5.

210   *Characters*

In passing, we note that unlike the left side of (17.16) the expression (52.10) does not directly follow from a single *L*-function; that is, the function

$$\prod_{p \equiv \ell \bmod q} \left(1 - \frac{1}{p^s}\right)^{-1} \tag{52.11}$$

is hard to handle if q is arbitrarily given. Dirichlet overcame this difficulty by means of the fundamental idea (52.1) supported by (51.1)–(51.3). Thus, to pick up a particular reduced residue class $\bmod q$, one needs to take into account the Fourier analysis over the set of all reduced residue classes mod q, a typical instance to appreciate the merit of the Fourier–Dirichlet invention, or rather that of Lagrange and Vandermonde, as has been mentioned at the end of §48.

Notes

[52.1]   The lower line in the definition (52.1) was suggested already by Dirichlet (1837a; Werke I, p.336), apparently for convenience. However, it has turned out later to be genuinely needed; see Note [55.2] below. By the way, the use of the symbol χ in the present context was started by Dedekind (Dirichlet (1871, p.341, footnote)).

[52.2]   The concept and the notation of *L*-functions originated with Dirichlet (1837a; Werke I, pp.317–318). The extensive use of them began with Landau (1908a, p.427; 1909, §102); he introduced the symbol $L(s, \chi)$ in the latter (p.482). Euler (1748b, Tomi Primi, Caput X, §176, etc.) seems to have considered special instances of *L*-functions.

[52.3]   To study the behavior of an arithmetic function f in an arithmetic progression, the Dirichlet series for f twisted by a character

$$\lfloor \chi f \rfloor(s) = \sum_{n=1}^{\infty} \chi(n) f(n) n^{-s}$$

is a natural means, when f is multiplicative. However, even then the Dirichlet series for f twisted by an additive character

$$\lfloor \psi_{a/q} f \rfloor(s) = \sum_{n=1}^{\infty} \psi_{a/q}(n) f(n) n^{-s}$$

may turn out to be a better choice, specifically when dealing with additive variations applied to multiplicative functions in the sense of Note [15.2]. For instance, $\lfloor \psi_{a/q} \sigma_\alpha \rfloor(s)$ with σ_α the divisor function (16.4) is such a case; see Estermann (1931) and Motohashi (1994).

§53

We shall now look into periods of functions $\xi \in \Xi_q$ and $\chi \in \mathcal{X}_q$:

$$\xi \in \Xi_q \text{ is of period } k \quad \Leftrightarrow$$
$$\xi(m) = \xi(n) \text{ if } m \equiv n \text{ mod } k, \langle mn, q \rangle = 1; \tag{53.1}$$

and

$$\text{the period of } \chi = \iota_q \xi \in \mathcal{X}_q \text{ is that of } \xi. \tag{53.2}$$

For instance, the unit character $J_q \in \mathcal{X}_q$ is of period 1. Note that (53.1) does not mean that k is uniquely determined by ξ. Because of this, we are led to the following discussion; it is, of course, possible to argue via Dirichlet's definition (51.1)–(51.3), but we shall take a little more direct procedure:

Theorem 47 (1) *Without loss of generality, we may suppose $k|q$ in (53.1). Then via (53.2), there exists a unique $\chi' \in \mathcal{X}_k$ such that*

$$\chi = J_{q/k}\chi', \tag{53.3}$$

with J_* as in the preceding section. Conversely, this decomposition implies that χ is of period k.
(2) *If $\chi_j \in \mathcal{X}_{q_j}, j = 1, 2$, are such that $\chi_1(n) = \chi_2(n)$, $\langle n, q_1 q_2 \rangle = 1$, then they have the same period $\langle q_1, q_2 \rangle$.*

Proof (1) First, if $f \in \mathbb{N}$ is such that $\chi(m) = \chi(n)$ whenever $m \equiv n$ mod f, $\langle mn, q \rangle = 1$, then χ is of period $\langle f, q \rangle = d$. To show this, let $d = fu + qv$, $u, v \in \mathbb{Z}$ by Euclid's algorithm. Suppose $\langle n(n + dl), q \rangle = 1$ for an l; then $\langle n(n + flu), q \rangle = 1$. Thus, $\chi(n + dl) = \chi(n + flu) = \chi(n)$. Namely, provided that $m \equiv n$ mod d, $\langle mn, q \rangle = 1$, we have $\chi(m) = \chi(n)$, which means that χ has $\langle f, q \rangle$ as its period. The first part of (1) has been confirmed. So, let k be a period of χ such that $k|q$. Then, for any n, $\langle n, k \rangle = 1$, we may choose m such that $m \equiv n$ mod k, $\langle m, q \rangle = 1$ because of Note [29.2]; we define $\chi'(n)$ to be $\chi(m)$. By the definition of period, the value of $\chi(m)$ is uniquely determined; that is, the values of χ' are fixed by those of χ. In addition, we set $\chi'(n) = 0$ if $\langle n, k \rangle \neq 1$. We see that $\chi' \in \mathcal{X}_k$, as the complete multiplicativity of χ' follows readily from that of χ. Also, if $\langle n, q \rangle = 1$, then we may take $m = n$, and $\chi'(n) = \chi(n)$, which is equivalent to (53.3). The converse statement is immediate. We end the proof of (1). As for (2), let $m \equiv n$ mod $\langle q_1, q_2 \rangle$, $\langle mn, q_1 \rangle = 1$. With $n\bar{n} \equiv 1$ mod q_1, we have $m\bar{n} \equiv 1$ mod $\langle q_1, q_2 \rangle$. There exist $r, u, v \in \mathbb{Z}$ such that $m\bar{n} = 1 + r(uq_1 + vq_2)$, again by Euclid's algorithm. Then, we have $1 = \langle m\bar{n}, q_1 \rangle = \langle 1 + rvq_2, q_1 \rangle$, so $\langle 1 + rvq_2, q_1 q_2 \rangle = 1$. By the assumption, we get $\chi_1(m\bar{n}) = \chi_1(1 + rvq_2) = \chi_2(1 + rvq_2) = 1$. Hence,

$\chi_1(m) = \chi_1(n)$. Therefore, χ_1 is of period $\langle q_1, q_2 \rangle$. Analogously, χ_2 is of the same period. We end the proof.

With this preparation, we introduce two fundamental concepts:

$$\begin{array}{c} \text{the least period of } \chi \in \mathcal{X}_q \text{ is called} \\ \text{the conductor of } \chi, \text{ and denoted by } q^* ; \\ \text{if } q = q^*, \text{ then } \chi \text{ is called a primitive character.} \end{array} \qquad (53.4)$$

Remark 1 A better notation alternative to q^* would be q_χ. Nevertheless, q^* is employed for the sake of notational simplicity; any relevant ambiguity should be avoided.

Theorem 48 (1) *For each* $\chi \in \mathcal{X}_q$, *there exists a unique conductor* q^* *and a unique primitive character* $\chi^* \in \mathcal{X}_{q^*}$ *such that*

$$\chi = J_{q/q^*} \chi^*. \qquad (53.5)$$

(2) *If two primitive characters* $\chi_j \in \mathcal{X}_{q_j}$ *satisfy* $\chi_1(n) = \chi_2(n)$, $\langle n, q_1 q_2 \rangle = 1$, *then* $q_1 = q_2$, $\chi_1 = \chi_2$.

Remark 2 With the decomposition (53.5) we call χ induced by the primitive character χ^*.

Proof (1) In the decomposition (53.3), take $k = q^*$ the conductor of χ. Then χ_1 is primitive. For inserting $\chi_1 = J_{q^*/(q^*)^*}(\chi_1)_1$, $(\chi_1)_1 \in \mathcal{X}_{(q^*)^*}$, into (53.3), we get $\chi = J_{q/(q^*)^*}(\chi_1)_1$. This implies that χ is of period $(q^*)^*$; and by the definition of q^* we find that $(q^*)^* = q^*$; namely χ_1 is primitive. The uniqueness of $\chi_1 = \chi^*$ is asserted in (1) of the last theorem. The assertion (2) follows immediately from (2) of the last theorem. This ends the proof.

Here are trivial remarks: A combination of the above proof of (53.3) and (53.5) implies that

$$\text{periods of a character are multiples of its conductor;} \qquad (53.6)$$

also,

$$\text{the principal character mod } q \text{ is primitive only if } q = 1. \qquad (53.7)$$

According to (53.5), we have, for Dirichlet L-functions,

$$L(s, \chi) = L(s, \chi^*) \prod_{p \mid q/q^*} \left(1 - \frac{\chi^*(p)}{p^s} \right). \qquad (53.8)$$

$\S 53$ 213

This implies that analytic properties of $L(s, \chi)$ are essentially the same as those of $L(s, \chi^*)$. In particular, if we adopt the procedure leading to (52.10), then the study of the distribution of primes in arithmetic progressions can be restricted to the discussion concerning L-functions associated with primitive characters.

Analogously to the Riemann Hypothesis (12.19) we set out the Generalized Riemann Hypothesis: For any χ,

$$\text{GRH}: \quad \text{If } L(\rho, \chi) = 0 \text{ with } 0 < \operatorname{Re} \rho < 1, \text{ then } \operatorname{Re} \rho = \tfrac{1}{2}. \qquad (53.9)$$

If correct, this grand hypothesis will yield a far more refined conclusion on the distribution of primes than RH. Note that because of (53.8) the line $\operatorname{Re} s = 0$ contains infinitely many zeros of $L(s, \chi)$ if χ is not primitive.

Remark 3 GRH is sometimes called instead the Extended Riemann Hypothesis (ERH). Note as well that the term GRH is also used in discussing the vast family of functions that admit Euler-product representations. Once GRH is resolved affirmatively, many core assertions, especially those related to prime numbers, in the present essays will become trivia. However, it should be known also that there exist plenty of problems against which GRH seems to be helpless, at least presently. To this category belong the asymptotic analysis of $\pi(x+x^c) - \pi(x)$ with an absolute constant $c < \tfrac{1}{2}$, and the divisor problem in arithmetic progressions which is mentioned in Note [104.5] below; both may appear quite natural issues.

Notes

[53.1] The above is rudimental. Thus, our reasoning is the same as most expositions in literature; see Montgomery–Vaughan (2006, Section 9.1), for instance. The notion of primitive characters can be said to have been perceived first by Kinkelin (1862, p.29) (see Note [55.2] below), and defined explicitly by de la Vallée Poussin (*caractère propre*: 1897, Deuxième partie, p.49) and by Landau (*eigentliche Charaktere*: 1909, p.479).

[53.2] By (53.5), in $\mathcal{X}_p$ all elements save for the principal character are primitive. Every $\chi \in \mathcal{X}_{p^\alpha}$, $\alpha \geq 2$, save for J_{p^α}, satisfies $\chi(n) = \chi^*(n)$ for any $n \in \mathbb{Z}$. However, with a generic χ, there may exist an n such that $\chi(n) = 0 \neq \chi^*(n)$, as can be seen from the discussion in the next section.

[53.3] The number of primitive characters mod q is equal to

$$q \prod_{p \| q} \left(1 - \frac{2}{p}\right) \prod_{p^2 | q} \left(1 - \frac{1}{p}\right)^2,$$

which is equal to 1 if $q = 1$, for then the products are empty. Because of Theorem 48(1), $\varphi(q)$ is the same as the sum over $d|q$ of the number of primitive

214 *Characters*

characters mod d; so the Möbius inversion yields this formula. We see that $\mathcal{X}_q$ contains a primitive character if and only if $q \not\equiv 2 \bmod 4$. Note that if $2\|q$, then any character mod q is of period $q/2$ since that $m \equiv n \bmod q/2$ and $\langle mn, q \rangle = 1$ implies $m \equiv n \bmod q$.

§54

We shall give the precise structure of each primitive Dirichlet character:

Theorem 49 *Let* $\chi = \iota_q \xi$, $\xi \in \Xi_q$. *Then,* χ *is primitive if and only if it holds under the definition* (51.1)–(51.3) *that for* $p \geq 3$

$$\begin{cases} (p-1) \nmid h_p & \text{if } q(p) = 1, \\ p \nmid h_p & \text{if } q(p) \geq 2, \end{cases} \tag{54.1}$$

as well as that

$$\begin{cases} 2 \nmid h_2 & \text{if } q(2) = 2, \\ 2 \nmid k_2 & \text{if } q(2) \geq 3. \end{cases} \tag{54.2}$$

Proof To begin with, we assert that if $\chi = \chi_1 \chi_2$, $\chi_\nu \in \mathcal{X}_{q_\nu}$, $q = q_1 q_2$, $\langle q_1, q_2 \rangle = 1$, then

$$\chi \bmod q_1 q_2 \text{ is primitive} \iff \chi_\nu \bmod q_\nu \text{ are both primitive.} \tag{54.3}$$

Supposing the left side, let the conductor of χ_ν be q_ν^*. Then $\chi_1 \chi_2$ is of period $q_1^* q_2^*$. Hence, if χ is primitive, then $q_1^* q_2^* = q_1 q_2$; and $q_\nu^* = q_\nu$, so χ_ν are both primitive. Conversely, suppose the right side, and let the conductor of χ be ν. Then, $\langle \nu, q_1 \rangle$ is a period of χ_1; that is, $q_1 = \langle \nu, q_1 \rangle$. To confirm this, let $m \equiv n \bmod \langle \nu, q_1 \rangle$, $\langle mn, q_1 \rangle = 1$, and let m_1, n_1 be such that $m_1 \equiv m$, $n_1 \equiv n \bmod q_1$, $m_1 \equiv 1$, $n_1 \equiv 1 \bmod q_2$. Then, $m_1 \equiv n_1 \bmod \nu$, $\langle m_1 n_1, q \rangle = 1$. For since $m_1 \equiv n_1 \bmod \langle \nu, q_1 \rangle$ and $m_1 \equiv n_1 \bmod q_2$, we have $m_1 \equiv n_1 \bmod \langle \nu, q_1 \rangle q_2$; moreover, $\nu = \langle \nu, q_1 \rangle \langle \nu, q_2 \rangle$. From these, it follows that $\chi_1(m) = \chi(m_1) = \chi(n_1) = \chi_1(n)$, as required. Similarly, $q_2 = \langle \nu, q_2 \rangle$. Therefore, $\nu = q$, and χ is primitive. We have confirmed (54.3).

We may now move to (51.1) and consider the period of

$$\xi_p(n) = e\big(h_p \mathrm{Ind}_{r_p}(n)/\varphi(p^\alpha)\big), \quad p \nmid n, \ p \geq 3. \tag{54.4}$$

Because of the first part of Note [53.2], we may suppose that $\alpha \geq 2$. According to the choice of r_p, we have $\mathrm{Ind}_{r_p}\big(1 + p^{\alpha-1}\big) \equiv 0 \bmod \varphi(p^{\alpha-1})$ but $\not\equiv 0 \bmod \varphi(p^\alpha)$. Thus, if ξ_p is of period $p^{\alpha-1}$, then $h_p \mathrm{Ind}_{r_p}\big(1 + p^{\alpha-1}\big) \equiv 0 \bmod \varphi(p^\alpha)$, so $p | h_p$. Conversely, if $p | h_p$, then for any $p \nmid n$ and t, we have

$h_p \text{Ind}_{r_p}(n + p^{\alpha-1}t) \equiv h_p \text{Ind}_{r_p}(n) \mod \varphi(p^\alpha)$ because of $(44.6)_{\alpha \mapsto \alpha-1}$. Hence, ξ_p is of period $p^{\alpha-1}$. The assertion (54.1) has been proved.

As for the modulus 2^α, if $\alpha = 1$, then there exists only the principal character. If $\alpha = 2$, then $\xi_2(n) = e(h_2 \text{Ind}_3(n)/2)$. Thus, if $2|h_2$, then ξ_2 is principal; if $2 \nmid h_2$, then the period of ξ_2 is neither 2 nor 1 but 4, which yields the first line of (54.2). We suppose that $\alpha \geq 3$. Under the definition (42.2), we note that $v(1 + 2^{\alpha-1}) \equiv 0 \mod 2$; also according to the discussion on (42.4), we have $w(1 + 2^{\alpha-1}) \equiv 0 \mod 2^{\alpha-3}$ as well as $\not\equiv 0 \mod 2^{\alpha-2}$. Hence, if $\xi^{(2)}$ is of period $2^{\alpha-1}$, then $k_2 w(1 + 2^{\alpha-1}) \equiv 0 \mod 2^{\alpha-2}$, which means $2|k_2$. Conversely, if $2|k_2$, then for any t and $2 \nmid n$ we have $k_2 w(n + 2^{\alpha-1}t) \equiv k_2 w(n) \mod 2^{\alpha-2}$ because of $(44.8)_{\alpha \mapsto \alpha-1}$. Also, $v(n + 2^{\alpha-1}t) \equiv v(n) \mod 2$, as $n + 2^{\alpha-1}t \equiv n \mod 4$. We see that $\xi^{(2)}$ is of period $2^{\alpha-1}$. We have got the second line of (54.2). We end the proof. See Note [55.5].

Here is an important corollary:

Theorem 50 *Primitive real characters $\chi \in \mathcal{X}_q$ exist if and only if*

$$q = 2^l q_0, \ l = 0, 2, 3, \text{ with } 2 \nmid q_0 \text{ being sqf.} \tag{54.5}$$

Remark This assertion includes the case $q = 1$ as well.

Proof We may assume that $q \neq 1$. Let ϱ_p be as in (51.6). We have $\chi(\varrho_p) = -1$, for if $\chi(\varrho_p) = 1$ with a certain $p|q$, then the period of χ would be smaller than q. Hence, for all $p|q$, $p \geq 3$, we have $-1 = \chi(\varrho_p) = e(h_p/\varphi(p^{q(p)}))$; that is, $h_p \equiv \varphi(p^{q(p)})/2 \mod \varphi(p^{q(p)})$. Because of (54.1), we find that $q(p) = 1$; hence q_0 is sqf. On the other hand, if $l \geq 4$, then let λ_2 be as in (51.9). We have $\chi(\lambda_2) = -1$; that is, $k_2 \equiv 2^{l-3} \mod 2^{l-2}$, so $2|k_2$, which means that (54.2) is violated. Hence, $l \leq 3$. Moreover, $l \neq 1$ because of Note [53.3]. We have obtained the necessity part of (54.5). The sufficiency part is an easy consequence of (54.3) and the following observation: For odd n, we have that

(i) if $l = 2$, then $h_2 = 1$; thus

$$\xi_2(n) = (-1)^{(n-1)/2}; \tag{54.6}$$

(ii) if $l = 3$, then $k_2 = 1$; thus

$$\xi^{(2)}(n) = \begin{cases} (-1)^{(n^2-1)/8} & \text{if } k_1 = 0, \\ (-1)^{(n-1)/2+(n^2-1)/8} & \text{if } k_1 = 1. \end{cases} \tag{54.7}$$

The assertion (54.6) is obvious. On the other hand, if $l = 3$, then $n \equiv (-1)^v 5^w$ mod 8 is equivalent to

$$\{n; v, w\} = \{1; 0, 0\}, \{3; 1, 1\}, \{5; 0, 1\}, \{7; 1, 0\}; \tag{54.8}$$

216   *Characters*

and (51.2) becomes

$$\xi^{(2)}(n) = (-1)^{k_1 v + w} = \begin{cases} 1, -1, -1, 1 & \text{if } k_1 = 0, \\ 1, 1, -1, -1 & \text{if } k_1 = 1. \end{cases} \tag{54.9}$$

This is expressed as (54.7) since in (54.8) it holds that

$$n \equiv 2v + 1 \bmod 4 \Rightarrow v \equiv (n-1)/2 \bmod 2,$$
$$n \equiv \pm(2w + 1) \bmod 8 \Rightarrow w \equiv (n^2 - 1)/8 \bmod 2. \tag{54.10}$$

We end the proof. Continued to Theorem 60.

Notes

[54.1]   The conductor of $\iota_p \xi_p \in \mathcal{X}_{p^\alpha}$ with ξ_p as in (54.4) is equal to $p^{\alpha-\gamma}$, provided $p^\gamma \| h_p$, $\gamma \le \alpha - 1$. For by $(44.6)_{\alpha \mapsto \alpha - \gamma}$, $h_p \mathrm{Ind}_{r_p}\left(n + p^{\alpha-\gamma}l\right) \equiv h_p \mathrm{Ind}_{r_p}(n) \bmod \varphi(p^\alpha)$ where l is arbitrary and $p \nmid n$; moreover, $h_p \mathrm{Ind}_{r_p}\left(1 + p^{\alpha-\gamma-1}\right) \not\equiv 0 \bmod \varphi(p^\alpha)$. One may deal with $\iota_2 \xi^{(2)} \in \mathcal{X}_{2^\alpha}$ in much the same way.

[54.2]   The set Ξ_{16} is composed of the functions

$$\tau_{k_1, k_2}(n) = e\left(k_1 v/2 + k_2 w/4\right),$$
$$k_1 \bmod 2, \ k_2 \bmod 4, \ n \equiv (-1)^v 5^w \bmod 2^4.$$

On noting that

$$\{n; v, w\} : \quad \begin{array}{l} \{1; 0, 0\}, \{3; 1, 3\}, \{5; 0, 1\}, \{7; 1, 2\}, \\ \{9; 0, 2\}, \{11; 1, 1\}, \{13; 0, 3\}, \{15; 1, 0\}, \end{array}$$

we have, for instance,

$$\tau_{0,1}(1) = 1, \ \tau_{0,1}(3) = -i, \ \tau_{0,1}(5) = i, \ \tau_{0,1}(7) = -1,$$
$$\tau_{0,1}(9) = -1, \tau_{0,1}(11) = i, \ \tau_{0,1}(13) = -i, \ \tau_{0,1}(15) = 1;$$

$$\tau_{1,1}(1) = 1, \ \tau_{1,1}(3) = i, \ \tau_{1,1}(5) = i, \ \tau_{1,1}(7) = 1,$$
$$\tau_{1,1}(9) = -1, \tau_{1,1}(11) = -i, \ \tau_{1,1}(13) = -i, \ \tau_{1,1}(15) = -1.$$

According to the lower line of (54.2), both of $\iota_2 \tau_{0,1}, \ \iota_2 \tau_{1,1} \in \mathcal{X}_{16}$ are primitive. Also, we have

$$\tau_{1,2}(1) = 1, \ \tau_{1,2}(3) = 1, \ \tau_{1,2}(5) = -1, \ \tau_{1,2}(7) = -1,$$
$$\tau_{1,2}(9) = 1, \ \tau_{1,2}(11) = 1, \ \tau_{1,2}(13) = -1, \ \tau_{1,2}(15) = -1.$$

Thus $\tau_{1,2}$ is of period 8. The same is true of $\tau_{0,2}$. Compare these with (54.7), (54.9). Further, $\tau_{1,0}$ is of period 4, corresponding to (54.6). The remaining $\tau_{0,3}, \tau_{1,3}$ yield primitive characters. Hence, there exist four primitive characters in $\mathcal{X}_{16}$ in accord with Note [53.3]. They are two pairs of complex conjugates; indeed, $\tau_{0,1}\tau_{0,3}$, $\tau_{1,1}\tau_{1,3} = \tau_{0,0}$. Non-trivial real characters come from $\tau_{0,2}$, $\tau_{1,2}$, $\tau_{1,0}$. Counting the principal character $\iota_2\tau_{0,0}$, in total we have $4 + 3 + 1 = \varphi(16)$ characters.

[54.3] The set Ξ_{15} is composed of the functions

$$\lambda_{a,b}(n) = e\big(aU(n)/2 + bV(n)/4\big), \quad \langle n, 15 \rangle = 1,$$

where, $n \equiv 2^{U(n)} \bmod 3$, $n \equiv 2^{V(n)} \bmod 5$, $0 \le a \le 1$, $0 \le b \le 3$. We have

$$\{n; U, V\} : \begin{array}{l} \{1; 0, 0\}, \{2; 1, 1\}, \{4; 0, 2\}, \{7; 0, 1\}, \\ \{8; 1, 3\}, \{11; 1, 0\}, \{13; 0, 3\}, \{14; 1, 2\}. \end{array}$$

According to the first line of (54.1), primitive characters in $\mathcal{X}_{15}$ are given by $\lambda_{1,1}$, $\lambda_{1,2}$, $\lambda_{1,3}$ in accord with Note [53.3]. Among these, $\lambda_{1,2}$ yields a real character, an instance of (54.5):

$$\lambda_{1,2}(1) = 1, \ \lambda_{1,2}(2) = 1, \ \lambda_{1,2}(4) = 1, \ \lambda_{1,2}(7) = -1,$$
$$r\lambda_{1,2}(8) = 1, \ \lambda_{1,2}(11) = -1, \ \lambda_{1,2}(13) = -1, \ \lambda_{1,2}(14) = -1.$$

The period is neither 3 or 5. Also, $\lambda_{1,1}$ and $\lambda_{1,3}$ are complex conjugates. Further, $\lambda_{0,1}$, $\lambda_{0,2}$, $\lambda_{0,3}$ are of period 5. For instance, from the table

$$\lambda_{0,2}(1) = 1, \ \lambda_{0,2}(2) = -1, \ \lambda_{0,2}(4) = 1, \ \lambda_{0,2}(7) = -1,$$
$$\lambda_{0,2}(8) = -1, \ \lambda_{0,2}(11) = 1, \ \lambda_{0,2}(13) = -1, \ \lambda_{0,2}(14) = 1,$$

we see that $\iota_{15}\lambda_{0,2} \in \mathcal{X}_{15}$ is induced by the primitive character $\iota_5\tilde{\lambda}_{0,2}$, $\tilde{\lambda}_{0,2} \in \Xi_5$, where $\tilde{\lambda}_{0,2}(1) = 1$, $\tilde{\lambda}_{0,2}(2) = \lambda_{0,2}(2) = -1$, $\tilde{\lambda}_{0,2}(3) = \lambda_{0,2}(8) = -1$, $\tilde{\lambda}_{0,2}(4) = \lambda_{0,2}(4) = 1$. That is, $\iota_{15}\lambda_{0,2} = J_3\iota_5\tilde{\lambda}_{0,2}$. In addition to these, $\lambda_{1,0}$ yields a real character of period 3. Including the principal character, we get $3 + 3 + 1 + 1 = \varphi(15)$ characters in total.

§55

We shall now consider Fourier expansions of characters $\chi \in \mathcal{X}_q$: These functions are of period q, and by (50.1)–(50.2) we have

$$\chi(n) = \frac{1}{\sqrt{q}} \sum_{a \bmod q} \widehat{\chi}(a) e(an/q),$$

$$\widehat{\chi}(a) = \frac{1}{\sqrt{q}} \sum_{n \bmod q} \chi(n) e(-an/q).$$

$$(55.1)$$

218   *Characters*

If $\langle a, q \rangle = 1$, then in the second line we may replace n by $-n\bar{a}$, $a\bar{a} \equiv 1 \bmod q$, and have

$$\widehat{\chi}(a) = \frac{1}{\sqrt{q}} \overline{\chi}(-a) G(\chi), \quad G(\chi) = \sum_{n \bmod q} \chi(n) e(n/q). \tag{55.2}$$

This $G(\chi)$ is called the Gauss sum associated with χ; see Remark following (61.1) below. In particular, by (18.18),

$$G(J_q) = c_q(1) = \mu(q). \tag{55.3}$$

A computation of $\widehat{\chi}(a)$, which is valid for any combination of a and χ, will be developed in the next section. Here we shall restrict ourselves to primitive characters. The following assertion will play a fundamental rôle throughout the rest of the present essays.

Theorem 51 *For any primitive character $\chi \in \mathcal{X}_q$ and for any $n \in \mathbb{Z}$, it holds that*

$$\chi(n) = \frac{G(\chi)}{q} \sum_{a \bmod q} \overline{\chi}(a) e(-na/q), \tag{55.4}$$

including the case $q = 1$. As a consequence we have

$$|G(\chi)| = \sqrt{q}. \tag{55.5}$$

Remark What makes the expansion (55.4) fundamental is in that no restriction is imposed upon n. Also, notice that (55.5) means that there exists an amazing amount of cancellation among the $\varphi(q)$ terms $\{\chi(n)e(n/q)\}$.

Proof We may suppose that $q > 1$. We note first that provided χ is primitive the following holds:

the vanishing theorem of Kinkelin: $\widehat{\chi}(a) = 0$ if $\langle a, q \rangle \neq 1$. $\tag{55.6}$

The case $q | a$ is the same as (52.4) with $\chi' = J_q$. Thus we suppose that $q \nmid a$; and $\langle a, q \rangle = d$ with $1 < d < q$. With $a_1 = a/d$, $q_1 = q/d$,

$$\widehat{\chi}(a) = \frac{1}{\sqrt{q}} \sum_{l \bmod q_1} e(-a_1 l/q_1) \sum_{t \bmod q/q_1} \chi(l + q_1 t). \tag{55.7}$$

To deal with the inner sum, we note that q_1 is not a period of χ since χ is primitive. Hence, there exist u, v such that $u \equiv v \bmod q_1$, $\langle uv, q \rangle = 1$, $\chi(u) \neq \chi(v)$; and we have a c such that $c \equiv 1 \bmod q_1$, $\chi(c) \neq 1, 0$. Then we make the change of variable $t = (c-1)l/q_1 + cw$. The sum under consideration becomes

$$\chi(c) \sum_{w \bmod q/q_1} \chi(l + q_1 w), \tag{55.8}$$

which confirms (55.6). With this, the first line of (55.1) is refined as

$$\chi(n) = \frac{1}{\sqrt{q}} \sum_{\substack{a \bmod q \\ \langle a,q \rangle = 1}} \widehat{\chi}(a) e(na/q). \tag{55.9}$$

Inserting (55.2) into this right side, we obtain (55.4). As for (55.5), it suffices to take $n = 1$ in (55.4). We end the proof.

As an immediate application of the Fourier expansion (55.4) we obtain the following extension of (12.7)–(12.8):

Theorem 52 *Let $q \geq 3$. Then the function $L(s, \chi)$ with a primitive character $\chi \bmod q$ continues analytically to an entire function, satisfying the functional equation*

$$L(1-s, \overline{\chi}) = \frac{2i^{-\delta_\chi}}{G(\chi)} \left(\frac{q}{2\pi} \right)^s \cos \left(\tfrac{1}{2}\pi(s + \delta_\chi) \right) \Gamma(s) L(s, \chi), \tag{55.10}$$

where $\delta_\chi = \tfrac{1}{2}(1 - \chi(-1))$. This is equivalent to

$$\left(\frac{q}{\pi} \right)^{(1-s)/2} \Gamma\left(\tfrac{1}{2}(1 - s + \delta_\chi) \right) L(1-s, \overline{\chi})$$
$$= \frac{i^{\delta_\chi} q^{1/2}}{G(\chi)} \left(\frac{q}{\pi} \right)^{s/2} \Gamma\left(\tfrac{1}{2}(s + \delta_\chi) \right) L(s, \chi). \tag{55.11}$$

Proof We use the integral representation

$$L(s, \chi) = \frac{-1}{q^s \Gamma(s)(e^{\pi i s} - e^{-\pi i s})} \sum_{1 \leq a < q} \chi(a) \int_C \frac{(-x)^{s-1} e^{-ax/q}}{1 - e^{-x}} dx, \tag{55.12}$$

which is an extension of (12.5). The integral is an entire function of s, and thus $L(s, \chi)$ is meromorphic over $\mathbb{C}$; see §94 for necessary facts on the Gamma function. Restricting s to the half plane $\operatorname{Re} s < 0$, we expand C to infinity as we did in §12. Computing the residues and invoking (55.4), we find that the sum over a in (55.12) equals

$$\frac{(2\pi)^s q}{G(\overline{\chi})} \left(e^{-\pi i s/2} - \chi(-1) e^{\pi i s/2} \right) L(1-s, \overline{\chi}), \tag{55.13}$$

which with (55.5) and (94.30) proves (55.10). Then an application to (55.10) of (94.25) and (94.30) yields (55.11). As for the regularity assertion, we note that (55.12) is, for any s,

$$L(s, \chi) = -\frac{\Gamma(1-s)}{2\pi i q^s} \sum_a \cdots . \tag{55.14}$$

220 *Characters*

It is implied that $L(s, \chi)$ is regular for $\operatorname{Re} s \leq \frac{3}{2}, s \neq 1$. Moreover, at $s = 1$ the integral in (55.12) is equal to $2\pi i$, and the sum over a vanishes, as $q \geq 3$. This ends the proof, for $L(s, \chi)$ is obviously regular for $\operatorname{Re} s \geq \frac{3}{2}$. See Note [60.3] below.

Notes

[55.1] The sum $G(\chi)$ for general χ was first introduced by Kinkelin (1862, §§XII–XV); as a matter of fact he dealt with only moduli which are either a prime power or a square-free integer, but his argument is general. He considered the functional equation for $L(s, \chi)$, and was led to $G(\chi)$; see Landau (1908a, p.428) and Hecke (1917b, p.77, footnote). However, a relation between L-functions and Gauss sums had been noticed by Dirichlet (1837a; Werke I, pp.324–325). The origin of $G(\chi)$ can be traced back to the Lagrange–Vandermonde resolvent in the theory of algebraic equations, as we shall explain in §70.

[55.2] The fact (55.6) can be said to have been pointed out by Kinkelin (1862, p.29) for the first time; see Note [55.5]. As far as primitive characters are concerned, the Fourier expansion (55.4) holds for $\langle n, q \rangle \neq 1$ as well. Because of this, the lower line of the definition (52.1) is indeed indispensable. The above proof of (55.6) is due to I. Schur; see Landau (1908a, pp.429–431).

[55.3] With the notation in Note [54.2], $\gamma = \iota_2 \tau_{0,2} \in \mathcal{X}_{16}$ is not primitive. According to the definition (55.1), we have

$$\widehat{\gamma}(2) = \frac{2}{\sqrt{16}} \left(e(-1/8) - e(-3/8) - e(-5/8) + e(-7/8) \right) = \sqrt{2},$$

since $\gamma(1) = \gamma(9) = 1, \gamma(3) = \gamma(11) = -1, \gamma(5) = \gamma(13) = -1, \gamma(7) = \gamma(15) = 1$; see (56.2) of the next section. On the other hand, we have

$$\widehat{\gamma^*}(2) = \frac{1}{\sqrt{8}} \left(e(-1/4) - e(-3/4) - e(-5/4) + e(-7/4) \right) = 0;$$

which is an example of (55.6).

[55.4] Gauss sums $\{G(\chi)\}$ are quasi-multiplicative: We have, for $\chi_j \in \mathcal{X}_{q_j}$, $\langle q_1, q_2 \rangle = 1$,

$$G(\chi_1\chi_2) = \chi_1(q_2)\chi_2(q_1)G(\chi_1)G(\chi_2), \quad \widehat{\chi}(a) = \chi_1(q_2)\chi_2(q_1)\widehat{\chi_1}(a)\widehat{\chi_2}(a)$$

by (32.3).

[55.5] Hence, to prove (55.6) one may restrict moduli to odd-prime powers; the case of powers of 2 can be omitted, as it is essentially the same: Kinkelin (1862, §XIV) considered, for $p \geq 3$,

$$\sum_{\substack{n=1 \\ p\nmid n}}^{p^\alpha} \xi_p(n)e\left(-an/p^\alpha\right), \quad p|a,$$

where ξ_p is as in (54.4). If $\alpha = 1$, this sum obviously vanishes, provided $p\nmid h_p$. Thus, suppose that $\alpha > 1$ and $a = pa_1$. We need to consider, instead,

$$\sum_{\substack{t=1 \\ p\nmid t}}^{p^{\alpha-1}} e\left(-a_1 t/p^{\alpha-1}\right) \sum_{\substack{n \bmod p^\alpha \\ n \equiv t \bmod p^{\alpha-1}}} \xi_p(n).$$

By $(44.6)_{\alpha \mapsto \alpha-1}$, we have $\mathrm{Ind}_{r_p}(n) \equiv \mathrm{Ind}_{r_p}(t) \bmod \varphi(p^{\alpha-1})$, so the inner sum is equal to

$$\xi_p(t) \sum_{f=0}^{p-1} e\left(h_p f/p\right).$$

This vanishes if $p\nmid h_p$, that is, if $\iota_p\xi_p$ is primitive. Therefore we have got (55.6) again. Kinkelin's recognition of the critical nature of the situation $p\nmid h_p$ appears to have eventually yielded the concept of primitive characters; nowadays, we argue in a reversed way as we did in §§53–54.

[55.6]   The evaluation of $G(\iota_{15}\lambda_{1,2})$ with $\lambda_{1,2}$ as in Note [54.3]: We have

$$G(\iota_{15}\lambda_{1,2}) = 2i\mathrm{Im}\,A, \quad A = e(1/15) + e(2/15) + e(4/15) + e(8/15).$$

With $B = e(7/15) + e(11/15) + e(13/15) + e(14/15)$, we have

$$A + B = \mu(15) = 1, \quad A \cdot B = \sum_{j=0}^{14} e(j/15) - e(5/15) - e(10/15) + 3 = 4.$$

Thus $A^2 - A + 4 = 0$, and $A = (1 + i\sqrt{15})/2$, for $\mathrm{Im}\,A$ is obviously positive. We have got $G(\iota_{15}\lambda_{1,2}) = i\sqrt{15}$. See Note [56.4] below.

§56

For the sake of completeness, we state the following explicit formula for $\widehat{\chi}(a)$ with a generic $\chi \bmod q$:

Theorem 53 *With the decomposition* (53.5),

$$q^* \nmid \frac{q}{\langle a,q \rangle} \;\Rightarrow\; \widehat{\chi}(a) = 0, \tag{56.1}$$

222 Characters

$$q^* \Big| \frac{q}{\langle a, q \rangle} \;\Rightarrow\; \widehat{\chi}(a) = \overline{\chi}^* \left(-\frac{a}{\langle a, q \rangle} \right) \mu \left(\frac{q}{q^* \langle a, q \rangle} \right) \chi^* \left(\frac{q}{q^* \langle a, q \rangle} \right)$$

$$\times \frac{\varphi(q)}{\sqrt{q} \varphi(q/\langle a, q \rangle)} G(\chi^*) . \tag{56.2}$$

Proof We may suppose that $q > 1$. First, we shall show that for $s | q$

$$\sum_{\substack{k \bmod q \\ k \equiv l \bmod s}} \chi(k) = \begin{cases} J_s(l) \chi^*(l) \varphi(q)/\varphi(s) & \text{if } q^* | s, \\ 0 & \text{if } q^* \nmid s. \end{cases} \tag{56.3}$$

If $q^* \nmid s$, then s is not a period of χ because of (53.6). As in the proof of the preceding theorem one may take a $c \equiv 1 \bmod s$ such that $\chi(c) \neq 1, 0$; multiplying the sum by $\chi(c)$ does not change it, which implies the lower line of (56.3). Also, if $\langle l, s \rangle \neq 1$ or $J_s(l) = 0$, then $\langle k, q \rangle \neq 1$ and this sum vanishes. Thus, we may suppose that $J_s(l) = 1$ as well as $q^* | s$. The sum is then

$$\sum_{u \bmod q/s} \chi(l + su) = \chi^*(l) \sum_{\substack{u \bmod q/s \\ \langle l + su, q \rangle = 1}} 1 = \chi^*(l) \varphi(q)/\varphi(s) \tag{56.4}$$

because of Notes [32.4]–[32.5], which confirms (56.3).

With this, we return to (55.7) and apply (56.3). We find that if $q^* \nmid q_1$, $q_1 = q/\langle a, q \rangle$, then $\widehat{\chi}(a) = 0$, which proves (56.1). On the other hand, if $q^* | q_1$, then with $a_1 = a/\langle a, q \rangle$

$$\widehat{\chi}(a) = \frac{\varphi(q)}{\sqrt{q} \varphi(q_1)} \sum_{l \bmod q_1} J_{q_1}(l) \chi^*(l) e(-a_1 l/q_1)$$

$$= \frac{\varphi(q)}{\sqrt{q} \varphi(q_1)} \overline{\chi}^*(-a_1) G(J_{q_1} \chi^*), \tag{56.5}$$

in which we have used the change of variable: $l \mapsto -\overline{l} \overline{a}_1$, $a_1 \overline{a}_1 \equiv 1 \bmod q_1$, as well as the fact that $J_{q_1} \chi^* \in \mathcal{X}_{q_1}$. Hence, we shall deal with $G(J_{q_1} \chi^*)$. With $\langle q^*, q_1/q^* \rangle = h$, we have

$$G(J_{q_1} \chi^*) = \sum_{\substack{f \bmod q_1/h \\ \langle f, q_1/h \rangle = 1}} \sum_{v \bmod h} \chi^*(f + vq_1/h) e\big((f + vq_1/h)/q_1\big). \tag{56.6}$$

Here we have applied the fact that if $J_{q_1}(f + vq_1/h) = 1$, then $\langle f, q_1/h \rangle = 1$, and if $\langle f, q_1/h \rangle = 1$ as well as $\chi^*(f + vq_1/h) \neq 0$, then $\langle f + vq_1/h, q^* q_1/h \rangle = 1$, so $J_{q_1}(f + vq_1/h) = 1$, as $h | q^*$. On noting that $\chi^*(f + vq_1/h) = \chi^*(f)$, as $q^* | (q_1/h)$, we have

$$G(J_{q_1}\chi^*) = \sum_{\substack{f \bmod q_1/h \\ \langle f, q_1/h \rangle = 1}} \chi^*(f)e(f/q_1) \sum_{v \bmod h} e(v/h)$$

$$= \sum_{\substack{f \bmod q_1/h \\ \langle f, q_1/h \rangle = 1}} \chi^*(f)e(f/q_1) \cdot \begin{cases} 1 & h = 1, \\ 0 & h > 1. \end{cases} \tag{56.7}$$

If $h = 1$, then we have the decomposition $f = uq_1/q^* + wq^*$ by $(30.4)_{K=2}$; and

$$G(J_{q_1}\chi^*) = \sum_{\substack{u \bmod q^*, \langle u, q^* \rangle = 1 \\ w \bmod q_1/q^*, \langle w, q_1/q^* \rangle = 1}} \chi^*\big(uq_1/q^*\big)e\big(u/q^* + w/(q_1/q^*)\big)$$

$$= \chi^*(q_1/q^*) \sum_{u \bmod q^*} \chi^*(u)e(u/q^*) \sum_{\substack{w \bmod q_1/q^* \\ \langle w, q_1/q^* \rangle = 1}} e\big(w/(q_1/q^*)\big)$$

$$= \mu(q_1/q^*)\chi^*(q_1/q^*)G(\chi^*)$$

$$\tag{56.8}$$

where we have used (18.18). This holds with $h > 1$ as well, for then $\chi^*(q_1/q^*) = 0$. We end the proof.

Notes

[56.1] The above theorem was first stated by Hasse (1964, pp.449–450). See also Montgomery and Vaughan (2006, Section 9.2).

[56.2] Take $a = 1$ in (56.8); then it follows that

$$G(\chi) = \mu(q/q^*)\chi^*(q/q*)G(\chi^*).$$

In particular, $G(\chi) = 0$ for any non-primitive $\chi \in \mathcal{X}_{p^\alpha}$, $\alpha \geq 2$.

[56.3] Returning to Note [54.3], we have

$$G(\iota_{15}\lambda_{0,2}) = \sum_{a=1}^{2} \sum_{b=1}^{4} \lambda_{0,2}(5a + 3b)e(a/3 + b/5)$$

$$= \sum_{b=1}^{4} \tilde{\lambda}_{0,2}(3b)e(b/5) \sum_{a=1}^{2} e(a/3)$$

$$= -\tilde{\lambda}_{0,2}(3) \sum_{b=1}^{4} \tilde{\lambda}_{0,2}(b)e(b/5) = G(\iota_5\tilde{\lambda}_{0,2}).$$

For the character $\iota_{15}\lambda_{0,2}$ is induced by the primitive $\iota_5\tilde{\lambda}_{0,2} \in \mathcal{X}_5$. This is an instance of the identity in the preceding Note. We put $C = e(1/5) + e(4/5)$, $D = e(2/5) + e(3/5)$, then $G(\iota_5\tilde{\lambda}_{0,2}) = C - D$. We note that

224   *Characters*

$C = 2\cos(2\pi/5) > 0$; also $C^2 + C - 1 = 0$, as $C + D = -1$, $C \cdot D = -1$. Hence $C = (\sqrt{5} - 1)/2$, $D = -(\sqrt{5} + 1)/2$; thus $G(\iota_5\tilde{\lambda}_{0,2}) = \sqrt{5}$. We find that $G(\iota_{15}\lambda_{0,2}) = \sqrt{5}$.

[56.4] Further, since $\iota_{15}\lambda_{1,2} = \iota_{15}\lambda_{1,0}\lambda_{0,2} = (\iota_3\tilde{\lambda}_{1,0}) \cdot (\iota_5\tilde{\lambda}_{0,2})$ with $\tilde{\lambda}_{1,0}$ being defined analogously to $\tilde{\lambda}_{0,2}$, Note [55.4] implies that

$$G(\iota_{15}\lambda_{1,2}) = (\iota_3\tilde{\lambda}_{1,0})(5)(\iota_5\tilde{\lambda}_{0,2})(3)G(\iota_3\tilde{\lambda}_{1,0})G(\iota_5\tilde{\lambda}_{0,2})$$
$$= (-1)(-1)\sqrt{5}G(\iota_3\tilde{\lambda}_{1,0}),$$
$$G(\iota_3\tilde{\lambda}_{1,0}) = e(1/3) - e(2/3) = i\sqrt{3}.$$

Hence, we get $G(J_{15}\lambda_{1,2}) = i\sqrt{15}$, alternatively to Note [55.6].

[56.5] The two sets Ψ_q and $\mathcal{X}_q$ have analogous orthogonal structures, as (50.3), (50.8), (52.3), and (52.4) indicate. However, the contents of Theorem 53 which should connect these structures via the Fourier expansion (55.1) appear to be too involved to immediately reveal the relation in question. Nevertheless, if we consider the set of all moduli not greater than a given D instead of a single modulus, then the quasi-orthogonality (Note [50.4]) in the system $\{\psi_\tau : \tau \in \mathcal{F}_D\}$ propagates to the system

$$\bigsqcup_{q \leq D} \{\text{primitive } \chi \in \mathcal{X}_q\}$$

via Theorem 51. Therefore, in view of (52.9) and (53.8), one may surmise that the distribution of primes in arithmetic progressions should become more tractable if it is considered in terms of the statistics with respect to varying differences (moduli). This viewpoint originated with Rényi (1948). It has eventually turned out to be one of the greatest innovations in the history of number theory. Details will be developed in Chapter 5.

[56.6] It will become essential in §105 to view each of the sets of all additive and all multiplicative characters as a system of distinct arithmetic functions. So we make here a trivial observation: When one proceeds with increasing modulus q, the genuinely new additive characters over the group $\mathbb{Z}/q\mathbb{Z}$ are those $\psi_{h/q}$ with $\langle h, q \rangle = 1$. Correspondingly, the genuinely new multiplicative characters over the group $(\mathbb{Z}/q\mathbb{Z})^*$ are the primitive characters mod q. Each primitive character is unique like each rational number in the unit interval.

§57

We now return to the beginning of §39 and enter into a detailed study of quadratic congruences:

$$x^2 \equiv d \bmod p, \quad p \nmid d, \ p \geq 3. \tag{57.1}$$

We have replaced the generic residue $a \bmod p$ by $d \bmod p$ for convenience. The discussion will continue to the end of the present chapter and beyond to some extent.

According to Euler's criterion (Theorem 39), there exists a solution to (57.1) if and only if the congruence $d^{(p-1)/2} \equiv 1 \bmod p$ holds. On the other hand, Fermat's theorem (28.11) implies that $d^{(p-1)/2} \equiv \pm 1 \bmod p$. With this, Legendre (1798, p.186) introduced the notation:

$$\left(\frac{d}{p}\right) = \begin{cases} 1 & d \bmod p \text{ is a quadratic residue,} \\ -1 & d \bmod p \text{ is a quadratic non-residue,} \end{cases} \quad p \nmid d. \tag{57.2}$$

In other words,

$$\left(\frac{d}{p}\right) \equiv d^{(p-1)/2} \bmod p, \quad p \nmid d. \tag{57.3}$$

Alternatively, with any primitive root $r_p \bmod p$,

$$\left(\frac{d}{p}\right) = (-1)^{\mathrm{Ind}_{r_p}(d)}, \quad p \nmid d. \tag{57.4}$$

For whether $d \bmod p$ is a quadratic residue or not is equivalent to whether $\mathrm{Ind}_{r_p}(d)$ is even or odd, respectively; besides, this parity does not depend on the choice of the primitive root, as is apparent from the remark on the base change in the definition (44.1) of the discrete logarithm. In particular, we see that

$$\begin{aligned} &\text{the number of residue classes } d \bmod p \\ &\text{with which (57.1) has a solution is } (p-1)/2. \end{aligned} \tag{57.5}$$

One may refer to (43.2) as well. Further, including the case $p \nmid d$, we define

$$\text{Legendre symbol}: \quad \left(\frac{d}{p}\right) = \iota_p e\big(\mathrm{Ind}_{r_p}(d)/2\big) \tag{57.6}$$

under the convention (52.5). Then,

$$\begin{aligned} &\text{Legendre symbol (57.6) is} \\ &\text{a unique non-trivial real element in } \mathcal{X}_p. \end{aligned} \tag{57.7}$$

This uniqueness assertion is contained in the discussion following (54.5) ($l = 0$, $q_0 = p$); see Note [57.3]. As it is completely multiplicative and of period p, we have

$$\left(\frac{ab}{p}\right) = \left(\frac{a}{p}\right)\left(\frac{b}{p}\right), \quad a,b \in \mathbb{Z},$$

$$\left(\frac{c}{p}\right) = \left(\frac{d}{p}\right), \quad c \equiv d \bmod p,$$

(57.8)

without any exception.

Notes

[57.1]   The definition (57.6) follows that of Dirichlet characters which were introduced historically later than Legendre's (57.2). In fact, the concept of characters themselves stemmed from Legendre's idea (1798, p.186): *Comme les quantités analogues à $N^{(c-1)/2}$ se rencontreront fréquemment..., nous emploierons le caractère abrégé $\left(\frac{N}{c}\right)$*... Previously, he (1785, art.IV) had used a crude abbreviation: $N^{(c-1)/2} = \pm 1$. By the way, although the term character should be more appropriate, we shall use symbol instead; this is to respect a famous tradition in number theory.

[57.2]   The chic symbol (57.2) is never a mere mathematical convention. By virtue of Legendre's esprit, we now have a beautiful far-reaching expression for one of the most fundamental structures in number theory. The lucidity of Dirichlet's *Vorlesungen* (1863/1894), which was actually compiled by Dedekind, is due not only to the typical transparency of their reasoning but also to their employment of the congruence notation of Gauss and the symbol of Legendre. Dirichlet (1863, p.83) justly admired the invention of (57.2): *... welches in allen folgenden Untersuchungen eine grosse Rolle spielt.* Indeed, behind the Legendre symbol is a deep algebraic structure; see Note [90.6].

[57.3]   The uniqueness stated in (57.7) is also a consequence of Note [51.6]; namely, $\alpha_2 = 1$ with $A = (\mathbb{Z}/p\mathbb{Z})^*, p \geq 3$.

[57.4]   According to Note [34.3], the congruence equation $x^2 \equiv d \bmod p^{\nu}$, $p \geq 3, \nu \geq 1, p \nmid d$, can be reduced to the case $\nu = 1$. Hence, the number of solutions equals $1 + \left(\frac{d}{p}\right)$ for all $\alpha \geq 1$. One may use (43.2) as well.

[57.5]   The congruence $x^2 \equiv d \bmod q$ with $q = 2^{\tau}q_1, 2 \nmid q_1$, does not have any solution either if $\tau = 2, d \not\equiv 1 \bmod 4$, or if $\tau \geq 3, d \not\equiv 1 \bmod 8$, or if $\left(\frac{d}{p}\right) = -1$ with a $p|q_1$. Otherwise, the number of solutions equals 2^J if $\tau \leq 1$, 2^{J+1} if $\tau = 2$, and 2^{J+2} if $\tau \geq 3$, where J is the number of different prime factors of q_1. To confirm this, use Theorem 39 ($\ell = 2$) for the modulus 2^{τ}, and the preceding Note for the modulus q_1.

[57.6]   Let $\tau \geq 3$ and $d \equiv 1 \bmod 8$. Let $\xi \bmod 2^{\tau}$ be a solution to $x^2 \equiv d \bmod 2^{\tau}$. Then, all the solutions are $\pm\xi, \pm\xi + 2^{\tau-1} \bmod 2^{\tau}$. See (43.3) (D).

§58

Legendre (1798, p.214) stated the truly amazing assertion (3) below, calling it *loi de réciprocité*:

Theorem 54 *It holds that for odd primes p, q*

$$
(1) \quad \left(\frac{-1}{p} \right) = (-1)^{(p-1)/2},
$$

$$
(2) \quad \left(\frac{2}{p} \right) = (-1)^{(p^2-1)/8}, \qquad\qquad (58.1)
$$

$$
(3) \quad \left(\frac{q}{p} \right) \left(\frac{p}{q} \right) = (-1)^{(p-1)(q-1)/4}, \quad p \neq q.
$$

Before entering into the proof, we give remarks: The assertions (1), (2) are called the supplementary laws, and (3) the reciprocity law; the set of these three is called collectively the quadratic reciprocity law. According to (1), $x^2 \equiv -1 \bmod p$ has a solution if and only if $p \equiv 1 \bmod 4$, for then $(p-1)/2$ is even, and otherwise odd. According to (2), $x^2 \equiv 2 \bmod p$ has a solution if and only if $p \equiv \pm 1 \bmod 8$, for then $(p^2 - 1)/8$ is even, and otherwise odd. According to (3),

$$
\begin{aligned}
&\text{the solubilities of } x^2 \equiv q \bmod p \text{ and that of } x^2 \equiv p \bmod q \\
&\text{are opposite to each other if } p \equiv q \equiv -1 \bmod 4, \\
&\text{and otherwise they are the same,}
\end{aligned} \qquad (58.2)
$$

as can be seen by checking the parity of $(p-1)(q-1)/4$. This is a really fundamental fact in number theory, which is not just amazing but might appear even counter-intuitive: prime numbers are not totally independent of each other. The statement (58.2) itself was first made by Legendre (1785, p.517); see §69[b].

The virtue of this theorem is in its ability to make far more precise what has been stated prior to Theorem 39, as far as quadratic congruences are concerned. The combination of (57.8) and (58.1) allows one to determine, by a routine computation, that is, without recourse to (57.3)–(57.4), whether a given integer modulo an odd prime is a quadratic residue or not. It should be stressed, in advance, that this practical aspect of (58.1) will be greatly enhanced when it is supplemented with Theorem 58, §62. Moreover, then a solution to the fundamental problem $(43.5)_{\ell=2}$ will be achieved with relative ease (Theorem 59, §62, and (63.11)–(63.12) below).

Euler published his consideration on (1) several times. Thus, it is appropriate to view (1) as a consequence of his basic assertion (43.2); note also that (1) has been proved in [35.4] and will be in Note [84.3], both by independent

228 *Characters*

arguments. As for (2), it was gradually perceived by Fermat, Euler, and Lagrange via their observations on prime factors of the integers $\{x^2 \pm 2y^2 : x, y \in \mathbb{N}\}$. We shall give a proof in this context in Note [58.3]; see also Remark following the proof of Theorem 58, §62. A statement equivalent to the reciprocity law (3) was first made by Euler (1772a, p.486; 1783, I, p.84) from his immense numerical examination of the prime factors of the integers $\{ax^2 \pm by^2 : x, y \in \mathbb{N}\}$. However, (3) itself was asserted by Legendre (1798, p.214) via his own investigation (1785, p.517) as has been indicated above. His approach to the proof of (3) will be detailed later.

The first rigorous proof of (3) was achieved by Gauss. Actually, he published five other proofs in his life time. The 1st proof [DA, arts.135–144] exploits faithfully the concept of quadratic residues, and is involved; later a simplification of it was given by Dirichlet (1854a; 1863/1894, §§48–51). However, we omit to deal with the 1st proof, save for a related digression made in Note [88.3]. The 2nd proof is in [DA, art.262]. It depends on the genus theory of quadratic forms; our account on it is in Note [91.6]. The 3rd proof (1808) is quite elementary; it appears to have stemmed from Lagrange's observation (Note [35.4]); see Note [58.10]. The 4th proof (1811) is closely related to the cyclotomic theory (equal dissections of the unit circle on $\mathbb{C}$) and relies specifically on the theory of Gauss sums; we shall give its account in §§59–61 but with an analytical twist as is explained below. The 5th proof (1818) is related to the 3rd; we omit to deal with it. The 6th proof (1818) is algebraically oriented; it contains, in a somewhat disguised form, an extremely fundamental scheme of expanding the realm of number theory; the details will be provided in §66.

There exist yet two other proofs due to Gauss, the 7th and the 8th, which are posthumously published in (1863a(2)). Their basis is placed in the arithmetics of polynomial rings over finite fields if expressed with modern terminology. The 7th is conceptually close to the 4th and reveals the algebraic structure behind the reciprocity law; and the 8th is essentially a greatly simplified version of the 6th. Both are breathtakingly beautiful; we shall present them in §67.

Remark As for the actual historical order of Gauss' eight proofs of the reciprocity law, see Note [67.2].

Thus, among Gauss' eight proofs of the law, we shall be concerned mainly with the 4th and the 7th. Describing the 4th proof, we shall adopt the procedure of Dirichlet (1835; 1863/1894, Supplement I) rather than Gauss' original. Dirichlet's approach to (58.1) is regarded as analytic in contrast to Gauss' algebraic reasoning since it relies on Fourier analysis, as will be detailed in §60. We shall augment it in (61.7)–(61.13) by showing Cauchy's approach.

$\S58$ 229

Another aim of ours in the present context is to develop a rectified and completed version of Legendre's proof of the reciprocity law by a combination of $\S69$ [d], $\S88$[A][B], and $\S89$[C_0]. His argument is strikingly different from any of other numerous proofs known today which are mostly reworks of Gauss' ideas. We shall be engaged in Legendre's reasoning, albeit it is rather involved: This is not only because it results in a genuinely alternative proof of the reciprocity law but also because it yields, in the course of discussion, two fundamental theorems in number theory, i.e., Dirichlet's prime number theorem (88.1) and Legendre's Theorem 66, $\S69$, or rather its extension Theorem 91, $\S89$, by Dedekind. These two assertions will, in fact, play a very essential rôle in our account of Gauss' genus theory of quadratic forms to be developed in $\S91$.

It should be added that a conceptually deeper proof of the reciprocity law will be indicated briefly in Note [90.6]; it relies on the theory of ideals of quadratic number fields.

Notes

[58.1]   Legendre (1798, p.214: the title of $\S$VI) called (58.1)(3) *Théorème contenant une loi de réciprocité qui existe entre deux nombres premiers quelconques*; it could not be more veracious. Gauss called it *theorema fundamentale* ([DA, art.130]), and *theorema aureum* in private correspondences.

[58.2]   Characters can be regarded as waves ($\S\S50$–51): a viewpoint that makes Dirichlet's analytic argument look natural. As a matter of fact, the 4th proof of Gauss is often said to be the first encounter of number theory and harmonic analysis, although one may actually go back a little further in history if the theory of algebraic equations is taken into account; see $\S70$.

[58.3]   A proof of the supplementary law (2) is, for instance, in [DA, arts.112–116]. Incorporating an argument of Dirichlet (1863/1894, $\S41$), we shall give its simplified account here: Suppose that $2 \bmod p$ is a quadratic residue with respect to a $p \equiv 3$ or $5 \bmod 8$; that is, there exists $0 < f < p$ such that $f^2 - 2 = pk$. We may assume that $2 \nmid f$, for if f is even, then we may take $p - f$ instead. Hence, $pk \equiv -1 \bmod 8$; and $k \equiv 3$ or $5 \bmod 8$. Then k has a prime factor q which is $\equiv 3$ or $5 \bmod 8$ since if all relevant prime factors are $\equiv \pm 1 \bmod 8$, then $k \equiv \pm 1 \bmod 8$. As $2 \bmod q$ is a quadratic residue, and $q < p$, we shall eventually reach a contradiction. Hence, $2 \bmod p$ is a quadratic non-residue, if $p \equiv 3$ or $5 \bmod 8$. Next, suppose that $-2 \bmod p$ is a quadratic residue with $p \equiv -1 \bmod 8$. Then there exists $0 < g < p$ such that $g^2 + 2 = pl$; we may assume that g is odd as before. We have $l \equiv -3 \bmod 8$, and l should have a prime factor $\equiv -1 \bmod 8$. If not, its prime

230 *Characters*

factors would be either 1, 3, or 5 mod 8. However, the case 5 mod 8 is rejected because of what we have shown above; indeed, if t is such a prime factor, then -2 mod t is a quadratic residue, and by the supplementary law (1) we have the contradiction that 2 mod t is a quadratic residue. Moreover, the prime factors of l can be neither $\equiv 1$ nor 3 mod 8 since product of such factors does not make -3 mod 8. Hence, -2 mod p with $p \equiv -1$ mod 8 is a quadratic non-residue; namely, via the supplementary law (1) we conclude that 2 mod p with $p \equiv -1$ mod 8 is a quadratic residue. What remains is the treatment of those $p \equiv 1$ mod 8. To this end, Dirichlet adopted Gauss' reasoning [DA, art.114]. Thus, take a primitive root ρ mod p, and note that $\rho^{(p-1)/2} \equiv -1$ mod p. This implies that $\left(\rho^{(p-1)/4}+1\right)^2 \equiv 2\rho^{(p-1)/4}$ mod p; and we find that 2 mod p with $p \equiv 1$ mod 8 is a quadratic residue, as $\rho^{(p-1)/4}$ is a square. We end the proof. See Notes [58.10] and [67.6]. Also, in the Remark attached to Theorem 58, §62, is given a proof of the fact that (2) is, in fact, a consequence of (1) and (3).

[58.4] An application of Theorem 54: the quadratic congruence $x^2 \equiv 31$ mod 430883. The modulus is a prime (Note [41.3]). Both the modulus and the residue are $\equiv -1$ mod 4, and the solubility in question is opposite to that of $y^2 \equiv 430883 \equiv 14$ mod 31. We apply the supplementary law (2) and then the reciprocity law (3), getting

$$\left(\frac{14}{31}\right) = \left(\frac{2}{31}\right)\left(\frac{7}{31}\right) = (+1) \cdot (-1)\left(\frac{3}{7}\right) = \left(\frac{1}{3}\right) = 1.$$

Namely, the original quadratic congruence equation does not have any solution. Returning to the criterion (43.2), we see that we should get $31^{215441} \equiv -1$ mod 430883; indeed, this has been observed in Note [41.3]. Dividing squares by 430883, we shall never get the residue 31.

[58.5] Another application: $x^2 \equiv f$ mod 430883, $f = 221129$. We need to decompose f, for the congruence $2^{f-1} \equiv 137695$ mod f means that f is composite. We apply Pollard's method (38.1) with $k_0 = 1$, $c = 1$; then $k_{12} \equiv 74745$, $k_{24} \equiv 136572$ mod f, $g_{12} = 557$. Hence, $221129 = 397 \cdot 557$; both factors are primes. We have now

$$\left(\frac{221129}{430883}\right) = \left(\frac{397}{430883}\right)\left(\frac{557}{430883}\right) = \left(\frac{2 \cdot 3 \cdot 23}{397}\right)\left(\frac{2 \cdot 7 \cdot 23}{557}\right)$$

$$= \left(\frac{3}{397}\right)\left(\frac{23}{397}\right)\left(\frac{7}{557}\right)\left(\frac{23}{557}\right)$$

$$= \left(\frac{6}{23}\right)\left(\frac{5}{23}\right) = \left(\frac{7}{23}\right) = -\left(\frac{2}{7}\right) = -1.$$

We should have $f^{215441} \equiv -1$ mod 430883, although we omit to give details. The process of the application of Theorem 54 is as follows. The first equality is due to the multiplicativity of the Legendre symbol. We move to the moduli 397 and 557 via the reciprocity; they are both $\equiv 1 \bmod 4$, and there is no change of sign. We then check $430883 \equiv 2 \cdot 3 \cdot 23 \bmod 397$ and $\equiv 2 \cdot 7 \cdot 23 \bmod 557$, which imply the second equality. With respect to these prime moduli, 2 is quadratic non-residue because of the supplementary law (2). Hence, we get the third equality. Next, we move to the moduli 3, 7, 23 without change of sign; and since $397 \equiv 1 \bmod 3$, $557 \equiv 4 \bmod 7$, we may ignore the relevant two factors; also, since $397 \equiv 6$, $557 \equiv 5 \bmod 23$, we get the fourth equality. Further, because of $6 \cdot 5 \equiv 7 \bmod 23$, we have the fifth equality. Moving to the modulus 7 causes change of sign. The final step is an application of the supplementary law (2). Later, in Note [62.2], we shall show a far quicker computation. Incidentally, one of the first applications of the reciprocity law in history is, naturally, due to Legendre (1798, pp.228–229). He got $\left(\frac{601}{1013}\right) = -1$, and then performed the confirmation $601^{506} \equiv -1 \bmod 1013$, precisely by means of the modular exponentiation (36.2).

[58.6]   This is an addition to Note [21.5]. As a matter of fact, prior to Smith (1855) a similar argument had been devised by Serret (1849b), although he needed the supplementary law (1) whereas Smith didn't. Serret's article has an addendum by C. Hermite reporting the following alternative argument: This is quite simple. First, we assume that $p \equiv 1 \bmod 4$; and choose an a such that $a^2 \equiv -1 \bmod p$, $1 < a < p/2$. In the continued fraction expansion of a/p, the denominators of the convergents increase from 1 to p. Hence, there exist adjacent convergents $d/c, v/u$ such that $c < \sqrt{p} < u$. We have $|a/p - d/c| \leq 1/cu$ (see (22.10)); that is, $|ac - pd| < \sqrt{p}$. Hence $(ac - pd)^2 + c^2 < 2p$. On noting that this left side is a multiple of p, we end the proof. Continued in Notes [78.3]–[78.4]. Obviously there is no necessity to restrict our consideration to primes; Hermite's argument is effective for any $n > 0$ such that $x^2 \equiv -1 \bmod n$ has a solution. See Note [81.7].

[58.7]   Prime numbers $p \equiv 1, 2, 4 \bmod 7$ have representations $x^2 + 7y^2 = p$, $x, y \in \mathbb{N}$. To show this, we note that $\left(\frac{-7}{p}\right) = \left(\frac{p}{7}\right)$ by (58.1) (1)(3). Thus, we see that there exists an a such that $a^2 \equiv -7 \bmod p$ for these p. We take s, t as in Note [29.3]. Then $7s^2 + t^2 \equiv (7 + a^2)s^2 \equiv 0 \bmod p$; there exists an f such that $7s^2 + t^2 = fp$, $0 < f < 8$. If $f = 1$, then the proof ends. If $f = 7$, then $7|t$, $s^2 + 7(t/7)^2 = p$. As for the remaining f, since $\left(\frac{fp}{7}\right) = 1$, we have $\left(\frac{f}{7}\right) = 1$; thus it suffices to consider $f = 2, 4$. The case in which s, t are both odd is to be discarded, for then $7s^2 + t^2 \equiv 0 \bmod 8$. Thus, s, t are both even; and $f = 4$, getting $7(s/2)^2 + (t/2)^2 = p$. For instance, $123^2 + 7 \cdot 124^2 = 122761$ (see Note [41.5]).

[58.8] Any prime factor s of the Mersenne number M_p satisfies $s \equiv 1$ mod $2p$, as is pointed out in Note [41.1]. This implies that $2^{(s-1)/2} \equiv 1$ mod s, and we see that 2 mod s is a quadratic residue. Hence, by the supplementary law (2), we also obtain $s \equiv \pm 1$ mod 8. As for a prime factor t of the Fermat number F_r, the order of 2 mod t is equal to 2^{r+1}. We suppose that $r \geq 2$. Then, we have $t \equiv 1$ mod 8; thus, the supplementary law (2) implies that 2 mod t is a quadratic residue, and $2^{(t-1)/2} \equiv 1$ mod t, which means that $2^{r+1}|(t-1)/2$ or $t \equiv 1$ mod 2^{r+2}. This observation is due to Lucas (1878a, pp. 280–283). Further, we have the criterion due to Pépin (1877):

$$F_r, r \geq 2, \text{ is a prime } \iff 5^{(F_r-1)/2} \equiv -1 \text{ mod } F_r.$$

For if F_r is a prime, then by the reciprocity law $\left(\frac{5}{F_r}\right) = \left(\frac{2}{5}\right) = -1$, which proves the necessity part; and the sufficiency part is due to the Legendre criterion (41.1). One may use the base 3 instead of 5.

[58.9] An observation of Chebyshev (1848: 1889, Anhang II) concerning primitive roots. If primes p, q are in the relation $p = 2q + 1 \equiv 3$ mod 8, then $2^{(p-1)/q} \not\equiv 1$ mod p, and $2^{(p-1)/2} \equiv -1$ mod p by the supplementary law (2). Hence, by (41.1), 2 mod p is a primitive root. Also, if $p \equiv 7$ mod 8, then q mod p is a primitive root. For otherwise it would follow that $q^q \equiv 1$ mod p; however, by the supplementary law (2) we would have $2^q \equiv 1$ mod p, and $(2q)^q \equiv 1$ mod p, which is a contradiction, as $2q \equiv -1$ mod p. For instance, the pairs $\{p = 20000243,\ q = 10000121\}$ and $\{p = 20000159,\ q = 10000079\}$ belong to the first and the second cases, respectively. As for p with primitive root 3 mod p, see the same article of Chebyshev.

[58.10] Following tradition, here we shall show the 3rd proof of Theorem 54 due to Gauss (1808), adopting its account by Dirichlet (1863/1894, §44).

Lemma Let $p > 2$, and $p \nmid d$. Let $dj \equiv r_j$ mod p with $|r_j| < p/2$, for $j = 1, 2 \ldots (p-1)/2$. Then we have $\left(\frac{d}{p}\right) = (-1)^s$ with $s = |\{r_j < 0\}|$.

Proof For $d^{(p-1)/2}((p-1)/2)! \equiv r_1 r_2 \cdots r_{(p-1)/2} \equiv (-1)^s((p-1)/2)!$ mod p, and we have (57.3).

The law (1): Take $d = -1$ in Lemma.

The law (2): Take $d = 2$ in Lemma, so $s = |\{j : p/2 < 2j < p\}| = [p/2] - [p/4]$, which is even and odd according as $p \equiv \pm 1$ and ± 3 mod 8, respectively.

The law (3): Take $d = q$ in Lemma. We note that if $r_j > 0$, then $qj = [qj/p]p + r_j$, and if $r_j < 0$ then $qj = ([qj/p]+1)p + r_j$. Thus

$$q(p^2 - 1)/8 = p \sum_{j=1}^{(p-1)/2} [qj/p] + \sum_{j=1}^{(p-1)/2} r_j + ps.$$

On noting that $(p^2 - 1)/8 = \sum_{j=1}^{(p-1)/2} |r_j|$, we get

$$(q+1)(p^2 - 1)/8 = p \sum_{j=1}^{(p-1)/2} [qj/p] + 2 \sum_{r_j>0} r_j + ps.$$

Hence

$$s \equiv \sum_{j=1}^{(p-1)/2} [qj/p] \bmod 2.$$

This implies that

$$\left(\frac{q}{p}\right)\left(\frac{p}{q}\right) = (-1)^{A(p,q)}, \quad A(p,q) = \sum_{j=1}^{(p-1)/2} [qj/p] + \sum_{k=1}^{(q-1)/2} [pk/q].$$

It now suffices to show that $A(p,q) = (p-1)(q-1)/4$. Thus, we consider the set $\{up - vq : 0 < u < q/2, 0 < v < p/2\}$; its size is $(p-1)(q-1)/4$. The number of the positive and the negative elements are $\sum_u [pu/q]$ and $\sum_v [qv/p]$, respectively, which ends the proof. By the way, there exists a claim that the 3rd proof is the one that Gauss himself liked most among his eight proofs. This appears to have originated in the remark by Smith (1859, art.19, the ending passage) on *Demonstrationem itaque genuinam* from Gauss (1808, p.4). It is true that the 3rd proof does not contain any *heterogeneis derivatæ*.

§59

We shall now describe Gauss' 4th proof of Theorem 54, relying on Dirichlet's idea (1835) that is to be given in the next section, which amounts to another milestone in the developments of analytic number theory.

Since the Legendre symbol is a primitive character, we have, by Theorem 51,

$$\left(\frac{d}{p}\right) = \frac{1}{p} G(-1,p)G(d,p),$$

$$G(d,p) = \sum_{u \bmod p} \left(\frac{u}{p}\right) e(du/p). \tag{59.1}$$

If $p \nmid d$, then denoting quadratic residues and non-residues mod p by s and t, respectively, we have

$$G(d,p) = \sum_s e(ds/p) - \sum_t e(dt/p) = 1 + 2\sum_s e(ds/p)$$
$$= \sum_{v=0}^{p-1} e(dv^2/p), \tag{59.2}$$

for $1 + \sum_s + \sum_t = 0$, and $v_1^2 \equiv v_2^2 \Leftrightarrow v_1 \equiv v_2, p - v_2 \bmod p$. Or rather apply Note [57.4], and proceed from the right side of (59.2) to the left. Since $G(cd,p) = \left(\frac{c}{p}\right)G(d,p)$ for any $c \bmod p$, taking $c = (p+1)/2$ we have

$$\left(\frac{d}{p}\right) = \frac{1}{p}H(-1,p)H(d,p), \quad p \nmid d, \tag{59.3}$$

where

$$H(a,b) = \sum_{w=0}^{|b|-1} \exp\left(\pi i a w^2/b + \pi i a w\right), \quad ab \ne 0, \ a,b \in \mathbb{Z}. \tag{59.4}$$

The introduction of the sum $H(a,b)$ might appear abrupt; in fact, it is to reveal a reciprocity/automorphic structure behind the sum $G(d,p)$:

Theorem 55 *For any non-zero $a,b \in \mathbb{Z}$ we have the functional equation*

$$H(a,b) = |b/a|^{1/2} \exp\left(\tfrac{1}{4}\pi i(\operatorname{sgn}(ab) - ab)\right)H(-b,a). \tag{59.5}$$

A proof will be given in the next section. We shall first apply this identity to prove Theorem 54. Thus, we take $a = -1$, $b = p$; then

$$H(-1,p) = p^{1/2} \exp\left(\tfrac{1}{4}\pi i(p-1)\right). \tag{59.6}$$

We get, by (59.3),

$$\left(\frac{-1}{p}\right) = \exp\left(\tfrac{1}{2}\pi i(p-1)\right), \tag{59.7}$$

which is the supplementary law (1). Also, since $H(-p,2) = 1 - \exp(-\tfrac{1}{2}\pi ip)$, we have

$$H(2,p) = (p/2)^{1/2} \exp\left(\tfrac{1}{4}\pi i(1 - 2p)\right)\left(1 - \exp(-\tfrac{1}{2}\pi ip)\right). \tag{59.8}$$

We get, again by (59.3),

$$\left(\frac{2}{p}\right) = (-1)^{(p-1)/2}\sqrt{2}\sin\left(\tfrac{1}{4}p\pi\right), \tag{59.9}$$

which is equivalent to the supplementary law (2). Further, combining (59.3), (59.5), and (59.6), we have

$$\left(\frac{q}{p}\right) = \frac{1}{\sqrt{q}} \exp\left(\tfrac{1}{4}\pi i(p-1) + \tfrac{1}{4}\pi i(1 - pq)\right)H(-p,q), \tag{59.10}$$

Also, applying the replacement $p \mapsto q, d \mapsto -p$ in (59.3) we have

$$\left(\frac{-p}{q}\right) = \frac{1}{\sqrt{q}} \exp\left(\tfrac{1}{4}\pi i(q-1)\right) H(-p,q). \qquad (59.11)$$

Therefore,

$$\left(\frac{q}{p}\right) = \exp\left(\tfrac{1}{4}\pi i(p-1) + \tfrac{1}{4}\pi i(q-1) + \tfrac{1}{4}\pi i(1-pq)\right)\left(\frac{p}{q}\right), \qquad (59.12)$$

which is equivalent to the reciprocity law (58.1) (3).

Notes

[59.1] An automorphic structure is evident in (59.5). The transformations $T^2 : {}^t\{a,b\} \mapsto {}^t\{a+2b,b\}$, and $W : {}^t\{a,b\} \mapsto {}^t\{-b,a\}$ do not alter the shape of $H(a,b)$, except for an explicit multiplier; here T, W are as in §5. Hence, the function $H(a,b)$ is automorphic with respect to the group generated by these elements of Γ according to the convention introduced after Theorem 4, in the extended sense stated there.

[59.2] The identity (59.5) is due not to Dirichlet but essentially to Schaar (1850). Dirichlet treated only $H(2,n) = \sum_{h=0}^{n-1} e(h^2/n)$, which suffices to prove (58.1) (3), though; see (61.4). A merit of (59.5) is in that it yields (58.1) (1)–(3) as a set.

§60

To prove Theorem 55, we shall employ the sum formula of Poisson (1823):

Theorem 56 *For any function f which is continuously differentiable in the closed interval $[A,B]$, $A < B$, $A, B \in \mathbb{Z}$, it holds that*

$$\tfrac{1}{2}f(A) + f(A+1) + \cdots + f(B-1) + \tfrac{1}{2}f(B)$$
$$= \sum_{n=-\infty}^{\infty} \int_A^B f(x)e(nx)\,dx, \qquad (60.1)$$

where $e(x) = \exp(2\pi i x)$ as usual.

Remark If $A + 1 = B$, then the sum $f(A+1) + \cdots + f(B-1)$ is empty, i.e., equal to 0. See (60.8) below. The application of Poisson's sum formula to arithmetic problems was started by Dirichlet.

236 *Characters*

Proof We shall verify first the expansion

$$[x] - x + \tfrac{1}{2} = \sum_{n=1}^{\infty} \frac{\sin(2\pi nx)}{\pi n}, \qquad x \in \mathbb{R} - \mathbb{Z}, \tag{60.2}$$

(see also (92.53)). Integrating the identity

$$2\sum_{n=1}^{N} \cos(2\pi n\theta) = \frac{\sin\left((2N+1)\pi\theta\right)}{\sin \pi\theta} - 1, \tag{60.3}$$

we have, for $0 < x < 1$,

$$\begin{aligned}\sum_{n=1}^{N} \frac{\sin(2\pi nx)}{\pi n} &= \int_{0}^{x} \sin\left((2N+1)\pi\theta\right)h(\theta)d\theta \\ &\quad + \int_{0}^{(2N+1)\pi x} \frac{\sin\theta}{\pi\theta}d\theta - x,\end{aligned} \tag{60.4}$$

where $h(\theta) = (\sin(\pi\theta))^{-1} - (\pi\theta)^{-1}$. Applying integration by parts to the first integral on the right side and taking N to $+\infty$, we get

$$\sum_{n=1}^{\infty} \frac{\sin(2\pi nx)}{\pi n} = \int_{0}^{\infty} \frac{\sin\theta}{\pi\theta}d\theta - x = \tfrac{1}{2} - x, \tag{60.5}$$

since setting $x = \tfrac{1}{2}$ on the left side, we find that this integral is equal to $\tfrac{1}{2}$. On noting the periodicity, we end the verification of (60.2).

It should be noted that (60.2) converges boundedly; that is, any partial sum of the infinite sum of (60.2) is uniformly bounded by an absolute constant. More precisely, there exists an absolute constant $c > 0$ such that for any integer $N \geq 0$

$$\left| \sum_{n=N+1}^{\infty} \frac{\sin(2\pi nx)}{\pi n} \right| \leq \frac{c}{1 + N\{x\}}, \qquad \{x\} = \min_{n \in \mathbb{Z}} |x - n|. \tag{60.6}$$

For assuming that $0 < x \leq \tfrac{1}{2}$, we subtract (60.4) from (60.5), and see that the sum on the right of (60.6) is equal to

$$\int_{(2N+1)\pi x}^{\infty} \frac{\sin\theta}{\pi\theta}d\theta - \int_{0}^{x} \sin\left((2N+1)\pi\theta\right)h(\theta)d\theta, \tag{60.7}$$

from which (60.6) follows via integration by parts.

Next, let $\rho(x)$ be the left side of (60.2); then we have, for each integer k such that $A \leq k < B$,

$$\tfrac{1}{2}\left(f(k) + f(k+1)\right) = \int_{k}^{k+1} f(x)dx - \int_{k}^{k+1} f'(x)\rho(x)dx. \tag{60.8}$$

By (60.2) and (60.6), this second integral is equal to

$$\sum_{n=1}^{N} \frac{1}{\pi n} \int_{k}^{k+1} f'(x) \sin(2n\pi x)dx + E_N \tag{60.9}$$

where $N \geq 1$ is arbitrary, and

$$|E_N| \leq c \int_{k}^{k+1} \frac{|f'(x)|}{1 + N\{x\}}dx. \tag{60.10}$$

We divide the last integral into two parts depending on whether $\{x\} \leq N^{-1/2}$ or not. We let $N \to \infty$; then E_N tends to 0, and (60.9) becomes

$$-2\sum_{n=1}^{\infty} \int_{k}^{k+1} f(x) \cos(2n\pi x)dx. \tag{60.11}$$

Inserting this into (60.8) and summing over k, we end the proof of Theorem 56.

Proof of Theorem 55: By the sum formula (60.1),

$$H(a,b) = \sum_{n=-\infty}^{\infty} \int_{0}^{|b|} \exp\left(\pi iax^2/b + \pi iax + 2\pi inx\right) dx. \tag{60.12}$$

For

$$H(a,b) = 1 + \sum_{w=1}^{|b|-1} \exp\left(\pi iaw^2/b + \pi iaw\right), \tag{60.13}$$

and this term 1 is equal to half of the sum of the values at $w = 0, |b|$ of the summand; if $|b| = 1$, then the sum is empty. The integral in (60.12) equals

$$\exp\left(-\tfrac{1}{4}\pi iab\right) \exp\left(-\pi i(b/a)n^2 - \pi ibn\right)$$
$$\times \int_{0}^{|b|} \exp\left(\pi i(a/b)(x + (n + a/2)(b/a))^2\right)dx. \tag{60.14}$$

We write $n = m + u|a|$, $m, u \in \mathbb{Z}$, i.e., $n \equiv m \mod |a|$. Then

$$\exp\left(-\pi i(b/a)n^2 - \pi ibn\right) = \exp\left(-\pi i(b/a)m^2 - \pi ibm\right); \tag{60.15}$$

also the integral of (60.14) is equal to

$$\int_{bm/a+b/2+(|a|b/a)u}^{bm/a+b/2+(|a|b/a)(u+|ab|/ab)} \exp\left(\pi i(a/b)x^2\right)dx. \tag{60.16}$$

238   *Characters*

We sum this over $|u| \le U$ and let $U \to \infty$; the convergence can be verified via integration by parts. Then we sum the result over $m \bmod |a|$. We get

$$
\begin{aligned}
H(a,b) &= \exp\left(-\tfrac{1}{4}\pi iab\right)H(-b,a)\int_{-\infty}^{\infty}\exp\left(\pi i(a/b)x^2\right)dx \\
&= |b/a|^{1/2}\exp\left(-\tfrac{1}{4}\pi iab\right)H(-b,a)\int_{-\infty}^{\infty}\exp\left(\pi i\,\mathrm{sgn}\,(ab)x^2\right)dx.
\end{aligned}
\tag{60.17}
$$

The last integral is equal to $e^{\pi i\,\mathrm{sgn}\,(ab)/4}$ for arbitrary a,b, $ab \ne 0$, as can be seen by considering the case $a = b = 1$. We have obtained the identity (59.5), which ends the proof of Theorem 54.

Notes

[60.1]   Another look, an important one, at (60.1) is to be given in Note [98.3]. Incidentally, Euler's famous sum $\sum_{n=1}^{\infty} 1/n^2 = \pi^2/6$, which we needed in Note [18.5], can easily be shown by (60.1): take $f(x) = (x - 1/2)^2$ and $A = 0, B = 1$.

[60.2]   This is a typical application of the Euler–Maclaurin sum formula mentioned in Note [13.8]: We elaborate (12.1). Let $\rho(x)$ be as in (60.8). Then, by (12.1),

$$
\zeta(s) = \frac{1}{s-1} + \tfrac{1}{2} + s\int_{1}^{\infty}\frac{\rho(x)}{x^{s+1}}dx, \quad \mathrm{Re}\,s > 0.
$$

On noting the expansion (60.2) and integrating twice by parts,

$$
\zeta(s) = \frac{1}{s-1} + \tfrac{1}{2} + \tfrac{1}{12}s + s(s+1)(s+2)\int_{1}^{\infty}\frac{\tilde{\rho}(x)}{x^{s+3}}dx, \quad \mathrm{Re}\,s > -2,
$$

with

$$
\tilde{\rho}(x) = -\frac{1}{4\pi^3}\sum_{n=1}^{\infty}\frac{\sin(2\pi nx)}{n^3}.
$$

The same procedure can be repeated indefinitely, and we find that $(s-1)\zeta(s)$ is entire and that $(s-1)\zeta(s) \ll |s|^{2A}$ for $\mathrm{Re}\,s \ge -A$ with any integer $A \ge 1$. However, a better bound follows from the combination of the functional equation (12.7)–(12.8) and Stirling's formula (94.37). Nevertheless, in many applications it suffices to have the fact that $\zeta(s)$ is of polynomial growth as $|\mathrm{Im}\,s|$ tends to infinity while $\mathrm{Re}\,s$ is bounded from below.

[60.3]   As for $L(s,\chi)$ with an arbitrary $\chi \bmod q$, we have, for $\mathrm{Re}\,s > 0$,

$$
L(s,\chi) = \sum_{h=1}^{q}\chi(h)\left\{\frac{1}{q(s-1)}h^{1-s} + \tfrac{1}{2}h^{-s} + \frac{s}{q^s}\int_{0}^{\infty}\frac{\rho(x)}{(x+h/q)^{s+1}}dx\right\}.
$$

This is regular at $s = 1$, if $\chi \neq J_q$, for then $\sum_{h=1}^{q} \chi(h) = 0$. Applying integration by parts repeatedly, we find that $L(s, \chi)$, $\chi \neq J_q$, is entire and of polynomial growth with respect to both s and q, provided $\operatorname{Re} s$ is bounded from below, a fact that will be applied in §101. One may, of course, use instead the combination of (53.8) and the functional equation (Theorem 52).

[60.4]   The values of the zeta and L-functions at non-positive integers can be discussed by means of either the preceding two Notes or by their functional equations. However, we skip it, since those assertions thus obtained do not have essential relevance to the purpose of the present essays.

§61

This is a digression. We shall add some consideration to the discussion of the previous sections. Thus, Theorem 55 gives

Theorem 57 *Let $H(a,b)$ be defined by (59.4). Then we have, for any $m \in \mathbb{N}$,*

$$H(2,m) = \sum_{h=0}^{m-1} e(h^2/m) = \tfrac{1}{2}(1 + i)(1 + i^{-m})\sqrt{m}, \qquad (61.1)$$

where $e(x) = \exp(2\pi i x)$ as before.

This is the prototype of various sums which are collectively called Gauss sums; in this context, the present sum is specifically called the quadratic Gauss sum.

Remark   In literature, the usage of the term Gauss sum is confused, as it denotes, on the one hand, the sum introduced at (55.2) and, on the other hand, the one defined in Note [61.2] below. We shall follow the same practice but with care.

If $\langle m, n \rangle = 1$, $m, n \in \mathbb{N}$, then

$$H(2n, m)H(2m, n) = \sum_{h=0}^{m-1}\sum_{k=0}^{n-1} e\big(((nh)^2 + (mk)^2)/mn\big) = H(2, mn), \quad (61.2)$$

since $(nh)^2 + (mk)^2 \equiv (nh + mk)^2 \bmod mn$ and $\{nh + mk\}$ is a residue system mod mn. Also, via (59.2), we have

$$G(d,p) = H(2d,p),$$
$$H(2d,p) = \left(\frac{d}{p}\right)H(2,p), \quad p \nmid d. \qquad (61.3)$$

240   *Characters*

Hence, for different odd primes p, q, we have

$$\left(\frac{q}{p}\right)\left(\frac{p}{q}\right) = \frac{H(2, pq)}{H(2, p)H(2, q)}. \tag{61.4}$$

Inserting (61.1) into this right side, we obtain (59.12), which is the (original) 4th proof of (58.1) (3) by Gauss. However, his proof of (61.1) is completely different from that we have given in §§59–60. He related once that he needed strenuous efforts to establish the explicit formula (61.1) (see Note [65.1]). Hence, one can say that Dirichlet (1835) discovered an accessible alternative argument to reach (61.1).

In passing, we remark for the sake of a later purpose that the explicit formula

$$G(1, p)^2 = (-1)^{(p-1)/2} p \tag{61.5}$$

can be proved independently of (61.1). For this sake it suffices to put $d = 1$ in (59.1); or rather by (61.3) we have

$$\left(\frac{-1}{p}\right) G(1, p)^2 = H(2, p)H(-2, p)$$

$$= \sum_{b \bmod p} \sum_{a \bmod p} e\big((a^2 - b^2)/p\big) \tag{61.6}$$

$$= \sum_{c \bmod p} \sum_{b \bmod p} e\big(c(2b + c)/p\big) = p,$$

in which the last line is due to the change of variables $a = b + c$. Gauss' 6th proof of (58.1) (3) is essentially the same as raising both sides of (61.5) to the power $(q - 1)/2$; details will be given in §66.

Additionally, we shall show Cauchy's proof (1840) of the explicit formula (61.1). To this end, we introduce the theta-function

$$\vartheta(z, \tau) = \sum_{n=-\infty}^{\infty} \exp\big(-\pi z(n + \tau)^2\big), \tag{61.7}$$

where $|\arg(z)| < \pi/2$, and $\tau \in \mathbb{C}$ is arbitrary. In the sum formula (60.1), we take $f(x) = \exp\big(-\pi z(x + \tau)^2\big)$, and let $A \to -\infty, B \to +\infty$. Then the right side converges uniformly; to see it, apply integration by parts twice to each integral. We get the theta-transformation formula:

$$\vartheta(z, \tau) = \frac{1}{z^{1/2}} \sum_{n=-\infty}^{\infty} \exp\big(-\pi n^2/z + 2\pi i n\tau\big), \quad |\arg(z^{1/2})| < \pi/4. \tag{61.8}$$

For the sake of a later purpose we note that differentiating both sides term-wise with respect to τ, which is obviously legitimate, one obtains

$$\sum_{n=-\infty}^{\infty} (n+\tau)\exp\left(-\pi z(n+\tau)^2\right) = \frac{1}{iz^{3/2}} \sum_{n=-\infty}^{\infty} n\exp\left(-\pi n^2/z + 2\pi in\tau\right),$$

$$(61.9)$$

with $|\arg(z^{3/2})| < 3\pi/4$.

We then observe that for non-zero integers a, b,

$$\vartheta\left(z - 2ia/b, 0\right) = \sum_{\ell=0}^{|b|-1} e\left(a\ell^2/b\right)\vartheta\left(b^2 z, \ell/b\right). \tag{61.10}$$

To the right side we apply (61.8), and get

$$\lim_{z\to 0} z^{1/2}\vartheta\left(z - 2ia/b, 0\right) = \frac{1}{|b|}H(2a, b), \tag{61.11}$$

where z approaches the origin while staying in the angular domain $|\arg(z)| \le \pi/2 - \varepsilon$. Since $1/(z - 2ia/b) = \xi + ib/2a$, $\xi = (b/2a)^2 z/(1 + (b/2a)iz)$, using (61.8) again, we get

$$z^{1/2}\vartheta\left(z - 2ia/b, 0\right) = \frac{(z/\xi)^{1/2}}{(z - 2ia/b)^{1/2}}\xi^{1/2}\vartheta\left(\xi + ib/2a, 0\right). \tag{61.12}$$

We let $z \to 0$ on both sides and obtain

$$\frac{1}{|b|}H(2a, b) = \frac{(-i\,\mathrm{sgn}\,(ab))^{-1/2}}{2(2|ab|)^{1/2}}H(-2b, 4a). \tag{61.13}$$

Taking $a = 1, b = m$ and noting here $(-i)^{-1/2} = \exp\left(\frac{1}{4}\pi i\right)$ because of the choice of the branch for $z^{1/2}$ in (61.8), we end an alternative proof of (61.1).

It should be worthwhile to give, at this point, a proof of the functional equation (12.8) as an application of (61.8). Thus, we have, for $\mathrm{Re}\,s > 1$,

$$\pi^{-s/2}\Gamma\left(\tfrac{1}{2}s\right)\zeta(s) = \tfrac{1}{2}\int_0^{\infty} \left(\vartheta(x, 0) - 1\right)x^{s/2-1}dx, \tag{61.14}$$

where a use of (94.21) has been made, and an exchange of the order of infinite summation and integration has been performed, which is verified readily by absolute convergence. We divide this integral at $x = 1$, and to the part with $0 < x < 1$ apply the change of variable $x \mapsto 1/x$. Then, via $(61.8)_{\tau=0}$, we get

$$\pi^{-s/2}\Gamma\left(\tfrac{1}{2}s\right)\zeta(s) = \frac{1}{s(s-1)} + \tfrac{1}{2}\int_1^{\infty} \left(x^{s/2} + x^{(1-s)/2}\right)\left(\vartheta(x, 0) - 1\right)\frac{dx}{x}, \tag{61.15}$$

242   *Characters*

where the integral is an entire function of s, which does not alter if s is replaced by $1 - s$. This is the second proof of (12.8) by Riemann (1860); his first proof is (12.4)–(12.6). Riemann might have extracted, from (12.8), the formula $(61.8)_{\tau=0}$, and returned to (12.8), although he attributed $(61.8)_{\tau=0}$ to Jacobi. With respect to this historical fact, it should be added that the functional equation for the zeta-function and the sum formula of Poisson are essentially equivalent; namely (60.1) follows from (12.7). See Note [98.3].

We can extend the above argument to Dirichlet L-functions; we omit the details, though. Thus, we can twist the theta-transformation formula by a character χ mod q in the following sense: provided that χ mod q is primitive,

$$\sum_{n=-\infty}^{\infty} \chi(n) \exp\left(-\pi z n^2/q\right) = \frac{G(\chi)}{(qz)^{1/2}} \sum_{n=-\infty}^{\infty} \overline{\chi}(n) \exp\left(-\pi n^2/qz\right);$$

$$(61.16)$$

$$\sum_{n=-\infty}^{\infty} \chi(n) n \exp\left(-\pi z n^2/q\right) = \frac{G(\chi)}{iq^{1/2}z^{3/2}} \sum_{n=-\infty}^{\infty} \overline{\chi}(n) n \exp\left(-\pi n^2/qz\right).$$

$$(61.17)$$

Via the Fourier expansion (55.4), the former follows from (61.8) and the latter from (61.9). As for an alternative proof of the functional equation (55.11), we employ (61.16) if $\chi(-1) = 1$, and (61.17) if $\chi(-1) = -1$.

Notes

[61.1]   The first proof of the beautiful explicit formula (61.1) is due to Gauss (1811). Various alternative proofs are known. Landau (1927, vierter Teil, Kapitel 6) collected those due to Kronecker (1889), Mertens (1896), and Schur (1921) in addition to Dirichlet's. Also, Cauchy (1840) gave two proofs. One of them is shown above; it is via an observation of a reflection property (61.8) of the theta-function, which can be regarded as a beginning of the theory of automorphic functions and forms. The extension to algebraic number fields is discussed in Hecke (1923).

[61.2]   One may wonder whether or not any analogue of the explicit formula (61.1) holds for the sum

$$g_\ell(p) = \sum_{m=0}^{p-1} e(m^\ell/p), \quad \ell \geq 3,$$

which we call the Gauss sum mod p of order ℓ. We shall give just an introductory account of this and related issues: Thus, we begin with the notion of the order of a $\chi \in \mathcal{X}_q$. That is, the least $\ell \in \mathbb{N}$ such that $\chi^\ell = {}_Jq$; such

$$\S 61 \qquad\qquad 243$$

a χ is called an ℓth power character mod q. We have of course $\ell | \varphi(q)$. In the case with $q = p \geq 3$, $\ell | (p-1)$, let

$$\chi(n) = \iota_p e\big(\mathrm{Ind}_r(n)/\ell\big),$$

with a primitive root $r \bmod p$. Then the set of all non-trivial ℓ^{th} power characters mod p is $\{\chi^j : 0 < j < \ell\}$. With this, we have, for the number $K_\ell(n,p)$ of solutions to $x^\ell \equiv n \bmod p$,

$$K_\ell(n;p) = 1 + \sum_{j=1}^{\ell-1} \chi^j(n),$$

since the right side is equal to 1 if $p | n$, to ℓ if $\ell | \mathrm{Ind}_r(n)$ or $n \bmod p$ is an ℓth power residue, and to 0 otherwise, which is the same as $K_\ell(n;p)$ in view of (43.2). This is a generalization of Note [57.4] and can be regarded as a spectral decomposition of the arithmetic function K_ℓ. The expression $(59.2)_{d=1}$ is now generalized to

$$\mathfrak{g}_\ell(p) = \sum_{n=0}^{p-1} K_\ell(n;p) e\big(n/p\big) = \sum_{j=1}^{\ell-1} G(\chi^j).$$

In particular, (55.5) yields the estimation

$$|\mathfrak{g}_\ell(p)| \leq (\ell-1)\sqrt{p}.$$

A great amount of cancellation takes place among $\{e(m^\ell/p) : 0 \leq m < p\}$, when $p \equiv 1 \bmod \ell$ tends to infinity while ℓ remains bounded.

[61.3]   Thus the above problem on $\mathfrak{g}_\ell(p)$ will be solved if we are able to give an explicit formula for $G(\chi^j)$ in the preceding Note. Closely related to this issue is the Jacobi sum (1837):

$$J(\chi_1, \chi_2) = \sum_{n \bmod p} \chi_1(n)\chi_2(1-n), \qquad \chi_1, \chi_2 \in \mathcal{X}_p \backslash \{J_p\},$$

where the condition can, of course, be dropped, but we shall keep it for the sake of simplicity. By (55.4)–(55.5), we have

$$J(\chi_1, \chi_2) = \begin{cases} G(\chi_1)G(\chi_2)/G(\chi_1\chi_2) & \text{if } \chi_1\chi_2 \neq J_p, \\ -\chi_1(-1) & \text{if } \chi_1\chi_2 = J_p, \end{cases}$$

which implies

$$|J(\chi_1, \chi_2)| \leq \sqrt{p}.$$

244 *Characters*

We get a generalization of (61.5): For χ as in the preceding Note,

$$G(\chi)^\ell = \chi(-1)p \prod_{j=1}^{\ell-2} J(\chi, \chi^j),$$

which holds for $\ell = 2$ as well; an empty product is equal to 1. It can be asserted at least that the right side is algebraically simpler than the left side, for the right side is an expression involving only integers and ℓ^{th} roots of unity while the left side involves p^{th} roots of unity. Numerical examples are in the next Note and in §70(xi). By the way, Gauss (1863b) can be regarded, in today's view, as a discussion on the relation between Gauss sums and Jacobi sums; details are in §70. A readable account of modern achievements on Gauss and Jacobi sums can be found in Berndt and Evans (1981).

[61.4]   We now consider the case $\ell = 4$, i.e., $\chi(n) = \iota_p e\big(\text{Ind}_r(n)/4\big)$, $p \equiv 1 \bmod 4$: We note first that $J(\chi, \chi) = a + bi$, $a, b \in \mathbb{Z}$, and by (55.5) we have $p = a^2 + b^2$. Namely, we have obtained a proof of Euler's theorem (78.3); see Note [78.2]. On the other hand, $g_4(p) = \sqrt{p} + G(\chi) + G(\overline{\chi})$ since χ^2 is the Legendre symbol $\bmod p$, and $G(\chi^2) = G(1, p) = H(2, p)$ in the notation of §59, so $G(\chi^2) = \sqrt{p}$ by (61.1). Also, by the preceding Note,

$$(G(\chi) + G(\overline{\chi}))^2 = 2G(\chi)G(\overline{\chi}) + \big(J(\chi, \chi) + J(\overline{\chi}, \overline{\chi})\big)G(\chi^2)$$
$$= 2\chi(-1)p + 2a\sqrt{p}$$

Thus, we have formally

$$g_4(p) = \sqrt{p} + \sqrt{2\chi(-1)p + 2a\sqrt{p}}, \quad \chi(-1) = i^{(p-1)/2} = \left(\frac{2}{p}\right).$$

The problem to fix a and the sign of this nested radical for general $p \equiv 1 \bmod 4$ had been a longstanding issue until finally solved by Matthews (1979, II). His method is too advanced to be explained in the present essays. However, if p is given explicitly, then we are able to proceed as follows: We note that $d^{(p-1)/4}$ with $p \nmid d$ is one of the four solutions $\{1, \eta, \eta^2, \eta^3\} \equiv \{1, \eta, -1, -\eta\} \bmod p$ of $x^4 \equiv 1 \bmod p$. For instance, when $p = 29$, a primitive root is $3 \bmod 29$, and we may take $\eta \equiv 3^7 \bmod 29$. This combination gives the table:

$d \bmod 29$	d^7	$\text{Ind}_3(d) \bmod 4$	$\chi(d)$
$1, 7, 16, 20, 23, 24, 25$	1	0	i^0
$2, 3, 11, 14, 17, 19, 21$	η	1	i^1
$4, 5, 6, 9, 13, 22, 28$	η^2	2	i^2
$8, 10, 12, 15, 18, 26, 27$	η^3	3	i^3

Hence, $J(\chi, \chi) = -5 - 2i$; that is, $a = -5$, $b = -2$, $\chi(-1) = -1$;

$$\big(G(\chi) + G(\overline{\chi})\big)^2 = -58 - 10\sqrt{29}, \quad \big(G(\chi) - G(\overline{\chi})\big)^2 = 58 - 10\sqrt{29}.$$

$\S61$ 245

A numerical computation gives $G(\chi) \doteqdot 1.0183751677 - 5.2879969760\,i$. Thus,

$$\big(G(\chi) + G(\overline{\chi})\big)/2i = \operatorname{Im} G(\chi) < 0, \quad \big(G(\chi) - G(\overline{\chi})\big)/2 = \operatorname{Re} G(\chi) > 0.$$

Therefore, we find that

$$G(\chi) = \tfrac{1}{2}\left(\sqrt{58 - 10\sqrt{29}} - i\sqrt{58 + 10\sqrt{29}}\,\right),$$

$$\mathfrak{g}_4(29) = \sqrt{29} - i\sqrt{58 + 10\sqrt{29}}\,.$$

One may show, in much the same way, that

$$\mathfrak{g}_4(37) = \sqrt{37} + i\sqrt{74 + 2\sqrt{37}}, \qquad \mathfrak{g}_4(41) = \sqrt{41} - \sqrt{82 - 10\sqrt{41}},$$

$$\mathfrak{g}_4(53) = \sqrt{53} - i\sqrt{106 - 14\sqrt{53}}, \qquad \mathfrak{g}_4(61) = \sqrt{61} + i\sqrt{122 + 10\sqrt{61}},$$

$$\mathfrak{g}_4(73) = \sqrt{73} + \sqrt{146 + 6\sqrt{73}}, \qquad \mathfrak{g}_4(89) = \sqrt{89} + \sqrt{178 - 10\sqrt{89}}, \ \ldots$$

[61.5] Here is a typical instance of applications of Jacobi sums: Let $p \geq 3$, and let $l_v | (p-1)$, $1 < l_v < p-1$. We consider

$$K_{l_1, l_2}(a, b; p) = \big|\{\{x, y\} : ax^{l_1} - by^{l_2} \equiv 1 \bmod p\}\big|,$$

where a, b, $p \nmid ab$, are constants, and x, y are variables $\bmod p$. This is, in other words, to count the number of points on the curve $ax^{l_1} - by^{l_2} = 1$ over the finite field $\mathbb{F}_p$. We have, with $K_\ell(u; p)$ as in Note [61.2],

$$K_{l_1, l_2}(a, b; p) = \sum_{u \bmod p} K_{l_1}(u; p) K_{l_2}(\overline{b}(au - 1); p).$$

Thus, we are led to

$$K_{l_1, l_2}(a, b; p) = p + \sum_{j_1=1}^{l_1-1} \sum_{j_2=1}^{l_2-1} \overline{\chi}_1^{j_1}(a) \overline{\chi}_2^{j_2}(-b) J(\chi_1^{j_1}, \chi_2^{j_2}),$$

with $\chi_v(n) = \jmath_p e(\operatorname{Ind}_r(n)/l_v)$. As a numerical example, take $p = 29$, $l_1 = 2$, $l_2 = 4$, $a = 2$, $b = 3$ together with $r = 3$, so we are evaluating the number of points on the curve $2x^2 - 3y^4 = 1$ over $\mathbb{F}_{29}$. Let $\chi_1 = \psi$ be the Legendre symbol mod 29, and $\chi_2 = \chi$ with χ as in the preceding Note. Then $\psi(2) = -1$ and $\chi(-3) = i^3$. Hence,

$$K_{2,4}(2, 3; 29) = 29 + 2\operatorname{Im} J(\psi, \chi) + J(\psi, \psi) = 32,$$

since $J(\psi, \psi) = -1$, and the explicit value of $G(\chi)$ evaluated in the preceding Note implies that

$$\operatorname{Im} J(\psi, \chi) = \operatorname{Im} \{G(\psi) G(\chi)/G(\chi^3)\} = \chi(-1)\sqrt{29}\operatorname{Im} \{G(\chi)/\overline{G(\chi)}\} = 2.$$

246   *Characters*

Indeed, the curve consists of the following 32 points:

$$\begin{array}{llll}
\{\pm 3, \pm 11\}, & \{\pm 3, \pm 13\}, & \{\pm 4, \pm 6\}, & \{\pm 4, \pm 14\}, \\
\{\pm 8, \pm 3\}, & \{\pm 8, \pm 7\}, & \{\pm 14, \pm 4\}, & \{\pm 14, \pm 10\},
\end{array} \quad \text{mod } 29,$$

with any combination of signs.

[61.6] The bound for Jacobi sums, which is stated in Note [61.3], yields the estimation

$$|K_{l_1,l_2}(a,b;p) - p| \le (l_1 - 1)(l_2 - 1)\sqrt{p}.$$

The number of points on the curve $ax^{l_1} - by^{l_2} - 1 = 0$ over $\mathbb{F}_p$ is asymptotically equal to p as $p \equiv 1 \bmod [l_1, l_2]$ increases indefinitely. Since the quality of the bound, i.e., that the error is $\ll \sqrt{p}$, is independent of local situations, it is suggested that this should be a geometric fact. A. Weil proved a similar assertion on curves over $\mathbb{F}_p$ in general; for that purpose he established first an analogue of RH concerning his curves over $\mathbb{F}_p$, appealing to an advanced theory of algebraic geometry. Later Stepanov (1969) discovered an alternative argument to prove a core part of Weil's achievement. His is a truly elementary method that can be said well within the reach of what we have so far developed in the present essays. W. Schmidt and E. Bombieri then extended Stepanov's method to deal fully with Weil's assertions; a thorough account can be found in Schmidt (2004).

[61.7] It should be appropriate to mention here the sum

$$S(m,n;c) = \sum_{\substack{1 \le d \le c \\ \langle c,d \rangle = 1}} e\big((m\overline{d} + nd)/c\big).$$

Ever since introduced by Kloosterman (1926, p.420), this has been one of the most exploited trigonometrical sums in analytic number theory. The basic reason for this is in the decomposition (29.9)–(29.10); thus, the sum is closely related to various subjects in the automorphic context. Methods referred to in the preceding Note yield the deep individual bound $S(m,n;c) \ll c^{1/2+\varepsilon}$, which is best possible. However, in applications it is more essential to have cancellation among $S(m,n;c)$ as c varies; see Linnik (1962). To this end, Selberg (1965, (3.10)) introduced the non-holomorphic Poincaré series, an important instance of Γ-automorphic functions: under Note [5.1] (2)

$$P_k(z,s) = \sum_{A \in \Gamma_\infty \backslash \Gamma} (\operatorname{Im} A(z))^s e(kA(z)),$$

where $\operatorname{Im} z > 0$, $\operatorname{Re} s > 1$, $k \in \mathbb{N} \cup \{0\}$. We have, via (29.9)–(29.10) and the sum formula (60.1),

$$P_k(z,s) = y^s e(kz) + y^{1-s} \sum_{l=-\infty}^{\infty} e(lx) \sum_{c=1}^{\infty} c^{-2s} S(k,l;c)$$

$$\times \int_{-\infty}^{\infty} \exp\left(-2\pi ly\xi i - \frac{2\pi k}{c^2 y(1-\xi i)}\right)(1+\xi^2)^{-s} d\xi.$$

Kuznetsov (1977/1981) demonstrated, by means of an ingenious inversion of the spectral expansion of inner products of $P_k(z,s)$'s, i.e., a specialization of (94.70) below, the astounding bound

$$\sum_{c\leq x} S(m,n;c)/c \ll x^{1/6}(\log x)^{1/3},$$

where the implied constant depends on $m, n \geq 1$. Therefore, in some cases, deep algebro-geometric bounds of trigonometrical sums can be superseded, though on average, by analytical means. It is indeed hard to surmise that any purely algebraic arguments would ever be able to detect such a massive cancellation. Thus it should appear quite natural that the spectral method of Kuznetsov in the automorphic context has been yielding tremendous impacts throughout analytic number theory since then.

[61.8] Specializing $P_m(z,s)$ by $m = 0$, we get the real-analytic Eisenstein series:

$$E(z,s) = \sum_{A\in\Gamma_\infty\backslash\Gamma} (\mathrm{Im}(A(z)))^s.$$

Since $S(0,n;c)$ is the Ramanujan sum, we have, via (18.19), the expansion

$$E(z,s) = y^s + \varphi_\Gamma(s)y^{1-s}$$
$$+ \frac{2\pi^s \sqrt{y}}{\Gamma(s)\zeta(2s)} \sum_{n\neq 0} |n|^{s-\frac{1}{2}} \sigma_{1-2s}(|n|)K_{s-\frac{1}{2}}(2\pi |n|y)e(nx)$$

Here $\varphi_\Gamma(s) = \sqrt{\pi}\Gamma(s - \frac{1}{2})\zeta(2s - 1)/(\Gamma(s)\zeta(2s))$, and K_ν is the K-Bessel function of order ν; one may call $E(z,s)$ a generating function of divisor sum functions. It will be indicated at (94.71) that $E(z,s)$ plays a fundamental rôle in the spectral theory of Γ-automorphic forms. Full accounts of these two Notes can be found, for instance, in Motohashi (1997, 2007).

§62

The effectiveness of Theorem 54 is limited, however. The reason for this is in that the computation of the symbol $\left(\frac{d}{p}\right)$ requires the prime power decomposition of d, which can be insurmountably hard if d is large (thus,

248 *Characters*

p as well). For instance, the computation in Note [58.5] is mostly occupied by the procedure to achieve the decomposition $221129 = 397 \cdot 557$. It is therefore extremely fortunate that we have an algorithm by Jacobi (1837) that totally eliminates this difficulty inherent in the computation of Legendre symbols. Jacobi's algorithm is akin to Euclid's, and offers a quick and effective procedure, as is to be shown below by means of examples. The determination of whether an arbitrary $d \bmod p$ is a quadratic residue or non-residue becomes literally a simple arithmetic. This particular liberation from the prime power decomposition is a rare and outstanding achievement in the whole of number theory.

We thus define the Jacobi symbol: For any odd $m \in \mathbb{N}$ and any $n \in \mathbb{Z}$, we put Jacobi symbol

$$\left(\frac{n}{m}\right) = \prod_{p^\alpha \| m} \left(\frac{n}{p}\right)^\alpha \quad \text{with Legendre symbols on the right.} \tag{62.1}$$

This does not cause any notational confusion, for if m is a prime, then it coincides with the Legendre symbol $\bmod m$. As a function of n it is completely multiplicative and of period m, i.e., a character $\bmod m$.

Theorem 58 *For any odd integers $m, n > 0$, it holds that*

$$(1) \quad \left(\frac{-1}{m}\right) = (-1)^{(m-1)/2},$$

$$(2) \quad \left(\frac{2}{m}\right) = (-1)^{(m^2-1)/8}, \tag{62.2}$$

$$(3) \quad \left(\frac{n}{m}\right)\left(\frac{m}{n}\right) = \begin{cases} (-1)^{(m-1)(n-1)/4} & \langle m, n \rangle = 1, \\ 0 & \langle m, n \rangle > 1. \end{cases}$$

Proof For any odd integers $m_1, m_2 > 0$, we have $4 | (m_1 - 1)(m_2 - 1)$, and

$$\tfrac{1}{2}(m_1 m_2 - 1) \equiv \tfrac{1}{2}(m_1 - 1) + \tfrac{1}{2}(m_2 - 1) \bmod 2. \tag{62.3}$$

We apply this identity repeatedly to compute the right side of $(62.1)_{n=-1}$ in conjunction with $(58.1)\,(1)$ and get $(62.2)\,(1)$. Also, $64 | (m_1^2 - 1)(m_2^2 - 1)$ implies that

$$\tfrac{1}{8}((m_1 m_2)^2 - 1) \equiv \tfrac{1}{8}(m_1^2 - 1) + \tfrac{1}{8}(m_2^2 - 1) \bmod 2. \tag{62.4}$$

This and $(58.1)\,(2)$ gives $(62.2)\,(2)$. As for $(62.2)\,(3)$, provided $\langle m, n \rangle = 1$ and p, q denote prime factors, including repetitions, of m, n, respectively, we have

$$\left(\frac{n}{m}\right)\left(\frac{m}{n}\right) = \prod_{p,\,q}\left(\frac{q}{p}\right)\left(\frac{p}{q}\right) = (-1)^{\left(\sum_p (p-1)/2\right)\left(\sum_q (q-1)/2\right)}$$

$$= (-1)^{\left(\left(\prod_p p-1\right)/2\right)\left(\left(\prod_q q-1\right)/2\right)} \tag{62.5}$$

$$= (-1)^{(m-1)(n-1)/4}.$$

The first line is due to (58.1)(3), and the second line to (62.3). We end the proof.

Remark In (62.2), the assertion (2) follows actually from (1) and (3): Note that for any odd $n \geq 5$

$$\left(\frac{2}{n}\right) = (-1)^{(n-1)/2}\left(\frac{n-2}{n}\right) = (-1)^{(n-1)/2+(n-1)(n-3)/4}\left(\frac{2}{n-2}\right), \tag{62.6}$$

and proceed by induction. This gives an alternative proof of (58.1)(2) as well.

It should be noted that if m is not a square, then the value of the Jacobi symbol $\left(\frac{n}{m}\right)$ is equal to $+1$ for half of reduced residues $n \bmod m$, and to -1 for the other half. To see this, take a and a prime power $p^\alpha \| m$ with an odd α, such that $\left(\frac{a}{p}\right) = -1$ and $a \equiv 1 \bmod m/p^\alpha$; see (57.5). Then observe the correspondence

$$\left(\frac{n}{m}\right) = \pm 1 \;\leftrightarrow\; \left(\frac{an}{m}\right) = \mp 1. \tag{62.7}$$

We are now able to make the very essence of the laws (58.1) explicit, that is, an answer to the fundamental problem $(43.5)_{\ell=2}$:

Theorem 59 *Let d be an arbitrary* sqf-*integer, and let*

$$d_0 = \begin{cases} |d| & d \equiv 1 \bmod 4, \\ 4|d| & d \equiv 2,3 \bmod 4. \end{cases} \tag{62.8}$$

Then there exists a set of $\varphi(d_0)/2$ reduced residues $\bmod\, d_0$, which depend only on d, such that the quadratic congruence equation $x^2 \equiv d \bmod p$ has a solution if and only if $p \geq 3$ is contained in one of the residue classes in this set.

Proof We compute the Legendre symbol $\left(\frac{d}{p}\right)$ by means of (62.2):

(1) $d \equiv 1 \bmod 4,\ d > 0 \Rightarrow \left(\frac{p}{d}\right)$,

(2) $d \equiv 1 \bmod 4,\ d < 0 \Rightarrow \left(\frac{-1}{p}\right)\left(\frac{|d|}{p}\right) = \left(\frac{p}{|d|}\right)$,

250 Characters

(3) $d \equiv 3 \bmod 4, d > 0 \Rightarrow (-1)^{(p-1)/2}\left(\frac{p}{d}\right)$,

(4) $d \equiv 3 \bmod 4, d < 0 \Rightarrow \left(\frac{-1}{p}\right)\left(\frac{|d|}{p}\right) = (-1)^{(p-1)/2}\left(\frac{p}{|d|}\right)$,

(5) $d \equiv 2 \bmod 8, d > 0 \Rightarrow \left(\frac{2}{p}\right)\left(\frac{d/2}{p}\right) = (-1)^{(p^2-1)/8}\left(\frac{p}{d/2}\right)$,

(6) $d \equiv 2 \bmod 8, d < 0 \Rightarrow \left(\frac{-2}{p}\right)\left(\frac{|d|/2}{p}\right) = (-1)^{(p^2-1)/8}\left(\frac{p}{|d|/2}\right)$,

(7) $d \equiv 6 \bmod 8, d > 0 \Rightarrow \left(\frac{2}{p}\right)\left(\frac{d/2}{p}\right) = (-1)^{(p^2-1)/8+(p-1)/2}\left(\frac{p}{d/2}\right)$,

(8) $d \equiv 6 \bmod 8, d < 0 \Rightarrow \left(\frac{-2}{p}\right)\left(\frac{|d|/2}{p}\right) = (-1)^{(p^2-1)/8+(p-1)/2}\left(\frac{p}{|d|/2}\right)$.

For instance, in the case (8), the sign factor is equal to $+1$ if $p \equiv 1, 3 \bmod 8$, and to -1 if $5, 7 \bmod 8$. The part of the Jacobi symbol is, by (62.7), equal to $+1$ for half of reduced classes modulo $|d|/2$ (odd and sqf) and to -1 for the other half. In total, $d \bmod p$ is a quadratic residue if and only if p is contained in one of $2\varphi(|d|/2)$ reduced residue classes modulo $8 \cdot |d|/2 = 4|d| = d_0$; and $2\varphi(|d|/2) = \varphi(d_0)/2$. Other seven cases are similar. We end the proof.

This is an addition to the statement of the last theorem:

$$\text{any of those particular reduced residue classes mod } d_0 \qquad (62.9)$$
$$\text{designated in Theorem 59 cannot be excluded,}$$

because any reduced residue class mod d_0 contains infinitely many primes by virtue of the prime number theorem of Dirichlet that will be established in §91 [A], independently of the quadratic reciprocity law. For each of odd primes p contained in those $\varphi(d_0)/2$ reduced residue classes mod d_0 the congruence equation $x^2 \equiv d \bmod p$ has a solution without any exception; and for any odd prime contained in the other $\varphi(d_0)/2$ reduced residue classes it has no solution, without any exception. In other words, the congruence equation $x^2 \equiv d \bmod q$, $\langle 2d, q \rangle = 1$, has a solution if and only if all prime factors of q are classified mod d_0 to the first type (apply Note [57.4] as well); note that one may include even moduli with a minor modification taking (43.3) into account. This grand view in number theory impressed immensely Euler, Lagrange, and Legendre, although they never achieved any complete verification. Eventually, Gauss grasped this structure in his genus theory of quadratic forms, as is to be made precise in the later part of §91.

In passing, we note that Theorem 50 detailed with (54.6)–(54.7) is identical to

Theorem 60 *Primitive real characters exist in $\mathcal{X}_q$ if and only if $2^l \| q$, $l = 0, 2, 3$, and $q/2^l$ is sqf. In terms of Jacobi symbol, they are represented as follows:*

$$l = 0 \;\Rightarrow\; \left(\frac{n}{q}\right),$$

$$l = 2 \;\Rightarrow\; j_2(n)(-1)^{(n-1)/2}\left(\frac{n}{q/4}\right),$$

$$l = 3 \;\Rightarrow\; \begin{cases} j_2(n)(-1)^{(n^2-1)/8}\left(\dfrac{n}{q/8}\right), \\[2ex] j_2(n)(-1)^{(n-1)/2+(n^2-1)/8}\left(\dfrac{n}{q/8}\right), \end{cases}$$

$$(62.10)$$

where j_2 is the principal character mod 2.

Notes

[62.1]   A discussion closely related to (62.1) is in [DA, art.133].

[62.2]   Returning to Note [58.5], we now apply (62.2). Then,

$$\left(\frac{221129}{430883}\right) = \left(\frac{11375}{221129}\right) = \left(\frac{1251}{11375}\right) = -\left(\frac{29}{1251}\right) = -1.$$

By (62.2)(3) we may move to the modulus 221129 without sign change. Since $430883 \equiv -11375 \bmod 221129$ and this negative sign can be ignored by (62.2)(1), getting the first equality. Moving to the modulus 11375, by (3) we get no sign change; also in $221129 \equiv 4 \cdot 1251 \bmod 11375$ we may, of course, ignore the factor 4, getting the second equality. Moving to the modulus 1251, by (62.2)(3) we do get sign change; also in $11375 \equiv 4 \cdot 29 \bmod 1251$ we may again ignore the factor 4, getting the third equality. Finally, moving to the modulus 29, by (62.2)(3) we get no sign change; and by the congruence $1251 \equiv 4 \bmod 29$ we end the computation. This procedure is different from that of Note [58.5] in that no consideration of prime factors, except for 2, is necessary, and solely by a repeated application of the division algorithm (2.1) we can proceed. This is a typical instance which indicates well that sometimes one can avoid prime power decomposition by means of Euclid's algorithm: the very essence of the number theory of Euclid.

[62.3]   Theorem 59 can be regarded as essentially due to Legendre (1798, Seconde Part., §X), and more explicitly to Gauss [DA, art.149], but the statement including (62.9) to Dirichlet (1863/1894, §§52, 132–137). Euler (1748a: Ann. 13–16) surmised from his huge numerical examination that prime factors of $ax^2 \pm by^2$, $\langle ax, by \rangle = 1$, should be contained in specific residue classes $\bmod 4|ab|$ and conjectured that the converse should hold. According to Theorem 59, he grasped correctly the essence of the reciprocity law (see Kronecker (1876, p.268)).

§63

Now, if a computation of the Legendre symbol signals that (57.1) has solutions, it may be asked how any one of them can be derived explicitly. This task has been discussed already in §45 (Tonelli's method), but because here we have the exponent $\ell = 2$, we are able to press the matter a little further, adding a simplification: This concerns the determination of the exponent k at (45.9); the corresponding part of the argument for $\ell = 2$ is replaced by the algorithm (63.7)–(63.10) below, which is due to the observation (63.8) that does not readily extend to the situation $\ell \geq 3$.

We shall deal with the congruence equation

$$x^2 \equiv d \bmod p, \quad p \nmid d; \quad p - 1 = 2^g t, \ g \geq 1, \ 2 \nmid t. \tag{63.1}$$

According to Theorem 39 or rather (28.11), we should have

$$d^{(p-1)/2} \equiv 1 \bmod p. \tag{63.2}$$

Then we put

$$X_0 \equiv d^{(t+1)/2}, \quad Y_0 \equiv d^t \bmod p. \tag{63.3}$$

We see that

$$X_0^2 \equiv dY_0 \bmod p; \text{ the order of } Y_0 \bmod p \text{ is } 2^{g_0}, \ g_0 < g, \tag{63.4}$$

where the second part is due to (63.2). Hence,

$$\text{if } g_0 = 0, \text{ then } \pm X_0 \bmod p \text{ is a root of (63.1).} \tag{63.5}$$

On the other hand,

$$\text{if } g_0 > 0, \text{ then take a quadratic non-residue } Z \bmod p. \tag{63.6}$$

The order of $Z^t \bmod p$ equals 2^g. We put

$$X_1 \equiv \left(Z^t\right)^{2^{g-g_0-1}} X_0, \quad Y_1 \equiv \left(Z^t\right)^{2^{g-g_0}} Y_0 \bmod p. \tag{63.7}$$

The order of $Y_1 \bmod p$ equals $2^{g_1}, 0 \leq g_1 < g_0$. For

$$Y_1^{2^{g_0-1}} \equiv \left(Z^t\right)^{2^{g-1}} Y_0^{2^{g_0-1}} \equiv (-1) \cdot (-1) \bmod p. \tag{63.8}$$

Since

$$X_1^2 \equiv Y_1 Y_0^{-1} X_0^2 \equiv dY_1, \tag{63.9}$$

we obtain the root $X_1 \bmod p$ of (63.1), if $g_1 = 0$. Otherwise, we iterate the procedure (63.7):

$$(X_0, Y_0, g_0) \mapsto (X_1, Y_1, g_1) \mapsto (X_2, Y_2, g_2) \mapsto \cdots, \tag{63.10}$$

$$\S63 \qquad 253$$

where $X_{v+1} \equiv (Z^t)^{2^{g-g_v-1}} X_v$, $Y_{v+1} \equiv (Z^t)^{2^{g-g_v}} Y_v \bmod p$, and $g_{v+1} (< g_v)$ can be determined by repeatedly squaring Y_{v+1} until getting 1 mod p; note that in the same way as (63.8) we have $Y_{v+1}^{2^{g_v-1}} \equiv 1 \bmod p$. We reach $g_* = 0$ eventually, and obtain the roots $\pm X_* \bmod p$ of (63.1).

Therefore, the discussion on the quadratic congruence equation $(39.1)_{\ell=2}$ will be completed, provided that

$$\text{both the prime power decomposition (33.4) and} \atop \text{the procedure (63.6) have been achieved.} \qquad (63.11)$$

This is because these two suppositions and the Hensel lifting (Note [34.3]) reduce our problem to the three fundamental tasks:

(1) fixing those d for which (63.1) has a solution;

(2) fixing those p for which (63.1) has a solution; $\qquad (63.12)$

(3) finding a solution of (63.1) in an effective way;

and (1) is solved with Theorems 54, 58; (2) with Theorem 59; (3) with the above procedure. It should be noted that the complexity of the present algorithm, i.e., whether it is polynomial time or not, is not taken into account here; see Note [68.6].

As for (3), an alternative argument due to Cipolla (1903; 1907) will be shown in §68. The step corresponding to (63.6) will become a key-point in his method as well.

Notes

[63.1] When $g = 1$, that is, $p \equiv -1 \bmod 4$, Lagrange (1769, p. 500) remarked that a root of (63.1)–(63.2) is given by $d^{(p+1)/4}$. This is obviously the same as $(45.12)_{\ell=2}$.

[63.2] Cipolla (1907) discussed how to express special solutions to (45.1) in terms of a polynomial of a. In particular, when $g = 2$, that is, $p \equiv 5 \bmod 8$, he (p. 59) showed that a root of (63.1)–(63.2) is

$$\rho \equiv \tfrac{1}{2} d^{(p+3)/8} \big((2^t - 1)(d^t - 1) - 2 \big) \bmod p.$$

To confirm this, note that as $d^{2^t} \equiv 1 \bmod p$

$$\rho \equiv \begin{cases} -d^{(t+1)/2} & \text{if } d^t \equiv 1 \\ -2^t d^{(t+1)/2} & \text{if } d^t \equiv -1 \end{cases} \bmod p.$$

Further observe that $2^{2t} \equiv 2^{(p-1)/2} \equiv -1 \bmod p$ because of (58.1) (2). See also Legendre (1798, pp. 230–233) and Cipolla (1903, p. 154).

[63.3] The congruence equation $x^2 \equiv 17 \bmod 101$. This fits to the specification in the preceding Note, but we shall rather apply Tonelli's algorithm: By the reciprocity law, 17 mod 101 is a quadratic residue, as $101 \equiv 16 \bmod 17$. Then, we note first that $101 - 1 = 4 \cdot 25$; thus $g = 2$, $t = 25$; $X_0 \equiv 17^{(25+1)/2} \equiv 65$, $Y_0 \equiv 17^{25} \equiv -1 \bmod 101$. Hence, $g_0 = 1$. According to the supplementary law $(58.1)(2)$, we may take $Z = 2$. We have then $X_1 \equiv 2^{25} \cdot 65 \equiv 44$, $Y_1 \equiv (2^{25})^2(-1) \equiv 1$. Therefore, $\pm 44 \bmod 101$ are the solutions. Indeed $44^2 = 1936 = 17 + 19 \cdot 101$.

[63.4] The congruence equation $x^2 \equiv 2 \bmod 19073$. This modulus is a prime. By the supplementary law $(58.1)(2)$, 2 mod 19073 is a quadratic residue. We then note that $19073 - 1 = 2^7 \cdot 149$; thus $g = 7$, $t = 149$; $X_0 \equiv 2^{(149+1)/2} \equiv -693$, $Y_0 \equiv 2^{149} \equiv 1712$. Repeating squaring 5 times, we get $1712^{2^5} \equiv -1$, so $g_0 = 6$. By means of the reciprocity law, we take the quadratic non-residue $Z = 3 \bmod 19073$. Then, starting the procedure (63.10), $X_1 \equiv -3^{149} \cdot 693 \equiv -6559$, $Y_1 \equiv (3^{149})^2 \cdot 1712 \equiv 5433$. We have $Y_1^{2^4} \equiv -1 \bmod 19073$, i.e., $g_1 = 5$. So $X_2 = -(3^{149})^2 \cdot 6559 \equiv 8140$, $Y_2 \equiv (3^{149})^{2^2} 5433 \equiv -1 \bmod 19073$, i.e., $g_2 = 1$. Further, $X_3 \equiv (3^{149})^{2^5} \cdot 8140 \equiv -8284$, $Y_3 \equiv -(3^{149})^{2^6} \equiv 1 \bmod 19073$. Hence, we find that $8284^2 \equiv 2 \bmod 19073$. Indeed, $8284^2 = 68624656 = 2 + 3598 \cdot 19073$. That is, dividing $1^2, 2^2, \ldots n^2, \ldots$ by 19073, we get the residue 2 first at $n = 8284$. By the way, to confirm that 19073 is a prime, apply (41.1) to see that the order of 3 mod 19073 is equal to 19072.

[63.5] The congruence equation $x^2 \equiv -5 \bmod p$ with $p = 404321$. According to the reciprocity law there exists a root. We note first that $p - 1 = 2^5 \cdot 5 \cdot 7 \cdot 19^2$; thus $g = 5$, $t = 12635$; $X_0 \equiv (-5)^{(t+1)/2} \equiv 366109$, $Y_0 \equiv (-5)^t \equiv 211830$, $Y_0^2 \equiv -1 \bmod p$, so $g_0 = 2$. We take the quadratic non-residue $Z \equiv 3 \bmod p$. Then $X_1 \equiv (Z^t)^{2^2} X_0 \equiv 103745$, $Y_1 \equiv (Z^t)^{2^3} Y_0 \equiv -1 \bmod p$; so $g_1 = 1$. We have further $X_2 \equiv (Z^t)^{2^3} X_1 \equiv -160284$, $Y_2 \equiv (Z^t)^{2^4} Y_1 \equiv 1 \bmod p$. In this way, we find the root $\pm 160284 \bmod p$. By the way, that 404321 is a prime can be confirmed by that the order of 6 mod 404321 is equal to 404320.

[63.6] In [DA, art.328, I] the congruence equation $x^2 \equiv -1365 \bmod k$ with $k = 5428681$ is discussed. Gauss applied the genus theory of quadratic forms and got four roots mod k. Here we shall show an alternative treatment: Note that the factorization $k = 307 \cdot 17683$ has been obtained in Note [38.5]; the primality of 17683 can be shown via the fact that the order of 5 mod 17683 is equal to $17682 = 2 \cdot 3 \cdot 7 \cdot 421$ by (41.1). So we deal with $x^2 \equiv -1365 \equiv 170 \bmod 307$. By using the Jacobi symbol,

$$\left(\frac{170}{307}\right) = \left(\frac{2}{307}\right)\left(\frac{85}{307}\right) = -\left(\frac{-33}{85}\right) = -\left(\frac{8}{11}\right) = -(-1)^3 = 1;$$

hence, 170 mod 307 is a quadratic residue. This fits to the specification of Note [63.1], and we get the two roots $\pm 170^{(307+1)/4} \equiv \pm 28$ mod 307. As for the congruence equation $x^2 \equiv -1365$ mod 17683, on noting that $1365 = 3 \cdot 5 \cdot 7 \cdot 13$, we have

$$\left(\frac{-1365}{17683}\right) = -\left(\frac{1365}{17683}\right) = -\left(\frac{1}{3}\right)\left(\frac{3}{5}\right)\left(\frac{1}{7}\right)\left(\frac{3}{13}\right) = 1;$$

so -1365 mod 17683 is a quadratic residue; the roots are $\pm(-1365)^{(17683+1)/4} \equiv \mp 861$ mod 17683. With this, we move to solving the system of congruences

$$x \equiv \begin{cases} \pm 28 & \mod 307, \\ \pm 861 & \mod 17683. \end{cases}$$

By Euclid's algorithm or rather (22.12),

$$\frac{17683}{307} = 57 + \frac{1}{1+} \frac{1}{1+} \frac{1}{2+} \frac{1}{61} \Rightarrow 17683 \cdot 5 - 307 \cdot 288 = -1.$$

Hence

$$x \equiv 17683 \cdot 5 \cdot (\pm 28) - 307 \cdot 288 \cdot (\pm 861)$$
$$\equiv \pm 2350978, \ \pm 2600262 \mod 5428681.$$

This set of four roots coincides, of course, with that given by Gauss.

[63.7]   In [DA, art.328, II] the congruence equation $x^2 \equiv -286$ mod q with $q = 4272943$ is discussed. Gauss used a method related to the theory of quadratic forms. Here we shall show an alternative argument. First, by using the Jacobi symbol

$$\left(\frac{-286}{q}\right) = -\left(\frac{143}{q}\right) = \left(\frac{103}{143}\right) = -\left(\frac{40}{103}\right) = -\left(\frac{5}{103}\right) = -\left(\frac{3}{5}\right) = 1.$$

Hence, if q is a prime, then -286 mod q is a quadratic residue; and by Note 63.1 the roots are $\pm(-286)^{(q+1)/4} \equiv \pm 1493445$ mod q. Although we may proceed to a confirmation by checking $1493445^2 \equiv -286$ mod q, we shall stay in the context of the present section and rather proceed to establish that q is a prime. First, we have $3^{(q-1)/2} \equiv -1$ mod q; thus 3 mod q can be a primitive root. Here we turn to (41.1); since $q - 1 = 2 \cdot 3 \cdot 712157$ and $3^{(q-1)/3} \not\equiv 1$ mod q, we need to see whether $q_1 = 712157$ is a prime or not. We have $2^{(q_1-1)/2} \equiv -1$ mod q_1; and thus 2 mod q_1 can be a primitive root. Since $q_1 - 1 = 2^2 \cdot 178039$, we need further to see if $q_2 = 178039$ is a prime. We have $q_2 - 1 = 2 \cdot 3^4 \cdot 7 \cdot 157$, and $3^{(q_2-1)/2}$, $3^{(q_2-1)/3}$, $3^{(q_2-1)/7}$, $3^{(q_2-1)/157}$ are all $\not\equiv 1$ mod q_2. The order of 3 mod q_2 is equal to $q_2 - 1$; hence, q_2 is a prime. Therefore, q_1 is a prime, and so is q.

256 *Characters*

[63.8] The congruence equation $ax^2 + bx + c \equiv 0 \bmod q$ is transformed into $X^2 \equiv b^2 - 4ac \bmod q$, $X = 2ax + b$. Thus, if we find a root $X \equiv \rho \bmod q$, then we need further to solve $2ax + b \equiv \rho \bmod q$. For instance,

$$34x^2 + 5x + 23 \equiv 0 \bmod q \implies (68x + 5)^2 \equiv -3103 \bmod p, \ \forall p | q.$$

Provided $\langle 3103, q \rangle = 1$, $2 \nmid q$, we need to have $\left(\frac{p}{29}\right)\left(\frac{p}{107}\right) = 1$ (see (2) in the proof of Theorem 59). As a numerical example, we take $q = p_1 p_2$ with $p_1 = 14563$, $p_2 = 188333$. We have to solve $y_v^2 \equiv -3103 \bmod p_v$. By Note 63.1,

$$y_1 \equiv \pm(-3103)^{(p_1+1)/4} \equiv \mp 1472 \bmod p_1.$$

Also, by Note [63.2],

$$\begin{aligned} y_2 &\equiv \pm\tfrac{1}{2}(-3103)^{(p_2+3)/8}\big((2^{(p_2-1)/4} - 1)((-3103)^{(p_2-1)/4} - 1) - 2\big) \\ &\equiv \mp 1735 \bmod p_2. \end{aligned}$$

Then, we solve $68x_v + 5 \equiv y_v \bmod p_v$ by Euclid's algorithm, getting $x_1 \equiv 3234, 4904 \bmod p_1$ and $x_2 \equiv 149584, 177229 \bmod p_2$. Another application of Euclid's algorithm gives $26550p_1 - 2053p_2 = 1$. Combining these, we obtain

$$34x^2 + 5x + 23 \equiv 0 \bmod 2742693479,$$
$$x \equiv 205809220, 803605807, 1374415485, 1972212072.$$

There exists no other solution.

§64

The discussion that we have developed so far since the proof of Euler's congruence (28.10) indicates well that there exists a distinctive resemblance between reduced residue classes and roots of unity. Gauss was evidently captivated by this analogy and proceeded to construct a beautiful theory of cyclotomy, as has been indicated in Note [40.4] and §58. Its core part was revealed in [DA, Section VII].

However, viewing from the reciprocity law, equally important or rather more essential is the unpublished Sectio VIII, i.e., *Caput octavum* that has been referred to in our Note [11.6] already. In its unfinished manuscript (1863a (2)) Gauss developed, in modern terminology, a theory of algebraic extensions of the finite field $\mathbb{F}_p = \mathbb{Z}/p\mathbb{Z}$ and gave, as noted in our §58, the 7th and the 8th proofs of the law, which despite of the inconsistency in numbering can, in fact, be regarded as forerunners of the 4th and the 6th proofs, respectively; the relevant historical details are to be given in our Notes [67.2]–[67.3].

$\S 64$ 257

Hence we shall first give, in the present and the next sections, a condensed account of [DA, Sectio VII] with some later twists and move to §66 on the 6th proof; further, in §67 we shall develop elements of the theory of finite fields and give an account of the 7th and the 8th proofs.

To begin with, we recall the basic arithmetic operations in $\mathbb{Q}[x]$, regarding them as obvious analogues of those in $\mathbb{Z}$. For instance, the divisibility $g(x)|h(x)$ of polynomials in $\mathbb{Q}[x]$ should not need any explanation. Also, an analogue of the division algorithm (2.1) holds with respect to the polynomial long division; the necessary order relation is provided by the comparison of degrees of polynomials. We readily get an analogue of Euclid's algorithm in $\mathbb{Q}[x]$.

Then, following Gauss we start our discussion on cyclotomy with [DA, art.42]:

Theorem 61 *Let $A(x)$, $B(x) \in \mathbb{Q}[x]$ be both monic; that is, their coefficients of the highest degree are equal to 1. Then $A(x)B(x) \in \mathbb{Z}[x]$ if and only if $A(x)$ and $B(x)$ are both in $\mathbb{Z}[x]$.*

Proof We shall show first that if $P_j(x) \in \mathbb{Z}[x]$ are such that $P_1(x)P_2(x) \in p\mathbb{Z}[x]$ with a certain prime p, then either $P_1(x)$ or $P_2(x)$ is in $p\mathbb{Z}[x]$. Namely, in the notation of Note [27.3], we have $P_1(x) \equiv 0 \bmod p$, if $P_1(x)P_2(x) \equiv 0$ and $P_2(x) \not\equiv 0 \bmod p$. To confirm this, let $P_1(x), P_2(x) \notin p\mathbb{Z}[x]$, and let $c_{l_j}x^{l_j}$ be the term of highest degree in $P_j(x)$ such that $p \nmid c_{l_j}$. Then the coefficient of $x^{l_1+l_2}$ in $P_1(x)P_2(x)$ is not divisible by p, which ends the confirmation. With this, let $[[A]]$, $[[B]]$ be the least common multiples of the denominators of the coefficients (irreducible fractions) of $A(x)$, $B(x)$, respectively. Then, if either $A(x)$ or $B(x)$ is not in $\mathbb{Z}[x]$, then there exists a prime p such that $[[A]][[B]] \equiv 0 \bmod p$. Hence, if $A(x)B(x) \in \mathbb{Z}[x]$, then $[[A]]A(x) \cdot [[B]]B(x) \equiv 0 \bmod p$, which implies that $[[A]]A(x) \equiv 0 \bmod p$, for instance. Since $[[A]]$ is the coefficient of the highest power of x in $[[A]]A(x)$, we have $[[A]] \equiv 0 \bmod p$. However, according to the definition of $[[A]]$, there should be a term in $[[A]]A(x)$ whose coefficient is not divisible by p, which is a contradiction. We end the proof.

Next, let ρ be a primitive q^{th} root of unity; so $\rho = e(k/q) = \exp(2\pi i k/q)$ with $\langle k, q \rangle = 1$. Then

$$\{\rho^h\} \text{ is the set of all primitive } q^{\text{th}} \text{ roots of unity}$$
$$\Leftrightarrow \{h\} \text{ is a reduced residue system mod } q. \tag{64.1}$$

This is equivalent to (6.3), for $\left(e(hk/q)\right)^a = 1 \Leftrightarrow q|ahk \Leftrightarrow (q/\langle h,q\rangle)|a$. On the other hand, $e(a/n)$ with any fraction a/n is a nth root of unity; and if $a/n = k/q$ in lowest terms, then $e(a/n) = e(k/q)$ is a primitive q^{th} root of

258 *Characters*

unity. Thus, an nth root of unity is a primitive q^{th} root of unity with a certain q dividing n; further, all primitive q^{th} roots of unity are n^{th} roots of unity for any n divisible by q. Hence, if $X_q(x)$ denotes the monic polynomial over $\mathbb{C}$ whose roots coincide exactly with all primitive q^{th} roots of unity, then

$$\prod_{q \mid n} X_q(x) = x^n - 1. \tag{64.2}$$

We call X_q the qth cyclotomic polynomial; note that this notion is different from cyclotomic equation (Note [33.2]). Applying Note [17.6], we get

$$X_q(x) = \frac{\displaystyle\prod_{u \mid q, \, \mu(u)=1} \left(x^{q/u} - 1\right)}{\displaystyle\prod_{v \mid q, \, \mu(v)=-1} \left(x^{q/v} - 1\right)}. \tag{64.3}$$

In particular, $X_q(x) \in \mathbb{Z}[x]$ is monic as required, and $\deg(X_q(x)) = \varphi(q)$. For the denominator and the numerator on the right side of (64.3) are both monic and in $\mathbb{Z}[x]$; the polynomial long division yields a monic polynomial in $\mathbb{Z}[x]$; also, the assertion on the degree is due to (64.1), and alternatively follows from (18.5) and the expression (64.3).

We are going to look into the algebraic nature of $X_q(x)$. We remark first that if $q = q_1 q_2$, $\langle q_1, q_2 \rangle = 1$, then by (32.3) and (64.1)

$$\text{a primitive } q^{\text{th}} \text{ root of unity} = \text{a primitive } q_1^{\text{th}} \text{ root of unity} \times \text{a primitive } q_2^{\text{th}} \text{ root of unity.} \tag{64.4}$$

Also, we shall occasionally use the following well-known fact without mentioning repeatedly (see Note [64.1]):

$$\text{any symmetric polynomial over } \mathbb{Z} \text{ is} \\ \text{a polynomial over } \mathbb{Z} \text{ of elementary symmetric polynomials.} \tag{64.5}$$

Theorem 62 *Any cyclotomic polynomial $X_q(x)$ is irreducible over $\mathbb{Q}$. That is, if $X_q(x) = a(x)b(x)$ with $a(x), b(x) \in \mathbb{Q}[x]$, then either $a(x)$ or $b(x)$ is of degree 0.*

Proof We shall deal first with the case $q = p^\alpha$. Thus, we assume that $X_{p^\alpha}(x) = X_p(x^{p^{\alpha-1}}) = a(x)b(x)$ with $a(x), b(x) \in \mathbb{Q}[x]$ such that $\deg(a(x)) \cdot \deg(b(x)) \neq 0$. We may suppose that $a(x), b(x)$ are both monic. By the preceding theorem, $a(x), b(x) \in \mathbb{Z}[x]$. With this, since $a(1)b(1) = X_{p^\alpha}(1) = X_p(1) = p$, we may suppose that $|a(1)| = 1$. Also, there should exist a primitive $(p^\alpha)^{\text{th}}$ root of unity such that $a(\rho) = 0$. We then put $W(x) = \prod_h a(x^h)$, $1 \le h \le p^\alpha - 1$, $p \nmid h$. If $X_{p^\alpha}(\xi) = 0$, then there exists an h such that $\xi^h = \rho$, which means that $W(\xi) = 0$, whence $X_{p^\alpha}(x)$ divides $W(x)$, that is,

$W(x) = X_{p^\alpha}(x)Y(x)$ with $Y(x) \in \mathbb{Z}[x]$. This implies that $W(1) = pY(1)$, which is a contradiction since $|W(1)| = 1$. We have shown that $X_{p^\alpha}(x)$ is irreducible over $\mathbb{Q}$.

As for the case with an arbitrary q, we shall proceed by induction with respect to the number of different prime factors of q. Thus, let $q = p^\alpha q'$, $p \nmid q'$, $q > 1$, and suppose that $X_{q'}(x)$ is irreducible over $\mathbb{Q}$. We assume that there exists a non-trivial decomposition $X_q(x) = a(x)b(x)$, where $a(x), b(x) \in \mathbb{Z}[x]$ are both monic. Let $A(x)$ and $B(x)$ be the polynomials whose zeros are $(p^\alpha)^{\text{th}}$ powers of zeros of $a(x)$, $b(x)$, respectively. By Note [28.6], we have

$$A(x) \equiv a(x), \ B(x) \equiv b(x) \bmod p. \tag{64.6}$$

Also, on noting (64.4), we decompose ρ, $a(\rho) = 0$, into the product $\rho = \lambda\xi$, where $X_{p^\alpha}(\lambda) = 0$, $X_{q'}(\xi) = 0$. Since $\rho^{p^\alpha} = \xi^{p^\alpha}$ is a root of $X_{q'}(x)$ as $p \nmid q'$, we see that $A(x)$ and $X_{q'}(x)$ have a common zero; then by the irreducibility of the latter we have $X_{q'}(x)|A(x)$, for if $X_{q'}(x) \nmid A(x)$, then by Euclid's algorithm for polynomials there exist $g(x), h(x) \in \mathbb{Q}[x]$ such that $X_{q'}(x)g(x) + A(x)h(x) = 1$, and the specialization $x = \rho^{p^\alpha}$ yields a contradiction. Just in the same way we also have $X_{q'}(x)|B(x)$. We next take an arbitrary root ω of $X_{q'}(x)$. We have $X_q(\omega) = (a(\omega) - A(\omega))(b(\omega) - B(\omega)) = p^2 w(\omega)$, $w(x) \in \mathbb{Z}[x]$, because of (64.6). On the other hand, since $X_q(x)|\left((x^q - 1)/(x^{q/p} - 1)\right)$ and $\omega^{q/p} = 1$, we have $p = X_q(\omega)z(\omega)$, $z(x) \in \mathbb{Z}[x]$. That is, $1 = pw(\omega)z(\omega)$. We divide $w(x)z(x)$ by $X_{q'}(x)$, and let the remainder be $\sum_{j=0}^{k} c_j x^j \in \mathbb{Z}[x]$, $k = \varphi(q') - 1$. Then $1 = p(c_0 + c_1\omega + \cdots + c_k\omega^k)$. We get $1 = pc_0$ and $c_j = 0$ for $j > 0$ since otherwise ω would satisfy an equation over $\mathbb{Q}$ whose degree is at most k, which contradicts the irreducibility of $X_{q'}(x)$. This ends the proof.

In our later discussion the following consequence of Theorem 62 ($q = p$) will become crucial:

$$e(h/p), \ h = 1, 2, \ldots, p - 1, \text{ are linearly independent over } \mathbb{Q}. \tag{64.7}$$

However, we shall require the following deeper assertion as well (see §70(ix)):

Theorem 63 *If $X_m(\xi) = 0$ and $p \nmid m$, then $X_p(x)$ is irreducible over $\mathbb{Q}(\xi)$. Hence*

$$e(h/p), \ h = 1, 2, \ldots, p - 1, \text{ are linearly independent over } \mathbb{Q}(\xi). \tag{64.8}$$

Remark As usual, $\mathbb{Q}(\xi)$ is composed of all $u(\xi)/v(\xi)$, $u(x), v(x) \in \mathbb{Q}[x]$. Obviously $\mathbb{Q}(\xi)$ is a field, but actually it coincides with the ring $\mathbb{Q}[\xi]$. For since $X_m(x)$ is irreducible over $\mathbb{Q}$ (preceding Theorem) and $v(\xi) \neq 0$, Euclid's algorithm for polynomials yields $s(x), t(x) \in \mathbb{Q}[x]$ such that

260 *Characters*

$v(x)s(x) + X_m(x)t(x) = 1$; and $1/v(\xi) = s(\xi)$. In particular, each element of $\mathbb{Q}(\xi)$ can be put in the form

$$\frac{1}{D}(d_0 + d_1\xi + \cdots + d_\ell\xi^\ell), \quad \langle D, d_0, d_1, \ldots, d_\ell \rangle = 1, \quad \ell = \varphi(m) - 1, \quad (64.9)$$

with $D > 0$. This expression is unique. For let $\{D', d_0', \ldots, d_\ell'\}$ be another sequence of the same nature. Then we have $d_j/D = d_j'/D'$ or $D'd_j = Dd_j'$ again by the irreducibility of $X_m(x)$. Thus, $D'\langle d_0, \ldots, d_\ell \rangle = D\langle d_0', \ldots, d_\ell' \rangle$. Then, via (7.3), we see that $D|D'$, $D'|D$; hence, $D = D'$, which ends the confirmation.

Proof Obviously we may suppose that $p \neq 2$. We assume that $X_p(x) = a(x)b(x)$ with monic $a(x), b(x) \in F_\xi[x]$, $\deg(a(x)) \cdot \deg(b(x)) \neq 0$. Here $F_\xi = \mathbb{Q}[\xi]$. We have $p = a(1)b(1)$. Also, $a(1) = A(\xi)/M$, $b(1) = B(\xi)/N$, where $A(x), B(x)$, as well as M, N are as in (64.9) with an obvious correspondence; these $A(x), B(x)$ are of course different from those in the proof of the last theorem. In particular, we have

$$pMN = A(\xi)B(\xi). \quad (64.10)$$

We shall deduce a contradiction from this. Note first that $a(1) = \prod(1 - \rho^h)$, $\rho = e(1/p)$, where $\{h\} \subset \{1, 2, \ldots, p-1\}$. Thus, $a(1)^p = pC(\rho)$, $C(x) \in \mathbb{Z}[x]$, since the binomial expansion gives $(1 - \rho^h)^p = pE(\rho^h)$, $E(x) \in \mathbb{Z}[x]$. Also, by (64.5) we have $\prod_{j=1}^p (x - pC(\rho^j)) = x^p - pG(x)$, $G(x) \in \mathbb{Z}[x]$. In particular, $a(1)^{p^2} = pG(a(1)^p)$, i.e., $A(\xi)^{p^2} = pM^{p^2}G((A(\xi)/M)^p)$. Since $\deg G \leq p - 1$, we find that $A(\xi)^{p^2} = pH(\xi)$ with $H(x) \in \mathbb{Z}[x]$. Then, we note that because of the basic condition $p \nmid m$ there exists an exponent $k \geq 2$ such that $p^k \equiv 1 \bmod m$ by Theorem 28. With this, we now consider the relation $A(\xi)^{p^k} = (pH(\xi))^{p^{k-2}}$. The left side equals $A(\xi^{p^k}) + pK(\xi) = A(\xi) + pK(\xi)$, $K(x) \in \mathbb{Z}[x]$. Hence, $A(\xi) = pV(\xi)$, $V(x) \in \mathbb{Z}[x]$. Just in the same way, we find that $B(\xi) = pW(\xi)$, $W(x) \in \mathbb{Z}[x]$. Therefore, by (64.10), $MN = pV(\xi)W(\xi)$. Then, following the ending argument of the proof of the last theorem, we get $p|MN$. However, $p|M$ is not consistent with $A(\xi) = pV(\xi)$. Also, $p|N$ yields a contradiction. We end the proof.

Notes

[64.1] One of the earliest references to the theory of symmetric polynomials is Weber (1895, Vierter Abschnitt). See also Niven et al (1991, §A.2) as well as our Note [97.4].

[64.2] Theorem 61 is called Gauss' lemma (i.e., one of his famous lemmas). The first assertion in the proof is obviously an extension of (11.2). A consequence of it is that if $P_1(x)P_2(x) \equiv 0 \bmod d$ as a polynomial, then

$\S64$ 261

products of any pair of coefficients of $P_1(x)$ and $P_2(x)$ are all divisible by d. This assertion plays a basic rôle in establishing the unique prime-ideal power decomposition of ideals, a genuine extension of (11.1) to algebraic number fields. See Hecke (1923, $\S\S24$–25).

[64.3] The case $q = p$ in Theorem 62 is the same as Gauss [DA, art.341]. He introduced the notion of irreducible polynomials and the linear independence (64.7), although adopting different terms. These are truly fundamental concepts in modern algebra. The notion of being linearly independent is an exceptionally powerful means; in the next section, at (65.8), we shall show its typical application. The term *irréductible* polynomials or equations was introduced by Abel (1829) and conveyed by Galois (1830/1831) to the present day; see Note [66.3] below. They were both inspired by Gauss [DA, Sectio VII]. By the way, a French edition of [DA] was published in 1807; it may be appropriate to mention here that [DA] is designed for self-study by those who have basic arithmetical technicalities.

[64.4] As for the above proof of Theorem 62, its first half is due to Kronecker (1845) and the second half to Arndt (1859c); we have adopted the historically earliest reasonings. Theorem 63, together with the proof, is due to Kronecker (1854); he also treated X_q with an arbitrary q in place of X_p. See also Dedekind (1857b), Bachmann (1872, Fünfte Vorlesung), Landau (1929), and Schur (1929). Their arguments are all applications of the very basic Note [28.6] due to Schönemann.

[64.5] Cauchy (1829, p.231): One may be curious about possible relations between primitive roots mod p and $(p - 1)^{st}$ roots of unity. Thus, we consider the relation
$$X_{p-1}(x)C(x) = D(x),$$
$$C(x) = \prod_{\substack{u|(p-1) \\ \mu(u)=-1}} \left(x^{(p-1)/u} - 1\right), \quad D(x) = \prod_{\substack{v|(p-1) \\ \mu(v)=1}} \left(x^{(p-1)/v} - 1\right),$$
and let x be a primitive root r mod p. Then $X_{p-1}(r) \equiv 0$ mod p, for $C(r) \not\equiv 0$ and $D(r) \equiv 0$ mod p. Because of Note [40.2], we find that

all of the $\varphi(p - 1)$ roots of $X_{p-1}(x) \equiv 0$ mod p

are primitive roots mod p.

For instance, when $p = 31$, primitive roots are $3, 11, 12, 13, 17, 21, 22, 24$ mod 31. Thus,
$$(x - 3)(x - 11)(x - 12)(x - 13)(x - 17)(x - 21)(x - 22)(x - 24)$$
$$= x^8 - 123x^7 + 6448x^6 - 187458x^5 + 3288541x^4 - 35372427x^3$$
$$+ 224940402x^2 - 755014392x + 970377408.$$

262   *Characters*

On the other hand, by (64.3)

$$X_{30}(x) = \frac{(x^2 - 1)(x^3 - 1)(x^5 - 1)(x^{30} - 1)}{(x - 1)(x^6 - 1)(x^{10} - 1)(x^{15} - 1)}$$
$$= x^8 + x^7 - x^5 - x^4 - x^3 + x + 1.$$

These two polynomials are congruent mod 31. Indeed, we have $123 \equiv -1$, $6448 \equiv 0$, $187458 \equiv 1$, $3288541 \equiv -1$, $35372427 \equiv 1$, $224940402 \equiv 0$, $755014392 \equiv -1$, $970377408 \equiv 1 \bmod 31$. As for the first and the last congruences, see Notes [44.4] and [44.3], respectively.

§65

In [DA, art.357] a quadratic decomposition of $X_p(x) = (x^p - 1)/(x - 1)$ is given; see also art.124. It is (65.3) below. We shall prove this attractive identity as a consequence of Theorem 62 or rather the assertion (64.7). It should be appropriate to stress here that the fundamental means exploited by Gauss in [DA, Sectio VII] are the following two facts:

(1) Let $r \bmod p$ be a primitive root, and let the convention (29.1)–(29.2) be effective concerning the notation $r^w \bmod p$. Then

$$\text{the roots of } X_p(x) = 0 \text{ are } \{e(r^w/p) : w \bmod (p - 1)\}. \tag{65.1}$$

(2) Thus, for each primitive root $r \bmod p$,

$$\{e(r^w/p) : w \bmod (p - 1)\} \text{ are linearly independent over } \mathbb{Q}. \tag{65.2}$$

The observation (65.1) is stated in [DA, art.343]; it is, of course, equivalent to the fact that $\{r^w \bmod p : w \bmod (p - 1)\}$ is a reduced residue system mod p; a relevant historical consideration is given in our §70.

Remark One may describe (65.1) as a parametrization of roots of $X_p(x) = 0$ which makes explicit the isomorphism between the cyclic group $(\mathbb{Z}/p\mathbb{Z})^*$ and the Galois group of X_p. An aim of ours in the present section is, in fact, to provide one of the origins of the Galois theory; see Artin (1971).

Theorem 64 *For each prime $p \geq 3$, there exist polynomials $Y(x), Z(x) \in \mathbb{Z}[x]$ such that*

$$4X_p(x) = Y^2(x) - (-1)^{(p-1)/2} p Z^2(x). \tag{65.3}$$

Proof Let r be as above, and let

$$\Xi_p(x,\theta) = \prod_{h=0}^{(p-3)/2} \left(x - \theta^{r^{2h}}\right) \in \mathbb{Z}[x,\theta]. \tag{65.4}$$

Dividing this by the polynomial $X_p(\theta)$, we get

$$\Xi_p(x,\theta) = \sum_{u=1}^{p-1} c_u(x)\theta^u + d(x,\theta)X_p(\theta),$$
$$c_u(x) \in \mathbb{Z}[x], \ d(x,\theta) \in \mathbb{Z}[x,\theta]. \tag{65.5}$$

We should have put $\sum_{u=0}^{p-2}$ instead, but we have made a modification using the fact $1 = -\sum_{u=1}^{p-1}\theta^u + X_p(\theta)$. Then, assuming $r > p$ without loss of generality, we let $u \equiv r^v \bmod p$, $1 \le v \le p-1$. So $\theta^{r^v} - \theta^u = \theta^u(\theta^{r^v-u} - 1)$ is divisible by $\theta^p - 1$ and thus by $X_p(\theta)$. Hence, (65.5) can be put in the form

$$\Xi_p(x,\theta) = \sum_{v=0}^{p-2} C_v(x)\theta^{r^v} + D(x,\theta)X_p(\theta),$$
$$C_v(x) \in \mathbb{Z}[x], \ D(x,\theta) \in \mathbb{Z}[x,\theta]. \tag{65.6}$$

In this we take $\theta = e(r^a/p)$. Then,

$$\Xi_p\big(x,e(r^a/p)\big) = \sum_{v \bmod (p-1)} C_v(x)e(r^{a+v}/p), \tag{65.7}$$

with an obvious convention on $\{C_v : v \bmod (p-1)\}$. If we take $a = 2$, then $\Xi_p\big(x,e(r^2/p)\big)$ is identical to $\Xi_p\big(x,e(1/p)\big)$, and

$$v \equiv v' \bmod 2 \ \Rightarrow \ C_v(x) = C_{v'}(x). \tag{65.8}$$

This is because (65.2) implies that the expression (65.7) is unique. We get

$$\Xi_p\big(x,e(1/p)\big) = C_0(x)\xi_0 + C_1(x)\xi_1,$$
$$\Xi_p\big(x,e(r/p)\big) = C_0(x)\xi_1 + C_1(x)\xi_0, \tag{65.9}$$

where

$$\xi_v = \sum_{h \bmod (p-1)/2} e(r^{v+2h}/p), \quad v \bmod 2. \tag{65.10}$$

Let s denote a quadratic residue mod p. Then, via (59.1)–(59.2), we have

$$\xi_0 = \sum_s e(s/p) = \tfrac{1}{2}\big(G(1,p) - 1\big),$$
$$\xi_1 = \sum_s e(rs/p) = -\tfrac{1}{2}\big(G(1,p) + 1\big), \tag{65.11}$$

264   *Characters*

for $\left(\frac{r}{p}\right) = -1$. Thus, denoting by t a quadratic non-residue mod p,

$$\Xi_p\big(x, e(1/p)\big) = \prod_s \big(x - e(s/p)\big) = -\tfrac{1}{2}\big(Y(x) - G(1,p)Z(x)\big),$$

$$\Xi_p\big(x, e(r/p)\big) = \prod_t \big(x - e(t/p)\big) = -\tfrac{1}{2}\big(Y(x) + G(1,p)Z(x)\big), \qquad (65.12)$$

where $Y(x) = C_0(x) + C_1(x)$, $Z(x) = C_0(x) - C_1(x)$. On noting (61.5) and $\Xi_p\big(x, e(1/p)\big)\Xi_p\big(x, e(r/p)\big) = X_p(x)$, we end the proof.

The polynomials C_0, C_1, and thus $Y(x), Z(x)$ are independent of the primitive root r mod p since ξ_ν are so. In passing, we remark that if $p \equiv 1$ mod 4, then (65.3) with $x = 1$ implies that there exist $u, v \in \mathbb{N}$ such that $u^2 - pv^2 = -4$. See Note [85.2].

Notes

[65.1]   By the identities (61.5) and (65.11), we have

$$(W - \xi_0)(W - \xi_1) = W^2 + W + \tfrac{1}{4}\big(1 - (-1)^{(p-1)/2}p\big) \in \mathbb{Z}[W].$$

One may ask which of the two roots of the quadratic polynomial of W on the right side is equal to ξ_0. The discriminant of the equation is $(-1)^{(p-1)/2}p$, whence if $p \equiv 1$ mod 4, then the roots are real, and otherwise imaginary. This problem is equivalent to determine the explicit formula for $G(1,p)$. To solve it, Gauss needed four years. At the end of [DA, art.356] he stated that the problem belonged to a higher phase of his investigation and would be treated in future; that is, he was yet unable to settle the problem. After a strenuous effort, he finally got the solution just before the end of August 1805. A few days later, he wrote a letter (Werke X-1, pp.24–25) to the astronomer Olbers, revealing his decent excitement: ... *die Bestimmung des Wurzelzeichens*, ... *Wie der Blitz einschlägt, hat sich das Räthsel gelöst* ... The details of his argument are contained in his article (1811). The result is

$$\xi_0 = \tfrac{1}{2}\left(-1 + \sqrt{(-1)^{(p-1)/2}p}\,\right),$$

with the radical part: $p \equiv 1$ mod 4 $\Rightarrow \sqrt{p}$, and $p \equiv -1$ mod 4 $\Rightarrow i\sqrt{p}$. By the way, if the prime p is given explicitly, then it is not needed to have (61.1) in order to determine ξ_0: instead one may rely on numerical computation. Indeed, Gauss employed such a procedure in [DA, art.353 ($p = 19$); art.354 ($p = 17$)]. See also Gauss (1811, pp.14–15) and our Notes [55.6], [56.3], [61.4] as well as (70.29).

$\S65$ 265

[65.2] If (65.10) is regarded as the case of $g = 2$ in the decomposition $p - 1 = fg$, then we are led to the definition

$$\psi_v^{[g]} = \sum_{h \bmod (p-1)/g} e\left(r^{v+gh}/p\right), \quad v \bmod g;$$

this ψ should not be confused with the one in $\S50$. Gauss [DA, art.343] called these sums f-term periods, and considered their algebraic properties; note that the g numbers $\{\psi_v^{[g]}\}$ are different from each other because of (64.8). We shall briefly follow his reasoning, specializing it to the famous case corresponding to the regular polygon of 17 edges. Thus, we introduce the polynomial

$$\Psi^{[g]}(\theta) = \sum_{h=0}^{(p-1)/g-1} \theta^{r^{gh}} \in \mathbb{Z}[\theta].$$

With $4|(p-1)$, we transform the polynomial

$$\left(W - \Psi^{[4]}(\theta)\right)\left(W - \Psi^{[4]}(\theta^{r^2})\right) \in \mathbb{Z}[W,\theta],$$

adopting the argument (65.5)–(65.6). Then, taking $\theta = e(r^2/p)$ in the resulting identity, we see, in much the same way as in (65.8)–(65.9), that the coefficients of the quadratic equation

$$\left(W - \psi_0^{[4]}\right)\left(W - \psi_2^{[4]}\right) = 0$$

are linear combinations over integers of the roots $\xi_0 = \psi_0^{[2]}$ and $\xi_1 = \psi_1^{[2]}$ of the quadratic equation given in the preceding Note. The equation

$$\left(W - \psi_1^{[4]}\right)\left(W - \psi_3^{[4]}\right) = 0$$

is of the same nature. Namely, with $4|(p-1)$, the four $\frac{1}{4}(p-1)$-term periods $\{\psi_v^{(4)} : v \bmod 4\}$ can be reached via solving two quadratic equations whose coefficients are obtainable by solving the first quadratic equation over $\mathbb{Z}$. Further, with $8|(p-1)$

$$\left(W - \Psi^{[8]}(\theta)\right)\left(W - \Psi^{[8]}(\theta^{r^4})\right) \in \mathbb{Z}[W,\theta]$$

can be treated similarly, and we find that the four quadratic equations

$$\left(W - \psi_j^{[8]}\right)\left(W - \psi_{j+4}^{[8]}\right) = 0, \quad j \bmod 4,$$

have coefficients which are linear combinations over $\mathbb{Z}$ of $\{\psi_v^{[4]} : v \bmod 4\}$. With this, we take $p = 17$. Then $\psi_0^{[8]} = 2\cos(2\pi/17)$, for $\psi_0^{[8]} = e(1/17) + e(r^8/17)$, and $r^8 = r^{(p-1)/2} \equiv -1 \bmod p$. Due to the same reason, the complex conjugate of $\psi_v^{[g]}$ is equal to $\psi_{8+v}^{[g]} = \psi_v^{[g]}$, and $\psi_v^{[g]}$ with $g = 2,4,8$ are all real numbers. From this, we find that the projection to the real line of the basic vertex $e(1/17)$ of the regular polygon of 17 edges can be reached by solving certain quadratic equations over $\mathbb{R}$ in succession. We need, as a matter of fact, to determine the resulting nested radicals explicitly, but this task can be resolved by numerical observations, as has been mentioned already in the preceding Note. In this way, Gauss established the possibility of the Euclidean construction (see Note [1.3]) of the regular polygon of 17 edges. It was March 1796, in the Spring of his nineteenth year. Two thousand and a few hundred years since the ancient Greek mathematics. Genuinely a stupendous discovery. In [DA, art.365] Gauss expresses his delight with similar words; of course, that is no exaggeration at all. We repeat that his idea is condensed in the observation (65.1)–(65.2). A thorough account is developed in Bachmann (1872, pp.63–75).

[65.3]   To make it certain, we give here an alternative proof that for Fermat primes $p = 2^{2^n} + 1$, the Euclidean construction of the p-sided regular convex polygon is possible. Thus we note that

$$e(1/p) = \frac{1}{p-1} \sum_{\chi \bmod p} G(\chi).$$

If the order of the character χ is equal to 2^r, $r \leq 2^n$, then $G(\chi) = -1$ for $r = 0$; $G^2(\chi) = \chi(-1)p$ for $r = 1$; and for $r \geq 2$

$$G(\chi)^{2^r} = \chi(-1)p \prod_{j=1}^{2^r-2} J(\chi, \chi^j)$$

by Note [61.3]. This right side is a polynomial over $\mathbb{Z}$ of $e(1/2^r)$. Hence, $e(1/p)$ can be expressed in terms of nested square roots. See Artin (1971, pp.80–82).

[65.4]   If viewed from the history of the theory of algebraic equations, then there had been an epoch-making discovery well prior to Gauss' construction of the 17-sided regular convex polygon. That was made by Vandermonde (1774; presented to the Paris Academy in 1770/1771) concerning the equation $x^{11} - 1 = 0$. Continued in §70.

[65.5]   In [DA, arts.119, 123–124] the cases $p = 3, 5, 7$ of (65.3) were quoted, and Gauss seems to have thought that they were due to Lagrange. In fact, they had been known to Euler (1772c, p.532, p.537), though stated in terms of homogeneous polynomials of two variables. In the single variable format, he should have got, for instance,

$$4X_5(x) = (2x^2 + x + 2)^2 - 5x^2.$$

Although Euler's motivation appears to have been elsewhere, he could have proceeded a little further and noticed the decomposition

$$4X_5(x) = \left(2x^2 + (1 + \sqrt{5})x + 2\right)\left(2x^2 + (1 - \sqrt{5})x + 2\right);$$

$$A: \ 2x^2 + (1 + \sqrt{5})x + 2 = 2(x - \alpha^2)(x - \alpha^3),$$
$$B: \ 2x^2 + (1 - \sqrt{5})x + 2 = 2(x - \alpha)(x - \alpha^4),$$

where

$$\alpha = \tfrac{1}{4}\left(-1 + \sqrt{5} + i\sqrt{10 + 2\sqrt{5}}\right) = e\left(\tfrac{1}{5}\right);$$

see (70.8) below. The exponents $2, 3$ to α appearing in A are quadratic non-residues mod 5, and the exponents $1, 4$ in B are quadratic residues mod 5. We have in particular $1 + 2\alpha + 2\alpha^4 = \sqrt{5}$. Thus, by (59.2), we have $G(1,5) = \sqrt{5}$. That is, the relation between (65.3) and $G(d,p)$ had already existed as a latent image in Euler's works; its development into a positive image needed only an effective fixing-agent: that is precisely the observation (65.1). In this way Gauss might have grasped the key for the general case. Continued in §70.

[65.6]   For $p \equiv 1 \bmod 3$, a discussion analogous to the first part of Note [65.1] is developed in [DA, art.358]. Thus, Gauss discovered, in particular, that the three $\frac{1}{3}(p - 1)$-term periods $\psi_v^{[3]}$, $v = 0, 1, 2$, satisfy the equation

$$W^3 + W^2 - \tfrac{1}{3}(p - 1)W - \tfrac{1}{9}\left(pa + \tfrac{1}{3}(p - 1)\right) = 0,$$

where a is an integer such that $4p = (3a - 2)^2 + 27b^2$ with another integer b. As for the extension of (65.3), Bachmann (1872) discussed a cubic decomposition of $3^3X_p(x)$, $p \equiv 1 \bmod 3$, and a quartic decomposition of $4^4X_p(x)$, $p \equiv 1 \bmod 4$. These assertions are closely related to the reciprocity laws for cubic and quartic power residues, as surmised naturally from the fact that quadratic reciprocity law (58.1) is hidden in (65.3).

§66

In this section, we shall show, with some modification, Gauss' 6th proof (1818, pp.55–59) of the quadratic reciprocity law. The essence of his argument is to view the set $\mathbb{Z}[x]$ with respect to the modulus $X_p(x)$; but see Note [66.1]: Simplifying notation slightly, we denote $X_p(x)$ by X_p, and extend the notion of congruence introduced in §27 to that $a(x) \equiv b(x) \bmod X_p$ means there exists a $c(x) \in \mathbb{Z}[x]$ such that $a(x) - b(x) = c(x)X_p(x)$. Since the polynomial $X_p(x)$ is monic, the polynomial long division by it can be performed without leaving

Z[x]; namely we are dealing with the ring $\mathbb{Z}[x]/(X_p(x)\mathbb{Z}[x])$. By the way, the division by $X_p(\theta)$ that was performed in the preceding section is the same as using modulo $X_p(\theta)$.

Remark The present section is independent from the fact proved in §64 that X_p is irreducible over $\mathbb{Q}$. Here the aim is to extend the notion of integers and congruences to integral polynomials. This is one of the major inventions due to Gauss.

If $p|m$, $m \geq 0$, then $x^m \equiv 1 \bmod X_p$, as $x^p - 1$ divides $x^m - 1$. Also, since $x \cdot x^{p-1} \equiv 1 \bmod X_p$, we may define $x^{-1} \bmod X_p$ to be the residue class $x^{p-1} \bmod X_p$. For, if $x \cdot a(x) \equiv 1 \bmod X_p$ with $a(x) \in \mathbb{Z}[x]$, then $x(a(x) - x^{p-1}) \equiv 0 \bmod X_p$. We have also $(X_p(x) - xr(x))(a(x) - x^{p-1}) = a(x) - x^{p-1}$, where $r(x) = 1 + x + \cdots + x^{p-2}$. Hence, $a(x) \equiv x^{p-1} \bmod X_p$; see paragraph (2) of the next section. With this, we define $x^{-l} \bmod X_p$, $l \geq 0$, as $(x^{-1})^l \bmod X_p$. It holds, for any $a, b \in \mathbb{Z}$, that $x^a \cdot x^b \equiv x^{a+b} \bmod X_p$; the verification is skipped, though. In particular,

$$c \equiv d \bmod p \;\Rightarrow\; x^c \equiv x^d \bmod X_p. \tag{66.1}$$

We now consider the polynomial

$$g_p(x) = \sum_{h=1}^{p-1} \left(\frac{h}{p}\right) x^h \equiv \sum_{k \bmod p} \left(\frac{k}{p}\right) x^k \bmod X_p. \tag{66.2}$$

We have

$$g_p(x)^2 \equiv (-1)^{(p-1)/2} p \bmod X_p, \tag{66.3}$$

which corresponds to (61.5). To show this, we note that for any $a \not\equiv 0 \bmod p$

$$\begin{aligned} g_p(x^a) &\equiv \left(\frac{a}{p}\right) g_p(x) \\ &\equiv \sum_{n \bmod p} x^{an^2} \bmod X_p. \end{aligned} \tag{66.4}$$

The first line follows immediately from (66.1)–(66.2). As for the second line, we follow (59.2) and denote by s, t quadratic residues, and non-residues mod p; then

$$\begin{aligned} g_p(x^a) &\equiv 1 + 2\sum_s x^{as} - \left(1 + \sum_s x^{as} + \sum_t x^{at}\right) \\ &\equiv 1 + 2\sum_s x^{as} \equiv \sum_{n \bmod p} x^{an^2} \bmod X_p. \end{aligned} \tag{66.5}$$

§66   269

Thus, in much the same way as in (61.6), we have

$$\left(\frac{-1}{p}\right)g_p(x)^2 \equiv g_p(x)g_p(x^{-1}) \equiv \sum_{l=0}^{p-1}\sum_{k=0}^{p-1} x^{k^2-l^2}$$

$$\equiv \sum_{m=0}^{p-1} x^{m^2} \sum_{l=0}^{p-1} x^{2lm} \equiv p \bmod X_p, \tag{66.6}$$

since if $p \nmid m$, then $\{2lm\}$ is a residue system mod p.

Next, with two different primes $p, q \geq 3$, we raise both sides of the congruence (66.3) to the power $(q-1)/2$ and multiply the result by $g_p(x)$. We get

$$g_p(x)^q \equiv (-1)^{(p-1)(q-1)/4} p^{(q-1)/2} g_p(x) \bmod X_p. \tag{66.7}$$

On the other hand, we have, by the congruence stated at the end of Note [28.6], with an obvious replacement,

$$g_p(x)^q \equiv g_p(x^q) + q \sum_{k=1}^{p-1} b_k x^k \bmod X_p, \tag{66.8}$$

where we have also used $\sum_{j=1}^{p-1} x^j \equiv -1 \bmod X_p$. Via (57.3), (66.4), (66.7), and (66.8), we find that

$$\left\{\left(\frac{q}{p}\right) - (-1)^{(p-1)(q-1)/4}\left(\frac{p}{q}\right)\right\} g_p(x) + q \sum_{k=1}^{p-1} c_k x^k \equiv 0 \bmod X_p. \tag{66.9}$$

This left side, in fact, vanishes identically since it is obviously an integer multiple of X_p, and the multiplier is equal to 0 as can be seen by taking $x = 0$. Consequently,

$$\left(\frac{q}{p}\right) - (-1)^{(p-1)(q-1)/4}\left(\frac{p}{q}\right) \equiv 0 \bmod q. \tag{66.10}$$

which proves the reciprocity law.

Notes

[66.1]   As a matter of fact, Gauss (1818, pp.55–59) did not use the congruence modulo X_p. He developed instead an explicit polynomial computation and deduced an identity which is essentially the same as (66.9). According to Kummer (1860, p.22), Jacobi (1827: Legendre–Jacobi (1875, p.213)) appears to be the first who perceived that the use of the modulus X_p should simplify Gauss' 6th proof a lot. In any event, Gauss (1818) did not use the modulus X_p,

270   *Characters*

most probably because he wanted to leave its proper use for Sectio VIII, as it transpires from his manuscript (1863a).

[66.2] The present and the next notes could be put in Notes attached to the next section: The existence of the manuscript (1863a) became widely known only after the publication of Werke II-1 in 1863. Nevertheless, extending the argument of Gauss' 6th proof had been undertaken independently of him. That is a creation of the theory of finite fields by Galois (1830) and independently by Schönemann (1846a/b), followed by Serret (1854, pp.343–370) and Dedekind (1857a); see Bachmann (1902, p.363, footnote). The aim of Sectio VIII was suggested in the introduction (art.335) to Sectio VII that it would be a thorough investigation of *congruentiis*. The theory of congruences mentioned by Gauss is exactly the study of $(33.1)_{q=p}$. Combined with the 6th proof, this minimum disclosure was more than enough for Galois and others to find out what Gauss intended to develop in Sectio VIII. Continued to §69[c].

[66.3] The theory of finite fields was initially formulated in terms of congruence in $\mathbb{Z}[x]$ with respect to double moduli: Let p be a prime, and let $u(x) \in \mathbb{Z}[x]$. Then, for $a(x), b(x) \in \mathbb{Z}[x]$, the notation $a(x) \equiv b(x) \bmod \{p, u(x)\}$ means that there exists $c(x), d(x) \in \mathbb{Z}[x]$ such that $a(x) - b(x) = u(x)c(x) + pd(x)$, which apparently goes back to Lagrange (1775); see Note [40.2]. Of course, this is the same as discussing the congruence $\bmod u(x)$ in $\mathbb{F}_p[x]$ or rather the ring $\mathbb{F}_p[x]/(u(x)\mathbb{F}_p[x])$, in the modern terms. After explaining the notion of irreducibility modulo p of integral polynomials, Galois (1830, p.399) proceeded to the concept of imaginary roots of congruence equations $\bmod p$. That was the birth of the modern concept of algebraic extensions; namely, the fact that with an irreducible $u(x)$ the above ring becomes a field (i.e., (4) in the next section). Independently of this, Schönemann (1846a, §14) also introduced double moduli and constructed a theory of finite fields that is essentially the same as Galois', and in fact more. We add that Gauss (1863a (2); Dedekind's comment, pp.241–242) also used double moduli, and his discussion turned out to essentially contain Schönemann's, as had been surmised in the preface of latter's article quoted above. Continued in Note [67.1].

§67

Now, we begin the discussion leading to Gauss' 7th and 8th proofs of the quadratic reciprocity law. We shall use the language of finite fields, as is customary today. This will not cause any essential deviation from the relevant thoughts developed in Gauss (1863a (2)).

$\S 67$ 271

(1) Let $\mathbb{F}$ be an arbitrary field. Let $\mathbb{F}[x]$ be the set of all polynomials in the indeterminate x over $\mathbb{F}$ (that is, their coefficients belong to $\mathbb{F}$). The degree deg of the polynomial $\sum_{j=0}^{n} a_j x^j \in \mathbb{F}[x]$, $a_n \neq 0$, is equal to n. With the ordinary polynomial arithmetic, $\mathbb{F}[x]$ is a commutative ring; its zero element 0 is the polynomial whose coefficients are all equal to $0 \in \mathbb{F}$, and whose degree is defined to be $-\infty$. By the way, the fact $\deg(k(x)) = 0$, $k(x) \in \mathbb{F}[x]$, is the same as that $k(x)$ is an element of $\mathbb{F}^* = \mathbb{F} \setminus \{0\}$. In the ring $\mathbb{F}[x]$ we are able to use the concept of divisors and multiples. Thus, that the polynomial $a(x)$ is divisible by a polynomial $b(x) \neq 0$ means that there exists a polynomial $c(x)$ such that $a(x) = b(x)c(x)$. We denote this fact by $b(x)|a(x)$, provided all relevant polynomials are in $\mathbb{F}[x]$. The order relation between polynomials is according to their degrees. With this, for instance (1.2) extends to $\mathbb{F}[x]$ with obvious changes; and we have the polynomial division algorithm in $\mathbb{F}[x]$. The inequality (2.1) is now read as $\deg(r(x)) < \deg(b(x))$. Moreover, the uniqueness of the quotient and the rest holds under the convention of ignoring the multipliers from $\mathbb{F}^*$, which we shall impose in the sequel. Then, Euclid's algorithm extends readily to $\mathbb{F}[x]$. The analogues of the greatest common divisor and the least common multiple can be considered in $\mathbb{F}[x]$; we shall continue using these terms and the notations. Thus, $d(x)$ is a common divisor of $a(x)$, $b(x)$, if $d(x)|a(x)$, $d(x)|b(x)$. The greatest common divisor $\langle a(x), b(x) \rangle$ is the one which has the greatest degree among common divisors; it is unique under the convention just introduced. As an analogue of (3.6), there exist $g(x)$, $h(x) \in \mathbb{F}[x]$ such that $a(x)g(x) + b(x)h(x) = \langle a(x), b(x) \rangle$. Naturally the analogue of (3.7) holds also. Further, that $a(x)$, $b(x)$ are coprime means that as an analogue of (3.5) $\langle a(x), b(x) \rangle = 1$ (or an element of $\mathbb{F}^*$). In particular, an analogue of Theorem 5 holds. What correspond to prime numbers are irreducible polynomials. Thus, a non-zero $t(x) \in \mathbb{F}[x]$ is called irreducible over the field $\mathbb{F}$ (or just irreducible), if $u(x)|t(x)$, $u(x) \in \mathbb{F}[x]$, implies that $\deg(u(x)) = \deg(t(x))$ or 0; in $\S 64$ we have already encountered examples. An obvious analogue of (11.2) holds, and we have then a self-evident extension of Theorem 10. Namely, each polynomial in $\mathbb{F}[x]$ can be expressed as a product of irreducible polynomials. Here it should be noted that each irreducible polynomial is unique save for multipliers from $\mathbb{F}^*$. Because of this, in order to avoid ambiguity, we impose that irreducible polynomials be all monic, i.e., their coefficients of the highest terms are equal to 1. Then, the following analogue of Theorem 10 holds in $\mathbb{F}[x]$. Any monic polynomial $g(x) \in \mathbb{F}[x]$ is uniquely expressed as a product of powers of polynomials $t_j(x)$ irreducible over $\mathbb{F}$:

$$g(x) = t_1^{e_1}(x)t_2^{e_2}(x) \cdots t_r^{e_r}(x) \tag{67.1}$$

(2) The analogue in $\mathbb{F}[x]$ of the notion (27.1)–(27.2) should be obvious. Thus, with an arbitrary $q(x) \in \mathbb{F}[x]$, $\deg(q(x)) \geq 1$, we are able to construct the residue class ring $\{\mathbb{F}[x] \bmod q(x)\}$ in just the same way as with $\mathbb{Z}$; in the preceding section we handled in fact the instance $\mathbb{F} = \mathbb{Q}$, $q(x) = X_p(x)$. Here we may suppose that $\mathbb{F} \subset \{\mathbb{F}[x] \bmod q(x)\}$ (the embedding of $\mathbb{F}$) since all the elements of $\mathbb{F}$ are incongruent to each other $\bmod q(x)$. Further, let $\{\mathbb{F}[x] \bmod q(x)\}^*$ be the set $\{a(x) \bmod q(x) : \langle a(x), q(x)\rangle = 1\}$. If $\{\mathbb{F}[x] \bmod q(x)\}^* \ni a(x) \bmod q(x)$, then the inverse residue class of $a(x) \bmod q(x)$ exists since Euclid's algorithm gives $a(x)A(x) + q(x)B(x) = 1$ with $A(x), B(x) \in \mathbb{F}[x]$, and $a(x)A(x) \equiv 1 \bmod q(x)$; this is a generalization of $x^{-1} \bmod X_p$ introduced in the preceding section, i.e., $\langle x, X_p \rangle = 1$.

(3) Extending the notion of congruence equations (33.1) over $\mathbb{Z}$, we consider the congruence equation $\sum_{k=0}^{K} a_k(x) X^k \equiv 0 \bmod q(x)$ over $\mathbb{F}[x]$. Then an analogue of Theorem 34, §34, readily holds. Namely, if $q(x)$ is irreducible over $\mathbb{F}$, then the number of different solutions $X \bmod q(x)$ is not greater than K. This assertion will play a basic rôle in our discussions below, in much the same way as Lagrange's Theorem 34 has done so far.

(4) If $t(x) \in \mathbb{F}[x]$, $\deg(t(x)) = m$, is irreducible, then the ring $\{\mathbb{F}[x] \bmod t(x)\}$ is in fact a field. In order to make this precise, let $t(x) = \sum_{j=0}^{m} c_j x^j \in \mathbb{F}[x]$, $c_m = 1$, and let $x \bmod t(x)$ be denoted by ρ. Then in $\{\mathbb{F}[x] \bmod t(x)\}$ we have $t(\rho) = 0$; and $\{\mathbb{F}[x] \bmod t(x)\} = \mathbb{F}[\rho]$. The set on this right side coincides with $\{a_1 \rho + \cdots + a_m \rho^m : a_j \in \mathbb{F}\}$. These coefficients $\{a_j\}$ are unique; in particular $\{\rho^j : j = 1, 2, \ldots m\}$ are linearly independent over $\mathbb{F}$. To confirm this assertion, let $a_1' \rho + \cdots + a_m' \rho^m$, $a_j' \in \mathbb{F}$, be another representation of the same element; and let $f(x) = \sum_{j=1}^{m} (a_j - a_j') x^{j-1}$. Then $f(\rho) = 0$. On the other hand $f(x)$ is a non-zero polynomial, $t(x)$ is irreducible, moreover $\deg(f(x)) < \deg(t(x))$. Hence $\langle f(x), t(x)\rangle = 1$. Thus in $\mathbb{F}[x]$ it holds that $f(x)A(x) + t(x)B(x) = 1$. This implies the contradiction $0 = 1$. On the other hand, if $\mathbb{F}[\rho] \ni g(\rho) \neq 0$, then $t(x) \nmid g(x)$ and $\langle g(x), t(x)\rangle = 1$, for otherwise $t(x) | g(x)$ and $g(\rho) = 0$. Again by Euclid's algorithm we have a $G(x) \in \mathbb{F}[x]$ such that $g(\rho)G(\rho) = 1$. Namely, any non-zero element of $\mathbb{F}[\rho]$ has its inverse in $\mathbb{F}[\rho]$. That is, the ring $\mathbb{F}[\rho]$ is in fact a field. It is called an algebraic extension of degree m of the basis field $\mathbb{F}$.

(5) This is a repetition of what we have already stated in the preceding paragraph: concerning an element $g(x)$ and an irreducible $t(x)$ in $\mathbb{F}[x]$, we have the equivalence

$$t(x) | g(x) \quad \Leftrightarrow \quad \begin{array}{l} \text{in a certain algebraic extension field of } \mathbb{F} \\ t(x) \text{ and } g(x) \text{ have a common root.} \end{array} \tag{67.2}$$

The necessity is trivial: use the field $\mathbb{F}[\rho]$. To show the sufficiency, suppose that $t(x) \nmid g(x)$. Then by Euclid's algorithm we have $t(x)A(x) + g(x)B(x) = 1$ in $\mathbb{F}[x]$.

This identity holds over any extended field, of course. Hence, the existence of such a common root yields the contradiction $0 = 1$. In particular, all the roots of $t(x)$ in any extension field are simple, for otherwise $t(x)|t'(x)$, which is obviously a contradiction.

(6) Now we specialize $\mathbb{F}$ to be $\mathbb{F}_p = \mathbb{Z}/p\mathbb{Z}$; then, Note [28.6] becomes relevant. We take an arbitrary $M \geq 2$ and let $t(x)$ be an irreducible factor of degree m of $x^{p^M} - x$ in $\mathbb{F}_p[x]$. For any $h(x) \in \mathbb{F}_p[x]$, we have $h(x)^{p^M} \equiv h(x^{p^M}) \equiv h(x) \bmod t(x)$; hence the congruence equation $X^{p^M} - X \equiv 0 \bmod t(x)$ has any $h(x) \bmod t(x)$ as its root. Here we note that $|\{\mathbb{F}_p[x] \bmod t(x)\}| = p^m$ since each coefficient a_j in (4) can take p different values. Hence, by (3) we have $p^m \leq p^M$ or $m \leq M$. On the other hand, if we put $M = mu + v, 0 \leq v < m$, then $x^{p^v} \equiv x \bmod t(x)$. For just analogously to Fermat's (28.13) we have $x^{p^m} \equiv x \bmod t(x)$, and $x^{p^{mu}} \equiv x \bmod t(x)$. If $v > 0$, then by what we have shown already we get $m \leq v$, which is a contradiction. Consequently, $m|M$. The converse holds trivially. Therefore, with $t(x)$, $\deg t = m$, which is irreducible over $\mathbb{F}_p$, it holds that

$$t(x)|\left(x^{p^M} - x\right) \Leftrightarrow m|M. \tag{67.3}$$

(7) In particular, let $\Phi_p(d)$ be the number of all irreducible polynomials of degree d in $\mathbb{F}_p[x]$. Then

$$p^m = \sum_{d|m} d\Phi_p(d), \quad \forall m \geq 1. \tag{67.4}$$

For provided $d|m$, the product of these $\Phi_p(d)$ irreducible polynomials divides $x^{p^m} - x$ and its degree is equal to $d\Phi_p(d)$; also, according to (1) the polynomial $x^{p^m} - x$ decomposes into a product of irreducible polynomials without any repetition as can be seen by differentiating $x^{p^m} - x$. Applying the Möbius inversion (Theorem 15) to (67.4), we get

$$\Phi_p(m) = \frac{1}{m} \sum_{d|m} \mu(d)p^{m/d}. \tag{67.5}$$

We have $\Phi_p(m) > 0$ because of Note [28.5]. Therefore, the following important assertion has been obtained: For each $m \geq 1$ there exist irreducible polynomials of degree m in $\mathbb{F}_p[x]$; hence,

$$\begin{gathered} \text{for any } m \geq 1 \text{ there exists} \\ \text{an algebraic extension of degree } m \text{ of } \mathbb{F}_p. \end{gathered} \tag{67.6}$$

We shall denote such an extension simply by $\mathbb{F}_{p^m}$; we shall omit to explicitly mention the relevant irreducible polynomial, as this convention should not cause any confusion in our later discussion.

274 Characters

(8) Let $t(x)$ be an irreducible polynomial of degree m in $\mathbb{F}_p[x]$. Then the decomposition

$$t(X) = \prod_{j=0}^{m-1} \left(X - \rho^{p^j} \right), \quad \rho = x \bmod t(x), \tag{67.7}$$

holds in the field $\mathbb{F}_p[\rho]$. To confirm this, note that $0 = t(\rho)^p = t(\rho^p)$, and thus $\rho^{p^j}, 0 \le j \le m-1$, are different zeros of $t(X) = 0$: if $\rho^{p^u} = \rho^{p^v}, 0 \le u < v$, then with $\xi = \rho^{p^u}$ we have $t(\xi) = 0$ as well as $\xi^{p^{v-u}} = \xi$. Then by (67.2) we have $t(x) | (x^{p^{v-u}} - x)$ in $\mathbb{F}_p[x]$. Hence, by (67.3) we have $m | (v-u)$.

(9) Further, there exists a primitive root $\bmod\, t(x)$. Namely, $\left(\mathbb{F}_{p^m}\right)^*$ is a cyclic group of order $p^m - 1$. This is an analogue of the second assertion of Theorem 36, §40. The proof is essentially the same as before; it suffices to note that an analogue of the consequence (40.1) holds, as has been observed in (3) above. Here one may invoke Note [30.6] as well.

(10) It should be noted that the fixed points of the Frobenius map $f \mapsto f^p$ in $\mathbb{F}_{p^m}$ are all contained in $\mathbb{F}_p$. For by (3) above the roots of the equation $X^p = X$ in $\mathbb{F}_{p^m}$ are exhausted by the elements of $\mathbb{F}_p$.

(11) Further, as an immediate consequence of the assertion (67.3), $f(x) \in \mathbb{F}_p[x]$, $\deg(f(x)) = m$, is irreducible if and only if $f(x) | \left(x^{p^m} - x \right)$ and $\langle f(x), x^{p^{m/\varpi}} - x \rangle = 1$ over $\mathbb{F}_p$ for any prime factor ϖ of m. This is an analogue of (41.1).

With these modern sophistications, we shall explain Gauss' 7th and 8th proofs of the reciprocity law (1863a (2), arts.365–366). Thus, take different primes $p, q \ge 3$, and let the order of $p \bmod q$ be equal to m. We take an irreducible polynomial $t(x)$ of degree m in $\mathbb{F}_p[x]$ and construct the field $F = \{\mathbb{F}_p[x] \bmod t(x)\} = \mathbb{F}_{p^m}$. According to (9) above, the multiplicative group F^* is cyclic; let ϱ be its generator, and let $\lambda = \varrho^{(p^m-1)/q}$. The order of λ is q. In a generalized sense, F contains a root of the cyclotomic equation $X^q - 1 = 0$ (that is, $\lambda \doteq \exp(2\pi i/q)$). We then introduce a Gauss sum in F:

$$\gamma(a) = \sum_{h \bmod q} \left(\frac{h}{q} \right) \lambda^{ah}. \tag{67.8}$$

We have

$$\gamma(a) = \left(\frac{a}{q} \right) \gamma(1),$$
$$\gamma(1)^2 = \left(\frac{-1}{q} \right) q = (-1)^{(q-1)/2} q. \tag{67.9}$$

In particular, $\gamma(1) \neq 0$. The proof of these facts is just the same as dealing with the original quadratic Gauss sum and the sum (66.2); it depends on the representation

$$\gamma(1) = \omega_0 - \omega_1 = \sum_{n \bmod q} \lambda^{n^2},$$

$$\omega_\nu = \sum_{l \bmod (q-1)/2} \lambda^{r^{\nu+2l}}. \tag{67.10}$$

where $r \bmod q$ is a primitive root.

The 7th proof:
From the relations $\omega_0 + \omega_1 = -1$, $\omega_0 = (\gamma(1) - 1)/2$, $\omega_1 = -(\gamma(1) + 1)/2$, we have

$$\omega_0 \neq \omega_1, \quad \omega_0 \omega_1 = \tfrac{1}{4}\left(1 - (-1)^{(q-1)/2} q\right). \tag{67.11}$$

Hence, the quadratic equation

$$(2W + 1)^2 = (-1)^{(q-1)/2} q \text{ over } \mathbb{F}_p \tag{67.12}$$

has two simple zeros ω_0, ω_1 in the extension field F. Also, on noting Note [28.2], we have, in F,

$$\omega_\nu^p = \sum_{l \bmod (q-1)/2} \lambda^{pr^{\nu+2l}} = \omega_{\nu'}, \quad \nu' \equiv \nu + \mathrm{Ind}_r p \bmod 2. \tag{67.13}$$

If $2 \mid \mathrm{Ind}_r(p)$, that is, if $p \bmod q$ is a quadratic residue, then we have $\omega_\nu^p = \omega_\nu$; and by (10) above we get $\omega_\nu \in \mathbb{F}_p$. That is, (67.12) has roots in $\mathbb{F}_p$, which means that $(-1)^{(q-1)/2} q \bmod p$ is a quadratic residue. On the other hand, if $p \bmod q$ is a quadratic non-residue, then $\omega_\nu^p = \omega_{\nu+1} \neq \omega_\nu$; and $\omega_\nu \notin \mathbb{F}_p$. Namely, (67.12) does not have roots in $\mathbb{F}_p$, and $(-1)^{(q-1)/2} q \bmod p$ is a quadratic non-residue. In other words,

$$\left(\frac{(-1)^{(q-1)/2} q}{p}\right) = \left(\frac{p}{q}\right), \tag{67.14}$$

which proves (58.1) (3), or (58.2).

The 8th proof:
By the definition (67.8) and the first line of (67.9), we have

$$\gamma(1)^p = \gamma(p) = \left(\frac{p}{q}\right) \gamma(1). \tag{67.15}$$

276 *Characters*

Also, by the second line of (67.9),

$$\gamma(1)^p = (-1)^{(p-1)(q-1)/4} q^{(p-1)/2} \gamma(1)$$
$$= (-1)^{(p-1)(q-1)/4} \left(\frac{q}{p}\right) \gamma(1), \tag{67.16}$$

since in F we have $q^{(p-1)/2} = \left(\frac{q}{p}\right)$ (see (57.3)). We have obtained another proof of (58.1) (3).

We now end our account of Gauss' 3rd, 4th, 6th, 7th, and 8th proofs of the quadratic reciprocity law, while the 1st and the 2nd are to be touched in Notes [88.3] and [91.6], respectively.

The proof of the reciprocity law itself has become an almost triviality as the 8th proof tells clearly. However, in mathematics in general, not always simple proofs are cherished by later generation. This applies to the reciprocity law as well. Proofs by Dirichlet and Legendre (see (69.3)) are not simple, but they are rich in content. Nonetheless, the simplicity of the 8th proof and the structural beauty of the 7th proof are indeed enthralling.

Notes

[67.1]   Schönemann (1846a) started his discussion by extending [DA, Sectio I] to $\mathbb{F}_p[x]$, if expressed in terms of the modern notation. Thus, he did not use Euclid's algorithm for polynomials but first introduced the concept of irreducible polynomial as objects corresponding to prime numbers. The explicit formula (67.5) should be attributed to Schönemann (1846a, §§46–48); it was stated later by Dedekind (1857a, p.21); see Gauss (1863a(2), p.222). The fundamental assertion (67.6) is in Schönemann (1846a, §44), although the concept of field is due to (Dedekind (1871, p.424)). Also, the well-known decomposition (67.7) appeared in Schönemann (1846a, §18); but see Gauss (1863a(2), p.229) and Galois (1830). In the introduction of Dedekind (1857a), the contributions of Gauss, Galois, Serret (1854, pp.343–370), and Schönemann are decently mentioned. In any event, Lagrange (1775), Legendre (1785), and specifically Gauss [DA, Sectio VII] together with his 6th proof of the reciprocity law must have inspired Galois, and a paradigm shift took place in algebra; but it had, in fact, occurred already to the young Gauss several decades earlier.

[67.2]   One may wonder when Gauss composed *Analysis residuorum* (1863a), especially the part *Caput octavum*. According to Dedekind (Gauss Werke II-1, p.240), the original seems to have been prepared sometime in 1797 or in 1798, but the manuscript itself must have been written after the publication of [DA]. See Bachmann (1911).

$\S 67$ 277

[67.3] A chronological confusion in the second paragraph of our §64 may be resolved if one takes into account the above comment by Dedekind. In short, Sectio VII was most likely written after a tentative text of Sectio VIII was compiled. More precisely, the 7th and the 8th proofs of the reciprocity law are called *tertia* and *quartam*, respectively, in art.366 of *Caput octavum*. From this fact, one may infer that Gauss got his eight proofs actually in the order 1st, 2nd, 7th, 8th, 3rd, 4th, 5th, and 6th in today's ordering; see Bachmann (1902, p.203), but see also Kronecker (1876, p.272, footnote). At any event, Gauss must have started his deeper investigation of the reciprocity law with the theory of $x^m - 1 \equiv 0 \bmod p$ (1863a (1)), that is, the cyclotomic polynomial over finite fields, and with its core results in background he proceeded to the cyclotomic polynomials over $\mathbb{C}$. This explanation might not appear to go well with *quartam* (1811, p.43) and *quinta, sexta* (1818, title), but there Gauss referred evidently to the publication order, and hence no new confusion should arise.

[67.4] Continuing Note [11.8], we confirm here that $f(x) = x^9 - x^3 + 2520$ is irreducible over $\mathbb{Q}$. For this sake, it suffices to show that $f(x)$ is irreducible over $\mathbb{F}_{31}$. Following the prescription (11) above, we need to show that $f(x) | (x^{31^9} - x)$ as well as $\langle f(x), x^{31^3} - x \rangle = 1$ over $\mathbb{F}_{31}$. We have, by an obvious extension of the modular exponentiation,

$$x^{31^3} \equiv 5x, \ x^{31^9} \equiv 5^3 x \equiv x \bmod f(x) \text{ in } \mathbb{F}_{31}[x].$$

This yields the irreducibility of f over $\mathbb{Q}$. That 504 is a common divisor of all values of $f(n), n \in \mathbb{N}$, can be seen by observing that $x^3(x^6 - 1) \equiv 0 \bmod q$ holds for $q = 2^3, 3^2, 7$, as $7 \nmid x \Rightarrow 7 | (x^6 - 1)$; $3 \nmid x \Rightarrow 9 | (x^{\varphi(9)} - 1)$; $2 \nmid x \Rightarrow 8 | (x^2 - 1)$.

[67.5] By the way, there exist polynomials in $\mathbb{Z}[x]$ which are irreducible over $\mathbb{Q}$ but reducible over $\mathbb{F}_p$ for all p: for instance, Lee (1969) showed that if $a \neq \pm 1$ is sqf, then $x^4 - 2(a - 1)x^2 + (a + 1)^2$ is such a polynomial.

[67.6] We gave, in Note [58.3] and at (62.6), two proofs of the supplement law (58.1)(2). Here we show an alternative way via the discussion of the present section: We take a quadratic extension $\mathbb{F}_p(\omega)$ of $\mathbb{F}_p$, and put $\gamma = \lambda - \lambda^3 - \lambda^5 + \lambda^7$, with $\lambda = \omega^{(p^2-1)/8}$; note that $p^2 \equiv 1 \bmod 8$. We have $\gamma^2 = 8$. Hence, $\gamma^p = (2^{(p-1)/2})^3 \gamma = (\frac{2}{p})\gamma$. Also, $\gamma^p = \lambda^p - \lambda^{3p} - \lambda^{5p} + \lambda^{7p}$ is equal to $\gamma, -\gamma, -\gamma, \gamma$, corresponding to $p \equiv 1, 3, 5, 7 \bmod 8$, respectively. This ends the proof.

[67.7] G. Frei's article [DA, (d), II.4] is a readable account of Gauss (1863a (2)).

§68

To explicitly construct a root of (57.1), Cipolla (1903) invented an algorithm, which is different from Tonelli's (§§45, 63):

Suppose that d mod p is a quadratic residue, and that there exists a $w \in \mathbb{Z}$ such that $w^2 - d$ mod p is a quadratic non-residue. With this, let S, T be such that in the complex number field

$$\left(w - (w^2 - d)^{1/2}\right)^{(p+1)/2} = S - T(w^2 - d)^{1/2}. \tag{68.1}$$

Then $S, T \in \mathbb{Z}$, and $S^2 \equiv d$ mod p. For instance, 7 mod 53 is, by (58.1)(3), a quadratic residue, and $2 = 3^2 - 7$ mod 53 is not by (58.1)(2). We have

$$(3 - \sqrt{2})^{27} = 128827982345121405 - 91095139922635829\sqrt{2}, \tag{68.2}$$

and $S \equiv 31$ mod 53; $31^2 = 7 + 18 \cdot 53$. By the way, $T \equiv 0$ mod 53.

We shall explain Cipolla's algorithm by means of the theory of finite fields, generalizing slightly the situation to (45.1) so that the conditions (45.2)–(45.4) are given; see also Cipolla (1907). We suppose that $w^\ell - d$ is an ℓth power non-residue mod p. Then, according to §67(11), the polynomial $q(x) = x^\ell - (w^\ell - d)$ is irreducible over $\mathbb{F}_p$. We need to verify this: On noting $p \equiv 1$ mod ℓ,

$$x^{p^\ell} = x^{\ell((p^\ell-1)/\ell)}x \equiv (w^\ell - d)^{(p^\ell-1)/\ell}x \equiv x \bmod q(x) \quad (\text{in } \mathbb{F}_p[x]), \tag{68.3}$$

for

$$(w^\ell - d)^{(p^\ell-1)/\ell} = \left\{(w^\ell - d)^{p-1}\right\}^{(1+p+\cdots+p^{\ell-1})/\ell} \equiv 1 \bmod p. \tag{68.4}$$

Also, as $q(x)$ does not have any zero in $\mathbb{F}_p$, we have $\langle q(x), x^p - x \rangle = 1$. We end the verification; note that ℓ is supposed to be a prime. Then let $\rho = x$ mod $q(x)$ in the field $\mathrm{F} = \{\mathbb{F}_p[x] \bmod q(x)\}$. We assert that

$$x_0 = (w - \rho)^{(1+p+\cdots+p^{\ell-1})/\ell} \in \mathbb{F}_p, \quad x_0^\ell \equiv d \bmod p, \tag{68.5}$$

which means that we have got an algorithm to construct a root of (45.1). To confirm this, we note that via (67.7)

$$x_0^\ell = (w - \rho)^{1+p+\cdots+p^{\ell-1}} = \prod_{j=0}^{\ell-1} (w - \rho^{p^j}) = q(w) = d; \tag{68.6}$$

moreover, applying the Frobenius map (§67(10)), we have

$$x_0^p = \left(x_0^\ell\right)^{(p-1)/\ell}x_0 = d^{(p-1)/\ell}x_0 \implies x_0^p = x_0 \implies x_0 \in \mathbb{F}_p. \tag{68.7}$$

In the case $\ell = 2$, we have $x_0 = S - T\rho$ because of the obvious isomorphism between the relevant fields. Hence, $x_0 \in \mathbb{F}_p$, and $S^2 \equiv d$, $T \equiv 0$ mod p.

Notes

[68.1] Legendre (1798, p.234) made a similar consideration, much earlier than Cipolla, although he was, in fact, concerned with the lifting of a solution $a \bmod p$ of (57.1) (see Note [34.3]). He exploited the following trivial relation:

$$\left(a - \sqrt{d}\,\right)^m = A - B\sqrt{d} \;\Rightarrow\; A^2 \equiv B^2 d \bmod p^m.$$

Here, $p \nmid B$, for

$$B = \sum_{j=0}^{[(m-1)/2]} \binom{m}{2j+1} a^{m-2j-1} d^j \equiv (2a)^{m-1} \not\equiv 0 \bmod p.$$

Hence, $AB^{-1} \bmod p^m$ is a lifting of the solution $a \bmod p$. For instance, a solution of $x^2 \equiv 7 \bmod 53^3$ is given by the relation

$$\left(31 - \sqrt{7}\,\right)^3 = 30442 - 2890\sqrt{7}, \quad 30442 \cdot (2890)^{-1} \equiv -33783 \bmod 53^3,$$

as (21.14) yields $1273 \cdot 53^3 - 65578 \cdot 2890 = 1$. Indeed $33783^2 = 7 + 7666 \cdot 53^3$.

[68.2] We apply Cipolla's method to the example treated in Note [63.4]: Thus, we consider $x^2 \equiv 2 \bmod p$, $p = 19073$. Here, $7 = 3^2 - 2 \bmod p$ is a quadratic non-residue. We take a ρ in the relevant extension field and compute $(3 - \rho)^{9537}$. We have $9537 = 1 + 2^6 + 2^8 + 2^{10} + 2^{13}$; and

$$(3 - \rho)^2 = 16 - 6\rho, \; (3 - \rho)^{2^2} = 508 - 192\rho, \; (3 - \rho)^{2^3} = 1141 - 4342\rho,$$

$$(3 - \rho)^{2^4} = 9578 - 9557\rho, \; (3 - \rho)^{2^5} = 4664 - 11238\rho,$$

$$(3 - \rho)^{2^6} = 5561 - 2856\rho, \; (3 - \rho)^{2^7} = -22 - 7887\rho,$$

$$(3 - \rho)^{2^8} = -2723 + 3714\rho, \; (3 - \rho)^{2^9} = 4378 - 9064\rho,$$

$$(3 - \rho)^{2^{10}} = 2095 - 1631\rho, \; (3 - \rho)^{2^{11}} = 8114 - 5756\rho,$$

$$(3 - \rho)^{2^{12}} = 9145 - 7887\rho, \; (3 - \rho)^{2^{13}} = 11786 - 4131\rho.$$

Hence,

$$(3 - \rho)^{9537}$$
$$= (3 - \rho)(5561 - 2856\rho)(-2723 + 3714\rho)(2095 - 1631\rho)(11786 - 4131\rho)$$
$$= -67658409161395108 + 25576083198222686\rho = -8284,$$

which coincides with the solution given in Note [63.4]. The computation has been done actually in $\mathbb{R}$ with the replacement $\rho \mapsto \sqrt{7}$ and the reduction $p = 0$ in handling coefficients resulting during the procedure.

[68.3] As with Tonelli's algorithm, Cipolla's is also probabilistic since the selection of w involves trial and error. Comparing two algorithms, one may

280   *Characters*

assert that Cipolla's is more direct than Tonell's since applications of the latter need to go through the procedure (45.9) and (63.7)–(63.10), while Cipolla's does not have the corresponding steps. The fact that it does not have any direct relation to the discrete logarithm is an advantage to Tonell's. However, this comparison does not seem to be applicable in general. Continued to the next Note.

[68.4]   We apply Cipolla's algorithm to the congruence equation $x^3 \equiv 5$ mod p with $p = 163$. Theorem 39 implies that 5 mod p is a cubic residue, and $2^3 - 5 = 3$ mod p is not. We compute $(2 - \rho)^{8911}$ in the relevant field F, where $\rho^3 = 3$ and $8911 = (1 + p + p^2)/3$. We have $8911 = 1 + 2 + 2^2 + 2^3 + 2^6 + 2^7 + 2^9 + 2^{13}$; and

$$(2 - \rho)^{2^2} = -8 - 29\rho + 24\rho^2, \ (2 - \rho)^{2^3} = -37 + 73\rho - 32\rho^2,$$

$$(2 - \rho)^{2^6} = 14 - \rho + 11\rho^2, \ (2 - \rho)^{2^7} = -33 + 9\rho - 17\rho^2,$$

$$(2 - \rho)^{2^9} = 71 + 89\rho + 52\rho^2, \ (2 - \rho)^{2^{13}} = -16 + 77\rho + 4\rho^2,$$

$$(2 - \rho)^{8911} = -127519950324 - 766195639272\rho + 592554715652\rho^2 = 68.$$

Indeed, $68^3 = 314432 = 5 + 1929p$. We may simplify the computation somewhat by noting $8911 = 54p + 109$, although we omit the details. On the other hand, with Tonell's algorithm (§45), we have $g = 4, t = 2, h = 1$; $X_0 = 5, Y_0 = 25, g_0 = 3; Z = 3, k = 7$. Hence a solution is given by $X_1 = (3^2)^{(3^3 - 7) \cdot 3^{4-3-1}} \cdot 5^{3/3} \equiv 32$ mod p. Indeed we have $32^3 = 5 + 201p$. Since the order of $Z^{(p-1)/3} \equiv 58$ mod p is equal to 3, other roots are $32 \cdot 58 \equiv 63$ and $32 \cdot 58^2 \equiv 68$ mod p. By the way, as for the example $x^3 \equiv 7$ mod 8101, which was treated in Note [45.3](1) with Tonelli's algorithm, Cipolla's requires to compute $(3 - \rho)^{21878101}$ with $\rho^3 = 20$. Even with the simplification via $21878101 = 2700p + 5401$, the computation is still involved. Thus, it appears that which algorithm to choose Tonelli's or Cipolla's depends very much on the combination of the size of the exponent ℓ, the nature of the residue d, and, of course, on the size of the modulus p.

[68.5]   Smith (1859, II, art.65) lamented by saying, *We now come to the problem of the actual solution of binomial congruences* (i.e., (39.1)) – *a subject upon which our knowledge is confined within very narrow limits.* The situation has not essentially changed since then. This is largely due to the three facts: that we have been unable to liberate ourselves from the premise (33.4), that any criterion other than Theorem 39 concerning power residues has not been discovered yet, and that any effective method to find power non-residues, relevant both to Tonelli's method and to Cipolla's, has not been attained yet.

$\S69$ 281

Number theory has made tremendous advancements since the time of Gauss. However, a variety of extremely fundamental problems have been left open.

[68.6]   It should be added that for the problem $(57.1)/(63.1)$ there exists a deterministic algorithm due to Schoof (1985) which is polynomial time with respect to the modulus but not to the residue. It depends on the theory of elliptic curves over finite fields.

§69

Here we collect some historical facts concerning the quadratic reciprocity law.

[a]   *Conclusio* of Euler (1772a) that was posthumously published in (1783, I, p.84) and in (1849c, I, p.486) is exactly the reciprocity law (58.1) (3). In the notation (57.2), it is equivalent to the following set of four assertions: For two different primes $p, q \geq 3$ we have, with respect to the modulus 4,

(1) $q \equiv 1:$ $\left(\frac{q}{p}\right) = 1 \Rightarrow \left(\frac{p}{q}\right) = 1, \left(\frac{-p}{q}\right) = 1,$

(2) $q \equiv -1:$ $\left(\frac{-q}{p}\right) = 1 \Rightarrow \left(\frac{p}{q}\right) = 1, \left(\frac{-p}{q}\right) = -1,$

(3) $q \equiv 1:$ $\left(\frac{q}{p}\right) = -1 \Rightarrow \left(\frac{p}{q}\right) = -1, \left(\frac{-p}{q}\right) = -1,$

(4) $q \equiv -1:$ $\left(\frac{-q}{p}\right) = -1 \Rightarrow \left(\frac{p}{q}\right) = -1, \left(\frac{-p}{q}\right) = 1.$

This is the same as (67.14).

[b]   What Legendre (1785, pp.516–517) stated, not yet using his own symbol, is equivalent to that for two different primes $p, q \geq 3$ we have, with respect to the modulus 4,

(1) $p \equiv 1, q \equiv -1:$ $\left(\frac{q}{p}\right) = 1 \Rightarrow \left(\frac{p}{q}\right) = 1,$

(2) $p \equiv -1, q \equiv 1:$ $\left(\frac{q}{p}\right) = -1 \Rightarrow \left(\frac{p}{q}\right) = -1,$

(3) $p, q \equiv 1:$ $\left(\frac{q}{p}\right) = 1 \Rightarrow \left(\frac{p}{q}\right) = 1,$

(4) $p, q \equiv 1:$ $\left(\frac{q}{p}\right) = -1 \Rightarrow \left(\frac{p}{q}\right) = -1,$

(5) $p \equiv -1, q \equiv 1:$ $\left(\frac{q}{p}\right) = 1 \Rightarrow \left(\frac{p}{q}\right) = 1,$

(6) $p \equiv 1, q \equiv -1:$ $\left(\frac{q}{p}\right) = -1 \Rightarrow \left(\frac{p}{q}\right) = -1,$

(7) $p, q \equiv -1:$ $\left(\frac{q}{p}\right) = 1 \Rightarrow \left(\frac{p}{q}\right) = -1,$

(8) $p, q \equiv -1:$ $\left(\frac{q}{p}\right) = -1 \Rightarrow \left(\frac{p}{q}\right) = 1.$

This set of eight assertions is also equivalent to (67.14). The pairs $\{(1), (2)\}$, $\{(3), (4)\}$, and $\{(5), (6)\}$ are contrapositions. Legendre noted that $\left(\frac{q}{p}\right) = \left(\frac{p}{q}\right)$

except for the case $p \equiv q \equiv -1 \bmod 4$, i.e., (7), (8), that is, the statement (58.2). What Gauss [DA, art.131, 1.–8.] asserted, with a different notation, coincides exactly with this set, albeit his arrangement is (3), (4), (5), (2), (1), (6), (7), (8).

[c]   From today's point of view, it is evident that the reciprocity law was hidden in a series of Euler's articles, and emerged publicly in Euler (1783). However, it appears to be a prevailing view that the existence of the above *conclusio* started to be known only when Chebyshev (1848, p.VIII) pointed it out. Chebyshev was one of the editors of the collection (1849c) of Euler's number theory works which includes the article under the present discussion; thus his revelation of Euler's discovery of the law was indeed proper and appropriate, although any editorial remark on *conclusio* is not made in the collection itself. See Smith (1859, art.16), Kummer (1860, p.19), Kronecker (1876, p.269), Baumgart (1885, pp.3–4), and Bachmann (1902, p.202).

Gauss [DA, art.151] credited the discovery of the reciprocity law to Legendre (1785). However, in a letter (May 1796: Schlesinger (1912, pp.20–21)) to his mentor E.A.W. Zimmermann, Gauss divulged that he had recently studied Legendre's article (1785) for the first time and investigated the proof of the reciprocity law asserted there (his conclusion of this study is given in [d] below) and continued that he himself had discovered the law and after efforts for almost a year achieved its proof. The same claim by him can be found in *Anhang* (Werke I, p.475, bottom lines), which is a collection of his handwritten notes to his own copy of [DA]; see also Werke II-1, p.4. Thus Gauss states that he discovered the law in March 1795 and proved it in April 1796.

Legendre (1785) stated Theorem 54 without using his character symbol; today's standard expression (58.1) (3) appeared in his treatise (1798, p.214) for the first time in history. In his writings, references to Euler's works are plenty but any mention to *conclusio* is not made by him. The same is with Gauss: Euler's *Opuscula analytica*, which contains the article under discussion, is referred to in [DA, arts.76, 93, 110, 151] but nowhere is *conclusio* mentioned.

There exists an extended discussion about to whom the reciprocity law should be ascribed properly. In Dirichlet's *Vorlesungen* there are some comments; compare those in the first edition (1863, p.100) and the fourth (1894, pp.95–96), of which the latter is apparently due to Dedekind. As for Landau, he compressed the whole of the relevant history into *quadratisches Reziprozitätsgesetz: zuerst von Euler vermutet, zuerst von Gauß bewiesen* (1927, Satz 86). This is correct but all too telegraphic.

[d]   Prior to starting this subsection, we stress that our interest here and in §§88–89 concerns strictly the idea of Legendre (1785) on the reciprocity law.

§69   283

Later developments (1798/1830) made by himself do not essentially interest us, but see Pieper (1997). In any event, Legendre (1785, pp.518–530) is certainly the first who really made efforts to prove the reciprocity law; see Note [91.2] below. However, his proof was imperfect: as he himself (1785, p.552) acknowledged, he had to rely on

Legendre's conjecture 1

$$\text{Any arithmetic progression with its initial term and difference being coprime contains infinitely many prime numbers.} \qquad (69.1)$$

Gauss [DA, art.297] stated that this must be extremely difficult to establish although it appears correct, and continued his criticism by pointing it out that Legendre (1785, p.520) actually required the following as well:

Legendre's conjecture 2

$$\text{For each prime } p \equiv 1 \bmod 4 \text{ there exists a prime } \varpi \equiv 3 \bmod 4 \text{ such that } \left(\frac{p}{\varpi}\right) = -1, \qquad (69.2)$$

which Gauss thought could not be proved without the reciprocity law. He repeated the same criticism in [DA, additamenta (arts.151, 296, 297)] which concerns the proof claimed by Legendre (1798).

However, even Gauss' insight had limitation. It turned out later that the conjecture (69.1) can be readily established independently of the reciprocity law, as Landau (1909, Zweiter Bd., §196) and Ingham (1930) showed. We shall give the details of this fact in §88[A]. Moreover, (69.2) can be proved by means of an elementary reasoning of Selberg (1950a, pp.71–72) again independently of the reciprocity law. We shall give the details in §88[B]; an additional remark is in Note [88.3]. Therefore, we assert:

Theorem 65

$$\text{The idea of Legendre (1785) for the proof of the reciprocity law is right.} \qquad (69.3)$$

In other words we are able to complete his proof by some augmentation: his idea has turned out to be well grounded as we shall explicitly assert at the end of §89[C_0].

Despite its flaws, Legendre's approach to the reciprocity law has been influential in number theory. There are two reasons for this: One is in that he pointed out the very basic importance in number theory general of the distribution of primes in arithmetic progressions, indeed for the first time in history. Since then, this viewpoint of his has ever been increasing its value.

284   *Characters*

An early instance is shown in §91 (see the reasoning immediately before
(91.14)); and much more are being accumulated via various applications of
sieve methods. Another reason is in his exploitation of the following theorem
(1785, p.513) which he himself proved independently of the reciprocity law:

Theorem 66 *Let the integers* $a, b, c \neq 0$ *satisfy the following conditions:*

    (i)   not all of a, b, c are of the same sign,

    (ii) $|abc|$ is sqf,

    (iii) $-ab$ mod $|c|$, $-bc$ mod $|a|$, $-ca$ mod $|b|$ are all quadratic residues.

$$(69.4)$$

Then the indefinite equation

$$ax^2 + by^2 + cz^2 = 0, \quad \langle ax, by, cz \rangle = 1, \tag{69.5}$$

has a solution.

A proof will be presented in §89[C_0], of course, independently of the
reciprocity law. As is well-known, this fine assertion is a basis of the local–
global principle due to H. Hasse.

For the sake of completeness, we shall prove [b](1)–(8) here, following
Legendre's original reasoning. Note that (58.1)(1) is regarded as having been
proved, that (58.1)(2) has nothing to do in the present discussion, and that in
view of the pairs of contrapositions we actually need to prove only (1), (3), (5),
(7), and (8).

(1): If $\left(\frac{p}{q}\right) = -1$, then $\left(\frac{-p}{q}\right) = 1$. We specialize Theorem 66 with $a = 1, b = p, c = -q$; then $x^2 + py^2 - qz^2 = 0$, $\langle x, py, qz \rangle = 1$, has a solution. However, $x^2 + py^2 - qz^2 \equiv x^2 + y^2 + z^2$ mod 4, and thus $x, y, z \equiv 0$ mod 2, which is a contradiction.

(3): If $\left(\frac{p}{q}\right) = -1$, then via (69.2) we take a prime $\varpi \equiv -1$ mod 4 such that $\left(\frac{q}{\varpi}\right) = -1$. We have $\left(\frac{\varpi}{q}\right) = -1$ by (2). Thus, $\left(\frac{p\varpi}{q}\right) = 1$. We specialize Theorem 66 with $a = 1, b = q, c = -p\varpi$, and argue as in (1).

(5): If $\left(\frac{p}{q}\right) = -1$, then via (69.1) we take a prime $\varpi \equiv 1$ mod 4 such that $\left(\frac{\varpi}{p}\right) = -1, \left(\frac{\varpi}{q}\right) = -1$. We have $\left(\frac{p}{\varpi}\right) = -1$ by (2) and $\left(\frac{q}{\varpi}\right) = -1$ by (4). We specialize Theorem 66 with $a = \varpi, b = q, c = -p$, and argue as in (1).

(7): If $\left(\frac{p}{q}\right) = 1$, then we specialize Theorem 66 with $a = p, b = q, c = -1$, and argue as in (1).

(8): If $\left(\frac{p}{q}\right) = -1$, then via (69.1) we take a prime $\varpi \equiv 1$ mod 4 such that $\left(\frac{\varpi}{p}\right) = -1, \left(\frac{\varpi}{q}\right) = -1$. By (2) we have $\left(\frac{p}{\varpi}\right) = -1, \left(\frac{q}{\varpi}\right) = -1$. We specialize Theorem 66 with $a = p, b = q, c = -\varpi$, and argue as in (1).

$$\S70 \qquad\qquad 285$$

Notes

[69.1] Kummer (1860, pp.19–20) asserted that (69.2) should follow from (69.1), so Legendre's proof of the reciprocity law could be completed with Dirichlet's analytic argument; see also Smith (1859/1869, art.16, Addition). However, it is not known yet how to derive (69.2) from (69.1) without using the law. Moreover, one should know also that Dirichlet (1837a, 1839) fully utilized the law to prove (69.1); details are in Note [88.2].

[69.2] Euler (1772e), Lagrange (1798, §V), and Gauss [DA, arts.294–295] are discussions relevant to Theorem 66. A sophisticated version of this assertion of Legendre is our Theorem 91, §89, which is due to Dedekind (Dirichlet (1871/1894, ending of §157)).

§70

In the present section we shall discuss the origin of Gauss sums. What we have commented in Note [65.5] with the same intention is our own surmise. Gauss himself actually indicated in his article (1863b, posth.) that the origin was in the resolvents of cyclotomic polynomials. We shall show that those resolvents and their innovative use, involving a prototype of the procedure explained in Note [65.2], can be recognized well in Vandermonde's article (1774: presented in 1770) on the equation $x^{11} - 1 = 0$.

Because of this historical fact, we shall take a roundabout, beginning our discussion with a brief account of the history of algebraic equations prior to [DA, Sectio VII]; see also the discussion [DA, (d), II.1] by O. Neumann. We shall use similar notations in proximity; this convention should not cause any confusion.

(i) It is said that the theory of algebraic equations originated in Babylonian mathematics. A typical specimen of mathematical clay tablets (teaching material, the third millennium BCE) contains problems like the following: When the sum of the long and the short sides of a rectangle is equal to l, and its area to s, find the length ρ_1, ρ_2 of the respective sides. According to the solutions remaining on tablets to various cases of the same sort of problems, we see that the solutions were obtained by this procedure (see Gandz (1937)): Subtract s from $(l/2)^2$, take the square root of the result, and add $l/2$; then one gets ρ_1. Namely, the roots of the equation $x(l - x) = s$ are $\rho_1 = l/2 + \sqrt{((l/2)^2 - s)}$ and $\rho_2 = l - \rho_1$. This is exactly the well-known formula for the roots of the general quadratic equation.

286 *Characters*

A little advanced reasoning to reach the same formula is as follows: With the equation $f(x) = x^2 + ax + b = (x - \rho_1)(x - \rho_2) = 0$ (so the existence of roots is presupposed), we put $\tau_\nu = \rho_1 + \omega^\nu \rho_2$, $\omega = e(1/2)$; thus $\rho_1 = \frac{1}{2}(\tau_0 + \tau_1)$, $\rho_2 = \frac{1}{2}(\tau_0 + \omega\tau_1)$. If the transposition $\rho_1 \leftrightarrow \rho_2$ is applied, then $\tau_1 \mapsto \omega\tau_1$; and we observe the relation

$$(X - \tau_1)(X - \omega\tau_1) = X^2 - \tau_1^2 = 0, \ \tau_1^2 = a^2 - 4b. \tag{70.1}$$

In other words, one can derive a linear equation, i.e., in terms of X^2, whose coefficient is obtained from those of f by means of (64.5) and via the square roots of whose root $(= \tau_1^2)$ one may reach the roots $\{\rho_1, \rho_2\}$ of $f(x) = 0$.

Next, with the cubic equation $f(x) = (x - \rho_1)(x - \rho_2)(x - \rho_3) = 0$, we put $\tau_\nu = \rho_1 + \omega^\nu \rho_2 + \omega^{2\nu} \rho_3$, $\omega = e(1/3)$, so $\rho_1 = \frac{1}{3}(\tau_0 + \tau_1 + \tau_2)$, $\rho_2 = \frac{1}{3}(\tau_0 + \omega^2\tau_1 + \omega\tau_2)$, $\rho_3 = \frac{1}{3}(\tau_0 + \omega\tau_1 + \omega^2\tau_2)$. If the transpositions $\rho_1 \leftrightarrow \rho_2$, $\rho_1 \leftrightarrow \rho_3$, $\rho_2 \leftrightarrow \rho_3$ are applied, then we get $\{\tau_1, \tau_2\} \mapsto \{\omega\tau_2, \omega^2\tau_1\}$, $\{\omega^2\tau_2, \omega\tau_1\}$, $\{\tau_2, \tau_1\}$, respectively. A computation shows that $\tau_1^3 + \tau_2^3$ and $\tau_1\tau_2$ are integral symmetric polynomials of $\{\rho_1, \rho_2, \rho_3\}$. Hence, the coefficients of the equation

$$\begin{aligned} &(X - \tau_1)(X - \omega\tau_1)(X - \omega^2\tau_1)(X - \tau_2)(X - \omega\tau_2)(X - \omega^2\tau_2) \\ &= (X^3 - \tau_1^3)(X^3 - \tau_2^3) = 0 \end{aligned} \tag{70.2}$$

are integral polynomials of the coefficients of f due to (64.5). Namely, $\{\tau_1, \tau_2\}$ are cubic roots of roots of a quadratic equation that can be derived from f rationally. In this way we are led to the well-known formula for the roots of the general cubic equation. In the case of the simplified cubic equation, into which any cubic equation can be readily transformed, we have

$$\begin{aligned} f(x) &= x^3 + ax + b = 0 \\ \Rightarrow \tau_1^3 + \tau_2^3 &= -27b, \ \tau_1\tau_2 = -3a. \end{aligned} \tag{70.3}$$

Thus, $(70.2)_{X^3 = W}$ takes the form

$$\begin{aligned} W^2 + 27bW - (3a)^3 &= 0 \\ \Rightarrow \tau_\nu^3 &= \tfrac{1}{2}(-27b \pm \sqrt{\Delta}), \ \Delta = 27(4a^3 + 27b^2), \end{aligned} \tag{70.4}$$

for $\nu = 1, 2$, with an appropriate choice of the branch of $\sqrt{\Delta}$. As is well-known,

$$\Delta = -27\Big((\rho_1 - \rho_2)(\rho_1 - \rho_3)(\rho_2 - \rho_3)\Big)^2. \tag{70.5}$$

Further, to the general equation of degree 4, one may extend the above reasoning; but this time a minor simplification can be made (i.e., the corresponding ω is not $e(1/4)$ but $e(1/2)$). With $f(x) = (x - \rho_1)(x - \rho_2)(x - \rho_3)(x - \rho_4) = 0$, we put $\tau_0 = \rho_1 + \rho_2 + \rho_3 + \rho_4$, $\tau_1 = \rho_1 - \rho_2 + \rho_3 - \rho_4$, $\tau_2 = -\rho_1 + \rho_2 + \rho_3 - \rho_4$, $\tau_3 = -\rho_1 - \rho_2 + \rho_3 + \rho_4$, so that $\rho_1 = \frac{1}{4}(\tau_0 + \tau_1 - \tau_2 - \tau_3)$, $\rho_2 = \frac{1}{4}(\tau_0 - \tau_1 + \tau_2 - \tau_3)$, $\rho_3 = \frac{1}{4}(\tau_0 + \tau_1 + \tau_2 + \tau_3)$, $\rho_4 = \frac{1}{4}(\tau_0 - \tau_1 - \tau_2 + \tau_3)$. Then, applications of any transposition over the set $\{\rho_1, \rho_2, \rho_3, \rho_4\}$ do not cause any change of the coefficients of the equation

$$\left(W - \tau_1^2\right)\left(W - \tau_2^2\right)\left(W - \tau_3^2\right) = 0. \tag{70.6}$$

Namely, by means of square roots of the roots of a cubic equation which is derived rationally from the original, one may express the roots of the general quartic equation in terms of radicals of rational functions of the coefficients of f.

(ii) The formula of the roots of the general cubic equation was discovered by S. del Ferro and N. Fontana (Tartaglia); the quartic case was discovered by L. de Ferrari. These formulas became widely known via *Artis Magnæ* (Cardano (1545)), one of the greatest science books in the Renaissance. This advance, the first since the Babylonian mathematics, came after 3000 years or more; in fact, specific cubic equations and others had been treated in ad hoc ways though. From today's point of view ((70.1)–(70.6)), Cardano's presentation or his classification of cases appears awfully complicated. However, that was inevitable, for the concept of negative numbers was yet to be generally accepted in Europe. Thus, for instance, with $a, b > 0$, the equations $x^3 + ax = b$ and $x^3 = ax + b$ were thought to be different issues to discuss. By the way, Cardano commenced his treatise by a praise for al-Khwarizmi (Mahomete, Mosis Arabis filio), and proceeded to the résumé of all the contributions to the subject of his book, candidly attributing them to the respective mathematicians; but see the footnote 8 on p.8 of Cardano (1993).

(iii) Some two centuries after Cardano, Euler (1740) wrote an account of the subject treated in (i) with a somewhat different argument. Then he applied his method to the equation $x^n - 1 = 0$, and for $n \leq 10$ he found that roots can be expressed in terms of radicals of rational numbers. For instance, in the case of $n = 5$, he did not apply Ferrari's (70.6) but set $y = x + 1/x$ and used

$$x^5 - 1 = x^2(x - 1)\left(y^2 + y - 1\right) = 0 \Rightarrow y = \tfrac{1}{2}\left(-1 \pm \sqrt{5}\right). \tag{70.7}$$

Thus, the 5th roots of unity are 1 and

$$\alpha = \tfrac{1}{4}\left(-1 + \sqrt{5} + i\sqrt{10 + 2\sqrt{5}}\right), \qquad \bar{\alpha} = \alpha^4,$$

$$\alpha^2 = \tfrac{1}{4}\left(-1 - \sqrt{5} + i\sqrt{10 - 2\sqrt{5}}\right), \qquad \bar{\alpha}^2 = \alpha^3; \tag{70.8}$$

288 *Characters*

see Note [65.5]. As for $n = 6$ he used the square roots of the roots for $n = 3$; for $n = 8$, he nested square roots three times; for $n = 9$, he nested cubic radicals; for $n = 10$, he used the square root of the roots for $n = 5$. In the remaining case $n = 7$, he set $y = x + 1/x$ again, and considered

$$x^7 - 1 = x^3(x-1)(y^3 + y^2 - 2y - 1), \qquad (70.9)$$

which is included in (i). An explicit representation of the roots of $x^7 - 1 = 0$ is thus obtained. It should be noted that he considered, in fact, the equations which become the same as (70.7), (70.9) if transformed by $y \mapsto -y$. Euler closed this fine article of his by saying the following: If the same procedure is applied to the equation $x^{11} - 1 = 0$, then an equation of 5th degree comes up, *hic subsistere debemus.*

(iv) It was the violinist mathematician Vandermonde (1774) who discovered the way out of this stalemate encountered by Euler. Admittedly his article is quite complicated, but the essence of his idea is, in fact, simple and daring. His discussion can be summarized as follows:

The equation of 5th degree in question is

$$y^5 + y^4 - 4y^3 - 3y^2 + 3y + 1 = 0 \qquad (70.10)$$

(in the original (p.415) it is set with $y \mapsto -y$ as Euler did). Vandemonde's clue was

$$\text{to exploit a particular configuration of the zeros.} \qquad (70.11)$$

To explain this, we follow the tradition (i) partly: let $\{\rho_j\}$ be the zeros of (70.10), and use $\alpha = e(1/5)$ to form

$$\tau_v = \sum_{j=1}^{5} \alpha^{(j-1)v} \rho_j; \quad \rho_k = \frac{1}{5} \sum_{v=0}^{4} \alpha^{v(1-k)} \tau_v. \qquad (70.12)$$

Then we infer from (i) that we should consider the set

$$\left\{ \tau_1^5, \tau_2^5, \tau_3^5, \tau_4^5 \right\}. \qquad (70.13)$$

For this sake we take (70.11) into account, and put

$$\rho_j = 2\cos(2^{j-1} \cdot 2\pi/11) = e(2^{j-1}/11) + e(-2^{j-1}/11); \qquad (70.14)$$

the significance of this configuration is to be revealed at (70.21). We note the following relations

$$\rho_j^2 = \rho_{j+1} + 2, \, j \bmod 5;$$

$$\rho_1\rho_2 = \rho_1 + \rho_4, \quad \rho_1\rho_3 = \rho_4 + \rho_5, \quad \rho_1\rho_4 = \rho_2 + \rho_3,$$

$$\rho_1\rho_5 = \rho_3 + \rho_5, \quad \rho_2\rho_3 = \rho_2 + \rho_5, \quad \rho_2\rho_4 = \rho_1 + \rho_5,$$

$$\rho_2\rho_5 = \rho_3 + \rho_4, \quad \rho_3\rho_4 = \rho_1 + \rho_3, \quad \rho_3\rho_5 = \rho_1 + \rho_2, \quad \rho_4\rho_5 = \rho_2 + \rho_4,$$

$$\tag{70.15}$$

which is the same as $a^2 = -b + 2$, $b^2 = -d + 2, \ldots$ on p.415 of the original. On noting $\rho_1 + \rho_2 + \cdots \rho_5 = -1$, we see that any polynomial of $\{\rho_j\}$ over $\mathbb{Z}$ can be expressed as a linear combination of $\{\rho_j : 1 \le j \le 5\}$ over $\mathbb{Z}$. Hence,

$$\tau_\nu^5 = \sum_{k=1}^{5} M_{\nu,k}(\alpha)\rho_k, \quad M_{\nu,k}(\alpha) \in \mathbb{Z}[\alpha]. \tag{70.16}$$

Vandermonde then observed that

the cyclic permutation $\sigma : \rho_j \mapsto \rho_{j+1}$, $j \bmod 5$,

commutes with (70.15); i.e., $(\rho_1\rho_2)^\sigma = \rho_1^\sigma \rho_2^\sigma$, etc. $\tag{70.17}$

This is probably the first instance in history of recognizing the algebraic conjugacy: rational relations between algebraic numbers are stable under the application of conjugacy mappings. Anyway, by induction $\left(\prod_{\nu=1}^{r} \rho_{j_\nu}\right)^\sigma = \prod_{\nu=1}^{r} \rho_{j_\nu}^\sigma$ for every $\{j_\nu\}$, so $\left(\tau_\nu^5\right)^\sigma = \left(\tau_\nu^\sigma\right)^5$, and

$$\left(\tau_\nu^5\right)^\sigma = \left(\alpha^{-\nu}\tau_\nu\right)^5 = \tau_\nu^5. \tag{70.18}$$

Hence,

$$\tau_\nu^5 = \frac{1}{5} \sum_{\ell=0}^{4} \left(\tau_\nu^5\right)^{\sigma^\ell} = \frac{1}{5} \sum_{\ell=0}^{4} \sum_{k=1}^{5} M_{\nu,k}(\alpha)\rho_{k+\ell} = -\frac{1}{5} \sum_{k=1}^{5} M_{\nu,k}(\alpha). \tag{70.19}$$

This implies that via the second formula in (70.12) we are able to express $\cos(2\pi/11)$ by means of the 5th roots of certain elements of $\mathbb{Q}[e(1/5)]$. Therefore, we reach

Vandermonde's discovery:

$e(1/11)$ can be expressed in terms of nested radicals $\tag{70.20}$

starting from rational numbers.

Further details will be given in (xi) below.

It should be stressed that the configuration (70.14) together with (70.15) makes it possible to restrict the set of permutations to be taken into account (that is, (70.17)); more on this point in the next subsection.

290   *Characters*

(v) Lagrange (1771a) investigated possible extensions of (i) at almost the same time as Vandermonde did; in particular, on p.254 he took up Euler's equation (70.10). However, he was unable to go so far as to enter into the core of (iv). The reason for this is in that he came to (70.12) but missed the configuration (70.14) since he seems to have never thought about (70.11). That is, he did not grasp anything similar to the invariance (70.18), and consequently had to deal with much more permutations, which resulted in uncontrollably many relations among $\{\tau_v\}$, a quagmire. This is the awful nature of the general 5th order equation which Lagrange, and probably Euler as well, tried in vain to cope with. What Vandermonde did was the discovery of a manageable set of permutations by observing (70.11). In retrospect, that is, as Galois would do half a century later, Vandrmonde identified the minimum set of permutations (in the above case, the cyclic group generated by σ, (70.17)) which are intimately related to the particular equation (70.10); and he achieved (70.20). This is no less than the birth of the modern theory of algebraic equations.

(vi) In his later years, Lagrange (1808, Notes XIII–XIV) analyzed the works on algebraic equations by Vandermonde and Gauss, while looking back on his own work (1771a); and he left this beautiful praise: *Vandermonde est le premier qui ait franchi les limites dans lesquelles la résolution des équations à deux termes se trouvait resserrée* (p.360). Exactly as he pointed out (p.350), *franchir* was enabled by the fact that

$$2 \bmod 11 \text{ is a primitive root.} \qquad (70.21)$$

Its consequences are the table (70.15) and the relations (70.17)–(70.19). However, Vandermonde himself stated nothing about the relation between (70.14) and (70.21). This can be regarded as quite natural because the concept of primitive roots modulo a prime was just being introduced by Euler (1772c; published in 1774) (see Note [39.1]). Nevertheless, it may be asserted that, even lacking the explicit notion of primitive roots, anybody who was inspired by the behavior of $2^j \bmod 11$ could have turned to the general cyclotomic equation $x^p - 1 = 0$ and tried to generalize Vandermonde's idea. Here it should be reminded that the clue $3^n \bmod 13$ left by Fermat (Note [28.1] above) inspired Euler (1755) deeply and led him to the later discovery of primitive roots $\bmod p$. Therefore, one may naturally surmise that the clue (70.14) must have stirred up genuine curiosity among specialists. It is, however, a real surprise that the history gives no relevant evidence except for Lagrange (1808).

(vii) Let $h(x) = 0$ be an equation of degree n. With its roots $\{\rho_j : j = 1, 2, \ldots n\}$ and ω an nth root of unity, we define

$$\tau_v = \sum_{j=1}^{n} \omega^{(j-1)v} \rho_j, \quad 0 \le v \le n-1. \tag{70.22}$$

This is called the Lagrange resolvent of h; see (1771a, p.331). From the point of view of §47(9), the consideration of the sum (70.22) can be regarded as a Fourier analysis of the distribution of the set of roots $\{\rho_j\}$. Through a particular configuration of $\{\rho_j\}$, one may achieve a kind of enhancement among the Fourier transforms $\{\tau_v\}$, perhaps an analogue of (70.19). There arises a possibility that one may effectively express each zeros by means of a combination of nested radicals of rational functions of coefficients of h. Here one recognizes the fundamental contribution of Vandermonde: in the case $h = X_{11}$ he discovered that such a configuration is (70.14). Therefore, if $\{\rho_j\}$ are arranged following (70.11), then we should call (70.22) instead the Vandermonde resolvent of h; see (1774, p.375).

(viii) Now, it can be asserted that Gauss [DA, art.360] must have introduced the configuration (65.1) in view of (70.11) in constructing the resolvents for $X_p(x) = 0$. One may confirm, by his posthumous article (1863b), that this was indeed the case. Namely, with $\rho = e(1/p)$, $\chi(n) = \iota_p e(\mathrm{Ind}_r(n)/g)$, $p-1 = fg$, $\omega = e(1/g)$, we are led to

$$\sum_{n \bmod p} \chi^v(n)e(n/p) = \sum_{j \bmod (p-1)} \omega^{jv}\rho^{r^j} = \sum_{u \bmod g} \omega^{uv}\psi_u^{[g]}. \tag{70.23}$$

On the left is the Gauss sum $G(\chi^v)$, in the middle a (Vandermonde) resolvent of X_p, and on the right a Fourier transform of the sequence of the f-term periods $\psi_u^{[g]}$ (Note [65.2]). Naturally, the right side can be inverted so that each f-term period is expressible as a linear combination of $G(\chi^v)$. The τ_v of Vandermonde is exactly a Gauss sum mod 11. Given the observation (70.21), it is no less easy to proceed to (70.23).

(ix) Continuing the preceding subsection, we consider

$$G(\chi^v)^g = \sum_{j \bmod (p-1)} R_{v,j}(\omega)\rho^{r^j}, \quad R_{v,j}(x) \in \mathbb{Q}[x]. \tag{70.24}$$

The mapping $\rho \mapsto \rho^r$ does not alter the left side (an extension of (70.18)). Then, by virtue of Kronecker's theorem (64.8), we get immediately $R_{v,j} = R_{v,j+1}$. That is,

$$G(\chi^v)^g = -R_{v,1}(\omega), \tag{70.25}$$

which is a refined extension of (70.19). Therefore, $G(\chi^v)$ is a gth root of an element of $\mathbb{Q}(\omega)$. According to the discussion of Note [61.3], we get the same conclusion by means of Jacobi sums (its special case is Note [65.3]). There

292 *Characters*

are assertions in [DA, art.360] and Lagrange (1808, p.334) which correspond to (70.25). In any event, taking $g = p - 1$, we see that $\rho = e(1/p)$ can be expressed as a $(p - 1)^{\text{st}}$ root of an element of $\mathbb{Q}(e(1/(p - 1)))$. Then, applying the decomposition (64.4) to $e(1/(p - 1))$, we may conclude, by induction, that $e(1/p)$ can be expressed in terms of radicals; here we need the prime power decomposition of $p - 1$, which can be a serious problem in practice, though. This is a typical instance of the theory of Abel (1829) and Galois (1831/1846) concerning the solubility of algebraic equations in terms of radicals. Hence, we are almost ready to assert that the origin of their theory is (70.20). However, we are unable to trace the possible historical route connecting Vandermonde and either Abel or Galois. Abel said, pointing to the cyclotomic equation, *La résolution de ces équations est fondée sur certaines relations qui existent entre les racines.* That is, (70.11). However, he got this view apparently through [DA, Sectio VII].

(x) Gauss' article (1863b) mentioned above seems to have been written with the aim to indicate how he had reached (70.23). As a numerical example, he treated the 2-term periods mod 11, that is, (70.14); see Sections 13, 17: there he employed exactly (70.21). Gauss referred to Lagrange (1808) which as detailed in (vi) contains a thorough discussion of Vandermonde (1774), stressing (70.21). However, Gauss' stance on Vandermonde's contribution is indefinite as far as we are able to learn from his article (1863b).

(xi) Finally, combining the ideas of Vandermonde, Lagrange, and Jacobi, we shall perform a practical computation (see also Bachmann (1872, pp.96–98)). Thus, let $p = 11$, and let $\chi(n) = \iota_{11}e(\text{Ind}_2(n)/5)$. Also, let α be as in (70.8). On noting that

$$\chi(-1) = 1, \chi(2) = \alpha, \chi(3) = \alpha^3, \chi(4) = \alpha^2, \chi(5) = \alpha^4,$$
$$\chi(6) = \alpha^4, \chi(7) = \alpha^2, \chi(8) = \alpha^3, \chi(9) = \alpha, \tag{70.26}$$

the τ_ν defined by (70.12) is the Gauss sum, or more correctly the Vandermonde sum

$$G(\chi^\nu) = \rho_1 + \alpha^\nu \rho_2 + \alpha^{2\nu} \rho_3 + \alpha^{3\nu} \rho_4 + \alpha^{4\nu} \rho_5, \tag{70.27}$$

where ρ_j is as in (70.14). Note the relations

$$G(\chi^3) = G(\overline{\chi}^2) = \overline{G(\chi^2)}, \ G(\chi^4) = G(\overline{\chi}) = \overline{G(\chi)} \tag{70.28}$$

and the numerical data

$$G(\chi) \doteqdot 2.6361055643 + 2.0126965628\, i,$$
$$G(\chi^2) \doteqdot 2.0701620998 + 2.5912215035\, i. \tag{70.29}$$

$$\S70 \qquad\qquad 293$$

We need to have explicit expressions for $\tau_\nu^5 = G(\chi^\nu)^5$, $\nu = 1,2$. To this end, as is prescribed in Note [61.3], we compute the Jacobi sums $J(\chi,\chi)$, $J(\chi,\chi^2)$, $J(\chi,\chi^3)$, $J(\chi^2,\chi^2)$, and $J(\chi^2,\chi^4)$. We have, by (70.26),

$$\begin{aligned} J(\chi,\chi) &= 2\alpha - 2\alpha^2 - \alpha^3, J(\chi,\chi^2) = -\alpha + 2\alpha^2 - 2\alpha^4, \\ J(\chi,\chi^3) &= J(\chi,\chi), \ J(\chi^2,\chi^2) = J(\chi,\chi^2), \ J(\chi^2,\chi^4) = \overline{J(\chi,\chi)}, \end{aligned} \qquad (70.30)$$

in which we have used $J(\chi^2,\chi^4) = J(\chi^{-3},\chi^{-1}) = \overline{J(\chi^3,\chi)}$. We find the following relations

$$\begin{aligned} G(\chi)^5 &= 11J(\chi,\chi)^2 J(\chi,\chi^2) = 11\big(6\alpha + 41\alpha^2 + 16\alpha^3 + 26\alpha^4\big), \\ G(\chi^2)^5 &= 11J(\chi,\chi^2)^2 \overline{J(\chi,\chi)} = 11\big(16\alpha + 6\alpha^2 + 26\alpha^3 + 41\alpha^4\big). \end{aligned} \qquad (70.31)$$

Therefore, we obtain, via (70.8),

$$\begin{aligned} G(\chi)^5 &= -\tfrac{11}{4}\left\{89 + 25\sqrt{5} + i\left(20\sqrt{10+2\sqrt{5}} - 25\sqrt{10-2\sqrt{5}}\right)\right\}, \\ G(\chi^2)^5 &= -\tfrac{11}{4}\left\{89 - 25\sqrt{5} + i\left(25\sqrt{10+2\sqrt{5}} + 20\sqrt{10-2\sqrt{5}}\right)\right\}. \end{aligned} \qquad (70.32)$$

Taking (70.29) into account, we fix the 5th roots of these right sides; then using the second formula in (70.12) together with the relations $G(\chi^3) = \overline{G(\chi^2)}$ and $G(\chi^4) = \overline{G(\chi)}$, we reach an explicit formula for ρ_1 in terms of nested radicals. The rest of the computation is the explicit formula for $e(1/11)$; that is, to solve a quadratic equation.

The formulas (70.32) coincide, respectively, with the θ' and θ'' in Lagrange (1808, the lower part of p.358). Vandermode's corresponding assertion can be confirmed by (70.32) and by $\sqrt{10 \pm 2\sqrt{5}} = \sqrt{5+2\sqrt{5}} \pm \sqrt{5-2\sqrt{5}}$ (signs correspond). As a matter of fact, his concluding result contains a minor error. Lagrange (1808, pp.359–360) indicated a correction, and expressed the praise mentioned in (vi). A beautiful retrospection. We note: Vandermonde (b.1735–d.1796) and Lagrange (b.1736–d.1813).

Notes

[70.1] With $p \equiv 1 \bmod 5$ the congruence equation $x^5 \equiv 1 \bmod p$ has five roots because of (40.1). It might be interesting to know that (70.8) can be used to construct these solutions $\bmod p$: For instance, if $p = 50051$, then $\sqrt{5}$ mod p, that is, a root of $x^2 \equiv 5$ is $x \equiv 5^{(p+1)/4} \equiv 10055 \bmod p$ by Note [63.1]. With the same abuse of notation, we have $\sqrt{-10 - 2\sqrt{5}} \equiv \sqrt{-20120} \equiv 24559$. Hence

$$\begin{aligned} \tfrac{1}{4}\left(-1 + \sqrt{5} + i\sqrt{10+2\sqrt{5}}\right) &\equiv \tfrac{1}{4}(-1 + 10055 + 24559) \\ &\equiv \tfrac{1}{4}(34613 + 50051) = 21166 = s. \end{aligned}$$

294 *Characters*

Indeed, $s^5 \equiv 1$; and other roots are $s^2 \equiv 43106$, $s^3 \equiv 1917$, and $s^4 \equiv 33912 \bmod p$. This somewhat bizarre way of computing roots of congruence equations should become quite natural if viewed through algebraic correspondences between fields. The argument can be generalized in many ways. Poinsot (1818; published in 1824) gave some examples. It is interesting that he devised a way to extract primitive roots from explicit formulas in radicals for roots of unity (see pp.151–164 there). However, Legendre's criterion (41.1) provides a simpler way to find primitive roots.

[70.2]   We shall translate the cubic-part of the subsection (i), specifically the transformation (70.2)–(70.4) with $a, b \in \mathbb{Z}$, into finite fields $\mathbb{F}_p$ ($p \geq 5$), and consider the possibility of full splitting: There exist $\rho_v \in \mathbb{Z}$ such that

$$(*) \qquad \begin{aligned} x^3 + ax + b &\equiv (x - \rho_1)(x - \rho_2)(x - \rho_3), \\ \rho_j &\not\equiv \rho_k \ (j \neq k) \ \Leftrightarrow \ \Delta \not\equiv 0 \end{aligned} \qquad \bmod p.$$

We shall omit considering the case of multiple zeros, i.e., $\Delta \equiv 0$; see Smith (1859, art.67). Hereafter we shall work in $\mathbb{F}_p$ and its algebraic extensions.

(I) $p \equiv 1 \bmod 3$. The key point is that $X_3 = 0$ is soluble in $\mathbb{F}_p$; $\omega = r^{(p-1)/3} \in \mathbb{F}_p$ is such a root, where r being a generator of $(\mathbb{F}_p)^*$, i.e., a primitive root mod p. Since we should have $\tau_v \in \mathbb{F}_p$, the congruence $(*)$ follows from (70.4) if and only if

$$\left(2W + 27b\right)^2 = \Delta \text{ with } \left(\tfrac{\Delta}{p}\right) = 1,$$

and the solution W is a cubic element in $\mathbb{F}_p$.

We take $p = 1117$ and consider $x^3 + 13x + 17 = 0$ in $\mathbb{F}_p$. As $\Delta = 447957 = 40$ is a quadratic residue, a root is $40^{140}\left((2^{279} - 1)(40^{279} - 1) - 2\right)/2 = 75$ by Note [63.2], that is, $2W + 27 \cdot 17 = 75$, and $W = -192$. This is a cubic residue, for $(-192)^{(p-1)/3} = 1$. So $\tau_1^3 = -192$. We compute τ_1 by Tonelli's method (§45). We have $g = 2, t = 124$, $h = 2$; $X_0 = (-192)^{83} = 285$, $Y_0 = (-192)^{248} = 120$; $Y_0^3 = 1 \Rightarrow g_0 = 1$. Following (45.8), we take $Z = 2$, a primitive root. That is, $Z^t = 529$; $Z^{6t} = Y_0 \Rightarrow k = 2$. Collecting these (see (45.10)), $\tau_1 = Z^t X_0 = 1087$. On noting $\tau_1 \tau_2 = -39$ (see (70.3)), we find that

$$\{\tau_1, \tau_2\} = \{1087, 113\}.$$

On the other hand, $\omega = 2^{(p-1)/3} = 996$; thus, $\omega\tau_1 = 279$, $\omega^2\tau_1 = 868$; $\omega\tau_2 = 848$, $\omega^2\tau_2 = 156$. Hence, the roots of the cubic congruence under consideration are

$$\rho_1 = \left(\tau_0 + \tau_1 + \tau_2\right)/3 = 1200/3 = 400,$$
$$\rho_2 = \left(\tau_0 + \omega^2\tau_1 + \omega\tau_2\right)/3 = 1716/3 = 572,$$
$$\rho_3 = \left(\tau_0 + \omega\tau_1 + \omega^2\tau_2\right)/3 = 435/3 = 145,$$

$$\S70 \qquad 295$$

since $\tau_0 = 0$. In this way, we obtain the full decomposition

$$x^3 + 13x + 17 \equiv (x - 145)(x - 400)(x - 572) \bmod 1117.$$

(II) $p \equiv -1 \bmod 3$. This time $X_3 = 0$ is not soluble in $\mathbb{F}_p$. By the reciprocity law, -3 is a quadratic non-residue; hence, $4X_3 = (2x + 1)^2 + 3$ is irreducible over $\mathbb{F}_p$. We then construct the algebraic extension $F = \{\mathbb{F}_p[x] \bmod X_3\}$ and shall work in it. Thus, $\omega = x \bmod X_3$ is a non-trivial cubic root of unity in F, i.e., $\omega = (-1 + \sqrt{-3})/2$ with the fancy of the preceding Note. Now, if the situation $(*)$ holds, then $\tau_1^3, \tau_2^3 \in F - \mathbb{F}_p$. For if $\tau_v^3 = d \in \mathbb{F}_p$, then with ℓ such that $3\ell \equiv 1 \bmod (p - 1)$ we have $\tau_v^3 = d^{3\ell}$, so $\tau_v = d^\ell \omega^c$ with an integer c. This implies that two of ρ_1, ρ_2, ρ_3 are equal to zero because of $\S67(4)$, i.e., the linear independence of $\{1, \omega, \omega^2\}$ over $\mathbb{F}_p$; hence, the initial assumption $\Delta \neq 0$ would be violated. We have shown the necessity part of the following assertion: The congruence $(*)$ follows from (70.4) if and only if

$$(2W + 27b)^2 = \Delta \text{ with } \left(\tfrac{\Delta}{p}\right) = -1,$$

and the solution W is a cubic element in $F \backslash \mathbb{F}_p$.

To confirm the sufficiency part, we need to show that, with our present translation, ρ_j given prior to (70.2) in terms of τ_v are indeed in $\mathbb{F}_p$. To this end, we apply $\S67(10)$, i.e., the Frobenius map. First, on noting $(2W + 27b)^2 - \Delta$ is irreducible over $\mathbb{F}_p$, we see that $(67.7)_{m=2}$ implies $\tau_1^{3p} = \tau_2^3$, that is, either $\tau_1^p = \tau_2$ or $\omega^{\pm 1}\tau_2$. The second case and the fact $\tau_1\tau_2 = -3a$ (see (70.3)) imply that

$$1 = \tau_1^{p^2-1} = \tau_1^{p(p-1)}\tau_1^{p-1} = \omega^{\pm(p-1)}(\tau_1\tau_2)^{p-1} = \omega^{\mp 2},$$

which is a contradiction. Hence, $\tau_1^p = \tau_2$, with which we get $\rho_v^p = \rho_v$ immediately, i.e., $\rho_v \in \mathbb{F}_p$. As an example, we consider $x^3 + 13x + 17 = 0$ in $\mathbb{F}_p$ with $p = 857$. In this case $(2W + 459)^2 = \Delta = (-3)(-201)$ is a quadratic non-residue, with -201 being a quadratic residue. Applying Tonelli's algorithm, we find that $-201 = 399^2$. Thus $W = (399\sqrt{-3} - 459)/2 = 399\omega - 30$ in F since $\sqrt{-3} = 2\omega + 1$. We observe that $W^{(p^2-1)/3} = 1$, and see, by an obvious extension of Theorem 39, that W is a cubic element of F; here we have used $(p^2 - 1)/3 = 285p + 571$. Next, on noting $p^2 - 1 = 3 \cdot t$, $3 \nmid t$, we extend (45.12) to F or rather argue directly; and find that a cubic root of W is equal to $\tau_1 = W^{(2t+1)/3}$. Thus,

$$\{\tau_1, \tau_2\} = \{\tau_1, -39/\tau_1\} = \{-13\omega - 258, 13\omega + 612\}.$$

Also,

$$\omega\tau_1 = -245\omega + 13, \qquad \omega^2\tau_1 = 258\omega + 245,$$
$$\omega\tau_2 = 599\omega - 13, \qquad \omega^2\tau_2 = -612\omega - 599.$$

In this way we obtain the full decomposition

$$x^3 + 13x + 17 \equiv (x - 118)(x - 363)(x - 376) \bmod 857.$$

4

Quadratic Forms

§71

Based on what we have developed so far, we now enter into a discussion on integral binary quadratic forms which are generically defined as

$$Q(x,y) = ax^2 + bxy + cy^2 = (x,y) \cdot Q \cdot {}^{\mathrm{t}}(x,y),$$
$$Q = \begin{pmatrix} a & b/2 \\ b/2 & c \end{pmatrix} = [|a,b,c|], \tag{71.1}$$

with three constant integers $\{a,b,c\}$, two independent variables $\{x,y\}$, and matrix multiplications on the rightmost side of the first line. To simplify the notation, we shall use interchangeably either the form $Q(x,y)$ or the matrix $Q = [|a,b,c|]$, writing often just Q, possibly with subscripts. Capital letters other than Q as well will occasionally appear for the same purpose. Also, we shall say simply a quadratic form or just a form instead of an integral binary quadratic form. It should be noted further that in the present chapter we shall often use matrix-notations to avoid any ambiguity which otherwise might creep in especially when handling products of quadratic forms.

We shall proceed, constantly attending either explicitly or implicitly to our fundamental task:

$$\text{to acquire an algorithm that yields all integer solutions}$$
$$\text{to the indefinite equation } Q(x,y) = m \text{ for a given integer } m \neq 0. \tag{71.2}$$

If such a solution $\{u,v\}$ exists, then we say that Q represents m; or we indicate the same by stating that $Q(u,v) = m$ is a representation. If $\langle u,v \rangle = 1$ in this, then Q is said to properly represent m or rather $Q(u,v) = m$ to be a proper representation. If $\langle u,v \rangle \neq 1$, then the indication improper is applied instead. It is also understood that

296

$$Q(u, v) = m \text{ and } Q(-u, -v) = m$$
are regarded as different representations. $\qquad$ (71.3)

We consider, by tradition, the modification

$$4aQ(x, y) = (2ax + by)^2 - Dy^2, \quad D = b^2 - 4ac, \qquad (71.4)$$

and make the definition

the discriminant of the form Q: $D = -4 \det Q$. $\qquad$ (71.5)

Our problem (71.2) is thus closely related to the congruence equation $X^2 \equiv DY^2 \mod |m|$. In the next section this connection will be made more precise. As our discussion develops, it will be gradually revealed that the notion of discriminants plays a principal rôle and the term is indeed an exquisite selection.

We collect here some basic facts concerning discriminants: Throughout the present chapter, the letter D with or without subscripts stands for the discriminant of a quadratic form; by the way, it is worth noting that in (71.4)

$$b \equiv D \mod 2. \qquad (71.6)$$

If D is a square, then (71.2) becomes a problem concerning a pair of linear indefinite equations, as is explained in Note [71.4]. Thus, it is assumed that

$$D \text{ is not a square of an integer} \qquad (71.7)$$

unless otherwise stated; any negative discriminant satisfies this. It is implied that we have always $ac \neq 0$ in $Q = [|a, b, c|]$, and that if $Q(x, y) = 0$ with $x, y \in \mathbb{Z}$, then $x, y = 0$ because of (71.4) and Note [13.1]. We then note that

a non-square d is a discriminant $\Leftrightarrow d \equiv 0, 1 \mod 4$. $\qquad$ (71.8)

The necessity is obvious, and the sufficiency is confirmed by

$$4|d \;\Rightarrow\; [|1, 0, -d/4|],$$
$$4|(d - 1) \;\Rightarrow\; [|1, 1, (1 - d)/4|]. \qquad (71.9)$$

These are said to belong to the set of principal forms. It is the collection of all forms which represent 1; see remark 2, §74. We shall look at principal forms rather collectively than individually; the reason for this will become obvious only in the proof of Theorem 93, §90.

The equation (71.4) implies also that if $D < 0$ and $a > 0$, then $Q(x, y) > 0$ over $\mathbb{Z}^2 \setminus \{0, 0\}$. With this situation, Q is called positive definite. Similarly, if D and a are both negative, then Q is called negative definite. These two types of forms can be treated completely analogously since if Q is negative

298   *Quadratic Forms*

definite, then $-Q$ is positive definite. On the other hand, if $D > 0$, then $Q(1,0)Q(b, -2a) = -a^2 D < 0$, so $Q(x,y)$ takes both positive and negative values over $\mathbb{Z}^2$, and Q is called indefinite. It will become apparent in the course of discussion that positive/negative definite forms are easier to deal with than indefinite forms. However,

$$\text{throughout §§71–76 we shall not distinguish} \atop \text{these three types of forms.} \tag{71.10}$$

Remark 1 Following a common practice, we shall use the adjective indefinite for problems of solving $Q(x,y) = m$ as well as for forms of positive discriminants, despite possible confusions.

We next note the trivial fact that if the form $[|a,b,c|]$ is of discriminant D and $\langle a,b,c \rangle = g$, then the form $[|a',b',c'|] = [|a,b,c|]/g$ is of discriminant D/g^2 and $\langle a',b',c' \rangle = 1$; representing $n \equiv 0 \bmod g$ by $[|a,b,c|]$ is the same as n/g by $[|a',b',c'|]$. Thus we introduce the notion of primitive forms: these are $[|a,b,c|]$ with $\langle a,b,c \rangle = 1$. As is easily seen, forms of discriminants with non-trivial square factors are not necessarily non-primitive.

We write

$$\mathcal{Q}(D) = \{\text{all primitive forms of discriminant } D\}. \tag{71.11}$$

Then, for each D,

$$|\mathcal{Q}(D)| = \infty, \tag{71.12}$$

since the principal forms $[|1, 2k, k^2 - D/4|]$ and $[|1, 2k+1, k(k+1)+(1-D)/4|]$, $k \in \mathbb{Z}$, corresponding to $D \equiv 0$ and $1 \bmod 4$, respectively, are all contained in $\mathcal{Q}(D)$.

It can happen that forms of a particular discriminant are all primitive. To specify such a situation we make the definition:

$$\text{D is a fundamental discriminant} \atop \Leftrightarrow \text{ all forms of discriminant } D \text{ are primitive.} \tag{71.13}$$

We have

Theorem 67 *The necessary and sufficient condition for D to be a fundamental discriminant is that either*

     (i) $D \equiv 1 \bmod 4$, $|D|$: sqf   or   (ii) $D/4 \equiv 2, 3 \bmod 4$, $|D/4|$: sqf

$$\tag{71.14}$$

holds. Also, for any discriminant D there exists a unique fundamental discriminant D_0 such that $D = D_0 R^2$ with an integer $R \geq 1$.

$$\S71 \qquad\qquad 299$$

Proof If $Q = [|a,b,c|]$ satisfies (i), then $\langle a,b,c\rangle = 1$ follows immediately. If it satisfies (ii) and $p|\langle a,b,c\rangle$, then with $p \neq 2$ we have the contradiction $p^2|D/4$, and with $p = 2$ we have $D/4 = (b/2)^2 - 4(a/2)(c/2) \equiv 0,1 \bmod 4$, which is again a contradiction. Therefore, if either (i) or (ii) holds, then D is a fundamental discriminant. Conversely, let D be a fundamental discriminant and $D \equiv 1 \bmod 4$. If $p^2|D$, then $D/p^2 \equiv 1 \bmod 4$, and the form $[|p,p,p(1 - D/p^2)/4|]$ yields a contradiction; that is, (i) is necessary. On the other hand, let D be a fundamental discriminant and $4|D$. If $p^2|D/4$, then the existence of the form $[|p,0, -p(D/4p^2)|]$ yields a contradiction. Hence, $|D|/4$ is sqf. With this, if $D/4 \equiv 1 \bmod 4$, then the existence of the form $[|2,2,(1 - D/4)/2|]$ yields a contradiction. This ends the proof of the first part of the theorem. As for the second part, we put $D = st^2$ with s being sqf and $t \in \mathbb{N}$. If $s \equiv 1 \bmod 4$, then we can take $D_0 = s, R = t$. Otherwise, $s \equiv 2,3 \bmod 4$ and $2|t$; thus we can take $D_0 = 4s, R = t/2$. To show the uniqueness of D_0, suppose $D = D_0'R'^2$ with a fundamental discriminant D_0'. Obviously $D_0 D_0' > 0$. If $D_0 \equiv D_0' \bmod 2$, then $D_0 = D_0'$, as the prime power decompositions of $|D_0|$ and $|D_0'|$ coincide. On the other hand, if $2|D_0$ and $2 \nmid D_0'$, then $D_0/4 = D_0' \equiv 1 \bmod 4$, which is a contradiction. This ends the proof.

If we restrict our concern to forms of fundamental discriminants, then we shall be able to enjoy some conveniences. However, we shall not generally impose this restriction since forms of non-fundamental discriminants are far more common in practice.

Nevertheless, we shall restrict our discussion throughout the present chapter to primitive forms:

$$\text{we shall work solely with } \mathcal{Q}(D) \text{ with various } D\text{'s} \qquad (71.15)$$

unless otherwise stated. This should not cause any loss of generality.

Notes

[71.1]   The plan of the present chapter in brief:

(0) We shall proceed appreciating Lagrange's pioneering and pleasantly practical ideas, Legendre's inspiring attempts, and Dirichlet's penetrating arguments, both algebraic and analytic, while being guided by Gauss' grand view. Note that assertions on quadratic forms of these authors and others will be quoted after making necessary changes due to the premise (71.15), which should not alter their mathematical contents in any essential way. See also Note [71.5].

(1) Gauss' theory of quadratic forms is divided into two parts: arts.153–222 and arts.223–307 of [DA]. The first part is largely a reconstruction of

300 *Quadratic Forms*

investigations by Lagrange and Legendre. It corresponds to our §§71–87, although we shall employ a reasoning which is closer to that of Lagrange and Legendre than Gauss', especially when discussing indefinite forms. Besides, we shall be attentive, in a restrictive manner, to the fact that behind Gauss' investigation the notion of algebraic number fields, or more specifically, quadratic number fields was forming; continued in Note [71.7].

(2) Thus the general structure of $\mathcal{Q}(D)$ is analyzed in §§71–76 via a classification of forms by means of the action of Γ, the use of which is a fundamental prescription by Gauss, as is to be explained in due course. This matrix group, which is introduced in §5, arises naturally from the notion of proper representations defined above through the mechanism that a pair of coprime integers involved in such a representation yields elements of the group, that is, Euclid's algorithm (3.6) plays a fundamental rôle in the present chapter again. The first glimpse of this classification, which we shall later, at (74.10)–(74.11), define precisely and allocate the notation $\mathcal{Q}(D)/\!\!/\Gamma$, appears already in the next section (paragraphs (b) (c)). It turns out to be a finite set (§76). With this, a recipe for the systematic resolution of the task (71.2) is briefly drawn in §76.

(3) The recipe is materialized in §§77–80 for $D < 0$ and in §§82–86 for $D > 0$. In the former we follow Lagrange's procedure but in the form streamlined by Gauss. In the latter, we shall rather return to the original method of Lagrange, as has been indicated above; in fact our aim is achieved by means of an overall application of the theory of continued fractions that Legendre elaborated greatly. Separately, two genuinely practical approaches to (71.2) are devised in §81 ($D < 0$) and §87 ($D > 0$), the origins of both of which can also be traced back to Lagrange. Besides, the later part of §85 is motivated by the thrilling existence of the Selberg zeta-function, whence a geometrical relation between the group Γ and the family of $\mathcal{Q}(D)$ with variable $D > 0$ will be touched on. Further, in §86 the dreamy *cakravâla* algorithm from classical Indian mathematics will be presented; it stands out as though a serendipitous discovery.

(4) That is part of the story, though. Our discussion will continue beyond §87 and become focused on an algebraic, that is, a group-theoretic structure of $\mathcal{Q}(D)/\!\!/\Gamma$, the existence of which Legendre discerned and Gauss established rigorously. One may view this development as a counterpart of the recognition of the group structure of reduced residue classes modulo an integer.

(5) We shall, however, have an intermezzo in §§88–89 beforehand. The discussion of §69[d] will be closed; namely, Legendre's incomplete proof of the quadratic reciprocity law will be completed, while simultaneously making preparations for the succeeding sections. Thus, we shall supply a quite simple

§71 301

proof of Dirichlet's prime number theorem for arithmetic progressions and also provide a proof of Legendre's Theorem 66, §69, or rather an extension of it by Dedekind. Further, a basic lemma of Selberg which resolves (69.2) is proved. Needless to say, these proofs are all independent of the reciprocity law.

(6) The second part of Gauss' theory of quadratic forms is devoted to the compositions and the genera of forms, that is, the grasp of the class group vaguely indicated in (4) above and a glimpse of its structure. Thus, in §90 we shall develop an account of Gauss' composition theory relying on the simplification devised by Arndt and Dirichlet–Dedekind. We shall close the section describing Legendre's attempt at a composition theory.

(7) Next, in §91 we shall develop an account of the genus theory, pursuing further the analogy touched on in (4). Thus, as a sort of extension of our third chapter, we shall try to make the first step into the Fourier analysis or the theory of the character group on each class group $\mathcal{Q}(D)/\!/\Gamma$. We shall ponder specifically its subgroup composed of real characters because of their decisive rôle in §93 which is analogous to that played by the real characters on reduced residue classes in Dirichlet's original proof (Note [88.2]) of his prime number theorem for arithmetic progressions. It turns out to be a synonym of developing the genus theory from scratch. Then, an approach due to Arndt and Dirichlet–Dedekind is again appreciated; and our reasoning is greatly facilitated by means of the theorems mentioned in (5) above. With such an experience in background, we shall make a visit, belated a little though, to the origin of the genus theory, that is, Lagrange's pioneering attempt and Legendre's more explicit aim at classifying those integers representable by forms in $\mathcal{Q}(D)$ in terms of the reduced residue classes modulo $|D|$ or the group $(\mathbb{Z}/|D|\mathbb{Z})^*$; there Legendre's argument reveals the essence of the reciprocity law (58.1). The core of the story will be succinctly summarized by means of an explicitly constructed isomorphism, i.e., the genus map (91.34). Then the section is closed with the construction of desired real characters on each class group, or the corresponding genus characters, in terms of elements in $\Xi_{|D|}$ defined in §51. Hence, the shadow of $\mathcal{Q}(D)/\!/\Gamma$ is visible inside $(\mathbb{Z}/|D|\mathbb{Z})^*$. Modern algebraic number theory has been built with the efforts to make this fascinating scene just more visible and more persuading.

(8) We then devote §92 to a proof of the class-number formula of Dirichlet, i.e., the explicit formula for the cardinality of the finite set $\mathcal{Q}(D)/\!/\Gamma$. Our argument belongs to the basic theory of automorphic functions and forms, another core impetus for the creation of modern number theory. Thus, we shall be leaving classical arithmetic for analytic number theory.

302 *Quadratic Forms*

(9) In §93, concluding the chapter, we shall give a proof of the Dirichlet–Weber prime number theorem: any primitive form represents infinitely many prime numbers, an analogue, for quadratic forms, of Dirichlet's prime number theorem. Thus, a conjecture of Legendre (Note [11.9]) is verified; as indicated above, the reasoning depends essentially on the notion of the genus characters. Then we shall make a return to the heart of the genus theory itself: An alternative/analytic approach originated with Dirichlet and Kronecker will be developed, or reworked. As a reward, we shall obtain our version of Gauss' 2nd proof of the reciprocity law. Sections 92–93 display classical instances indicating the power of analytic methods in discussing arithmetical problems.

(10) We are naturally well aware that this field of number theory has been cultivated by a great number of authors. However, we shall refer to only quite a limited number of core treatises and articles; we shall start from and proceed faithfully on the very basic ideas devised by the four great pioneers cited in (0) above, in just the same way as the present author did in his solitary student days of early 1960s without being able to access to wide literature. Thus, our principal aim in the present chapter is to provide ample experiences with which one may acquire the ability of explicitly solving (71.2) and closely related problems by means of moderately theoretical approaches while discerning the historical background of the rise of analytic number theory and algebraic number theory. A familiarity with these episodes and skills is a minimum requirement for a further study of number theory.

[71.2]   In the above we did not mention works by Fermat or Euler either, although we shall do so in the course of our discussion. The reason for this is in that it is Lagrange who really started the investigation of the structure of the set $\mathcal{Q}(D)$ with the aim to solve (71.2). The theory of quadratic forms originated with his works; they are all marvelously readable.

[71.3]   Occasionally we shall have to deal with solutions to $Q(x,y) = m$ which are improper. Such an m ought to have a non-trivial square factor. One should, however, note that even then m is possibly represented properly by the same form. The simplest case takes place with the form $[|1,0,1|] \in \mathcal{Q}(-2^2)$; it offers the example $3^2+4^2 = 5^2$. A little involved case occurs, for instance, with the form $[|7^3,119,9|] \in \mathcal{Q}(1813)$; it represents 7^5 properly and improperly with $\{x,y\} = \{37,-147\}$ and $\{7,0\}$, respectively.

[71.4]   The explicit attention to the quantity D associated with each form originated in Lagrange (1773); but he left D unnamed, an attitude typical of him. Also, Legendre (1798, p.70) made only this description of D without naming it: *celle qui détermine la nature de la formule*. It is Gauss [DA, art.154] who named D as *determinantem*; and Dirichlet (1863, p.139) adopted it, that is, *Determinante*, although our description of this historical fact is slightly incorrect, as is explained in the next Note. Today's discriminant appears to be a diversion from the theory of algebraic equations. By the way, if $Q = [|a,b,c|] \in \mathcal{Q}(D)$ with $D = f^2, f \in \mathbb{Z}$, then $aQ(x,y) = \left(ax + \frac{1}{2}(b+f)y\right)\left(ax + \frac{1}{2}(b-f)y\right)$, $b \equiv f \bmod 2$. With $d_1 = \langle a, \frac{1}{2}(b+f)\rangle$, $d_2 = \langle a, \frac{1}{2}(b-f)\rangle$, we have

$$aQ(x,y) = d_1 d_2 (\alpha x + \beta y)(\alpha' x + \beta' y), \quad \langle \alpha, \beta \rangle = 1, \quad \langle \alpha', \beta' \rangle = 1$$
$$\Rightarrow \quad a = a\langle a,b,c \rangle = d_1 d_2 \langle \alpha\alpha', \alpha\beta' + \beta\alpha', \beta\beta' \rangle = d_1 d_2.$$

Hence $Q(x,y) = (\alpha x + \beta y)(\alpha' x + \beta' y)$; $D = \det \left(\begin{smallmatrix} \alpha & \alpha' \\ \beta & \beta' \end{smallmatrix}\right)^2$. The issue (71.2) becomes the same as solving $\alpha x + \beta y = m_1$, $\alpha' x + \beta' y = m_2$, $m_1 m_2 = m$ with integers x, y, m_1, m_2. When m is huge, this can be a hard problem because of the fact stated in Note [2.6]. See also Gauss' extensive discussion [DA, arts.206–212] on forms with square discriminants; he (art.215) considered further the case $D = 0$.

[71.5]   As a matter of fact, for more than several decades following the publication of [DA], the form $[|a, 2b, c|]$ with $a,b,c \in \mathbb{Z}$ was mainly treated, and $D/4 = b^2 - ac$ was called its determinant. Namely, the situation $D \equiv 0 \bmod 4$ was mostly discussed. Nowadays, (71.1) is the standard form in number theory. This change from Gauss' format was proposed by Dedekind (Dirichlet (1879, p.388, footnote)), which he made with his ideal theory in mind, as is indicated at (72.24) below; the common use of (71.1) is said to have begun with Kronecker (1885, p.768), and Weber (1908, §103) made it standard. Lagrange and Legendre treated forms in either format equally but separately. One ought to take care when translating accounts of quadratic forms in the style of Gauss and Dirichlet into the modern version. This is one of the annoying complications caused by the prime number 2 that have been touched upon in Note [10.4]. Throughout the present chapter, we shall quote older investigations and results on quadratic forms after translating them into the modern format, mostly without mention. See also the last lines of Note [87.7] below.

[71.6]   By means of (71.4) the issue (71.2) can be simplified so that one may treat instead the indefinite equation

$$\text{pell}_d(n) : \quad x^2 - dy^2 = n \quad x, y \in \mathbb{Z}.$$

This is called a Pell equation; as for the notation pell_d indicating the form $[|1, 0, -d|]$, $d \neq 0$, see Notes [84.13]–[84.14] below. The equations $\text{pell}_d(\pm 1)$ and $\text{pell}_d(\pm 4)$, $d > 0$, are discussed at several separate places because of their fundamental importance in the theory of indefinite quadratic forms.

304 *Quadratic Forms*

[71.7]   The notion of *Fundamental-Discriminante* due to Kronecker (1885, pp.768–769) seems to have stemmed from [DA, art.226]; he employed (71.14) for its definition. Weber (1908, p.321) proposed the term *Stammdiskriminante* but niceties are often unable to survive. There exists no known algorithm of polynomial time to determine whether or not a given D is a fundamental discriminant; this is of course related to Note [17.3]. Each fundamental discriminant coincides with the discriminant of a quadratic number field; if it is positive, then the field is called real quadratic, and otherwise imaginary quadratic. What we develop in the present chapter is largely equivalent to the basic theory of quadratic number fields. We shall, however, keep ourselves from general algebraic number theory, save for short discussions such as the relation between quadratic forms and the notion of ideals within a particular quadratic number field, as we shall indicate in the later part of the next section and in Note [72.7]. In doing this, we shall employ actually an isomorphic setting expressed in terms of matrices, by which we shall be able to adhere to the arithmetic in $\mathbb{Z}$. Naturally, a comparative study of the present chapter and the theory of quadratic number fields is certainly worthwhile; Weber (1908, §§84–113) and Hecke (1923, §§44–53) are recommended for this purpose. By the way, a remarkable effect of using ideal theory in the investigation of binary quadratic forms will be acutely felt at Note [90.5]. Nevertheless, one should be aware that in order to explicitly solve (71.2) with a particular indefinite form the old method of Lagrange (§87) is still superbly effective.

§72

The most basic clue about the task (71.2) is the following assertion due to Lagrange (1773); see Note [72.1]:

Theorem 68 *Provided*

$$m \in \mathbb{N}, \tag{72.1}$$

the necessary and sufficient condition for m to be properly represented by a form of discriminant D is that the congruence equation

$$X^2 \equiv D \bmod 4m \tag{72.2}$$

has a solution.

Proof   Let $Q = [|a, b, c|]$ be of discriminant D, and A an integral matrix $\left(\begin{smallmatrix} \alpha & \kappa \\ \beta & \lambda \end{smallmatrix}\right)$. We have

$$a' = a\alpha^2 + b\alpha\beta + c\beta^2,$$

$${}^t A[|a,b,c|]A = [|a',b',c'|], \qquad b' = 2a\alpha\kappa + b(\alpha\lambda + \beta\kappa) + 2c\beta\lambda, \qquad (72.3)$$

$$c' = a\kappa^2 + b\kappa\lambda + c\lambda^2,$$

which implies, via determinants,

$$4Q(\alpha,\beta)Q(\kappa,\lambda) = \big(2a\alpha\kappa + b(\alpha\lambda + \beta\kappa) + 2c\beta\lambda\big)^2 - (\alpha\lambda - \beta\kappa)^2 D; \quad (72.4)$$

see (72.20) for an alternative derivation of this identity. Assuming that $Q(u,v) = m$ is a proper representation, we may take k,l such that $U = \left(\begin{smallmatrix} u & k \\ v & l \end{smallmatrix}\right) \in \Gamma$ by Euclid's algorithm (3.6). Letting A be U, the relation (72.4) implies that (72.2) has a solution. Conversely, if $\xi \bmod 4m$ is a solution to (72.2), then the form

$$M_{m,\xi} = [|m,\xi,c|], \quad c = (\xi^2 - D)/4m, \qquad (72.5)$$

represents m properly. We end the proof. By the way, this notation $M_{m,\xi}$ will be used frequently in our discussion; the relevant discriminant will be indicated contextually.

One may call this theorem the Lagrange principle. Indeed, it will be seen to govern the whole theory of integral binary quadratic forms. Thus, a few remarks are in order:

(a) The theorem does not require that forms be primitive. The positivity condition (72.1) is imposed merely for the sake of convenience. It will in fact cause no real restriction. If $D < 0$, then (72.1) is relevant only to positive definite forms, which of course implies no loss of generality; on the other hand, if $D > 0$, then see Note [74.6] below. A more essential aspect to be observed, especially from practical point of view, is in that since we base our whole discussion on a quadratic congruence mod $4m$ it is required, in general, to have or rather suppose the prime power decomposition of m a priori; see (33.4) and the beginning of §46. On the other hand, the necessity of the prime power decomposition of D depends on individual situations, while in the genus theory (§91) it becomes indispensable.

(b) If a proper representation $Q(u,v) = m$ is found by some means, then the corresponding solution $\xi \bmod 4m$ to (72.2) is obtained from the non-diagonal part of the matrix ${}^t UQU$; thus, it is fixed, save for a little ambiguous aspect caused by the choice of the right column of U that is to be clarified in §74. Conversely, we try to discover a way that starts at a solution to (72.2) and leads to a $U \in \Gamma$ with which the upper-left element of ${}^t UQU$ becomes m, that is, ${}^t UQU = M_{m,\xi}$. With Q being fixed, such a U is not always guaranteed to exist. Here emerges a principal issue of the present chapter, namely, to characterize the set $\{{}^t F M_{m,\xi} F : \forall F \in \Gamma\}$ so as to be able to see whether the particular Q is contained in this set or not, i.e., the clarification of the structure of the set $\mathfrak{Q}(D) /\!/ \Gamma$ to be defined at (74.11).

(c) Any integer pair $\{k, l\}$ such that $ul - vk = -1$ (i.e., $\det A = -1$ in (72.3)) works equally well as far as the proof of the last theorem is concerned; Lagrange (1773, p.724) stated in fact only $(ul - vk)^2 = 1$, or $ul - vk = \pm 1$. Gauss must have analyzed closely the effect of taking A with $\det A = -1$ and concluded that one should fix, as above, $A = U \in \Gamma$ in order to eliminate the ambiguity which would arise if A's with $\det A = \pm 1$ were indiscriminately used in (72.3). The nature of this ambiguity and the real merit of taking $U \in \Gamma$ will be clearly perceived only after having studied the present chapter throughly; Note [90.8] should be one of the best persuading numerical examples for this sake. Gauss himself expressed the situation by stating that *ceterum usus harum distinctionum mox innotescet* ([DA, art.158]); see also the end of the first paragraph of art.222. In any event, by means of a higher resolution thus gained, Gauss was able to discover a beautiful algebraic, i.e., group-theoretic, structure of $\mathfrak{Q}(D) /\!/ \Gamma$, which is regarded as one of the greatest events in the history of number theory. Hence we shall follow this prescription of Gauss throughout the present chapter.

We are, of course, mainly concerned with the issue of properly representing an integer with a primitive form, as has been indicated in the preceding section. Thus, we introduce the arithmetic function q_D defined as

$$q_D(m) = \begin{cases} 1 & \text{if } m \text{ is properly representable by a form in } \mathfrak{Q}(D), \\ 0 & \text{otherwise.} \end{cases} \tag{72.6}$$

Under (71.10), we have $q_D(-m) = q_D(m)$, so it suffices to consider q_D on $\mathbb{N}$. Then, we shall show that

$$q_D \text{ is a multiplicative function on } \mathbb{N}. \tag{72.7}$$

To verify this, we let $U \in \Gamma$ be as in the above proof and write

$$\begin{aligned} {}^t U Q U &= M_{m,\xi}, \\ \xi = 2auk + b(ul + vk) + 2ckl; \quad \xi^2 &\equiv D \bmod 4m. \end{aligned} \tag{72.8}$$

We observe first that

$$q_D(m) = 1 \implies M_{m,\xi} \in \mathfrak{Q}(D). \tag{72.9}$$

For, we see in (72.3) that $\langle a, b, c \rangle$ divides $\langle a', b', c' \rangle$, and that if $A = U \in \Gamma$, then $U^{-1} \in \Gamma$, and ${}^t U^{-1} ({}^t U Q U) U^{-1} = Q$ implies the opposite divisibility as well, whence (72.9) follows. With this, suppose that $q_D(m_1m_2) = 1$, $\langle m_1,m_2\rangle = 1$. By (72.9), we have $M_{m_1m_2,\xi} = [|m_1m_2,\xi,c|] \in \mathcal{Q}(D)$ with a certain ξ and $c = (\xi^2 - D)/4m_1m_2$. Then both the forms $M_{m_1,\xi} = [|m_1,\xi,m_2c|]$, $M_{m_2,\xi} = [|m_2,\xi,m_1c|]$ are obviously in $\mathcal{Q}(D)$ and properly represent m_1, m_2, respectively; that is, $q_D(m_1) = 1, q_D(m_2) = 1$. Conversely, suppose that $q_D(m_1) = 1$, $q_D(m_2) = 1$ with $\langle m_1,m_2\rangle = 1$. There exist ξ_1,ξ_2 such that $M_{m_1,\xi_1}, M_{m_2,\xi_2} \in \mathcal{Q}(D)$. We take a ξ satisfying $\xi \equiv \xi_j \bmod 2m_j$, $j = 1,2$. The existence of such a $\xi \bmod 2m_1m_2$ follows readily from (31.2) because of (71.6). We have $\xi^2 \equiv D \bmod 4m_1m_2$. Then, ${}^t T^{r_j} M_{m_j,\xi_j} T^{r_j} = M_{m_j,\xi}$, where T is as in (5.1) and $r_j = (\xi - \xi_j)/2m_j$. By the assumption, $M_{m_j,\xi}$, $j = 1,2$, are both in $\mathcal{Q}(D)$, from which it follows readily that $M_{m_1m_2,\xi} \in \mathcal{Q}(D)$. Hence, $q_D(m_1m_2) = 1$. This ends the proof of (72.7). Note that not ξ itself but the residue class $\xi \bmod 2m_1m_2$ is fixed, which will not cause any ambiguity in our discussion.

Incidentally, we have employed, in the above, a way to produce a new form from two given forms. Thus, we introduce a chart in $\mathcal{Q}(D)$:

$$M_{n_1,\eta_1} \times M_{n_2,\eta_2} \longmapsto M_{n_1n_2,\,\eta_3}, \qquad (72.10)$$

where $\langle n_1,n_2\rangle = 1$, and $\eta_3 \bmod 2|n_1n_2|$ is to be fixed in terms of $\eta_j \bmod 2|n_j|$, $j = 1,2$. This view point will play a basic rôle in our account (§90) of Gauss' theory of compositions of quadratic forms; Theorem 92, the central assertion in his theory, is a maximal extension of (72.10).

The assertion (72.7) implies that we may confine ourselves to the two separate cases:

$$\begin{aligned} &\text{(A)} \quad m > 0, \ \langle m,D\rangle = 1, \\ &\text{(B)} \quad m > 0, \quad m|D^\infty, \end{aligned} \qquad (72.11)$$

as far as we are concerned with the possibility of properly representing m by not a particular form but a certain form in $\mathcal{Q}(D)$. It holds that for any $\mu \geq 1$

$$p \nmid D \implies q_D(p^\mu) = \begin{cases} 1, & p = 2, D \equiv 1 \bmod 8, \\ 0, & p = 2, D \equiv 5 \bmod 8, \\ \frac{1}{2}\left(1 + \left(\dfrac{D}{p}\right)\right), & p \geq 3; \end{cases} \qquad (72.12)$$

see Notes [57.4]–[57.5]. This gives an overall description of the case (A); an extensive discussion will be developed in later sections. As for the case (B), in view of (72.10) our task is equivalent to finding all $M_{p^\mu,\xi}$ under

$$\mu \geq 1, \quad p|D. \qquad (72.13)$$

However, detailed discussion is omitted, as it is rudimental. Indeed, it depends only on the simple additional observation: If $q_D(p^\mu) = 1$, then there exists a

308   *Quadratic Forms*

form $[|p^\mu, \xi, c|]$ in $\mathcal{Q}(D)$ such that $\xi^2 \equiv D \bmod 4p^\mu$ and $c = (\xi^2 - D)/4p^\mu$. We have $p|\xi$, so $p \nmid c$, which yields restrictions on the powers of p dividing ξ, D when μ is given.

Returning to the basic identity (72.4), which is due to Lagrange (1773, p.724), we make here an essential observation: It is said in some literature that the origin of the formula can be traced back to that of Brahmagupta (628: Datta–Singh (1935: 1962, Chap. III, §16))

$$(\alpha^2 - \beta^2 d)(\kappa^2 - \lambda^2 d) = \begin{cases} (\alpha\kappa + \beta\lambda d)^2 - (\alpha\lambda + \beta\kappa)^2 d & (1), \\ (\alpha\kappa - \beta\lambda d)^2 - (\alpha\lambda - \beta\kappa)^2 d & (2). \end{cases} \qquad (72.14)$$

We apply, via some rearrangement and (71.4), these identities with $d = D$ to $Q(\alpha, \beta)Q(\kappa, \lambda)$. Then we find that (2) yields (72.4) but (1) does not readily accomplish it, i.e., we need an additional rearrangement to reach the same conclusion. As a matter of fact, Lagrange considered the linear transformation ${}^t(x, y) \mapsto A \cdot {}^t(x, y)$ and proceeded to (72.4) by computing $b'^2 - 4a'c'$, which is, of course, the same as we did but via the notion of determinants of matrices; Lagrange never mentioned (72.14), as far as his article (1773) is concerned. On the other hand, Legendre (1798, p.422) used (72.14) in constructing his theory of composing forms while being apparently annoyed by *l'ambiguité du signe* which (72.14) bears; see Note [72.5].

The ambiguity in (72.14) is in that the replacement $\lambda \mapsto -\lambda$, for instance, on the left side does not cause any change but on the right side exchanges (1) and (2). A way to eliminate this inconvenience in (72.14) is to consider, instead, the matrix identity

$$\begin{pmatrix} \alpha & \beta d \\ \beta & \alpha \end{pmatrix} \begin{pmatrix} \kappa & \lambda d \\ \lambda & \kappa \end{pmatrix} = \begin{pmatrix} \alpha\kappa + \beta\lambda d & (\alpha\lambda + \beta\kappa)d \\ \alpha\lambda + \beta\kappa & \alpha\kappa + \beta\lambda d \end{pmatrix}, \qquad (72.15)$$

which corresponds equally to the two formulas above, for the effect of $\lambda \mapsto -\lambda$ takes place now on both sides. As a matter of fact, a fundamental stance is induced here, that is, to view quadratic forms in the light of ideal theory:

To make the matter explicit, we introduce the notation

$$\begin{pmatrix} x & yd \\ y & x \end{pmatrix} = x\mathfrak{e} + y\mathfrak{d} = [\mathfrak{e}, \mathfrak{d}] \cdot {}^t(x, y), \quad \mathfrak{e} = \begin{pmatrix} 1 & 0 \\ 0 & 1 \end{pmatrix}, \ \mathfrak{d} = \begin{pmatrix} 0 & d \\ 1 & 0 \end{pmatrix}; \quad (72.16)$$

note that $\mathfrak{d}^2 = d\mathfrak{e}$, which is, of course, a generalization of the matrix equivalent of the imaginary unit, i.e., $\left(\begin{smallmatrix} 0 & -1 \\ 1 & 0 \end{smallmatrix}\right)^2 = -\left(\begin{smallmatrix} 1 & 0 \\ 0 & 1 \end{smallmatrix}\right)$; see (5.1). Also, $[\mathfrak{e}, \mathfrak{d}]$ denotes an ordered pair of matrices, i.e., a row vector whose elements are matrices. One should be aware that

$$\begin{gathered} \text{the multiplication of matrices} \\ \text{of the type (72.16) is commutative,} \end{gathered} \qquad (72.17)$$

provided $\mathfrak{d}$ is fixed appropriately.

Let $Q = [|a,b,c|] \in \mathfrak{Q}(D)$. Then, we have the correspondence

$$Q(x,y) \mapsto \begin{pmatrix} ax + \frac{1}{2}by & \frac{1}{2}yD \\ \frac{1}{2}y & ax + \frac{1}{2}by \end{pmatrix} = ax\mathfrak{e} + \frac{1}{2}y(b\mathfrak{e} + \mathfrak{d}) \tag{72.18}$$
$$= \left[a\mathfrak{e}, \tfrac{1}{2}(b\mathfrak{e} + \mathfrak{d})\right] \cdot {}^t(x,y),$$

where $\mathfrak{d}^2 = D\mathfrak{e}$. Note that

$$aQ(x,y) = \det\left(ax\mathfrak{e} + \tfrac{1}{2}y(b\mathfrak{e} + \mathfrak{d})\right),$$
$$aQ(x,y)\mathfrak{e} = \left(ax\mathfrak{e} + \tfrac{1}{2}y(b\mathfrak{e} + \mathfrak{d})\right) \cdot \left(ax\mathfrak{e} + \tfrac{1}{2}y(b\mathfrak{e} - \mathfrak{d})\right); \tag{72.19}$$

also, we get (72.4) by taking the determinants of both sides of the matrix identity

$$\left(a\alpha\mathfrak{e} + \tfrac{1}{2}\beta(b\mathfrak{e} + \mathfrak{d})\right) \cdot \left(a\kappa\mathfrak{e} + \tfrac{1}{2}\lambda(b\mathfrak{e} - \mathfrak{d})\right)$$
$$= \tfrac{1}{2}a\left(2(\alpha,\beta) \cdot Q \cdot {}^t(\kappa,\lambda)\mathfrak{e} - (\alpha\lambda - \beta\kappa)\mathfrak{d}\right). \tag{72.20}$$

The thought here is to view the form Q via the matrix-module:

$$\mathbf{Q} = a\mathfrak{e}\mathbb{Z} + \tfrac{1}{2}(b\mathfrak{e} + \mathfrak{d})\mathbb{Z}, \tag{72.21}$$

which is structurally deeper by one stratum or more than dealing with Q as it is. Then, let $\mathbf{I}_D = \mathfrak{e}\mathbb{Z} + \tfrac{1}{2}(\delta\mathfrak{e} + \mathfrak{d})\mathbb{Z}$ correspond to the principal form $(71.9)_{d=D}$ with $D \equiv \delta \bmod 2$, $\delta = 0, 1$. This is a ring, as indicated by (72.22)–(72.23) below with $b = \delta$. We have $\mathbf{Q} \subset \mathbf{I}_D$ because of (71.6). Besides, $\mathbf{Q}$ is mapped into itself under the multiplication of any element of $\mathbf{I}_D$ as

$$\left(u\mathfrak{e} + \tfrac{1}{2}v(\delta\mathfrak{e} + \mathfrak{d})\right) \cdot \left(ax\mathfrak{e} + \tfrac{1}{2}y(b\mathfrak{e} + \mathfrak{d})\right) = ax_1\mathfrak{e} + \tfrac{1}{2}y_1(b\mathfrak{e} + \mathfrak{d}), \tag{72.22}$$

with

$$\begin{pmatrix} x_1 \\ y_1 \end{pmatrix} = \begin{pmatrix} u + \tfrac{1}{2}(\delta - b)v & -cv \\ av & u + \tfrac{1}{2}(\delta + b)v \end{pmatrix} \begin{pmatrix} x \\ y \end{pmatrix}; \tag{72.23}$$

elements of this matrix are integers. That is,

$$\mathbf{Q} \text{ is an ideal of the ring } \mathbf{I}_D. \tag{72.24}$$

In what follows we shall use the terms module and ideal interchangeably in describing (72.21).

Hence, one may interpret the theory of quadratic forms into that of matrix modules and ideals. We shall, however, take reasonings in terms of matrix

310 *Quadratic Forms*

modules only when it becomes advantageous. By the way, the more traditional expression (Lagrange (1773, p.730))

$$aQ(x,y) = \left(ax + \tfrac{1}{2}(b + \sqrt{D})y\right)\left(ax + \tfrac{1}{2}(b - \sqrt{D})y\right) \qquad (72.25)$$

for $Q \in \mathfrak{Q}(D)$, which corresponds to the lower line of (72.19), i.e., $1 \leftrightarrow \mathfrak{e}$ and $\sqrt{D} \leftrightarrow \mathfrak{d}$, will also be employed, but be aware that then an imprecision can be caused by the way of choosing the branch of square roots.

Incidentally, (72.25) implies that if we confine ourselves to the ring $\mathbf{I}_D$, then $\mathbf{I}_{\ell^2 D}$, $\ell \in \mathbb{N}$, is to be identified with the subring

$$\mathfrak{e}\mathbb{Z} + \ell\mathbf{I}_D. \qquad (72.26)$$

Namely, $\mathfrak{d} \mapsto \ell\mathfrak{d}$ should be applied throughout the above discussion.

Notes

[72.1] Theorem 68, or rather the proof is due to Lagrange (1773, pp.723–724). Gauss [DA, art.154] started his great theory of quadratic forms with this theorem.

[72.2] The above seminal article of Lagrange begins with the following observation (Théorème I): If $r > 0$ is such that the congruence equation of two unknowns

$$ax^2 + bxy + cy^2 \equiv 0 \bmod r$$

has a coprime solution $\{u, v\}$, then there exist $A, B, C, u_1, v_1 \in \mathbb{Z}$ such that

$$r = Au_1^2 + Bu_1v_1 + Cv_1^2, \quad \langle u_1, v_1 \rangle = 1, \quad B^2 - 4AC = b^2 - 4ac.$$

This follows immediately from the last theorem. Lagrange's original argument is given in the next Note; interestingly it does not depend on (72.4). In any event, the key point is the assumption $\langle u, v \rangle = 1$, which induces the use of Euclid's algorithm or the rôle played by the group $GL(2, \mathbb{Z})$. The heart of Lagrange's discovery is that any positive integer which is properly represented by a particular quadratic form is a product of primes, each of which can be represented by a certain form of the same discriminant as that of the original form. The historical background of this fascinating fact can be traced back at least to Diophantus (Alexandria, the 3rd century), who observed the case of $65 = 5 \cdot 13$; see Note [78.2]. However, the genuine origin is in the following theorem due to Euler (1749a, 1772d, etc.): If $m|(a^2+b^2)$, $\langle a, b \rangle = 1$, then there exist c, d such that $m = c^2 + d^2$, $\langle c, d \rangle = 1$. See Notes [78.9]–[78.10] below. These prototypes belong to the case $D = -4$.

[72.3] Lagrange argued as follows: Suppose that $au^2 + buv + cv^2 = rs$, $\langle u, v \rangle = 1$. Let $s = ds_1, v = dv_1, \langle s_1, v_1 \rangle = 1$. Then $a = da_1$, for $d|au^2$. Hence, $rs_1 = a_1u^2 + buv + cdv_1^2$. Further, let $s_1u_1 + v_1w = u$ by Euclid's algorithm; note that $\langle u_1, v_1 \rangle = 1$. We have

$$rs_1 = a_1 s_1^2 u_1^2 + (2a_1 w + b)s_1 u_1 v_1 + (a_1 w^2 + bw + cd)v_1^2.$$

We take $A = a_1 s_1$, $B = 2a_1 w + b$, $C = (a_1 w^2 + bw + cd)/s_1$, and end the proof.

[72.4] We see, via Note [72.2], that a proper representation $Q(u, v) = m_1 m_2$ implies that there exist two forms Q_j and $\{u_j, v_j\}$ such that

$$Q(u, v) = Q_1(u_1, v_1)Q_2(u_2, v_2) \text{ with } \begin{cases} \det Q_j = \det Q, \\ Q_j(u_j, v_j) = m_j, \ \langle u_j, v_j \rangle = 1. \end{cases}$$

Then, reversing the situation, one may ask what can be asserted about the product of two positive integers which are properly representable by respective forms whose discriminants are equal to each other. If the two integers are coprime, then Theorem 68 ensures that the product can be properly represented by a third form of the same discriminant; indeed, we have the procedure following (72.9). Hence, more drastically, when $Q_j \in \mathcal{Q}(D)$, $j = 1, 2$, are arbitrarily given, it may be proposed to compose $Q \in \mathcal{Q}(D)$ such that $Q_1(x_1, y_1)Q_2(x_2, y_2) = Q(x, y)$ where the coefficients of Q are the functions of those of Q_j only, and $\{x, y\}$ a function of $\{x_j, y_j\}$ and the coefficients of Q_j. It was Gauss who rigorously established the existence of such an algorithm; its version due to Arndt (1859a) is our Theorem 92, §90, where as Gauss asserted the discriminants of Q_j, can, in fact, be different, though in a restricted way.

[72.5] Earlier than Gauss, Legendre commenced an investigation on this multiplicative structure of $\mathcal{Q}(D)$ and discussed his findings in his treatise (1798, pp.421–434). The chart (72.10) can be attributed to him: For Legendre took two forms $M_{a_j, b_j} = [|a_j, b_j, c_j|] \in \mathcal{Q}(D)$, $\langle a_1, a_2 \rangle = 1$, and viewed the matter in the scheme (72.10). He needed to fix the variables $\{x, y\}$; he tried to solve this by transforming the forms into $4a_j M_{a_j, b_j}(x_j, y_j) = (2a_j x_j + b_j y_j)^2 - D y_j^2$ and taking the matter into (72.14) of Brahmagupta. He faced, however, an annoying complexity because of the above *l'ambiguité* in (72.14). Continued in Note [90.8].

[72.6] To the discussion in Note [72.4] we add that one cannot, in general, be certain about whether the same Q represents either m_1 or m_2 or neither. The last theorem implies only that both can be represented but by forms possibly different from Q. Indeed, there exists a possibility that neither m_1 nor m_2 is representable by Q itself. Such a situation is shown in Note [81.4]. Behind this subtlety is a fascinating algebraic structure of $\mathcal{Q}(D)$ or rather $\mathcal{Q}(D)/\!/\Gamma$ which is to be revealed only in §§90–91.

[72.7] Provided D is a fundamental discriminant, the ring $\mathbf{I}_D$ corresponds to the set of all algebraic integers of the field $\mathbb{Q}(\sqrt{D})$ (see Hecke (1923, Kap.VII)) and $\mathbf{Q}$ can then be regarded as equivalent to an ideal of this field. Hence it should be possible to deal with quadratic forms by means of the

312   *Quadratic Forms*

theory of quadratic number fields. This is an idea of Dedekind (Dirichlet (1879, §181)); see also Hecke (1923, §53). However, we deal generally with non-fundamental D's, as premised above. Then, our $\mathbf{Q}$ is an ideal of the ring $\mathbf{I}_D$ but does not, in general, correspond to an ideal of the integer ring of $\mathbb{Q}(\sqrt{D_0})$, where $D = D_0 R^2$ is as in Theorem 67. The notion for $\mathbb{Q}(\sqrt{D_0})$ which corresponds to our $\mathbf{I}_D$ is called the quadratic order of the conductor R; see (72.26) with $\ell = R$ and $D = D_0$. It should be expedient to quote here elements of the theory of orders. However, our direct treatment of primitive quadratic forms of arbitrary discriminants is quite adequate for the purpose of the present essays. Nevertheless, it is very true that the use of quadratic number fields and their orders has a definitive merit, as it extends to higher algebraic domains.

[72.8] Actually, from the ideal theory we need to adopt only very basic notions and facts: Let D be arbitrary. The product $\mathbf{AB}$ of $\mathbf{I}_D$-ideals $\mathbf{A}$ and $\mathbf{B}$ is defined as the $\mathbf{I}_D$-ideal generated by all finite sums of products $\alpha\beta$ with $\alpha \in \mathbf{A}$, $\beta \in \mathbf{B}$. Then, let $N(\mathbf{A}) = \mathbf{A}\overline{\mathbf{A}}$, where the bar is the conjugation induced by $\mathfrak{d} \mapsto -\mathfrak{d}$. Obviously, this conjugation commutes with multiplication in $\mathbf{I}_D$. Since $\overline{\mathbf{I}}_D = \mathbf{I}_D$, both $\overline{\mathbf{A}}$ and $N(\mathbf{A})$ are $\mathbf{I}_D$-ideals. We have $N(\mathbf{A})N(\mathbf{B}) = N(\mathbf{AB})$. Also, if $N(\mathbf{A}) = \mathbf{I}_D$, then $\mathbf{A} = \mathbf{I}_D$. Specifically, we have, for $Q = [|a,b,c|] \in \mathcal{Q}(D)$,

$$N(\mathbf{Q}) = a\mathbf{I}_D,$$

since

$$\begin{aligned}
\mathbf{Q}\overline{\mathbf{Q}} &= \left[a^2\mathfrak{e}, \tfrac{1}{2}a(b\mathfrak{e} + \mathfrak{d}), \tfrac{1}{2}a(b\mathfrak{e} - \mathfrak{d}), \tfrac{1}{4}(b^2 - D)\mathfrak{e}\right]_{\mathbb{Z}} \\
&= a\left[a\mathfrak{e}, c\mathfrak{e}, \tfrac{1}{2}(b\mathfrak{e} + \mathfrak{d}), \tfrac{1}{2}(b\mathfrak{e} - \mathfrak{d})\right]_{\mathbb{Z}} \\
&= a\left[a\mathfrak{e}, b\mathfrak{e}, c\mathfrak{e}, \tfrac{1}{2}(b\mathfrak{e} + \mathfrak{d})\right]_{\mathbb{Z}} \\
&= a\left[\mathfrak{e}, \tfrac{1}{2}(b\mathfrak{e} + \mathfrak{d})\right]_{\mathbb{Z}} = a\left[\mathfrak{e}, \tfrac{1}{2}(\delta\mathfrak{e} + \mathfrak{d})\right]_{\mathbb{Z}},
\end{aligned}$$

because of $\langle a, b, c \rangle = 1$ and (71.6); here $[\cdots]_{\mathbb{Z}}$ denotes the module generated by $[\cdots]$ over $\mathbb{Z}$. This yields a typical instance related to the divisibility of ideals, an important extension of the observation made in Note [4.2] about the divisibility in $\mathbb{Z}$: with an $\mathbf{I}_D$-ideal $\mathbf{A}$,

if $\mathbf{A} \subset \mathbf{Q}$, then there exists an $\mathbf{I}_D$-ideal $\mathbf{B}$

such that $\mathbf{A} = \mathbf{QB}$.

For the verification, it suffices to note that $\mathbf{A}\overline{\mathbf{Q}} \subset a\mathbf{I}_D$, so $\mathbf{A}\overline{\mathbf{Q}}/a$ is an $\mathbf{I}_D$-ideal. An application will be made in Note [90.5].

[72.9] The theory of quadratic orders originated in Dedekind (1877c). He extended the notion to general algebraic number fields; Hilbert (1897, §§31–35) named it *Zahlring* and made it a fundamental algebraic device. On the other hand, the use of matrices (72.16) as an isomorphic replacement of quadratic irrationals originated with Poincaré (1880, p.239). We have employed matrix modules in order to stay strictly inside $\mathbb{Z}$ or rather proceed internally.

§73

To deal with the case (72.11) (A) or

$$m > 0, \ \langle m, D \rangle = 1, \tag{73.1}$$

we return to (72.2).

Remark 1 Since it stems from the basic fact (72.7) as remarked at (72.11), the condition (73.1) is quite a natural restriction and makes our discussion simpler considerably if imposed. Nevertheless, for the sake of generality, we shall proceed not always supposing (73.1) but invoking it only when necessary to do so. One should attend to this aspect throughout the present section.

Thus, we consider, on (73.1), the congruence equation $x^2 \equiv D \bmod p$ for each $p|m$. Because of Theorem 68, Theorem 59 implies that all the prime factors of m belong to a special set of reduced residue classes modulo $|D|$, which strongly restrict the nature of integers representable by forms in $\mathfrak{Q}(D)$. Also, we find in (72.4) that the four times of the product of two values of the form Q is a quadratic residue modulo $|D|$. These facts indicate well that the set of all integers which are properly representable by Q and satisfy (73.1) has a certain multiplicative structure that can be analyzed by means of the quadratic reciprocity law, which we suggested in Note [71.1](7) was indeed an aim of Legendre (1785, pp.524–531 and the tables I–III). We shall closely investigate this aspect in §91 and obtain an in-depth assertion. In essence, the structure of the group $(\mathbb{Z}/|D|\mathbb{Z})^*$ will be found to be a clue to this fundamental matter. We shall repeat the same observation in Remark 2 below, with a little more details. At all events, discriminants will turn out to be exactly as described in Note [71.4].

For the sake of a very basic preparation towards that destination, i.e., §91, here we consider the number of solutions of (72.2) under (73.1). To this end, we introduce the Kronecker symbol κ_D associated with a discriminant D. It is initially defined as a completely multiplicative function over $\mathbb{N}$ such that

Quadratic Forms

$$\kappa_D(2) = \begin{cases} 0 & D \equiv 0 \bmod 4, \\ 1 & D \equiv 1 \bmod 8, \\ -1 & D \equiv 5 \bmod 8; \end{cases} \tag{73.2}$$

$$2 \nmid a \implies \kappa_D(a) = \left(\frac{D}{a}\right) \text{ (Jacobi symbol)};$$

compare this with (72.12) and note that $\kappa_D(2) = (-1)^{(D^2-1)/8}$ if $2 \nmid D$.

Theorem 69 *The number of solutions* $\bmod 4m$ *of* (72.2) *under* (73.1) *is equal to*

$$2 \sum_{f \mid m} \mu^2(f) \kappa_D(f), \tag{73.3}$$

where μ is the Möbius function.

Proof This expression is equal to

$$2\left(1 + \kappa_D(2)\right)^\eta \prod_{\substack{p \mid m \\ p \neq 2}} \left(1 + \left(\frac{D}{p}\right)\right), \quad \eta = \begin{cases} 1 & 2 \mid m, \\ 0 & 2 \nmid m, \end{cases} \tag{73.4}$$

where the convention $0^0 = 1$ is observed. The product part is the consequence of Note [57.4]. As for $p = 2$, in the case $\eta = 0$ the issue is the same as to solve $X^2 \equiv D \bmod 4$; thus, if $D \equiv 0 \bmod 4$, then $X \equiv 0, 2 \bmod 4$; and if $D \equiv 1 \bmod 4$, then $X \equiv \pm 1 \bmod 4$. Hence, the number of solutions is equal to 2. In the case $\eta = 1$, we need to consider $X^2 \equiv D \bmod 2^\lambda$, $\lambda \geq 3$. Here we have $D \equiv 1 \bmod 4$ because of (73.1); so $D \equiv 1, 5 \bmod 8$. Then by Note [57.5] the number of solutions is equal to $2(1 + \kappa_D(2))$. We end the proof.

We shall collect a few facts concerning the Kronecker symbol. The most basic is the following:

Theorem 70 *The function κ_D is a Dirichlet character* $\bmod |D|$.

Remark 2 We are making an extremely essential step, as here the reciprocity law comes into the theory of quadratic forms: The merit of using the Kronecker symbol stems from this assertion. For instance, the formula (73.3), which itself has been obtained independently of the law, now states that the congruence (72.2) under (73.1) is a matter inside $(\mathbb{Z}/|D|\mathbb{Z})^*$, and it already gives a hint about a close relation between $\mathfrak{Q}(D)$ and this group; we repeat that we shall show in §91 that indeed it is, in a quite deep sense. Another instance is in the fact that the Dirichlet series associated with the arithmetic function κ_D is identical to a Dirichlet L-function, namely

$$\lfloor \kappa_D \rfloor(s) = L(s, \kappa_D), \tag{73.5}$$

which has far-reaching implications, as we shall show in §§92–93. One may assert that the real development of analytic number theory started with the recognition of (73.5). Anyway, the emergence of the Dirichlet character κ_D means that the congruence equation (72.2) can be viewed not only from the modulus $4m$ but also from the residue D. This is a consequence of the reciprocity law, as the proof below shows. In other words, the theory of quadratic forms, especially the theory of representations of integers by forms, is built on the law, in a way indispensably related to the discriminants. Thus, an appearance, in what follows, of Kronecker's symbol signals most always a moment when the law enters into the relevant discussion. It should be worth remarking that because of this dependency on the law from the outset, our approach to the theory of quadratic forms is different from Gauss' to a considerable extent; a relevant discussion is to be given in Notes [73.1] and [91.4]–[91.6]. We shall apply the present theorem while keeping it in mind that the reciprocity law is actually the prime mover in the present chapter. Precisely because of this, we shall make it explicit when the crude expression (73.3) is employed without being processed by the reciprocity law. Important instances such as will take place in §92 and §93[C]; indeed, they are related to a proof of Gauss' 2nd proof of the reciprocity law.

Proof   We first extend the domain of κ_D to $\mathbb{Z}$ by the definition

$$a < 0 \;\Rightarrow\; \kappa_D(a) = \operatorname{sgn}(D)\kappa_D(|a|). \tag{73.6}$$

As we shall show in a moment, we have the following facts:

$$\begin{aligned}
&(1)\; a \equiv b \bmod |D|,\; a,b > 0 \;\Rightarrow\; \kappa_D(a) = \kappa_D(b),\\
&(2)\; \kappa_D(|D| - 1) = \operatorname{sgn}(D).
\end{aligned} \tag{73.7}$$

For instance, if $a \equiv b \bmod |D|$, $a > 0, b < 0$, then $a \equiv (|D| - 1)|b| \bmod |D|$; thus by (73.6), $\kappa_D(a) = \kappa_D(|D| - 1)\kappa_D(|b|) = \operatorname{sgn}(D)\kappa_D(|b|) = \kappa_D(b)$, where the last identity is by the definition (73.7). Other cases are treated in just the same way. Also, it is obvious that (73.7) yields a completely multiplicative function of $a \in \mathbb{Z}$ which vanishes if $\langle a, D \rangle \neq 1$. Hence, this extension of κ_D satisfies $(52.1)_{q=|D|}$. In passing, we mention that the extension is unique because (73.7) is necessary.

We shall prove the assertion (73.6) (1). If $D \equiv 1 \bmod 4$, then we write $a = 2^\alpha s$, $2 \nmid s$. We have, by Theorem 58,

$$\begin{aligned}
\kappa_D(a) &= (\kappa_D(2))^\alpha \kappa_D(s)\\
&= \left(\frac{2}{|D|}\right)^\alpha \left(\frac{D}{s}\right) = \left(\frac{2}{|D|}\right)^\alpha \left(\frac{s}{|D|}\right) = \left(\frac{a}{|D|}\right),
\end{aligned} \tag{73.8}$$

316   *Quadratic Forms*

which gives (73.6)(1) for any odd D. If $D \equiv 0 \bmod 4$, then with $D = 2^\beta f$, $2 \nmid f$, we have, again by Theorem 58,

$$\kappa_D(a) = (-1)^{(f-1)(a-1)/4+\beta(a^2-1)/8} \left(\frac{a}{|f|} \right), \quad 2 \nmid a > 0. \tag{73.9}$$

Thus, on noting that if $2 \mid \beta$, then $a \equiv b \bmod 4$, and that if $2 \nmid \beta$, then $a \equiv b \bmod 8$, we get (73.6)(1) again. Therefore, the assertion has been verified.

As for (73.6)(2), we consider, for instance, the case $D \equiv 1 \bmod 4$, $D < 0$. We put $D = 1 - 2^\gamma d$, $2 \nmid d$, where $\gamma \geq 2$ and $d \geq 1$. Then $\kappa_D(|D| - 1) = \kappa_D(2)\kappa_D(2^{\gamma-1}d - 1)$. Here $\kappa_D(2)$ is equal to -1 if $\gamma = 2$, and to 1 if $\gamma \geq 3$. Also, since $D \equiv -1 \bmod (2^{\gamma-1}d - 1)$, we have

$$\kappa_D(2^{\gamma-1}d - 1) = \left(\frac{-1}{2^{\gamma-1}d - 1} \right), \tag{73.10}$$

which is equal to 1 if $\gamma = 2$, and to -1 if $\gamma \geq 3$. Other cases are easier, and details can be omitted. We end the proof.

Remark 3 The formulas (73.8)–(73.9) hold also for negative a as well.

We have the following crucial relation between the Kronecker symbol associated with a fundamental discriminant and the notion of primitive real characters:

Theorem 71 *If D is a fundamental discriminant, then κ_D is a primitive real character mod $|D|$. Conversely, any primitive real character χ mod q is identical to κ_D with the fundamental discriminant $D = \chi(-1)q$.*

Proof For any fundamental discriminant $D = 2^\beta f$, $2 \nmid f$, the combination of (62.10), (73.8), (73.9), and the last Remark readily yields the first part, on noting that if $\beta = 2$, then $f \equiv 3 \bmod 4$, and if $\beta = 3$, then $f \equiv 1, 3 \bmod 4$. As for the second part, let χ be a primitive real character mod q. On noting (54.5)–(54.7), we arrange as follows:

$$
\begin{aligned}
l = 0 &\Rightarrow D = (-1)^{(q-1)/2}q, \\
l = 2 &\Rightarrow D = (-1)^{(q/4+1)/2}q, \\
l = 3, (54.7)_{k_1=0} &\Rightarrow D = (-1)^{(q/8-1)/2}q, \\
l = 3, (54.7)_{k_1=1} &\Rightarrow D = (-1)^{(q/8+1)/2}q.
\end{aligned}
\tag{73.11}
$$

These D's are fundamental discriminants, and the combination of (62.10) and the pair (73.8)–(73.9) implies $\kappa_D = \chi$. By the last Remark and $D = \mathrm{sgn}\,(D) \cdot q$ it follows that $D = \kappa_D(-1)q = \chi(-1)q$. We end the proof.

$$\S 73 \qquad\qquad 317$$

With the notation $\Delta(t) = \prod_{p|t} \left((-1)^{(p-1)/2} p\right)$, $p \geq 3$, we see that D's in (73.11) are respectively equal to $\Delta(q), -4\Delta(q/4), 8\Delta(q/8), -8\Delta(q/8)$, where we have used (62.3). We then let

$$p^* = \begin{cases} (-1)^{(p-1)/2} p, & p \geq 3, \\ \text{either } -4 \text{ or } 8 \text{ or } -8, & p = 2. \end{cases} \tag{73.12}$$

These are called prime discriminants. Any fundamental discriminant D is a product of prime discriminants, and accordingly κ_D is decomposed:

$$D\text{: fundamental} \implies \begin{array}{c} \text{independently of the reciprocity law} \\ D = \prod_{p|D} p^*, \quad \kappa_D = \prod_{p|D} \kappa_{p^*}. \end{array} \tag{73.13}$$

Also,

$$\kappa_{p^*}(a) = \left(\frac{a}{p}\right), \quad p \geq 3, \ p \nmid a, \ \text{(by the reciprocity law)} \tag{73.14}$$

$$\kappa_{2^*}(a) : \begin{cases} \kappa_{-4}(a) = (-1)^{(a-1)/2}, & \kappa_8(a) = (-1)^{(a^2-1)/8}, \\ \kappa_{-8}(a) = (\kappa_{-4} \cdot \kappa_8)(a) = (-1)^{(a-1)/2 + (a^2-1)/8}, \end{cases} \quad 2 \nmid a, \tag{73.15}$$

which are special cases of (73.8) and (73.9), respectively. Each κ_{p^*} is called a prime character.

Remark 4 Here is an obvious consequence of the above discussion: Any real character is uniquely expressible as a product of prime characters and a trivial character because of Theorem 71 and (73.13). This decomposition will be exploited in the second half of $\S 91$.

Further, for an arbitrary discriminant D we have the following:

Theorem 72 *Let $D = D_0 R^2$ be as in Theorem 67. Then we have the decomposition*

$$\kappa_D = J_R \kappa_{D_0}. \tag{73.16}$$

In other words, the conductor of the character κ_D is equal to $|D_0|$.

Remark 5 Here is a conflict of terms: in Note [72.7] R is called the conductor of the ring $\mathbf{I}_D$, while here $|D_0|$ is the conductor of κ_D. Dedekind (Dirichlet (1879, p.613)) called R *Index.*

Proof If $2 \nmid D$, then (73.16) follows immediately from (73.8). If $2|D$, $2 \nmid D_0$, then we have, in (73.9), $2|\beta$ and $f = D_0(R/2^{\beta/2})^2 \equiv 1 \bmod 4$; thus

$$\kappa_D(a) = J_2(a)\left(\frac{a}{|f|}\right) = J_R(a)\left(\frac{a}{|D_0|}\right) = J_R(a)\kappa_{D_0}(a). \tag{73.17}$$

318   *Quadratic Forms*

If $2|D$, $D_0/4 \equiv 3 \bmod 4$, then $2|\beta$ and $f = (D_0/4)(R/2^{\beta/2-1})^2 \equiv D_0/4 \bmod 4$; thus, by (73.9),

$$\kappa_D(a) = J_R(a)(-1)^{(D_0/4-1)(a-1)/4}\left(\frac{a}{|D_0/4|}\right) = J_R(a)\kappa_{D_0}(a). \qquad (73.18)$$

Finally, if $2|D$, $D_0/4 \equiv 2 \bmod 4$, then $2 \nmid \beta$ and $f = (D_0/8)(R/2^{(\beta-3)/2})^2 \equiv D_0/8 \bmod 4$; thus, by (73.9),

$$\kappa_D(a) = J_R(a)(-1)^{(D_0/8-1)(a-1)/4+(a^2-1)/8}\left(\frac{a}{|D_0/8|}\right) = J_R(a)\kappa_{D_0}(a).$$
$$(73.19)$$

We end the proof.

In passing, we remark that

$$D: \text{an even fundamental discriminant}$$
$$\Rightarrow \kappa_D(a+|D|/2) = -\kappa_D(a); \qquad (73.20)$$

that is, κ_D is skew-symmetric. This can be deduced from either (73.18) or (73.19).

Notes

[73.1]   As we shall gradually make it clear, the earlier efforts to solve various issues in the theory of quadratic forms ushered in the law (58.1)(3). Now that we have the law, we should exploit it in the investigation of quadratic forms; and indeed we shall do so in the present chapter, save for some sensitive places as those indicated in Remark 2 above. The first explicit application of this strategy was made by Legendre (1785, pp.528–531) as noted above. On the other hand, Gauss seems to have tried to develop as far as possible his theory of quadratic forms independently of the reciprocity law; that stance led him to the 2nd proof of the law ([DA, art.262]) as an application of the initial part of his genus theory of quadratic forms. Then, he started using the law in arts.263–264. See Note [91.6] below.

[73.2]   The explicit formula (73.3) originated in Dirichlet (1839; Théorèmes I, II). It became accessible by the introduction of κ_D by Kronecker (1885, pp.769–770). This symbol is usually defined as a specific extension of Jacobi's so that the denominators in (62.1) can be even but the numerators are restricted to discriminants; see for instance Landau (1927, pp.51–52). However, as far as the aim of the present essays is concerned, such a modification of the Jacobi symbol does not provide any notable advantage; so we employ (73.2).

§74

We now take a $Q \in \mathcal{Q}(D)$ and discuss under the premise that

$$Q(u,v) = m > 0 \text{ is a proper representation;}$$
$$\text{however, the coprimality condition in (73.1) is not imposed.} \tag{74.1}$$

With this, we compare (72.8) to the chain of relations: ${}^t\{u,v\} \mapsto U = \begin{pmatrix} u & k \\ v & l \end{pmatrix} \mapsto \xi$. It is not a chain of mappings, for although the second arrow is a mapping, that is, ξ is fixed by U, the first is not because the right column of U is not unique but only known to be in the set ${}^t\{k + ru, l + rv\}$, $r \in \mathbb{Z}$, with a particular ${}^t\{k, l\}$ fixed by means of Euclid's algorithm, an aspect which has been indicated in §72, (b). Namely, with T as in (5.1), all UT^r have the same left column, and vice versa. Then, with the sets

$$\mathcal{S}_Q(m) = \big\{ \text{all } {}^t\{u,v\} \text{ satisfying (74.1)} \big\},$$
$$\mathcal{U}_Q(m) = \big\{ \text{all } U \in \Gamma \text{ whose left column is in } \mathcal{S}_Q(m) \big\}, \tag{74.2}$$

we have the bijection

$$\mathcal{S}_Q(m) \to \mathcal{U}_Q(m)/\Gamma_\infty, \quad \Gamma_\infty = \big\{ T^r : r \in \mathbb{Z} \big\}. \tag{74.3}$$

Remark 1 As a matter of fact, there exists a way to hide the right column of U behind discussion. See Note [74.3].

As for the mapping $U \mapsto \xi$, replacing U by UT^r in (72.8), the image ξ is shifted to $\xi + 2mr$, for ${}^t T^r M_{m,\xi} T^r = M_{m,\xi+2mr}$. Hence, if viewed from the right side of (74.3), then the image ξ is uniquely determined with respect to modulo $2m$. To make this situation precise, we introduce the canonical system of forms concerning proper representations of m by a certain form in $\mathcal{Q}(D)$:

$$\mathcal{M}_m(D) = \big\{ M_{m,\xi} : \xi \bmod 2m \big\} \cap \mathcal{Q}(D). \tag{74.4}$$

Here the congruence condition means that a representative system $\{\xi\}$ of solutions to (72.2) is fixed modulo $2m$. The restriction to $\mathcal{Q}(D)$ is necessary, as we have (72.9) while $M_{m,\xi}$ defined by (72.5) cannot be automatically primitive without assuming side-condition like (73.1) which is currently not in effect; however, be aware that there can be situations $M_{m,\xi} \in \mathcal{Q}(D)$ even if $\langle m, \xi \rangle \neq 1$. Anyway, the mapping $U \mapsto \xi$ is now read as the mapping $U\Gamma_\infty \mapsto M_{m,\xi} \in \mathcal{M}_m(D)$. The representation (74.1) takes place only if there exists a $U \in \Gamma$ such that ${}^t UQU \in \mathcal{M}_m(D)$.

We have thus the chain of mappings:

$$\mathcal{S}_Q(m) \to \mathcal{U}_Q(m)/\Gamma_\infty \to \mathcal{M}_m(D). \tag{74.5}$$

320 *Quadratic Forms*

The second mapping is not necessarily injective, though. In order to modify it into an injection, we consider when the images of the two cosets $U_1\Gamma_\infty$ and $U_2\Gamma_\infty$ with $U_1, U_2 \in \mathcal{U}_Q(m)$ coincide. We have then ${}^t U_1 Q U_1 = {}^t(U_2 T^s)Q(U_2 T^s)$ with an $s \in \mathbb{Z}$; that is, $U_2 T^s U_1^{-1} = V \in \mathrm{Aut}_Q$, where

$$\mathrm{Aut}_Q = \{Y \in \Gamma : {}^t YQY = Q\}, \tag{74.6}$$

which is called the automorphism group of Q in the matrix group Γ. We have $U_2 = VU_1 T^{-s}$, so U_1 and U_2 are in the same double coset with respect to Aut_Q on the left and Γ_∞ on the right of an element of Γ. In this way we obtain the injective mappings:

$$\mathrm{Aut}_Q \backslash \mathcal{S}_Q(m) \to \mathrm{Aut}_Q \backslash \mathcal{U}_Q(m)/\Gamma_\infty \to \mathcal{M}_m(D). \tag{74.7}$$

Since the rightmost set is finite, we see in particular that

$$\mathrm{Aut}_Q \backslash \mathcal{S}_Q(m) \text{ is a finite set.} \tag{74.8}$$

Therefore, the task (71.2) will be settled if we attain a method to determine Aut_Q and these finitely many seed solutions, i.e., the minimal set of solutions that generate $\mathcal{S}_Q(m)$ together with the action of Aut_Q.

We shall re-reconsider the above discussion in a wider perspective. We have been concerned with proper representations of a particular integer by an individual form in $\mathfrak{Q}(D)$. Instead, we shall now deal with all proper representations of all eligible integers by arbitrary forms in $\mathfrak{Q}(D)$. In view of (72.8), this is equivalent to considering the following motion inside $\mathfrak{Q}(D)$:

$$Q \mapsto {}^t UQU, \quad \mathfrak{Q}(D) \ni \forall Q, \ \Gamma \ni \forall U, \tag{74.9}$$

which is called the transformation of a form by an element of Γ; the upper-left coefficient of the matrix ${}^t UQU$ is properly represented by Q and vice versa. Since (74.9) does not take forms outside $\mathfrak{Q}(D)$, we may introduce the new notion

$$Q_1 \equiv Q_2 \bmod \Gamma \quad \text{in } \mathfrak{Q}(D) \tag{74.10}$$

to indicate that there exists a $U \in \Gamma$ such that ${}^t UQ_1 U = Q_2$. We shall say simply that the forms Q_1 and Q_2 are equivalent; we shall see the matter more precisely in the next section.

Remark 2 Following Lagrange, Gauss [DA, art.157] defined forms Q_j, $j = 1, 2$, as equivalent if ${}^t VQ_1 V = Q_2$ with $V \in \mathrm{GL}(2, \mathbb{Z})$. Then, he (art.158) called the two cases with $\det V = +1$ and $\det V = -1$ as proper and improper, respectively. Thus our notion (74.10) of equivalence is the proper one in his definition. This simplification of terms should not cause any inconvenience in our discussion. More in the next section.

$$\S74 \qquad\qquad 321$$

The classification axiom is obviously valid with (74.10). We then make an important definition:

$$\mathcal{K}(D) = \mathcal{Q}(D)/\!/\Gamma = \{\text{all equivalence classes in } \mathcal{Q}(D)\}. \qquad (74.11)$$

We shall assert and prove later, in Theorem 74, §76, that $\mathcal{K}(D)$ is a finite set: It is one of the most fundamental discoveries in the theory of numbers, and is due to Lagrange (1773).

Remark 3 A form is equivalent to the respective principal form (71.9) if and only if it represents 1. For if it is equivalent to $[|1,b,c|]$, then apply (76.3) below so that $2|b \Rightarrow [|1,0, -D/4|]$ and $2\nmid b \Rightarrow [|1,1,(1-D)/4|]$, which are $M_{1,\xi}$ with $\xi^2 \equiv D \bmod 4$. Thus, principal forms of a particular determinant are all equivalent to each other. This fact has an important application in §87. Anyway, all the principal forms of discriminant D form a single class of $\mathcal{K}(D)$.

Proper representations and the classification (74.10) are closely related to each other; it is obvious that the sets of integers properly represented by equivalent forms coincide with each other. Whether (74.1) is possible or not is relevant not to an individual form but rather to the class to which the form belongs. It should be appropriate to introduce here

$$\text{convention :} \qquad \begin{array}{l} \text{a class } \mathcal{C} \in \mathcal{K}(D) \text{ represents } m \text{ properly} \\ \Leftrightarrow \text{ a form } Q \in \mathcal{C} \text{ represents } m \text{ properly.} \end{array} \qquad (74.12)$$

Note that it is possible that two inequivalent forms, i.e., two different classes represent properly the same set of integers, as we shall show in the next section.

It transpires now that in order to describe proper representations in general we need to investigate the following two issues: to determine

(1) whether or not two given forms in $\mathcal{Q}(D)$ are equivalent;

(2) all elements in Γ connecting equivalent forms each other. $\qquad (74.13)$

The issue (1) is the same as to devise a method or an algorithm with which one can construct all Γ-orbits arising from the motion (74.9) that do not cross each other; Q represents m properly if and only if the orbit of Q crosses the set $\mathcal{M}_m(D)$; then, of course, every form on the orbit has the same property. As for the issue (2), it is equivalent to construct Aut_Q for each Q: For the relation ${}^t U Q U = {}^t U' Q U'$ is equivalent to $U' U^{-1} \in \mathrm{Aut}_Q$, and the coset decomposition

$$\mathrm{Aut}_Q \backslash \Gamma : \quad \Gamma = \bigsqcup_v \mathrm{Aut}_Q \cdot U_v \qquad (74.14)$$

gives that

$$\text{the points on the } \Gamma\text{-orbit of } Q \text{ are } \{{}^t U_v Q U_v\}. \qquad (74.15)$$

322   *Quadratic Forms*

Remark 4 The geometric flavor one may discern here is closely related to a variety of fascinating developments in modern analytic number theory, a typical instance among which we shall dwell on in the later parts of §§85 and 94.

We then return to the discussion (74.1)–(74.8). Let $\mathcal{M}_m(\mathcal{C}, D)$ be the subset of $\mathcal{M}_m(D)$ that is composed of $M_{m,\xi}$ in a class $\mathcal{C} \in \mathcal{K}(D)$ so that

$$\mathcal{M}_m(D) = \bigsqcup_{\mathcal{C} \in \mathcal{K}(D)} \mathcal{M}_m(D, \mathcal{C}); \tag{74.16}$$

in other words $\mathcal{M}_m(D, \mathcal{C})$ is the collection of all $M_{m,\xi}$, ξ mod $2m$, at which the Γ-orbit (74.15) meats $\mathcal{M}_m(D)$. With this, let $M_{m,\xi} \in \mathcal{M}_m(D, \mathcal{C})$ and $Q \in \mathcal{C}$ be taken arbitrarily. Then there exists a $U \in \Gamma$ such that ${}^t U Q U = M_{m,\xi}$, i.e., $U \in \mathcal{U}_Q(m)$. Hence, the configuration (74.7) is now refined: we have a chain of bijections

$$\mathrm{Aut}_Q \backslash \mathcal{S}_Q(m) \;\to\; \mathrm{Aut}_Q \backslash \mathcal{U}_Q(m)/\Gamma_\infty \;\to\; \mathcal{M}_m(D, \mathcal{C}), \tag{74.17}$$

for each $Q \in \mathcal{C}$; note that $\mathcal{M}_m(D, \mathcal{C})$ does not depend on the choice of $Q \in \mathcal{C}$. This diagram will yield a theoretical algorithm (76.10) below to solve (71.2). Also to be indicated in §76, the classification $\mathcal{S}_Q(m)$ modulo Aut_Q becomes more interesting with $D > 0$; a useful criterion for this purpose will be given in Note [76.4].

Pasting (74.17) according to (74.16), we get another chain of bijections

$$\bigsqcup_Q \mathrm{Aut}_Q \backslash \mathcal{S}_Q(m) \;\to\; \bigsqcup_Q \mathrm{Aut}_Q \backslash \mathcal{U}_Q(m)/\Gamma_\infty \;\to\; \mathcal{M}_m(D), \tag{74.18}$$

where Q runs over a set of all representative forms of classes in $\mathcal{K}(D)$. Naturally the pair of the two mappings in (74.18) depends on the choice of representative forms.

Finally, we consider the effect of (73.1) upon the above argument. It is immediate: (74.4) is replaced by

$$(73.1) \;\Rightarrow\; \mathcal{M}_m(D) = \{ M_{m,\xi} : \xi \text{ mod } 2m \}, \tag{74.19}$$

since $\langle m, \xi \rangle = 1$. Consequently, we have, via Theorem 69 and (74.18), thus independently of the reciprocity law,

$$(73.1) \;\Rightarrow\; \sum_Q \big| \mathrm{Aut}_Q \backslash \mathcal{S}_Q(m) \big| = \big| \mathcal{M}_m(D) \big| = \sum_{f \mid m} \mu^2(f) \kappa_D(f), \tag{74.20}$$

on noting that if $\xi^2 \equiv D$ mod $4m$, then $(\xi + 2m)^2 \equiv D$ mod $4m$, so the set $\{\xi$ mod $4m\}$ contains the set $\{\xi$ mod $2m\}$ in double. In the case where (73.1)

$$\S 74 \qquad\qquad 323$$

does not hold, one ought to return to (72.2); a typical situation is discussed in Notes [87.4]–[87.5] below.

In passing, we make here a trivial but crucial observation: with (74.9)

$$U^{-1}\mathrm{Aut}_Q U = \mathrm{Aut}_{{}^t UQU}. \qquad (74.21)$$

On the left side, the Γ-conjugacy of subgroups and on the right side, the equivalence of forms mod Γ are visible. This identity is a key to reveal a geometrical structure of the group $\Gamma/\{\pm 1\}$, as we shall discuss in the later part of §85; in fact, we have already suggested this in Remark 4 above. It will be needed in §85 to be aware about the nature of the Γ-centralizer C_Q of Aut_Q, that is,

$$C_Q = \{C \in \Gamma : C^{-1}\mathrm{Aut}_Q C = \mathrm{Aut}_Q\}; \qquad (74.22)$$

thus

the different members among $\{U^{-1}\mathrm{Aut}_Q U : U \in \Gamma\}$ is

$\{H^{-1}\mathrm{Aut}_Q H : H \in C_Q\backslash\Gamma\}.$ $\qquad (74.23)$

We remark that $\mathrm{Aut}_Q \subset C_Q$, or

the mapping $\mathcal{K}(D) \ni \mathcal{C} \ni Q \mapsto \mathrm{Aut}_Q$

is not injective, in general. $\qquad (74.24)$

This should be compared with (74.14)–(74.15). See Note [74.8].

Notes

[74.1] The present section has its origin in Lagrange (1773). In his view were not only individual forms but also the whole set of quadratic forms, as we remarked already in [71.2]. He relied on (72.3) and observed that if $\det A = \pm 1$, then the new coefficient a' is properly represented by $[|a,b,c|]$. This and the notion of the equivalence (74.10) appeared first in Lagrange (1773, Problémes III–IV), although we have refined his original classification by following Gauss' prescription that U be in Γ. The discussion [DA, arts.155–170] is essentially covered with the present and the succeeding two sections. A small dose of matrix algebra and group theory has made Gauss' massive argument accessible.

[74.2] Gauss [DA, art.155] expressed the situation (72.8) by stating that the proper representation $Q(u,v) = m$ belongs to ξ mod $2m$. Naturally, $Q \equiv Q'$ mod Γ if and only if both have proper representations of a certain integer m which belong to the same ξ mod $2m$, for then both of them are equivalent to $M_{m,\xi}$. The convention (74.12) leads us to the statement: a proper representation of m by a class $\mathcal{C}$ belongs to ξ mod $2m$ if and only if $M_{m,\xi} \in \mathcal{C}$.

324   *Quadratic Forms*

[74.3]   Dirichlet (1851) (see Smith (1862, art.111) and Dedekind (Dirichlet (1871/1894, §60, footnote)) recasted Gauss' thoughts in the preceding Note as follows: That the proper representation $[|a, b, c|](u, v) = m$ belongs to ξ mod $2m$ is equivalent to the congruence

$$\begin{pmatrix} \xi - b & -2c \\ 2a & \xi + b \end{pmatrix} \begin{pmatrix} u \\ v \end{pmatrix} \equiv \begin{pmatrix} 0 \\ 0 \end{pmatrix} \text{ mod } 2m.$$

To confirm, let the right side be $2m \cdot {}^t\{k, l\}$. Then

$$2m(ul - vk) = \big(2au + (\xi + b)v\big)u - \big((\xi - b)u - 2cv\big)v = 2m.$$

Hence, $\binom{u\ k}{v\ l} \in \Gamma$; in particular, $\langle u, v \rangle = 1$, and the representation is proper. Also,

$$0 = \big(2au + (\xi + b)v\big)k - \big((\xi - b)u - 2cv\big)l = 2auk + (vk + ul)b + 2cvl - \xi,$$

which implies (72.8). Conversely, (72.8) gives that

$$(\xi - b)u - 2cv = 2mk, \quad 2au + (\xi + b)v = 2ml,$$

which ends the confirmation. The merit of this congruence is in that it makes possible to discuss without mentioning explicitly the right column of $\binom{u\ k}{v\ l}$, whence it will turn out to be a handy device at the extremely crucial steps (iii)–(v) of §90; the corresponding discussion by Gauss [DA, arts.237–239] is awfully involved. One should be aware, however, that there exists yet a simpler argument depending on Note [74.7] below, as we shall indicate in Note [90.5].

[74.4]   [DA, art.228]. Let $N \neq 0$ be arbitrary, and let $Q = [|a, b, c|] \in \mathfrak{Q}(D)$. Then there exists a form $[|a', b', c'|] \in \mathfrak{Q}(D)$ equivalent to Q such that $\langle a', N \rangle = 1$. This can be shown readily by choosing an r such that $\langle Q(r, 1), N \rangle = 1$ and transforming Q by $\binom{r\ r-1}{1\ 1}$. The detection of such an r is not problematic in practice. However, in order to make the assertion rigorous, we use the decomposition $N = N_1 N_2 N_3$ with

$$N_1 = \frac{N}{\langle N, c^\infty \rangle}, \ N_2 = \frac{\langle N, c^\infty \rangle}{\langle N, \langle a, c \rangle^\infty \rangle}, \ N_3 = \langle N, \langle a, c \rangle^\infty \rangle.$$

Then use $a' = Q(N_1, N_2)$. Indeed, we have $\langle N_j, N_k \rangle = 1, j \neq k$, and $\langle a', N_j \rangle = 1$ (the case $j = 3$ is due to $\langle a, b, c \rangle = 1$). By the way, (Legendre, 1798, p.421) made a suggestion of the same effect; so this useful device seems to have been commonly known before [DA]. It has evolved into a standard device in the ideal theory of algebraic number fields.

[74.5]   Continuing the discussion, we may take a form $[|a', b', c'|] \in \mathfrak{Q}(D)$ with $a' > 0$ and $\langle a', N \rangle = 1$: We remarked in §71 that for each $Q \in \mathfrak{Q}(D)$ there exists a $\{u, v\}$, $\langle u, v \rangle = 1$, such that $Q(u, v) > 0$. This means that we have

$$\S74 \qquad 325$$

a $Q_1 \equiv Q \bmod \Gamma$ with which $Q_1(1,0) > 0$; that is, we may suppose $a > 0$. Returning to the above argument, if $Q(N_1,N_2) < 0$, then in the identity

$$4aQ(N_1 + Nk, N_2) = \big(2a(N_1 + Nk) + bN_2\big)^2 - DN_2^2$$

we let k be so large that the right side becomes positive, i.e., $Q(N_1 + Nk, N_2) > 0$. The rest of the proof is only to observe $Q(N_1 + Nk, N_2) \equiv Q(N_1, N_2) \bmod N$.

[74.6]   With a practical purpose in mind, we make it precise, though trivial, that the problem $Q(x,y) = m < 0$ with $Q = [|a,b,c|] \in \mathfrak{Q}(D)$, $a > 0, D > 0$, can be discussed in the following way: Consider $-Q(x,y) = |m|$ instead. Then take $Q' = [|a',b',c'|] \equiv -Q \bmod \Gamma$ with $a' > 0$ by means of the preceding Note. Consider $Q'(x,y) = |m|$, and return to $Q(x,y) = m$. Also, if $\mathrm{pell}_D(-4)$ is solvable, then there is an alternative way to take; namely, then it becomes enough to consider $Q(x,y) = |m|$. See (85.18).

[74.7]   In terms of the matrix–module representation (§72) of quadratic forms, the equivalence (74.10) is rendered in the following way: Let $Q = [|a,b,c|]$, $Q' = [|a',b',c'|] \in \mathfrak{Q}(D)$ correspond to the $\mathbf{I}_D$-ideals $\mathbf{Q}, \mathbf{Q}'$, respectively. Then, with $\mathfrak{d}^2 = D\mathfrak{e}$,

$$Q \equiv Q' \bmod \Gamma \quad \Leftrightarrow \quad \begin{array}{l} \exists\, \mathfrak{r} \in [\mathfrak{e}, \mathfrak{d}]_{\mathbb{Q}} \text{ such that } \det \mathfrak{r} = a/a' \text{ and} \\ \mathbf{Q} = \mathfrak{r} \cdot \mathbf{Q}', \end{array}$$

where $\mathfrak{r}$ is a linear combination of $\mathfrak{e}, \mathfrak{d}$ with coefficients in $\mathbb{Q}$. We show the sufficiency first. Thus, suppose that the equality of the two modules holds. Then, the pair $\{a'\mathfrak{r}, \frac{1}{2}(b'\mathfrak{e} + \mathfrak{d})\mathfrak{r}\}$ is a basis of the Abelian group $\mathbf{Q}$. Hence, there exists an integral matrix A such that $\det A = \pm 1$ and

$$\left[a\mathfrak{e}, \tfrac{1}{2}(b\mathfrak{e} + \mathfrak{d}) \right] \cdot A = \mathfrak{r} \cdot \left[a'\mathfrak{e}, \tfrac{1}{2}(b'\mathfrak{e} + \mathfrak{d}) \right].$$

We write this as $[\mathfrak{a}_1, \mathfrak{b}_1] = [\mathfrak{a}_2, \mathfrak{b}_2]$, and compute the both sides of $\mathfrak{a}_1 \cdot \overline{\mathfrak{b}}_1 - \overline{\mathfrak{a}}_1 \cdot \mathfrak{b}_1 = \mathfrak{a}_2 \cdot \overline{\mathfrak{b}}_2 - \overline{\mathfrak{a}}_2 \cdot \mathfrak{b}_2$, where the bar is as in Note [72.8]. We find that $-a(\det A)\mathfrak{d} = -a'(\det \mathfrak{r})\mathfrak{d}$, that is, $\det A = +1$. Hence,

$$\left[a\mathfrak{e}, \tfrac{1}{2}(b\mathfrak{e} + \mathfrak{d}) \right] \cdot U \cdot {}^{\mathrm{t}}(x,y) = \mathfrak{r} \cdot \left[a'\mathfrak{e}, \tfrac{1}{2}(b'\mathfrak{e} + \mathfrak{d}) \right] \cdot {}^{\mathrm{t}}(x,y), \quad \exists\, U \in \Gamma.$$

Taking determinants on both sides, we find that

$$a({}^{\mathrm{t}}UQU)(x,y) = (\det \mathfrak{r})a'Q'(x,y),$$

which verifies the sufficiency. Conversely, suppose that we have $(72.3)_{A=U}$, $U = \begin{pmatrix} u & k \\ v & l \end{pmatrix} \in \Gamma$, so that ${}^{\mathrm{t}}UQU = Q'$. We have

$$\left[a\mathfrak{e}, \tfrac{1}{2}(b\mathfrak{e} + \mathfrak{d}) \right] \cdot U = \left[\mathfrak{a}, \mathfrak{b} \right],$$
$$\mathfrak{a} = au\mathfrak{e} + \tfrac{1}{2}v(b\mathfrak{e} + \mathfrak{d}), \quad \mathfrak{b} = ak\mathfrak{e} + \tfrac{1}{2}l(b\mathfrak{e} + \mathfrak{d}).$$

326   *Quadratic Forms*

A computation gives that $\mathfrak{b}/\mathfrak{a} = \mathfrak{b}\bar{\mathfrak{a}}/aa' = (1/2a')(b'\mathfrak{e} + \mathfrak{d})$, and we find that

$$\left[a\mathfrak{e}, \tfrac{1}{2}(b\mathfrak{e} + \mathfrak{d})\right] \cdot U = \frac{\mathfrak{a}}{a'} \cdot \left[a'\mathfrak{e}, \tfrac{1}{2}(b'\mathfrak{e} + \mathfrak{d})\right],$$

which ends the verification, as $\det(\mathfrak{a}/a') = a/a'$. See Hecke (1923, Theorem 154).

[74.8]   For instance, the form $Q = [|2, 7, -2|] \in \mathfrak{Q}(65)$ has the property:

$$^t WQW = -Q,$$

with $W \in \Gamma$ as in (5.1). Hence, two different members in a class of $\mathcal{K}(65)$ share the same set of automorphisms. In the later part of §85, we shall apply a refinement to the map (74.24) so that an injective map becomes available.

§75

We ought to be constantly attentive to what is stated in §72(c) and Note [74.1], although the reason for this will be fully revealed only in §§90–91.

Thus, making precise Remark 2 of the preceding section, Lagrange considered the transformation (74.9) not by the elements of $\mathrm{SL}(2, \mathbb{Z}) = \Gamma$ but, in fact, by those of $\mathrm{GL}(2, \mathbb{Z}) = \Gamma \sqcup \diamond \Gamma$, $\diamond = \left(\begin{smallmatrix} 1 & 0 \\ 0 & -1 \end{smallmatrix}\right)$. The same was practiced by Legendre. Gauss refined their thought by calling forms Q_1, Q_2 properly equivalent if there exists a $U \in \Gamma$ such that $^t U Q_1 U = Q_2$, and forms Q'_1, Q'_2 improperly equivalent if there exists a $K \in \diamond \Gamma$ such that $^t K Q'_1 K = Q'_2$. The discussion in §72(c) can be rephrased by means of these two terms. In the preceding section, we said just equivalent in place of Gauss' term properly equivalent. We shall keep this practice of ours throughout the rest of the present chapter since it suffices for us to make it known only when dealing with improper equivalence.

With this, if $Q(u, v) = m$ is a proper representation, then $(\diamond Q \diamond)(u, -v) = m$ is so; thus, the forms Q and $\diamond Q \diamond$, which are of the same determinant, properly represent the same set of integers. However, these two improperly equivalent forms are not equivalent in general. More precisely, with (72.8), let $\tilde{Q} = {}^t S Q S$, $S \in \diamond \Gamma$, $S^{-1} U = \left(\begin{smallmatrix} u' & k' \\ v' & l' \end{smallmatrix}\right)$. Then $\tilde{Q} \in \mathfrak{Q}(D)$, and we get the proper representation $\tilde{Q}(u', v') = m$. This representation of m by $\tilde{Q}$ belongs to $-\xi \bmod 2m$ in the sense of Note [74.2] since

$$^t(S^{-1}U\diamond)\tilde{Q}S^{-1}U\diamond = \diamond M_{m,\xi}\diamond = M_{m,-\xi}, \quad S^{-1}U\diamond \in \Gamma. \tag{75.1}$$

Thus, if $M_{m,\xi}$ and $M_{m,-\xi}$ are not equivalent, then Q and $\tilde{Q}$ are not either. On the other hand, if $M_{m,\xi}$ and $M_{m,-\xi}$ are equivalent, then Q and $\tilde{Q}$ are so, and there

§75   327

should exist an $L \in \diamond\Gamma$ such that ${}^t LQL = Q$. This L is like an automorphism of Q but not in Γ, i.e., not in Aut_Q; see Note [76.5] below. The form Q is improperly equivalent to itself. Thus, there can exist situations in which the improper equivalence coincides with the ordinary one: If Q, L are as above and $Q_1 = {}^t UQU$ with any $U \in \Gamma$, then $Q_1 = {}^t(LU)Q(LU)$ with $LU \in \diamond\Gamma$. Namely, all forms in the class to which Q belongs are improperly equivalent to Q as well; moreover, each form in the class is improperly equivalent to itself, i.e., ${}^t(U^{-1}LU)Q_1(U^{-1}LU) = Q_1$.

Gauss' own view [DA, art.164] on this intricacy concerning equivalence of forms is as follows: He [DA, art.163] begins with the definition

$$[|a, b, c|] \text{ is an ambiguous form } (\textit{forma anceps}) \iff a | b. \tag{75.2}$$

Theorem 73 *The following two definitions* (i) *and* (ii) *are equivalent:*

$$\begin{array}{ll} & \text{(i) containing an ambiguous form,} \\ \text{ambiguous class} & \\ : & \text{(ii) containing a form improperly} \quad\quad (75.3) \\ (\textit{classis anceps}) & \\ & \text{equivalent to itself.} \end{array}$$

Proof   We shall show first that (i) implies (ii). Thus, let $Q = [|a, b, c|]$ be such that $b = ka$, $k \in \mathbb{Z}$; we take $K = \left(\begin{smallmatrix} 1 & k \\ 0 & -1 \end{smallmatrix}\right) \in \diamond\Gamma$; then ${}^t KQK = Q$, which implies (ii). Conversely, suppose (ii); and let $Q = [|a, b, c|]$ be such that ${}^t AQA = Q$, $A \in \diamond\Gamma$. We have

$$A = \begin{pmatrix} s & t \\ u & -s \end{pmatrix}, \quad s^2 + tu = 1. \tag{75.4}$$

To confirm this, let $A = \left(\begin{smallmatrix} s & t \\ u & v \end{smallmatrix}\right)$; then

$$(1) \quad sv - tu = -1,$$
$$(2) \quad a = as^2 + (bs + cu)u, \tag{75.5}$$
$$(3) \quad b = 2ast + b(sv + tu) + 2cuv.$$

From (1) and (3) follows $ast + (bs + cu)v = 0$. Combining this and (2), we get $(s^2 - 1)v - stu = 0$, which and (1) give $v = -s$. Hence, (75.4) has been confirmed. In particular, if $u = 0$, then $s = \pm 1$; and (3) gives $b = \pm at$, so Q is an ambiguous form. As for the case $u \neq 0$, let $R = \left(\begin{smallmatrix} \sigma & \tau \\ \lambda & \mu \end{smallmatrix}\right) \in \Gamma$. Then

$$B = R^{-1}AR = \begin{pmatrix} * & * \\ \ell & * \end{pmatrix}, \quad \ell = u\sigma^2 - 2s\sigma\lambda - t\lambda^2. \tag{75.6}$$

We see that $B \in \diamond\Gamma$ satisfies ${}^t B({}^t RQR)B = {}^t RQR$. Thus, if $\ell = 0$, then the reduction to the case $u = 0$ will be attained. For this sake, note that (75.4)

328 *Quadratic Forms*

yields $u\ell = (u\sigma - s\lambda)^2 - \lambda^2$. We choose coprime σ, λ such that $u\sigma = (s+1)\lambda$. We end the proof.

Let $Q \in \mathfrak{Q}(D)$ belong to an ambiguous class, and let the representation (74.1) hold. Then in (72.8) we have $M_{m,-\xi} \equiv M_{m,\xi}$ mod Γ as asserted above. Hence, the class $\mathcal{C}$ represented by Q has both $M_{m,\xi}$ and $M_{m,-\xi}$ as elements of $\mathcal{M}_m(D, \mathcal{C})$. Here it should be noted that if $m|\xi$, then only $M_{m,\xi}$ is to be taken, i.e., appears in $\mathcal{M}_m(D)$; however, note also that under (73.1) such a situation never takes place. If Q is not in an ambiguous class, then $M_{m,\xi}$ and $M_{m,-\xi}$ are in different classes. If $\mathrm{GL}(2,\mathbb{Z})$ were employed, in (74.10), instead of $\Gamma = \mathrm{SL}(2,\mathbb{Z})$, then this resolution, i.e., the ability of watching at $M_{m,\xi}$ and $M_{m,-\xi}$ as being well separated, would be lost, and thence the algebraic structure of $\mathfrak{Q}(D)$ to be described in §§90–91 would not be captured, albeit certainly premature to state this.

Nevertheless, we stress that transformations of forms by matrices in $\mathrm{GL}(2,\mathbb{Z})$ will play a basic rôle in our account of the theory of indefinite forms which we shall develop in §§82–87 following closely the tradition established by Lagrange and Legendre.

Notes

[75.1]   The set of all ambiguous classes of $\mathcal{K}(D)$ will turn out to constitute a core part of the genus theory (§91). The above proof of Theorem 73 is due to Dirichlet (1857; 1863, §58).

[75.2]   The origin of the term ambiguous form/class: In the first edition of Dirichlet's *Vorlesungen* (1863, p.148) the term *anceps* is employed, following [DA, art.163] and Dirichlet's original lecture. In the second edition (1871, p.136, text and footnote) it is changed into *ambig* adopting Kummer's opinion; it remains the same in the third edition (1879, p.138). In the final (fourth) edition (1894, p.139, text and footnote) the author Dedekind proposes *zwei-seitig*. Also, in the French edition (1807, p.132) of [DA] the term *ambiguë* is employed. By the way, Weber (1882, p.309) called real characters on finite Abelian groups *ambige Charaktere*. This came apparently from the theory of quadratic forms or more precisely, from the genus theory.

[75.3]   A detection of an ambiguous form in $\mathfrak{Q}(D)$ can imply a non-trivial factorization of D since (75.2) means $a|D$. See Note [92.7].

[75.4]   With $Q=[|3,-19,-11|] \in \mathfrak{Q}(493)$,   we   take $U_1 = \begin{pmatrix} 300 & -73 \\ 37 & -9 \end{pmatrix}$, $U_2 = \begin{pmatrix} 324 & 79 \\ 41 & 10 \end{pmatrix}$, and have ${}^tU_1QU_1 = M_{m,-\xi}$, ${}^tU_2QU_2 = M_{m,\xi}$, $m = 44041$ (a prime), $\xi = 21455$. Since $U_1, U_2 \in \Gamma$, we find that $M_{m,-\xi}$ and $M_{m,\xi}$ are equivalent. Hence, Q belongs to an ambiguous class of $\mathcal{K}(493)$; ${}^tVQV = Q$

with $V = \left(\begin{smallmatrix} 7 & -48 \\ 1 & -7 \end{smallmatrix}\right)$, $V \in \diamond \Gamma$. Alternatively, with $C = \left(\begin{smallmatrix} 673 & 103 \\ 98 & 15 \end{smallmatrix}\right) \in \Gamma$ we have ${}^t CQC = [|17, -17, -3|]$, which is an ambiguous form according to (75.2); note that $17|493$. See Note [83.1].

[75.5]   In the notation of the later part of §72,

$$Q \text{ is an ambiguous form } \Leftrightarrow \overline{\mathbf{Q}} = \mathbf{Q}.$$

To see this, we note the relation

$$\left[a\mathfrak{e}, \tfrac{1}{2}(b\mathfrak{e} - \mathfrak{d})\right] = \left[a\mathfrak{e}, \tfrac{1}{2}(b\mathfrak{e} + \mathfrak{d})\right] \cdot A, \quad A = \begin{pmatrix} 1 & b/a \\ 0 & -1 \end{pmatrix}.$$

Thus, if Q is ambiguous, i.e., (75.2), then A is integral, and $\overline{\mathbf{Q}} = \mathbf{Q}$. Conversely, if this module identity holds, then

$$\left[a\mathfrak{e}, \tfrac{1}{2}(b\mathfrak{e} - \mathfrak{d})\right] = \left[a\mathfrak{e}, \tfrac{1}{2}(b\mathfrak{e} + \mathfrak{d})\right] \cdot B$$

with an integral B. It follows that $B = A$, and thus $a|b$. The definition (75.2) is made rather abruptly in [DA]; however, if viewed from modules, it becomes quite natural, as we have seen here. This suggests that Gauss was, in fact, dealing with integral modules behind his characteristically heavy discussion of [DA, Sectio V]. See Note [90.7].

§76

We return to (74.13). An answer to (1) stems from the following fundamental discovery by Lagrange (1773, Théorème III):

Theorem 74 *The $\mathcal{K}(D)$ defined by (74.11) is a finite set.*

Remark 1 We thus introduce

$$\text{the class number: } h_D = |\mathcal{K}(D)|. \tag{76.1}$$

Proof Let $[|a,b,c|]$ be such that $|a|$ is the minimum among all forms in a particular class. Then we may suppose that

$$\text{Lagrange's reduction condition: } |b| \leq |a| \leq |c|. \tag{76.2}$$

is fulfilled. If we have ever $|b| > |a|$, then the transformation by $T^v = \left(\begin{smallmatrix} 1 & v \\ 0 & 1 \end{smallmatrix}\right)$ gives

$$[|a,b,c|] \mapsto [|a,b + 2av, av^2 + bv + c|]. \tag{76.3}$$

330 *Quadratic Forms*

Thus, one may take v so that $|b + 2av| \leq |a|$. Then it can be supposed that $|av^2 + bv + c| \geq |a|$ as well; otherwise transform further by $W = \begin{pmatrix} 0 & -1 \\ 1 & 0 \end{pmatrix}$. It now follows from (76.2) that

$$D < 0 \;\Rightarrow\; |a| \leq \sqrt{|D|/3} \tag{76.4}$$

as well as

$$D > 0 \;\Rightarrow\; |a| \leq \sqrt{D/4}. \tag{76.5}$$

To confirm (76.4), we note that if $D < 0$, then $|D| = 4ac - b^2 \geq 3a^2$. As for (76.5), we have $|ac| \geq b^2 = D + 4ac > 4ac$, so $ac < 0$, and $4a^2 \leq 4|ac| = -4ac = D - b^2 \leq D$. The number of the tuples $\{a,b,c\}$ satisfying either (76.4) or (76.5) together with (76.2) is obviously finite. This ends the proof.

As for (74.15)(2), the answer is contained in [DA, art.162]. The equation $\mathrm{pell}_D(4)$ becomes salient:

Theorem 75 *For any* $Q = [|a,b,c|] \in \mathfrak{Q}(D)$,

$$\mathrm{Aut}_Q = \left\{ \begin{pmatrix} \frac{1}{2}(t - bu) & -cu \\ au & \frac{1}{2}(t + bu) \end{pmatrix} : t^2 - Du^2 = 4 \right\}. \tag{76.6}$$

Proof Since $(t - bu)(t + bu) \equiv t^2 - Du^2 \equiv 0 \bmod 4$, we have $t \pm bu \equiv 0 \bmod 2$; thus, each matrix in (76.6) is in Γ. The confirmation that it is an automorphism of Q is just a routine computation. Conversely, let $V = \begin{pmatrix} \alpha & \beta \\ \gamma & \delta \end{pmatrix}$ be an automorphism of Q. Then by the formulas in (72.3) we have

$$a = a\alpha^2 + b\alpha\gamma + c\gamma^2. \tag{76.7}$$

Also, $b = 2a\alpha\beta + b(1 + 2\beta\gamma) + 2c\gamma\delta$ since $V \in \Gamma$, and thus

$$0 = a\alpha\beta + b\beta\gamma + c\gamma\delta. \tag{76.8}$$

Eliminating b from these identities, we get $a\beta = -c\gamma$. Also, eliminating c, we get $a(\alpha - \delta) = -b\gamma$. Thus $a | \langle \gamma b, \gamma c \rangle$. On noting that $Q \in \mathfrak{Q}(D)$ implies $\langle a, \langle b, c \rangle \rangle = 1$, we find that $a | \gamma$. Hence, with $\gamma = au$, we have $\beta = -cu$, $\alpha - \delta = -bu$. Further, $(\alpha + \delta)^2 = (bu)^2 + 4\alpha\delta = (bu)^2 + 4(1 + \beta\gamma) = Du^2 + 4$. We put $\alpha + \delta = t$, and end the proof.

Remark 2 This proof is an adaptation of Dirichlet (1863, §61); see Theorem 87, §85, for an alternative proof, which as being constructive is, in fact, more satisfactory than the above, especially in the case of positive determinants; moreover, in the same section, we shall show an intrinsic, or geometric relation between $\mathrm{pell}_D(4)$ with variable $D > 0$ and the group $\Gamma/\{\pm 1\}$ itself. In this context, Notes [76.3]–[76.6] below will become crucial.

By the way, in (76.6) it is not supposed that $\langle t, u \rangle = 1$ (see Note [85.1]); also, both of t, u can be either positive or negative.

Namely, the issue (73.13) (2) is equivalent to solving $\mathrm{pell}_D(4)$. When $D < 0$, this is immediate:

$$D = -3 : \{\{\pm 1, \pm 1\}, \{\pm 2, 0\}\}, \quad |\mathrm{Aut}_\varrho| = 6;$$
$$D = -4 : \{\{t, u\} = \{0, \pm 1\}, \{\pm 2, 0\}\}, \quad |\mathrm{Aut}_\varrho| = 4; \qquad (76.9)$$
$$D < -4 : \{\{\pm 2, 0\}\}, \quad |\mathrm{Aut}_\varrho| = 2,$$

counting all combinations of signs. On the other hand, with $D > 0$ the problem becomes highly interesting, as will be observed in §§84–85.

Summing up:

In view of (74.17) and the contents of the present section, it is inferred that (71.2) will be settled, if the following five steps are accomplished: the global issues

(a) to find a system of h_D inequivalent forms, i.e., $\mathcal{K}(D)$,

(b) to check whether two given forms are equivalent or not, $\qquad$ (76.10)

(c) to solve $\mathrm{pell}_D(4)$,

and the local issues under (74.1)

(d) to solve (72.2) and compose $\mathcal{M}_m(D)$,

(e) to identify the seeds of $\mathcal{S}_Q(m)$, i.e., $\mathrm{Aut}_\varrho \backslash \mathcal{S}_Q(m)$. $\qquad$ (76.11)

We shall see:

(a) is solved in §77 for $D < 0$ and in §83 for $D > 0$.

(b) is solved by transforming any given form into one of the representative forms which are identified in step (a). Two forms are equivalent if and only if they are transformed into the same representative. For the case $D < 0$, an explicit procedure to transform a given form in such a way is stated as pre-reduction [+] in §77; and for the case $D > 0$ as pre-reduction [±] in §83.

(c) need not be discussed for $D < 0$ because of (76.9). With $D > 0$, a discussion is developed in §85. The problem will be settled in an explicitly constructive way.

(d) can be hard in general. We are, anyway, able to try at least the algorithms of Tonelli and Cipolla, although there persist the issue of the prime power decomposition of m and the difficulty of finding quadratic non-residues modulo primes.

332 *Quadratic Forms*

(e) will be solved if the elements of $\mathcal{M}_m(D, \mathbb{C})$ in (74.17) are identified. For this sake, we apply (b) to transform Q to a unique form among the representatives fixed by (a). We do the same with each element of $\mathcal{M}_m(D)$ in place of Q, which amounts to making the right side of (74.16), i.e., each $\mathcal{M}_m(D, \mathbb{C})$ explicit. Then, via the transformations involved in this procedure we get simultaneously a set of matrices in $\mathcal{U}_Q(m)$, the left columns of which comprise all the seeds for $\mathrm{Aut}_Q \backslash \mathcal{S}_Q(m)$. This way of solving (e) is theoretical; the existence of a far more practical method is indicated in the next paragraph.

(æ) will make it possible to dispense with the above reduction method, if one is concerned not with the whole structure of $\mathfrak{Q}(D)$ but solely with solving the issue (71.2). Thus, as has been indicated already in Note [71.1] (3), we shall show less demanding methods in §81 for $D < 0$ and in §87 for $D > 0$, which yield special solutions to a particular $Q(x, y) = m$ in a direct way: direct in the sense that (a)–(e) is skipped basically, yet in place of (d) or rather (72.2) the congruence $x^2 \equiv D \bmod 4|a|m$ needs to be solved; observe that the modulus is $4|a|m$, not $4m$. Moreover, in the case $D > 0$ we shall be able to identify all seed solutions mod Aut_Q in §87.

In passing, we augment the last theorem as follows: Suppose that $D > 0$ and

$$\mathrm{pell}_D(-4) \text{ has solutions.} \tag{76.12}$$

With $Q = [|a, b, c|] \in \mathfrak{Q}(D)$ let

$$\mathrm{Aut}_Q^* = \left\{ \begin{pmatrix} \frac{1}{2}(t - bu) & -cu \\ au & \frac{1}{2}(t + bu) \end{pmatrix} : t^2 - Du^2 = -4 \right\} \subset \diamond \Gamma; \tag{76.13}$$

be aware that Aut_Q^* is not a group. Then we have a counterpart of (74.6):

$$^t Y Q Y = -Q, \quad \forall\, Y \in \mathrm{Aut}_Q^*. \tag{76.14}$$

Obviously

$$\mathrm{Aut}_Q^* = Y \cdot \mathrm{Aut}_Q, \quad \forall\, Y \in \mathrm{Aut}_Q^*. \tag{76.15}$$

This plays a core rôle in the theory of indefinite forms, as we shall show in later sections.

Notes

[76.1] Lagrange (1773) did not explicitly state Theorem 74 but indicated a way of transforming given forms so that the resulting forms satisfy (76.2). The reduction method (76.3) is Théorème II, and the combination of (76.2), (76.4), (76.5) corresponds to the corollaries of Théorème III.

§76   333

[76.2] Euler (1733a, p.6; 1758b, p.298) recognized the rôle of $\mathrm{pell}_D(4)$ in the mechanism of the automatic generation of solutions to quadratic indefinite equations; however, see Note [86.1].

[76.3] In terms of the representation (72.25), the action on Q of the matrix in (76.6) is expressed as

$$ax + \tfrac{1}{2}(b + \sqrt{D})y \;\mapsto\; \tfrac{1}{2}(t + u\sqrt{D})\left(ax + \tfrac{1}{2}(b + \sqrt{D})y\right),$$

where $\sqrt{D} > 0$ if $D > 0$, and $\sqrt{D} = i\sqrt{|D|}$ if $D < 0$. The right side is equal to

$$ax' + \tfrac{1}{2}(b + \sqrt{D})y',$$
$$x' = \tfrac{1}{2}(t - bu)x - cuy, \quad y' = aux + \tfrac{1}{2}(t + bu)y;$$

that is, $Q(x', y') = Q(x, y)$. On the other hand, with the notion of matrix module, i.e., (72.21),

$$t^2 - Du^2 = 4 \;\Rightarrow\; \tfrac{1}{2}(t\varepsilon + u\eth) \cdot \mathbf{Q} = \mathbf{Q}.$$

For the verification, note that (72.23), after a change of variables, gives

$$\tfrac{1}{2}(t\varepsilon + u\eth) \cdot [a\varepsilon, \tfrac{1}{2}(b\varepsilon + \eth)] = [a\varepsilon, \tfrac{1}{2}(b\varepsilon + \eth)] \cdot V_Q(u, v),$$

where $V_Q(u, v)$ is the matrix in (76.6).

[76.4] Here is a useful way to express the automorphism relation in $S_Q(n)$, $Q = [\,|a, b, c|\,]$, in terms of congruence: We let

$$\langle a, n \rangle = 1,$$

but do not suppose $(73.1)_{m=n}$. Let $S_Q(n) \ni {}^t\{\alpha, \beta\}, {}^t\{\alpha', \beta'\}$, so in particular $\langle \beta\beta', n \rangle = 1$. Then,

$$(\dagger) \qquad \alpha\beta^{-1} \equiv \alpha'\beta'^{-1} \bmod |n| \;\Leftrightarrow\; {}^t\{\alpha, \beta\} \equiv {}^t\{\alpha', \beta'\} \bmod \mathrm{Aut}_Q,$$

where the equivalence on the right side is in the sense of (74.8). For the verification, we invoke the relation

$$\left(a\alpha\varepsilon + \tfrac{1}{2}\beta(b\varepsilon + \eth)\right) \cdot \left(a\alpha'\varepsilon + \tfrac{1}{2}\beta'(b\varepsilon + \eth)\right)^{-1} = \frac{1}{2n}(k\varepsilon + \ell\eth),$$
$$k = 2(\alpha, \beta) \cdot Q \cdot {}^t(\alpha', \beta'), \quad \ell = \alpha\beta' - \beta\alpha', \quad \eth^2 = D\varepsilon,$$

which follows from (72.19) and (72.20). If the left side of $(\dagger)$ holds, then $\ell \equiv 0$ as well as $k \equiv 2\beta'\beta^{-1}Q(\alpha, \beta) \equiv 0 \bmod |n|$. Thus, the right side of the first line of the last expression is equal to $\tfrac{1}{2}(t\varepsilon + u\eth)$, $t, u \in \mathbb{Z}$. Taking the determinants of both sides, we find that $t^2 - Du^2 = 4$. Hence, by the last identity of the preceding Note we get

$${}^t\{\alpha, \beta\} = V_Q \cdot {}^t\{\alpha', \beta'\}.$$

334   *Quadratic Forms*

Conversely, if the right side of (†) holds, then multiply the last equation of the preceding Note by ${}^t\{\alpha, \beta\}$ from the right, and see that $k, \ell \equiv 0 \bmod |n|$ in the first line of the last expression, which ends the verification.

[76.5]   It should be added that if Aut^*_Q is as in (76.13) and if ${}^t\{\alpha, \beta\} \in S_Q(n)$ and ${}^t\{\alpha', \beta'\} \in S_Q(-n)$ with $\langle a, n \rangle = 1$ as above, then we have

$$(\dagger\dagger) \qquad \alpha\beta^{-1} \equiv \alpha'\beta'^{-1} \bmod |n| \ \Leftrightarrow \ {}^t\{\alpha, \beta\} \equiv {}^t\{\alpha', \beta'\} \bmod \mathrm{Aut}^*_Q,$$

with an obvious abuse of notation on the right side, as Aut^*_Q is not a group. Note that the left side of (††) gives, in fact, the condition for the existence of Aut^*_Q or rather the solutions to $\mathrm{pell}_D(-4)$. Continued to the later part of §85. This and the preceding Notes have important applications in §§87 and 93.

[76.6]   Let $V_Q(t, u)$ be as in Note [76.3]. Then we have, for $U \in \Gamma$,

$$U^{-1} V_Q(t, u) U = V_{{}^t U Q U}(t, u),$$

which is, of course, a refinement of (74.21). To prove this, we use the expression

$$V_Q(t, u) = \tfrac{1}{2} t \mathfrak{e} + u \mathbf{i} Q, \quad \mathbf{i} = \mathfrak{d}, \ d = -1,$$

where $\mathbf{i}Q$ is a matrix product; $\mathbf{i} = W$ in the notation of (5.1). Since $U^{-1}\mathbf{i} = \mathbf{i}\,{}^t U$, we get

$$U^{-1} V_Q(t, u) U = \tfrac{1}{2} t \mathfrak{e} + u \mathbf{i}\,{}^t U Q U.$$

§77

Now, in the case of positive definite forms, we shall make the reduction argument (76.10)–(76.11) practical in order to constructively solve (71.2). In the present section, the convention Note [5.1] is in effect.

Thus, in the present and the next four sections, we shall be concerned solely with the set

$$\mathcal{Q}_+(D) = \Big\{ [|a, b, c|] \in \mathcal{Q}(D) : a > 0 \Big\}, \text{ with } D < 0. \tag{77.1}$$

As is readily seen, the restriction to $\mathcal{Q}_+(D)$ of the discussion in the last six sections including those associated Notes induces no essential changes. For instance, the transformation (74.9) and consequently the equivalence (74.10) can be thus restricted; specifically, we have $\mathcal{M}_m(D) \subset \mathcal{Q}_+(D)$ more precisely than (74.4), which implies that (74.18) holds as it is.

However, we need to modify (74.11) and consequently (76.1) as well. On noting that $\mathcal{Q}(D) = \mathcal{Q}_+(D) \sqcup (-\mathcal{Q}_+(D))$ and that (72.3) implies that if $Q \in \mathcal{Q}_+(D)$, then ${}^t U Q U \in \mathcal{Q}_+(D)$ for any $U \in \Gamma$, we introduce

$$\mathcal{K}_+(D) = \mathcal{Q}_+(D)/\!\!/\Gamma, \quad \mathfrak{h}_+(D) = |\mathcal{K}_+(D)|. \tag{77.2}$$

We have

$$D < 0 \implies h_D = 2\mathfrak{h}_+(D). \tag{77.3}$$

With this, we shall first solve (76.10) (a) and (b) by showing a method to identify a representative form for each class in $\mathcal{K}_+(D)$. It is possible to achieve this with Lagrange's criterion (76.2) or rather by following its proof. We shall, however, show an alternative criterion which is of a higher resolution. Indeed it is sharp, that is, it enables one to exactly fix the full set of representative forms. It is due to Gauss [DA, art.171]. However, in our discussion we shall follow later thoughts by Dedekind and others.

Thus, we consider the correspondence

$$\mathcal{Q}_+(D) \ni Q = [|a,b,c|] \;\mapsto\; \omega(Q) = \frac{-b + i\sqrt{|D|}}{2a}; \tag{77.4}$$

note that $|\omega(Q)|^2 = c/a$. This complex number or point is on

$$\text{the upper-half plane: } \mathcal{H} = \{z \in \mathbb{C} : \operatorname{Im} z > 0\}. \tag{77.5}$$

We note that

$$(77.4) \text{ is injective.} \tag{77.6}$$

To confirm, take two forms $Q_v = [|a_v, b_v, c_v|] \in \mathcal{Q}_+(D)$, $v = 1,2$, and suppose that $\omega(Q_1) = \omega(Q_2)$. Comparing the imaginary parts of these points, we get $a_1 = a_2$; and the real parts give $b_1 = b_2$, so $c_1 = c_2$ as well, which ends the confirmation. Also, with $U = \begin{pmatrix} \alpha & \beta \\ \gamma & \delta \end{pmatrix} \in \Gamma$, we have

$$U^{-1}\big(\omega(Q)\big) = \omega\big({}^{\mathrm{t}}UQU\big); \tag{77.7}$$

Here the convention Note [5.1] (3) becomes relevant; so

$$U(z) = \frac{\alpha z + \beta}{\gamma z + \delta}, \quad z \in \mathbb{C}; \tag{77.8}$$

and

$$U^{-1}(z) = \frac{\delta z - \beta}{-\gamma z + \alpha}. \tag{77.9}$$

To see (77.7), it suffices to note that ${}^{\mathrm{t}}\tilde{\omega}Q\tilde{\omega} = 0$ with $\tilde{\omega} = {}^{\mathrm{t}}\{\omega(Q), 1\}$, and

$$
{}^{\mathrm{t}}\tilde{\eta}\big({}^{\mathrm{t}}UQU\big)\tilde{\eta} = 0, \quad \tilde{\eta} = U^{-1}\tilde{\omega}, \tag{77.10}
$$

where U^{-1} is the matrix inverse of U.

336   *Quadratic Forms*

As is well-known, the elements of Γ, in fact $\Gamma/\{\pm 1\}$, can be regarded as a motion in $\mathcal{H}$, for (77.8) implies that

$$\operatorname{Im} U(z) = \frac{\operatorname{Im} z}{|\gamma z + \delta|^2}. \tag{77.11}$$

The most basic fact about this action of Γ on $\mathcal{H}$ is that it is discontinuous; that is, it induces a tessellation of $\mathcal{H}$. To make it precise, we introduce the fundamental domain $\mathcal{F}$ of Γ:

$$\mathcal{F} = \left\{ z \in \mathcal{H} : |z| > 1, |x| < \tfrac{1}{2} \right\} \sqcup \partial,$$

$$\partial = \left\{ \operatorname{Re} z = -\tfrac{1}{2}, \ \operatorname{Im} z \geq \tfrac{1}{2}\sqrt{3} \right\} \cup \left\{ |z| = 1, \ -\tfrac{1}{2} \leq \operatorname{Re} z \leq 0 \right\}. \tag{77.12}$$

Also, following (74.9)–(74.10), we make the definition:

$$z_1 \equiv z_2 \mod \Gamma \ \Leftrightarrow \ U(z_1) = z_2, \ \exists U \in \Gamma. \tag{77.13}$$

Theorem 76 *The family of sets $\left\{ U\mathcal{F} : \forall U \in \Gamma \right\}$ covers $\mathcal{H}$ completely*:

$$\mathcal{H} = \bigcup_{U \in \Gamma} U\mathcal{F}, \text{ i.e., } \mathcal{F} = \Gamma \backslash \mathcal{H}. \tag{77.14}$$

Overlapping occurs only at the Γ images of the points $\left\{ i, \ \tfrac{1}{2}(-1 + i\sqrt{3}) \right\}$.

Proof Let $z \in \mathcal{H}$ be arbitrary. We consider the set $\left\{ \operatorname{Im} U(z) : \forall U \in \Gamma \right\}$, where $U(z)$ is generically expressed as in (77.8). There should exist a maximum in this set, for the set $\{\gamma z + \delta\}$ is contained in the lattice generated by z and 1, and it is obvious that the denominator in (77.11) attains a positive minimum. Let $z_0 = x_0 + iy_0$ have the maximum imaginary part in this context. Applying T^ν appropriately we may assume that $-\tfrac{1}{2} \leq x_0 < \tfrac{1}{2}$. Also, on noting that $\operatorname{Im} W(z_0) = y_0/|z_0|^2 \leq y_0$, we see that we can assume $|z_0| \geq 1$ as well. With this, if $|z_0| > 1$, then $z_0 \in \mathcal{F}$. If $|z_0| = 1$, then either z_0 or $W(z_0)$ is on ∂. We have shown (77.14). Next, assume that both $z_1 = x_1 + iy_1$ and $U(z_1) = x_2 + iy_2$ are contained in $\mathcal{F}$. We may suppose that $\gamma \geq 0$ and $y_2 \geq y_1$. Since this means that $|\gamma z_1 + \delta| \leq 1$, we have $\gamma y_1 \leq 1$. We have also $y_1 \geq \tfrac{1}{2}\sqrt{3}$, so $\gamma = 0$ or 1. If $\gamma = 0$, then we may suppose that $\alpha = \delta = 1$, and $U(z) = z + \beta$. If $\beta \neq 0$, then $U(z_1) \notin \mathcal{F}$. Hence we have $\beta = 0$, and U is the identity transformation. As for the case $\gamma = 1$, we have $|\delta| \leq 1$, for $|z_1 + \delta| \leq 1$. Since the case $\delta = -1$ is excluded obviously, we have either $\delta = 0$ or 1. If $\delta = 0$, then $|z_1| = 1$ and $U(z) = \alpha - 1/z$. If $\alpha \neq 0$, then $U(z_1) \notin \mathcal{F}$, and this is a contradiction. Hence $z_2 = -1/z_1$; that is $z_1 = i$, and $z_1 = z_2 \in \partial$. On the other hand, if $\delta = 1$, then $z_1 = \tfrac{1}{2}\left(-1 + i\sqrt{3} \right)$ and $U(z) = \alpha - 1/(z + 1)$, which gives $\alpha = 0$, and thus $z_1 = z_2 \in \partial$. This ends the proof.

$$\S 77 \qquad\qquad 337$$

We are now able to refine Theorem 74 under (77.1). We note first that (77.6)–(77.7) implies that in $\mathfrak{Q}_+(D)$

$$Q_1 \equiv Q_2 \bmod \Gamma \iff \omega(Q_1) \equiv \omega(Q_2) \bmod \Gamma, \qquad (77.15)$$

where Note [5.1] is in effect. Therefore, as a corollary of the preceding theorem, we have:

Theorem 77

$$\text{Each class in } \mathcal{K}_+(D) \text{ is represented} \atop \text{by a unique form } Q \in \mathfrak{Q}_+(D) \text{ such that } \omega(Q) \in \mathcal{F}. \qquad (77.16)$$

Namely, in the case of positive definite forms, the condition (76.2) of Lagrange can be replaced by the sharper

$$\text{Gauss' condition for reduced forms of } \mathfrak{Q}_+(D):$$
$$\text{either } -a < b \le a < c \text{ or } 0 \le b \le a = c, \qquad (77.17)$$
$$\text{as well as } \langle a,b,c \rangle = 1.$$

That is, we shall call a form $[|a,b,c|]$ reduced if it satisfies this.

Remark The condition $\langle a,b,c \rangle = 1$ should not be dropped from (77.17). For there exists the possibility that a particular $\{a,b,c\}$, $D = b^2 - 4ac$, satisfies the required inequalities but not the coprimality.

We have obtained

Theorem 78 *A solution to the step* (76.10) (a) *for $D < 0$:*

$$\text{the } \mathfrak{h}_+(D) \text{ equals the number} \atop \text{of all reduced forms of discriminant } D. \qquad (77.18)$$

In other words, for any positive definite form, there must exist an element of Γ that transforms it into a unique reduced form. In order to find such an element, it is expedient to use Theorem 4, while noting

$$W : [|a,b,c|] \mapsto [|c, -b, a|],$$
$$T^v : [|a,b,c|] \mapsto [|a, 2av + b, av^2 + bv + c|]. \qquad (77.19)$$

Pre-reduction [+]:
Take an arbitrary $Q = [|a,b,c|] \in \mathfrak{Q}_+(D)$.

(1) Suppose first that $a \ne c$. Applying W if necessary, we may assume that $0 < a < c$; thus, $|\omega(Q)| > 1$. If $\omega(Q) \notin \mathcal{F}$, then $b \notin (-a,a]$. Applying T^v to Q, we have $Q' = [|a,b',c'|]$, $b' \in (-a,a]$. If $\omega(Q') \notin \mathcal{F}$, then $|\omega(Q')| \le 1$. If the equality holds, then apply W. If $|\omega(Q')| < 1$, then $c' < a$. We repeat the procedure.

(2) With the case $a = c$, if $b \in (-a, a]$ but $\omega(Q) \notin \mathcal{F}$, then apply W. If $b \notin (-a, a]$, then applying either T or T^{-1}, the resulting Q' satisfies $|\omega(Q')| < 1$. We proceed to (1).

(3) Within a finite number of steps, we should reach a point in $\mathcal{F}$. To see this, we note that the coefficient of x^2 is not increasing with the above procedure while not less than 1; and it becomes a minimum eventually. Thus, we may assume that $Q = [|a, b, c|]$ is such that a is a minimum. If we still have $\omega(Q) \notin \mathcal{F}$, then in the case of (1) we have $a < c$, i.e., no need to apply W. The form Q' satisfies $|\omega(Q')| \geq 1$ because of the minimality of a. If $|\omega(Q')| > 1$, then the procedure ends. Otherwise, we have $|\omega(Q')| = 1$ and shall need at most one application of W to finish. In the case of (2) we do not even need to apply $T^{\pm 1}$.

Therefore, in the case of positive definite forms, task (71.2) can be solved completely by the combination of Theorem 78 (or rather (77.17)) and the pre-reduction [+], of course provided the solutions to (72.2) are known. Details will be inferred from numerical examples to be discussed in the next three sections.

Notes

[77.1] The notion of the fundamental domain of Γ is due to Dedekind (1877b, §2), including the above proof. A similar geometric idea is in Gauss (Werke III, pp.477–478; VIII, pp.102–105 (noted by Fricke)). As for the drawing of the tessellation (77.14), which has become familiar since then, see Klein (1890, Zweiter Abschnitt, Zweites Kapitel).

[77.2] The fundamental domain $\mathcal{F}$ has a highly interesting relevance to the classification of indefinite forms as well. See the closing paragraph of §82.

[77.3] The condition (77.17) and the consequence (77.18) are contained in Gauss' discussion [DA, arts.171–182].

[77.4] The pre-reduction [+] is essentially due to Lagrange (1773, Probléme III). By this we do not mean that he used the fundamental domain $\mathcal{F}$ but that the idea of searching for a minimum leading coefficient a as indicated in the step (3) is indeed due to him (see also Lagrange (1798, pp.61–63)). If we describe his original reduction procedure in terms of a combination of Gauss' equivalence notion (74.10) and its visualization by Dedekind (the proof of Theorem 76), then we are led to the pre-reduction [+].

[77.5] The essence of the pre-reduction [+] is to find non-trivial minimum values of positive definite quadratic forms, as indicated above. In relation to this, Legendre (1798, pp.69–76) found that if $Q = [|a, b, c|]$ satisfies the condition (76.2), then the first, the second, and the third minimums of the set $\{Q(x, y) : \langle x, y \rangle = 1\}$ are equal to a, c, $a - |b| + c$, respectively, including equalities among these; see also Humbert (1915). It should be noted that with any positive definite form, even if of a quite simple shape, it is extremely hard to exactly locate the kth minimum for a given k. Continued in Note [78.12].

[77.6] Theorem 76 yields the following useful assertion on the basis of subgroups of rank 2 in the free Abelian group $\mathbb{Z}^2$: In the situation of Theorem 30, §29, we may take the basis $\{\mathbf{u}, \mathbf{v}\}$ in such a way that

$$\text{the acute angle between } \mathbf{u}, \mathbf{v} \text{ is in the interval } \left[\tfrac{1}{3}\pi, \tfrac{2}{3}\pi\right].$$

Let $\{\mathbf{u}, \mathbf{v}\} = \{{}^{\text{t}}\{a, b\}, {}^{\text{t}}\{c, d\}\}$ be an arbitrary basis. Let $u = a + bi, v = c + di$. We may assume naturally that $u/v \in \mathcal{H}$. Then take a $U = \begin{pmatrix} \alpha & \beta \\ \gamma & \delta \end{pmatrix} \in \Gamma$ so that $U(u/v) \in \mathcal{F}$. We have $\tfrac{1}{3}\pi \leq \arg(U(u/v)) \leq \tfrac{2}{3}\pi$. The acute angle between $\alpha \mathbf{u} + \beta \mathbf{v}$ and $\gamma \mathbf{u} + \delta \mathbf{v}$ is equal to $\arg(U(u/v))$. See Siegel (1957, Chapter 2, §7).

[77.7] Rabinowicz (1913). Let k be a prime. The values $f_k(j)$ for $1 \leq j \leq k - 1$ of the polynomial $f_k(x) = x^2 - x + k$ are all primes if and only if $\mathfrak{h}_+(1 - 4k) = 1$. Hence, Euler's observation given in Note [11.7] is equivalent to $\mathfrak{h}_+(-163) = 1$. As for the discussion on the general case, we may suppose that $k \geq 7$ since the cases $k = 2, 3, 5$ are easy to verify. With this, we assume that $\mathfrak{h}_+(1 - 4k) = 1$. If $f_k(n)$ is composite for an $n \in [1, k - 1]$, then it has a prime factor $p \leq \sqrt{f_k(n)} < k$ (Note [10.6]). Also, since $f_k(n)$ can be regarded as being properly represented by the form $[|1, -1, k|]$, according to Lagrange's theorem (Note [72.2]) this p is represented by a form in $\mathcal{Q}_+(1 - 4k)$. By the current assumption, the form $[|1, -1, k|]$ itself represents p, so there exist u, v such that $(2u - v)^2 + (4k - 1)v^2 = 4p < 4k - 1$. This holds only with $v = 0, u = \pm 1$; and we are led to the contradiction $p = 1$. Next, assume that $\mathfrak{h}_+(1 - 4k) \geq 2$. Then, there exists a form $[|a, b, c|] \in \mathcal{Q}_+(1 - 4k)$ which is not equivalent to $[|1, -1, k|]$ and satisfies $a \geq 2$ (see Remark 2, §74); thus, by (76.4) (or the first line of (77.17)), we may assume that $2 \leq a \leq ((4k - 1)/3)^{1/2}$. Further, $X^2 \equiv 1 - 4k \bmod 4a$ should have a solution. If $2|a$, then we would have $X^2 \equiv -3 \bmod 8$, which is a contradiction. If $2 \nmid a$, then since X should be odd there exists a g such that $(2g - 1)^2 \equiv 1 - 4k \bmod 4a$ and $1 \leq g \leq a$. It is implied that $f_k(g) \equiv 0 \bmod a$. On noting that $g \leq ((4k - 1)/3)^{1/2} < k$, if $f_k(g)$ is composite, then we end the proof. If $f_k(g)$ is a prime, then $a = f_k(g)$, and $f_k(a + g) = a(a + 2g)$ is composite; moreover $a + g \leq 2((4k - 1)/3)^{1/2} < k$. This ends the proof. In his original proof, Rabinowicz employed the theory of quadratic number fields; we are unable to find to whom the present proof should be attributed. See Granville and Mollin (2000) for a further discussion.

340　　　　　　　　　　　　　*Quadratic Forms*

§78

We shall now apply the concluding assertion at the end of the text of the preceding section to the problem of expressing a given prime p as a sum of two squares. Despite its outward simplicity, this problem contains most of the essential features of the task (71.2) in the case of positive definite forms. We shall proceed with the reciprocity law, although its use is not totally indispensable as is to be remarked in Note [78.3].

We may assume that $p \geq 3$. The relevant quadratic form is $[|1,0,1|] \in \mathfrak{Q}_+(-4)$. If $Q \in \mathfrak{Q}_+(-4)$ is such that $\omega(Q) \in \mathcal{F}$, then by (77.17) we have $a = 1$; so b is equal to either 0 or 1, while b is even because of (71.6). Hence, $a = c = 1$, $b = 0$. According to Theorem 78, all forms in $\mathfrak{Q}_+(-4)$ are equivalent to $\left(\begin{smallmatrix} 1 & 0 \\ 0 & 1 \end{smallmatrix}\right)$. Therefore,

$$\mathfrak{h}_+(-4) = 1 :$$
$$Q \in \mathfrak{Q}_+(-4) \;\Leftrightarrow\; Q = {}^t U U, \; \exists\, U \in \Gamma. \tag{78.1}$$

Then, turning to (72.2), we need to consider the congruence equation $X^2 \equiv -4 \mod 4p$; note that any representation of p, if ever exists, is proper. By the supplementary law (58.1)(1) we should have $p \equiv 1 \mod 4$. Then there exist r, s such that $r^2 - 4ps = -4$, and the primitive form $[|p, r, s|]$ is contained in $\mathfrak{Q}_+(-4)$. Hence, by (78.1) there exists $U = \left(\begin{smallmatrix} u_1 & u_2 \\ u_3 & u_4 \end{smallmatrix}\right) \in \Gamma$ such that

$$({}^t U U)(x,y) = px^2 + rxy + sy^2. \tag{78.2}$$

This gives rise to the representation $p = u_1^2 + u_3^2$.

In short, first solve $X^2 \equiv -1 \mod p$, and compose the right side of (78.2), to which apply the pre-reduction [+] to find a U satisfying (78.2), as we shall do with the example given in Note [78.1]. The application of (æ), §76, to the present situation will be discussed in Note [81.7]. Also, see Note [78.3] about the use of the supplementary law (58.1)(1).

In this way, we have obtained Euler's theorem; for the history of this famous assertion, see Note [78.2].

Theorem 79

$$p = x^2 + y^2 \text{ has a solution} \quad\Leftrightarrow\quad \text{either } p = 2 \text{ or } p \equiv 1 \mod 4. \tag{78.3}$$

We consider the representations of an arbitrary positive integer m in place of a prime. We shall show that (78.3) gives the assertion:

$$n = f^2 + g^2, \quad f, g \in \mathbb{Z},$$
$$\Leftrightarrow\; n = 2^\alpha \prod_{p \equiv 1 \bmod 4} p^\beta \prod_{p' \equiv -1 \bmod 4} p'^{2\gamma}, \tag{78.4}$$

in which the exponents $\alpha, \beta, \gamma \geq 0$ are arbitrary: The right side gives the left side, as is readily seen by the relation:

$$n = (ac)^2 + (bc)^2, \quad c = 2^{[\alpha/2]} \prod_{p'} p'^{\gamma} \tag{78.5}$$

with

$$a + bi = (1 + i)^{\alpha - 2[\alpha/2]} \prod_{u^2 + v^2 = p} (u + vi)^{\beta}, \quad i = \sqrt{-1}. \tag{78.6}$$

Conversely we shall show that (78.5) follows from the left side of (78.4). Thus, if $p' \equiv -1 \bmod 4$ is such that $n = f^2 + g^2 \equiv 0 \bmod p'$, then $f^2 \equiv -g^2 \bmod p'$, and $p' | fg$. For if $p' \nmid fg$, then $(f\overline{g})^2 \equiv -1 \bmod p'$, $g\overline{g} \equiv 1 \bmod p'$, which contradicts (58.1)(1). Or more directly, raising both sides of $f^2 \equiv -g^2 \bmod p'$ to the power $(p' - 1)/2$, we get the contradiction $1 \equiv -1 \bmod p'$. Hence, $p' | f$, $p' | g$; that is, $p'^2 | n$, and $(f/p')^2 + (g/p')^2 = n/p'^2$. Repeating the same procedure, we conclude that $2 | n(p')$ in the notation of (13.1). Further, if $2^{\alpha} \| n$ and $2^{\delta} \| \langle f, g \rangle$, then $2^{\alpha - 2\delta}(n/2^{\alpha}) = (f/2^{\delta})^2 + (g/2^{\delta})^2$. Since one of $f/2^{\delta}$ and $g/2^{\delta}$ is odd, $2^{\alpha - 2\delta}(n/2^{\alpha})$ is divisible by the first power of 2 at most. So $\alpha - 2\delta = 0$ or 1, as $n/2^{\alpha}$ is odd; that is, $\delta = [\alpha/2]$. We have obtained (78.5) and proved (78.4).

We proceed to the issue of the number of representations; so we introduce the following arithmetic function on $\mathbb{N}$:

$$r(n) = \left| \left\{ \{x, y\} \in \mathbb{Z}^2 : n = x^2 + y^2 \right\} \right|. \tag{78.7}$$

We are going to prove an explicit formula for this. Note that the counting concerns all possible combinations of x, y; see Note [78.5] for unordered and unsigned representations. Here the premise (71.3) becomes relevant.

To this end we consider the function $r^*(m) = \left| \left\{ m = u^2 + v^2 : \langle u, v \rangle = 1 \right\} \right|$. If $2 \nmid m$, then (73.1) is satisfied with $D = -4$; and $r^*(m)$ is the number of different proper representations of m by the form $Q = [|1, 0, 1|]$. Thus, by (74.3), $r^*(m) = |\mathcal{U}_Q(m)/\Gamma_{\infty}|$. According to (74.20)

$$2 \nmid m \implies r^*(m) = |\text{Aut}_Q| |\mathcal{M}_m(-4)|, \tag{78.8}$$

as all forms in $\mathcal{M}_m(-4)$ are equivalent to Q. In addition, we note that for any $n \in \mathbb{N}$

$$r(2n) = r(n), \tag{78.9}$$

since we have the bijective correspondence

$$n = a^2 + b^2 \iff 2n = k^2 + l^2$$
$$a = \tfrac{1}{2}(k + l), b = \tfrac{1}{2}(k - l); \quad k = a + b, l = a - b, \tag{78.10}$$

which is related to (78.6).

Theorem 80 *We have, for each $n \in \mathbb{N}$,*

$$r(n) = 4 \sum_{d|n} \kappa_{-4}(d), \tag{78.11}$$

where κ_{-4} is a Kronecker symbol associated with the discriminant -4, i.e., the unique non-principal character mod 4.

Proof Here Theorem 70 becomes relevant. On noting the relation (78.9) as well as $\kappa_{-4}(2) = 0$, we see that it suffices to consider the case $2 \nmid n$. With this, the classification according to $\langle x, y \rangle = \ell$ gives

$$r(n) = \sum_{\ell^2|n} r^*(n/\ell^2). \tag{78.12}$$

Combining (74.20), (76.9)$_{D=-4}$, and (78.8), we get

$$\begin{aligned}
r(n) &= 4 \sum_{h|n/\ell^2} \mu(h)^2 \kappa_{-4}(h) \\
&= 4 \sum_{d|n} \kappa_{-4}(d) \sum_{d=\ell^2 h} \mu^2(h) = 4 \sum_{d|n} \kappa_{-4}(d).
\end{aligned} \tag{78.13}$$

We have used the facts that $\kappa_{-4}(\ell^2) = 1$ because $2 \nmid \ell$ and the combination $\{\ell, h\}$ such that $d = \ell^2 h$, $\mu^2(h) = 1$, exists uniquely for each d. We end the proof.

The sum on the right side of (78.11) is multiplicative (see (16.2)), so

$$\begin{aligned}
r(n) &= 4 \prod_p \sum_{v=0}^{n(p)} \kappa_{-4}(p^v) \\
&= 4 \prod_{p \equiv 1 \bmod 4} (n(p) + 1) \prod_{p' \equiv -1 \bmod 4} \tfrac{1}{2}((-1)^{n(p')} + 1).
\end{aligned} \tag{78.14}$$

It is perhaps worth remarking that the representation (78.3) of $p \equiv 1 \bmod 4$ is essentially unique, for $r(p) = 8$; but this particular fact can be proved independently of (78.11): it is implied by the argument of Euler given in Note [78.11].

In much the same way, one may discuss the representations by the form $[|1, 0, d|]$ provided $d > 0$ and $\mathfrak{h}_+(-4d) = 1$; see Note [78.14]. However, the problem that is far more interesting is to consider whether or not any explicit formula analogous to (78.11) exists on $\mathfrak{h}_+(-4d) \geq 2$. See (79.6) which concerns the case $d = 5$.

Notes

[78.1] For $p = 430897$ (see Note [41.4]), we shall find integers u, v such that $p = u^2 + v^2$. We have first to solve $X^2 \equiv -1 \bmod p$. By the reciprocity law, we see that 5 mod p is a quadratic non-residue; hence, $u^2 \equiv -1 \bmod p$ with $u = 5^{(p-1)/4}$. By the device (36.2),

$$\tfrac{1}{4}(p - 1) = 2^2 + 2^3 + 2^6 + 2^7 + 2^{10} + 2^{13} + 2^{15} + 2^{16}$$
$$\Rightarrow 5^{(p-1)/4} \equiv 76715 \bmod p,$$

so that $76715^2 + 1 = 13658 \cdot 430897$. Namely, the quadratic form

$$Q(x, y) = 430897x^2 + 153430xy + 13658y^2$$

is in $\mathcal{Q}_+(-4)$, and there should exist a $U \in \Gamma$ such that $Q = {}^t\!UU$; of course $Q = M_{p,\xi}, \xi = 153430$. The pre-reduction [+] works as follows:

$$[|430897, 153430, 13658|] \xRightarrow{W} [|13658, -153430, 430897|]$$

$$\xRightarrow{T^6} [|13658, 10466, 2005|] \xRightarrow{W} [|2005, -10466, 13658|] \xRightarrow{T^3} [|2005, 1564, 305|]$$

$$\xRightarrow{W} [|305, -1564, 2005|] \xRightarrow{T^3} [|305, 266, 58|] \xRightarrow{W} [|58, -266, 305|]$$

$$\xRightarrow{T^2} [|58, -34, 5|] \xRightarrow{W} [|5, 34, 58|] \xRightarrow{T^{-3}} [|5, 4, 1|] \xRightarrow{W} [|1, -4, 5|] \xRightarrow{T^2} [|1, 0, 1|].$$

Therefore,

$$U = \left(WT^6 WT^3 WT^3 WT^2 WT^{-3} WT^2\right)^{-1} = \begin{pmatrix} -601 & -107 \\ 264 & 47 \end{pmatrix}$$

yields the representation

$$430897 = 601^2 + 264^2.$$

[78.2] An observation relevant to Theorem 79 is said to have been made by Diophantus (ca 250 CE). Bachet (1621, Diophantus, p.173) raised interest in this historical fact; thence Girard (Stevin (1625, p.622)) and Fermat (1640: Oeuvres II, pp.214–215) gave correct assertions. Much later, Euler (1749a; 1751; 1772d) established Theorem 79. His proof is given in Note [78.9]; as for Fermat's, a mention is in the same Note. Among several known proofs, the one by Thue (1902), which has been given in Note [29.3] (A), is ingenious; but it is an existential argument. The same can be said about the proof by Smith (1855) which is presented in Note [21.5]. Compared with these, the method of C. Hermite (Serret (1849b, Note)) given in Notes [29.3] (B) and [58.6] is quite practical; its extension will be shown in §81; and see Note [81.7]. As for the argument via Jacobi sums that was indicated in Note [61.4], it depends on a primitive root r mod p and the computation of Ind_r, so it should rather

344   *Quadratic Forms*

be regarded as an existential argument. Further, Legendre (1808, pp.59–60) found a method that exploits the continued fraction expansion of $\sqrt{p}$; see Notes [84.3] and [84.6]. See also Gauss' discussion in [DA, art.265].

[78.3]   It is said widely that the supplementary law (58.1)(1) is indispensable to prove Theorem 79. However, this is a fallacy. Smith's proof mentioned above is independent of the notion of quadratic residues.

[78.4]   As a matter of fact, Smith remarked at the end of the article just referred to that his method yields a solution to $x^2 \equiv -1 \bmod p$. Indeed the matrix representation of the case (1) in Note [21.5] implies that

$$\left(A_h^2 + A_{h-1}^2\right)\left(B_h^2 + B_{h-1}^2\right) - \left(A_h B_h + A_{h-1} B_{h-1}\right)^2 = 1,$$

so

$$\left(A_h B_h + A_{h-1} B_{h-1}\right)^2 \equiv -1 \bmod p,$$

provided $A_h^2 + A_{h-1}^2 = p$. In Smith's proof the reasoning proceeds in a way opposite to that in conventional expositions on (78.3); namely, after a direct proof of (78.3) the assertion (58.1)(1) is obtained. In the case $p = 430897$ the symmetric continued fraction expansion in question is

$$\frac{430897}{76715} = 5 + \frac{1}{1+} \frac{1}{1+} \frac{1}{1+} \frac{1}{1+} \frac{1}{1+} \frac{1}{1+} \frac{1}{3+} \frac{1}{2+} \frac{1}{2+} \frac{1}{3}$$
$$+ \frac{1}{1+} \frac{1}{1+} \frac{1}{1+} \frac{1}{1+} \frac{1}{1+} \frac{1}{1+} \frac{1}{5};$$

$$\begin{pmatrix} 5 & 1 \\ 1 & 0 \end{pmatrix} \begin{pmatrix} 1 & 1 \\ 1 & 0 \end{pmatrix}^6 \begin{pmatrix} 3 & 1 \\ 1 & 0 \end{pmatrix} \begin{pmatrix} 2 & 1 \\ 1 & 0 \end{pmatrix} = \begin{pmatrix} 601 & 264 \\ 107 & 47 \end{pmatrix} \in \diamond \Gamma.$$

Thus we get the same decomposition of p as in Note [78.1]. Also

$$(601 \cdot 107 + 264 \cdot 47)^2 = 76715^2 = -1 + 13658 \cdot 430897.$$

On the other hand, relevantly to Note [58.6], we note that

$$\frac{76715}{430897} = 0 + \frac{1}{5+} \frac{1}{1+} \frac{1}{1+} \frac{1}{1+} \frac{1}{1+} \frac{1}{1+} \frac{1}{1+} \frac{1}{3+} \frac{1}{2+} \frac{1}{2+} \cdots.$$

The 9th and the 10th convergents are $\frac{107}{601}$ and $\frac{261}{1466}$, respectively. On noting $[\sqrt{p}] = 656$, we find that $a = 76715$, $c = 601$, $d = 107$, $ac - pd = -264$, which gives the result of Note [78.1] again. See Brillhart (1972).

[78.5]   The explicit formula (78.11) is commonly attributed to Dirichlet (1840a, p.3; 1863, p.245) but see the next Note. An alternative proof is given in Landau (1927, III. Teil, Kapitel 2). It should be known that (78.11) is not of elementary nature: the determination of divisors of n has to be taken into

$§78$ 345

account, which is a hard problem in general, as we have mentioned repeatedly. This is just the same as with the fundamental formula (74.20). As for the number $r_0(n)$ of unordered and unsigned representations $n = x^2 + y^2$, $xy \neq 0$, we find readily that

$$
8r_0(n) = \begin{cases} r(n) - 4(-1)^\alpha & n/2^\alpha \text{ is a square,} \\ r(n) & \text{otherwise,} \end{cases}
$$

where 2^α as in (78.4). See Fermat (Oeuvres II, pp.214–215) and Gauss [DA, art.182, p.162, footnote].

[78.6] While developing his theory of elliptic functions, Jacobi (1828: Legendre and Jacobi (1875, p.242)) encountered the following identity, which is equivalent to (78.11):

$$
\left(\sum_{n=-\infty}^{\infty} \xi^{n^2} \right)^2 = 1 + 4 \sum_{n=1}^{\infty} (-1)^{n-1} \frac{\xi^{2n-1}}{1 - \xi^{2n-1}}, \quad |\xi| < 1.
$$

The sum on the left side is the origin of the ϑ-function introduced at (61.7).

[78.7] As a corollary of (78.11) we shall prove the well-known expansion:

$$
\frac{\pi}{4} = \sum_{j=1}^{\infty} \frac{(-1)^{j-1}}{2j - 1}.
$$

We observe first that the number of ordinary lattice points on the disk of radius $\sqrt{M}$ centered at the origin equals

$$
1 + \sum_{1 \leq m \leq M} r(m) = 1 + 4 \sum_{1 \leq m \leq M} \sum_{d \mid m} \chi(d), \quad \chi = \kappa_{-4}.
$$

The left side equals $\pi M + O(\sqrt{M})$; this error term corresponds to the number of those lattice points on the ring between the two circles of radius $\sqrt{M} \pm 2$. On the other hand, the double sum on the right side equals

$$
\sum_{d \leq \sqrt{M}} \chi(d) \sum_{f \leq M/d} 1 + \sum_{f \leq \sqrt{M}} \sum_{d \leq M/f} \chi(d) - \sum_{d \leq \sqrt{M}} \chi(d) \sum_{f \leq \sqrt{M}} 1.
$$

The absolute values of the sums over d in the second and the third double sums are not greater than 1. The first double sum is equal to

$$
\sum_{d \leq \sqrt{M}} \chi(d) \left(\frac{M}{d} + u_d \right), \quad |u_d| \leq 1.
$$

346   *Quadratic Forms*

Further,

$$\sum_{d \le \sqrt{M}} \frac{\chi(d)}{d} = \left(\sum_{d=1}^{\infty} - \sum_{d > \sqrt{M}} \right) \frac{\chi(d)}{d}.$$

On the right side, the absolute value of the second sum is less than $1/\sqrt{M}$, as it is an alternating series. Summing up, we find that

$$\pi M = 4M \sum_{d=1}^{\infty} \frac{\chi(d)}{d} + O(\sqrt{M}).$$

This ends the proof. The dissecting argument applied to the initial double sum is the same as (16.27).

[78.8]   If an integer n has the representation $n = u^2 + v^2$ with $u, v \in \mathbb{Q}$, then there exist integers $a, b \in \mathbb{Z}$ such that $n = a^2 + b^2$; an observation due to Diophantus or earlier. For confirmation, we note that then there exist $q, s, t \in \mathbb{Z}$ such that $nq^2 = s^2 + t^2$, so the prime power decomposition of nq^2 is of the form (78.4), and n as well.

[78.9]   The proof of the sufficiency part of (78.3) by Euler (1772d, Theorema 1); his earlier proofs are essentially the same: It begins with an application of the supplement law (58.1)(1); there exists an integer a such that $a^2 + 1 \equiv 0 \bmod p$. Hence, it suffices to prove that from the existence of u, v such that $u^2 + v^2 \equiv 0 \bmod n$ follows the existence of u_0, v_0 such that $n = u_0^2 + v_0^2$. For this sake, let $|u_1|, |v_1| \le n/2$ be such that $u \equiv u_1, v \equiv v_1 \bmod n$. Then, $nn_1 = u_1^2 + v_1^2, 0 < n_1 \le n/2$. Let $u_1 = \alpha n_1 + r, v_1 = \beta n_1 + s, |r|, |s| \le n_1/2$, and put $w = \alpha r + \beta s$. From the identity $nn_1 = n_1^2(\alpha^2 + \beta^2) + 2n_1 w + r^2 + s^2$ it follows that $n_1 n_2 = r^2 + s^2, 0 < n_2 \le n_1/2$, so $n = n_1(\alpha^2 + \beta^2) + 2w + n_2$. At this step, an application of the identity $((72.14)_{d=-1})$

$$\left(a^2 + b^2\right)\left(f^2 + g^2\right) = (af + bg)^2 + (ag - bf)^2$$

implies that $n_1 n_2 (\alpha^2 + \beta^2) = w^2 + z^2, z = r\beta - s\alpha$. So $nn_2 = u_2^2 + v_2^2$, $u_2 = w + n_2, v_2 = z$. By induction, we conclude the proof. As for the argument of Fermat, it is vaguely indicated in Œuvres II, p.432, and appears to be similar to Euler's.

[78.10]   The identity in the preceding Note is trivial if viewed as a relation between absolute values of complex numbers. However, if viewed from the theory of numbers, then it states clearly that the set of all integers which are sums of two squares is closed with respect to multiplication, a phenomenon that has been noticed since great antiquity as indicated in Note [78.1]. It implies that the representation of a prime as a sum of two squares is the core issue, that is, the statement of Theorem 79. This structural fact is said to be the origin of the theory of compositions of quadratic forms, as we have mentioned a few times already. Also the condition $p \equiv 1 \bmod 4$ in (78.3) is the earliest glimpse of the genus theory of quadratic forms; see (71.1)(6). Indeed, Lagrange and Legendre tried hard to extend this fact beyond $D = -4$.

[78.11]   The formula (78.14) implies obviously that if n can be expressed in two genuinely different ways as sums of two squares, then n is composite; and a factorization of such an n becomes an issue. Euler (1749a) gave a method for this purpose. We shall employ an argument somewhat different from his, and prove that if $n = a^2 + b^2 = c^2 + d^2$ with $a,b,c,d > 0$ and $\{a,b\} \neq \{c,d\}, \{d,c\}$, then $\langle n, ad - bc \rangle$ is a non-trivial factor of n. To show this, we note that $a^2 \equiv -b^2, d^2 \equiv -c^2 \Rightarrow (ad)^2 \equiv (bc)^2 \bmod n$ (see §49 (iv)). Hence, $n = \langle n, (ad - bc)(ad + bc) \rangle$. Suppose that $\langle n, ad - bc \rangle = 1$; then $ad + bc \equiv 0 \bmod n$, and $n = ad + bc$ since $0 < a,b,c,d < \sqrt{n} \Rightarrow ad + bc < 2n$. Applying the identity quoted in Note [78.9], $n^2 = (ad + bc)^2 \Rightarrow ac - bd = 0 \Rightarrow c^2(a^2 + b^2) = b^2(c^2 + d^2) \Rightarrow b = c$, and $a = d$, which is a contradiction. Also, $ad - bc = 0$ would imply $a = c, b = d$, a contradiction again. Therefore, $1 < \langle n, ad - bc \rangle < n$, in which the second inequality follows from the bounds for a,b,c,d. This ends the proof. With $n = 1000009$ that was treated in Note [38.4], Euler (1749a, p.170) showed $n = 1000^2 + 3^2 = 972^2 + 235^2$ and got the decomposition $n = 293 \cdot 3413$. By the present argument, $ad - bc = 232084$, and the factor $3413 = \langle n, ad - bc \rangle$ is obtained. By the way, $293 = \langle n, ad + bc \rangle$. Euler (1765, p.390) proved that 10091401 is a prime by means of the fact that $10091401 = 2920^2 + 1251^2$ is the sole representation. Continued in Note [81.8].

[78.12]   Continuing Note [77.5], if the form is $[|1,0,1|]$, then the problem becomes equivalent to arranging, in the order of magnitude, the integers which are expressible as a sum of two squares. In spite of its elementary appearance, this is, in fact, one of the most difficult unsolved problems in mathematics. In order to indicate the reason for this claim of ours, we introduce the function

$$B(s) = \sum_{m=1}^{\infty} \frac{b(m)}{m^s}, \quad b(m) = \begin{cases} 1 & r(m) \neq 0, \\ 0 & r(m) = 0. \end{cases}$$

By (78.4), we find readily that for $\operatorname{Re} s > 1$

$$B(s) = \left(1 - \frac{1}{2^s}\right)^{-1} \prod_{p \equiv 1 \bmod 4} \left(1 - \frac{1}{p^s}\right)^{-1} \prod_{p \equiv 3 \bmod 4} \left(1 - \frac{1}{p^{2s}}\right)^{-1}.$$

348   *Quadratic Forms*

Thus

$$C(s) = \left(1 - \frac{1}{2^s}\right)^{-1} \prod_{p \equiv 3 \bmod 4} \left(1 - \frac{1}{p^{2s}}\right)^{-1},$$

$$B(s) = \{C(s)R(s)\}^{1/2},$$

$$R(s) = \tfrac{1}{4}\sum_{n=1}^{\infty} r(n)n^{-s} = \zeta(s)L(s,\kappa_{-4}).$$

Here the branch of the square root is such that $\lim_{s \to +\infty} B(s) = 1$. From this follows

$$\sum_{m \le x} b(m) \sim K\frac{x}{\sqrt{\log x}}, \quad K = \tfrac{1}{2}\sqrt{C(1)} = 0.76422\ldots$$

See Landau (1908b); or one may proceed in much the same way as the proof of (95.18) below. This K is called the Landau–Ramanujan constant. Because of the appearance of the zeta and L-functions, it is conceivable to make an appeal to GRH. However, it is far beyond the reach of GRH to replace this asymptotic formula by an expression so sharp that it can give a solution to our present ordering problem.

[78.13]   We omit the discussion of the two traditional subjects: expressing positive integers as sums of four squares (Lagrange's theorem (1770c)) and three (Legendre's theorem (1798, p.202 and pp.398–399)). Landau (1927, Dritter Teil) gave a lucid account of their famous results. He adopted the alternative argument of Euler (1772d, pp.543–544) and partly that of Gauss [DA, arts.288–293], respectively; notably, in the latter (p.124) Landau used Dirichlet's prime number theorem, i.e., (88.1) below. More interesting is, of course, the number of expressing an integer as sums of squares, i.e., $|\{n = r_1^2 + r_2^2 + \cdots + r_s^2 : r_j \in \mathbb{Z}\}|$. The issue is to find any analogue of (78.11) for arbitrary $s \ge 3$, or more correctly to see whether such a formula exists or not. Bateman (1951) is an informative account of this problem. The basic device is the theory of powers of the Jacobi ϑ-function, i.e., the transformation properties of the function against the action of Γ.

[78.14]   Gauss [DA, art.303] gave a table of the values of $\mathfrak{h}_+(-4k)$. In particular, he conjectured that $\mathfrak{h}_+(-4k) = 1$ should hold only for $k = 1,2,3,4,7$. This was settled by Landau (1903b).

§79

We next consider the case $D = -20$. Let $[|a,b,c|] \in \mathcal{Q}_+(-20)$ be as in (77.17). We have $2|b$ by (71.6); also, $2\sqrt{5/3} \ge a$; thus, $a = 2$ or 1. The point $\omega(Q) \in \mathcal{F}$ is equal either to $i\sqrt{5}$ or to $(-1+i\sqrt{5})/2$. That is, the representative forms are

$$Q_1 = [|1,0,5|] \text{ and } Q_2 = [|2,2,3|]. \text{ We have } \mathfrak{h}_+(-20) = 2 \text{ by Theorem 78.}$$
Thus,

$$Q \in \mathcal{Q}_+(-20) \iff Q = \text{either } {}^tUQ_1U \text{ or } {}^tUQ_2U, \exists\, U \in \Gamma. \qquad (79.1)$$

The application of pre-reduction [+] will be detailed in the examples given in Notes below.

The congruence equation $X^2 \equiv -20 \bmod 4p$ is equivalent to $X^2 \equiv -5 \bmod p$, provided $p \geq 7$. We should have $\left(\frac{-5}{p}\right) = 1$. Hence, by the reciprocity law (58.1),

$$(-1)^{(p-1)/2}, \; \left(\frac{p}{5}\right) = \begin{cases} (1) : +1, +1, \\ (2) : -1, -1. \end{cases} \qquad (79.2)$$

The primes satisfying either condition can be expressed by either Q_1 or Q_2. To make the situation a little more precise, we observe that

$$\begin{aligned} p = Q_1(x,y) = x^2 + 5y^2 &\implies \left(\frac{p}{5}\right) = 1, \\ 2p = 2Q_2(x,y) = (2x+y)^2 + 5y^2 &\implies \left(\frac{2p}{5}\right) = 1. \end{aligned} \qquad (79.3)$$

Therefore, via (58.1)(2), we conclude that

$$(-1)^{(p-1)/2}, \; \left(\frac{p}{5}\right) = \begin{cases} (1) \; +1, +1 \implies Q_1, \\ (2) \; -1, -1 \implies Q_2. \end{cases} \qquad (79.4)$$

In terms of congruence, this can be expressed as

$$\begin{aligned} &(1) \; p \equiv 1, 3^2 \bmod 20 : \quad \{n^2 \bmod 20 : \langle n, 20 \rangle = 1\} \implies Q_1, \\ &(2) \; p \equiv 3, 3^3 \bmod 20 : \; 3 \cdot \{n^2 \bmod 20 : \langle n, 20 \rangle = 1\} \implies Q_2. \end{aligned} \qquad (79.5)$$

Legendre (1785, p.529) was the first who grasped this congruence condition from the viewpoint of the reciprocity law; namely, he combined (79.2) with (79.3) to deduce (79.5). Here is a glimpse of the genus theory: a precise condition for the representability of a prime by forms in $\mathcal{Q}_+(-20)$ can be described in terms of the reduced residue class mod 20, $|D| = 20$, to which the prime belongs; more on this aspect in Note [80.4] and in §91.

The explicit formula (78.11) has the following analogue:

Theorem 81 *Let $t(n)$ be the number of all integer solutions of the indefinite equation $x^2 + 5y^2 = n = 2^\alpha 5^\beta m$, $\langle m, 10 \rangle = 1$. Then we have*

$$t(n) = \left(1 + (-1)^\alpha \kappa_5(m)\right) \sum_{d|m} \kappa_{-20}(d) \qquad (79.6)$$

in the notation (73.2).

Proof Let $t_j^*(m)$ be the number of proper representations of m by Q_j. Then, by (74.18) and (76.9), we have $t_1^*(m) + t_2^*(m) = 2|\mathcal{M}_m(-20)|$. In order to separate the summands on this left side, we observe, as in (79.3), that

$$t_2^*(m) = 0 \Leftrightarrow \kappa_5(m) = 1, \quad t_1^*(m) = 0 \Leftrightarrow \kappa_5(m) = -1, \qquad \text{provided } |\mathcal{M}_m(-20)| \neq 0, \tag{79.7}$$

where we have used $(73.14)_{p=5}$. In particular, $t_1^*(m)t_2^*(m) = 0$. Hence, by (74.20), which is applicable as $\langle m, 10 \rangle = 1$,

$$t_j^*(m) = \left(1 - (-1)^j \kappa_5(m)\right) \sum_{d|m} \mu^2(d)\kappa_{-20}(d) \tag{79.8}$$

including the case $|\mathcal{M}_m(-20)| = 0$. We then classify the cases according to the parities of α, β and obtain an analogue of (78.9) as follows:

· $2|\beta \Rightarrow 5^{\beta/2}\|x, 5^{\beta/2}\|y \Rightarrow t(n) = t(2^\alpha m)$;

· $2\nmid\beta \Rightarrow 5^{(\beta+1)/2}|x, 5^{(\beta-1)/2}\|y \Rightarrow t(n) = t(2^\alpha m)$;

· $2|\alpha \Rightarrow$ either $2^{\alpha/2}\|x, 2^{\alpha/2+1}|y$ or $2^{\alpha/2+1}|x, 2^{\alpha/2}\|y$. Hence, $t(2^\alpha m) = t(m)$;

· $2\nmid\alpha \Rightarrow 2^{(\alpha-1)/2}\|x, 2^{(\alpha-1)/2}\|y$. Hence, $t(2^\alpha m) = t(2m)$. Then, on noting that $x^2 + 5y^2 = 2m \Rightarrow 2\nmid xy$, we put $x = 2u + v, y = v$, and find that $Q_2(u, v) = m$. Therefore, $2|\alpha \Rightarrow t(n) = \sum_{g^2|m} t_1^*(m/g^2)$, and $2\nmid\alpha \Rightarrow t(n) = \sum_{g^2|m} t_2^*(m/g^2)$. Into these we insert (79.8) while observing that $j \equiv \alpha - 1 \bmod 2$. We end the proof. See Dickson (1929, p.84).

The formula (79.6) itself does not depend on the reciprocity law; however, when deriving the condition for $t(n) > 0$ from it, the law comes in, as then we need to appeal to Theorem 70.

Notes

[79.1] According to (79.5), $p = 404321$ should have the representation $p = x^2 + 5y^2$. We shall find such integers x, y. We need to solve $X^2 \equiv -5 \bmod p$; we have done it already in Note [63.5]: $160284^2 + 5 = 63541p$. Hence, there should exist a $U \in \Gamma$ such that

$$Q = \begin{pmatrix} 404321 & 160284 \\ 160284 & 63541 \end{pmatrix} = {}^tUQ_1U.$$

Applying the pre-reduction [+],

$$[|404321, 320568, 63541|] \xrightarrow{W} [|63541, -320568, 404321|]$$

$$\xrightarrow{T^3} [|63541, 60678, 14486|] \xrightarrow{W} [|14486, -60678, 63541|]$$

$$\xrightarrow{T^2} [|14486, -2734, 129|] \xrightarrow{W} [|129, 2734, 14486|] \xrightarrow{T^{-11}} [|129, -104, 21|]$$

$$\xrightarrow{W} [|21, 104, 129|] \xrightarrow{T^{-2}} [|21, 20, 5|] \xrightarrow{W} [|5, -20, 21|] \xrightarrow{T^2} [|5, 0, 1|]$$

$$\xrightarrow{W} [|1, 0, 5|].$$

Hence,

$$U = \left(WT^3WT^2WT^{-11}WT^{-2}WT^2W\right)^{-1} = -\begin{pmatrix} 111 & 44 \\ 280 & 111 \end{pmatrix}.$$

Indeed we have

$$Q = \begin{pmatrix} 111 & 280 \\ 44 & 111 \end{pmatrix}\begin{pmatrix} 1 & 0 \\ 0 & 5 \end{pmatrix}\begin{pmatrix} 111 & 44 \\ 280 & 111 \end{pmatrix}.$$

Therefore,

$$404321 = 111^2 + 5 \cdot 280^2.$$

In Note [81.2] we shall give an alternative argument to prove this decomposition of 404321, which is an example for (æ), §76.

[79.2]   As for $p = 430883$, we see, by (79.5), that $p = 2x^2 + 2xy + 3y^2$ should have a solution. We need first to solve $X^2 \equiv -5 \bmod p$. We may apply Lagrange's observation (Note [63.1]) and get $X \equiv \pm(-5)^{(p+1)/4} \equiv \mp 95402 \bmod p$. Hence, with a $U \in \Gamma$, we should have

$$\begin{pmatrix} 430883 & 95402 \\ 95402 & 21123 \end{pmatrix} = {}^tUQ_2U.$$

By the pre-reduction [+],

$$U = \left(WT^5WT^2WT^{-15}WT^{-3}W\right)^{-1} = \begin{pmatrix} 140 & 31 \\ -411 & -91 \end{pmatrix}.$$

Therefore,

$$430883 = 2 \cdot 140^2 + 2 \cdot 140 \cdot (-411) + 3 \cdot (-411)^2.$$

[79.3]   The assertion (79.5) has a long history similar to that of Euler's theorem (78.3). We omit it, however. More important is to observe the following multiplicative structure of the set of integers which can be represented by forms in $Q_+(-20)$: We have the three identities

(1) $\left(x^2 + 5y^2\right)\left(u^2 + 5v^2\right) = (xu - 5yv)^2 + 5(xv + yu)^2,$

(2) $\left(2x^2 + 2xy + 3y^2\right)\left(2u^2 + 2uv + 3v^2\right)$
$$= \left(2xu + 3yv + xv + yu\right)^2 + 5(xv - yu)^2,$$

(3) $\left(x^2 + 5y^2\right)\left(2u^2 + 2uv + 3v^2\right) = 2(xu - yu - 3yv)^2$
$$+ 2(xu - yu - 3yv)(xv + 2yu + yv) + 3(xv + 2yu + yv)^2.$$

These correspond respectively to the symbolic identities

$$Q_1 \times Q_1 = Q_1, \ Q_2 \times Q_2 = Q_1, \ Q_1 \times Q_2 = Q_2,$$

352 *Quadratic Forms*

with Q_1, Q_2 as above. That is, the set $\{Q_1, Q_2\}$ appears like a group of order 2. It is said that Fermat (1658: Œuvres II, p.405(3°)) noticed this structure, but the first explicit mention of the identity (2) was made by Lagrange (1775, pp.788–789).

[79.4]   It should be appropriate to remark here that the identity

$$\left(f_1 x_1^2 + g x_1 y_1 + f_2 k y_1^2\right)\left(f_2 x_2^2 + g x_2 y_2 + f_1 k y_2^2\right)$$
$$(*) \qquad = f_1 f_2 x_3^2 + g x_3 y_3 + k y_3^2,$$
$$x_3 = x_1 x_2 - k y_1 y_2, \quad y_3 = f_1 x_1 y_2 + f_2 x_2 y_1 + g y_1 y_2,$$

holds with any $\{f_1, f_2, g, k, x_1, x_2, y_1, y_2\} \in \mathbb{C}^8$; in particular (2) above is a consequence of the case with $f_1 = 2, g = 2, f_2 = 3, k = 1$. For the proof of $(*)$, one may use the matrix representation (72.18) of forms so that with $\mathfrak{d}^2 = (g^2 - 4 f_1 f_2 k)\mathfrak{e}$

$$\left(f_1 x_1 \mathfrak{e} + \tfrac{1}{2} y_1 (g \mathfrak{e} + \mathfrak{d})\right)\left(f_2 x_2 \mathfrak{e} + \tfrac{1}{2} y_2 (g \mathfrak{e} + \mathfrak{d})\right) = f_1 f_2 x_3 \mathfrak{e} + \tfrac{1}{2} y_3 (g \mathfrak{e} + \mathfrak{d}).$$

The well-poised identity $(*)$ plays a basic rôle in our account (§90) of Gauss' composition theory of quadratic forms. Since Dirichlet (1851) applied $(*)$ explicitly and quite effectively in his simplification of Gauss' theory, it can be attributed to him; however, see Note [90.9].

[79.5]   As for the separation argument applied at (79.7), we note that κ_5 is an instance of the genus characters which are to be discussed in §91, specifically at (91.36) together with the foregoing paragraph (iii). The character κ_5 is used to pick up a particular genus (for this term see the explanation following (80.6) below), i.e., either Q_1 or Q_2 in the present situation; anyway, this operation becomes possible because of the group structure suggested in Note [79.3], and thus extends to all discriminants, although the effect is limited since a genus is not composed of a single class in general. A closely related discussion is in the next section, as indicated above, and in Note [91.10].

[79.6]   Analogously to $R(s)$ defined in Note [78.12], let

$$T(s) = \sum_{n=1}^{\infty} t(n) n^{-s}.$$

Then (79.6) implies that

$$T(s) = \zeta(s) L(s, \kappa_{-20}) + L(s, \kappa_{-4}) L(s, \kappa_5)$$

for $\operatorname{Re} s > 1$; note that the last term is related to the later part of Note [93.4]. Thus, $T(s)$ continues analytically to the whole $\mathbb{C}$ in much the same way as

$\zeta(s)$. Moreover, it satisfies a functional equation similar to (12.8) or rather to that for $\zeta^2(s)$: under the change $s \mapsto 1 - s$, the function

$$(20/\pi)^{s/2}\Gamma(\tfrac{1}{2}s)\Gamma(\tfrac{1}{2}(s+1))T(s)$$

is invariant, as can be seen via (55.11) and Note [55.4], i.e., $G(\kappa_{-20}) = G(\kappa_{-4}) \cdot G(\kappa_5) = 2i \cdot 5^{1/2}$ (see Note [56.3]); note that $\Gamma(\tfrac{1}{2}s)\Gamma(\tfrac{1}{2}(s+1)) = 2\pi^{1/2}2^{-s}\Gamma(s)$ by (94.25) below. Yet $T(s)$ is known to have infinitely many zeros off the critical line $\mathrm{Re}\,s = \tfrac{1}{2}$. This is strikingly different from what is expected concerning $R(s)$, i.e., an analogue of RH should hold with it. See Voronin (1976) for a more general and precise assertion.

§80

We next consider the case $D = -231 \equiv 1 \bmod 4$. The assertion (77.17)–(77.18) yields the following system of forms representing classes of $\mathcal{K}_+(-231)$:

$$Q_1 = [|1, 1, 58|], \; Q_2^{\pm} = [|2, \pm 1, 29|], \; Q_3 = [|3, 3, 20|],$$
$$Q_4^{\pm} = [|4, \pm 3, 15|], \; Q_5^{\pm} = [|5, \pm 3, 12|], \; Q_6^{\pm} = [|6, \pm 3, 10|], \qquad (80.1)$$
$$Q_7 = [|7, 7, 10|], \; Q_8 = [|8, 5, 8|].$$

Thus, $\mathfrak{h}_+(-231) = 12$. According to Theorem 68, a prime number $p \nmid 231$, $p > 2$, is represented by one of these forms if and only if

$$\left(\frac{-231}{p}\right) = \left(\frac{p}{231}\right) = +1, \qquad (80.2)$$

where the reciprocity law (58.1) is applied. Since $231 = 3 \cdot 7 \cdot 11$, we have the following four cases:

$$\left(\frac{p}{3}\right), \left(\frac{p}{7}\right), \left(\frac{p}{11}\right) = \begin{cases} (1) \; +1, +1, +1, \\ (2) \; +1, -1, -1, \\ (3) \; -1, +1, -1, \\ (4) \; -1, -1, +1. \end{cases} \qquad (80.3)$$

Then, one may ask by which form in (80.1) p is represented actually. In the case $D = -20$, the assertion (79.4)–(79.5) gives a perfect answer to the analogous problem. Thus, imitating (79.3), we make the following observation:

$$4p = 4Q_1 = (2x+y)^2 + 231y^2 \Rightarrow \left(\tfrac{p}{3}\right), \left(\tfrac{p}{7}\right), \left(\tfrac{p}{11}\right) = 1,$$

$$8p = 8Q_2^{\pm} = (4x \pm y)^2 + 231y^2 \Rightarrow \left(\tfrac{2p}{3}\right), \left(\tfrac{p}{7}\right), \left(\tfrac{2p}{11}\right) = 1,$$

$$4p = 4Q_3 = 3(2x+y)^2 + 77y^2 \Rightarrow \left(\tfrac{2p}{3}\right), \left(\tfrac{5p}{7}\right), \left(\tfrac{p}{11}\right) = 1,$$

$$16p = 16Q_4^{\pm} = (8x \pm 3y)^2 + 231y^2 \Rightarrow \left(\tfrac{p}{3}\right), \left(\tfrac{p}{7}\right), \left(\tfrac{p}{11}\right) = 1,$$

$$20p = 20Q_5^{\pm} = (10x \pm 3y)^2 + 231y^2 \Rightarrow \left(\tfrac{2p}{3}\right), \left(\tfrac{5p}{7}\right), \left(\tfrac{p}{11}\right) = 1,$$

$$8p = 8Q_6^{\pm} = 3(4x \pm y)^2 + 77y^2 \Rightarrow \left(\tfrac{p}{3}\right), \left(\tfrac{5p}{7}\right), \left(\tfrac{2p}{11}\right) = 1,$$

$$4p = 4Q_7 = 7(2x+y)^2 + 33y^2 \Rightarrow \left(\tfrac{p}{3}\right), \left(\tfrac{5p}{7}\right), \left(\tfrac{2p}{11}\right) = 1,$$

$$32p = 32Q_8 = (16x + 5y)^2 + 231y^2 \Rightarrow \left(\tfrac{2p}{3}\right), \left(\tfrac{p}{7}\right), \left(\tfrac{2p}{11}\right) = 1.$$

$$(80.4)$$

Thus, we find that

$$\left(\tfrac{p}{3}\right), \left(\tfrac{p}{7}\right), \left(\tfrac{p}{11}\right) = \begin{cases} (1) \ +1, +1, +1 \ \Rightarrow \ Q_1, Q_4^{\pm}, \\ (2) \ -1, +1, -1 \ \Rightarrow \ Q_2^{\pm}, Q_8, \\ (3) \ -1, -1, +1 \ \Rightarrow \ Q_3, Q_5^{\pm}, \\ (4) \ +1, -1, -1 \ \Rightarrow \ Q_6^{\pm}, Q_7. \end{cases} \qquad (80.5)$$

Or rather in terms of residue classes,

$$\begin{array}{llll} (1) & p \equiv 2^{2r} & \Rightarrow & Q_1, Q_4^{\pm}, \\ (2) & p \equiv 2 \cdot 2^{2r} & \Rightarrow & Q_2^{\pm}, Q_8, \\ (3) & p \equiv 5 \cdot 2^{2r} & \Rightarrow & Q_3, Q_5^{\pm}, \\ (4) & p \equiv 10 \cdot 2^{2r} & \Rightarrow & Q_6^{\pm}, Q_7, \end{array} \qquad (80.6)$$

where the congruences are with respect to the modulus $231 = |D|$, and also $r \bmod 15$; in just the same way as (79.5) we see here, i.e., in (1), the appearance of the set $\{n^2 \bmod |D| : \langle n, D\rangle = 1\}$. Each of the families (1)–(4) of classes in $\mathcal{K}_+(-231)$ is called a genus ([DA, art.231]). In order that p, $p \nmid 231$, be represented by one of the forms contained in a prescribed genus, it is necessary and sufficient that p satisfies the set of signs allocated to the genus according to (80.5) or rather the corresponding congruence condition listed in (80.6). The resolution has become higher than (80.3). However, (80.5)–(80.6) do not yet provide a criterion sharp enough with which one may identify the form actually representing a given prime satisfying (80.2). We shall continue the discussion on this point in the Notes below.

Notes

[80.1] The prime number $p_1 = 402767 \equiv 2^{21}$ mod 231 corresponds to the genus (2). The solutions to the congruence equation $X^2 \equiv -231$ mod $4p_1$ satisfy $X \equiv \pm(-231)^{(p_1+1)/4} \equiv \pm 40933$ mod p_1 by Note [63.1] (Lagrange), so $40933^2 + 231 = 1040 \cdot 4p_1$. Thus, there should exist a $U \in \Gamma$ such that

$$Q = \begin{pmatrix} 402767 & 40933/2 \\ 40933/2 & 1040 \end{pmatrix} = \text{either } {}^tUQ_2^{\pm}U \text{ or } {}^tUQ_8U.$$

By the pre-reduction [+],

$$U = \left(WT^{20}WT^3WT^{-6}\right)^{-1} = -\begin{pmatrix} 374 & 19 \\ 59 & 3 \end{pmatrix}, \quad Q = {}^tUQ_2^+U.$$

Hence, p_1 is represented by Q_2^+:

$$402767 = 2 \cdot 374^2 + 374 \cdot 59 + 29 \cdot 59^2.$$

Naturally, Q_2^- represents p_1 also:

$$402767 = 2 \cdot 374^2 - 374 \cdot (-59) + 29 \cdot (-59)^2.$$

[80.2] The prime $p_2 = 378179 \equiv 2^5$ mod 231 as well corresponds to the genus (2). In much the same way as above,

$$Q = \begin{pmatrix} 378179 & 40977/2 \\ 40977/2 & 1110 \end{pmatrix} = \text{either } {}^tUQ_2^{\pm}U \text{ or } {}^tUQ_8U.$$

By the reduction [+],

$$U = \left(WT^{19}WT^2WT^6W\right)^{-1} = -\begin{pmatrix} 37 & 2 \\ 203 & 11 \end{pmatrix}, \quad Q = {}^tUQ_8U.$$

Hence, p_2 is represented by Q_8:

$$378179 = 8 \cdot 37^2 + 5 \cdot 37 \cdot 203 + 8 \cdot 203^2.$$

By the way Q_8 belongs to an ambiguous class; see Note [91.7].

[80.3] The prime number $p_3 = 5807$ which is congruent to p_2 mod 231 is represented not by Q_8 but by $Q_2^{\pm}$: $p_3 = Q_2^+(54, -1) = Q_2^-(54, 1)$. This implies the crucial fact, an annoying one, that it cannot be determined, by means of the classification mod 231 only, which class in a particular genus listed in (81.6) contains the form that actually represents the prime in question.

[80.4] Concerning the tables (79.5) and (80.6), group structures become evident if one views them through $(\mathbb{Z}/20\mathbb{Z})^*$ and $(\mathbb{Z}/231\mathbb{Z})^*$, respectively. However, observe that not the whole of these groups but their respective

356   *Quadratic Forms*

subgroups of order $\varphi(|D|)/2$ actually come up, the reason for which is in the Lagrange principle (Theorem 68). Efforts to extend this observation to general $\mathcal{K}(D)$ resulted in the genus theory (§91).

§81

Here we shall give details to (æ), §76, in the case of positive definite forms. We shall proceed without assuming (73.1) until Remark below.

Thus, let $Q = [|a,b,c|] \in \mathfrak{Q}_+(D)$ and $m > 0$ be given arbitrarily. Our basic problem is to algorithmically construct, from the data $\{a,b,c,m\}$, a solution to the indefinite equation $Q(x,y) = m$, if any. It is true that the problem can be solved in general by means of (76.10)–(76.11) reinforced with the pre-reduction [+], §77. However, such a procedure should become extremely involved as $|D|$ increases indefinitely. In order to ease this difficulty to a certain extent, we shall show an argument which stems directly from the viewpoint of Note [71.6]; directly, in the sense that we are able to dispense with (76.10)–(76.11) and the pre-reduction [+].

We begin with a simplification procedure: Taking Note [74.4] into account, we may suppose, without loss of generality, that Q is such that

$$\langle a, mD \rangle = 1. \tag{81.1}$$

Let $Q(u,v) = m$ is a proper representation, so $(2au + bv)^2 + |D|v^2 = 4am$. We have $\langle 2au + bv, v \rangle = \langle 2a, v \rangle = \langle 2, v \rangle$ since (81.1) implies that $\langle a, v \rangle = 1$. If $2|v$, then we consider instead the relation $(au + bv/2)^2 + |D|(v/2)^2 = am$. Thus $Q(u,v) = m$ is transformed into $u_1^2 + |D|v_1^2 = m_1$, $\langle u_1, v_1 \rangle = 1$, where m_1 is equal to either $4am$ or am. We apply Note [14.3] to the pair $\{|D|, m_1\}$ so that $|D| = \alpha_0\alpha_2\gamma(\alpha_3/\gamma)$ and $m_1 = \beta_0\beta_2\gamma(\beta_3/\gamma)$. Then α_3/γ, β_3/γ ought to be coprime squares and their product divides u_1^2. To confirm this claim, suppose that $p^\tau \| (\alpha_3/\gamma)$ with an odd τ. Then, since $p^{\tau+1} \nmid \beta_2$ by definition, we have in particular $p | u_1$, so $p \nmid v_1$. Namely, $p^\tau \| Dv_1^2$ and $p^{\tau+1} | m_1$, whence the contradiction $p^\tau \| u_1^2$ follows. Hence, α_3/γ is a square. In just the same way, we see that β_3/γ is a square. Since $(\alpha_3/\gamma)|\beta_2$, $(\beta_3/\gamma)|\alpha_2$, the square integer $\alpha_3\beta_3/\gamma^2$ divides u_1^2. Thus, there exist $u_2|u_1$, $\alpha_2^*|\alpha_2$ and $\beta_2^*|\beta_2$ such that $u_2^2 + \alpha_0\alpha_2^*\gamma v_1^2 = \beta_0\beta_2^*\gamma$. Dividing this identity by γ, we get $\gamma^*u_3^2 + \alpha_0\alpha_2^*v_1^2 = \beta_0\beta_2^*$, where $u_3 = u_2/\langle u_2, \gamma^\infty \rangle$ and $\gamma^* = \langle u_2, \gamma^\infty \rangle^2/\gamma$. We have $\langle \gamma^*u_3, \alpha_0\alpha_2^*v_1, \beta_0\beta_2^* \rangle = \langle \gamma^*u_3, v_1, \beta_0\beta_2^* \rangle = 1$, as $\langle \alpha_0\alpha_2^*, \beta_0\beta_2^* \rangle = 1$ and $\langle \gamma^*u_3, v_1 \rangle = 1$.

§81 357

With this, we see that we may discuss instead the indefinite equation

$$fx^2 + hy^2 = n,$$
$$\langle fx, hy, n \rangle = 1, \quad x, y \geq 1, \tag{81.2}$$

on the condition

$$1 \leq f < h, \quad f + h < n; \tag{81.3}$$

note that if $n \leq f + h$, then the discussion of (81.2) becomes trivial. We have, anyway,

$$\langle f, h \rangle = 1, \quad \langle x, y \rangle = 1, \quad \langle fhxy, n \rangle = 1, \quad 6 \leq 4f + h \leq n, \tag{81.4}$$

where the last is due to the fact that $\min\{fx^2 + hy^2\} = 4f + h$ provided $x, y \in \mathbb{N}$ while $xy \neq 1$. Also the case $fh = 1$ will be treated separately in Note [81.7].

Each proper representation of m by Q implies that $\gamma^* x^2 + \alpha_0 \alpha_2^* y^2 = \beta_0 \beta_2^*$, which is the type of (81.2), has solutions satisfying the coprimality condition. Thus, our strategy is to solve this derived equation first and feed the acquired data into the original problem with Q, although one should be aware that desired solutions do not necessarily follow via this argument.

Now, if $\{\sigma, \tau\}$ is a solution to (81.2)–(81.3), then there exists a ξ such that $\sigma \equiv \tau \xi \bmod n$ and

$$f\xi^2 \equiv -h \bmod n, \quad 1 < \xi < \tfrac{1}{2}n; \tag{81.5}$$

note that $1 < \xi$ because of the second condition in (81.3) and that $2\xi = n$ is obviously impossible. We apply Euclid's algorithm to the coprime pair $\{\xi, n\}$, and let $\{r_j : -1 \leq j \leq k\}$ be the residues thus arising; here $k \geq 2$, because of the inequality in (81.5). Let A_j/B_j be the convergent of the continued fraction expansion of ξ/n. Then we have

Theorem 82 *With $\{r_j\}$ as above, there exists a unique v such that*

$$r_{v+1} < \sqrt{n/f} < r_v, \quad 0 \leq v \leq k - 1; \tag{81.6}$$

and it holds that

$$\sigma = r_{v+1}, \quad \tau = B_v. \tag{81.7}$$

Proof Since $r_{-1} = \xi, r_0 = n, r_1 = \xi, r_k = 1$ by (21.10), the existence of such a v is immediate. The formula (21.12) gives $r_{j+1} = (-1)^j (\xi B_j - nA_j)$, so

$$fr_{j+1}^2 + hB_j^2 \equiv (f\xi^2 + h)B_j^2 \equiv 0 \bmod n, \quad -1 \leq j \leq k. \tag{81.8}$$

If $v = 0$ in (81.6), i.e., $\xi < \sqrt{n/f}$, then $f\xi^2 + h < n + h < 2n - 1$ because of the second condition in (81.3). Hence $f\xi^2 + h = n$; that is

$$v = 0 \implies fr_1^2 + hB_0^2 = n, \tag{81.9}$$

as $B_0 = 1$. We assume next that $\nu \geq 1$ in (81.6). Then, we exploit the fact that there exists an integer ℓ such that $\sigma = \tau\xi - \ell n$. We apply Theorem 23 or rather the contraposition of (23.11), with $a = \xi$, $b = n$, $s = \ell$, $t = \tau$, $V = \tau$, and find that

$$\text{if } \sigma < r_j \text{ with a } j, \ 1 \leq j \leq k, \text{ then } B_j \leq \tau. \tag{81.10}$$

This procedure is legitimate since the range of ξ stated in (81.5) means that the third case in (23.13) holds. Hence, on noting $\sigma < \sqrt{n/f} < r_\nu$, we have $hB_\nu^2 \leq h\tau^2 < n$, so $fr_{\nu+1}^2 + hB_\nu^2 < 2n$ because of (81.6). By the congruence $(81.8)_{j=\nu}$, we are led to the assertion that

$$(81.6) \ \Rightarrow \ fr_{\nu+1}^2 + hB_\nu^2 = n, \tag{81.11}$$

where we have taken (81.9) into account.

It remains to show (81.7): If $\sigma > r_{\nu+1}$, then $n = f\sigma^2 + h\tau^2 > fr_{\nu+1}^2 + hB_\nu^2 = n$, a contradiction. On the other hand, if $\sigma < r_{\nu+1}$, then we get $fr_{\nu+2}^2 + hB_{\nu+1}^2 = n$ just analogously as (81.11). It is implied that

$$\begin{aligned}
n^2 &= \left| r_{\nu+1}\sqrt{f} + iB_\nu\sqrt{h} \right|^2 \left| r_{\nu+2}\sqrt{f} + iB_{\nu+1}\sqrt{h} \right|^2 \\
&= \left(r_{\nu+1}r_{\nu+2}f - B_\nu B_{\nu+1}h \right)^2 + fh\left(r_{\nu+1}B_{\nu+1} + r_{\nu+2}B_\nu \right)^2.
\end{aligned} \tag{81.12}$$

However, $(21.11)_{j=\nu+1}$ gives $n = r_{\nu+1}B_{\nu+1} + r_{\nu+2}B_\nu$, and we are led to a contradiction again since $fh \geq 2$. Hence, $\sigma = r_{\nu+1}$ and $\tau = B_\nu$. This ends the proof.

We are now led to an algorithm to solve (81.2): Let f, h, n satisfying (81.3) be such that $\langle f, h \rangle = 1$, $\langle fh, n \rangle = 1$. Suppose that a solution ξ to (81.5) has been found by some means. Expand ξ/n into a continued fraction, and identify ν as in (81.6). Then check whether the identity (81.11) holds or not. If affirmative, then we have obtained a proper solution to (81.2), that is, $\langle r_{\nu+1}, B_\nu \rangle = 1$. Indeed, with $f\xi^2 + h = kn$, we have, by $(21.12)_{j=\nu}$,

$$1 = kB_\nu^2 - 2f\xi A_\nu B_\nu + fnA_\nu^2, \tag{81.13}$$

so $\langle B_\nu, n \rangle = 1$, and thus $\langle r_{\nu+1}, B_\nu \rangle = 1$ again by $(21.12)_{j=\nu}$. Therefore, in view of the last theorem we conclude that exactly all the solutions to (81.2) should be found in this way.

Or more drastically, i.e., without computing the residues $\{r_j\}$,

$$\text{Cornacchia's algorithm :}$$
$$\text{check if there exists a } j \text{ such that } ((n - hB_j^2)/f)^{1/2} \in \mathbb{N}. \tag{81.14}$$

This is because those y satisfying (81.2) ought to be in the set $\{B_j\}$, although the step concerning the coprimality remains.

$$\S81 \qquad\qquad 359$$

Remark If (73.1) is imposed, then we have $\langle am, D\rangle = 1$ in addition to (81.1), and the initial simplification gives $u_1^2 + |D|v_1^2 = m_1$ with $\langle D, m_1\rangle = 1$ or 4. Thus we are led to the situation $f = 1$. With this, the theorem is modified as follows:

$$\text{if there exists an integer } s \text{ such that } r_{\nu+1}^2 + hs^2 = n, \qquad (81.15)$$
$$\text{then } s = B_\nu \text{ and } \langle r_{\nu+1}, B_\nu\rangle = 1.$$

To confirm this, let k be such that $\xi^2 + h = kn$. Then by $(21.12)_{j=\nu}$

$$n = r_{\nu+1}^2 + hs^2 = n\bigl(ks^2 - 2\xi A_\nu B_\nu + nA_\nu^2\bigr) + (B_\nu^2 - s^2)\xi^2, \qquad (81.16)$$

so $n|(B_\nu^2 - s^2)$. On the other hand, (23.17) gives $B_\nu \le n/r_\nu$, and $B_\nu \le \sqrt{n}$ by $(81.6)_{f=1}$. Hence $|B_\nu^2 - s^2| < n$, as $s^2 < n$. Thus $B_\nu^2 - s^2 = 0$, that is, $s = B_\nu$. With this, we are led to (81.13) and finish the confirmation.

Notes

[81.1] The present section is a rework of Cornacchia (1908); see also Hardy et al (1990). The application of continued fractions or rather Euclid's algorithm to the problem (81.2) but with $f = 1$, $h = 2, 3$ is due to Lucas (1891, pp.455–456). This restriction on the coefficients was caused by that he followed the argument of Hermite (Note [58.6]). Cornacchia, still with $f = 1$, removed the restriction on h. Compared with the standard reduction argument, i.e., (76.10)–(76.11) accompanied by the pre-reduction [+], the advantage of Cornacchia's algorithm is, of course, in that one may reach a conclusion swiftly by means of continued fraction expansions, without making a possibly big excursion such as constructing $\mathcal{K}_+(-4fh)$. However, it should be noted that although the integer solutions to the original problem $Q(x, y) = m$ yield those for the derived equation of the type (81.2), there remains a step to check if any of the solutions to the latter recovers a desired $\{x, y\}$, as indicated already in the paragraph between (81.4) and (81.5). Nevertheless, in most cases we do not need to thoroughly perform the above simplification, as the numerical examples in the Notes below should indicate.

[81.2] We treated the indefinite equation $x^2 + 5y^2 = 404321$ in Note [79.1] and got the solution $\{111, 280\}$. Alternatively we proceed as follows: We apply Euclid's algorithm to the pair $\{\xi, n\} = \{160284, 404321\}$, and find that the relevant residues are

$$r_1 = 160284, \ r_2 = 83753, \ r_3 = 76531, \ r_4 = 7222,$$
$$r_5 = 4311, \ r_6 = 2911, \ r_7 = 1400, \ r_8 = 111, \ \ldots$$

360   *Quadratic Forms*

Thus, $r_v = 1400$, $r_{v+1} = 111$ ($v = 7$), and $\{(n - 111^2)/5\}^{1/2} = 280$. By the
way, we have the expansion

$$\frac{\xi}{n} = 0 + \frac{1}{2} + \frac{1}{1} + \frac{1}{1} + \frac{1}{10} + \frac{1}{1} + \frac{1}{1} + \frac{1}{2} + \frac{1}{12} + \cdots + \frac{1}{3}.$$

The 7th convergent is $\frac{111}{280}$, or $B_7 = 280$. Or rather by (81.14) it suffices to see
that $(n - 5 \cdot B_7^2)^{1/2} = 111$.

[81.3]   According to (79.5), the indefinite equation $x^2 + 5y^2 = 430883 = n$,
a prime, does not have any solution. By Note [63.1] (Lagrange) we have $\xi \equiv$
$95402 \bmod n$. Applying Euclid's algorithm, we have the candidate $r_v = 962$,
$r_{v+1} = 131$, $B_v = 411$ ($v = 7$). However, $r_{v+1}^2 \not\equiv n \bmod 5$, and moreover
$n - 5B_v^2 < 0$. Or rather by (81.14) it suffices to verify $(n - 5 \cdot B_j^2)^{1/2} \notin \mathbb{N}$ for
any j such that $B_j < (n/5)^{1/2}$.

[81.4]   Consider the indefinite equation $x^2 + 5y^2 = 435629 = n$. In the
notation of §37 we have $n \notin \text{pp}(2)$, so n is composite. By the ρ algorithm
(38.1) we find the decomposition $n = p_1 p_2$, $p_1 = 367$, $p_2 = 1187$. Both of
these prime factors satisfy (79.5) (2) and they can be represented by Q_2 but
neither by Q_1. Yet, by Note [79.3], n can be represented by Q_1, so the present
indefinite equation should have a solution. We shall confirm this by means
of Cornacchia's algorithm without relying on Note [79.3]: By Note [63.1]
(Lagrange), $t_l \equiv \pm(-5)^{(p_l+1)/4} \bmod p_l$, $l = 1, 2$, satisfy $t_l^2 \equiv -5 \bmod p_l$.
Thus, $t_1 \equiv \pm 27 \bmod p_1$, $t_2 \equiv \pm 282 \bmod p_2$. Then, we note that

$$\frac{367}{1187} = 0 + \frac{1}{3} + \frac{1}{4} + \frac{1}{3} + \frac{1}{1} + \frac{1}{2} + \frac{1}{1} + \frac{1}{5} \Rightarrow 207 \cdot 367 - 64 \cdot 1187 = 1.$$

Via $\xi \equiv 207 \cdot 367 \cdot t_2 - 64 \cdot 1187 \cdot t_1 \bmod n$, we have $\xi^2 \equiv -5 \bmod n$. There
are four candidates 49572, 204446, 231183, 386057 for ξ. Obviously we may
discard the last two. We find that

$$\xi = 49572 \ : \ (n - 5 \cdot B_6^2)^{1/2} = 123,$$
$$\xi = 204446 : \ (n - 5 \cdot B_7^2)^{1/2} = 228;$$

that is,

$$435629 = \begin{cases} 123^2 + 5 \cdot 290^2, \\ 228^2 + 5 \cdot 277^2. \end{cases}$$

[81.5]   (1) The indefinite equation $2x^2 + xy + 29y^2 = 402767 = p$ has
the solution $\{374, 59\}$; it was obtained already in Note [80.1]. Here we shall
apply Cornacchia's algorithm: We have $(4x + y)^2 + 231y^2 = 8p = n$. Solving

$X^2 \equiv -231 \bmod n$, we get eight roots $X_j^{\pm} \equiv 2jp \pm 40933 \bmod n$, $0 \le j \le 3$. Among these, we take $\xi = 764601 \equiv X_1^- \bmod n$. We have then

$$\frac{\xi}{n} = 0 + \frac{1}{4} + \frac{1}{4} + \frac{1}{1} + \frac{1}{2} + \frac{1}{34} + \frac{1}{1} + \frac{1}{3} + \frac{1}{4} + \frac{1}{45} + \frac{1}{2}.$$

We find that $(n - 231 \cdot B_4^2)^{1/2} = 1555$ with $B_4 = 59$. Further $(1555 - 59)/4 = 374$. The same solution is obtained with X_3^+, $v = 5$ since $X_1^- \equiv -X_3^+ \bmod n$. Any of other $X_j^{\pm}$ does not provide solution.

(2) The indefinite equation $2x^2 + xy + 29y^2 = 201119 = p'$ has no solution: Roots of $Y^2 \equiv -231 \bmod n'$ with $n' = 8p'$ are $Y_j^{\pm} \equiv 2jp' \pm 56013 \bmod n'$, $0 \le j \le 3$. We take $\xi' = 346225 \equiv Y_1^- \bmod n'$.

$$\frac{\xi'}{n'} = 0 + \frac{1}{4} + \frac{1}{1} + \frac{1}{1} + \frac{1}{1} + \frac{1}{5} + \frac{1}{49} + \cdots + \frac{1}{4}.$$

The fifth convergent is $\frac{17}{79}$, and $409^2 + 231 \cdot 79^2 = n'$. However, $409 \not\equiv 79 \bmod 4$. The same holds with Y_3^+. Any of other $Y_j^{\pm}$ does not yield solution either.

[81.6] We shall quickly solve the indefinite equation $43x^2 + 97y^2 = 8329187 = n$: That n is a prime can be proved by a repeated use of the criterion (41.1): $n - 1 = 2 \cdot 43 \cdot n'$, where $n' = 96851$ is a prime, for the order of 2 mod n' is equal to $n' - 1 = 2 \cdot 5^2 \cdot 13 \cdot 149$; and the order of 2 mod n is $n - 1$. Then, we apply Cornacchia's algorithm, with $f = 43$, $h = 97$. First, we need to have ξ such that $\xi^2 \equiv -97 \cdot 43^{-1} \equiv 2130720 \bmod n$. By Note [63.1] (Lagrange), $\xi = 2851144$ since $2130720^{(n+1)/4} \equiv -2851144 \bmod n$. We have

$$\frac{\xi}{n} = 0 + \frac{1}{2} + \frac{1}{1} + \frac{1}{11} + \frac{1}{1} + \frac{1}{2} + \frac{1}{1} + \frac{1}{1} + \frac{1}{157} + \cdots + \frac{1}{3},$$

with which $((n - 97 \cdot B_7^2)/43)^{1/2} = 203$, $B_7 = 260$, that is,

$$43 \cdot 203^2 + 97 \cdot 260^2 = 8329187.$$

[81.7] That (81.5) has a solution does not guarantee that (81.2) has indeed a solution, as observed in Note 81.3. Nevertheless, with $fh = 1$, one obtains a solution whenever there exists a ξ such that $\xi^2 \equiv -1 \bmod n$. In particular, one does not need to assume the existence of s in $(81.14)_{h=1}$; it suffices to have $\xi \bmod n$ only. To verify this claim, let v be as in $(81.6)_{f=1}$. Then, $B_v \le n/r_v < \sqrt{n}$. Thus, $r_{v+1}^2 + B_v^2 < 2n$. Via (81.8) with $fh = 1$, we get $r_{v+1}^2 + B_v^2 = n$. That is, we have proved the existence of s required in $(81.15)_{h=1}$; and consequently $\langle r_{v+1}, B_v \rangle = 1$.

[81.8] An extension of Note [78.11]: If (81.2), $fh \ge 2$, has two or more genuinely different solutions, then n is composite. More precisely, assume that

362   *Quadratic Forms*

$n = fa^2 + hb^2 = fc^2 + hd^2$ with $a, b, c, d > 0$ and $a \neq c$. Then $\langle n, ad - bc \rangle$ is a non-trivial factor of n: $fa^2 \equiv -hb^2, hd^2 \equiv -fc^2 \Rightarrow fh(ad)^2 \equiv fh(bc)^2$; and $(ad)^2 \equiv (bc)^2 \bmod n$, as $\langle fh, n \rangle = 1$. We have $n = \langle n, (ad - bc)(ad + bc) \rangle$. Besides, $|ad \pm bc| < n$, for $n^2 = (fac \mp hbd)^2 + fh(ad \pm bc)^2$. If $\langle n, ad - bc \rangle = 1$, then $ad + bc \equiv 0 \bmod n$, so $ad + bc = 0$, which is a contradiction. Also, if $ad - bc = 0$, then $a = c, b = d$, which is a contradiction again. Hence $1 < \langle n, ad - bc \rangle < n$. In Note [81.4], we have $n = 123^2 + 5 \cdot 290^2 = 228^2 + 5 \cdot 277^2$. So $ad - bc = -32049$. The factor $1187 = \langle n, ad - bc \rangle$ is retrieved; note that $\langle n, ad + bc \rangle = 367$. See Euler (1778a) as well as our §49 (iv).

§82

Now we turn to indefinite forms. We shall make the reduction argument (76.10)–(76.11) practical in order to constructively solve (71.2) in the case of positive discriminants. We shall proceed under the convention Note [5.1](3).

Thus, in the present and the succeeding five sections, we shall deal solely with the set

$$\mathcal{Q}_\pm(D) = \Big\{ [|a, b, c|] \in \mathcal{Q}(D) : D = b^2 - 4ac > 0 \Big\}, \tag{82.1}$$

where $a \leqslant 0$, but see Remark 1 below. This is, of course, the same as $\mathcal{Q}(D)$, and all the assertions in §§71–76 hold without change. Nevertheless, we attach the subscript $\pm$ in order to stress that the values of forms can now be either positive or negative; thus, for instance, $Q \in \mathcal{Q}_\pm(D) \Leftrightarrow -Q \in \mathcal{Q}_\pm(D)$, an aspect which is not shared by $\mathcal{Q}_+(D)$. Accordingly, we write

$$\mathcal{K}_\pm(D) = \mathcal{Q}_\pm(D) /\!\!/ \Gamma, \quad \mathfrak{h}_\pm(D) = |\mathcal{K}_\pm(D)|; \tag{82.2}$$

so $\mathcal{K}(D) = \mathcal{K}_\pm(D)$, and by (76.1)

$$D > 0 \Rightarrow \mathfrak{h}_D = \mathfrak{h}_\pm(D). \tag{82.3}$$

Remark 1 Because of Note [74.5], the sign of the first coefficient a is generally immaterial in dealing with indefinite forms; we may restrict ourselves to the situation $a > 0$. However, such a restriction can cause inconvenience as well, so we shall leave sgn a without being fixed and make it precise only if necessary to do so.

Our issue (71.2) becomes the task to devise an algorithm to find all integral points $\{x, y\}$ on a particular hyperbola $Q(x, y) = m > 0$, which is the same as fixing $\mathcal{S}_Q(m)$ defined by (74.2). Since hyperbolas extend to infinity, (71.2) becomes far more interesting than before; in the preceding five sections we

§82   363

dealt with ellipses which are of course bounded. The overall conclusion of the discussion to be developed hereafter is expressed by this statement: If $\mathcal{S}_Q(m) \neq \varnothing$, then it is a disjoint union of a finite number of infinite sequences; they are generated all in the same way, i.e., each by means of a special solution, or a seed solution and an infinite cyclic group, while the seed and the generator of the group can be computed explicitly.

To attain this grand assertion, we shall begin with a new approach to the reduction theory; it is based on the criterion (82.22) below or rather the combination of (82.13)–(82.14) which is of a higher resolution than (76.2) and enables us to analyze quite precisely the structure of $\mathcal{Q}_\pm(D)$ by means of the theory of continued fractions.

Thus, modifying (77.4), we associate each $Q = [\,|a,b,c|\,] \in \mathcal{Q}_\pm(D)$ with the pair $\{\omega^+(Q), \omega^-(Q)\}$ of irrational real numbers:

$$\omega^+(Q) = \frac{-b + \sqrt{D}}{2a}, \quad \omega^-(Q) = \frac{-b - \sqrt{D}}{2a},$$
$$a > 0 \Leftrightarrow \omega^+(Q) > \omega^-(Q), \quad a < 0 \Leftrightarrow \omega^+(Q) < \omega^-(Q). \tag{82.4}$$

The allocation of the signs of the radicals is essential; so

$$\omega^+(-Q) = \omega^-(Q), \quad \omega^-(-Q) = \omega^+(Q). \tag{82.5}$$

That is, we regard the two roots of the quadratic equation $ax^2 + bx + c = 0$ not as independent real numbers but as a conjugate pair:

$$\begin{aligned} &\text{if } \omega^+(Q) \text{ is a root of a quadratic equation over } \mathbb{Q}, \\ &\text{then the other root is } \omega^-(Q), \text{ and vice versa,} \end{aligned} \tag{82.6}$$

for such a quadratic polynomial is a constant multiple of $ax^2 + bx + c$.

The counterpart of (77.6) holds obviously with (82.4): in $\mathcal{Q}_\pm(D)$

$$\text{the mapping } Q \mapsto \omega^+(Q) \text{ is injective;} \tag{82.7}$$

and that of (77.7) is

$$U^{-1}(\omega^+(Q)) = \omega^+({}^tUQU), \quad U^{-1}(\omega^-(Q)) = \omega^-({}^tUQU), \quad \forall U \in \Gamma. \tag{82.8}$$

This is readily confirmed for $U = T, W$, and it suffices to quote Theorem 4. In particular,

$$V(\omega^\pm(Q)) = \omega^\pm(Q), \quad \forall V \in \mathrm{Aut}_Q, \tag{82.9}$$

with signs corresponding. Moreover, we have

$$\begin{aligned} P^{-1}(\omega^+(Q)) &= \omega^-({}^tPQP) = \omega^+({}^tP(-Q)P), \\ P^{-1}(\omega^-(Q)) &= \omega^+({}^tPQP) = \omega^-({}^tP(-Q)P), \end{aligned} \quad \forall P \in \diamond\Gamma, \tag{82.10}$$

364   *Quadratic Forms*

since $J(\omega^+(Q)) = \omega^-(JQJ)$ with $J = \left(\begin{smallmatrix} 0 & 1 \\ 1 & 0 \end{smallmatrix}\right)$, and

$$P^{-1}\big(\omega^+(Q)\big) = P^{-1}J^{-1}\big(\omega^-(JQJ)\big)$$
$$= \omega^-\big({}^t(JP)JQJ(JP)\big) = \omega^-\big({}^tPQP\big). \tag{82.11}$$

In the case of positive definite forms, we were able to interpret the equivalence (74.10) between forms into the equivalence (77.13) between points on the upper half-plane $\mathcal{H}$; and the discontinuity of the action of Γ on $\mathcal{H}$ was exploited in an essential manner. In the case of indefinite forms, we are obviously concerned with the action of Γ on $\mathbb{R} \cup \{\infty\}$. Then, a difficulty arises: this action is not discontinuous; more precisely, for any irreducible fraction a/b we may compose $U = \left(\begin{smallmatrix} a & * \\ b & * \end{smallmatrix}\right) \in \Gamma$ by Euclid's algorithm, and we have $U(\infty) = a/b$, whence any pair of rational numbers are equivalent mod Γ to each other. We have here an enormous maze which is tightly packed.

Nevertheless, if we consider the action (82.8) of Γ on the set of pairs

$$\Big\{ \{\omega^+(Q), \omega^-(Q)\} : Q \in \mathcal{Q}_\pm(D) \Big\}, \tag{82.12}$$

then fortunately we are able to attain a resolution that is to some extent comparable to Theorem 77. In order to make this fact precise and provide a proof to it, we make the following two basic definitions:

$$\omega^+(Q) \text{ is a reduced quadratic irrational}$$
$$\Leftrightarrow 0 < -\omega^-(Q) < 1 < \omega^+(Q), \tag{82.13}$$

and

$$Q \text{ is a reduced form}$$
$$\Leftrightarrow \omega^+(Q) \text{ is a reduced quadratic irrational.} \tag{82.14}$$

We shall call both situations reduced. It is not an empty definition since for each discriminant $D > 0$ there exists a reduced form: for $D = 4d + 1$, let $f = [(\sqrt{D} - 1)/2]$ or $(2f + 1)^2 < D < (2f + 3)^2$, and let

$$Q = [|1, -(2f + 1), f^2 + f - d|]. \tag{82.15}$$

Then the pair

$$\omega^+(Q) = f + \tfrac{1}{2}\big(1 + \sqrt{D}\big), \quad \omega^-(Q) = f + \tfrac{1}{2}\big(1 - \sqrt{D}\big) \tag{82.16}$$

satisfies (82.13). As for $D = 4d$, we let

$$Q = [|\,1, -2[\sqrt{d}\,], [\sqrt{d}\,]^2 - d\,|]. \tag{82.17}$$

Then the pair

$$\omega^+(Q) = [\sqrt{d}\,] + \sqrt{d}, \quad \omega^-(Q) = [\sqrt{d}\,] - \sqrt{d} \tag{82.18}$$

satisfies (82.13).

Relying on (82.13)–(82.14), we shall give an alternative proof to the finiteness of $\mathfrak{h}_{\pm}(D)$. The new proof, which is, in fact, contained (at (82.31) below) in that of the next theorem, offers a practical method to select the forms representing classes in $\mathcal{K}_{\pm}(D)$. It exploits the nature of the continued fraction expansion of $\omega^+(Q)$.

Remark 2 The concepts of reduced forms for negative and positive discriminants can, in fact, be discussed in a partly unified way: There exists a geometric argument classifying pairs $\{\omega^+(Q), \omega^-(Q)\}$ by means of Theorem 76, §77. It should be particularly stressed that this alternative argument opens a truly fascinating landscape in mathematics. See the closing paragraph of the present section. Nevertheless, for our immediate aim to solve (71.2) it is definitely realistic to use the theory of continued fractions, at least presently.

Thus, we have the following basic theorem concerning the relation between reduced quadratic irrationals and their continued fraction expansions: We call an irrational real number periodic if its continued fraction expansion is periodic, that is, a fixed block of a finite number of consecutive terms repeats from a certain step onward. If this repetition starts at the first term, then such a real number is called purely periodic.

Theorem 83

$$\text{The number of reduced forms in } \mathfrak{Q}_{\pm}(D) \text{ is finite;} \tag{82.19}$$

$$\text{Any } \omega^+(Q) \text{ is periodic;} \tag{82.20}$$

$$Q \text{ is reduced } \Leftrightarrow \omega^+(Q) \text{ is purely periodic.} \tag{82.21}$$

Proof These fundamental assertions are essentially due to Lagrange and Legendre, although the definition (82.13), or rather (82.22) below, is basically due to Gauss; see Notes [82.1]–[82.3] for historical comments. First we shall confirm (82.19). For any reduced $Q = [[a, b, c]] \in \mathfrak{Q}_{\pm}(D)$, we have $\omega^+(Q) - \omega^-(Q) > 0$ by the definition (82.13)–(82.14), so in particular $a > 0$, and by $\omega^+(Q) + \omega^-(Q) > 0$ we have $b < 0$. Hence, $|b| < \sqrt{D}$, as $\omega^-(Q) < 0$. Then, from $0 < D - b^2 = -4ac$ follows $c < 0$. Collecting these, we get

$$\text{Gauss's reduction condition concerning the set } \mathfrak{Q}_{\pm}(D) \quad : \quad \begin{array}{c} a > 0,\, b < 0,\, c < 0;\ |b| < \sqrt{D}; \\ \tfrac{1}{2}\left(\sqrt{D} - |b|\right) < a,\ |c| < \tfrac{1}{2}\left(\sqrt{D} + |b|\right); \\ \langle a, b, c \rangle = 1. \end{array} \tag{82.22}$$

This is an analogue of (77.17). In the middle line on the right, the inequality for a follows from $-\omega^-(Q) < 1 < \omega^+(Q)$, and that for c from $a|c| = \tfrac{1}{4}(D - b^2)$.

366   *Quadratic Forms*

The number of possible tuples $\{a, b, c\}$ is within a limit which depends solely on D. Hence (82.19) holds.

In order to confirm (82.20), we take an arbitrary $Q \in \mathcal{Q}_\pm(D)$ and write $\eta_0 = \omega^+(Q)$. As in (22.4)–(22.5), we have the expansion

$$
\begin{aligned}
\eta_0 &= s_0 + \cfrac{1}{s_1 + } \cfrac{1}{s_2 + } \cdots + \cfrac{1}{s_j + } \cfrac{1}{\eta_{j+1}} \\
&= \frac{F_j \eta_{j+1} + F_{j-1}}{G_j \eta_{j+1} + G_{j-1}}, \quad \eta_{j+1} = \mathrm{R}^{j+1}(\eta_0).
\end{aligned}
\tag{82.23}
$$

Accordingly, we introduce the cf-orbit, or rather the Legendre orbit $\{Q_j : j \geq 0\}$ of Q:

$$
Q_j = {}^t H_{j-1}\big((\det H_{j-1})Q\big)H_{j-1}, \quad H_j = \begin{pmatrix} F_j & F_{j-1} \\ G_j & G_{j-1} \end{pmatrix},
\tag{82.24}
$$

where $Q_0 = Q$, and H_{-1} is identity; note that $\det H_j = (-1)^{j+1}$. We have $H_{j-1}^{-1}(\eta_0) = \eta_j$, so via (82.8)–(82.10) we find that

$$
\begin{aligned}
\omega^+(Q_j) &= \mathrm{R}^j(\omega^+(Q)), \\
Q_{2j} &\equiv Q \bmod \Gamma,
\end{aligned} \quad j \geq 0.
\tag{82.25}
$$

On the other hand, with $Q_j = [|a_j, b_j, c_j|]$ we have

$$
\begin{aligned}
a_{j+1} &= (-1)^{j+1}\big(aF_j^2 + bF_j G_j + cG_j^2\big), \\
b_{j+1}^2 &= D + 4a_{j+1}c_{j+1}, \\
c_{j+1} &= (-1)^{j+1}\big(aF_{j-1}^2 + bF_{j-1}G_{j-1} + cG_{j-1}^2\big).
\end{aligned}
\tag{82.26}
$$

The absolute values of these coefficients are within the limit depending solely on Q. To see this, we note that

$$
\begin{aligned}
a_{j+1} = (-1)^{j+1}\, G_j^2\big(&a(F_j/G_j - \eta_0)(F_j/G_j - \eta_0 + 2\eta_0) \\
&+ b(F_j/G_j - \eta_0)\big),
\end{aligned}
\tag{82.27}
$$

where we have used $a\eta_0^2 + b\eta_0 + c = 0$ as well as $G_j \neq 0$, $j \geq 0$. Applying (22.10), we have $|a_{j+1}| < |a|(2|\eta_0| + 1) + |b|$. The same holds for $|c_{j+1}|$, and $|b_{j+1}|^2 < D + 4(|a|(2|\eta_0| + 1) + |b|)^2$, which ends the confirmation. Therefore, there exists $0 \leq k < l$ such that $Q_k = Q_l$. In particular,

$$
\eta_k = \omega^+(Q_k) = \omega^+(Q_l) = \eta_l.
\tag{82.28}
$$

This means that in the expansion (82.23) we have

$$
\eta_k = s_k + \cfrac{1}{s_{k+1} + } \cfrac{1}{s_{k+2} + } \cdots + \cfrac{1}{s_{l-1} + } \cfrac{1}{\eta_k}.
\tag{82.29}
$$

Namely, $\eta_0 = \omega^+(Q)$ is periodic, and (82.20) has been confirmed. Moreover, these $\eta_k > 1$ are reduced in the sense of (82.13). To see this, we put

$$\begin{pmatrix} s_k & 1 \\ 1 & 0 \end{pmatrix}\begin{pmatrix} s_{k+1} & 1 \\ 1 & 0 \end{pmatrix}\cdots\begin{pmatrix} s_{l-1} & 1 \\ 1 & 0 \end{pmatrix} = \begin{pmatrix} A & A' \\ B & B' \end{pmatrix}, \tag{82.30}$$

which is not the identity matrix. With $g(x) = Bx^2 + (B' - A)x - A'$, we have $g(\eta_k) = 0$, $g(0) = -A' < 0$, $g(-1) = B - B' + A - A' > 0$. Thus, the two roots $\{\eta_k, \eta_k'\}$ of $g(x) = 0$ satisfy $0 < -\eta_k' < 1 < \eta_k$; moreover, $\eta_k' = \omega^-(Q_k)$ by (82.6). We have found that η_k is reduced, and so is Q_k. Also, all $\eta_j, j > k$, are analogous to η_k; and $Q_j, j \geq k$, are all reduced. Therefore, via (82.24) we see that

$$\text{for each } Q \text{ there exists a reduced } S$$
$$\text{such that } Q \equiv S \bmod \Gamma; \text{ in particular, } \mathfrak{h}_\pm(D) < +\infty. \tag{82.31}$$

For if k is even, then $H_{k-1} \in \Gamma$, and $Q \equiv Q_k \bmod \Gamma$; if k is odd, then $H_k \in \Gamma$, and $Q \equiv Q_{k+1} \bmod \Gamma$.

It remains for us to confirm (82.21). The sufficiency part has already been proved in the discussion following (82.30). Thus, suppose that Q is reduced. We use the expansion (82.23) again; and the discussion up to (82.29) does not need any change, naturally. If $k = 0$, then the confirmation ends. Thus we suppose that $k \geq 1$. We then observe that not only $\omega^+(Q) = \eta_0$ but also $\omega^+(Q_1) = \eta_1 = (\eta_0 - s_0)^{-1} > 1$ is reduced. For we have $H_0^{-1}(\omega^-(Q)) = \omega^-(Q_1)$ by (82.10) and (82.24), and $\omega^-(Q_1) = (\omega^-(Q) - s_0)^{-1}$ which implies that $-1 < \omega^-(Q_1) < 0$. In the same way, we find that $\omega^+(Q_j) = \eta_j, j \geq 0$, are all reduced; and using the relation $\omega^-(Q_j) = (\omega^-(Q_{j-1}) - s_{j-1})^{-1}$, $j \geq 1$, we see that $0 < -s_{j-1} - 1/\omega^-(Q_j) < 1$, or $s_{j-1} = [-1/\omega^-(Q_j)]$ (the integral part). Consequently, in (82.29) we have $s_{l-1} = [-1/\omega^-(Q_l)] = [-1/\omega^-(Q_k)] = s_{k-1}$. That is, if $k \geq 1$, then $\eta_{k-1} = \eta_{l-1}$. Inductively, we get $\eta_0 = \eta_{l-k}$, which confirms (82.21). We end the proof of Theorem 83.

Following the general notion of periods, we make the definition:

$$\text{the period of } Q \text{ is the size of the least periodic block}$$
$$\text{in the continued fraction expansion of } \omega^+(Q). \tag{82.32}$$

The size of any periodic block of $\omega^+(Q)$ is a multiple of the period, which can be shown routinely by the division algorithm (2.1).

In passing, we remark a trivial fact: For any $Q \in \mathcal{Q}_\pm(D)$ and $j \geq 0$, there exist $v_j, w_j \in \mathbb{Z}$ such that

$$R^j(\omega^+(Q)) = \frac{v_j + \sqrt{D}}{2w_j}, \qquad \begin{array}{l} D \equiv 0 \bmod 4 \Rightarrow 2 \mid v_j, \\ D \equiv 1 \bmod 4 \Rightarrow 2 \nmid v_j. \end{array} \tag{82.33}$$

368   *Quadratic Forms*

Notice that the sign of the square root is plus. This formula follows from (82.25) and will have an application later in §87.

An alternative definition due to Smith (1877: written in 1874) of being reduced, concerning indefinite forms:

Take an arbitrary $Q = [|a,b,c|] \in \mathcal{Q}_\pm(D)$, and let $\mathfrak{l}_Q$ be the upper half of the circle in $\mathbb{C}$, starting at $\omega^+(Q)$ and ending at $\omega^-(Q)$; it is called a hyperbolic geodesic. When transforming Q by elements of Γ (that is, (74.9)), according to Theorem 76 and (82.8) there should be a corresponding image of $\mathfrak{l}_Q$ which passes through the fundamental domain $\mathcal{F}$. With this observation, we make the definition:

$$Q \text{ is Smith-reduced } \Leftrightarrow \mathfrak{l}_Q \cap \mathcal{F} \neq \varnothing. \tag{82.34}$$

If $Q = [|a,b,c|]$ satisfies this, then the geometric relation between the points $(\pm 1 + i\sqrt{3})/2$ and $\mathfrak{l}_Q = \{x,y : a(x^2 + y^2) + bx + c = 0, y > 0\}$ implies that $(4a \pm b)^2 + 3b^2 \leq 4D$ (either sign); here Remark 1 above is irrelevant. The number of such tuples $\{a,b,c\}$ is evidently finite with each $D > 0$. Therefore, we get $\mathfrak{h}_\pm(D) < \infty$. See Klein (1890, pp.250–260; especially the footnote for §6 by R. Fricke). Continued to the later parts of §§85 and 94. There we shall connect each class in $\mathcal{K}_\pm(D)$ with a closed geodesic on the Riemann surface $\mathcal{F}^*$ resulting from the identification, i.e., gluing, of edges of $\mathcal{F}$ which are Γ-congruent.

Notes

[82.1]   The above reduction method concerning indefinite quadratic forms via the theory of continued fractions is based essentially on the accounts by Lagrange (1773, Probléme IV) and Legendre (1798, Première partie, §XIII).

[82.2]   The reduction condition (82.22) is actually a modification of the version due to Dirichlet (1863, §74) of Gauss's original condition [DA, art.183]. Dirichlet based his argument on the correspondence $[|a,2b,c|] \leftrightarrow \left(-b \pm \sqrt{b^2 - ac}\right)/c$, and moreover, following Gauss he restricted himself to the use of the proper equivalence, that is, the use of elements of Γ. Thus our argument is different from that of Gauss and Dirichlet. We have adopted, from Gauss' theory, the notion of reduced forms, i.e., (82.14). Otherwise, we closely follow the method of Lagrange and Legendre; as mentioned already they relied, in a crucial way, on the theory of continued fractions which Gauss was apparently reluctant to fully use in [DA], which made his theory (arts.183–205) extremely hard to follow. It should be known, however, that he had a decent reason to do so: He must have hunted for arguments which would extend to situations higher than quadratic; for that purpose the theory of continued fractions appears to be of limited use. Nevertheless, one will find in the later sections, especially in §87, that the practical power of continued fractions is indeed irresistible despite its inherent limitation.

[82.3]   It is obvious that periodic continued fractions represent roots of quadratic equations over $\mathbb{Q}$. Its converse assertion (82.20) is a famous discovery due to Lagrange (1770a, pp.614–615). However, it is Euler (1759) who grasped the relation between the issue (71.2) for indefinite forms and periodic continued fractions. The above proof of (82.20) is due to Charves (1877). The explicit use of the cf-orbit $\{Q_j\}$ defined in (82.24) originated with Legendre (1798, p.126).

[82.4]   Galois (1828). Let r be the period of a reduced $\omega^+(Q)$. Then

$$\omega^+(Q) = s_0 + \cfrac{1}{s_1} + \cfrac{1}{s_2} + \cdots + \cfrac{1}{s_{r-1}} + \cfrac{1}{\omega^+(Q)}$$

$$\Leftrightarrow \quad \frac{-1}{\omega^-(Q)} = s_{r-1} + \cfrac{1}{s_{r-2}} + \cfrac{1}{s_{h-3}} + \cdots + \cfrac{1}{s_1} + \cfrac{1}{s_0} + \frac{1}{-1/\omega^-(Q)}.$$

To show this, we use (82.23) with $j = r - 1$. On noting (82.6), we have

$$\omega^+(Q) = \frac{F_{r-1}\omega^+(Q) + F_{r-2}}{G_{r-1}\omega^+(Q) + G_{r-2}} \quad \Leftrightarrow \quad \frac{-1}{\omega^-(Q)} = \frac{F_{r-1}(-1/\omega^-(Q)) + G_{r-1}}{F_{r-2}(-1/\omega^-(Q)) + G_{r-2}}.$$

On the right side, we have $-1/\omega^-(Q) > 1$ as well as

$$\begin{pmatrix} F_{r-1} & G_{r-1} \\ F_{r-2} & G_{r-2} \end{pmatrix} = {}^t\left\{ \begin{pmatrix} s_0 & 1 \\ 1 & 0 \end{pmatrix} \begin{pmatrix} s_1 & 1 \\ 1 & 0 \end{pmatrix} \cdots \begin{pmatrix} s_{r-1} & 1 \\ 1 & 0 \end{pmatrix} \right\}.$$

§83

Thus, each class in $\mathcal{K}_\pm(D)$ is represented by a reduced form satisfying (82.22). We shall now introduce a sort of visualization of Theorem 83 to indicate a practical way to select a representative set of reduced forms. Here the convention Note [5.1] is extended to $\diamond\Gamma$.

To this end, we observe that the theorem induces, in fact, a classification of forms in $\mathcal{Q}_\pm(D)$ with respect to $\mathrm{GL}(2,\mathbb{Z}) = \Gamma \sqcup \diamond\Gamma$: As is depicted by (82.21), cf-orbits of reduced forms look like circles perpetually rotating, and the converse holds true. According to (82.19), the number of these circles is finite. Moreover, after a finite number of steps, the cf-orbit of each form in $\mathcal{Q}_\pm(D)$ reaches one of these circles and starts winding, which is the assertion (82.20). If the cf-orbits of two different forms $Q, Q' \in \mathcal{Q}_\pm(D)$ come to a particular circle, possibly at its two different points, then obviously there exists a $Y \in \Gamma \sqcup \diamond\Gamma$ such that $Q' = {}^tY((\det Y)Q)Y$. Conversely, if this identity holds,

370   *Quadratic Forms*

then the cf-orbits of Q, Q' coincide from a certain point on, so they come to the same circle. For under this condition $Y^{-1}(\omega^+(Q)) = \omega^+(Q')$ by (82.8)–(82.10), and by Theorem 26, §25, there exist k, k' such that $R^k(\omega^+(Q)) = R^{k'}(\omega^+(Q'))$, that is, $\omega^+(Q_k) = \omega^+(Q'_{k'})$ because of (82.25), which with (82.7) implies $Q_k = Q'_{k'}$. Hence, we see that the circles of reduced forms stand for a classification of elements in $\mathfrak{Q}_\pm(D)$ with respect to the action of $GL(2, \mathbb{Z}) = \Gamma \sqcup \diamond \Gamma$.

Looking into this configuration, one may determine the class number $\mathfrak{h}_\pm(D)$, but only after further examining the circles, for if two forms are equivalent in the sense of (74.10), then their cf-orbits certainly come to the same circle, but the converse does not hold true in general. The correct picture is drawn in (83.1) below:

Pre-reduction [$\pm$]:

(1) find all reduced forms $\{Q\} \subset \mathfrak{Q}_\pm(D)$ according to (82.22). Expand quadratic irrationals $\{\omega^+(Q)\}$ into continued fractions; they ought to be purely periodic.

(2) examine the cf-orbit of a reduced form Q defined by (82.24). If its period r is

$$\begin{aligned} \text{odd} &\Rightarrow \text{ the orbit represents a single class,} \\ \text{even} &\Rightarrow \text{ the orbit represents two inequivalent classes.} \end{aligned} \tag{83.1}$$

(3) Then enter into (76.10).

Verification of (83.1): If r is odd, then the move of one step following rounding once is the same as an application of a transformation by an element in Γ, so all forms on a circle are equivalent to each other mod Γ. On the other hand, if r is even, then the assumption $Q_0 \equiv Q_1$ mod Γ induces a contradiction. To show this, we again invoke Theorem 26 but with $\nu \equiv 0$ mod 2. On choosing $J \equiv j$ mod 2 appropriately and noting (82.25), we are led to

$$R^{J+2}(\omega^+(Q_0)) = R^{j+2}(\omega^+(Q_1)) = R^{j+3}(\omega^+(Q_0)) \tag{83.2}$$
$$\Rightarrow r|(J - j - 1).$$

This is a contradiction, and we end the confirmation.

As a matter of fact, with a particular positive discriminant, only one of the two cases given in (83.1) takes place: More precisely we have

Theorem 84 *If a form of discriminant $D > 0$ turns out to be of an even or an odd period, then all forms in $\mathfrak{Q}_\pm(D)$ are of even or odd periods, respectively.*

$$\S83 \qquad\qquad 371$$

The proof will be given in the middle part of §85; see also (85.20). The examples for odd periods are in Notes [83.1], [83.3]; for the case of an even period see Note [83.2].

With this, we may resume the algorithm (76.10). How to proceed then to the solutions to (71.2) is indicated in the typical example Note [83.4]. One may think that it remains yet to solve (74.13)(2) or rather (76.10)(c), that is, to find all the solutions to $\mathrm{pell}_D(4)$. However, this task has, in fact, been resolved already in the proof of Theorem 83, though implicitly. Thus, if we take a reduced form Q in (82.24), then there exists an f such that $Q = {}^t\!H_f Q H_f$, $H_f \in \Gamma$, by repeating the rounding appropriately. Those H_f are all in Aut_Q; hence, Theorem 75 implies that infinitely many solutions to $\mathrm{pell}_D(4)$ have been obtained. Moreover, the scene taking place on the hyperbolas indicated in the second paragraph of §82 has become more visible. Details will be provided in §85. However, it is appropriate to discuss $\mathrm{pell}_d(\pm 1)$ beforehand, as we shall do in the next section.

Notes

[83.1] We shall prove $\mathfrak{h}_{\pm}(493) = 2$. With $D = 493 \equiv 1 \bmod 4$, the pairs $\{|b|, a|c|\}$ such that $a > 0$ and $b^2 + 4a|c| = D$ are $\{1, 3\cdot 41\}$, $\{3, 11^2\}$, $\{5, 3^2 \cdot 13\}$, $\{7, 3\cdot 37\}$, $\{9, 103\}$, $\{11, 3\cdot 31\}$, $\{13, 3^4\}$, $\{15, 67\}$, $\{17, 3\cdot 17\}$, $\{19, 3\cdot 11\}$, $\{21, 13\}$. Among these, $|b| = 1, 7, 9, 11, 15$ are to be discarded since they do not satisfy the middle line in the condition (82.22). Corresponding to the remaining values of b, we obtain the following set of reduced forms:

(1) : $[|11, -3, -11|]$, (2) : $[|9, -5, -13|]$, (3) : $[|13, -5, -9|]$, (4) : $[|9, -13, -9|]$,
(5) : $[|3, -17, -17|]$, (6) : $[|17, -17, -3|]$, (7) : $[|3, -19, -11|]$,
(8) : $[|11, -19, -3|]$, (9) : $[|1, -21, -13|]$, (10) : $[|13, -21, -1|]$.

Among these, (1) and (2) are not equivalent mod Γ. For

$$(1) : \quad \frac{3 + \sqrt{D}}{22} = 1 + \frac{1}{6} + \frac{1}{1} + \frac{1}{6} + \frac{1}{1} + \frac{1}{(3 + \sqrt{D})/22},$$

$$(2) : \quad \frac{5 + \sqrt{D}}{18} = 1 + \frac{1}{1} + \frac{1}{1} + \frac{1}{21} + \frac{1}{1} + \frac{1}{(5 + \sqrt{D})/18}.$$

Via continued fraction expansions, we find that (1), (5), (6), (7), (8) belong to the same circle of reduced forms; and (2), (3), (4), (9), (10) form another circle. The number of reduced forms in either circle is 5; hence, by (83.1) the class number equals 2. Further, denoting the form (1) by Q, we find that ${}^t\!HQH = -Q$ with

$$H = \begin{pmatrix} 1 & 1 \\ 1 & 0 \end{pmatrix} \begin{pmatrix} 6 & 1 \\ 1 & 0 \end{pmatrix} \begin{pmatrix} 1 & 1 \\ 1 & 0 \end{pmatrix} \begin{pmatrix} 6 & 1 \\ 1 & 0 \end{pmatrix} \begin{pmatrix} 1 & 1 \\ 1 & 0 \end{pmatrix} = \begin{pmatrix} 63 & 55 \\ 55 & 48 \end{pmatrix} \in \diamond \Gamma.$$

In particular, we see that

$$H^2 = \begin{pmatrix} 6994 & 6105 \\ 6105 & 5329 \end{pmatrix} \in \mathrm{Aut}_Q.$$

According to (76.6), we have $u = 6105/11 = 555$, $t = 2 \cdot 6994 - 3 \cdot 555 = 12323$: so

a special solution to $\mathrm{pell}_{493}(4)$: $12323^2 - 493 \cdot 555^2 = 4$.

The form (2) yields the same, as is indicated by Theorem 84. By the way, H gives

a special solution to $\mathrm{pell}_{493}(-4)$: $111^2 - 493 \cdot 5^2 = -4$.

[83.2]   We shall prove $\mathfrak{h}_\pm(268) = 2$. We have $D = 268 = 4d$, $d = 67$. The continued fraction expansion of the reduced quadratic irrational $\eta_0 = 8 + \sqrt{67}$ is as follows:

$$\eta_0 = 16 + \cfrac{1}{5+} \cfrac{1}{2+} \cfrac{1}{1+} \cfrac{1}{1+} \cfrac{1}{7+} \cfrac{1}{1+} \cfrac{1}{1+} \cfrac{1}{2+} \cfrac{1}{5+} \cfrac{1}{\eta_0} = \frac{96578\eta_0 + 17901}{5967\eta_0 + 1106}.$$

Adopting the notation (82.25),

$$\eta_1 = \frac{8+\sqrt{d}}{3}, \quad \eta_2 = \frac{7+\sqrt{d}}{6}, \quad \eta_3 = \frac{5+\sqrt{d}}{7}, \quad \eta_4 = \frac{2+\sqrt{d}}{9}, \quad \eta_5 = \frac{7+\sqrt{d}}{2},$$

$$\eta_6 = \frac{7+\sqrt{d}}{9}, \quad \eta_7 = \frac{2+\sqrt{d}}{7}, \quad \eta_8 = \frac{5+\sqrt{d}}{6}, \quad \eta_9 = \frac{7+\sqrt{d}}{3}.$$

The corresponding reduced forms and the cf-orbit composed of them are:

$$\eta_0 : [|1,-16,-3|] \;\mapsto\; \eta_1 : [|3,-16,-1|] \;\mapsto\; \eta_2 : [|6,-14,-3|] \;\mapsto$$

$$\eta_3 : [|7,-10,-6|] \;\mapsto\; \eta_4 : [|9,-4,-7|] \;\mapsto\; \eta_5 : [|2,-14,-9|] \;\mapsto$$

$$\eta_6 : [|9,-14,-2|] \;\mapsto\; \eta_7 : [|7,-4,-9|] \;\mapsto\; \eta_8 : [|6,-10,-7|] \;\mapsto$$

$$\eta_9 : [|3,-14,-6|] \;\mapsto\; \eta_{10} = \eta_0 : \cdots\cdots$$

According to (82.22), these exhaust all reduced forms; there exists only one circle. The period is 10, thus by (83.1) the class number equals 2. The above orbit sprits into two classes which are represented by $[|1,-16,-3|]$ and $[|3,-16,-1|]$, respectively. By the way, in $\mathrm{pell}_{268}(4)$: $t^2 - 268u^2 = 4$ we have $2|t$, and get a solution to $\mathrm{pell}_{67}(1)$. Namely, $\mathrm{pell}_{268}(4)$ and $\mathrm{pell}_{67}(1)$ are equivalent (see §85 (I)). In particular, from the expansion of η_0, we are led to an automorphism of the form $[|1,-16,-3|]$, and consequently we find, via (76.6),

$$\text{a special solution to pell}_{67}(1) : 48842^2 - 67 \cdot 5967^2 = 1.$$

By the law (58.1) (1), we see that $\text{pell}_{67}(-1)$ has no solution.

[83.3] We shall show that $\mathfrak{h}_\pm(628) = 1$. We have $D = 4d$, $d = 157$. The continued fraction expansion of the reduced quadratic irrational $\eta_0 = 12 + \sqrt{157}$ is as follows:

$$\eta_0 = 24 + \cfrac{1}{1} + \cfrac{1}{1} + \cfrac{1}{7} + \cfrac{1}{1} + \cfrac{1}{5} + \cfrac{1}{2} + \cfrac{1}{1} + \cfrac{1}{1}$$
$$+ \cfrac{1}{1} + \cfrac{1}{1} + \cfrac{1}{2} + \cfrac{1}{5} + \cfrac{1}{1} + \cfrac{1}{7} + \cfrac{1}{1} + \cfrac{1}{1} + \cfrac{1}{\eta_0} = \frac{9459858\eta_0 + 5013385}{385645\eta_0 + 204378}.$$

As before,

$$\eta_1 = \frac{12 + \sqrt{d}}{13}, \quad \eta_2 = \frac{1 + \sqrt{d}}{12}, \quad \eta_3 = \frac{11 + \sqrt{d}}{3}, \quad \eta_4 = \frac{10 + \sqrt{d}}{19},$$
$$\eta_5 = \frac{9 + \sqrt{d}}{4}, \quad \eta_6 = \frac{11 + \sqrt{d}}{9}, \quad \eta_7 = \frac{7 + \sqrt{d}}{12}, \quad \eta_8 = \frac{5 + \sqrt{d}}{11},$$
$$\eta_9 = \frac{6 + \sqrt{d}}{11}, \quad \eta_{10} = \frac{5 + \sqrt{d}}{12}, \quad \eta_{11} = \frac{7 + \sqrt{d}}{9}, \quad \eta_{12} = \frac{11 + \sqrt{d}}{4},$$
$$\eta_{13} = \frac{9 + \sqrt{d}}{19}, \quad \eta_{14} = \frac{10 + \sqrt{d}}{3}, \quad \eta_{15} = \frac{11 + \sqrt{d}}{12}, \quad \eta_{16} = \frac{1 + \sqrt{d}}{13}.$$

The corresponding reduced forms and the cf-orbit composed of them are:

$$\eta_0 : [|1, -24, -13|] \mapsto \eta_1 : [|13, -24, -1|] \mapsto \eta_2 : [|12, -2, -13|] \mapsto$$
$$\eta_3 : [|3, -22, -12|] \mapsto \eta_4 : [|19, -20, -3|] \mapsto \eta_5 : [|4, -18, -19|] \mapsto$$
$$\eta_6 : [|9, -22, -4|] \mapsto \eta_7 : [|12, -14, -9|] \mapsto \eta_8 : [|11, -10, -12|] \mapsto$$
$$\eta_9 : [|11, -12, -11|] \mapsto \eta_{10} : [|12, -10, -11|] \mapsto \eta_{11} : [|9, -14, -12|] \mapsto$$
$$\eta_{12} : [|4, -22, -9|] \mapsto \eta_{13} : [|19, -18, -4|] \mapsto \eta_{14} : [|3, -20, -19|] \mapsto$$
$$\eta_{15} : [|12, -22, -3|] \mapsto \eta_{16} : [|13, -2, -12|] \mapsto \eta_{17} = \eta_0 : \cdots \cdots .$$

According to (82.22), these are all reduced forms; in fact, $[|6, -14, -18|]$, $[|18, -14, -6|]$, $[|6, -22, -6|]$, $[|2, -22, -18|]$, $[|18, -22, -2|]$ arise as well, but these are not primitive. The period is 17, and by (83.1) we find that the class number equals 1. In order to choose a convenient representative form, we observe, for instance,

$$[|1, -24, -13|] \xRightarrow{T^{12}} Z = [|1, 0, -157|], \quad \eta_0 = T^{12}(\omega^+(Z)).$$

Thus, all forms in $\mathcal{Q}_\pm(628)$ are equivalent to $x^2 - 157y^2$, which is not reduced but almost.

[83.4] Continuing the discussion, we shall consider the representations of the prime number $p = 6781$ by the form Z. We proceed as prescribed in (76.10)–(76.11) accompanied by the pre-reduction [$\pm$]. First, following

374   *Quadratic Forms*

Theorem 68, we solve the congruence equation $X^2 \equiv 628 \bmod 4p$, that is, $X_1^2 \equiv 157 \bmod p$. By Note [63.1] (Legendre), we get $X_1 \equiv \pm 4719 \bmod p$. Hence the discriminant of $Z_1 = [|6781, -9438, 3284|]$ is equal to $628 = 4d$, and Z_1 should be equivalent to Z with respect to Γ. We shall confirm this. Because of the expansion

$$\omega^+(Z_1) = \frac{4719 + \sqrt{d}}{6781} = 0 + \frac{1}{1} + \frac{1}{2} + \frac{1}{3} + \frac{1}{4} + \frac{1}{\eta_{12}},$$

$$\omega^+(Z_1) = S(\eta_{12}), \quad S = \begin{pmatrix} 30 & 7 \\ 43 & 10 \end{pmatrix} \in \diamond\Gamma,$$

we see that the cf-orbit of Z_1 enters into the circle at η_{12}, with an obvious abuse of notation. Taking into account the rounding movement, we have, adopting the notation of (82.24)–(82.25),

$$\eta_{12} = H_{11}^{-1}(\eta_0), \quad \eta_0 = \eta_{17} = H_{16}^{-1}(\eta_0) \implies \eta_{12} = (H_{16}^{\kappa}H_{11})^{-1}(\eta_0),$$

$$H_{11} = \begin{pmatrix} 88823 & 33974 \\ 3621 & 1385 \end{pmatrix}, \quad H_{16} = \begin{pmatrix} 9459858 & 5013385 \\ 385645 & 204378 \end{pmatrix}.$$

Hence,

$$\omega^+(Z_1) = S(H_{16}^{\kappa}H_{11})^{-1}T^{12}(\omega^+(Z));$$

and

$$Z_1 = {}^t K_{\kappa} Z K_{\kappa}, \quad K_{\kappa} = T^{-12}H_{16}^{\kappa}H_{11}S^{-1} \in \Gamma, \quad 2 \nmid \kappa,$$

as $T, H_{11} \in \Gamma$, and $S, H_{16} \in \diamond\Gamma$. For instance,

$$K_1 = \begin{pmatrix} 2826905056841 & -1962062224509 \\ 225611584950 & -156589612789 \end{pmatrix},$$

and

$$2826905056841^2 - 157 \cdot 225611584950^2 = 6781.$$

Similarly, K_{-1} yields the representation $56009^2 - 157 \cdot 4470^2 = 6781$. By the way, K_0 gives $292512^2 - 157 \cdot 23345^2 = -6781$.

[83.5] In the example in Note [83.1], the periods of the two representative forms are the same, which does not hold in general, however. For instance, let $D = 377 \equiv 1 \bmod 4$. The relevant reduced quadratic irrationals are

$$\eta_0 = \frac{19 + \sqrt{D}}{2}, \quad \eta_1 = \frac{19 + \sqrt{D}}{8}, \quad \eta_2 = \frac{13 + \sqrt{D}}{26}, \quad \eta_3 = \frac{13 + \sqrt{D}}{8};$$

$$\xi_0 = \frac{19 + \sqrt{D}}{4}, \quad \xi_1 = \frac{17 + \sqrt{D}}{22}, \quad \xi_2 = \frac{5 + \sqrt{D}}{16},$$

$$\xi_3 = \frac{11 + \sqrt{D}}{16}, \ \xi_4 = \frac{5 + \sqrt{D}}{22}, \ \xi_5 = \frac{17 + \sqrt{D}}{4};$$

$$\eta_0 = 19 + \frac{1}{4} + \frac{1}{1} + \frac{1}{4} + \frac{1}{\eta_0}; \quad \xi_0 = 9 + \frac{1}{1} + \frac{1}{1} + \frac{1}{1} + \frac{1}{1} + \frac{1}{9} + \frac{1}{\xi_0}.$$

The periods of η_0 and ξ_0 are 4 and 6, respectively; they are different but their parities are the same, which is a situation indicated by Theorem 84. Thus, by (83.1), we get the classification

$$\eta_0 \equiv \eta_2, \ \eta_1 \equiv \eta_3, \ \xi_0 \equiv \xi_2 \equiv \xi_4, \ \xi_1 \equiv \xi_3 \equiv \xi_5 \text{ mod } \Gamma.$$

Therefore, $\mathfrak{h}_\pm(377) = 4$. By the way, we have $\eta_0 = (461\eta_0 + 96)/(24\eta_0 + 5)$ and $\xi_0 = (461\xi_0 + 48)/(48\xi_0 + 5)$. Both identities yield $466^2 - 377 \cdot 24^2 = 4$, namely $233^2 - 377 \cdot 12^2 = 1$ a solution to $\text{pell}_{377}(1)$. Neither $\text{pell}_{377}(-4)$ nor $\text{pell}_{377}(-1)$ has any solution. See Note [84.4] and (85.20) below. Also see Note [87.3], which contains the situation of two different odd periods.

§84

We now enter into a discussion on the Pell equations $\text{pell}_d(\pm 1)$, $d > 0$. It is said that the origin of the problem can be traced back to the approximations to $\sqrt{2}$ by irreducible fractions. People in great antiquity noticed that $\sqrt{2}$ is incommensurable, i.e., not rational, and sought for special fractions t/u ([Σ.II, Prop. 10]) which approximate $\sqrt{2}$ exceptionally well. Their findings were concealed in the plans of sacred temples and alters. Eventually someone realized that

> the fraction $2u/t$ as well ought to be a good approximation; thus, $|t/u - 2u/t| = |t^2 - 2u^2|/tu$ has to be exceptionally small. $\hspace{1em}$ (84.1)

The numerator $|t^2 - 2u^2|$ is to be as small as possible. Since this expression cannot be zero, the best possible situation will take place with those integers $\{t, u\}$ such that $t^2 - 2u^2 = \pm 1$, whence the problem $\text{pell}_2(\pm 1)$ emerges. On the other hand, it was apparently well-known that the fractions which give exceptionally good approximation to $\sqrt{2}$ arise from the application to it of the infinite version of Euclid's algorithm ([Σ.X, Prop. 2]: Note [23.4]), or the continued fraction expansion. It is then quite natural to surmise that all the solutions to $\text{pell}_2(\pm 1)$ would be obtained from the continued fraction expansion of $\sqrt{2}$. Lagrange (1768) made a fundamental contribution by giving a rigorous proof to this conjecture, indeed, in a far generalized and precise fashion.

376   *Quadratic Forms*

Thus, with an arbitrary non-square $d \geq 2$, the same argument as (84.1) leads us to the problem $\text{pell}_d(\pm 1)$. The continued fraction expansion of the irrational number $\sqrt{d}$ comes to our attention. As for $\text{pell}_d(1)$, certainly we are able to reach the desired conclusion via (82.18) and (82.21) together with Theorem 75, $D = 4d$, as we did in Note [83.2]. However, because of the particular importance of the subject and also in order to prepare for a beautiful encounter in §86 with classic Indian mathematics, we shall develop instead a direct reasoning.

To begin with, we assert that with $\theta_j = \mathrm{R}^j([\sqrt{d}] + \sqrt{d})$

$$\text{there exist integers } \{\alpha_j, \beta_j : j \geq 0\}, \text{ such that}$$
$$1 < \theta_j = (\alpha_j + \sqrt{d})/\beta_j, \ 1 \leq \alpha_j < \sqrt{d}, \ 1 \leq \beta_j < 2\sqrt{d}. \tag{84.2}$$

To prove this save for the first inequality which is trivial, we note that the relation $\theta_j = [\theta_j] + 1/\theta_{j+1}$ implies anyway that there exist tuples $\{\alpha_j, \beta_j : j \geq 0\}$ of rational numbers such that $\theta_j = (\alpha_j + \sqrt{d})/\beta_j$, and

$$(0) \ \alpha_0 = [\sqrt{d}], \ \beta_0 = 1,$$
$$(1) \ \alpha_j + \alpha_{j+1} = [\theta_j]\beta_j, \tag{84.3}$$
$$(2) \ \beta_j\beta_{j+1} = d - \alpha_{j+1}^2.$$

The assumption that $\alpha_j, \beta_j \in \mathbb{Z}$ for all $j \leq J$ implies $\alpha_{J+1}, \beta_{J+1} \in \mathbb{Z}$; note that $\beta_j\beta_{j+1} = d - ([\theta_j]\beta_j - \alpha_j)^2$. Thus, by induction $\alpha_j, \beta_j \in \mathbb{Z}$ for all $j \geq 0$. Let $\theta_j^* = (\alpha_j - \sqrt{d})/\beta_j$. Since (84.3) yields $\theta_j^* = (\theta_{j-1}^* - [\theta_{j-1}])^{-1}$ for $j \geq 1$, we get $-1 < \theta_j^* < 0$ for $j \geq 0$, again by induction. We have $1 < \theta_j - \theta_j^* = 2\sqrt{d}/\beta_j$ and $0 < \theta_j + \theta_j^* = 2\alpha_j/\beta_j$, so $0 < \beta_j < 2\sqrt{d}, 0 < \alpha_j$; further $\alpha_j < \sqrt{d}$, as $\theta_j^* < 0$. These hold for $j \geq 0$. We end the proof of (84.2).

For each d the number of possible integer pairs $\{\alpha_j, \beta_j\}$ in (84.2) is finite, whence there exist $0 \leq j_1 < j_2$ such that $\theta_{j_1} = \theta_{j_2}$. We claim that $\theta_{j_2-j_1} = \theta_0$: We have $\theta_j^* = [\theta_j] + 1/\theta_{j+1}^*, j \geq 0$, as noted above; hence, $0 < -[\theta_j] - 1/\theta_{j+1}^* < 1$, so $[\theta_j] = [-1/\theta_{j+1}^*]$ for $j \geq 0$. We see that $[\theta_{j_2-1}] = [-1/\theta_{j_2}^*] = [-1/\theta_{j_1}^*] = [\theta_{j_1-1}]$ for $j_1 \geq 1$. That is, $\theta_{j_2-1} = [\theta_{j_2-1}] + 1/\theta_{j_2} = [\theta_{j_1-1}] + 1/\theta_{j_1} = \theta_{j_1-1}$ for $j_1 \geq 1$, which verifies the claim. Therefore, we find that there exists an $l \geq 1$ such that $\mathrm{R}^l(\theta_0) = \theta_0$. The reasoning in the later part of §82 is a minor generalization of the present one.

We then introduce

$$r = \min\left\{l \geq 1 : \mathrm{R}^l(\sqrt{d}) = \theta_0 = [\sqrt{d}] + \sqrt{d}\right\}; \tag{84.4}$$

note that here we have $\mathrm{R}^l(\sqrt{d})$ in place of $\mathrm{R}^l(\theta_0)$, which makes a subtle difference in the discussion to be developed.

$$\S84 \qquad\qquad 377$$

Thus, starting from the square root $\sqrt{d}$, we have been led to the crucial notion (84.4). We shall show its relation to the equation $\mathrm{pell}_d(1)$. First, we note the fact that

$$\beta_j = 1 \text{ for a } j \geq 1 \iff \theta_j = \theta_0 \iff r|j. \qquad (84.5)$$

To verify this, let $\beta_j = 1$. Then $\theta_j^* = \alpha_j - \sqrt{d}$, and since $-1 < \theta_j^* < 0$ we have $\alpha_j = [\sqrt{d}]$ or $\theta_j = \theta_0$. Via (2.1) we have $r|j$. Conversely, if $j \geq 1$ is such that $r|j$, then $\theta_0 = \mathrm{R}^j(\sqrt{d}) = \theta_j$, so $\beta_j = 1$.

We then let

$$\sqrt{d} = a_0 + \cfrac{1}{a_1 +} \cfrac{1}{a_2 +} \cfrac{1}{a_3 +} \cdots + \cfrac{1}{a_{r-1} +} \cfrac{1}{2a_0 +} \cfrac{1}{a_1 +} \cfrac{1}{a_2 +} \cdots, \qquad (84.6)$$

be the continued fraction expansion of $\sqrt{d}$ so that $a_j = [\mathrm{R}^j(\sqrt{d})]$, i.e., $a_0 = [\sqrt{d}]$ and $a_j = [\theta_j]$ for $j \geq 1$; the set of periods of this expansion is, of course, given by (84.4). With this, let $t, u \in \mathbb{N}$ be supposed to be such that $t^2 - du^2 = 1$. Then,

$$\sqrt{d} - \frac{t}{u} = -\frac{1}{u^2(t/u + \sqrt{d})} > -\frac{1}{2u^2}, \qquad (84.7)$$

for $1 < \sqrt{d} < t/u$, so $|\sqrt{d} - t/u| < 1/2u^2$. Legendre's criterion (Theorem 24) implies that the irreducible fraction t/u coincides with a convergent A_k/B_k of (84.6) and

$$\sqrt{d} - \frac{A_k}{B_k} = -\frac{1}{B_k(A_k + B_k\sqrt{d})}. \qquad (84.8)$$

In view of (22.8), we find that k is odd and $A_k + B_k\sqrt{d} = B_{k-1} + B_k\mathrm{R}^{k+1}(\sqrt{d})$. Since $\mathrm{R}^{k+1}(\sqrt{d}) = \theta_{k+1}$, we have $\beta_{k+1} = 1$. Hence, $r|(k+1)$ by (84.5). In other words, we have $t = A_{l-1}, u = B_{l-1}$ with an even $l \geq 2$ contained in the set (84.4).

In addition, we consider $\mathrm{pell}_d(-1)$. Since $-1 \bmod d$ ought to be a quadratic residue, d cannot be an arbitrary non-square. We suppose anyway that $t, u \in \mathbb{N}$ are such that $t^2 - du^2 = -1$. Then, instead of (84.7), we have

$$\frac{1}{\sqrt{d}} - \frac{u}{t} = -\frac{1}{t^2 d(u/t + 1/\sqrt{d})} > -\frac{1}{2t^2}. \qquad (84.9)$$

Thus, $u/t = B_{k_1-1}/A_{k_1-1}$ with a certain odd integer k_1; here A_j/B_j is as above, and (25.8) has been invoked. A rearrangement gives

$$\sqrt{d} - \frac{A_{k_1-1}}{B_{k_1-1}} = \frac{1}{B_{k_1-1}(A_{k_1-1} + B_{k_1-1}\sqrt{d})}. \qquad (84.10)$$

As before, we find that $\beta_{k_1} = 1$, and $r|k_1$. Thus, in the present case, r has to be odd; and $t = A_{l-1}$, $u = B_{l-1}$ with an odd $l \geq 1$ contained in the set (84.4).

Conversely, take an l in the set (84.4). Then the identity (22.5) with $\eta = \sqrt{d}$ gives

$$\left(B_{l-1}([\sqrt{d}] + \sqrt{d}) + B_{l-2}\right)\sqrt{d} = A_{l-1}([\sqrt{d}] + \sqrt{d}) + A_{l-2}, \qquad (84.11)$$

which implies

$$A_{l-1} = B_{l-2} + B_{l-1}[\sqrt{d}], \quad dB_{l-1} = A_{l-2} + A_{l-1}[\sqrt{d}]. \qquad (84.12)$$

Eliminating $[\sqrt{d}]$, we find, via (21.6), that

$$A_{l-1}^2 - dB_{l-1}^2 = A_{l-1}B_{l-2} - B_{l-1}A_{l-2} = (-1)^l. \qquad (84.13)$$

Namely, $\{A_{l-1}, B_{l-1}\}$ is a solution to $\mathrm{pell}_d((-1)^l)$.

Hence, the solutions to $\mathrm{pell}_d(\pm 1)$ are intrinsically related to the notion (84.3). To make the situation more precise, we shall show that

$$A_{l-1} + B_{l-1}\sqrt{d} = (A_{r-1} + B_{r-1}\sqrt{d})^{l/r}. \qquad (84.14)$$

To this end, we remark that

$$\begin{aligned}
\begin{pmatrix} 2[\sqrt{d}] & 1 \\ 1 & 0 \end{pmatrix} \begin{pmatrix} a_1 & 1 \\ 1 & 0 \end{pmatrix} \cdots \begin{pmatrix} a_{r-1} & 1 \\ 1 & 0 \end{pmatrix} \\
= \begin{pmatrix} 1 & [\sqrt{d}] \\ 0 & 1 \end{pmatrix} \begin{pmatrix} A_{r-1} & A_{r-2} \\ B_{r-1} & B_{r-2} \end{pmatrix},
\end{aligned} \qquad (84.15)$$

since $a_0 = [\sqrt{d}]$ and $\begin{pmatrix} 2c & 1 \\ 1 & 0 \end{pmatrix} = \begin{pmatrix} 1 & c \\ 0 & 1 \end{pmatrix} \begin{pmatrix} c & 1 \\ 1 & 0 \end{pmatrix}$ for any c. Hence,

$$\begin{aligned}
\begin{pmatrix} A_{l+r-1} & A_{l+r-2} \\ B_{l+r-1} & B_{l+r-2} \end{pmatrix} &= \begin{pmatrix} A_{l-1} & A_{l-2} \\ B_{l-1} & B_{l-2} \end{pmatrix} \begin{pmatrix} 1 & [\sqrt{d}] \\ 0 & 1 \end{pmatrix} \begin{pmatrix} A_{r-1} & A_{r-2} \\ B_{r-1} & B_{r-2} \end{pmatrix} \\
&= \begin{pmatrix} A_{l-1} & dB_{l-1} \\ B_{l-1} & A_{l-1} \end{pmatrix} \begin{pmatrix} A_{r-1} & A_{r-2} \\ B_{r-1} & B_{r-2} \end{pmatrix};
\end{aligned} \qquad (84.16)$$

the second line is due to (84.12). Applying the vector $(1, \sqrt{d})$ from the right, we get

$$A_{l+r-1} + B_{l+r-1}\sqrt{d} = (A_{l-1} + B_{l-1}\sqrt{d})(A_{r-1} + B_{r-1}\sqrt{d}), \qquad (84.17)$$

which gives (84.14).

To state the concluding assertion, we introduce the convention:

$$\begin{aligned}
&\text{via the correspondence } t^2 - du^2 = \pm 1 \;\mapsto\; t + u\sqrt{d}, \\
&\text{the latter is identified with a solution to } \mathrm{pell}_d(\pm 1).
\end{aligned} \qquad (84.18)$$

Changing the signs of t, u we still have solutions to $\text{pell}_d(\pm 1)$; thus we may make the definition:

$$\text{the least solution of } \text{pell}_d(\pm 1):$$
$$\min\left\{t + u\sqrt{d} : t^2 - du^2 = \pm 1; \ t, u > 0\right\}, \tag{84.19}$$

where signs correspond.

We have obtained the famous theorem of Lagrange (1768):

Theorem 85 *For an arbitrary non-square $d \in \mathbb{N}$, expand $\sqrt{d}$ into an infinite continued fraction* (84.6) *with r the period and A_j/B_j the convergents. Under the convention* (84.18) *it holds that*
(i) *The indefinite equation* $\text{pell}_d(-1)$ *has solutions if and only if $2 \nmid r$. The least solution is $A_{r-1} + B_{r-1}\sqrt{d}$. The set of all solutions is*

$$\left\{\pm \left(A_{r-1} + B_{r-1}\sqrt{d}\right)^{2v+1} : v \in \mathbb{Z}\right\}. \tag{84.20}$$

(ii) *The indefinite equation* $\text{pell}_d(1)$ *has solutions always. The least solution is $A_{\ell-1} + B_{\ell-1}\sqrt{d}$ where $2|r \Rightarrow \ell = r$ and $2 \nmid r \Rightarrow \ell = 2r$. The set of all solutions is*

$$\left\{\pm \left(A_{\ell-1} + B_{\ell-1}\sqrt{d}\right)^{v} : v \in \mathbb{Z}\right\}. \tag{84.21}$$

In passing, we mention for the sake of later purposes that comparing the coefficients of $\sqrt{d}$ in the relation

$$\theta_{j+1} = \frac{\alpha_{j+1} + \sqrt{d}}{\beta_{j+1}} = \mathrm{R}^{j+1}\left(\sqrt{d}\right) = -\frac{A_{j-1} - B_{j-1}\sqrt{d}}{A_j - B_j\sqrt{d}}, \quad j \geq 0, \tag{84.22}$$

one obtains

$$A_j^2 - dB_j^2 = (-1)^{j+1}\beta_{j+1}, \quad j \geq -1, \tag{84.23}$$

where the case $j = -1$ follows from the definition $A_{-1} = 1, B_{-1} = 0$, $\beta_0 = 1$. The equation (84.23) is of course a generalization of (84.13). It should be remarked that (84.22)–(84.23) implies

$$A_{j-1}A_j - dB_{j-1}B_j = (-1)^j \alpha_{j+1} \tag{84.24}$$

as well. These will have an application at (86.2) below.

Further, we add the following assertion due to Legendre (1798, p.457):

Theorem 86 *Let $t_1 + u_1\sqrt{d}$ be the least solution to* $\text{pell}_d(1)$*, and let t_m, u_m be such that $(t_1 + u_1\sqrt{d})^m = t_m + u_m\sqrt{d}$ for $m \in \mathbb{Z}$. Then, for any $p \geq 3$, $p \nmid d$, we have*

$$t_{p-\omega} \equiv 1, \ u_{p-\omega} \equiv 0 \bmod p, \quad \omega = \left(\frac{d}{p}\right). \tag{84.25}$$

Quadratic Forms

Proof Using the matrix representation of forms (§72), we have, with $\eth^2 = d\mathfrak{e}$,

$$(t_1\mathfrak{e} + u_1\eth)^p \equiv t_1^p\mathfrak{e} + u_1^p\eth^p \equiv t_1\mathfrak{e} + \omega u_1\eth \bmod p, \tag{84.26}$$

since $\eth^p = d^{(p-1)/2}\eth$. Hence,

$$(t_1\mathfrak{e} + u_1\eth)^{p-\omega} \equiv (t_1\mathfrak{e} + \omega u_1\eth)(t_1\mathfrak{e} + \omega u_1\eth)^{-1} \equiv \mathfrak{e} \bmod p, \tag{84.27}$$

which ends the proof.

The congruence (84.27) can be regarded as an analogue of Fermat's theorem (28.11). Thus, more generally, corresponding to Euler's theorem (28.10), we have, for any $g \in \mathbb{Z}$, $\langle g, d \rangle = 1$,

$$(t_1\mathfrak{e} + u_1\eth)^{\varphi(g;\,d)} \equiv \mathfrak{e} \bmod |g|, \quad \varphi(g;d) = g \prod_{\substack{p|g \\ p>2}} \left(1 - \left(\frac{d}{p}\right)\frac{1}{p}\right). \tag{84.28}$$

The proof is an immediate extension of (28.12) together with the observation $(t_1\mathfrak{e} + u_1\eth)^2 \equiv \mathfrak{e} \bmod 2$. In particular, we find that

$$t_h \equiv 1, \ u_h \equiv 0 \bmod |g| \text{ for an } h|\varphi(g;d). \tag{84.29}$$

Notes

[84.1] The theoretical analysis of the relation between $\text{pell}_d(1)$ and the continued fraction for $\sqrt{d}$ was started by Euler (1759, published in 1767). However, the first rigorous proof of the existence of solutions to general $\text{pell}_d(1)$ and the algorithm to construct the solutions was achieved by Lagrange (1768). Their investigations are both based on the work of W. Brouncker (Wallis (1685, Chapter XCVIII)). The above reasoning leading to Theorem 85 is an adoption of Lagrange (1769: a revision of his (1768) based on Euler (1759)); we have referred to Legendre (1798, pp.50–57) as well. The use of matrices is of course modern.

[84.2] An observation on the size of the least solution to $\text{pell}_d(1)$:

(1) $d = 419: r = 18$.

$$\sqrt{419} = 20 + \cfrac{1}{2} + \cfrac{1}{7} + \cfrac{1}{1} + \cfrac{1}{2} + \cfrac{1}{3} + \cfrac{1}{1} + \cfrac{1}{2} + \cfrac{1}{1} + \cfrac{1}{19} + \cfrac{1}{1}$$
$$+ \cfrac{1}{2} + \cfrac{1}{1} + \cfrac{1}{3} + \cfrac{1}{2} + \cfrac{1}{1} + \cfrac{1}{7} + \cfrac{1}{2} + \cfrac{1}{20 + \sqrt{419}};$$

$$\text{pell}_{419}(1): \quad 270174970^2 - 419 \cdot 13198911^2 = 1.$$

(2) $d = 420$: $r = 2$.

$$\sqrt{420} = 20 + \cfrac{1}{2 + \cfrac{1}{20 + \sqrt{420}}} \; ;$$

$$\text{pell}_{420}(1): \; 41^2 - 420 \cdot 2^2 = 1.$$

(3) $d = 421$: $r = 37$.

$$\sqrt{421} = 20 + \cfrac{1}{1+} \cfrac{1}{1+} \cfrac{1}{13+} \cfrac{1}{5+} \cfrac{1}{1+} \cfrac{1}{3+} \cfrac{1}{1+} \cfrac{1}{2+} \cfrac{1}{1+} \cfrac{1}{1+} \cfrac{1}{1+}$$
$$+ \cfrac{1}{2+} \cfrac{1}{9+} \cfrac{1}{1+} \cfrac{1}{7+} \cfrac{1}{3+} \cfrac{1}{3+} \cfrac{1}{2+} \cfrac{1}{2+} \cfrac{1}{3+} \cfrac{1}{3+} \cfrac{1}{7+} \cfrac{1}{1+} \cfrac{1}{9}$$
$$+ \cfrac{1}{2+} \cfrac{1}{1+} \cfrac{1}{1+} \cfrac{1}{1+} \cfrac{1}{2+} \cfrac{1}{1+} \cfrac{1}{3+} \cfrac{1}{1+} \cfrac{1}{5+} \cfrac{1}{13+} \cfrac{1}{1+} \cfrac{1}{1+} \cfrac{1}{20 + \sqrt{421}} \; ;$$

$$\text{pell}_{421}(-1): \; 4404244569682141 8^2 - 421 \cdot 214649746353078 5^2 = -1 \, ;$$

$$\text{pell}_{421}(1): \; 387947404591492687946821716706144 9^2$$
$$- 421 \cdot 1890739959518390208804997807062 60^2 = 1.$$

As is indicated by these examples, close neighbors of those d which yield relatively small least solutions to $\text{pell}_d(1)$ often produce huge least solutions. This phenomenon is well depicted in Euler's table (1759, pp.335–336) and its extension by Legendre (1798, Table XII) for $2 \le d \le 1003$. According to Jacobson–Williams (2000), with ξ_d, ξ_{d+1} being the least solutions of $\text{pell}_d(1)$, $\text{pell}_{d+1}(1)$, respectively, we have $\log \xi_{d+1} / \log \xi_d \gg d^{1/6} / \log d$ for infinitely many d, i.e., ξ_{d+1}/ξ_d can be just huge. As for the size of ξ_d itself, the same as Remark 3, §85, holds. By the way, the above example (3) will be treated in Note [86.3] by an ancient Indian method which is far more efficient than the one applied here.

[84.3] Legendre (1785, p.549). If $p \equiv 1 \bmod 4$, then $\text{pell}_p(-1)$ admits solutions; so the period of $\sqrt{p}$ is odd in view of Theorem 85 (i). Incidentally, here the law (58.1) (1) is proved. To directly prove this solubility assertion, let $a^2 - pb^2 = 1$, $a, b \in \mathbb{N}$, by Theorem 85 (ii). We have $2 \nmid a$ and $2 | b$ since $a \equiv b \bmod 2$ is impossible, and $2 | a$ would imply $p \equiv -1 \bmod 4$. We put $a = 1 + 2A$, $b = 2B$, so $A(A + 1) = pB^2$. Via prime power decomposition, either $A = pU_1^2$, $A + 1 = V_1^2$ or $A = U_2^2$, $A + 1 = pV_2^2$. In the former case, we have $V_1^2 - pU_1^2 = 1$; but the same cannot repeat indefinitely, for $0 < U_1 < b$. Hence, we may suppose that the latter case takes place; and we get a solution to $\text{pell}_p(-1)$. If $p \equiv -1 \bmod 4$, then the period of $\sqrt{p}$ is even, of course.

[84.4] If $\text{pell}_d(-1)$ has a solution for a composite $d \equiv 1 \bmod 4$, then $p | d \Rightarrow p \equiv 1 \bmod 4$. But the converse assertion does not hold in general. For instance,

382   *Quadratic Forms*

$\mathrm{pell}_{377}(-1)$ does not have any solution because the period of $\sqrt{377}$ is 4, albeit $377 = 13 \cdot 29$. On the other hand, for $d = 481 = 13 \cdot 37$ we have $964140^2 - 481 \cdot 43961^2 = -1$; see also the ending part of Note [84.6]. One may wonder whether it is possible or not to determine the solubility of $\mathrm{pell}_d(-1)$ without examining the continued fraction expansion of $\sqrt{d}$. This problem is unsolved. It is commonly attributed to Dirichlet (1834) but actually can be traced back to Lagrange (1768, pp.721–723).

[84.5]   Legendre (1798, p.56). For any non-square $d \in \mathbb{N}$, the period block of the continued fraction of $\sqrt{d}$ is palindromic; that is, in (84.6) we have

$$\{a_1, a_2, \ldots, a_{r-1}\} = \{a_{r-1}, a_{r-2}, \ldots, a_1\}.$$

To confirm this, it suffices to show $-1/\theta^*_{r-j+1} = \theta_j$ in the discussion following (84.3). Thus, for $j = 1$ we have $-1/\theta^*_r = -1/\theta^*_0 = R(\theta_0) = \theta_1$ since $\theta_r = \theta_0$. Then, by induction, $-1/\theta^*_{r-j} = -\big([\theta_{r-j}] + 1/\theta^*_{r-j+1}\big)^{-1} = -\big([\theta_j] - \theta_j\big)^{-1} = R(\theta_j) = \theta_{j+1}$, where the first equality follows from (84.3). We end the confirmation. Alternatively, one may use the comparison of the two continued fraction expansions of $(\sqrt{d} - [\sqrt{d}])^{-1}$ which arise from Note [82.4] and (84.6). Yet another proof is given by the identity

$$\begin{pmatrix} dB_{r-1} & A_{r-1} \\ A_{r-1} & B_{r-1} \end{pmatrix} = \begin{pmatrix} A_{r-1} & A_{r-2} \\ B_{r-1} & B_{r-2} \end{pmatrix} \cdot \begin{pmatrix} [\sqrt{d}] & 1 \\ 1 & 0 \end{pmatrix}$$
$$= \begin{pmatrix} [\sqrt{d}] & 1 \\ 1 & 0 \end{pmatrix} \begin{pmatrix} a_1 & 1 \\ 1 & 0 \end{pmatrix} \cdots \begin{pmatrix} a_{r-1} & 1 \\ 1 & 0 \end{pmatrix} \begin{pmatrix} [\sqrt{d}] & 1 \\ 1 & 0 \end{pmatrix},$$

which is due to $(84.12)_{l=r}$. This phenomenon of palindrome had been observed by Euler (1759, p.319); see his table (pp.322–324) as well.

[84.6]   Legendre (1808, pp.59–60). If $r = 2g + 1$ in the expansion (84.6), then we have

$$d = \alpha_{g+1}^2 + \beta_{g+1}^2$$

with α_j, β_j as in (84.2). To confirm this, we note that the preceding Note implies that

$$\begin{pmatrix} dB_{r-1} & A_{r-1} \\ A_{r-1} & B_{r-1} \end{pmatrix} = Y \cdot {}^t Y, \quad Y = \begin{pmatrix} A_g & A_{g-1} \\ B_g & B_{g-1} \end{pmatrix}.$$

In particular, $d(B_g^2 + B_{g-1}^2) = A_g^2 + A_{g-1}^2$, and via (84.23) we get $\beta_{g+1} = \beta_g$. Then by (84.3) (2) we end the confirmation. In particular, in view of Note [84.3], we have obtained an alternative proof of Euler's (78.3). See Smith (1859/1869, art.123). For instance, with $d = 135013$, we have $r = 95$ and

$$Y = \begin{pmatrix} 104339167378203282234448346438 & 993471357530390289741091695551 \\ 283961572842415046800812777 & 270375637784849622321617756 \end{pmatrix},$$

which is in Γ; and $\theta_{48} = Y^{-1}(\sqrt{d}) = (18 + \sqrt{d})/367$. Hence, $135013 = 18^2 + 367^2$. Note that $d = 37 \cdot 41 \cdot 89$. The periods of the square-roots of its factors $d_1 = 37 \cdot 41$, $d_2 = 37 \cdot 89$, and $d_3 = 41 \cdot 89$ are 6, 5, and 43, respectively, so $\mathrm{pell}_d(-1)$, $\mathrm{pell}_{d_2}(-1)$, and $\mathrm{pell}_{d_3}(-1)$ have solutions, while $\mathrm{pell}_{d_1}(-1)$ does not.

[84.7]   Märcker (1840, pp.355–359). If $r = 2g$ in (84.6), then by Note [84.5] we see that $-1/\theta_{g+1}^* = \theta_g$, so $\alpha_g = \alpha_{g+1}$. Thus, by (84.3) (1) we get $\alpha_g = a_g \beta_g/2$ as $[\theta_g] = a_g$, and by (84.3) (2) $d = \beta_g(\beta_{g+1} + a_g^2 \beta_g/4)$. Here we have $\beta_g > 1$ because of (84.5). From this, we find that

$$\text{if } d \text{ is a prime} \equiv 3 \bmod 4, \text{ then } \beta_g = 2.$$

To confirm this, we may suppose that $2 \nmid a_g$, for otherwise β_g would be a proper divisor of d. So $2|\beta_g$, and $d = (\beta_g/2)(2\beta_{g+1} + a_g^2 \beta_g/2)$; thus, $\beta_g/2$ is a factor of d and not equal to d, that is, $\beta_g/2 = 1$. For instance, in the case of Note [83.2] ($d = 67$, a prime) we have $r = 10, \beta_5 = 2$. On the other hand, in the case of a composite d, we see that either β_g or $\beta_g/2$ is a proper divisor of d, whence we may factor d via the continued fraction expansion of $\sqrt{d}$. For instance, with $d = 473903$,

$$\sqrt{d} = 688 + \cfrac{1}{2+} \cfrac{1}{2+} \cfrac{1}{6+} \cfrac{1}{7+} \cfrac{1}{1+} \cfrac{1}{105+} \cfrac{1}{32+} \cfrac{1}{105+} \cfrac{1}{1+} \cfrac{1}{7+} \cfrac{1}{6+} \cfrac{1}{2+} \cfrac{1}{2+} \cfrac{1}{688 + \sqrt{d}},$$

so $r = 14$ and $\mathrm{R}^7(\sqrt{d}) = (688 + \sqrt{d})/43$, and $d = 43 \cdot 11021$. Further, with $d' = 11021$, we have

$$\sqrt{d'} = 104 + \cfrac{1}{1+} \cfrac{1}{51+} \cfrac{1}{2+} \cfrac{1}{51+} \cfrac{1}{1+} \cfrac{1}{104 + \sqrt{d'}},$$

so $r = 6$ and $\mathrm{R}^3(\sqrt{d'}) = (103 + \sqrt{d'})/103$. In this way, we get $d = 43 \cdot 103 \cdot 107$.

[84.8]   An alternative proof by Dirichlet (1863, §§141–142) of the fact that $\mathrm{pell}_d(1)$ always has a non-trivial solution: We return to (20.7); thus there exist an infinitely many $a, b \in \mathbb{N}$ such that $|a - b\sqrt{d}| < 1/b$. Since $0 < a + b\sqrt{d} \leq (1 + 2\sqrt{d})b$, we have $0 < |a^2 - db^2| < 1 + 2\sqrt{d}$; that is, $|a^2 - db^2|$ is less than a quantity which depends only on d. We see that there exists an m such that $\mathrm{pell}_d(m)$ has infinitely many solutions $\{a, b\}$. Consider $a \bmod |m|$, $b \bmod |m|$. There must be different pairs $\{a_1, b_1\}, \{a_2, b_2\} \in \mathbb{N}^2$ such that $a_1 \equiv a_2, b_1 \equiv b_2 \bmod |m|$. With this, we see that

$$\begin{aligned}(a_1 - b_1\sqrt{d})(a_2 + b_2\sqrt{d}) &= (A + B\sqrt{d})m \\ (a_1 + b_1\sqrt{d})(a_2 - b_2\sqrt{d}) &= (A - B\sqrt{d})m\end{aligned} \quad \Rightarrow \quad \begin{aligned}A^2 - dB^2 &= 1, \\ A, B &\in \mathbb{Z}.\end{aligned}$$

If $B = 0$, then $A = \pm 1$, and

$$a_1 - b_1\sqrt{d} = Am/(a_2 + b_2\sqrt{d}) = A(a_2 - b_2\sqrt{d}),$$

which yields the contradiction $a_1 = a_2$, $b_1 = b_2$. This ends the proof. The idea of using the classification modulo $|m|$ is due to Lagrange (1768, pp.676–678). Dirichlet devised a way to dispense with continued fractions. See also Landau (1927, Erster Teil, Kapitel 7).

[84.9] Among mathematical stories from ancient world, one of the most fascinating is *Problema Bovinum* (ca 250 BCE). It is a problem of Archimedes (Siracusa) with which he challenged mathematicians in Alexandria (contained in his letter to Eratosthenes; a relevant document was discovered in 1773 by the dramatist G.E. Lessing). The problem is said to be equivalent to

$$\text{pell}_A(1), \quad A = 410286423278424.$$

The least solution to this was obtained by Amthor (Krumbiegel–Amthor (1880, p.162)):

$$\left(T_0 + U_0\sqrt{d}\right)^{2329}, \ d = 4729494,$$
$$T_0 = 109931986732829734979866232821433543901088049,$$
$$U_0 = 50549485234315033074477819735540408986340.$$

By virtue of Legendre's congruence (84.25), we are able to quickly explain the reasoning of Amthor: We begin with the prime power decomposition $A = 2^3 \cdot 3 \cdot 7 \cdot 11 \cdot 29 \cdot 353 \cdot q^2$, $q = 4657$; thus $\sqrt{A} = 2q\sqrt{d}$. Theorem 85 yields that the least solution to $\text{pell}_d(1)$ is equal to $T_0 + U_0\sqrt{d}$ ($r = 92$). Hence, our task is equivalent to finding the least $K > 0$ such that $(T_0\mathfrak{e} + U_0\mathfrak{d})^K \equiv V\mathfrak{e} \bmod 2q$ with a $V \in \mathbb{N}$; however, since U_0 is even, we may replace the modulus by q. Also, considering the binomial expansion, we see that $(T_0\mathfrak{e} - U_0\mathfrak{d})^K \equiv V\mathfrak{e} \bmod q$ should hold as well. Hence, $V^2 \equiv 1 \bmod q$; that is, $(T_0\mathfrak{e} + U_0\mathfrak{d})^K \equiv \pm\mathfrak{e} \bmod q$. On the other hand, by the reciprocity law we have $\left(\frac{d}{q}\right) = -1$; and $(T_0\mathfrak{e} + U_0\mathfrak{d})^{q+1} \equiv \mathfrak{e} \bmod q$ by (84.25). Hence, $K|(q + 1)$ by the definition of K. Since $q + 1 = 2 \cdot 17 \cdot 137$, our K has to be equal to one of $2, 17, 34, 137, 274, 2329, 4658$. We need to compute the powers of $T_0\mathfrak{e} + U_0\mathfrak{d} \equiv -251\mathfrak{e} - 1606\mathfrak{d} \bmod q$. This can readily be executed by imitating the procedure developed in Note [70.2] (II), that is, by considering the powers of $-251 - 1606\rho$ in the quadratic extension field $\mathbb{F}_q(\rho)$ with $\rho^2 = d \equiv 2639 \bmod q$. We find that none of the exponents $2, 17, 34, 137, 274$ is appropriate, as they correspond to $262\mathfrak{e} + 551\mathfrak{d}$, $-1411\mathfrak{e} + 1933\mathfrak{d}$, $106\mathfrak{e} - 1579\mathfrak{d}$, $1686\mathfrak{e} + 2334\mathfrak{d}$, $-1006\mathfrak{e} - 82\mathfrak{d} \bmod q$, respectively. Then we reach

$$(T_0\mathfrak{e} + U_0\mathfrak{d})^{2329} \equiv -\mathfrak{e} \bmod q \;\Rightarrow\; K = 2329.$$

[84.10]   The algorithm of Lucas (1878b, pp.314–316) to test the primality of Mersenne numbers M_p. We shall explain it by means of (84.28): We begin with the fact that the least solution to $\mathrm{pell}_3(1)$ is $2 + \sqrt{3}$. We consider the powers of $2\mathfrak{e} + \mathfrak{d}$, $\mathfrak{d}^2 = 3\mathfrak{e}$. Let

$$L_n\mathfrak{e} = (2\mathfrak{e} + \mathfrak{d})^{2^n} + (2\mathfrak{e} - \mathfrak{d})^{2^n}, \quad n \geq 0.$$

Then,

$$\text{Lucas' test}: \quad L_{p-2} \equiv 0 \bmod M_p \;\Rightarrow\; M_p \text{ is a prime.}$$

For

$$(2\mathfrak{e} + \mathfrak{d})^{2^{p-2}} \equiv -(2\mathfrak{e} - \mathfrak{d})^{2^{p-2}} \;\Rightarrow\; (2\mathfrak{e} + \mathfrak{d})^{2^{p-1}} \equiv -\mathfrak{e} \bmod M_p,$$

which according to an obvious extension of (41.1) means that the order of $2\mathfrak{e} + \mathfrak{d} \bmod q$ is equal to 2^p for any prime factor q of M_p. We have $2^p | \varphi(q; 3)$, so $2^p \leq q + 1$; hence, $M_p = q$. This test is easy to apply: Squaring both sides of the identity defining L_n we see that $L_{n+1} = L_n^2 - 2$. Hence, starting from $L_0 = 4$, one may compute $L_n \bmod M_p$ quickly. For instance, in the case of $M_{31} = 2147483647$, which Fermat claimed to be a prime and Euler (1772f) confirmed, we have $L_0 = 4, L_1 = 14, L_2 = 194, L_3 = 37634, \ldots, L_{26} \equiv 211987665, L_{27} \equiv 1181536708, L_{28} \equiv 65536 \bmod M_{31}$; and $65536^2 - 2 = 4294967294 = 2M_{31}$, i.e., $L_{29} \equiv 0 \bmod M_{31}$. Euler's own argument is different and depends on the observation made in Note [58.8]. As a matter of fact, the above congruence condition of Lucas is necessary as well. To show this, we note that if M_p is a prime, then by the reciprocity law

$$\left(\frac{3}{M_p}\right) = -\left(\frac{(-1)^p - 1}{3}\right) = -1;$$

and with respect to mod M_p we have $(\mathfrak{e} + \mathfrak{d})^{M_p} \equiv \mathfrak{e} - \mathfrak{d} \Rightarrow (\mathfrak{e} + \mathfrak{d})^{2^p} \equiv -2\mathfrak{e} \Rightarrow (2(2\mathfrak{e} + \mathfrak{d}))^{2^{p-1}} \equiv -2\mathfrak{e}$. Then, note that 2 is a quadratic residue by (58.1)(2), that is, $2^{2^{p-1}} \equiv 2 \cdot 2^{(M_p-1)/2} \equiv 2$. Hence, $\left((2\mathfrak{e} + \mathfrak{d})^{2^{p-2}}\right)^2 \equiv -\mathfrak{e}$, from which follows $L_{p-2} \equiv 0$. See Lehmer (1930, p.443).

[84.11]   Legendre (1798, Seconde partie, §XV) described an argument which can be regarded as one of the origins of the theory of integer factorization. It is the method (49.1)(iii), and rests on the fact that the continued fraction expansion of $\sqrt{d}$ or the algorithm (84.3) (as well as its refinements (86.1), (86.3), (86.10) given below) do not require the prime power decomposition of d in general: Thus, prepare the continued fraction (84.6) and let β_{j+1} be as in

386   *Quadratic Forms*

(84.23). Search for those cases where $\beta_{j+1} = tu^2$ holds with a relatively small factor t (sqf). Since $\beta_{j+1} < 2\sqrt{d}$ by (84.2), the prime power decomposition of β_{j+1} can be expected to be less troublesome than that of d. If $p|d$ and $p \nmid \beta_{j+1}$, then $(-1)^{j+1}t \bmod p$ is a quadratic residue; and $p \bmod 4t$ is under the restriction imposed by Theorem 59; so Legendre correctly exploited the reciprocity law, although his proof of the law was not complete. Collect such t's and fix those $\varphi(4t)/2$ residue classes $\bmod 4t$ as is indicated in the proof of Theorem 59; Legendre did not have the Jacobi symbol, but that does not matter here. Then, by a simple sifting procedure, one may narrow the set of primes which can be candidates for prime factors of d. He applied his argument to $d = d_0 = 10091401$ (p.318) and reached a set of three primes (p.320); however, the relevant sifting procedure is cumbersome and does not seem appropriate to reproduce here. Anyway, these three primes do not divide d_0, and Legendre could verify Euler's assertion (Note [78.11]) that d_0 is a prime. Obviously, this argument of Legendre can yield prime factors of d if it is composite. He also remarked that one may try to factor hd instead of d with a relatively small h.

[84.12]   When applying the method (49.1) (iv) of Kraïtchik, one may wonder how to effectively collect the pairs $\{g_j, h_j\}$ with which (49.3) works well. The method of Legendre in the preceding Note suggests the use of the identity (84.23). See Pomerance (1985).

[84.13]   A digression about the eponym, John Pell, of the equation $\text{pell}_d(m)$: It has been asserted often that the equation should have been attributed instead to Fermat. He (1657: Œuvres II, pp.334–335) made a challenge towards mathematicians in England or rather Europe by proposing to find non-trivial solutions to $\text{pell}_d(1)$, $d = 109, 149, 433$, etc.; the fact that he selected specifically these integers indicates his possession of a relevant general theory, for the corresponding least solutions are all huge. According to Wallis (1685, bottom lines of p.364), Brouncker responded to the challenge with a set of correct answers. Euler mentioned, in his letter to Goldbach of August 1730, that according to Wallis' book $109xx + 1$ *quadratum ... solvendis excogitavit D. Pell Anglus peculiarem methodum* (see Fuss (1843, I, p.37)). It is said that the young Euler did not carefully read Wallis' book, and erroneously took the argument there for Pell's. Indeed, although Wallis attributed his Chapters LIX–LXIII to Pell, they have nothing to do with $\text{pell}_d(1)$, and the very Chapter XCVIII that is devoted to Brouncker's work contains Fermat's name but not Pell's. Brouncker exploited the periodicity of the continued fraction expansion of $\sqrt{d}$, if expressed in modern terms. This feature was observed later by Euler (1733a); notably, there (§§15–16) he mentioned the names of Pell, Fermat, and Wallis but not of Brouncker. Euler's basic investigations (1733a, 1759) led Lagrange to the complete resolution (1769) of the problem $\text{pell}_d(m)$; actually, his former work (1768), which is apparently independent of Euler's works, is the first resolution, but the work (1769) is far more thorough. In the context of the present Note, it should be pointed out that Lagrange referred to Fermat, Brouncker, Wallis, and Euler but not to Pell. Then, Gauss came in and devoted [DA, arts.198–201] to the discussion of $\text{pell}_d(m^2)$ (not $\text{pell}_d(m)$). In art.202 he related some historical facts about $\text{pell}_d(1)$; after mentioning Fermat and Brouncker, he continued that the equation was called *Pellianum* by some people because Euler (1733a, 1759, 1771, 1773b) attributed the solution in Wallis' treatise to Pell. Indeed, although *Pellius et Fermatius* is in Euler (1733a, §15), as remarked above, in the other three articles there are mentions solely to Pell, as if stressing. On the other hand, Gauss did not remark the fact that Pell's name was missing in Lagrange's articles (1768, 1769), although he referred to them in art.202. In any event, Gauss' stance about the eponym is indefinite. In contrast, his successor Dirichlet (1863, Contents: in the titles for §§83–84) explicitly employed *Pell'sche Gleichung*, which indicates well that in the mean time the name had already become an established term in number theory.

[84.14]   However, nowadays it is widely known that an amazingly effective algorithm to solve $\text{pell}_d(\pm 1)$ was actually developed by Brahmagupta (628) and gradually completed to the cakravâla (cyclic) method of Bhaskara II (1150); details are to be given in §86. In particular, the fact that $\text{pell}_d(1)$ has infinitely many solutions was known already in classic Indian mathematics (Datta–Singh (1935: 1962, Chap.III, §16)). More than a thousand years earlier than Lagrange, a marvelous number theoretical achievement had been done. This outstanding historical fact makes the use of the notation $\text{pell}_d(m)$ only less justifiable. Be that as it may, one ought to count yet the fact that so many investigations have already been conducted under the term Pell equations.

§85

We next develop a method to solve $\text{pell}_D(4)$, $D > 0$, i.e., the issue (76.10) (c). This task has actually been settled already as indicated in the ending paragraph of §83. Thus, we shall show an alternative way, which is more efficient because of the use of Theorem 25 instead of Theorem 24; the latter played a basic rôle in the preceding section. Also, in the later part of the present section, we shall describe the geometrical nature of the family $\{\text{pell}_D(4) : \forall D > 0\}$, and indicate how the Selberg zeta-function, a true mathematical marvel, emerges. It will

388   *Quadratic Forms*

transpire, even from our limited description presented here and in §94, that behind the very traditional subject $\text{pell}_D(4)$, $D > 0$, extends a deep, gorgeous world.

Remark 1 With any non-discriminant d, $\text{pell}_d(4)$ need not be discussed, for either if $d \equiv -1 \bmod 4$ or if $2\|d$, then each solution to $\text{pell}_d(4)$ is equal to $\{2u, 2v\}$ with a $\{u, v\}$ solving $\text{pell}_d(1)$.

(I) $D = 4d$.

Each solution to $\text{pell}_D(4)$ is equal to $\{2t, u\}$ with $\{t, u\}$ solving $\text{pell}_d(1)$; so the discussion of the preceding section should appear to suffice. We shall, however, show that there exists a simpler argument: Let $t^2 - du^2 = 1$, $t, u > 0$. This is equivalent to

$$\sqrt{d} = \frac{t\sqrt{d} + du}{u\sqrt{d} + t}, \quad \begin{pmatrix} t & du \\ u & t \end{pmatrix} \in \Gamma. \tag{85.1}$$

Theorem 25 implies that $t/u = A_k/B_k$; here A_k/B_k is the kth convergent of $\sqrt{d}$, and k is odd. There exists an integer $\tau \geq 0$, such that

$$\begin{pmatrix} t & du \\ u & t \end{pmatrix} = \begin{pmatrix} A_k & A_{k-1} \\ B_k & B_{k-1} \end{pmatrix} \begin{pmatrix} 1 & \tau \\ 0 & 1 \end{pmatrix}. \tag{85.2}$$

It is implied that with $a_j = [R^j(\sqrt{d})]$

$$\sqrt{d} = a_0 + \cfrac{1}{a_1 +} \cfrac{1}{a_2 + \cdots +} \cfrac{1}{a_k +} \cfrac{1}{\sqrt{d} + \tau}. \tag{85.3}$$

The continued fraction expansion of $\sqrt{d} + \tau$ is purely periodic. Then, by the argument following (82.30) we see that $-1 < -\sqrt{d} + \tau < 0$. Hence, we get $\tau = [\sqrt{d}] = a_0$. The rest of the argument can be omitted.

(II) $D = 4d + 1$.

Let $t^2 - Du^2 = 4$, $t, u > 0$. With $\eta = \frac{1}{2}(1 + \sqrt{D})$, this is equivalent to

$$\eta = \frac{\frac{1}{2}(t + u)\eta + du}{u\eta + \frac{1}{2}(t - u)}, \quad \begin{pmatrix} \frac{1}{2}(t + u) & du \\ u & \frac{1}{2}(t - u) \end{pmatrix} \in \Gamma. \tag{85.4}$$

Since $t > u$, Theorem 25 implies that $(t + u)/2u = A_k/B_k$; here A_k/B_k is the kth convergent of η and k is odd. There exists an integer $\tau' \geq 0$ such that

$$\begin{pmatrix} \frac{1}{2}(t + u) & du \\ u & \frac{1}{2}(t - u) \end{pmatrix} = \begin{pmatrix} A_k & A_{k-1} \\ B_k & B_{k-1} \end{pmatrix} \begin{pmatrix} 1 & \tau' \\ 0 & 1 \end{pmatrix}. \tag{85.5}$$

Thus, with $a_j = [R^j(\eta)]$,

$$\eta = a_0 + \cfrac{1}{a_1} + \cfrac{1}{a_2} + \cdots + \cfrac{1}{a_k} + \cfrac{1}{\eta + \tau'}, \tag{85.6}$$

and $\eta + \tau'$ is expanded into a purely periodic continued fraction. Hence, as before we find that

$$-1 < \tfrac{1}{2}(1 - \sqrt{D}) + \tau' < 0 \Rightarrow \tau' = f, \tag{85.7}$$

with f as in (82.15).

Theorem 87 *For any positive discriminant $D \equiv 1 \bmod 4$, non-trivial solutions to the indefinite equation $\mathrm{pell}_D(4)$ are given by*

$$t = 2A_k - B_k, \quad u = B_k, \quad 2 \nmid k. \tag{85.8}$$

Here A_k/B_k is the kth convergent of the continued fraction expansion (85.6).

Remark 2 If the expansion (85.6) holds with an even k, then take a further period so that k increases to $2k + 1$.

Proof It remains to show that (85.8) is a solution to $\mathrm{pell}_D(4)$. Thus, we note that

$$\eta + f = \begin{pmatrix} 1 & f \\ 0 & 1 \end{pmatrix} \begin{pmatrix} A_k & A_{k-1} \\ B_k & B_{k-1} \end{pmatrix} (\eta + f), \tag{85.9}$$

with the f as above. A quadratic equation for $\eta + f$ follows from this. Comparing it with that for $\omega^+(Q)$ where Q is given by (82.15), we have

$$B_{k-1} = A_k - (f + 1)B_k, \quad A_{k-1} = dB_k - fA_k. \tag{85.10}$$

Hence,

$$\begin{aligned} 1 &= A_k B_{k-1} - B_k A_{k-1} = A_k^2 - A_k B_k - dB_k^2 \\ &\Rightarrow \left(2A_k - B_k\right)^2 - DB_k^2 = 4. \end{aligned} \tag{85.11}$$

We end the proof.

With this, we return to $(76.6)_{D>0}$. We note that

$$\begin{aligned} &\begin{pmatrix} \tfrac{1}{2}(t_1 - bu_1) & -cu_1 \\ au_1 & \tfrac{1}{2}(t_1 + bu_1) \end{pmatrix} \begin{pmatrix} \tfrac{1}{2}(t_2 - bu_2) & -cu_2 \\ au_2 & \tfrac{1}{2}(t_2 + bu_2) \end{pmatrix} \\ &= \begin{pmatrix} \tfrac{1}{2}(t_3 - bu_3) & -cu_3 \\ au_3 & \tfrac{1}{2}(t_3 + bu_3) \end{pmatrix}. \end{aligned} \tag{85.12}$$

Here

$$\begin{aligned} &t_3 = \tfrac{1}{2}\left(t_1 t_2 + Du_1 u_2\right), \quad u_3 = \tfrac{1}{2}\left(u_1 t_2 + u_2 t_1\right), \\ &t_3^2 - Du_3^2 = \tfrac{1}{4}\left(t_1^2 - Du_1^2\right)\left(t_2^2 - Du_2^2\right) = 4; \end{aligned} \tag{85.13}$$

Quadratic Forms

or rather

$$\tfrac{1}{2}\left(t_1 + u_1\sqrt{D}\right) \cdot \tfrac{1}{2}\left(t_2 + u_2\sqrt{D}\right) = \tfrac{1}{2}\left(t_3 + u_3\sqrt{D}\right). \tag{85.14}$$

To sum up the above, we introduce

$$\epsilon_D = \min\left\{\tfrac{1}{2}\left(t + u\sqrt{D}\right) : t^2 - Du^2 = 4;\, t, u > 0\right\}, \tag{85.15}$$

and call it the least solution to $\mathrm{pell}_D(4)$, following (84.19). In the later part of the present section is a geometrical characterization of ϵ_D.

Remark 3 When $D > 0$ tends to infinity, the size of ϵ_D can probably become as huge as $D^{cD^{1/2}}$ with an absolute constant $c > 0$. See Hua (1942).

We have obtained

Theorem 88 *For any positive discriminant D, the set of all solutions to* $\mathrm{pell}_D(4)$ *can be identified with the multiplicative group*

$$\mathcal{E}_D = \left\{\pm\,\epsilon_D^\nu : \nu \in \mathbb{Z}\right\}. \tag{85.16}$$

Also, for any $Q \in \mathfrak{Q}_\pm(D)$,

$$\mathrm{Aut}_Q \cong \mathcal{E}_D. \tag{85.17}$$

Remark 4 In the context of $\Gamma/\{\pm 1\}$ (see Note [5.1]), the relation (85.17) is translated into $\mathrm{Aut}_Q/\{\pm 1\} \cong \{\epsilon_D^\nu : \nu \in \mathbb{Z}\}$. Then, that $\nu > 0$ is the same as $\epsilon_D^\nu = \tfrac{1}{2}(t + u\sqrt{D})$ with $t, u > 0$. For $t = \epsilon_D^\nu + \epsilon_D^{-\nu}$, $u = (\epsilon_D^\nu - \epsilon_D^{-\nu})/\sqrt{D}$.

Therefore, for any discriminant $D > 0$, $\mathrm{pell}_D(4)$ has infinitely many solutions. We have obtained a practical algorithm to compute them all. Also, we have determined the structure of Aut_Q for any $Q \in \mathfrak{Q}_\pm(D)$. With this, we return to (74.8) or equivalently to (74.17). Then, we see that the set of all solutions to $(71.2)_{D>0}$ splits into a finite number of orbits, each of which can be arranged by means of (85.17), confirming what has been stated in the paragraph following (82.3). That is, there exists a finite number of special solutions, or seeds of solutions to $(71.2)_{D>0}$ from which all the remaining solutions are generated by explicitly constructed automorphisms. As has been noticed already, this scheme for $(71.2)_{D>0}$ originated in Euler (1733a). We shall add, in §87, a practical way to identify the seeds. See also Note [92.5].

We may now close the discussion of §83:

Proof of Theorem 84.

If $t^2 - Du^2 = -4$, then for any $Q = [|a,b,c|] \in \mathfrak{Q}_\pm(D)$ we have

$$\,^tCQC = -Q, \quad C = \begin{pmatrix} \frac{1}{2}(t-bu) & -cu \\ au & \frac{1}{2}(t+bu) \end{pmatrix} \in \diamond\Gamma, \tag{85.18}$$

as remarked already at (76.14). On noting (82.10), we have $C(\omega^+(Q)) = \omega^+(Q)$. Then, by Theorem 26, there exist $h,h' \in \mathbb{N}$ such that $h - h' \equiv 1 \bmod 2$ and $R^h(\omega^+(Q)) = R^{h'}(\omega^+(Q))$. That is, the period of $\omega^+(Q)$ is odd. Conversely, if there exists a reduced form $Q' \in \mathfrak{Q}_\pm(D)$ whose circular cf-orbit is of odd period, then (82.24) implies that there exists a $C' \in \diamond\Gamma$ such that $\,^tC'Q'C' = -Q'$. Following faithfully the proof of Theorem 75, we reach a solution to $\mathrm{pell}_D(-4)$. We conclude that

$$\mathrm{pell}_D(-4) \text{ has a solution } \iff \text{ any form in } \mathfrak{Q}_\pm(D) \text{ is of odd period.} \tag{85.19}$$

We end the proof.

Hence, we have obtained the following alternative expression of (83.1):

$$\text{if } \mathrm{pell}_D(-4) \text{ has a solution, then}$$
$$\mathfrak{h}_\pm(D) \text{ is equal to the number of the circular cf-orbits;} \tag{85.20}$$
$$\text{otherwise, it is equal to twice the same number.}$$

Also, it should be added that

$$\mathrm{pell}_D(-4), D \equiv 1 \bmod 4, \text{ is solvable}$$
$$\text{if and only if (85.6) holds with even } k\text{'s;} \tag{85.21}$$
$$\text{the solutions are } t = 2A_k - B_k, \ u = B_k,$$

where A_k, B_k are as in (85.8) but with $2|k$ instead of $2 \nmid k$. The proof can be skipped, as it is analogous to that of Theorem 87. As for the case $D \equiv 0 \bmod 4$, the corresponding assertion follows readily from Theorem 85(i). We add that the counterpart of (ii) there implies that with Aut_Q^* defined by (76.13)

$$\mathrm{Aut}_Q = \left\{ \pm 1, \ \pm V_*^2 : \forall V_* \in \mathrm{Aut}_Q^* \right\}, \quad \mathrm{Aut}_Q^* = V_* \cdot \mathrm{Aut}_Q; \tag{85.22}$$

the latter holds with any fixed V_*. The verification is immediate.

We have thus almost finished the reduction theory of quadratic forms that we commenced in §74; almost, because further practical approaches to Pell equations will be presented in the next section, and the details of (æ), §76, for indefinite forms remain to be developed in §87. It should, however, be highly appropriate to have here a glimpse into a spectacular idea of Selberg (1956), with which he discovered a way to view all primitive indefinite forms, or rather all of their automorphisms, as an entity:

392   *Quadratic Forms*

Selberg's zeta-function for Γ (part 1)

The main rôle of the story is played by

$$\zeta_\Gamma(s) = \prod_{D>0} \prod_{n=0}^{\infty} \left(1 - \frac{1}{\epsilon_D^{2s+2n}}\right)^{\mathfrak{h}_\pm(D)}, \quad \operatorname{Re} s > 1. \tag{85.23}$$

Here D runs over all positive discriminants, with ϵ_D as above; the product is known to converge absolutely in the indicated range. Actually, Selberg defined his zeta-function in a quite wide framework; but the treatment of the case of Γ suffices to reveal the heart of his idea. Anyway, it turns out that

$$\zeta_\Gamma^*(s) = \frac{\zeta_\Gamma(s)}{\zeta(2s)\Gamma(2s)(2s-1)} \tag{85.24}$$

is an entire function which satisfies a functional equation similar to (12.7); and most astoundingly,

$$\zeta_\Gamma^*(s) \text{ satisfies an analogue of the Riemann hypothesis.} \tag{85.25}$$

The set of all complex zeros of $\zeta_\Gamma^*(s)$ is exactly the same as $\left\{\frac{1}{2} \pm i\omega, \ \omega > 0\right\}$ with $\left\{\frac{1}{4} + \omega^2\right\}$ being the whole discrete spectrum of the Laplace–Beltrami operator $\Delta = -y^2(\partial_x^2 + \partial_y^2)$ on $\mathcal{F}$ or rather on the surface $\mathcal{F}^*$ defined at the end of the text of §82.

We ought to give more details of course, but it can be done only in the later part of §94 because of a necessity of basics from the theory of the Riemann zeta-function. Thus, we shall beforehand make a preparation concerning a geometric structure of Γ that is proper to the present section. Be aware that we shall discuss under the convention Note [5.1](3). Let $D > 0$ be an arbitrary positive discriminant. Let $\mathcal{K}_\pm(D)$ be as before; and let a class $\mathcal{C} \in \mathcal{K}_\pm(D)$ be represented by a form Q. Then we have, by (74.14),

$$\mathcal{C} = \{{}^t UQU : U \in \operatorname{Aut}_Q\backslash\Gamma\}, \tag{85.26}$$

which induces the map

$$\mathcal{C} \mapsto \{\operatorname{Aut}_{{}^t UQU} = U^{-1}\operatorname{Aut}_Q U : U \in \operatorname{Aut}_Q\backslash\Gamma\}. \tag{85.27}$$

However, (74.24) and Note [74.9] imply that this is not injective; so we shall make a refinement. Combining (76.6) and Theorem 88, we introduce the map

$$Q = [|a,b,c|] \mapsto V_Q^k = \begin{pmatrix} \frac{1}{2}(t_k - bu_k) & -cu_k \\ au_k & \frac{1}{2}(t_k + bu_k) \end{pmatrix}, \tag{85.28}$$

where $\epsilon_D^k = \frac{1}{2}(t_k + u_k\sqrt{D}\,)$; in the notation of Note [76.3], $V_Q^k = V_Q(t_k, u_k)$. For each k this is injective; that is, if $V_{Q_1}^k = V_{Q_2}^k$, then $Q_1 = Q_2$. Hence (85.28) can be assembled into a single injective map:

$$Q \mapsto [V_Q]^+ = \{V_Q^n : n \in \mathbb{N}\}; \tag{85.29}$$

note that the exponent n runs over $\mathbb{N}$. Since $V_{-Q} = V_Q^{-1}$, we have $\mathrm{Aut}_Q = [V_Q] = [V_Q]^+ \sqcup \{1\} \sqcup [V_{-Q}]^+$. We now replace (85.27) by

$$\mathcal{C} \mapsto \{[V_{^tUQU}]^+ = U^{-1}[V_Q]^+U : U \in \mathrm{Aut}_Q\backslash\Gamma\}; \tag{85.30}$$

see Note [76.6]. This is a family of disjoint sets: we observe that with $\omega^\pm(Q)$ as in (82.4)

$$\phi_Q V_Q \phi_Q^{-1} = \begin{pmatrix} \epsilon_D & 0 \\ 0 & \epsilon_D^{-1} \end{pmatrix}, \quad \phi_Q = \begin{pmatrix} 1 & -\omega^+(Q) \\ 1 & -\omega^-(Q) \end{pmatrix}; \tag{85.31}$$

that is, the eigenvalues of V_Q are $\epsilon_D^{\pm 1}$. Thus, if $U_1^{-1}V_Q^{n_1}U_1 = U_2^{-1}V_Q^{n_2}U_2$ in (85.30), then a comparison of eigenvalues yields $n_1 = n_2 = k$, so $V_{^tU_1QU_1}^k = V_{^tU_2QU_2}^k$, which implies $^tU_1QU_1 = {}^tU_2QU_2$ since (85.28) is injective. We end the confirmation.

Next we shall view (85.30) from Γ. We introduce the following classification of non-trivial linear fractional transformations $\begin{pmatrix} \alpha & \beta \\ \gamma & \delta \end{pmatrix} \in \mathrm{PSL}(2,\mathbb{R})$:

elliptic: one fixed point on $\mathcal{H}$ $\Leftrightarrow$ $|\alpha + \delta| < 2$,
parabolic: only one fixed point on $\mathbb{R} \cup \{\infty\}$ $\Leftrightarrow$ $|\alpha + \delta| = 2$,
hyperbolic: two different fixed points on $\mathbb{R}$ $\Leftrightarrow$ $|\alpha + \delta| > 2$.
$$\tag{85.32}$$

Let $\Gamma^{(h)}$ be the set of all hyperbolic elements in $\Gamma/\{\pm 1\}$. We shall show that

$$\bigcup_Q \mathrm{Aut}_Q\backslash\{1\} = \Gamma^{(h)}, \tag{85.33}$$

where Q runs over all indefinite forms regardless of discriminants. The inclusion of all $\mathrm{Aut}_Q\backslash\{1\}$ in $\Gamma^{(h)}$ is obvious. To show the opposite inclusion, take an arbitrary $V = \begin{pmatrix} v_1 & v_2 \\ v_3 & v_4 \end{pmatrix} \in \Gamma^{(h)}$. Since we are presently dealing with an element in $\mathrm{PSL}(2,\mathbb{Z})$, we may assume, without loss of generality, that $v_1 + v_4 > 0$. With this, let

$$\langle v_3, v_1 - v_4, v_2 \rangle = u, \quad v_3 = au, \quad v_1 - v_4 = -bu, \quad v_2 = -cu, \tag{85.34}$$

so that $\langle a, b, c \rangle = 1$; note that $u > 0$, for $V \neq 1$. With $v_1 + v_4 = t$, we are led to $t^2 - Du^2 = 4$, where $b^2 - 4ac = D > 0$, a discriminant; the positivity of D is implied by that V has two fixed points on $\mathbb{R}$. We have got a map $V \mapsto Q = [|a,b,c|] \in \mathcal{Q}_\pm(D)$ such that

$$V = \begin{pmatrix} \frac{1}{2}(t - bu) & -cu \\ au & \frac{1}{2}(t + bu) \end{pmatrix} = V_Q(t,u). \tag{85.35}$$

394   *Quadratic Forms*

Hence, (85.33) has been verified. We have found, more essentially, that any hyperbolic V is identical to a unique V_Q^n because in (85.35) we have $t, u > 0$; see Remark 4 above. Classifying these Q with respect to modulo Γ and taking (85.30) into account, we obtain the decomposition

$$\Gamma^{(h)} = \bigsqcup_{D > 0} \bigsqcup_{\mathcal{C} \in \mathcal{K}(D)} \bigsqcup_{U \in [V_Q] \backslash \Gamma} U^{-1} [V_Q]^+ U \backslash \{1\} \tag{85.36}$$

where Q is a representative of $\mathcal{C}$ and $V_Q = V_Q(u_1, v_1)$. Continued to Part 2 in §94.

Notes

[85.1]   Since $\langle 2A_k - B_k, B_k \rangle = \langle 2, B_k \rangle$, if $2 | B_k$ for the first k satisfying (85.9), then $\mathrm{pell}_D(4)$, $D \equiv 1 \bmod 4$, does not have any proper solution. Examples are in Notes [83.5] and [87.2].

[85.2]   If $p \equiv 1 \bmod 4$, then $\mathrm{pell}_p(-4)$ has a solution; hence, periods of all forms in $\mathcal{Q}_\pm(p)$ are odd. To confirm this assertion, we combine Note [84.3] with Theorem 87. Thus, by $a^2 - pb^2 = -1$, $u^2 - pv^2 = 4$, and by the Brahmagupta identity (72.14),

$$(au + pbv)^2 - p(av + bu)^2 = (a^2 - pb^2)(u^2 - pv^2) = -4.$$

By the way, as noted at the end of §65, if we take $x = 1$ in the Euler–Gauss identity (65.3), then a solution to $\mathrm{pell}_p(-4)$ follows. Moreover, by another identity (Note 86.2(4) below) of Brahmagupta, we are led to a solution to $\mathrm{pell}_p(1)$ as well. In particular, solutions to these indefinite equations can be expressed in terms of pth roots of unity: a fascinating relation between $\mathrm{pell}_p(1)$ and the cyclotomy. See Bachmann (1872, pp.294–299).

[85.3]   An example for the preceding Note: $D = 421$, $\eta = \frac{1}{2}(1 + \sqrt{421})$;

$$\eta = 10 + \cfrac{1}{1} + \cfrac{1}{3} + \cfrac{1}{6} + \cfrac{1}{1} + \cfrac{1}{1} + \cfrac{1}{2} + \cfrac{1}{2}$$
$$+ \cfrac{1}{1} + \cfrac{1}{1} + \cfrac{1}{6} + \cfrac{1}{3} + \cfrac{1}{1} + \cfrac{1}{\eta + 9} = \frac{A_{12}(\eta + 9) + A_{11}}{B_{12}(\eta + 9) + B_{11}}.$$

The period is 13, so by (85.19) $\mathrm{pell}_{421}(-4)$ has solutions. The least one is, by (85.21),

$$(2A_{12} - B_{12})^2 - 421 \cdot B_{12}^2 = 444939^2 - 421 \cdot 21685^2 = -4.$$

This implies

$$\left(\tfrac{1}{2}(444939 + 21685\sqrt{421})\right)^2 = \tfrac{1}{2}\left(197970713723 + 9648502215\sqrt{421}\right),$$

which is the least solution ϵ_{421} to $\mathrm{pell}_{421}(4)$ in the sense of (85.15). Alternatively, via the expansion

$$\eta = 10 + \cfrac{1}{1} + \cfrac{1}{3} + \cfrac{1}{6} + \cfrac{1}{1} + \cfrac{1}{1} + \cfrac{1}{2} + \cfrac{1}{2} + \cfrac{1}{1} + \cfrac{1}{1} + \cfrac{1}{6}$$
$$+ \cfrac{1}{3} + \cfrac{1}{1} + \cfrac{1}{19} + \cfrac{1}{1} + \cfrac{1}{3} + \cfrac{1}{6} + \cfrac{1}{1} + \cfrac{1}{1} + \cfrac{1}{2} + \cfrac{1}{2}$$
$$+ \cfrac{1}{1} + \cfrac{1}{1} + \cfrac{1}{6} + \cfrac{1}{3} + \cfrac{1}{1} + \cfrac{1}{\eta + 9} = \frac{A_{25}(\eta + 9) + A_{24}}{B_{25}(\eta + 9) + B_{24}}$$

and (85.8), we get

$$2A_{25} - B_{25} = 197970713723, \quad B_{25} = 9648502215.$$

See Note [86.3]. By the way, $\mathfrak{h}_{\pm}(421) = 1$.

[85.4]   The representation (85.23) for $\zeta_\Gamma(s)$ was stated in Hejahl (1983, p.518), for instance. The same devices as (85.28) and (85.34) appeared in Sarnak (1982, p.232) and in Siegel (1957, Chap.3, §5), respectively. Thus one may call (85.36) the Sarnak–Siegel decomposition. Also, see Milnor (1982) and Arcozzi (2012) for historical accounts of the non-Euclidean (hyperbolic) geometry and its model due to Beltrami (1868).

§86

In the present section, we shall develop a discussion on an ancient Indian method for $\mathrm{pell}_d(1)$.

We begin with the observation that expanding $\sqrt{d}$ into an ordinary continued fraction is equivalent to fixing positive integers $\{\alpha_j, \beta_j : j \geq 0\}$ according to the cyclic algorithm

$$\begin{aligned}
&(0)\ \alpha_0 = [\sqrt{d}\,], \ \beta_0 = 1, \\
&(1)\ \alpha_j + \alpha_{j+1} \equiv 0 \bmod \beta_j, \\
&(2)\ \alpha_{j+1} < \sqrt{d} < \alpha_{j+1} + \beta_j, \\
&(3)\ \beta_j \beta_{j+1} = d - \alpha_{j+1}^2.
\end{aligned} \tag{86.1}$$

With this we invoke (84.24) and assert that

$$\begin{pmatrix} A_j & dB_j \\ B_j & A_j \end{pmatrix} = \prod_{u=0}^{j} \frac{1}{\beta_u} \begin{pmatrix} \alpha_{u+1} & d \\ 1 & \alpha_{u+1} \end{pmatrix}, \quad j \geq 0. \tag{86.2}$$

Taking determinants of both sides, we get $A_j^2 - dB_j^2 = (-1)^{j+1}\beta_{j+1}$ as noted at (84.23). Therefore the algorithm (86.1) gives rise to solutions to $\mathrm{pell}_d(1)$ since

Quadratic Forms

there should exist a j such that $\beta_{j+1} = 1$, $2 \nmid j$, because of the periodicity of the continued fraction expansion of $\sqrt{d}$.

We then look into the condition (86.1) (2). Suppose that $\theta_j = (\alpha_j + \sqrt{d})/\beta_j$ is such that $\frac{1}{2} < \theta_j - [\theta_j]$ or $[\theta_{j+1}] = 1$; this corresponds to the situation discussed in Note [24.4]. Such a θ_j is nearer to the integer $[\theta_j] + 1$ than to $[\theta_j]$. Equivalently, $\alpha_{j+1} + \frac{1}{2}\beta_j < \sqrt{d}$, as $\alpha_j + \alpha_{j+1} = [\theta_j]\beta_j$; that is, in (2) we have $\alpha_{j+1} + \beta_j - \sqrt{d} < \sqrt{d} - \alpha_{j+1}$. Via this observation, we are led to the following modification of (86.1):

$$
\begin{aligned}
&(0) \ \mu_0 = 0, \ \nu_0 = 1, \\
&(1) \ \mu_k + \mu_{k+1} \equiv 0 \bmod \nu_k, \\
&(2) \ |\sqrt{d} - \mu_{k+1}| \text{ is minimum,} \\
&(3) \ \nu_k \nu_{k+1} = |d - \mu_{k+1}^2|.
\end{aligned}
\tag{86.3}
$$

Starting form (0) or $k = 0$, we inductively identify μ_{k+1} as to be the μ such that $\mu \equiv -\mu_k \bmod \nu_k$ and $|\sqrt{d} - \mu|$ is the least. Then $d - \mu_{k+1}^2$ is divisible by ν_k, and ν_{k+1} satisfying (3) is obtained uniquely. With $\xi_k = (\mu_k + \sqrt{d})/\nu_k$, the criterion (2) means that

$$
\begin{aligned}
\xi_k < [\xi_k] + \tfrac{1}{2} &\Rightarrow \xi_k = [\xi_k] + \frac{1}{\xi_{k+1}}, \ \mu_k + \mu_{k+1} = [\xi_k]\nu_k; \\
\xi_k > [\xi_k] + \tfrac{1}{2} &\Rightarrow \xi_k = [\xi_k] + 1 - \frac{1}{\xi_{k+1}}, \ \mu_k + \mu_{k+1} = ([\xi_k] + 1)\nu_k.
\end{aligned}
\tag{86.4}
$$

Since $b_k = (\mu_k + \mu_{k+1})/\nu_k$ is equal to either $[\xi_k]$ or $[\xi_k] + 1$, we are led to the following half-regular (see Note [21.1]) continued fraction expansion of $\sqrt{d}$:

$$
\begin{aligned}
\sqrt{d} &= b_0 + \frac{\rho_1}{b_1 +} \frac{\rho_2}{b_2 +} \cdots + \frac{\rho_k}{b_k +} \frac{\rho_{k+1}}{\xi_{k+1}} \\
&= \frac{E_k \xi_{k+1}/\rho_{k+1} + E_{k-1}}{F_k \xi_{k+1}/\rho_{k+1} + F_{k-1}}, \quad \rho_k = \mathrm{sgn}\left(\sqrt{d} - \mu_k\right).
\end{aligned}
\tag{86.5}
$$

Here, we have, in view of Note [21.1],

$$
\begin{pmatrix} E_k & E_{k-1} \\ F_k & F_{k-1} \end{pmatrix} = \prod_{u=0}^{k} \begin{pmatrix} b_u & 1 \\ \rho_u & 0 \end{pmatrix}, \quad k \geq 0.
\tag{86.6}
$$

In much the same way as (84.22)–(84.23), we find that

$$
E_k^2 - dF_k^2 = (-1)^{k+1} \nu_{k+1} \prod_{u=0}^{k} \rho_{u+1}
\tag{86.7}
$$

and

$$\begin{pmatrix} E_k & dF_k \\ F_k & E_k \end{pmatrix} = \prod_{u=0}^{k} \frac{1}{v_u} \begin{pmatrix} \mu_{u+1} & d \\ 1 & \mu_{u+1} \end{pmatrix}, \quad k \geq 0. \tag{86.8}$$

Since (86.1) yields the expansion (86.5) which is a modification of (84.6) by means of the invertible transformation given in Note [21.4], the algorithm (86.3) as well should resolve $\text{pell}_d(1)$.

Further, we look into the condition (86.3) (2). We observe that

$$\begin{aligned} (\mu_{k+1} + v_k)^2 - d &< d - (\mu_{k+1})^2 \\ &\Rightarrow \mu_{k+1} + \tfrac{1}{2} v_k < \sqrt{d} \ \Rightarrow \ [\xi_k] + \tfrac{1}{2} < \xi_k, \end{aligned} \tag{86.9}$$

which is the situation stated in the lower line of (86.4). With this observation, we modify (86.3) so that $\{\mu_k, v_k : k \geq 0\}$ is replaced by the following $\{\sigma_l, \tau_l : l \geq 0\}$:

the *cakravâla* algorithm :
$$\begin{aligned} &\text{(0) } \sigma_0 = 0, \ \tau_0 = 1, \\ &\text{(1) } \sigma_l + \sigma_{l+1} \equiv 0 \bmod \tau_l, \\ &\text{(2) } \left| d - \sigma_{l+1}^2 \right| \text{ is minimum,} \\ &\text{(3) } \tau_l \tau_{l+1} = \left| d - \sigma_{l+1}^2 \right|, \end{aligned} \tag{86.10}$$

where σ_* is, of course, a local notation different from the divisor function (16.4). Let $\psi_l = (\sigma_l + \sqrt{d})/\tau_l$, and let $c_l = (\sigma_l + \sigma_{l+1})/\tau_l$, which is equal to either $[\psi_l]$ or $[\psi_l] + 1$. Then we have

$$\begin{aligned} \sqrt{d} &= c_0 + \frac{\pi_1}{c_1} + \frac{\pi_2}{c_2} + \cdots + \frac{\pi_l}{c_l} + \frac{\pi_{l+1}}{\psi_{l+1}} \\ &= \frac{G_l \psi_{l+1}/\pi_{l+1} + G_{l-1}}{H_l \psi_{l+1}/\pi_{l+1} + H_{l-1}}, \quad \pi_l = \text{sgn}\,(d - \sigma_l^2). \end{aligned} \tag{86.11}$$

As before, we have

$$\begin{pmatrix} G_l & G_{l-1} \\ H_l & H_{l-1} \end{pmatrix} = \prod_{j=0}^{l} \begin{pmatrix} c_j & 1 \\ \pi_j & 0 \end{pmatrix}, \quad l \geq 0; \tag{86.12}$$

and

$$\begin{pmatrix} G_l & dH_l \\ H_l & G_l \end{pmatrix} = \prod_{u=0}^{l} \frac{1}{\tau_u} \begin{pmatrix} \sigma_{u+1} & d \\ 1 & \sigma_{u+1} \end{pmatrix}, \tag{86.13}$$

$$G_l^2 - dH_l^2 = (-1)^{l+1} \tau_{l+1} \prod_{u=0}^{l} \pi_{u+1}. \tag{86.14}$$

398   *Quadratic Forms*

Since (86.11) is the same as the result of applying the transformation given in Note [21.4] to (84.6) not fully as is done in (86.5) but partly, that is, only when the inequality on the upper line of (86.9) holds, we are able to assert that

$$\text{the cakravâla algorithm (86.10) resolves pell}_d(1). \tag{86.15}$$

It should be stressed here that (86.3) and (86.10) are not identical because the reversal of the reasoning (86.9) does not work well in general; namely, it can happen that $(\mu_{k+1} + \nu_k)^2 - d \geq d - (\mu_{k+1})^2$ while $[\xi_k] + \frac{1}{2} < \xi_k$; an example is given in Note [86.4].

Since the expansions (86.5) and (86.11) are usually shorter than (84.6), both the algorithms (86.3) and (86.10) can be quicker than (86.1) to reach a solution of $\text{pell}_d(1)$. In literature, (86.1), (86.3), and (86.10) are sometimes called, respectively, the ordinary, the nearest integer, and the nearest square algorithms for continued fraction expansions of $\sqrt{d}$. This naming comes obviously from the step (2) of each algorithm. One may notice that among these three the easiest to apply is (86.10) (2) since it is simpler to identify the square closest to d than to approximate, on $\mathbb{R}$, the square-root of d, on the respective premise (1). Yet, Brahmagupta stated a remarkable enhancement, Note [86.2], which is applicable to any of these three algorithms.

Notes

[86.1]   The cakravâla algorithm is attributed to Brahmagupta (628: Colebrooke (1817, pp.363–372)) but usually stated in the form (86.10) due to Bhaskara II (1150: Strachey (1813, pp.36–53), Colebrooke (1817, pp.170–184), Datta–Singh (1935: 1962: II, 1938, §17)). In the explanation preceding (86.10) we adopted partly the argument of Ayyangar (1941). It does not seem that the general equation $\text{pell}_d(m)$ was systematically considered in classical Indian mathematics, although Brahmagupta knew already that if it has a solution, then there are infinitely many; indeed he (Note [72.5]) composed them by means of the solutions to $\text{pell}_d(1)$ just the same way as Euler did more than a thousand years later (Datta–Singh (1935: 1962, p.174)).

[86.2]   The composition formulas of Brahmagupta (Datta–Singh (1935: 1962, §16)):

(1) if $t^2 - du^2 = \pm 1$, then

$$\left(t^2 + du^2\right)^2 - d(2tu)^2 = 1.$$

(2) if $t^2 - du^2 = \pm 2$, then

$$\left(\tfrac{1}{2}(t^2 + du^2)\right)^2 - d(tu)^2 = 1.$$

(3) if $t^2 - du^2 = 4$, then

$$2 \mid t \Rightarrow \left(\tfrac{1}{2}(t^2 - 2) \right)^2 - d \left(\tfrac{1}{2} tu \right)^2 = 1,$$

$$2 \nmid t \Rightarrow \left(\tfrac{1}{2} t(t^2 - 3) \right)^2 - d \left(\tfrac{1}{2} u(t^2 - 1) \right)^2 = 1.$$

(4) if $t^2 - du^2 = -4$, then

$$\left(\tfrac{1}{2}(t^2 + 2)((t^2 + 2)^2 - 3) \right)^2 - d \left(\tfrac{1}{2} tu((t^2 + 2)^2 - 1) \right)^2 = 1.$$

Comparing the results in Note [84.2](3) and Note [85.3], we see that the difference between them is huge, indeed in the order of 6th power as (4) indicates. Interestingly, Euler (1773b) stated and applied these four identities.

[86.3]   We apply (86.10) to $\mathrm{pell}_{421}(1)$:

$$
\begin{array}{llllll}
\sigma_0 = 0, & \tau_0 = 1 & \Rightarrow & \sigma_1 = 21, & \tau_1 = 20, & \pi_1 = -1, & c_0 = 21, \\
\sigma_2 \equiv -21 \bmod 20 & & \Rightarrow & \sigma_2 = 19, & \tau_2 = 3, & \pi_2 = +1, & c_1 = 2, \\
\sigma_3 \equiv -19 \bmod 3 & & \Rightarrow & \sigma_3 = 20, & \tau_3 = 7, & \pi_3 = +1, & c_2 = 13, \\
\sigma_4 \equiv -20 \bmod 7 & & \Rightarrow & \sigma_4 = 22, & \tau_4 = 9, & \pi_4 = -1, & c_3 = 6, \\
\sigma_5 \equiv -22 \bmod 9 & & \Rightarrow & \sigma_5 = 23, & \tau_5 = 12, & \pi_5 = -1, & c_4 = 5, \\
\sigma_6 \equiv -23 \bmod 12 & & \Rightarrow & \sigma_6 = 25, & \tau_6 = 17, & \pi_6 = -1, & c_5 = 4, \\
\sigma_7 \equiv -25 \bmod 17 & & \Rightarrow & \sigma_7 = 26, & \tau_7 = 15, & \pi_7 = -1, & c_6 = 3, \\
\sigma_8 \equiv -26 \bmod 15 & & \Rightarrow & \sigma_8 = 19, & \tau_8 = 4, & \pi_8 = +1, & c_7 = 3. \\
\end{array}
$$

By (86.12) and (86.14), we obtain

$$G_7 = 444939, \ H_7 = 21685 : \ G_7^2 - 421 H_7^2 = -4 \, ;$$

see Note [85.3]. By Note [86.2](4), we get the assertion Note [84.2](3) again:

$$\tfrac{1}{2}(G_7^2 + 2)\left((G_7^2 + 2)^2 - 3\right) = 3879474045914926879468217167061449,$$

$$\tfrac{1}{2} G_7 H_7 \left((G_7^2 + 2)^2 - 1\right) = 189073995951839020880499780706260.$$

Also we have, by (86.11),

$$\sqrt{421} = 21 + \frac{-1}{2} + \frac{1}{13} + \frac{1}{6} + \frac{-1}{5} + \frac{-1}{4} + \frac{-1}{3} + \frac{-1}{3} + \frac{1}{\psi_8} = \frac{G_7 \psi_8 + G_6}{H_7 \psi_8 + H_6},$$

with $G_6 = 168886$, $H_6 = 8231$, $\psi_8 = (19 + \sqrt{421})/4$. In terms of the regular continued fraction expansion, we have

$$\sqrt{421} = 20 + \frac{1}{1} + \frac{1}{1} + \frac{1}{13} + \frac{1}{5} + \frac{1}{1} + \frac{1}{3} + \frac{1}{1} + \frac{1}{2} + \frac{1}{1} + \frac{1}{1} + \frac{1}{1} + \frac{1}{2} + \frac{1}{\psi_8}.$$

In view of Note [21.4] this is equivalent to the last half-regular expansion. Since the period of $\sqrt{421}$ is 37 (Note [84.2](3)), the ordinary method to solve $\mathrm{pell}_{421}(1)$ requires 74 terms.

400   *Quadratic Forms*

[86.4]   In (86.5) the integer b_k is the closest one to ξ_k, but c_l in (86.11) is not always the closest to ψ_l. That is, (86.3) and (86.10) are indeed different algorithms, as has been remarked already. For instance, we have, by (86.10),

$$\sqrt{133} = 12 + \cfrac{-1}{2} + \cfrac{1}{7} + \cfrac{1}{6} + \cfrac{-1}{3} + \cfrac{-1}{4} + \cfrac{-1}{3} + \cfrac{-1}{6} + \cfrac{1}{7} + \cfrac{1}{2} + \cfrac{-1}{\psi_{10}}$$
$$= \frac{2588599\psi_{10} - 1210008}{224460\psi_{10} - 104921}, \quad \psi_{10} = 12 + \sqrt{133},$$

which solves $\mathrm{pell}_{133}(1)$. Here

$$\sigma_8 = 11, \tau_8 = 3; \ \sigma_9 = 10, \tau_9 = 11 \ \Rightarrow \ c_8 = 7.$$

Since $\psi_8 = (11 + \sqrt{133})/3 = 7.51085\ldots$, this value of c_8 is not the integer closest to ψ_8. By the way,

$$12 + \cfrac{-1}{2} + \cfrac{1}{7} = \frac{173}{15} \ \Rightarrow \ 173^2 - 133 \cdot 15^2 = 4;$$

hence, by Note [86.2](3), we get

$$2588599^2 - 133 \cdot 224460^2 = 1,$$

while the ordinary continued fraction expansion of $\sqrt{133}$ requires 16 terms to produce the same. The classic Indian method to solve general $\mathrm{pell}_d(1)$ is indeed a marvel of algorithm. And it was discovered more than 1300 years ago.

§87

We shall now detail (æ), §76, in the case of indefinite forms; here the coprimality condition in (73.1) is not supposed. A pleasantly practical method will be presented that makes it possible to discuss the indefinite equation $Q(x,y) = m > 0$ for any $Q \in \mathfrak{Q}_\pm(D)$ without fully relying on the reduction argument. This stems from an idea of Lagrange (1770b, Problème V, 29; 1798, §VII) to transform a form into a principal form; see also Legendre (1798, Première Partie, §XI).

Let $Q = [|a,b,c|] \in \mathfrak{Q}_\pm(D)$. Without loss of generality, we may suppose that $a > 0$ and $\langle a,m \rangle = 1$ because of Note [74.5]. Let $Q(\alpha,\beta) = m$ be a proper representation; so, in particular, $\langle \beta,m \rangle = 1$. Let $\mu \equiv \alpha\beta^{-1} \bmod m$, i.e., $\alpha = \mu\beta + \lambda m$ with an integer λ. Then we get

$$am\lambda^2 + (2a\mu + b)\lambda\beta + \gamma\beta^2 = 1,$$
$$\gamma = \big((2a\mu + b)^2 - D\big)/4am = Q(\mu,1)/m, \tag{87.1}$$

$$\S 87 \qquad\qquad 401$$

where γ is an integer, for $Q(\mu, 1) \equiv \beta^{-2}Q(\alpha, \beta) \equiv 0 \bmod m$. Namely,

$$\mathfrak{Q}_\pm(D) \ni [|am, 2a\mu + b, \gamma|] \text{ represents } 1. \qquad (87.2)$$

By Remark 2, §74, this is a principal form and is equivalent mod Γ either to the form (82.15) or to the form (82.17) according as D is odd or even, respectively. Then, the combination of Theorem 26 and (82.8) implies that the continued fraction expansion of

$$\xi = \frac{v + \sqrt{D}}{2w}, \quad v = -2a\mu - b, \ w = am, \qquad (87.3)$$

has the same period block as

$$\omega_D = \begin{cases} [(\sqrt{D}-1)/2] + (1+\sqrt{D})/2 & \text{if } 4 \nmid D, \\ [\sqrt{D/4}] + \sqrt{D/4} & \text{if } 4 \mid D. \end{cases} \qquad (87.4)$$

More precisely, there should exist $h, h' \in \mathbb{N}$ such that $h \equiv h' \bmod 2$ and $R^h(\xi) = R^{h'}(\omega_D)$. With this, let $r \geq h'$, $2|r$, be a multiple of the period of ω_D so that $R^{h-h'+r}(\xi) = \omega_D$; there exists an odd J, i.e., $J = h - h' + r - 1$, with which $R^{J+1}(\xi) = \omega_D$. In other words, according to (82.25) the cf-orbit of the derived form (87.2) reaches the corresponding basic principal form in an even number of steps. On the other hand, in much the same way as the derivation of (84.23) we have

$$R^{j+1}(\xi) = -\frac{A_{j-1} - B_{j-1}\xi}{A_j - B_j\xi} \qquad (87.5)$$
$$\Rightarrow \quad (2A_j w - B_j v)^2 - DB_j^2 = 4(-1)^{j+1}ww_{j+1},$$

where $R^{j+1}(\xi) = (v_{j+1} + \sqrt{D})/2w_{j+1}$ and A_j/B_j is the jth convergent of ξ.

Consequently, if Q represents m properly under the condition imposed above, then there should exist an odd J such that $w_{J+1} = 1$, i.e.,

$$(2A_J w - B_J v)^2 - DB_J^2 = 4w, \qquad (87.6)$$

which corresponds to (85.11). This is equivalent to

$$amA_J^2 + (2a\mu + b)A_J B_J + \gamma B_J^2 = 1 \qquad (87.7)$$

or

$$Q(s, t) = m \text{ with } s = A_J m + B_J \mu, \ t = B_J. \qquad (87.8)$$

Since $\langle s, t \rangle = \langle m, B_J \rangle = 1$ as well as $\langle m, t \rangle = 1$ by (87.7), we have got a way to map an arbitrary ${}^t\{\alpha, \beta\} \in \mathcal{S}_Q(m)$ to a certain ${}^t\{s, t\} \in \mathcal{S}_Q(m)$ in such a way that $st^{-1} \equiv \alpha\beta^{-1} \bmod m$; that is, by Note [76.4], ${}^t\{s, t\} \equiv {}^t\{\alpha, \beta\} \bmod \text{Aut}_Q$. This map does not affect the classification of $\mathcal{S}_Q(m)$ by Aut_Q. Namely, it is the

402   *Quadratic Forms*

identical map on $\mathrm{Aut}_Q \backslash \mathcal{S}_Q(m)$, i.e., does not essentially alter the set of the seed solutions.

Reversing the reasoning, we obtain the following algorithm:

Theorem 89 *Let $Q = [|a,b,c|] \in \mathcal{Q}_{\pm}(D)$, $a > 0$. Let $m > 0$, $\langle a,m \rangle = 1$, be such that there exists a ρ mod m satisfying*

$$v_0^2 \equiv D \bmod 4am, \quad v_0 = -(2a\rho + b). \tag{87.9}$$

With

$$\xi_0 = \frac{v_0 + \sqrt{D}}{2w_0}, \quad w_0 = am; \quad R^j(\xi_0) = \frac{v_j + \sqrt{D}}{2w_j}, \tag{87.10}$$

suppose that there exists a v satisfying

$$2 \nmid v, \quad w_{v+1} = 1. \tag{87.11}$$

Then

$$Q(\sigma, \tau) = m, \quad \sigma = A_v m + B_v \rho, \quad \tau = B_v, \tag{87.12}$$

is a proper representation. Moreover, we have a bijection

$$\{\rho \bmod m\} \to \mathrm{Aut}_Q \backslash \mathcal{S}_Q(m). \tag{87.13}$$

Proof We have, by (87.5) with $v \mapsto v_0$, $w \mapsto w_0$, and $j \mapsto v$,

$$(2a\sigma + b\tau)^2 - D\tau^2 = 4am \implies Q(\sigma, \tau) = m, \tag{87.14}$$

where the left side of the arrow is equivalent to

$$w_0 A_v^2 - v_0 A_v B_v + (v_0^2 - D)B_v / 4w_0 = 1. \tag{87.15}$$

Thus, $\langle m, B_v \rangle = 1$, so ${}^t\{\sigma, \tau\} \in \mathcal{S}_Q(m)$. The assertion (87.13) is an immediate consequence. We end the proof.

Some remarks are in order:

(1) This algorithm yields all the seeds of $\mathcal{S}_Q(m)$ exactly. The assertion (87.13) sates that essential is the residue class mod m to which ρ belongs: as far as staying in the same residue class, any change of the value of ρ affects neither the crucial (87.11) nor the corresponding class in $\mathrm{Aut}_Q \backslash \mathcal{S}_Q(m)$. Also, there are actually infinitely many v satisfying (87.11) if there exists one; they all yield solutions belonging to the same class mod Aut_Q.

(2) There exist situations where (87.9) does not necessarily guarantee (87.11); see the second part of Note [87.4] as well as Note [87.5].

§87   403

(3) In practice, one may suppose that $Q = [|a,b,c|]$ is such that

$$a \nmid b, \tag{87.16}$$

since otherwise Q can be transformed into an equivalent form $[|a',b',c'|]$ with $\langle a',D \rangle = 1$, and thus $a' \nmid b'$; see Note [74.5]. This has a subtle consequence: If $X^2 \equiv D \bmod 4am$ has solutions $\pm \xi \bmod 2am$, then of course we have $\xi^2 \equiv b^2 \bmod 4a$; and only one of them possibly satisfies another requirement $X \equiv -b \bmod 2a$, for if both do, then it would follow that $a|b$, contradicting (87.16). Namely, (87.16) is to remove an ambiguity in choosing $(\bmod\, 2am)$ the solutions of $X^2 \equiv D \bmod 4am$. Relevant remarks are in Notes below.

(4) The condition (87.11) does not necessarily require that $R^{\nu+1}(\xi_0)$ is exactly equal to ω_D. It suffices to have $R^{\nu+1}(\xi_0) = \omega_D + f, f \in \mathbb{Z}$, since continuing the expansion we reach $R^{\nu'+1}(\xi_0) = \omega_D, 2 \nmid \nu'$, anyway. Relevant numerical examples are in Notes [87.2] and [87.4].

(5) We are, of course, able to deal with negative m as well; but then we need to have instead an even ν in (87.11). Such an example is contained in Note [87.7].

Notes

[87.1]   Although the above discussion suffices for our purpose, we remark that it is possible to make Lagrange's algorithm less dependent on the reduction theory by means of Theorem 25, §25. It is an extension of the argument of §85(I)–(II); see Matthews, K.R. (2002). In any event, the application of Theorem 89 does not require the knowledge of the reduction theory. Nevertheless, it is needed in order to explain the phenomenon mentioned in (2) above; see Note [87.3].

[87.2]   Let $Q = [|13,11,-10|] \in \mathcal{Q}_\pm(D), D = 641$. With $m = 140561$ consider $Q(x,y) = m$: We need to solve $X \equiv -11 \bmod 26, X^2 \equiv D \bmod 52m$; here $X = -(26\rho + 11)$ with ρ to be fixed later. Pollard's method (§38) gives $m = 367 \cdot 383$; both factors are prime. Thus, we consider $x_1^2 \equiv D \bmod 52$, $x_2^2 \equiv D \bmod 367$, and $x_3^2 \equiv D \bmod 383$. We take $x_1 = 15$, for x_1 should be $\equiv -11 \bmod 26$; this is related to the point touched upon in (3) above. As for the moduli 367, 383, we apply Note [63.1] (Lagrange), finding $x_2 \equiv \pm 170 \bmod 367, x_3 \equiv \pm 32 \bmod 383$. Hence, following Note [30.2], we have

$$X \equiv 367 \cdot 383 \cdot 21 \cdot 15 + 26 \cdot 383 \cdot 15 \cdot x_2 + 26 \cdot 367 \cdot 58 \cdot x_3 \bmod 26m,$$

so that the possible values of v_0 are

$$47907, 795459, 1173863, 3324089 \quad \text{(mutually different mod } 26m\text{)},$$

while the corresponding values of ρ are, respectively,

$$-1843, \; -30595, \; -45149, \; -127850 \quad \text{(mutually different mod } m\text{)}.$$

In particular,

$$\frac{47907 + \sqrt{D}}{26m} = 0 + \cfrac{1}{76} + \cfrac{1}{4} + \cfrac{1}{11} + \cfrac{1}{1} + \cfrac{1}{1} + \cfrac{1}{2} + \cfrac{1}{1} + \cfrac{1}{1}$$
$$+ \cfrac{1}{1} + \cfrac{1}{4} + \cfrac{1}{2} + \cfrac{1}{3} + \cfrac{1}{6} + \cfrac{1}{\omega_D}$$
$$= \frac{A_{13}\omega_D + A_{12}}{B_{13}\omega_D + B_{12}}; \quad A_{13} = 200297, \; B_{13} = 15271588,$$

where

$$\omega_D = \tfrac{1}{2}(25 + \sqrt{D}) = 25 + \cfrac{1}{6} + \cfrac{1}{3} + \cfrac{1}{2} + \cfrac{1}{4} + \cfrac{1}{1} + \cfrac{1}{1} + \cfrac{1}{1} + \cfrac{1}{2} + \cfrac{1}{1}$$
$$+ \cfrac{1}{1} + \cfrac{1}{12} + \cfrac{1}{12} + \cfrac{1}{1} + \cfrac{1}{1} + \cfrac{1}{2} + \cfrac{1}{1} + \cfrac{1}{1} + \cfrac{1}{1} + \cfrac{1}{4} + \cfrac{1}{2} + \cfrac{1}{3} + \cfrac{1}{6} + \cfrac{1}{\omega_D}.$$

Hence, we get the special solution

$$\sigma = A_{13}m + B_{13} \cdot (-1843) = 8409933, \quad \tau = 15271588.$$

Similarly, with $v_0 = 795459$, we use $\frac{A_7}{B_7} = \frac{377}{1732}$; with $v_0 = 1173863$, $\frac{A_{17}}{B_{17}} = \frac{42078063}{130998748}$. Further,

$$\frac{3324089 + \sqrt{D}}{26m} = 0 + \cfrac{1}{1} + \cfrac{1}{10} + \cfrac{1}{17} + \cfrac{1}{27} + \cfrac{1}{6} + \cfrac{1}{3} + \cfrac{1}{2} + \cfrac{1}{4}$$
$$+ \cfrac{1}{1} + \cfrac{1}{1} + \cfrac{1}{1} + \cfrac{1}{2} + \cfrac{1}{1} + \cfrac{1}{12} + \cfrac{1}{12} + \cfrac{1}{1} + \cdots$$

The term $\frac{1}{27}$ corresponds to $\omega_D + 2$; thus, $w_4 = 1$ with $\frac{A_3}{B_3} = \frac{171}{188}$. In this way, we get all the seeds:

$$\mathcal{S}_Q(m) = \text{Aut}_Q \cdot \Big\{ {}^t\{131, 188\}, \; {}^t\{957, 1732\},$$
$${}^t\{8409933, 15271588\}, \; {}^t\{72139891, 130998748\} \Big\}.$$

As a matter of fact, here any ρ satisfying (87.9) fulfills (87.11) automatically because ω_D is of odd period, and it yields a solution to $Q(x, y) = m$. Since $\mathfrak{h}_\pm(641) = 1$, all the elements in $\mathfrak{Q}_\pm(641)$ are principal forms; thus, returning to (74.20), which is applicable since (73.1) holds presently, we should have $|\text{Aut}_Q \backslash \mathcal{S}_Q(m)| = \sum_{f|m} \mu^2(f)\kappa_D(f)$. Indeed, we have the value 4 on both sides.

[87.3] Let $Q = [|39, 4, -26|] \in \mathfrak{Q}_\pm(D)$, $D = 4072$. With $m = 271167$ consider $Q(x, y) = m$: Since $m = 3 \cdot 13 \cdot 17 \cdot 409$ and $\langle a, m \rangle \neq 1$, we use the fact that $Q(3,1) = 337$ is a prime not dividing m. Transforming Q by $U = \begin{pmatrix} 3 & 2 \\ 1 & 1 \end{pmatrix}$, we move to $Q_1 = [|337, 436, 138|]$. We need to find X mod $674m$ such that $X \equiv -436 \equiv 238$ mod 674 and $X^2 \equiv D$ mod $1348m$. The latter is equivalent to the system of congruences $x_1^2 \equiv 28$ mod 1348, $x_2^2 \equiv 1$ mod 3, $x_3^2 \equiv 3$ mod 13, $x_4^2 \equiv 9$ mod 17, and $x_5^2 \equiv 391$ mod 409. We take $x_1 = 238$; see (3) above. Also, $x_2 = \pm 1$, $x_3 = \pm 4$, $x_4 = \pm 3$, $x_5 = \pm 105$; the last is obtained via Tonelli's algorithm. We have

$$X \equiv 317 \cdot 3 \cdot 13 \cdot 17 \cdot 409 \cdot 238 + 13 \cdot 17 \cdot 409 \cdot 674 \cdot x_2 - 3 \cdot 17 \cdot 409 \cdot 674 \cdot x_3$$
$$+ 13 \cdot 3 \cdot 13 \cdot 409 \cdot 674 \cdot x_4 + 208 \cdot 3 \cdot 13 \cdot 17 \cdot 674 \cdot x_5 \text{ mod } 674m.$$

There exist 16 solutions to (87.9) or rather ρ's with respect to mod m. We have to check each to see whether (87.11) is satisfied or not. For instance, with $\{x_2 = 1, x_3 = 4, x_4 = 3, x_5 = 105\}$, we get $v_0 = 119890684$, $\rho = -177880$. Then, $w_{42} = 1$, $\frac{A_{41}}{B_{41}} = \frac{1117747189349765013}{1703941381962094325}$. Thus, it holds that

$$m = Q_1(-940929009609250829, 1703941381962094325)$$
$$= Q(585095735096436163, 763012372352843496),$$

Also, $\{x_2 = 1, x_3 = 4, x_4 = -3, x_5 = -105\}$ gives $v_0 = 49470490$, $\rho = -73399$, $w_8 = 1$, $\frac{A_7}{B_7} = \frac{36}{133}$. Hence,

$$m = Q_1(-55, 133) = Q(101, 78),$$

which stands for the least seed. In this way, we are able to identify all the seeds representing the classes of $\mathrm{Aut}_Q \backslash S_Q(m)$, the number of which is 16. On the other hand, the right side of (74.20), which is applicable presently, has the value 16, as is readily computed. Since $\mathfrak{h}_{\pm}(4072) = 2$, the left side of (74.20) has two terms, and it follows that one of them is non-zero, i.e., 16, and the other zero. Then, take $Q_2 = [|2, 0, -509|] \in \mathfrak{Q}(D)$. It is not equivalent to Q, as can be seen by comparing the continued fraction expansion of $\omega^+(Q_2)$ with that of $\omega^+(Q)$. Thus we have $S_{Q_2}(m) = \varnothing$. We are able to alternatively verify this by means of the present algorithm. However, since (87.16) is not fulfilled, we first transform Q_2 into $Q_3 = [|139, 206, 69|]$ by $U_1 = \begin{pmatrix} 18 & 17 \\ 1 & 1 \end{pmatrix}$. Then we proceed just as before: The last theorem applied to the problem $Q_3(x, y) = m$ yields 16 forms in $\mathfrak{Q}(D)$, but all turn out to be congruent to Q_2, i.e., none of them is the principal form; hence, (87.11) is never attained this time. By the way, Q ought to be a principal form; indeed we have

$$Q(908633108775, 1184931392143) = 1.$$

By the way, the periods of Q and Q_2 are 17 and 15, respectively

[87.4] In the present section the coprimality condition in (73.1) is not assumed but $\langle a, m \rangle = 1$ is imposed. Here is a relevant instance: Let

406   *Quadratic Forms*

$Q = [|1630, -5247, 4222|] \in Q_{\pm}(D)$, $D = 3569$. Consider $Q(x, y) = 57233$ $= m$. Thus, $\langle D, m \rangle = 43$, yet $\langle 1630, m \rangle = 1$. We need to solve $X \equiv 5247$ mod 3260, $X^2 \equiv D$ mod $6520m$. There exist only two solutions 57987607, 150226047 mod $3260m$; more on this point in the next Note. We have

$$\frac{57987607 + \sqrt{D}}{3260m} = 0 + \frac{1}{3} + \frac{1}{4} + \frac{1}{1} + \frac{1}{1} + \frac{1}{2} + \frac{1}{9} + \frac{1}{1} + \frac{1}{\omega_D + 2},$$

with $\frac{A_7}{B_7} = \frac{239}{769}$. Hence ${}^t\{1253, 769\} \in S_Q(m)$. On the other hand, $v_0 = 150226047$ gives the solution ${}^t\{8770701965, 5387956153\}$. Since $\text{pell}_D(4)$ has the least solution $\{85003950, 1422872\}$, $\text{Aut}_Q = \{\pm V^k : k \in \mathbb{Z}\}$ with

$$V = \begin{pmatrix} 3775406667 & -6007365584 \\ 2319281360 & -3690402717 \end{pmatrix}.$$

Applying $-V^{-1}$ to the last solution we get ${}^t\{553, 349\}$. Therefore,

$$S_Q(m) = \text{Aut}_Q \cdot \left\{ {}^t\{553, 349\}, {}^t\{1253, 769\} \right\}.$$

We continue the discussion with $Q(x, y) = 28321 = m_1$. This time there exist four solutions $X \equiv 9469027, 48135887, 61221527, 88709847$ mod $3260m_1$. However, only the first two, i.e., the third and the fifth convergents of respective continued fractions yield the solutions:

$$S_Q(m_1) = \text{Aut}_Q \cdot \left\{ {}^t\{67, 39\}, {}^t\{193, 117\} \right\}.$$

[87.5] An analysis of the facts found in the preceding Note: Thus, let $D = 3569$. The criterion (82.22) yields a set of 46 reduced forms in $Q_{\pm}(D)$. It consists of one cf-cycle (I) of period 14 and two cycles (II, III) of period 16. Hence, by (83.1) we get $\mathfrak{h}_{\pm}(D) = 6$. The representative forms of these cf-cycles are

$$\text{I} : [|1, -59, -22|]; \quad \text{II} : [|2, -59, -11|]; \quad \text{III} : [|11, -59, -2|].$$

The six classes are

$$\mathfrak{s}_1 = \{S_1 = [|1, -59, -22|]\}, \quad \mathfrak{s}_2 = \{S_2 = [|22, -59, -1|]\},$$
$$\mathfrak{s}_3 = \{S_3 = [|2, -59, -11|]\}, \quad \mathfrak{s}_4 = \{S_4 = [|40, -57, -2|]\},$$
$$\mathfrak{s}_5 = \{S_5 = [|11, -59, -2|]\}, \quad \mathfrak{s}_6 = \{S_6 = [|22, -51, -11|]\}.$$

The form Q in the preceding Note belongs to $\mathfrak{s}_2$ since

$$Q = {}^t U S_2 U, \quad U = \begin{pmatrix} 8687 & -14141 \\ -515752 & 839559 \end{pmatrix} \in \Gamma.$$

Thus the above procedure applied to Q can be said to map $\mathfrak{s}_2$ to $\mathfrak{s}_1$, the principal class. With $m = 57233 = 11^3 \cdot 43$ and $D = 3529 = 43 \cdot 83$, we are unable to utilize (74.20). Nevertheless, returning to (72.2) we see that

$X^2 \equiv D \bmod 4m$ has two solutions $\xi_1 \equiv -46655$, $\xi_2 \equiv -67811 \bmod 2m$; that is, in the notation (72.5), M_{m,ξ_1} and M_{m,ξ_2}, which are both in $\mathfrak{Q}_\pm(D)$, represent m properly. These forms are equivalent to S_2, as the continued fractions of $(-\xi_\nu + \sqrt{D})/2m$, $\nu = 1, 2$, indicate. Therefore, we find that m is represented solely by the class $\mathfrak{s}_2$. On the other hand, with $m_1 = 28321 = 127 \cdot 223$, the right side of (74.20), which is now applicable, becomes 4. The above procedure, with 9469027, 48135887 mod $3260m_1$, maps $\mathfrak{s}_2$ to $\mathfrak{s}_1$; with 61221527, 88709847 mod $3260m_1$, $\mathfrak{s}_2$ is moved to $\mathfrak{s}_5$ and $\mathfrak{s}_4$, respectively. Since the left side of (74.20) presently has six terms, m_1 ought to be represented by classes other than $\mathfrak{s}_2$ as well. Indeed, Theorem 89 leads us to the conclusion that $\mathfrak{s}_3, \mathfrak{s}_6$ do:

$$S_3(546967764, 18425557) = m_1, \quad S_6(1153111236, 458157727) = m_1;$$

that is, $|\mathrm{Aut}_{S_3} \backslash \mathcal{S}_{S_3}(m_1)|$ and $|\mathrm{Aut}_{S_6} \backslash \mathcal{S}_{S_6}(m_1)|$ are both equal to 1, while neither $\mathfrak{s}_1$ nor $\mathfrak{s}_4$ nor $\mathfrak{s}_5$ represents m_1.

[87.6] As for the computation of $\{w_j\}$ in (87.10), we have the following procedure: for $j \geq 0$

$$(0)\ v_0^2 \equiv D \bmod 4am,\ w_0 = 2am,$$
$$(1)\ v_j + v_{j+1} \equiv 0 \bmod 2w_j,$$
$$(2)\ v_{j+1} < \sqrt{D} < v_{j+1} + 2w_j$$
$$(3)\ 4w_j w_{j+1} = D - v_{j+1}^2;$$

thus,

$$\begin{pmatrix} 2A_K w_0 - B_K v_0 & DB_K \\ B_K & 2A_K w_0 - B_K v_0 \end{pmatrix} = 2w_0 \prod_{j=0}^{K} \frac{1}{2w_j} \begin{pmatrix} v_{j+1} & D \\ 1 & v_{j+1} \end{pmatrix}.$$

One may add cakravâla flavor to this by imitating (86.10), which should make the algorithm just more efficient.

[87.7] We shall discuss briefly the problem to solve, in integers $\{x, y\}$, the equation $F(x, y) = 0$, with

$$F(x, y) = ax^2 + bxy + cy^2 + dx + ey + f$$
$$= Q(x, y) + dx + ey + f,$$

where $d, e, f \in \mathbb{Z}$, and Q is of discriminant $D > 0$ but not initially supposed to be primitive. Thus, first a reduction argument: We have, for variables x and y,

$$F(x, y) = (x, y, 1) \cdot A \cdot {}^t(x, y, 1), \quad A = \begin{pmatrix} a & b/2 & d/2 \\ b/2 & c & e/2 \\ d/2 & e/2 & f \end{pmatrix}.$$

408 *Quadratic Forms*

Let $\alpha = be - 2cd$, $\beta = bd - 2ae$. Then, with $D^* = \det A$,

$$Q(\xi,\theta) = D^2 F(x,y) + 4DD^*, \quad {}^t\{\xi,\theta\} = {}^t\{x,y\}D + {}^t\{\alpha,\beta\}.$$

To verify this identity, we expand $Q(\xi,\theta)$ noting the relations $2a\alpha + b\beta = Dd$, $2c\beta + b\alpha = De$; so it becomes $D^2(F(x,y) - f) + Q(\alpha,\beta)$. On the other hand

$$(\alpha,\beta,-D) \cdot A \cdot {}^t(\alpha,\beta,-D) = (0,0,4D^*) \cdot {}^t(\alpha,\beta,-D) = -4DD^*,$$

and this left side is equal to $Q(\alpha,\beta) - D(d\alpha + e\beta - Df)$, while $d\alpha + e\beta - Df = 8D^* + Df$; so $Q(\alpha,\beta) = 4DD^* + D^2 f$, which ends the verification. Hence,

$$F(x,y) = 0 \text{ in integers } x,y \quad \Leftrightarrow \quad \begin{aligned} Q(\xi,\theta) &= 4DD^*, \\ {}^t\{\xi,\theta\} &\equiv {}^t\{\alpha,\beta\} \bmod D. \end{aligned}$$

We remark that if a suitable ${}^t\{\xi,\theta\}$ exists, then there are actually infinitely many solutions to $F(x,y) = 0$. This follows from the fact that

$$(\dagger) \qquad V^2 \cdot {}^t\{\alpha,\beta\} \equiv {}^t\{\alpha,\beta\} \bmod D,$$

where $V \in \mathrm{Aut}_Q$; indeed,

$$\left(V^2 - \begin{pmatrix} 1 & 0 \\ 0 & 1 \end{pmatrix}\right) \cdot {}^t\{\alpha,\beta\} = {}^t\left\{\tfrac{1}{2}e(bu - t) - cdu, \tfrac{1}{2}d(bu + t) - aeu\right\} uD,$$

with $t^2 - Du^2 = 4$. Thus, for each integer l, there exists an integral ${}^t\{x_l, y_l\}$ such that

$$F(x_l, y_l) = 0, \quad V^{2l} \cdot {}^t\{\xi,\theta\} = {}^t\{x_l, y_l\}D + {}^t\{\alpha,\beta\}.$$

With this, we now restrict ourselves, for the sake of simplicity, to the situation where Q is primitive and ${}^t\{\xi,\theta\}$ is proper; note that the latter is possible only if Q is primitive, in general $\langle a,b,c \rangle$ divides α,β,D, so ξ,θ as well. Then, apply Theorem 89 to the problem to find integral vectors ${}^t\{\xi,\theta\} \equiv {}^t\{\alpha,\beta\} \bmod D$ such that $Q(\xi,\theta) = 4DD^*$, $\langle \xi,\theta \rangle = 1$. Here, if $D^* < 0$, then (5) at the end of the above text becomes relevant. It should be stressed that $(\dagger)$ means also that in selecting suitable members from the set of all proper solutions ${}^t\{\xi,\theta\}$, we need to test only the four members $\pm{}^t\{\xi_0,\theta_0\}$, $\pm V \cdot {}^t\{\xi_0,\theta_0\}$ with each seed ${}^t\{\xi_0,\theta_0\}$. See Lagrange (1768, pp.725–731); the reduction argument applied to $F(x,y)$ is a modern adaptation of that by Legendre (1798, pp.451–452). See also [DA, arts.216–221]. Incidentally, Gauss dealt with the case $2|\langle b,d,e \rangle$, but, interestingly, neither Lagrange nor Legendre had imposed such a restriction; see Note [71.5] above.

[87.8] An example:

$$\begin{aligned} G(x,y) &= 22x^2 - 59xy - y^2 + 173x + 237y - 35 \\ &= S_2(x,y) + 173x + 237y - 35, \end{aligned}$$

where S_2 is as in Note [87.5]. We have $D = 3569, \alpha = -13637, \beta = -20635$, $D^* = -3499933/4$. We need to solve

$$S_2(\xi,\theta) = -n; \quad \xi = 3569x - 13637, \ \theta = 3569y - 20635,$$

with $n = 12491260877$. We appeal to Theorem 89 but looking for even v. Though this n appears somewhat awful, it is in fact of relatively simple nature, for $n = 19 \cdot 23 \cdot 43 \cdot 83 \cdot 8009$. One may proceed as in Note [87.4]. Solving the congruence $\{X \equiv 59 \bmod 44, X^2 \equiv 3569 \bmod 88n\}$, we get eight solutions mod $44n$; four among them give seed solutions to $S_2(\xi,\theta) = -n$. However, the criterion indicated at the end of the preceding Note together with

$$V = \begin{pmatrix} 84476699 & 1422872 \\ 31303184 & 527251 \end{pmatrix},$$

implies that there exist only two suitable: $X \equiv 12166315195, \ 141819536379$ mod $44n$ yield that $S_2(132692, 50745), S_2(38755132, 14360877) = -n$; and

$$G(41, 20) = 0, \quad G(-10855, -4018) = 0,$$

respectively. Then, via the action of V^2, we are led to a faraway solution

$$G(268696932991993305, 99566740098166460) = 0,$$

and infinitely many more.

§88

Intermezzo: In the present and the next sections we shall complete the Legendre proof (§69[d]) of the quadratic reciprocity law (58.1)(3). We shall, in fact, develop more than what is actually needed for the sake of a later purpose. The discussion is composed of the parts [A]+[B] and [C]=[C$_0$]+[C$_1$]+[C$_\varpi$] in the present and the next sections, respectively. For the proof of the reciprocity law, the combination of [A], [B], and [C$_0$] suffices.

[A] We shall prove the following assertion independently of the reciprocity law, using only a very basic fact in the theory of functions:

Theorem 90

$$\begin{array}{c} \text{Dirichlet's prime number theorem:} \\ \text{the statement (69.1) of Legendre is valid.} \end{array} \qquad (88.1)$$

410   *Quadratic Forms*

Proof Returning to (52.6)–(52.10), we shall prove that for any coprime pair $\{\ell, q\}$

$$\lim_{s \to 1+0} \sum_{p \equiv \ell \bmod q} \frac{\log p}{p^s} = +\infty. \tag{88.2}$$

This is the same as

$$-\lim_{s \to 1+0} \sum_{\chi \bmod q} \overline{\chi}(\ell) \frac{L'}{L}(s, \chi) = +\infty. \tag{88.3}$$

We have

$$-\frac{L'}{L}(s, J_q) = -\frac{\zeta'}{\zeta}(s) + \sum_{p|q} \frac{\log p}{p^s - 1}, \tag{88.4}$$

and because of (12.2) the right side diverges to $+\infty$ as $s \to 1 + 0$. Thus it suffices to show that

$$\lim_{s \to 1+0} \sum_{\chi \neq J_q \bmod q} \overline{\chi}(\ell) \frac{L'}{L}(s, \chi) \neq \infty, \tag{88.5}$$

or rather

$$\chi \neq J_q \implies L(1, \chi) \neq 0, \tag{88.6}$$

since we have $L(1, \chi) \neq \infty$ as it follows readily from the combination of Note [13.8] and the remark after Theorem 46.

Thus, we proceed to a proof of (88.6): We introduce the function

$$M(s, \chi) = \sum_{n=1}^{\infty} |a(n, \chi)|^2 n^{-s}, \quad a(n, \chi) = \sum_{d|n} \chi(d). \tag{88.7}$$

Following closely the reasoning of Note [18.7], or rather replacing u^α, v^β there by $\chi(u), \overline{\chi}(v)$, respectively, we get, for $\operatorname{Re} s > 1$,

$$M(s, \chi) = \frac{\zeta(s) L(s, J_q) L(s, \chi) L(s, \overline{\chi})}{L(2s, J_q)} = \frac{\zeta^2(s) L(s, \chi) L(s, \overline{\chi})}{\zeta(2s) \prod_{p|q}(1 + p^{-s})}. \tag{88.8}$$

Hence, we see that since $L(1, \overline{\chi}) = \overline{L(1, \chi)}$,

$$L(1, \chi) = 0 \implies M(s, \chi) \text{ is regular at } s = 1. \tag{88.9}$$

Namely, if we suppose that (88.6) does not hold, then in the expression (88.8) the double pole at $s = 1$ of $\zeta^2(s)$ will be canceled out. We shall show that this yields a contradiction.

We observe first that under the present supposition the function $M(s, \chi)$ is regular for $\operatorname{Re} s > \frac{1}{2}$ and in a neighborhood of the segment of $0 < s \leq \frac{1}{2}$.

For the numerator in (88.8) is regular there as we have just noted, and as for the factor $1/\zeta(2s)$ it suffices to invoke the observation (12.21). Then we consider the Taylor expansion

$$M(s,\chi) = \sum_{k=0}^{\infty} \frac{M^{(k)}(2,\chi)}{k!}(s-2)^k. \tag{88.10}$$

Because of the above regularity of $M(s,\chi)$, the convergence radius is greater than $\frac{3}{2}$, so this series converges absolutely at a certain $s_1 < \frac{1}{2}$. Inserting the result of the term-wise differentiation into $M^{(k)}(2,\chi)$, the right side with $s=s_1$ becomes a double series whose terms are all non-negative; and we may exchange the order of summation. We find that

$$M(s_1,\chi) = \sum_{n=1}^{\infty} |a(n,\chi)|^2 n^{-s_1}. \tag{88.11}$$

Hence, the series (88.7) converges for $\operatorname{Re} s \geq s_1$. In particular, since we have $1/\zeta(2s) = 0$ at $s = \frac{1}{2} > s_1$,

$$0 = \sum_{n=1}^{\infty} |a(n,\chi)|^2 n^{-1/2}. \tag{88.12}$$

This is a contradiction, for $a(1,\chi) = 1$. We end the proof (88.1).

[B] We shall prove, independently of the reciprocity law, that

$$\text{Legendre's statement (69.2) is valid.} \tag{88.13}$$

To this end we shall prove the asymptotic formula

$$\sum_{\substack{2<p\leq x \\ \left(\frac{d}{p}\right)=1}} \frac{\log p}{p} = \tfrac{1}{2}\log x + O(1) \quad (d \neq 1 \text{ and } |d| : \text{sqf}); \tag{88.14}$$

hereafter within [B], all implicit constants may depend on d. Provided (88.14) is valid, we have, via (13.7),

$$\sum_{\substack{2<p\leq x \\ \left(\frac{d}{p}\right)=-1}} \frac{\log p}{p} = \tfrac{1}{2}\log x + O(1). \tag{88.15}$$

Hence, by means of the supplementary law (58.1)(1), which can, of course, be applied here, we have

Quadratic Forms

$$\sum_{\substack{p \le x,\, p \equiv 1 \bmod 4 \\ \left(\frac{d}{p}\right)=1}} + \sum_{\substack{p \le x,\, p \equiv 1 \bmod 4 \\ \left(\frac{d}{p}\right)=-1}} +2 \sum_{\substack{p \le x,\, p \equiv -1 \bmod 4 \\ \left(\frac{d}{p}\right)=-1}}$$

$$= \sum_{\substack{2 < p \le x \\ \left(\frac{d}{p}\right)=-1}} + \sum_{\substack{2 < p \le x \\ \left(\frac{-d}{p}\right)=1}} = \log x + O(1), \tag{88.16}$$

with an obvious abbreviation. This lower line is due to (88.15) and $(88.14)_{d \mapsto -d}$. The sum of the first two sums in the upper line is equal to $\sum_{p \le x,\, p \equiv 1 \bmod 4}$ with an error $O(1)$, and it is $(88.14)_{d=-1}$. Therefore,

$$\sum_{\substack{p \le x,\, p \equiv -1 \bmod 4 \\ \left(\frac{d}{p}\right)=-1}} = \tfrac{1}{4} \log x + O(1), \tag{88.17}$$

which implies (88.13); indeed, more than that.

In order to confirm (88.14) without recourse to the reciprocity law, we shall prove first the asymptotic formula

$$\sum_{\{m,n\} \in S(x)} \log \left| m^2 - dn^2 \right| = \frac{2}{\sqrt{|d|}} x \log x + O(x), \tag{88.18}$$

as x increases indefinitely; here $S(x)$ is the set of all pairs $\{m, n\}$ of integers such that

$$|m| \le (x/2)^{1/2}, \quad |n| \le (x/2|d|)^{1/2}, \quad \{m, n\} \ne \{0, 0\}. \tag{88.19}$$

We observe that for any $1 \le K \le x$

$$S_K(x) = \left\{ \{m, n\} \in S(x) : K/2 < |m^2 - dn^2| \le K \right\}$$
$$\Rightarrow |S_K(x)| = O\left((xK)^{1/2}\right). \tag{88.20}$$

This is trivial if $d < 0$. If $d > 0$, then the estimation of the part with $dn^2 < 2K$ is also trivial. As for the part with $dn^2 \ge 2K$, one may use either $(dn^2 + K/2)^{1/2} < |m| \le (dn^2 + K)^{1/2}$ or $(dn^2 - K)^{1/2} < |m| \le (dn^2 - K/2)^{1/2}$; the number of such m is not greater than $K^{1/2}$. Thus, we have confirmed (88.20). On noting that $\log |m^2 - dn^2| = \log x + O(j)$ on the set $S_{x/2^j}(x)$, we see that (88.20) gives the following asymptotic formula for the sum in (88.18):

$$|S(x)| \log x + \sum_j O(jx/2^{j/2}) = \frac{2}{\sqrt{|d|}} x \log x + O(x). \tag{88.21}$$

We shall next deduce (88.14) from (88.18). Our reasoning is similar to that of Note [13.5]: We consider the prime power decomposition

$$\prod_{\{m,n\}\in S(x)} \left|m^2 - dn^2\right| = \prod_{p\leq x} p^{T(p)}, \quad T(p) = \sum_j |t(p^j; x)|, \tag{88.22}$$

$$t(p^j; x) = \left\{\{m,n\} \in S(x) : m^2 \equiv dn^2 \bmod p^j\right\}, \quad p^j \leq x. \tag{88.23}$$

(1) The treatment of $T(p)$ with $p \geq 3$:

In order to estimate $|t(p^j; x)|$ for each $j \geq 1$, take a sufficiently large L, and let the tuple $\{A, B, p\}$ be such that

$$\left(\frac{d}{p}\right) = 1, \quad \begin{array}{l} A^2 \equiv dB^2 \bmod p^L, \\ p\nmid AB, \ \langle A, B\rangle = 1, \end{array} \tag{88.24}$$

while invoking the Hensel lifting (Note [34.3]). On noting that $m^2 \equiv dn^2 \bmod p^j$ is equivalent to $(aB + bA)(aB - bA) \equiv 0 \bmod p^j$, we consider the set

$$\tau(j_1, j_2; x) = \left\{\{m,n\} \in S(x) : \begin{array}{l} Bm + An \equiv 0 \bmod p^{j_1} \\ Bm - An \equiv 0 \bmod p^{j_2} \end{array}\right\}, \tag{88.25}$$

so that

$$t(p^j; x) = \bigcup_{j_1+j_2=j} \tau(j_1, j_2; x). \tag{88.26}$$

The condition (88.25) implies that $m, n \equiv 0 \bmod p^{\min\{j_1,j_2\}}$; thus,

$$j_1 \geq j_2 \Rightarrow \tau(j_1, j_2; x) = p^{j_2} \cdot \tau(j_1 - j_2, 0; x/p^{2j_2}), \tag{88.27}$$

in which the right side indicates $\{m,n\} = p^{j_2}\{m', n'\}$ as vectors. We now employ the argument for (29.5)–(29.8) with the modulus $q = p^{j_1 - j_2}$. The set $\tau(j_1, j_2; x)$ is contained in the lattice $\mathbb{Z}\mathbf{u} + \mathbb{Z}\mathbf{v}$ (row vectors), the area of the basic parallelogram is equal to $qp^{2j_2} = p^j = |\mathbf{u}||\mathbf{v}| \sin\theta$, where $|\mathbf{u}|, |\mathbf{v}|$ are the lengths of $\mathbf{u}, \mathbf{v}$, and θ is the acute angle between $\mathbf{u}, \mathbf{v}$, so by Note [77.6] (together with the remark following Theorem 30) we may suppose that $\frac{1}{3}\pi \leq \theta \leq \frac{2}{3}\pi$. In particular, $|\mathbf{u}||\mathbf{v}| \leq 2p^j/\sqrt{3}$. On the other hand, the end point $\{g, h\} \neq \underline{0}$ of $\mathbf{u}$ satisfies $g^2 \equiv dh^2 \bmod p^j$, and $|d||\mathbf{u}|^2 = |d|(g^2 + h^2) \geq |g^2 - dh^2| \geq p^j$; so $|\mathbf{u}| \geq (p^j/|d|)^{1/2}$. Likewise, $|\mathbf{v}| \geq (p^j/|d|)^{1/2}$. Hence, $|\mathbf{u}|, |\mathbf{v}| \approx p^{j/2}$. We see that the difference between $|\tau(j_1, j_2; x)|$ and $|S(x)|/p^j$ (the approximate number of the parallelograms in $S(x)$) is not greater than the number of points in $\mathbb{Z}\mathbf{u} + \mathbb{Z}\mathbf{v}$ which are within the distance $O(p^{j/2})$ from the boundary of $S(x)$, so in (88.26) we have $|\tau(j_1, j_2; x)| = |S(x)|/p^j + O\left((x/p^j)^{1/2}\right)$. In particular, $|t(p^j; x)| = O(jx/p^j)$, and

$$T(p) = |t(p; x)| + O(x/p^2). \tag{88.28}$$

Then we note that

$$|t(p; x)| = |\tau(1, 0; x)| + |\tau(0, 1; x)| - |\tau(1, 1; x)|. \tag{88.29}$$

414   *Quadratic Forms*

Thus we have found that

$$\left(\frac{d}{p}\right) = 1 \;\Rightarrow\; T(p) = \frac{4x}{p\sqrt{|d|}} + O\big((x/p)^{1/2}\big) + O\big(x/p^2\big). \tag{88.30}$$

Also, we assert that

$$\text{(i)}\;\; \left(\frac{d}{p}\right) = -1 \;\Rightarrow\; T(p) \ll x/p^2; \quad \text{(ii)}\;\; \left(\frac{d}{p}\right) = 0 \;\Rightarrow\; T(p) \ll x/p. \tag{88.31}$$

To verify these additional bounds for $T(p)$, we consider $t(p^j;x)$. In (88.23), let $p^k\|m$, $p^l\|n$, and $p^f\|(m^2 - dn^2)$, so $f \geq j$. Then in case (i) we have $2\min\{k,l\} = f$. Thus $m, n \equiv 0 \bmod p^{\lfloor (j+1)/2 \rfloor}$, which yields assertion (i). In case (ii), on noting that $p\|d$ as d is sqf, we have $\min\{2k, 2l+1\} = f$, so $m \equiv 0 \bmod p^{\lfloor (j+1)/2 \rfloor}$, which yields assertion (ii). This ends the discussion of $T(p)$, $p \geq 3$.

(2) The treatment of $T(2)$:

In dealing with $t(2^j;x)$, j can be supposed to be sufficiently large; otherwise we may utilize the bound $t(2^j;x) \ll x$. If $d \equiv 1 \bmod 8$, then by (43.3) an analogue of (88.24) is available, and the argument (88.25)–(88.30) works for $p = 2$ without any essential change. Let $d \not\equiv 1 \bmod 8$. If $2 \nmid d$ additionally, then the situation is analogous to that of (88.31)(i), and we find that with an obvious diversion of notation $2\min\{k,l\} \geq f - 2$. Hence $m, n \equiv 0 \bmod 2^{\lfloor (j-1)/2 \rfloor}$. Further, if $2|d$, then the situation is analogous to that of (88.31)(ii), so $m \equiv 0 \bmod 2^{\lfloor (j+1)/2 \rfloor}$. From these, we find that

$$T(2) \ll x. \tag{88.32}$$

Collecting (88.22), (88.30)–(88.32) while noting that (13.7) and the summation by parts give $\sum_{p \leq x} \log p/\sqrt{p} \ll \sqrt{x}$, we obtain

$$\sum_{\substack{\{m,n\}\in S(x)}} \log|m^2 - dn^2| = \frac{4x}{\sqrt{|d|}} \sum_{\substack{2<p\leq x \\ \left(\frac{d}{p}\right)=1}} \frac{\log p}{p} + O(x). \tag{88.33}$$

This and (88.18) imply (88.14). We end the proof (88.13).

Notes

[88.1]   The use of the function (88.7) is due to Ingham (1930). The slick application, i.e., (88.11), of the theory of functions is originally due to (Landau, 1909, Zweiter Bd., §196), which is based on an elementary observation by du Bois-Reymond (1883) concerning the radius of convergence of power series. See also the argument of de la Vallée Poussin (1896/1897, Deux. part, Chap.IV, §2), especially its part concerning real characters. Combined with a sieve method, Ingham's method can be extended considerably; see Motohashi (1983,

§88   415

Part II; 2015). It should be added that there exists a conceptually much simpler proof of (88.6), and thus that of Theorem 90, which is due to Landau (1927, Satz 152).

[88.2] Dirichlet (1837a) gave a complete proof of (88.2) only when the modulus q is a prime. In order to prove the core assertion (88.6) he needed to separate the cases of the complex and the real characters; the former was treated in much the same way as what we shall show below, while the latter required a use of Theorem 64, §65, somewhat unexpectedly. The situation with general moduli is discussed in Dirichlet (1839; 1863/1894, Supplmente VI), again dealing with the complex and real characters separately. Note that Ingham (1930) did not need such a separation. We stress specifically that with real characters Dirichlet needed the reciprocity law in an essential way. Although this aspect of his proof of (88.2) has been noted already in Note [69.1], we shall here make it more precise but with a tint of complex function theory to which Dirichlet did not appeal: His idea for dealing with complex characters is to exploit the trivial inequality

$$F(s) = \prod_{\chi \bmod q} L(s, \chi) \geq 1, \quad s > 1.$$

To confirm this, we consider the logarithm of the Euler product for each factor, and see that

$$F(s) = \prod_p \exp\left(\sum_{j=1}^{\infty} \frac{1}{jp^{js}} \sum_{\chi \bmod q} \chi(p^j) \right) = \prod_{p \nmid q} \exp\left(\varphi(q) \sum_{p^j \equiv 1 \bmod q} \frac{1}{jp^{js}} \right);$$

the sums on the right side are taken over all $j \geq 1$ that satisfy the condition indicated. These sums are all non-negative; and $F(s) \geq 1$ as claimed. On the other hand, for $s \in \mathbb{C}$ the function $F(s)$ is regular possibly except for the pole of order one at $s = 1$ due to the factor $L(s, J_q)$. If $L(1, \chi) = 0$ for a complex χ, then $L(s, \overline{\chi}) = \overline{L(\overline{s}, \chi)}$ also has a zero at $s = 1$. Hence, we get $F(1) = 0$. This contradicts the fact that $\lim_{s \to 1+0} F(s) \geq 1$. So (88.6) holds for complex characters. Then, as for a real character $\chi \neq J_q$, Dirichlet argued completely differently: we may suppose, by (53.8), that χ is primitive, so χ coincides with a Kronecker symbol associated with a fundamental discriminant by Theorem 71. Therefore, by a combination of (73.5) and Dirichlet's identities (92.32) and (92.48) below, we are able to conclude that $L(1, \chi) > 0$. This ends Dirichlet's proof of (88.6). What is essential in his reasoning is, of course, the characterization of real characters, for which Dirichlet proceeded in much the same way as we did in §73. Hence his proof of his prime number theorem depends indispensably on the reciprocity law.

416   *Quadratic Forms*

[88.3] Part [B] is a detailed version of Selberg (1950a, pp.71–72); see also Rogers (1974). We note that (88.15) proves a fact closely related to the assertion stated at the beginning of [DA, art.125], which Gauss needed in his 1st proof of the reciprocity law:

$$p \equiv 1 \bmod 4 \;\Rightarrow\; \exists\, \varpi \text{ a prime such that}$$
$$2 < \varpi < p \text{ and } \left(\frac{p}{\varpi}\right) = -1.$$

Gauss (art.129) argued as follows (see also Dirichlet (1863/1894, §50)): We may assume that $p \geq 13$. First let $p \equiv 5 \bmod 8$. Then, since $(p+1)/2 \equiv 3 \bmod 4$, there is a prime ϖ such that $\varpi \equiv 3 \bmod 4$ and $\varpi \,|\, ((p+1)/2)$. So $p \equiv -1 \bmod \varpi$, and $\left(\frac{p}{\varpi}\right) = \left(\frac{-1}{\varpi}\right) = -1$ by the supplement law (58.1) (1). Hence, we move to the case $p \equiv 1 \bmod 8$. Let $N = [\sqrt{p}]$, and assume that $\left(\frac{p}{q}\right) = 1$ for all odd prime $q \leq 2N + 1$. Then, there is an $R > N$ such that $R^2 \equiv p \bmod (2N+1)!$, as it follows immediately from Theorem 39. Thus, we have

$$\prod_{n=1}^{N}(p - n^2) \equiv \prod_{n=1}^{N}(R^2 - n^2) = \frac{1}{R}\binom{N+R}{2N+1}(2N+1)! \bmod (2N+1)! \,.$$

Since $\langle R, (2N+1)! \rangle = 1$, the left side should be a multiple of $(2N+1)!$. However,

$$\frac{1}{(2N+1)!}\prod_{n=1}^{N}(p - n^2) = \frac{1}{N+1}\prod_{n=1}^{N}\frac{p - n^2}{(N+1)^2 - n^2} < 1.$$

§89

[C_0] Next, we shall prove Theorem 66, §69, independently of the reciprocity law. The argument below is completely elementary, although somewhat intricate.

We begin with the following definition: with a, b, c as in (69.4),

$\varrho(a,b,c)$ is the term among $\{|ab|, |bc|, |ca|\}$ that is the middle in size. (89.1)

If there are equal terms in these three, then their value is $\varrho(a,b,c)$. In particular, if $\varrho(a,b,c) = 1$, then (69.5) is equivalent to $x^2 + y^2 = z^2$, which has been treated already in Note [13.7]. Thus we assume that $\varrho(a,b,c) \geq 2$, and proceed by induction with respect to the value of the function ϱ.

Because of symmetry, we may assume that $|a| \leq |b| \leq |c|$. Further, we may assume that $|b| < |c|$, for if $|b| = |c|$, then the condition (69.4) (ii) implies that $|bc| = 1$ which contradicts the current assumption on ϱ. Thus, $|c| \geq 2$ and $\varrho(a,b,c) = |ca|$. By the condition (69.4) (iii), we take a λ such that $a\lambda^2 + b \equiv 0 \bmod |c|$, $0 < |\lambda| \leq |c|/2$, so

$$a\lambda^2 + b = cc_0. \tag{89.2}$$

We may assume that $c_0 \neq 0$ since if $c_0 = 0$, then (69.4) (ii) implies that $|\lambda| = 1$, $a = -b = \pm 1$, which is the same as dealing with $x^2 - y^2 = cz^2$, and $\{1,1,0\}$ is an admissible solution. Hence, we have

$$0 < |c_0| \leq |ac|(|\lambda|/|c|)^2 + |b|/|c| < \tfrac{1}{4}|ac| + 1 < \varrho(a,b,c). \tag{89.3}$$

With this, let $a_1 = \langle a\lambda^2, b, cc_0 \rangle$; then, $a_1 = \langle a\lambda^2, b \rangle = \langle b, cc_0 \rangle = \langle cc_0, a\lambda^2 \rangle$. Because of (69.4) (ii), we may write $\lambda = a_1\alpha$, $b = a_1\beta$, $c_0 = a_1c_1\gamma^2$, where c_1 is sqf. We see that

$$aa_1\alpha^2 + \beta = cc_1\gamma^2, \tag{89.4}$$

$$1 = \langle aa_1\alpha^2, \beta \rangle = \langle \beta, cc_1\gamma^2 \rangle = \langle cc_1\gamma^2, aa_1\alpha^2 \rangle. \tag{89.5}$$

Then, via the change of variables

$${}^t\{x,y,z\} = \frac{1}{c_1\gamma^2} \begin{pmatrix} a_1\alpha & \beta & 0 \\ 1 & -a\alpha & 0 \\ 0 & 0 & c_1\gamma \end{pmatrix} \cdot {}^t\{x_1, y_1, z_1\}, \tag{89.6}$$

we are led to

$$ax^2 + by^2 + cz^2 = \frac{c}{c_1\gamma^2}\left(a_1x_1^2 + b_1y_1^2 + c_1z_1^2\right), \quad b_1 = a\beta. \tag{89.7}$$

Here we note, though trivial, that all coefficients are non-zero.

We shall show that the tuple $\{a_1, b_1, c_1\}$ satisfies the three conditions (69.4), and moreover $\varrho(a_1, b_1, c_1) < \varrho(a,b,c)$: If $ab < 0$, then $ab = a_1b_1 < 0$. If $ab > 0$, then $ac < 0$, $bc < 0$ and by $c(a\lambda^2 + b) = a_1c_1(c\gamma)^2$, we have $a_1c_1 < 0$. Hence, (69.4) (i) is satisfied by $\{a_1, b_1, c_1\}$. As for (69.4) (ii), we note that a_1, b_1, c_1 are all sqf according to their definitions. Also, because of $a_1|b$, we have $\langle a_1, a \rangle = 1$, and we see that $\langle a_1, b_1 \rangle = \langle a_1, \beta \rangle = 1$ via (89.5). Further, (89.5) implies that $\langle a, c_1 \rangle = 1$, so $\langle b_1, c_1 \rangle = \langle \beta, c_1 \rangle = 1$. Moreover, (89.5) gives $\langle c_1, a_1 \rangle = 1$ as well. Hence, $\{a_1, b_1, c_1\}$ satisfies (69.4) (ii). It remains to see whether (69.4) (iii) is satisfied or not. By (89.4) we have $a_1(a\alpha)^2 + b_1 = acc_1\gamma^2$, so $-a_1b_1 \bmod |c_1|$ is a quadratic residue. We note that $aca_1c_1 \bmod |\beta|$ and $-ac \bmod |a_1\beta|$ are quadratic residues; the latter is the same as the original condition on $-ac \bmod |b|$. Hence, $-a_1c_1 \bmod |\beta|$ is a

418 Quadratic Forms

quadratic residue. On the other hand, $ac(a_1\alpha)^2 + bc = a_1c_1(c\gamma)^2$ because of (89.2), and $-bc \bmod |a|$ is a quadratic residue originally; so $-a_1c_1 \bmod |a|$ is a quadratic residue. These imply that $-a_1c_1 \bmod |b_1|$ is a quadratic residue. Similarly, $b_1c_1 \equiv ac(c_1\gamma)^2 \bmod |a_1|$ as $a_1c_1(a\alpha)^2 + b_1c_1 = ac(c_1\gamma)^2$, and $-ac \bmod |a_1\beta|$ is a quadratic residue, so $-b_1c_1 \bmod |a_1|$ is a quadratic residue. We have confirmed that $\{a_1, b_1, c_1\}$ satisfies (69.4).

Furthermore, we have $|a_1b_1| = |ab| < |ac| = \varrho(a,b,c)$ as well as $|c_1a_1| \leq |c_1a_1|\gamma^2 = |c_0| < \varrho(a,b,c)$ by (89.3). Hence, $\varrho(a_1,b_1,c_1) < \varrho(a,b,c)$. Our problem has been transformed into that to show the existence of a non-trivial integral tuple $\{x_1, y_1, z_1\}$ with which the right side (89.7) vanishes. This will solve our original equation: The linear transformation (89.6) is, of course, non-degenerate, and such a tuple yields a non-trivial solution $\{x,y,z\} \in \mathbb{Q}^3$ to the equation of (69.5). Eliminating denominators of these rationals in an obvious way, we may suppose $\{x,y,z\} \in \mathbb{Z}^3$, $\langle x,y,z \rangle = 1$. With this, let $p|\langle ax, by, cz \rangle$. If $p|a$, then $p \nmid bc$ by (69.4)(ii), so $p|y, p|z$, and $p^2|ax^2$, i.e., $p|x$, which is impossible. Thus, by symmetry, we should have $p \nmid abc$. This implies the contradiction $p|\langle x,y,z \rangle$. Hence, the coprimality in (69.5) holds. We end the proof of Theorem 66.

Proof of Theorem 65

We have verified (69.1), (69.2) as well as (69.5) without recourse to the reciprocity law. Therefore our assertion (69.3) has now been fully established.

$[C_1]$ We shall extend Theorem 66: one may discard the condition (69.4)(ii). The proof will be completed in the next part $[C_\varpi]$.

For this sake, we shall show a way to derive another solution to (69.5) from

$$au^2 + bv^2 + cw^2 = 0, \quad \langle au, bv, cw \rangle = 1. \tag{89.8}$$

It should be made explicit here that we assume none of (69.4) but only that there exist integers u, v, w satisfying (89.8).

We note that one of au, bv, cw ought to be even. If $2|au$ for instance, then $\langle 2au, bv, cw \rangle = 1$, so by Euclid's algorithm we have a tuple $\{k,l,m\}$ such that

$$auk + bvl + cwm = 1, \quad 2|k. \tag{89.9}$$

We put $ak^2 + bl^2 + cm^2 = h$, and $\{u', v', w'\} = 2\{k,l,m\} - h\{u,v,w\}$ as vectors. Then,

$$au'^2 + bv'^2 + cw'^2 = 0, \quad \langle au', bv', cw' \rangle = 1, \tag{89.10}$$

$$auu' + bvv' + cww' = 2, \tag{89.11}$$

$$\{u', v', w'\} \equiv \{u, v, w\} \bmod 2. \tag{89.12}$$

$$\S 89 \qquad\qquad 419$$

To have (89.12), we need to verify that $2 \nmid h$. To show this, we may suppose that $2|bvl$ and $2 \nmid cwm$ in (89.9). If $2 \nmid bl$, then $2|v$, which contradicts (89.8), as $2|au$. That is, $2|bl$, and $h \equiv cm^2 \equiv 1 \bmod 2$. Hence (89.12) holds. On the other hand, the confirmation of the equations in (89.10) and (89.11) is immediate. It remains to show the coprimality assertion in (89.10). Thus, suppose that $p|\langle au', bv', cw'\rangle$. Then $p = 2$ by (89.11). Assume that $2|a$. If $2 \nmid v'w'$, then $2|\langle a, b, c\rangle$, and if $2|v'$ and $2 \nmid w'$, then $2|c$ as well as $2|v$ by (89.12); the case $2 \nmid v'$ and $2|w'$ is analogous. Whichever situation contradicts (89.8). Hence, $2|\langle v', w'\rangle$, and $2|\langle v, w\rangle$ by (89.12); this contradicts (89.8) again. In other words, $2 \nmid a$, and by symmetry $2 \nmid b, 2 \nmid c$. Thus, $2|\langle u', v', w'\rangle$, and (89.12) yields the contradiction $2|\langle u, v, w\rangle$. We have confirmed (89.10)–(89.12).

Further, let

$$2 \cdot {}^{\text{t}}\{u'', v'', w''\} = {}^{\text{t}}\{u, v, w\} \times {}^{\text{t}}\{u', v', w'\} \quad \text{(cross product).} \qquad (89.13)$$

Then u'', v'', w'' are integers, as this right side is divisible by 2 because of (89.12). We consider the transformation

$$^{\text{t}}\{r', s', t'\} = L \cdot {}^{\text{t}}\{r, s, t\}, \quad L = \tfrac{1}{2} \begin{pmatrix} u & u' & -2bcu'' \\ v & v' & -2cav'' \\ w & w' & -2abw'' \end{pmatrix}. \qquad (89.14)$$

We have

$$^{\text{t}}L \begin{pmatrix} a & 0 & 0 \\ 0 & b & 0 \\ 0 & 0 & c \end{pmatrix} L = \tfrac{1}{2} \begin{pmatrix} 0 & 1 & 0 \\ 1 & 0 & 0 \\ 0 & 0 & -2abc \end{pmatrix}. \qquad (89.15)$$

To show this, we need the relations

$$auu' = 1 + bcu''^2, \quad bvv' = 1 + cav''^2, \quad cww' = 1 + abw''^2, \qquad (89.16)$$

in which the first, for instance, is a consequence of

$$\left(bv^2 + cw^2\right)\left(bv'^2 + cw'^2\right) = \left(bvv' + cww'\right)^2 + bc\left(vw' - wv'\right)^2 \qquad (89.17)$$

(an identity of the Brahmagupta type) and the three identities in (89.8), (89.10), (89.11).

Hence, (89.14) yields

$$ar'^2 + bs'^2 + ct'^2 = rs - abct^2. \qquad (89.18)$$

However, r', s', t' are not necessarily integers even if r, s, t are. Because of this, we let r, s, t be integers such that

$$\begin{gathered} (1)\ rs = abct^2, \quad (2)\ r \equiv s \bmod 2, \\ (3)\ \text{if } p|\langle r, s\rangle, \text{ then } p = 2 \text{ and } r + s \equiv 2 \bmod 4. \end{gathered} \qquad (89.19)$$

420 *Quadratic Forms*

Here it is understood that if r, s satisfy (1), (2) but $\langle r, s \rangle = 1$, then (3) becomes void. We assert that r', s', t' are integers and satisfy

$$ar'^2 + bs'^2 + ct'^2 = 0, \quad \langle ar', bs', ct' \rangle = 1. \tag{89.20}$$

That r', s', t' are integers is a consequence of (89.12) and (89.19)(2). On the other hand, to verify this coprimality assertion, we note that from (89.10), (89.11), (89.13), and (89.16), it follows that

$$ {}^t\{r, s, t\} = L^{-1} \cdot {}^t\{r', s', t'\}, \quad L^{-1} = \begin{pmatrix} au' & bv' & cw' \\ au & bv & cw \\ u'' & v'' & w'' \end{pmatrix}. \tag{89.21}$$

In particular, if $p \mid \langle ar', bs', ct' \rangle$, then $p \mid \langle r, s \rangle$. Namely, if we have in fact $\langle r, s \rangle = 1$, then the coprimality in question holds. Otherwise, by (89.19)(3) we should have $p = 2$ and $(r+s) \equiv 2 \bmod 4$. However, (89.21) implies also that $(r+s) = ar' \cdot (u+u') + \cdots \equiv 0 \bmod 4$, which is a contradiction. Hence, the coprimality holds. Therefore, we have shown that by means of (89.14) and (89.19), we are able to derive (89.20) solely from (89.8).

$[C_\varpi]$ We shall prove that if a prime ϖ is such that $\varpi \nmid bc$ and

$$-bc \bmod |a|\varpi^2 \text{ is a quadratic residue}, \tag{89.22}$$

then the procedure $[C_1]$ yields that $\varpi \mid r'$ in (89.20); note that we have $\langle a, bc \rangle = 1$ because of (89.8). Namely, provided (89.8) and (89.22) there exists a tuple $\{r_1, s_1, t_1\}$ such that

$$a\varpi^2 r_1^2 + bs_1^2 + ct_1^2 = 0, \quad \langle a\varpi r_1, bs_1, ct_1 \rangle = 1. \tag{89.23}$$

To prove this we may suppose that

$$\varpi \nmid uu', \tag{89.24}$$

for if $\varpi \mid uu'$, then either (89.8) or (89.10) gives already (89.23).

(I) $\varpi \geq 3$:

Let α be such that $\alpha^2 \equiv -bc$ but $bcu'' + \alpha \not\equiv 0 \bmod \varpi$. If the latter congruence does not hold, then change the sign of α. Let $\varpi \varpi^{-1} \equiv 1 \bmod |u|$ and

$$\eta = (bcu'' + \alpha)(1 - \varpi \varpi^{-1})/u + \varpi k, \tag{89.25}$$

where we take k so that $\langle \eta, 2abc \rangle = 1$ (see Note [29.2]). Let

$$r = \tau \eta^2, \ s = \tau abc, \ t = \tau \eta, \ \tau = \begin{cases} 1 & abc \equiv 1 \bmod 2, \\ 2 & abc \equiv 0 \bmod 2, \end{cases} \tag{89.26}$$

which fulfills (89.19). Then, by (89.14), (89.16), and (89.24), we get

$$2ur' = \tau\{(u\eta - bcu'')^2 + bc\} \equiv 0 \bmod \varpi \Rightarrow \varpi \mid r'. \tag{89.27}$$

(II) $\varpi = 2$:

(i) If $2 \mid a$, then $-bc \bmod 8$ is a quadratic residue; thus, $bc \equiv -1 \bmod 8$, and $b \equiv -c \bmod 8$, so, by (89.8), we have $au^2 \equiv 0 \bmod 8$, as $2 \nmid vw$. With this, if $8 \nmid a$, then $2 \mid u$, which is excluded by (89.24). Thus, we may suppose that $8 \mid a$. Then, we take $r = abc/2$, $s = 2$, $t = 1$. The requirement (89.19) is fulfilled. Also, by (89.16), we have $1 - u''^2 \equiv 1 + bcu''^2 \equiv 0 \bmod 8$; thus, $2 \nmid u''$. Hence, via (89.14), $2r' = ur + 2u' - 2bcu'' \equiv 2(u' + u'') \equiv 0 \bmod 4$, as $2 \nmid u'$ because of (89.24). Therefore, $2 \mid r'$.

(ii) If $2 \nmid a$, then we may take $r = abc$, $s = 1$, $t = 1$, as $2 \nmid bc$. The condition (89.19) is satisfied; in fact, (3) is irrelevant now. With this, if $2 \nmid u''$, then (89.16) gives $auu' \equiv 1 + bc \equiv 0 \bmod 4$ because of (89.22). Thus, $2 \mid uu'$, contradicting (89.24). Hence, $2 \mid u''$, and $auu' \equiv 1 \bmod 4$, which implies that $uu'r = auu'bc \equiv -1 \bmod 4$ or $ur \equiv -u' \bmod 4$. By (89.14) we have $2r' = ur + u' - 2bcu'' \equiv 0 \bmod 4$. Therefore, $2 \mid r'$.

We have obtained (89.23) on the assumption (89.8) and (89.22). Starting from $[C_0]$ (Theorem 66), we repeatedly apply the procedure $[C_\varpi]$, and obtain, independently of the reciprocity law, the following assertion on diagonal ternary quadratic forms, which we attribute to Legendre and Dedekind:

Theorem 91 *Let* A, B, C *be mutually prime, and such that not all of them are of the same sign, and* $-BC \bmod |A|$, $-CA \bmod |B|$, $-AB \bmod |C|$ *are all quadratic residues. Then the indefinite equation*

$$Ax^2 + By^2 + Cz^2 = 0 \tag{89.28}$$

has a solution which satisfies

$$\langle Ax, By, Cz \rangle = 1. \tag{89.29}$$

Notes

[89.1] Part $[C_0]$ is the proof by Dedekind (Dirichlet (1871/1894, §157)) of Legendre's Theorem 66. An alternative argument can be found in Gauss [DA, arts.294–295] as well as in Stepanov (1994, pp.116–118). The parts $[C_1]$ and $[C_\varpi]$ are adoptions from Dedekind Dirichlet (1871/1894, §156). The discussion in $[C_1]$–$[C_\varpi]$ is essential for our purpose since it yields (89.28)–(89.29) which will play a core rôle in our proof of Gauss' famous theorem

420   *Quadratic Forms*

(91.14) below. Bounds for the least solution to (89.28) were claimed by some authors; however, they seem to have left the aspect (89.29) undiscussed. This coprimality will become quite essential in our approach to the genus theory to be developed in §91; see (91.16).

[89.2]   By means of $[C_0]$, we shall solve the indefinite equation

$$233x^2 + 337y^2 - 797z^2 = 0.$$

These coefficients satisfy the condition (69.4); note that $233, 337, 797$ are primes. According to (89.2), we need to take a λ such that $\lambda^2 \equiv -193$ mod 797. By Note 63.1 (Legendre), $\lambda = 396$; we have $233 \cdot 396^2 + 337 = (-797) \cdot (-45845)$. We move to the indefinite equation

$$x_1^2 + 233 \cdot 337y_1^2 - 45845z_1^2 = 0, \quad 45845 = 5 \cdot 53 \cdot 173,$$
$$Y_1: \quad x = 396y - 797y_1, \; x_1 = -45845y + 233 \cdot 396y_1, \; z_1 = z.$$

In (89.2) we take $a = 1, b = -45845, c = 233 \cdot 337$, that is, $\{x, y, z\} = \{x_1, z_1, y_1\}$. Then, we have to solve $\lambda_1^2 \equiv 45845$ mod $233 \cdot 337 \Rightarrow \lambda_1^2 \equiv -56$ mod 233, $\lambda_1^2 \equiv 13$ mod 337. By Tonelli's method (§63), we get $\lambda_1 \equiv \pm 115$ mod 233, $\lambda_1 \equiv \pm 32$ mod 337. Via Euclid's algorithm, we take $\lambda_1 = 4076$; we have $4076^2 - 45845 = 233 \cdot 337 \cdot 211$. We move to the indefinite equation

$$x_2^2 - 45845y_2^2 + 211z_2^2 = 0,$$
$$Y_2: \quad x_1 = 4076z_1 + 233 \cdot 337y_2, \; x_2 = 211z_1 + 4076y_2, \; z_2 = y_1.$$

As before we take $\{x, y, z\}$ for $\{x_2, z_2, y_2\}$. Solving $\lambda_2^2 + 211 \equiv 0$ mod 45845, that is, $\lambda_2^2 \equiv -1$ mod 5, $\lambda_2^2 \equiv 1$ mod 53, $\lambda_2^2 \equiv -38$ mod 173, we get $20298^2 + 211 = 45845 \cdot 8987$. With this, we move to

$$x_3^2 + 211y_3^2 - 8987z_3^2 = 0, \quad 8987 = 11 \cdot 19 \cdot 43,$$
$$Y_3: \quad x_2 = 20298z_2 - 45845y_3, \; x_3 = -8987z_2 + 20298y_3, \; z_3 = y_2.$$

Solving $\lambda_3^2 + 211 \equiv 0$ mod 8987, that is, $\lambda_3^2 \equiv -2$ mod 11, $\lambda_3^2 \equiv -2$ mod 19, $\lambda_3^2 \equiv 4$ mod 43, we get $3528^2 + 211 = 8987 \cdot 1385$. With this, we move to

$$x_4^2 + 211y_4^2 - 1385z_4^2 = 0, \quad 1385 = 5 \cdot 277,$$
$$Y_4: \quad x_3 = 3528y_3 - 8987y_4, \; x_4 = -1385y_3 + 3528y_4, \; z_4 = z_3.$$

Solving $\lambda_4^2 + 211 \equiv 0$ mod 1385, we get $627^2 + 211 = 1385 \cdot 2^2 \cdot 71$. We move to

$$x_5^2 + 211y_5^2 - 71z_5^2 = 0,$$
$$Y_5: \quad x_4 = 627y_4 - 1385y_5, \; x_5 = -284y_4 + 627y_5, \; z_5 = 2z_4.$$

We take $\{x,y,z\}$ for $\{x_5,z_5,y_5\}$. Solving $\lambda_5^2 - 71 \equiv 0 \bmod 211$, we get $55^2 - 71 = 211 \cdot 14$. We move to

$$x_6^2 - 71y_6^2 + 14z_6^2 = 0,$$

$$Y_6: \quad x_5 = 55z_5 + 211y_6, \ x_6 = 14z_5 + 55y_6, \ z_6 = y_5.$$

We take $\{x,y,z\}$ for $\{x_6,z_6,y_6\}$. Solving $\lambda_6^2 + 14 \equiv 0 \bmod 71$, we get $25^2 + 14 = 71 \cdot 3^2$. Finally, we are led to the trivial equation:

$$x_7^2 + 14y_7^2 - z_7^2 = 0,$$

$$Y_7: \quad x_6 = 25z_6 - 71y_7, \ x_7 = 9z_6 + 25y_7, \ z_7 = 3y_6.$$

Applying the transformations $Y_7^{-1}, Y_6^{-1}, \ldots, Y_2^{-1}$, we get

$$\begin{aligned}
\{x_7, y_7, z_7\} &= \{1, 0, 1\} \mapsto \{25, 3, 1\} \mapsto \{83, 1, -10\} \\
&\mapsto \{-13063, 136, -355\} \mapsto \{46057568, 492871, -491675\} \\
&\mapsto \{-21743732815, -102760075, 231586930\} \\
&\mapsto \{-397708148415, 231586930, 1882020535\} = \{x_1, y_1, z_1\}.
\end{aligned}$$

Finally, via the action of Y_1^{-1}, we reach the special solution

$$233 \cdot 3433621594^2 + 337 \cdot 474768699^2 - 797 \cdot 1882020535^2 = 0.$$

[89.3] While $\{a,b,c,u,v,w\}$ is fixed, the derived solution $\{s_1,r_1,t_1\}$ given in (89.23) is determined solely by the choice of $\eta \bmod \varpi$ defined by (89.25), provided $\varpi \geq 3$. Starting from the simple

$$5 \cdot 1^2 + 7 \cdot 4^2 - 13 \cdot 3^2 = 0,$$

we are led, for instance, to

$$5^3 \cdot 30131^2 + 7 \cdot 581573^2 - 13 \cdot 436866^2 = 0,$$
$$\langle 5 \cdot 30131, 7 \cdot 581573, 13 \cdot 436866 \rangle = 1, \qquad \varpi = 5, \ \eta = -3,$$

from which one may further proceed to

$$5^3 x^2 + 7^3 y^2 - 13^3 z^2 = 0, \quad \langle 5x, 7y, 13z \rangle = 1.$$

A solution exists; however, what $[C_1]$–$[C_\varpi]$ yields is huge.

§90

We now return to the investigation of the structure of the set of all integral binary quadratic forms: The aim of the present section is to develop an account of Gauss' *compositione formarum* or the composition theory of forms [DA, arts.234–251]. Note that this section is independent of the reciprocity law.

424   *Quadratic Forms*

We stress first of all that unlike Gauss we shall deal solely with primitive forms, i.e., we adhere to the premise (71.15). This makes our discussion considerably simpler than otherwise, yet the core of the composition theory will be adequately presented. We shall work thus with the notation

$$\mathcal{Q}(D) = \begin{cases} \mathcal{Q}_+(D) \\ \mathcal{Q}_\pm(D) \end{cases} \qquad \mathcal{K}(D) = \begin{cases} \mathcal{K}_+(D) \\ \mathcal{K}_\pm(D) \end{cases} \qquad \mathfrak{h}(D) = \begin{cases} \mathfrak{h}_+(D) \\ \mathfrak{h}_\pm(D) \end{cases} \tag{90.1}$$

where the upper parts are for $D < 0$ ((77.1)–(77.3)) and the lower for $D > 0$ ((82.1)–(82.3)).

The origin of the composition theory can be traced back to the issue of finding the converse of Lagrange's theorem (Note [72.2]), as mentioned in Note [72.4]. Repeating what is stated there, we are concerned not only with the multiplicative nature of the set of all integers which are expressible by quadratic forms but also with a possible multiplicative structure or composition among forms themselves, the existence of which is strongly suggested by (72.10) and Notes [79.3]–[79.4].

Following Gauss [DA, arts.235–236], we understand that the problem of composing forms Q_1 and Q_2, possibly of different discriminants, is to find a form Q and the transformation coefficients $\{s_\nu, s'_\nu\} \in \mathbb{Z}^8$ such that

$$Q_1(x_1, y_1) \cdot Q_2(x_2, y_2) = Q(x_3, y_3),$$

$$\{x_3, y_3\} = \{x_1, y_1\} \cdot \left[\begin{pmatrix} s_1 & s_2 \\ s_3 & s_4 \end{pmatrix}, \begin{pmatrix} s'_1 & s'_2 \\ s'_3 & s'_4 \end{pmatrix} \right] \cdot {}^{\mathrm{t}}\{x_2, y_2\}, \tag{90.2}$$

where the lower line indicates that $x_3 = \{x_1, y_1\} \cdot \begin{pmatrix} s_1 & s_2 \\ s_3 & s_4 \end{pmatrix} \cdot {}^{\mathrm{t}}\{x_2, y_2\}$, and y_3 is analogous. For instance, Dirichlet's identity $(*)$ of Note [79.4] can be restated as follows: For any $\{f_1, f_2, g, k, x_1, y_1, x_2, y_2\} \in \mathbb{C}^8$,

$$[|f_1, g, f_2 k|](x_1, y_1) \cdot [|f_2, g, f_1 k|](x_2, y_2) = [|f_1 f_2, g, k|](x_3, y_3),$$

$$\{x_3, y_3\} = \{x_1, y_1\} \cdot \left[\begin{pmatrix} 1 & 0 \\ 0 & -k \end{pmatrix}, \begin{pmatrix} 0 & f_1 \\ f_2 & g \end{pmatrix} \right] \cdot {}^{\mathrm{t}}\{x_2, y_2\}. \tag{90.3}$$

We shall demonstrate that the essentials of Gauss' theory of compositions, including its reworks and simplifications by Arndt and Dirichlet–Dedekind (see Note [90.3]), can be derived all from this identity.

One may instead start with a given Q, and regard (90.2) as its decomposition into the factor forms Q_1, Q_2. In [DA, art.235] Gauss developed an argument of a chain of literally heavy algebraic computation and was led to a complicated conclusion on the task to proceed from the right side to the left of (90.2), even without supposing the primitivity of forms. This aspect of the composition theory is, however, not genuinely related to our purpose, as we proceed from a

$$\S90 \qquad\qquad 425$$

given pair of forms to their composition, relying solely on (90.3). Nevertheless, it should be noted that we shall deal with a special case of this reversion problem in the subsection [C] of §93.

Anyway, provided that D is even and forms are primitive, Gauss' discovery can be summarized as follows:

[0] [art.235] If (90.2) holds, then there should exist integers m_1, m_2 such that $Q_j \in \mathcal{Q}(m_j^2 D)$ with D the discriminant of Q.

[1] [art.236] Conversely, if $Q_j \in \mathcal{Q}(\ell_j^2 D)$ with $\langle \ell_1, \ell_2 \rangle = 1$ are arbitrarily given, then there exist coefficients $\{s_v, s_v'\}$ and a form $Q \in \mathcal{Q}(D)$ that satisfy (90.2); Q is called the composition of Q_1, Q_2.

[2] [arts.237–239] The composition thus attained commutes with the Γ-equivalence, in the sense stated in (90.12) below.

[3] [arts.240–241] This composition is commutative and associative.

[4] [arts.242–244] The coefficients of Q are determined in a practical way, although Gauss dealt with only the case $\ell_1 = \ell_2 = 1$. He commented on the aspect that was later discussed by Dirichlet under the notion of concordance between forms; see the paragraph following (90.37) below.

[5] [arts.249–251] In terms of modern terminology, the set $\mathcal{K}(D)$ is a finite Abelian group under the composition as its operation. By the way, in arts.245–248 basic relations between genus and composition are discussed.

We shall generalize [1]–[5] to any D, either even or odd, and ℓ_j as in (90.6) below. In particular, each pair of forms in $\mathcal{Q}(D)$ can be composed in the sense of (90.2), while conforming to the Γ-equivalence: this is not only a desired answer to the problem raised in Note [72.4] but also leads to Theorem 93 below, a completion of [5]. The principal assertion of the present section is this great theorem.

Remark 1 With (90.2) it holds that if $Q_j, j = 1, 2$, are primitive, then Q ought to be primitive. To confirm this, let $Q_j = [|a_j, b_j, c_j|]$. Suppose that Q is not primitive. Then, there exists a p such that for any integers u_1, v_1, u_2, v_2 we have $p | Q_1(u_1, v_1) Q_2(u_2, v_2)$. In particular, $p | a_1 a_2$. If $p \nmid a_2$, then $p | a_1$. Also, we have $p | c_1 a_2$ and $p | (a_1 + b_1 + c_1) a_2$, which implies the contradiction $p | \langle a_1, b_1, c_1 \rangle$. Namely, we should have $p | a_1, p | a_2$. In exactly the same way, we should have $p | c_1, p | c_2$. Moreover, $p | (a_1 + b_1 + c_1)(a_2 + b_2 + c_2)$. Hence, $p | b_1 b_2$ which induces a contradiction.

Although the assertion [0] is irrelevant to our purpose, we give here an explanation on it (discriminants can now be odd as well): Suppose that Q be such that thedecomposition (90.2) with primitive Q_j holds. Since

$$^t\{x_3,y_3\} = \Psi \cdot {}^t\{x_2,y_2\}, \quad \Psi = \begin{pmatrix} s_1x_1 + s_3y_1 & s_2x_1 + s_4y_1 \\ s_1'x_1 + s_3'y_1 & s_2'x_1 + s_4'y_1 \end{pmatrix}, \tag{90.4}$$

the identity (90.2) is a relation between two quadratic forms Q, Q_2; that is, ${}^t\Psi Q\Psi = Q_1(x_1,y_1)Q_2$. Taking the determinants on both sides, we have

$$(\det \Psi)^2 \det Q = (Q_1(x_1,y_1))^2 \det Q_2. \tag{90.5}$$

Let $\det \Psi = Ex_1^2 + Fx_1y_1 + Gy_1^2$, and let $\psi_1 = \langle E,F,G\rangle$; naturally $\{s_v, s_v'\}$ ought to be such that $\det \Psi$ is a non-trivial form so that $\psi_1 \neq 0$. Specializing (90.5) by taking $\{x_1,y_1\} = \{1,0\},\{0,1\},\{1,1\}$, we find readily that since $(\det \Psi)/\psi_1$ and Q_1 are both primitive forms, $\psi_1^2 \det Q = \det Q_2$. Analogously, we have $\psi_2^2 \det Q = \det Q_1$ with another integer ψ_2. This ends the confirmation of [0].

We shall now expand on the core part of Gauss' composition theory; hereafter D is either even or odd in general while temporary restrictions may be imposed, with mention at each occasion:

Theorem 92 *Let*

$$2 \nmid \ell_1\ell_2, \quad \langle \ell_1, \ell_2 \rangle = 1. \tag{90.6}$$

Let $Q_j = [|a_j, b_j, c_j|] \in \mathfrak{Q}(\ell_j^2 D)$ *and*

$$\rho = \langle \ell_2 a_1, \ell_1 a_2, (\ell_2 b_1 + \ell_1 b_2)/2 \rangle. \tag{90.7}$$

Then the system of congruences in $\mathbb{Z}$

$$\begin{aligned} (\ell_2 a_1/\rho)x &\equiv a_1 b_2/\rho, \quad (\ell_1 a_2/\rho)x \equiv a_2 b_1/\rho, \\ ((\ell_2 b_1 + \ell_1 b_2)/2\rho)x &\equiv (b_1 b_2 + \ell_1\ell_2 D)/2\rho \end{aligned} \quad \mod 2|a_3|, \tag{90.8}$$

with $a_3 = a_1 a_2/\rho^2$, *has a unique solution* $x \equiv b_3 \mod 2|a_3|$. *The Gauss–Arndt composition of* Q_j *is thereby defined by*

$$Q_1 \circ Q_2 = [|a_3, b_3, c_3|], \quad c_3 = (b_3^2 - D)/4a_3, \tag{90.9}$$

so that $Q_1(x_1,y_1)Q_2(x_2,y_2) = (Q_1 \circ Q_2)(x_3,y_3)$ *with*

$$\begin{aligned} \frac{x_3}{\rho} &= x_1 x_2 + (b_2 - \ell_2 b_3)\frac{x_1 y_2}{2a_2} + (b_1 - \ell_1 b_3)\frac{y_1 x_2}{2a_1} \\ &\quad + \left(b_1 b_2 + \ell_1\ell_2 D - (\ell_2 b_1 + \ell_1 b_2)b_3\right)\frac{y_1 y_2}{4a_1 a_2}, \end{aligned} \tag{90.10}$$

$$\rho y_3 = \ell_2 a_1 x_1 y_2 + \ell_1 a_2 y_1 x_2 + \tfrac{1}{2}(\ell_2 b_1 + \ell_1 b_2)y_1 y_2.$$

We have

$$Q_1 \circ Q_2 \in \mathfrak{Q}(D) \tag{90.11}$$

as well as

$$Q_j \equiv Q_j', j = 1,2 \;\Rightarrow\; Q_1 \circ Q_2 \equiv Q_1' \circ Q_2' \mod \Gamma. \tag{90.12}$$

$$\S90 \qquad\qquad 427$$

Remark 2 The assertion (90.11) need not be proved in view of Remark 1. Our proof of the other parts of the present theorem consists of the five steps (i)–(v). We shall show in (ii) that (90.8) is indeed a system of congruences in $\mathbb{Z}$ and it has a unique solution b_3 mod $2|a_3|$ so that (90.10) is a transform with integral coefficients, i.e., (90.2) holds; the relevant $\{s_\nu, s'_\nu\}$ can be read off from (90.16) below. To this end, we shall adopt the reasoning of Arndt (1859a). However, save for part (ii), our argument is independent of either Gauss' or Arndt's. See Note [90.3].

Remark 3 This theorem is more than what will actually be needed later in the present chapter; our interest will then be focused on the situation of the same discriminant, i.e., $\ell_1 = \ell_2 = 1$, restricting (90.6). As a matter of fact, the condition $2 \nmid \ell_1 \ell_2$ can be dispensed with provided D is even, a fact which can be readily inferred from the discussion in part (ii) below. That is indeed the situation Arndt treated; following Gauss he did not consider the case of odd D's.

Remark 4 The treatment of compositions by means of the ideal theory of quadratic number fields originated with Dedekind (Dirichlet (1894, §187); see also Weber (1908, §§101–102). However, it is, of course, definitely preferable to go through the classical approach like the present one as at least a background knowledge or experience. Nevertheless, one should be aware of a clear advantage of the approach via matrix modules (see Note [90.5]). In any event, we shall require the contents of the present section to proceed to §91 and then to §93. We add that Bhargava (2001) developed a new and highly inspiring approach to Gauss' composition theory. Classical number theory is a rich land.

Proof (i) Supposing that a solution b_3 to (90.8) has been found, let

$$W_j = \begin{pmatrix} \rho^{-1/2} & \rho^{-1/2}\delta_j/\ell_j \\ 0 & \rho^{1/2}/\ell_j \end{pmatrix}, \qquad \ell_j b_3 = b_j + 2a_j\delta_j/\rho, \qquad (90.13)$$

with integers δ_j, and let $Q_j^* = {}^t W_j Q_j W_j$. Then

$$Q_1^* = [|a_1/\rho, b_3, a_2 c_3/\rho|], \qquad Q_2^* = [|a_2/\rho, b_3, a_1 c_3/\rho|]. \qquad (90.14)$$

By means of (90.3), we have

$$Q_1^*(w_1, z_1) Q_2^*(w_2, z_2) = [|a_3, b_3, c_3|](w_3, z_3),$$

$$\{w_3, z_3\} = \{w_1, z_1\} \cdot \left[\begin{pmatrix} 1 & 0 \\ 0 & -c_3 \end{pmatrix}, \begin{pmatrix} 0 & a_1/\rho \\ a_2/\rho & b_3 \end{pmatrix} \right] \cdot {}^t\{w_2, z_2\}. \qquad (90.15)$$

Via the transformation ${}^t\{w_j, z_j\} = W_j^{-1} \cdot {}^t\{x_j, y_j\}, j = 1, 2$, we are led to the relations

$$
{}^t W_1^{-1} \begin{pmatrix} 1 & 0 \\ 0 & -c_3 \end{pmatrix} W_2^{-1} = \begin{pmatrix} \rho & -\delta_2 \\ -\delta_1 & (\delta_1\delta_2 - \ell_1\ell_2 c_3)/\rho \end{pmatrix},
$$

$$
{}^t W_1^{-1} \begin{pmatrix} 0 & a_1/\rho \\ a_2/\rho & b_3 \end{pmatrix} W_2^{-1} = \frac{1}{\rho} \begin{pmatrix} 0 & a_1\ell_2 \\ a_2\ell_1 & \ell_1\ell_2 b_3 - (a_1\delta_1\ell_2 + a_2\delta_2\ell_1)/\rho \end{pmatrix}.
$$
(90.16)

After some rearrangements, we find that $\{w_3, z_3\} = \{x_3, y_3\}$, which is the same as (90.9)–(90.10).

(ii) We ought to verify that the coefficients of (90.8) and the modulus are all integers. Thus, we shall show first that

$$
\rho \mid \langle a_1 b_2, a_2 b_1, (b_1 b_2 + \ell_1 \ell_2 D)/2 \rangle, \quad \rho^2 | a_1 a_2; \tag{90.17}
$$

and a_3 is an integer. To this end, note that as ℓ_j are odd by (90.6), $b_1, b_2 \equiv D \bmod 2$, so $(\ell_2 b_1 + \ell_1 b_2)/2$ and $(b_1 b_2 + \ell_1 \ell_2 D)/2$ are both integers. Then, let

$$
\begin{aligned}
R &= \ell_2 a_1, \quad S = \ell_1 a_2, \quad T = (\ell_2 b_1 + \ell_1 b_2)/2, \\
U &= (\ell_2 b_1 - \ell_1 b_2)/2, \quad V = \ell_2 c_1, \quad W = \ell_1 c_2;
\end{aligned} \tag{90.18}
$$

and observe that

$$
\begin{aligned}
&(1) \quad 2 \nmid D \;\Rightarrow\; \langle R, S, T, U, 2V, 2W \rangle = 1, \\
&(2) \quad 2 | D \;\Rightarrow\; \langle R, S, T, U, V, W \rangle = 1.
\end{aligned} \tag{90.19}
$$

The first line is due to the fact that $\langle R, T + U, 2V \rangle = \ell_2$ and $\langle S, T - U, 2W \rangle = \ell_1$ since $2 \nmid b_j$ because $2 \nmid \ell_j^2 D$, and Q_j are primitive. The second line is analogous. By the coprimality condition in (90.6) we get (90.19) for both cases. Then, regardless of the parity of D, we have $b_2 R = \ell_2 a_1 b_2$ and $2a_1 T - b_1 R = \ell_1 a_1 b_2$, so $\rho | a_1 b_2$ again by (90.6). Similarly $\rho | a_2 b_1$. Also, $b_2 T - 2c_2 S = \frac{1}{2}\ell_2(b_1 b_2 + \ell_1 \ell_2 D)$ and $b_1 T - 2c_1 R = \frac{1}{2}\ell_1(b_1 b_2 + \ell_1 \ell_2 D)$, so $\rho | (b_1 b_2 + \ell_1 \ell_2 D)/2$. Hence, we have verified the first divisibility assertion in (90.17). We shall next confirm the second assertion there. Note first that $\rho | a_1 a_2$ since by Euclid's algorithm there exists $\mu_j \in \mathbb{Z}$ such that $a_1 a_2 = \mu_1 a_2 R + \mu_2 a_1 S$. In particular, $\rho^2 | a_1 a_2 R$, $\rho^2 | a_1 a_2 S$, $\rho^2 | a_1 a_2 T$. Further, $a_1 a_2 U = Ra_2 b_1 - a_1 a_2 T$, $2a_1 a_2 V = a_2 b_1 T - \frac{1}{2}(b_1 b_2 + \ell_1 \ell_2 D)R$, $2a_1 a_2 W = a_1 b_2 T - \frac{1}{2}(b_1 b_2 + \ell_1 \ell_2 D)S$; we have used $b_j^2 - 4a_j c_j = \ell_j^2 D$. If $2 \nmid D$, then combining these with (90.19)(1), we get $\rho^2 | a_1 a_2$. On the other hand, if $2 | D$, then $2 | b_j$, and $a_1 b_2, a_2 b_1, (b_1 b_2 + \ell_1 \ell_2 D)/2 \equiv 0 \bmod 2\rho$, so $a_1 a_2 V, a_1 a_2 W \equiv 0 \bmod \rho^2$. Hence, by (90.19)(2), we get $\rho^2 | a_1 a_2$ again. We have verified (90.17).

Without restricting the parity of D, we shall show that a special solution b_3 to (90.8) is given by

$$
\rho b_3 = \alpha a_1 b_2 + \beta a_2 b_1 + \gamma (b_1 b_2 + \ell_1 \ell_2 D)/2, \tag{90.20}
$$

$$\S90 \qquad\qquad 429$$

where the integers $\{\alpha, \beta, \gamma\}$ are fixed by Euclid's algorithm so that

$$\rho = \alpha a_1 \ell_2 + \beta a_2 \ell_1 + \gamma (\ell_2 b_1 + \ell_1 b_2)/2; \qquad (90.21)$$

see (90.7). Eliminating α from the last two identities, we get

$$\begin{aligned}
\rho a_1 (\ell_2 b_3 - b_2) &= \beta a_1 a_2 (\ell_2 b_1 - \ell_1 b_2) + a_1 \ell_1 \gamma (b_2^2 - \ell_2^2 D)/2 \\
&\equiv 0 \bmod 2|a_1 a_2|.
\end{aligned} \qquad (90.22)$$

Since the left side and the modulus are both divisible ρ^2, the first congruence of (90.8) follows; the second congruence is shown analogously. Eliminating γ, we see that

$$\begin{aligned}
\rho\big((\ell_2 b_1 &+ \ell_1 b_2) b_3 - b_1 b_2 - \ell_1 \ell_2 D\big)/2 \\
&= \alpha \ell_1 a_1 (b_2^2 - \ell_2^2 D)/2 + \beta \ell_2 a_2 (b_1^2 - \ell_1^2 D)/2 \equiv 0 \bmod 2|a_1 a_2|.
\end{aligned} \qquad (90.23)$$

Namely, the third congruence follows as well. Hence, (90.20) indeed gives a special solution to (90.8). On the other hand, to show the uniqueness of $b_3 \bmod 2|a_3|$, let B be a solution to (90.8) so that $\ell_2 a_1 B/\rho \equiv a_1 b_2/\rho$, $\ell_1 a_2 B/\rho \equiv a_2 b_1/\rho$, $(\ell_2 b_1 + \ell_1 b_2) B/2\rho \equiv (b_1 b_2 + \ell_1 \ell_2 D)/2\rho \bmod 2|a_3|$. Multiply these by the α, β, γ of (90.21), respectively, and sum; then we find that $B \equiv b_3 \bmod 2|a_3|$.

Further, we have

$$\begin{aligned}
(R/\rho)^2 (B^2 - D) &\equiv (a_1 b_2/\rho)^2 - (a_1/\rho)^2 \ell_2^2 D \\
&= 4a_1^2 a_2 c_2/\rho^2 \equiv 0 \bmod 4|a_3|,
\end{aligned} \qquad (90.24)$$

where we have used the fact that $m \equiv n \bmod 2k \Rightarrow m^2 \equiv n^2 \bmod 4k$. Similarly $(S/\rho)^2 (B^2 - D) \equiv 0 \bmod 4|a_3|$. Also,

$$\begin{aligned}
(T/\rho)^2 (B^2 - D) &\equiv \big((b_1 b_2 + \ell_1 \ell_2 D)/2\rho\big)^2 - \big((\ell_2 b_1 + \ell_1 b_2)/2\rho\big)^2 D \\
&= (b_1^2 - \ell_1^2 D)(b_2^2 - \ell_2^2 D)/4\rho^2 = 4a_1 a_2 c_1 c_2/\rho^2 \equiv 0 \bmod 4|a_3|
\end{aligned} \qquad (90.25)$$

From these we conclude that $B^2 \equiv D \bmod 4|a_3|$; in particular, c_3 is indeed an integer, and $[|a_3, b_3, c_3|] \in \mathfrak{Q}(D)$.

(iii) Our proof of the core assertion (90.12) is developed in the present and the next two subsections: Let $Q_j' = [|a_j', b_j', c_j'|] \equiv Q_j \bmod \Gamma$, $j = 1, 2$. We then construct two supplementary forms $K_j = [|r_j, \ell_j s, t_j|] \equiv Q_j \bmod \Gamma$, $j = 1, 2$, under the condition

$$\begin{aligned}
\langle r_1 r_2, N \rangle = 1, \ \langle r_1, r_2 \rangle = 1, \ r_1, r_2 > 0, \\
N = 2\ell_1 \ell_2 a_1 a_2 c_3 a_1' a_2' c_3'.
\end{aligned} \qquad (90.26)$$

To have this, we apply Notes [74.5]–[74.6] to get $L_1 = [|r_1, s_1, (s_1^2 - \ell_1^2 D)/4r_1|] \equiv Q_1 \bmod \Gamma$, $\langle r_1, N \rangle = 1$, $r_1 > 0$, and then $L_2 = [|r_2, s_2, (s_2^2 - \ell_2^2 D)/4r_2|] \equiv Q_2 \bmod \Gamma$, $\langle r_2, r_1 N \rangle = 1$, $r_2 > 0$. We take k_j by Euclid's

430   *Quadratic Forms*

algorithm so that $\ell_2(s_1 + 2r_1k_1) = \ell_1(s_2 + 2r_2k_2)$; note that $2|(\ell_2s_1 - \ell_1s_2)$. Let $\ell_1\ell_2s$ be the value of this identity. Then $K_j = {}^tT^{kj}L_jT^{kj} = [|r_j, \ell_j s, t_j|]$; here $T = \begin{pmatrix} 1 & 1 \\ 0 & 1 \end{pmatrix}$. Since $D = s^2 - 4r_1t_1/\ell_1^2 = s^2 - 4r_2t_2/\ell_2^2$ and $\langle r_1\ell_1, r_2\ell_2 \rangle = 1$, there exists a t such that

$$K_1 = [|r_1, \ell_1 s, \ell_1^2 r_2 t|], \quad K_2 = [|r_2, \ell_2 s, \ell_2^2 r_1 t|], \quad D = s^2 - 4r_1r_2t. \qquad (90.27)$$

We have, by (90.3) with the modification $\{x_j, y_j\} \mapsto \{x_j, \ell_j y_j\}$,

$$K_1(x_1, y_1)K_2(x_2, y_2) = K(x_3, y_3), \quad K = [|r_1r_2, s, t|]$$
$$x_3 = x_1x_2 - \ell_1\ell_2ty_1y_2, \quad y_3 = \ell_2r_1x_1y_2 + \ell_1r_2x_2y_1 + \ell_1\ell_2sy_1y_2. \qquad (90.28)$$

Or following (90.7)–(90.10) we get the same, i.e., $K = K_1 \circ K_2$; it is easy to see that $\rho = 1$ and $s \bmod 2r_1r_2$ is a solution to (90.8) on the present specification.

(iv) Let $r_j = Q_j(u_j, v_j)$ be the proper representation which corresponds to $r_j = K_j(1,0)$ in terms of a transformation belonging to Γ. Then, by Note [74.3], we have

$$2a_ju_j + (\ell_j s + b_j)v_j \equiv 0 \bmod 2r_j, \quad j = 1, 2. \qquad (90.29)$$

This is due to the fact that the proper representation $r_j = K_j(1,0)$ belongs to $\ell_j s \bmod 2r_j$ of course, and so does $r_j = Q_j(u_j, v_j)$. On the other hand, defining $\{u_3, v_3\}$ by (90.10) with $\{x_j, y_j\} = \{u_j, v_j\}, j = 1, 2$, we have obviously

$$r_1r_2 = a_3u_3^2 + b_3u_3v_3 + c_3v_3^2,$$
$$K(1,0) = Q_1 \circ Q_2(u_3, v_3). \qquad (90.30)$$

If it turns out that the right side is a proper representation of r_1r_2 belonging to $s \bmod 2r_1r_2$ in the same way as the left side, then $K \equiv Q_1 \circ Q_2 \bmod \Gamma$.

(v) Thus, yet by Note [74.3], it should be examined whether or not the following system of congruences holds:

$$(s - b_3)u_3 - 2c_3v_3 \equiv 0 \bmod 2r_1r_2, \qquad (90.31)$$
$$2a_3u_3 + (s + b_3)v_3 \equiv 0 \bmod 2r_1r_2. \qquad (90.32)$$

For this sake, we note the identity

$$\big(2a_1u_1 + (\ell_1 s + b_1)v_1\big)\big(2a_2u_2 + (\ell_2 s + b_2)v_2\big)$$
$$= 2\rho\left(2\frac{a_1a_2}{\rho^2}u_3 + (s + b_3)v_3\right) + \ell_1\ell_2(s^2 - D)v_1v_2, \qquad (90.33)$$

as can be verified by expanding the left side and rearranging by means of (90.10) with the present specification. By (90.29), the first term on this right side is divisible by $4r_1r_2$. Since $\langle 2\rho, r_1r_2 \rangle = 1$ by (90.26), we obtain (90.32);

$$\S 90 \qquad\qquad 431$$

note that $2|(s + b_3)$. Then, we multiply both sides of (90.32) by $2c_3$, and on noting that (90.27) implies $4a_3c_3 = b_3^2 - s^2 + 4r_1r_2t$, we get

$$(s + b_3)\big((s - b_3)u_3 - 2c_3v_3\big) \equiv 0 \bmod r_1r_2. \qquad (90.34)$$

Here $\langle s + b_3, r_1r_2 \rangle = 1$ since $\langle s^2 - b_3^2, r_1r_2 \rangle = \langle 4a_3c_3, r_1r_2 \rangle = 1$ by (90.26). We obtain (90.31); note that $2|(s - b_3)$. Therefore, $Q_1 \circ Q_2 \equiv K \bmod \Gamma$, indeed. In exactly the same way, we find that $Q_1' \circ Q_2' \equiv K \bmod \Gamma$. We have completed the proof of Theorem 92.

Continued to Note [90.5], where a proof, alternative to (iii)–(v), of (90.12) is given. It depends on the ideal theory of matrix modules and is far simpler than the above.

Now, we shall discuss the structure of the set $\mathcal{K}(D)$ in the light of the composition just obtained. Thus we shall confine ourselves to $\mathcal{Q}(D)$, that is, hereafter we shall always have

$$\ell_1 = \ell_2 = 1. \qquad (90.35)$$

We take classes $\mathcal{C}_j \in \mathcal{K}(D)$ and make the definition: with any $Q_j \in \mathcal{C}_j$,

$$\text{the product } \mathcal{C}_1\mathcal{C}_2 \text{ is the class in } \mathcal{K}(D) \text{ containing } Q_1 \circ Q_2 \qquad (90.36)$$

or rather denoting by (Q) the class containing Q, we introduce the operation

$$(Q_1)(Q_2) = (Q_1 \circ Q_2). \qquad (90.37)$$

By virtue of (90.12) this is well-defined.

The composition (90.9) is applicable to any pair of forms in $\mathcal{Q}(D)$. However, for the sake of computing the multiplication (90.36) of classes in $\mathcal{K}(D)$, it suffices to have a way to select a pair of forms which are easy to compose; such a specialization of forms in (90.37) entails no difference in (90.36). The exploitation of this simple observation originated with Dirichlet (1851). What Dirichlet was initially concerned with was, in fact, the relation between proper representations and Gauss' composition of forms. Apparently quoting the ending lines of [DA, art.244], he stated a sufficiency condition that the product of two integers which are properly represented by forms in $\mathcal{Q}(D)$ but not necessarily coprime is also properly represented by the form composed of these forms. He called this condition *radices concordantes*: Under the condition $m_1, m_2 > 0$, $\langle m_1m_2, D \rangle = 1$, that is, under (73.1), suppose that the proper representation $Q_j(u_j, v_j) = m_j$ belongs to $\xi_j \bmod 2m_j$ or $Q_j \equiv M_{m_j, \xi_j} \bmod \Gamma$. Then, on noting $\xi_1 \equiv \xi_2 \bmod 2$, we make the definition:

$$\{m_j, \xi_j\}, j = 1, 2, \text{ are concordant} \;\Leftrightarrow\; \langle m_1, m_2, (\xi_1 + \xi_2)/2 \rangle = 1; \qquad (90.38)$$

432   *Quadratic Forms*

as a matter of fact, this is different from Dirichlet's original but equivalent to. The composition (90.9) with (90.35) and $\rho = 1$ gives

$$M_{m_1, \xi_1} \circ M_{m_2, \xi_2} \equiv M_{m_1 m_2, \eta} \mod \Gamma, \tag{90.39}$$

where η is specified by (90.8). Hence, via (90.12) we conclude that $m_1 m_2$ is properly represented by $Q_1 \circ Q_2$. Namely, that $q(m_j) = 1$, $j = 1, 2$, $\Rightarrow q_D(m_1 m_2) = 1$, in the notation (72.6), can hold sometimes even if $\langle m_1, m_2 \rangle \neq 1$. Of course, one may replace M_{m_j, ξ_j} by $Q_j = [|a_j, b_j, c_j|]$ such that $\langle a_1, a_2, (b_1 + b_2)/2 \rangle = 1$ since $Q_j = M_{a_j, b_j}$. Hence, we are led further to the definition (Dedekind (Dirichlet (1871/1894, §146)):

$$\mathfrak{Q}(D) \ni Q_1, Q_2 \text{ are } einig/\text{concordant}$$
$$\Leftrightarrow \ Q_1 = [|a_1, b, a_2 c|], \ Q_2 = [|a_2, b, a_1 c|]; \tag{90.40}$$

indeed, $b = b_3$ defined by (90.8) with $\rho = 1$ works fine. This is but exactly the same as going back to the composition (90.3). In any event,

$$(90.40) \ \Rightarrow \ Q_1 \circ Q_2 = [|a_1 a_2, b, c|],$$
$$Q_1(x_1, y_1) Q_2(x_2, y_2) = (Q_1 \circ Q_2)(x_3, y_3), \tag{90.41}$$
$$x_3 = x_1 x_2 - c y_1 y_2, \ y_3 = a_1 x_1 y_2 + a_2 x_2 y_1 + b y_1 y_2.$$

The contribution of Dirichlet–Dedekind is in the recognition of the fact that the relation (90.41) provides a remarkable simplification of the computation of the composition or the multiplication of classes in $\mathcal{K}(D)$. Namely, imposing the concordance relation in considering the multiplication of classes causes no restriction within the framework of the Γ-equivalence. More precisely, given two classes $\mathcal{C}_j \in \mathcal{K}(D)$,

(A) One may choose a form from each of $\mathcal{C}_1, \mathcal{C}_2$ so that the resulting pair of forms is concordant in the sense of (90.40);
(B) The class to which the composition (90.41) belongs depends only on $\mathcal{C}_1, \mathcal{C}_2$,

where (A) is (iii), and (B) is (v) in the above proof, provided (90.35).

Therefore, in a somewhat drastic expression, the theory of the composition of quadratic forms begins with Note [79.4] ($*$), or (90.3) and requires in fact nothing more. Of course, this can be asserted only because of (90.12), a decisive discovery by Gauss [DA, arts.237–239]. His proof of (90.12) is extremely involved, however; and its simplification shown in (iii)–(v) as well is accomplished by means of none other than (90.3). However, see Note [90.5].

In any event, we have reached a completion of Gauss' fundamental theorem [DA, arts.249–251]:

$$§90 \qquad\qquad 433$$

Theorem 93 *There exists an arithmetic operation with which the set $\mathcal{K}(D)$ becomes a finite Abelian group.*

Remark 5 Gauss did not have the notion of groups, but what he discovered can be expressed as this statement. As a matter of fact, he came quite close even to the structure theorem of finite Abelian groups (Note [30.4]); see [DA, arts.305–307].

Proof The operation has been defined by (90.36). The unit element is the class (the principal class) that contains the principal forms defined by (71.9) and the adjacent paragraph. That is,

$$\text{(forms representing 1)} = 1 \text{ in } \mathcal{K}(D). \tag{90.42}$$

When $4|D$, we have $2|b$ in any form $[|a,b,c|]$. Thus, $[|1,0,-D/4|]$ is equivalent to $[|1,b,(b^2-D)/4|]$; and $[|a,b,c|] \circ [|1,b,(b^2-D)/4|] = [|a,b,c|]$ by (90.41). The case $D \equiv 1 \bmod 4$ is analogous. This confirms (90.42). Also, the inverse relation is inferred from $[|c,b,a|] \circ [|a,b,c|] = [|ac,b,1|]$. Since $[|c,b,a|] \equiv [|a,-b,c|] \bmod \Gamma$,

$$([|a,-b,c|]) = ([|a,b,c|])^{-1}. \tag{90.43}$$

Further, the operation (90.37) is associative. To confirm this by means of (90.41), we note that $\{[|a_1,b_1,c_1|] \circ [|a_2,b_2,c_2|]\} \circ [|a_3,b_3,c_3|]$ is easy to compute, provided $b_1 = b_2 = b_3 = b$, $c_1 = a_2a_3c$, $c_2 = a_1a_3c$, $c_3 = a_1a_2c$; and then the result of composition is equal to $[|a_1a_2a_3,b,c|]$. Namely, it suffices to show that it is possible to take respective forms from arbitrary three classes $\mathcal{C}_j$. Thus, applying Note [74.6] twice, we take $[|a_j,b_j,c_j|] \in \mathcal{C}_j$ such that $\langle a_1,a_2 \rangle = 1$, $\langle a_1a_2,a_3 \rangle = 1$. Then, let $b \equiv b_j \bmod 2a_j$ so that $[|a_j,b,(b^2-D)/4a_j|] \in \mathcal{C}_j$. The rest of the argument is omitted. We end the proof.

In passing, we note that (90.43) implies that

$$\mathcal{C} \in \mathcal{K}(D) \text{ is an ambiguous class } \Leftrightarrow \mathcal{C}^2 = 1. \tag{90.44}$$

This is a succinct summary of the contents of §75. Indeed, the equivalence between (75.3)(ii) and (90.44) is immediate.

The class group $\mathcal{K}(D)$ can, of course, be described by the structure theorem of finite Abelian groups. Naturally it should have special features due to particularities of the set $\mathcal{Q}(D)$ as well. For instance, its part concerning the 2-rank, i.e., the structure of the factor group $\mathcal{K}(D)/\mathcal{K}^2(D)$, will be described in the next section as the genus theory which connects $\mathcal{K}(D)$ with $(\mathbb{Z}/|D|\mathbb{Z})^*$ in a fascinating way.

434   *Quadratic Forms*

Returning to the subject mentioned in the third paragraph of the present section, we add the following: Take an arbitrary form $Q = [|a,b,c|] \in \mathfrak{Q}(D)$, $a > 1$, $\langle a, D \rangle = 1$. Then via the prime power decomposition $a = \prod_{j \leq J} p_j^{\nu_j}$ it follows that

$$Q = [|p_1^{\nu_1}, b, ac/p_1^{\nu_1}|] \circ [|p_2^{\nu_2}, b, ac/p_2^{\nu_2}|] \circ \cdots \circ [|p_J^{\nu_J}, b, ac/p_J^{\nu_J}|], \qquad (90.45)$$

and

$$[|p_j^{\nu_j}, b, ac/p^{\nu_j}|] = [|p_j, b, ac/p_j|] \circ \cdots \circ [|p_j, b, ac/p_j|] \quad (\nu_j \text{ times}). \quad (90.46)$$

Gauss [DA, art.243 (4)] must have been deeply content with this harmony between the two fundamental decompositions in arithmetic. Continued to Note [90.6].

Notes

[90.1] The composition theory of quadratic forms is one of the greatest achievements in the history of number theory. It is at the same time a typical instance of the cliché that great things have small beginnings since it began in an ancient, seemingly meager, observations indicated in Note [78.10] and in Note [72.5] (in the historical order). After isolated observations by Fermat, Euler, and especially by Lagrange (Note [79.3]), Legendre (1798, p.421) inaugurated an investigation that can be regarded as a theoretical analysis of compositions of quadratic forms.

[90.2] Legendre set out §III of Quatrième partie of his *Essai* with *Problême I: Étant donnés deux diviseurs quadratique* Δ, Δ', *d'une même formule* $t^2 + au^2$, *trouver le diviseur quadratique qui renferme leur produit* $\Delta\Delta'$. In modern terms, this is exactly the same as the problem of composing quadratic forms of the same discriminant $D = -4a$. He grasped (90.42)–(90.43), and devised multiplication tables for $a = 41, 89$ (1798, pp.432–434). However, they do not indicate the existence of any decent operation, the cause of which is to be detailed in Note [90.7].

[90.3] Gauss [DA, arts.236, 242–243] solved (90.2) in the sense that if restricted to our setting, then he found the coefficients of $Q_1 \circ Q_2$. However, the solution to the congruence (90.8) and the composition coefficients $\{s_j, s_j'\}$ were, both explicitly, given not by himself but by Arndt later. Because of this historical fact, (90.8) with (90.9) is customarily attributed to Arndt. Our reasoning in (ii) is precisely that of Arndt. The argument in the rest of our proof is different from the corresponding parts of Gauss'; Arndt did not discuss (90.12). Thus, while the procedure (i) does not seem to have been indicated before, (iii)–(v) are basically depend on the thoughts of Dirichlet (1851).

§90   435

In his discussion of the issue mentioned prior to (90.38), he introduced and used effectively the congruence relation stated in Note [74.3]; it was later adopted by Dedekind (Dirichlet (1871/1894, §146)) as a basic device. The fact that Gauss' composition theory can be accessibly reworked by means of Dirichlet's identity (90.3) and his congruence relation had been stated by Smith (1859/1869, art.111), together with a decent mention to Arndt's contribution. Repeating the remark made in Note 74.3, we stress here that Dirichlet's congruence relation means indeed the equivalence in terms of the group Γ. Therefore, the above proof of Theorem 92 is valid only under the proper equivalence of forms. Gauss' adoption of Γ-equivalence is penetrating indeed.

[90.4] In denoting their notions of composition, Legendre (1798, p.421) and Gauss [DA, p.273] used the multiplication and the addition symbols, respectively. In the French edition of [DA] (1807, p.274) the latter is replaced by the multiplicative one. Since Dirichlet's *Vorlesungen* (1871, p.385), the multiplicative notation has been common, although there have been a few exceptions till recently.

[90.5] This should have been stated earlier: under (90.6) the assertion (90.9)–(90.10) is equivalent to the matrix identity

$$
\text{(†)} \qquad \begin{aligned}
&\left(a_1 x_1 \mathfrak{e} + \tfrac{1}{2}(b_1 \mathfrak{e} + \ell_1 \mathfrak{d}) y_1\right)\left(a_2 x_2 \mathfrak{e} + \tfrac{1}{2}(b_2 \mathfrak{e} + \ell_2 \mathfrak{d}) y_2\right) \\
&\qquad = \rho\left(a_3 x_3 \mathfrak{e} + \tfrac{1}{2}(b_3 \mathfrak{e} + \mathfrak{d}) y_3\right),
\end{aligned}
$$

where $\mathfrak{e}, \mathfrak{d}$ are as in (72.16) with $d = D$; see Arndt (1857, pp.69–70). In particular, with $Q = Q_1 \circ Q_2$, $Q_j \in \mathfrak{Q}(\ell_j^2 D)$, it holds that

$$
\mathbf{Q}_1 \mathbf{Q}_2 = \rho \mathbf{Q},
$$

which means specifically that the finite sums of products of the elements of $\mathbf{Q}_j$ form an $\mathbf{I}_D$-ideal. For a confirmation, we note first that from (72.26) follows $\mathbf{I}_{\ell_1^2 D} \mathbf{I}_{\ell_2^2 D} = \mathbf{I}_D$ readily; see [DA, art.245]. Then, $\mathbf{I}_D \mathbf{Q}_1 \mathbf{Q}_2 = \mathbf{I}_{\ell_1^2 D} \mathbf{Q}_1 \mathbf{I}_{\ell_2^2 D} \mathbf{Q}_2 = \mathbf{Q}_1 \mathbf{Q}_2$, i.e., an $\mathbf{I}_D$-ideal. By (†) $\mathbf{Q}_1 \mathbf{Q}_2 \subseteq \rho \mathbf{Q}$, and by Note [72.8] $\mathbf{Q}_1 \mathbf{Q}_2 = \rho \mathbf{Q} \mathbf{B}$ with an $\mathbf{I}_D$-ideal $\mathbf{B}$. On the other hand, we have, with $N(\cdot)$ defined there, $N(\mathbf{Q}_1 \mathbf{Q}_2) = a_1 a_2 \mathbf{I}_{\ell_1^2 D} \mathbf{I}_{\ell_2^2 D} = \rho^2 a_3 \mathbf{I}_D = N(\rho \mathbf{Q})$, where the conjugation is as before, i.e., according to $\mathfrak{d} \mapsto -\mathfrak{d}$. Hence, $N(\mathbf{B}) = \mathbf{I}_D$, or $\mathbf{B} = \mathbf{I}_D$. One may now prove quickly the core assertion (90.12): By Note [74.7], there exist $\mathfrak{r}_j \in [\mathfrak{e}, \mathfrak{d}]_{\mathbb{Q}}, j = 1, 2$, such that $\det \mathfrak{r}_j = a_j / a_j'$ and $\mathbf{Q}_j = \mathfrak{r}_j \mathbf{Q}_j'$. Hence, $\mathbf{Q} = (\rho'/\rho) \mathfrak{r}_1 \mathfrak{r}_2 \mathbf{Q}'$, where Q', ρ' are analogous to Q, ρ, respectively; and

$$
\det((\rho'/\rho)\mathfrak{r}_1 \mathfrak{r}_2) = (\rho'/\rho)^2 (a_1/a_1')(a_2/a_2') = a_3/a_3'.
$$

436   Quadratic Forms

[90.6] We take a fundamental discriminant D for the sake of simplicity. The combination of (90.45)–(90.46) and the preceding Note yields

$$(\sharp) \qquad \mathbf{Q} = \mathbf{P}_1^{\nu_1} \mathbf{P}_2^{\nu_2} \cdots \mathbf{P}_J^{\nu_J},$$

where $\mathbf{P}_j$ is the $\mathbf{I}_D$-ideal corresponding to $[|p_j, b, ac/p_j|]$. This induces us to regard $\mathbf{P}_j$ as an analog of a prime number. In fact, it corresponds to a prime ideal of the filed $\mathbb{Q}(\sqrt{D})$, and $(\sharp)$ is an analogue of the prime power decomposition of integers. Also, one may readily see: If $\kappa_D(p) = +1$ (Kronecker symbol), then $p\mathbf{I}_D$ is decomposed into a product of two different prime ideals $\mathbf{P}, \overline{\mathbf{P}}$ (p splits); if $\kappa_D(p) = 0$, then $p\mathbf{I}_D$ is a square of a prime ideal, i.e., $\overline{\mathbf{P}} = \mathbf{P}$ (p ramifies); if $\kappa_D(p) = -1$, then $p\mathbf{I}_D$ itself is maximal (p is inert). These facts suggest us to ponder the problem of the general reciprocity law, which is indicated in (43.5), in the light of the mode of decomposition of prime numbers into prime ideals in appropriate algebraic number fields. See Hecke (1923, §§16, 29, 30, 46), for instance; specifically in §46, the quadratic reciprocity law (58.1) is proved by means of the ideal theory, albeit it depends on auxiliary assertions whose proofs are not straightforward. This approach to the law should be compared with Gauss' 2nd proof, i.e., his original version, to be mentioned in Note [91.6] below. There exists an apparent similarity between them, as both depend indispensably on the evaluation of the number of ambiguous classes. In this aspect, our version of the 2nd proof differs; see CONCLUDING REMARK of §91. Anyway, here is a historical impetus for the development of algebraic number theory: clarify the prime ideal splitting mode of primes and establish the general reciprocity law.

[90.7] It is claimed by some historians that Gauss must have developed his theory of compositions of forms first via multiplication of modules and then translated it into the literally awesome arithmetic computation as published in [DA, Sectio V]. See [DA, (d), II.3, footnote 3]. In any event, the use of the symbol $\sqrt{D}$ in the discussion on integral quadratic forms was a common practice before and after Gauss; it played a rôle of an operational convenience. To move to the integral module generated by $\{\mathfrak{e}, \mathfrak{d}\}$ or to quadratic algebraic integers does not appear to have required any special élan. It should be added that Dedekind (Dirichlet (1871, §§145–170; 1879, §§145–181; 1894, §§145–187)) discussed the composition of quadratic forms by means of the products of modules, or ideals, and reached essentially the same as Note [90.5](†).

[90.8] A table by Legendre (1798, p.432): Let $A = [|1, 2, 42|]$, $B = [|2, 2, 21|]$, $C = [|5, 6, 10|]$, $D = [|3, 2, 14|]$, $E = [|6, 2, 7|]$; note that we follow Legendre's notation, so D stands temporally for a form. These are in $\mathcal{Q}(-164)$. The right half of the table has the entries:

$$(\Diamond) \qquad CC = \begin{cases} A \\ B \end{cases} \quad CD = \begin{cases} D \\ E \end{cases} \quad CE = \begin{cases} D \\ E \end{cases}$$

$$DD = \begin{cases} A \\ C \end{cases} \quad DE = \begin{cases} B \\ C \end{cases} \quad EE = \begin{cases} A \\ C \end{cases}$$

Legendre was perplexed or rather embarrassed by this. To resolve his uneasiness, we appeal to the Γ-equivalence instead of the $\mathrm{GL}(2, \mathbb{Z})$-equivalence used by Lagrange and Legendre. Thus, $\mathcal{K}(-164)$ is represented by A, B, $C^{\pm} = [|5, \pm 6, 10|]$, $D^{\pm} = [|3, \pm 2, 14|]$, $E^{\pm} = [|6, \pm 2, 7|]$; namely, $\mathfrak{h}(-164) = 8$. With this, $(\Diamond)$ becomes

$$CC : \begin{cases} (A) = (C)(C^-) \\ (B) = (C)^2 \end{cases} \quad CD : \begin{cases} (D) = (C^-)(D^-) \\ (E) = (C)(D^-) \end{cases} \quad CE : \begin{cases} (D) = (C)(E^-) \\ (E) = (C^-)(E^-) \end{cases}$$

$$DD : \begin{cases} (A) = (D)(D^-) \\ (C) = (D^-)^2 \end{cases} \quad DE : \begin{cases} (B) = (D)(E^-) \\ (C) = (D)(E) \end{cases} \quad EE : \begin{cases} (A) = (E)(E^-) \\ (C) = (E^-)^2 \end{cases}$$

where in the notation (90.37), we have $(C^-) = (C)^{-1}$, $(D^-) = (D)^{-1}$, $(E^-) = (E)^{-1}$, of course; also $(B) = (B)^{-1}$, an ambiguous class. As an example we shall verify $(C) = (D)(E)$. Thus, $D \equiv [|3, 14, 30|]$, $E \equiv [|6, 14, 15|]$ by applications of powers of $\left(\begin{smallmatrix} 1 & 1 \\ 0 & 1 \end{smallmatrix} \right)$. These are concordant, so $D \circ E \equiv [|18, 14, 5|] \equiv [|5, -14, 18|] \equiv C$. As for $(B) = (D)(E^{-1})$, we have $D \equiv [|19, 30, 14|]$ and the pair $D \equiv [|19, 106, 6 \cdot 25|]$, $E^{-1} \equiv [|6, 106, 19 \cdot 25|]$. So $D \circ E^{-1} \equiv [|6 \cdot 19, 106, 25|] \equiv [|25, -106, 6 \cdot 19|] \equiv [|25, -6, 2|] \equiv [|2, 6, 25|] \equiv B$. By the way, the group $\mathcal{K}(-164)$ is cyclic, generated by (D):

$$(D)^2 = (C)^{-1}, \ (D)^3 = (E)^{-1}, \ (D)^4 = (B), \ (D)^5 = (E), \ (D)^6 = (C).$$

It is now evident what Legendre missed: the resolving power of the lens Γ. Thus, classes $(C), (C)^{-1}$, etc., were not split in his view. Or rather he did not notice the fact that the ambiguity in (72.14) is caused by the action of $\Diamond$, §75. This is the reason why he was led to $(\Diamond)$. Nonetheless, Legendre had, in his view, multiplication of (Lagrange's) classes of forms, beyond doubt.

[90.9] We have attributed the identity $(*)$ of Note [79.4] to Dirichlet. However, his proof of it (1851, p.159; Werke II, p.111) raises an interest, for his argument appears to be an adaptation of Legendre's mentioned in Note [72.5]. Namely, to the forms $[|f_1, g, f_2 k|]$, $[|f_2, g, f_1 k|]$ Dirichlet applied the identity

438   *Quadratic Forms*

(72.14)(1) and derived (∗). In other words, Legendre could have come to (∗) and used it in his initial theory of composition, only if he had not been bothered by (72.14)(2). Indeed, it transpires from the discussion developed in Legendre (1798, pp.421–422) that he was transforming pairs of his forms into those of the type (90.40) in order to multiply them. Constructing his theory of compositions, Dirichlet was undoubtedly inspired by Legendre.

§91

In the present section we shall develop an account of the genus theory of quadratic forms, which was started by Legendre (1785) and completed by Gauss [DA]. The genus theory enhances the fundamental Theorem 93 by connecting the groups $\mathcal{K}(D)$ and $(\mathbb{Z}/|D|\mathbb{Z})^*$ in an explicit way. However, our main concern is focused rather on the consequential notion of the genus characters associated with the discriminant D, as has been indicated in Note [71.1](7). We shall first make precise the reason for this preference of ours and then proceed to a construction of the genus theory. The original motivation for the genus theory itself will be related in Note [91.1].

Our argument depends, in an essential but subtle way, on the reciprocity law, mainly because of our application of Theorem 70; see the remark attached to Theorem 94 below, the main assertion of the present section. Besides, we shall employ, for the sake of technical ease, the theory of characters on Abelian groups, i.e., Notes [51.2]–[51.6], an origin of which is in fact in the genus theory. Hence, our approach is substantially different from Gauss', as is to be explained in Notes [91.3]–[91.6].

Thus, we point out an important practical feature of the genus theory: it provides a device that plays a principal rôle in our proof of the Dirichlet–Weber prime number theorem (Theorem 99, §93). We shall need there a means to take a particular element in the group $\mathcal{K}(D)$. The task is quite analogous to the first step in the proof of Dirichlet's prime number theorem for arithmetic progressions (88.1), where a particular element in the group $(\mathbb{Z}/q\mathbb{Z})^*$ was to be taken; for that sake, Dirichlet (1839) invented the character group Ξ_q, §51. It was extended by himself (1840b) to the group $\mathcal{K}(D)$, though in a quite special situation only; see Note [93.1]. To deal with the general case, Weber (1882) devised the notion of the character group of any finite Abelian group and applied it to $\mathcal{K}(D)$, that is, he reached the dual group $\widehat{\mathcal{K}}(D)$. Then, a situation similar to the core part of Note [88.2] emerged: to identify the subgroup

$\mathcal{R}(D)$ of $\widehat{\mathcal{K}}(D)$ composed of all real characters of $\mathcal{K}(D)$. Translated into modern formulation, Weber encountered a typical moment to try the duality theory concerning finite Abelian groups: With D an arbitrary discriminant, the combination of Notes [51.2], 51.5(1), and [51.6]$_{p=2}$ implies that

$$\mathcal{R}(D) \cong \mathcal{K}(D)/\mathcal{K}^2(D), \quad \mathcal{K}^2(D) = \left\{ \mathcal{C}^2 : \mathcal{C} \in \mathcal{K}(D) \right\}, \tag{91.1}$$

since $\mathcal{R}(D) = (\mathcal{K}^2(D))^{\perp}$. However, this formality is unable to yield anything practical. The real issue is obviously

$$\begin{array}{c} \text{to construct a homomorphism of } \mathcal{K}(D) \\ \text{whose kernel is } \mathcal{K}^2(D). \end{array} \tag{91.2}$$

If its image has a tangible structure, then such a mapping should well describe the nature of the elements of $\mathcal{R}(D)$. Weber could solve this problem by means of a constructive argument, but we proceed with an alternative way, as is indicated in Remark 2 below. Weber found anyway that $\mathcal{R}(D)$ is generated by means of the prime characters (§73) in $\Xi_{|D|}$ modulo a classification coming from a basic nature of integers represented by classes in $\mathcal{K}(D)$, that is, the Lagrange principle (Theorem 68 above), thereby gaining the concept of genus characters and achieving a proof of Theorem 99.

With these historical facts in the background, we begin an account of the genus theory: Thus, in view of Theorems 67 and 72 there exists a fundamental discriminant $D_0|D$ such that

$$D = D_0 R^2, \quad \kappa_D = J_R \kappa_{D_0}. \tag{91.3}$$

We let $\varpi = (-1)^{(p-1)/2} p$ with $\varpi|D, p \geq 3$, and

$$2^{\beta} \| D; \ \tau = \left| \{ \varpi : \varpi |D \} \right|; \ E_0 = \prod_{\varpi | D_0} \varpi. \tag{91.4}$$

Here and within the present section the restriction (15.3) is suspended, i.e., negative divisors are admitted; also, be aware that β and τ depend on D while E_0 on D_0 only. The quotient D_0/E_0 equals one of $1, -4, 8, -8$, for ϖ is the same as the prime discriminant p^* in (73.12)–(73.13). Our discussion is separated into eleven parts (i)–(xi) according to the values of β and D_0/E_0. We shall first deal with the case $\beta = 0$, i.e., $2 \nmid D$, in detail, to which the treatment of other cases is more or less analogous.

Remark 1 Hereafter we shall deal with various Kronecker symbols. Our operation on them will not take us outside the framework $\left\{ (\mathbb{Z}/|D|\mathbb{Z})^*, \Xi_{|D|} \right\}$, as is guaranteed by Theorem 70. In particular, our Kronecker symbols act solely on integers coprime to D.

440 *Quadratic Forms*

Remark 2 The prime number theorem (88.1) for arithmetic progressions, Theorem 91 on diagonal ternary quadratic forms, and the quadratic reciprocity law (58.1), mostly in the form (73.14) or rather Theorem 70, will be the principal means in our reasoning.

(i) $\beta = 0$; thus, $D_0 = E_0$, $D = E_0 R^2$, $2 \nmid R$.

Define Θ to be the map of $\mathcal{K}(D)$ into the multiplicative group $\{\{\eta_\varpi : \varpi | D\}\} = \{\{\pm 1\}^\tau\}$ with τ as above, in the following way: Assume (73.1), and let m be properly represented by a given class $\mathcal{C}$; the convention (74.12) is in use. Then, consider

$$\mathcal{C} \mapsto \Theta(\mathcal{C}) = \left\{ \kappa_\varpi(m) : \varpi | D \right\} \in \left\{ \{\eta_\omega : \varpi | D\} : \prod_{\varpi | E_0} \eta_\omega = 1 \right\}; \quad (91.5)$$

here the reason that not only $\varpi | E_0$ but all $\varpi | D$ are let participate will become apparent only when (91.16) comes into play. We ought to verify that Θ is well-defined. Thus, if $2|m$, then (72.12) implies that $E_0 \equiv 1 \bmod 8$, so $\kappa_{E_0}(2) = 1$; moreover, $\kappa_{E_0}(p) = 1$ for all odd $p|m$. Thus, $\kappa_{E_0}(m) = 1$ in (91.5), and via (73.13), $\prod_{\varpi | E_0} \kappa_\varpi(m) = 1$, which confirms the inclusion required in (91.5). We next take two forms $Q_1, Q_2 \in \mathcal{C}$, and suppose that $Q_j(u_j, v_j) = m_j > 0$, $\langle m_1 m_2, D \rangle = 1$. Since Q_1 properly represents m_2 as well, we may suppose $Q_1 = Q_2$. Then, by (72.4), there exist integers s, t derived from $\{u_j, v_j\}$ such that

$$4 m_1 m_2 = s^2 - t^2 D, \quad (91.6)$$

and $4 m_1 m_2 \equiv s^2 \bmod |\varpi|$ for $\varpi | D$. On noting that $\langle s, \varpi \rangle = 1$ here and κ_ϖ is a character mod $|\varpi|$ (Theorem 70), we get $\kappa_\varpi(m_1 m_2) = 1$. That is, $\kappa_\varpi(m_1) = \kappa_\varpi(m_2)$ for each $\varpi | D$, which implies that Θ is a function on $\mathcal{K}(D)$. We end the verification.

Remark 3 The observation following (91.6) originated in Legendre (1798, p.279). Gauss [DA, art.229] stated that integers representable by a form in $\mathcal{Q}(D)$ do not have such a fixed character property as above modulo any prime not dividing D; so the restriction to prime factors or ϖ's of D becomes our concern. In retrospect, this is natural because of the Dirichlet–Weber prime number theorem (Theorem 99) but extended to arithmetic progressions. Thus, let $q > 1$, $\langle q, D \rangle = 1$, be arbitrary; then there exist infinitely many primes in each reduced residue class $\bmod q$ which are represented by a given $\mathcal{C} \in \mathcal{K}(D)$. Hence, all the values of any non-principal $\chi \in \Xi_q$ are taken by integers represented by $\mathcal{C}$ while the corresponding values of κ_ϖ are definitely fixed for each $\varpi | D$.

In particular,

$$|\Theta(\mathcal{K}(D))| \leq 2^{\tau - 1}. \quad (91.7)$$

§91   441

For the fundamental theorem on homomorphisms applied to the map
$\{\{\pm 1\}^\tau\} \ni \{\eta_\varpi : \varpi | D\} \mapsto \prod_{\varpi | E_0} \eta_\varpi \in \{-1, 1\}$ gives

$$\left| \left\{ \{\eta_\varpi : \varpi | D\} : \prod_{\varpi | E_0} \eta_\varpi = 1 \right\} \right| = 2^{\tau - 1}, \tag{91.8}$$

as $E_0 \neq 1$ here. In fact, Θ is surjective, that is, equality holds in (91.7): By
virtue of Dirichlet's prime number theorem (88.1) there exists a sufficiently
large prime ℓ such that $\kappa_\varpi(\ell) = \eta_\varpi$ for each $\varpi | D$; and

$$\left(\frac{D}{\ell} \right) = \left(\frac{E_0}{\ell} \right) = \kappa_{E_0}(\ell) = \prod_{\varpi | E_0} \kappa_\varpi(\ell) = \prod_{\varpi | E_0} \eta_\varpi = 1; \tag{91.9}$$

Then, by Theorem 68 the prime ℓ is represented, of course properly, by a
certain class $\mathcal{C} \in \mathcal{K}(D)$, and $\Theta(\mathcal{C}) = \{\kappa_\varpi(\ell) : \varpi | D\} = \{\eta_\varpi : \varpi | D\}$. Hence,

$$|\Theta(\mathcal{K}(D))| = 2^{\tau - 1}. \tag{91.10}$$

This application of the prime number theorem (88.1) is a salient point of our
argument. Also, note that although the theorem (88.1) is independent of the
reciprocity law, the selection of ℓ in (91.9) becomes possible only because of
Theorem 70, i.e., $\kappa_\varpi(\ell) = \left(\frac{\varpi}{\ell} \right) = \left(\frac{\ell}{p} \right)$.

Further, Θ is a homomorphism into the group $\{\{\pm 1\}^\tau\}$. For the confirmation,
let $n_j > 0$, $\langle n_1, n_2 \rangle = 1$, $\langle n_1 n_2, D \rangle = 1$, be properly represented by the classes
$\mathcal{C}_1, \mathcal{C}_2 \in \mathcal{K}(D)$, respectively; here the condition $\langle n_1, n_2 \rangle = 1$ does not cause
any loss of generality because of Note [74.5], as usual. Then, (90.39) implies
that $n_1 n_2$ is properly represented by the product class $\mathcal{C}_1 \mathcal{C}_2$. Since κ_ϖ are
multiplicative, we get $\Theta(\mathcal{C}_1 \mathcal{C}_2) = \Theta(\mathcal{C}_1) \cdot \Theta(\mathcal{C}_2)$, as required to show.

We now define the notion of genus/genera as follows:

each coset of the kernel $\mathcal{P}(D)$ of Θ is called a genus of $\mathcal{K}(D)$;
$\mathcal{G}(D)$ stands for the collection of all genera;
$\mathfrak{g}(D) = |\mathcal{G}(D)|$, the genus number of the discriminant D. (91.11)

Specially, $\mathcal{P}(D)$ is called the principal genus. We have of course

$$\mathcal{G}(D) \cong \mathcal{K}(D)/\mathcal{P}(D) \tag{91.12}$$

as well as the basic assertions

$$\mathfrak{g}(D) = 2^{\tau - 1},$$
the number of classes in each genus is equal to $\mathfrak{h}(D)/2^{\tau - 1}$; (91.13)

Remark 4 As indicated in the last Remark, apparently following Legendre,
Gauss introduced the concept of genus in [DA, arts.228–231], which is but
different from ours; the details are in Note [91.5]. Also, one should be aware
that the definition of genus, either Legendre–Gauss' or ours, works well under

442 *Quadratic Forms*

the Lagrange–Legendre classification of forms by the action of $GL_2(\mathbb{Z}) = \Gamma \sqcup \diamond\Gamma$. This is due to the fact that the bare identity (91.6) or (72.4) with an arbitrary $\{\alpha, \beta, \kappa, \lambda\}$ is used to define genera by both procedures; see the ending lines of Note [91.1].

Then, we return to the task (91.2). Obviously $\mathcal{K}^2(D) \subseteq \mathcal{P}(D)$. A famous discovery of Gauss [DA, arts.286–287], after an appropriate interpretation (see Note [91.5]), states that here we have equality actually:

$$\text{the principal genus theorem}: \quad \mathcal{P}(D) = \mathcal{K}^2(D). \qquad (91.14)$$

In other words,

$$\mathcal{R}(D) \cong \mathcal{G}(D);$$
$$\text{the 2-rank of } \mathcal{K}(D) \text{ is equal to } \tau - 1; \qquad (91.15)$$
$$\text{in particular, } 2^{\tau-1} \| \mathfrak{h}(D),$$

according to the notion introduced in Note [30.6]. For the verification of (91.14), we note first that if $\Theta(\mathcal{C}) = $ unit, then every a, $\langle a, D\rangle = 1$, which $\mathcal{C}$ properly represents is a quadratic residue mod $|D|$, for $\kappa_{\varpi}(a) = 1$ implies that $a \bmod |\varpi|^{\nu}$, $\nu \geq 1$, are all quadratic residues; here we need to invoke the reciprocity law, i.e., (73.14) as well as the lifting (Note [34.3]). We have thus a set of integers all of which are quadratic residues mod $|D|$. Then, it is surmised, naïvely though, that such a set should contain a square; see Dedekind (Dirichlet (1871/1894, §155)), although he dealt with even D's only. We shall show that this is indeed the case; thence $\mathcal{C} \in \mathcal{K}^2(D)$ will follow immediately. Here one should be aware that quadratic residues and squares are different notions; the latter makes a far smaller set than the former. Hence, capturing these squares is never a trivial task, and we shall settle it by means of Theorem 91: We may assume additionally that $a > 0$ because of Note [74.5]. By the Lagrange principle, $D \bmod 4a$ is a quadratic residue. Thus, by virtue of (89.28)–(89.29), there exists a tuple $\{u, v, w\}$ such that

$$4au^2 + Dv^2 - w^2 = 0, \quad \langle 2au, Dv, w\rangle = 1. \qquad (91.16)$$

We have $u \neq 0$, for D is not a square presently. Since $2 \nmid vw$, we have $w = v + 2z$ with an integer z, and

$$au^2 + \tfrac{1}{4}(D-1)v^2 - vz - z^2 = 0. \qquad (91.17)$$

Namely, the principal form $[|1, 1, (1-D)/4|]$ represents au^2. Hence, the form $Q \circ [|1, 1, (1-D)/4|]$ with any $Q \in \mathcal{C}$ represents the square $(au)^2$, $\langle au, D\rangle = 1$, and consequently a certain square properly. The form thus composed is, of course, equivalent to Q, so $\mathcal{C}$ itself indeed represents properly a square α^2, $\langle \alpha, D\rangle = 1$; see what we have mentioned in Note [89.1]. The Lagrange

§91   443

principle ensures then that there exists a ξ such that $\xi^2 \equiv D \bmod 4\alpha^2$, $\langle \xi, \alpha \rangle = 1$; and in $\mathcal{K}(D)$

$$\mathcal{C} = ([\,|\alpha^2, \xi, (\xi^2 - D)/4\alpha^2|\,]) = ([\,|\alpha, \xi, (\xi^2 - D)/4\alpha|\,])^2, \qquad (91.18)$$

by (90.3). Therefore, we have established (91.14) for every odd discriminant.

(ii) $\beta = 2$ and $D_0 = E_0$; thus, $D = E_0(2R)^2, 2 \nmid R$.

We retain the definition (91.5). There is none to alter up to (91.13). However, as for the proof of (91.14), we must impose $a \equiv 1 \bmod 4$ in order to have (91.19) below. To this end, let $\mathcal{C} \ni [\,|a,b,c|\,]$ and observe that $ac = (b/2)^2 - D/4 \equiv (b/2)^2 - 1 \bmod 4$, and thus if $4|b$, then either a or $c \equiv 1 \bmod 4$. Because of this, we take the form $[\,|a, 2ka + b, c_1|\,] \in \mathcal{C}$ in such a way that $2ka + b = 4f$, $\langle f, D \rangle = 1$ with a sufficiently large f. Since $\langle a, D \rangle = 1$ implies $\langle a, (a+1)b/4 \rangle = 1$, we see, by Note [29.2], that there exist infinitely many h such that $\langle D, ah + (a+1)b/4 \rangle = 1$; hence one may take $k = 2h + b/2$, i.e., $f = ah + (a+1)b/4$. Then $ac_1 = 4f^2 - D/4 \equiv -1 \bmod 4$, that is, $\langle ac_1, D \rangle = 1$ and either a or c_1 is congruent to 1 mod 4, while they are both positive and represented properly by $\mathcal{C}$. That is, without loss of generality, we may suppose that $a \equiv 1 \bmod 4$, so we replace (91.16) by

$$au^2 + Dv^2 - w^2 = 0, \quad \langle au, Dv, w \rangle = 1, \ a > 0, a \equiv 1 \bmod 4. \qquad (91.19)$$

Namely, $au^2 = w^2 - (D/4)(2v)^2$, and the principal form $[\,|1, 0, -D/4|\,]$ represents au^2, $\langle au, D \rangle = 1$. The rest of the argument can be omitted. The reason for the relative intricacy of the present part is given in the later paragraph of (iv) below.

(iii) $\beta = 2$ and $D_0 = -4E_0$; thus $D = -4E_0R^2, 2 \nmid R$.

We replace the definition (91.5) by

$$\mathcal{C} \mapsto \Theta(\mathcal{C}) = \big\{ \kappa_{-4}(m), \{\kappa_{\varpi}(m) : \varpi | D\} \big\}$$
$$\in \Big\{ \{\eta_{-4}, \{\eta_{\varpi} : \varpi | D\}\} : \ \eta_{-4} \prod_{\varpi | E_0} \eta_{\varpi} = 1 \Big\} \subseteq \{\{\pm 1\}^{\tau+1}\}. \qquad (91.20)$$

The insertion of κ_{-4} and η_{-4} is required since $\kappa_{-4E_0}(p) = 1$ for all odd $p|m$. The identity (91.6) presently gives $m_1 m_2 = (s/2)^2 - t^2(D/4)$. We have $D/4 \equiv -1 \bmod 4$, and either $2|(s/2), 2 \nmid t$ or $2 \nmid (s/2), 2|t$. Hence, $m_1 m_2 \equiv 1 \bmod 4$, so $\kappa_{-4}(m_1) = \kappa_{-4}(m_2)$. The argument leading to (91.13) holds with τ being replaced by $\tau + 1$ and with other minor changes like that E_0 in (91.9) is to be replaced by $-4E_0$. As for the proof of (91.14), we use (91.19); note that $a \equiv 1 \bmod 4$ since $\kappa_{-4}(a) = 1$ by definition. The rest of the argument can be omitted.

444   *Quadratic Forms*

(iv) $\beta = 3$ and $D_0 = 8E_0$; thus, $D = 8E_0R^2$, $2\nmid R$.

We replace the definition (91.5) by

$$\mathcal{C} \mapsto \Theta(\mathcal{C}) = \{\kappa_8(m), \{\kappa_\varpi(m) : \varpi|D\}\}$$
$$\in \left\{\{\eta_8, \{\eta_\varpi : \varpi|D\}\} : \eta_8 \prod_{\varpi|E_0} \eta_\varpi = 1\right\} \subseteq \{\{\pm 1\}^{\tau+1}\}. \tag{91.21}$$

The insertion of κ_8 and η_8 is required since $\kappa_{8E_0}(p) = 1$ for all odd $p|m$. The identity (91.6) gives $m_1m_2 = (s/2)^2 - 2t^2(D/8)$. We have $2 \nmid (s/2)$, $D/8 \equiv 1 \bmod 4$. Hence, $m_1m_2 \equiv 1 - 2t^2 \equiv \pm 1 \bmod 8$. Thus, $\kappa_8(m_1) = \kappa_8(m_2)$. As for the proof of (91.14), we use

$$au^2 + (D/4)v^2 - w^2 = 0, \ \langle au, (D/4)v, w\rangle = 1, \ a > 0. \tag{91.22}$$

The rest of the argument can be omitted. By the way, one might wonder why (91.22) was not used in (ii). The reason is in that the fact $\langle au, D\rangle = 1$ which is necessary to have $\langle \alpha, D\rangle = 1$ at the step corresponding to (91.18) does not follow from (91.22) if $2 \nmid D/4$, the case of (ii). Thus (91.19) was instead applied. However, then $a \equiv 1 \bmod 4$ is required, and to attain it the intricacy was caused.

(v) $\beta = 3$ and $D_0 = -8E_0$; thus, $D = -8E_0R^2$, $2\nmid R$.

We replace the definition (91.5) by

$$\mathcal{C} \mapsto \Theta(\mathcal{C}) = \{\kappa_{-8}(m), \{\kappa_\varpi(m) : \varpi|D\}\}$$
$$\in \left\{\{\eta_{-8}, \{\eta_\varpi : \varpi|D\}\} : \eta_{-8} \prod_{\varpi|E_0} \eta_\varpi = 1\right\} \subseteq \{\pm 1\}^{\tau+1}. \tag{91.23}$$

The identity (91.6) presently gives $m_1m_2 = (s/2)^2 - 2t^2(D/8)$. We have $2 \nmid (s/2)$, $D/8 \equiv -1 \bmod 4$. Hence, $m_1m_2 \equiv 1 + 2t^2 \equiv 1, 3 \bmod 8$. Thus, $\kappa_{-8}(m_1) = \kappa_{-8}(m_2)$. As for the proof of (91.14), we use (91.22). The rest of the argument can be omitted.

(vi) $\beta = 4$ and $D_0 = E_0$; thus $D = E_0(4R)^2$, $2\nmid R$.

We replace the definition (91.5) by

$$\mathcal{C} \mapsto \Theta(\mathcal{C}) = \{\kappa_{-4}(m), \{\kappa_\varpi : \varpi|D\}\}$$
$$\in \left\{\{\eta_{-4}, \{\eta_\varpi : \varpi|D\}\} : \prod_{\varpi|E_0} \eta_\varpi = 1\right\} \subseteq \{\{\pm 1\}^{\tau+1}\}. \tag{91.24}$$

The identity (91.6) gives $m_1m_2 = (s/2)^2 - 4t^2(D/16)$. Since $2\nmid(s/2)$, we have $m_1m_2 \equiv 1 \bmod 4$. Thus, $\kappa_{-4}(m_1) = \kappa_{-4}(m_2)$. As for the proof of (91.14), we use (91.22); note that this time $a \equiv 1 \bmod 4$ is required in applying (91.22), which is handled by the presence of κ_{-4}. The rest of the argument can be omitted.

$$\S 91 \qquad\qquad 445$$

(vii) $\beta = 4$ and $D_0 = -4E_0$; thus, $D = -4E_0(2R)^2, 2 \nmid R$.

We replace the definition (91.5) by

$$\mathcal{C} \mapsto \Theta(\mathcal{C}) = \{\kappa_{-4}(m), \{\kappa_{\varpi}(m) : \varpi | D\}\}$$
$$\in \left\{\{\eta_{-4}, \{\eta_{\varpi} : \varpi | D\}\} : \eta_{-4} \prod_{\varpi | E_0} \eta_{\varpi} = 1\right\} \subseteq \{\{\pm 1\}^{\tau+1}\}; \qquad (91.25)$$

note the difference between this and (91.24). As for the proof of (91.14), we use (91.22). The rest of the argument can be omitted.

(viii) $\beta \geq 5$ and $D_0 = E_0$; thus, $D = E_0(2^\lambda R)^2, \lambda \geq 3, 2 \nmid R$.

We replace the definition (91.5) by

$$\mathcal{C} \mapsto \Theta(\mathcal{C}) = \{\kappa_{-4}(m), \kappa_8(m), \{\kappa_{\varpi} : \varpi | D\}\}$$
$$\in \left\{\{\eta_{-4}, \eta_8, \{\eta_{\varpi} : \varpi | D\}\} : \prod_{\varpi | E_0} \eta_{\varpi} = 1\right\} \subseteq \{\{\pm 1\}^{\tau+2}\}. \qquad (91.26)$$

The identity (91.6) presently gives $m_1 m_2 = (s/2)^2 - 8t^2(D/32) \equiv 1 \bmod 8$, so $\{\kappa_{-4}(m_1 m_2), \kappa_8(m_1 m_2)\} = \{1, 1\}$. The argument leading to (91.13) holds with τ being replaced by $\tau + 2$ and with other minor changes. As for the proof of (91.14), we use (91.22). Because of (43.3), we need to have $a \equiv 1 \bmod 8$ which is but equivalent to $\{\kappa_{-4}(a), \kappa_8(a)\} = \{1, 1\}$. The rest of the argument can be omitted.

(ix) $\beta \geq 5$ and $D_0 = -4E_0$; thus, $D = -4E_0(2^\lambda R)^2, \lambda \geq 2, 2 \nmid R$.

$$\mathcal{C} \mapsto \Theta(\mathcal{C}) = \{\kappa_{-4}(m), \kappa_8(m), \{\kappa_{\varpi}(m) : \varpi | D\}\}$$
$$\in \left\{\{\eta_{-4}, \eta_8, \{\eta_{\varpi} : \varpi | D\}\} : \eta_{-4} \prod_{\varpi | E_0} \eta_{\varpi} = 1\right\} \subseteq \{\{\pm 1\}^{\tau+2}\}. \qquad (91.27)$$

(x) $\beta \geq 5$ and $D_0 = 8E_0$; thus, $D = 8E_0(2^\lambda R)^2, \lambda \geq 1, 2 \nmid R$.

$$\mathcal{C} \mapsto \Theta(\mathcal{C}) = \{\kappa_{-4}(m), \kappa_8(m), \{\kappa_{\varpi}(m) : \varpi | D\}\}$$
$$\in \left\{\{\eta_{-4}, \eta_8, \{\eta_{\varpi} : \varpi | D\}\} : \eta_8 \prod_{\varpi | E_0} \eta_{\varpi} = 1\right\} \subseteq \{\{\pm 1\}^{\tau+2}\}. \qquad (91.28)$$

(xi) $\beta \geq 5$ and $D_0 = -8E_0$; thus, $D = -8E_0(2^\lambda R)^2, \lambda \geq 1, 2 \nmid R$.

$$\mathcal{C} \mapsto \Theta(\mathcal{C}) = \{\kappa_{-4}(m), \kappa_8(m), \{\kappa_{\varpi}(m) : \varpi | D\}\}$$
$$\in \left\{\{\eta_{-4}, \eta_8, \{\eta_{\varpi} : \varpi | D\}\} : \eta_{-4}\eta_8 \prod_{\varpi | E_0} \eta_{\varpi} = 1\right\} \subseteq \{\{\pm 1\}^{\tau+2}\}. \qquad (91.29)$$

Summing up, we obtain

Theorem 94 *Classify discriminants D into the eleven types (i)–(xi), and accordingly define the homomorphism Θ as above. Then, under the definitions*

446 *Quadratic Forms*

(91.11), *the assertion* (91.14) *holds. Also,* (91.13) *holds but with* τ *being replaced by*

$$\tau^* = \begin{cases} \tau & \text{(i)–(ii)}, \\ \tau + 1 & \text{(iii)–(vii)}, \\ \tau + 2 & \text{(viii)–(xi)}; \end{cases} \quad \text{i.e., } \mathfrak{g}(D) = 2^{\tau^* - 1}. \tag{91.30}$$

Remark 5 Our proof of this fundamental assertion depends on the reciprocity law. In case (i), this has been made clear after (91.6) and (91.10), i.e., an application of Theorem 70 but the mention has been omitted in all other cases.

We then return to the first line following (91.2). Although Θ satisfies (91.2), the image $\Theta(\mathcal{K}(D))$ is not tangible enough, and it ought to be further transformed: For this purpose, we prepare the configuration

$$K(D) = \left\{ n \bmod |D| : \kappa_D(n) = 1 \right\} \subset (\mathbb{Z}/|D|\mathbb{Z})^*, \\ G(D) = K(D)/P(D), \tag{91.31}$$

with

$$P(D) = \Big\{ n \bmod |D| : \\ \{\kappa_{-4}^{\delta_1}(n), \kappa_8^{\delta_2}(n), \kappa_{-8}^{\delta_3}(n), \{\kappa_{\varpi}(n) : \varpi | D\}\} = \{1\}^{\tau^*} \Big\}; \tag{91.32}$$

Here $\delta_j = 0, 1$ are chosen according to the types (i)–(xi) of D, and it is understood that if $\delta_j = 0$, then the corresponding term does not appear; that is,

$$\{\delta_1, \delta_2, \delta_3\} = \begin{cases} \{0, 0, 0\} & \text{(i), (ii)}, \\ \{1, 0, 0\} & \text{(iii), (vi), (vii)}, \\ \{0, 1, 0\} & \text{(iv)}, \\ \{0, 0, 1\} & \text{(v)}, \\ \{1, 1, 0\} & \text{(viii), (ix), (x), (xi)}; \end{cases} \tag{91.33}$$

so $\tau^* = \tau + \delta_1 + \delta_2 + \delta_3$. Note that in (91.31) the condition $\kappa_D(n) = 1$ is the same as $\kappa_{D_0}(n) = 1$ and $\langle n, D \rangle = 1$ because of (73.16); also note that the residue classes modulo $|D|$ in $P(D)$ can be determined solely with the discriminant D, provided the prime power decomposition of $|D|$ is available.

Now, let m satisfy (73.1) and be represented properly by $\mathcal{C} \in \mathcal{K}(D)$. Then, we have the bijection, the genus map:

$$\mathcal{C}\mathcal{K}^2(D) = \mathcal{C}\mathcal{P}(D) \mapsto mP(D), \\ \text{i.e., } \mathcal{R}(D) \cong \mathcal{G}(D) \cong G(D). \tag{91.34}$$

The verification is skipped, for it is just a consequence of the above discussion on the map Θ leading to the last theorem. What is essential here is that $G(D)$ can be constructed without any knowledge about $\mathcal{K}(D)$; only the prime power decomposition of $|D|$ is required. Namely, (91.34) amounts to a desired practical solution to (91.2).

It is stated via the inverse of the mapping in (91.34), which is a core of the genus theory, that a prime number ℓ can be represented properly by a form in a given genus associated with a discriminant D if and only if ℓ belongs to a particular set of residue classes $\mathrm{mod}\,|D|$, that is, a coset of $K(D)$ modulo $P(D)$; see (91.9). Namely, by knowing only the residue class $\ell \bmod |D|$, one can make it exact which genus contains the form that represents ℓ, including whether such a genus exists or not. In other words,

$$\text{the factor group } \mathcal{K}(D)/\mathcal{K}^2(D)$$
$$\text{can be seen via congruence mod } |D| \text{ or the group } \left(\mathbb{Z}/|D|\mathbb{Z}\right)^*. \tag{91.35}$$

A glimpse of the structure of the class group $\mathcal{K}(D)$ can be captured in a mechanism working in its basement $\left(\mathbb{Z}/|D|\mathbb{Z}\right)^*$. More precisely, with any m which satisfies (73.1) and the Lagrange principle, i.e., (72.2), we are able to pinpoint the genus which contains the form that represents m, just by computing

$$\left\{\kappa_{-4}^{\delta_1}(m), \kappa_8^{\delta_2}(m), \kappa_{-8}^{\delta_3}(m), \{\kappa_{\varpi}(m) : \varpi|D\}\right\}. \tag{91.36}$$

Since the values of the prime characters at a particular reduced residue $m \bmod |D|$ can be computed rather immediately, (91.36) is a more practical means than (91.34), albeit the prime power decomposition of $|D|$ is needed. We have already encountered such instances at (78.3), (79.5), and (80.6).

However, it is also true that the genus theory does not provide us with effective means to solve our fundamental problem (71.2), where the form Q is given firsthand. This is obvious, for $\mathfrak{h}(D)$ should be divisible by odd primes in general, so the 2-rank of $\mathcal{K}(D)$ is too modest a clue to examine the whole nature of the group. This inherent limitation of the genus theory has already been indicated in Note [80.3]. As a matter of fact, splitting a given genus in order to identify the class or form that actually represents the prime number in question can be discussed but with an appeal to a deep realm of algebraic number theory and automorphic function theory. Needless to say, any account of such a method is beyond the scope of the present essays; moreover, its practicability is another matter to be seriously examined, although the formation of the method itself is indeed a great success of relevant algebraic and analytic theories.

It is amazing, nevertheless, that the genus theory is able to exhibit a part of the structure of the group $\mathcal{K}(D)$ by means of congruence classes $\bmod |D|$.

448   *Quadratic Forms*

This fact gave tremendous stimuli to the formation of advanced algebraic number theory, as a result of which have been brought about generalizations of the structure (91.35) in the context of algebraic extensions of fields.

In any event, if one opts for the practical efficiency in considering the problem (71.2), then the algorithms described in §81 and §87 are most probably still the best to try first. With any pair of a form Q and an integer m, the method makes it possible to precisely construct the set of seeds, or $\mathrm{Aut}_Q \backslash S_Q(m)$, including its emptiness. However, this requires at least the prime power decompositions of m and quadratic non-residues which are needed if either Tonelli's or Cipolla's methods are to be applied; moreover, checking the relevant continued fraction expansions can be tedious, especially if the involved parameters are huge.

We now return to the original aim of the present section and construct the real characters of the class group $\mathcal{K}(D)$, or the subgroup $\mathcal{R}(D)$ of $\widehat{\mathcal{K}}(D)$. We shall show, via the inverse of the map (91.34), that the prime characters actually appearing in (91.32) can be regarded as the seeds generating $\mathcal{R}(D)$:

Theorem 95 *Let* w *and* W *be the subgroups of* $\Xi_{|D|}$ *which are generated, respectively, by* $\{1, \kappa_D\}$ *and by the prime characters appearing in* (91.32) *with the specification* (91.33). *Then,*

$$\mathcal{R}(D) \cong \mathrm{W/w}. \tag{91.37}$$

Proof In order to identify the elements of $\mathcal{R}(D)$, it suffices to consider the same concerning $\widehat{G}(D)$ since $\mathcal{R}(D) \cong \widehat{\mathcal{G}}(D) \cong \widehat{G}(D)$ by (91.15) and (91.34). We apply Note [51.5](4) to the combination $\{A = \Xi_{|D|},\ B = \mathrm{W},\ C = \mathrm{w}\}$. On noting that $\mathrm{W}^{\perp} = P(D)$ and $\mathrm{w}^{\perp} = K(D)$, i.e., $\mathrm{W} = P(D)^{\perp}$ and $\mathrm{w} = K(D)^{\perp}$, we find that

$$\widehat{G}(D) \cong P(D)^{\perp}/K(D)^{\perp} = \mathrm{W/w}. \tag{91.38}$$

This should end the discussion. However, more explicitly one may argue as follows: Let $\kappa \in \mathrm{W}$, and let

$$[\kappa](\mathcal{C}) = \kappa(m) \text{ with } m \text{ as in (91.34).} \tag{91.39}$$

The map $\kappa \mapsto [\kappa]$ is a homomorphism from W into $\mathcal{R}(D)$. Let $[\kappa'] = [\kappa]$ with $\kappa, \kappa' \in \mathrm{W}$. Reversing (91.39), take an arbitrary $m \bmod |D| \in K(D)$, and find $\mathcal{C}$ corresponding to it by just following (91.9). From $[\kappa'](\mathcal{C}) = [\kappa](\mathcal{C})$ it follows that $\kappa'\kappa^{-1}$ is trivial on $K(D)$, or it is contained in $K(D)^{\perp} = \mathrm{w}$. Namely, $|\text{image of } \mathrm{W}| = |\mathrm{W/w}|$. On noting $|\mathrm{W/w}| = 2^{\tau^*-1} = |\mathcal{R}(D)|$, we end the proof.

$§91$ 449

The assertion (91.37) may be termed the genus character theorem. The genus characters or the real characters on $\mathcal{K}(D)$ are obtained by classifying, inside $\Xi_{|D|}$, the characters κ generated by W in terms of modulo w, that is, κ and $\kappa\kappa_D$ which are different in $\Xi_{|D|}$ are regarded as the same character of $\mathcal{K}(D)$ via the correspondence (91.39): Hence, the elements of W/w are the genus characters associated with the discriminant D.

Notes

[91.1] The origin of the genus theory was in the investigation of the set of the residue classes mod $|D|$ which contain prime numbers represented by a form in $\mathcal{Q}(D)$. An initial instance was Fermat's observation on the forms $[|1,0,5|]$ and $[|2,2,3|]$ which we mentioned in Note [79.3]; Euler also made numerous findings of the same nature on prime numbers represented by the forms $[|a,0,b|]$. More systematic studies were made by Lagrange (1773; 1775); he composed four tables I–IV where I, III are for negative discriminants $-4d$, and II, IV for positive $4d$. For instance, he gave the following at the entry $d = 26$:

$$\text{I} : \quad 1,2,3,5 \quad | \quad 0,0,1,2 \quad | \quad 26,13,9,6 ;$$

$$\text{III} : \quad \begin{cases} 1,3 & 1,3,9,17,25,27,35,43,49,51,-23,-29, \\ 2,5 & 5,7,15,21,31,37,45,47,-11,-19,-33,-41. \end{cases}$$

In modern terms, this is the same as the structure of $\mathcal{Q}(-4d)$ viewed via the equivalence mod $GL(2,\mathbb{Z})$. In terms of mod Γ, $\mathfrak{h}(-4d) = 6$ and $\mathfrak{g}(-4d) = 2$, and the table can be refined as follows:

$$\text{I}^* : \quad Q_1 = [|1,0,26|], \, Q_2 = [|2,0,13|], \, Q_3^{\pm} = [|3, \pm 2,9|], \, Q_5^{\pm} = [|5, \pm 4,6|];$$

$$\text{III}^* : \quad \begin{cases} \{Q_1,Q_3^{\pm}\} & \Theta : (+1,+1) \\ \{Q_2,Q_5^{\pm}\} & \Theta : (-1,-1) \end{cases} \quad \text{type (v).}$$

Lagrange's table I corresponds to the set I^* of representative forms, and III shows the two genera III^* in terms of $2 \times 12 = 24$ reduced residue classes mod $4d$; note that $24 = \varphi(4d)/2 = |K(-4d)|$. This is an instance of (91.34). However, Lagrange himself seems to have been unable to discern any general structure from his tables. By the way, by applying the classification in terms of Γ, table I is refined to I^*, but III and III^* are not essentially different, which is an example for the second part of Remark 4 above.

[91.2] We attribute the origin of the theoretical study of genera of forms to Legendre (1785). The reason for this is as follows: The article contains four tables at its end, which are similar to Lagrange's in the preceding Note. Legendre's tables I, II correspond to the above type (iii) and tables III, IV to (ii). However, there exists a genuine difference between Lagrange's and Legendre's investigations. Legendre combined his own recently discovered reciprocity law (58.1) (3) with the Lagrange principle and found that the rôles of D and m in (72.2) can be interchanged (i.e., our Theorem 59, §62), in a definitive manner. This argument implies quite analogously that the 24 residue classes in Lagrange's III exhaust all the reduced residue classes mod $4d$ representable by the forms in his table I; thus table III means that just by checking the residue class of a prime number in question one can say whether or not a selected genus contains the form that represents it. Lagrange's difficulty was resolved by Legendre. If the reciprocity law had been known so widely as nowadays, then Legendre's assertion could have appeared to be a rudimental matter. However, in his time, i.e., 1780s, the power of the reciprocity law just started to be felt among specialists; see what Legendre (1798, p.286) said about his own discovery and compare our (91.30) with his new tables III–VII in the treatise and their respective explanations (pp.266–286) to find their perfect coincidence (even though Legendre's tables are subsets of (91.30) because of his restriction of discriminants). For the first time in history, he firmly grasped the heart of the quadratic reciprocity law as a structure among the integers represented by quadratic forms, albeit his proof of the law was incomplete. Later this subject was given the term genus theory and completed by Gauss.

[91.3] Gauss' assertion (91.14) played a Merkmal rôle in the development of algebraic number theory; see for instance [DA, (d), VIII.3] by F. Lemmermeyer. Because of this historical fact, it has been given the special terms like *Hauptgeschlechtssatz*. For the sake of its proof, Gauss developed a theory of ternary quadratic forms (*Digressio*: [DA, arts.266–285]) and established the theorem in arts.286–287. To avoid Gauss' involved discussion, we followed instead Arndt (1859a) with minor changes; anyway it also depends in an essential manner on a special instance of the theory of ternary quadratic forms, i.e., the Legendre–Dedekind theorem (89.28). By the way, Gauss' theory of ternary quadratic forms contains Legendre's theorem (see Note [89.1]); see the relevant comment in Dirichlet (1871/1894, §158, footnote by Dedekind). However, in §93[C] an alternative proof that originated with Kronecker (1865) will be presented; it dispenses, in particular, with the theory of the ternary quadratic forms, although we shall restrict our discussion to the case of odd discriminants for the sake of simplicity.

[91.4] By the general duality consideration concerning finite Abelian groups, i.e., Note [51.2], we have obviously

$$\mathcal{R}(D) \cong \mathcal{A}(D) = \{\mathcal{C} \in \mathcal{K}(D) : \mathcal{C}^2 = 1\},$$

where

$$\mathcal{A}(D) \text{ is the kernel of the map } \mathcal{C} \mapsto \mathcal{C}^2;$$

it is called the ambiguous group associated with the discriminant D, as $\mathcal{C}^2 = 1$ is another way to say that $\mathcal{C}$ is an ambiguous class. We have

$$|\mathcal{A}(D)| = |\mathcal{K}(D)/\mathcal{K}^2(D)| \geq |\mathcal{K}(D)/\mathcal{P}(D)| = |\mathcal{G}(D)| = 2^{\tau^* - 1},$$

for $\mathcal{K}^2(D) \subseteq \mathcal{P}(D)$. This is, of course, the situation where we have not applied the principal genus theorem yet. In other words, if we are able to show by some way that $|\mathcal{A}(D)| \leq 2^{\tau^* - 1}$, then the surjectivity of Θ, which we have established with Dirichlet's prime number theorem, implies as well that $\mathcal{K}^2(D) = \mathcal{P}(D)$ the principal genus theorem; that is, here is a possibility to dispense with the theory of the ternary quadratic forms. In Gauss' discussion on genera, $\mathcal{A}(D)$ plays a major rôle. He [DA, arts.257–259] computed explicitly the value of $|\mathcal{A}(D)|$ by an independent but involved argument. However, to prove the lower bound stated above, he needed the theory of ternary quadratic forms. Namely, it can be said that if Gauss had had the prime number theorem for arithmetic progressions, then he could have dispensed with the theory of ternary quadratic forms. Yet, he had other motivations to develop such a roundabout reasoning; continued in the next two Notes.

[91.5] Our definition of genera is not entirely the same as the one that is due to Legendre and Gauss; although Legendre did not invent the term genera, he must have been aware of their existence. Thus, ours is via the map Θ, i.e., (91.5), and his (1798, pp.278–286) is via

$$\Theta^*(\mathcal{C}) = \left\{ \left(\frac{m}{p} \right) : p | D \right\},$$

which Gauss [DA, art.229] apparently adopted. Here m is as in the definition of Θ, and D is supposed to be odd or the case (i) for the sake of simplicity. Note also Remark 4 above. Anyway, the reciprocity law (58.1)(3) or rather (73.14) implies that

$$\Theta = \Theta^*,$$

which is implicit in Legendre's argument indicated in Note [91.2]; with this, (91.14) is to be understood, although in retrospect. However, Gauss proceeded with Θ^* without relying on the law, and it apparently caused him a considerable extra work, i.e., the estimation of $|\mathcal{A}(D)|$ in order to reach the inequality

$$(\dagger) \qquad |\Theta^*(\mathcal{K}(D))| \leq 2^{\tau - 1}$$
$$\text{independently of the reciprocity law,}$$

450   *Quadratic Forms*

([DA, art.261]), which becomes, of course, equivalent to our simple inequality (91.7), only if the law is employed. Further, to establish the surjectivity of Θ^*, i.e., the equality in (†), Gauss proved the principal genus theorem in art.286 and reached the equality, finally in art.287, after a great digression to the theory of ternary quadratic forms. Our proof of the same concerning Θ is facilitated, indeed greatly, by the use of the reciprocity law as well as the prime number theorem (88.1); moreover, we did not need the principal genus theorem to be proved beforehand. It is an idea of Smith (1859/1869, p.207); see also Dirichlet (1871/1894, §125, the ending footnote by Dedekind).

[91.6]   Nevertheless, Gauss' approach via Θ^* to the genus theory has its own merit. In particular, it led him to his 2nd proof [DA, art.262] of the reciprocity law, although it is involved, as it depends on the explicit computation of $|\mathcal{A}(D)|$. However, in §93[C], we shall prove, independently of both the reciprocity law and the notion of ambiguous classes, that

$$(\dagger\dagger) \qquad D \equiv 1 \bmod 4 \;\Rightarrow\; |\mathcal{K}(D)/\mathcal{K}^2(D)| \le 2^{\tau - 1}.$$

This implies (†) for odd D since $|\Theta^*(\mathcal{K}(D))| \le |\mathcal{K}(D)/\mathcal{K}^2(D)|$, as the kernel of Θ^* obviously contains $\mathcal{K}^2(D)$. Anyway, the inequality (†) or rather (††) yields a proof of the law. We reproduce it below with a minor change, following Dedekind (Dirichlet (1871, §154)). Thus, with different odd primes p, q, we proceed as follows, supposing that (58.1) (1) is available:

(a) Let $p \equiv 1 \bmod 4$ and $\left(\frac{p}{q}\right) = 1$. There exists a ξ such that $\xi^2 \equiv p \bmod 4q$. Then $Q_1 = [|q, \xi, (\xi^2 - p)/4q|] \in \mathcal{Q}(p)$. From either (†) or (††) it follows that $|\Theta^*(\mathcal{K}(p))| = 1$. Since $Q_1(1,0) = q$, we obtain $\left(\frac{q}{p}\right) = 1 = \left(\frac{p}{q}\right)$.

(b) Let $p \equiv 1 \bmod 4$ and $\left(\frac{q}{p}\right) = 1$. There exists an η such that $\eta^2 \equiv q^* \bmod 4p$ with $q^* = (-1)^{(q-1)/2} q$. Then, $Q_2 = [|p, \eta, (\eta^2 - q^*)/4p|] \in \mathcal{Q}(q^*)$, and $Q_2(1,0) = p$. Hence, $\left(\frac{p}{q}\right) = 1 = \left(\frac{q}{p}\right)$.

(c) Let $p, q \equiv -1 \bmod 4$ and $p > q$. Take the forms $Q_3 = [|1, -1, (1-pq)/4|]$, $Q_4 = [|(pq-1)/4, 1, -1|] \in \mathcal{Q}(pq)$. On noting $Q_3(1,0) = 1$, $Q_4(1,0) = (pq-1)/4$, we see that

$$\Theta^*(Q_3) = \{+1, +1\}, \quad \Theta^*(Q_4) = \{-1, -1\},$$

since

$$\left(\frac{(pq-1)/4}{p}\right) \cdot \left(\frac{4}{p}\right) = \left(\frac{pq-1}{p}\right) = \left(\frac{-1}{p}\right) = -1,$$

and the same holds with p, q being exchanged. On the other hand, we have $|\Theta^*(\mathcal{K}(pq))| \le 2$, so $\Theta^*(Q)$ with any $Q \in \mathcal{Q}(pq)$ is equal to either $\Theta^*(Q_3)$ or $\Theta^*(Q_4)$. We take $Q_5 = [|(p-q)/4, p, p|]$. Then

$$\left(\left(\frac{-q}{p} \right), \left(\frac{p}{q} \right) \right) = \text{either } (+1, +1) \text{ or } (-1, -1).$$

Collecting (a), (b), (c), we end the proof (58.1) (3). The proof of the supplementary laws can be omitted. After this proof (i.e., his own version of it), Gauss finally started using the reciprocity law in his discussion of quadratic forms; see arts.263–264. It thus transpires that Gauss wanted to include the theory of quadratic residues completely inside the arithmetic theory of quadratic forms. He had a good reason to take such a way, for he certainly longed for a proof of the law which should extend to the analogous situations higher than quadratic. It was the starting point of the long way towards modern algebraic number theory as we indicated already in the above. See also Note [93.4].

[91.7] The case of the fundamental discriminant $D = -231$. This belongs to (i). Thus,

$$D = (-3)(-7)(-11) \equiv 1 \bmod 4; \ \tau = \tau^* = 3;$$
$$|\mathcal{K}| = 12; \ |\mathcal{G}| = |\mathcal{A}| = 2^2; \ |\mathcal{P}| = |\mathcal{K}^2| = |\mathcal{K}|/|\mathcal{A}| = 3.$$

The representative forms of $\mathcal{K}(-231)$ are given in (80.1). Writing $Q_2^+ = Q_2$, etc.,

$$\mathcal{K} = \left\{ (Q_1), (Q_2)^{\pm 1}, (Q_3), (Q_4)^{\pm 1}, (Q_5)^{\pm 1}, (Q_6)^{\pm 1}, (Q_7), (Q_8) \right\};$$
$$\mathcal{P} = \mathcal{K}^2 = \left\{ (Q_1), (Q_4)^2, (Q_4)^4 \right\} = \left\{ (Q_1), (Q_4)^{-1}, (Q_4) \right\};$$
$$\mathcal{A} = \left\{ (Q_1), (Q_3), (Q_7), (Q_8) \right\};$$

on noting $(Q_8) = (Q_8)^{-1}$. The four entries in the third line are inequivalent ambiguous classes, and because of their number there exist no other representatives belonging to $\mathcal{A}(-231)$. As for the second line, we have $(Q_4)^3 = 1$ since with $T = \left(\begin{smallmatrix} 1 & 1 \\ 0 & 1 \end{smallmatrix} \right)$ we have ${}^t T^{-1} Q_4 T^{-1} = [|4, -5, 16|]$, and $[|4, -5, 16|] \circ [|4, -5, 16|] \equiv [|16, -5, 4|] \bmod \Gamma$, which gives $(Q_4)^2 = (Q_4)^{-1}$. Thus, (Q_4) is of order 3. With this, we obtain the elements of $\mathcal{K}^2, \mathcal{P}$ and find that they are the same as (80.6) (1). Further,

$$(Q_3)\mathcal{K}^2 = \left\{ (Q_3), (Q_5)^{-1}, (Q_5) \right\} : \ (80.6) \ (3),$$

since $(Q_3)(Q_4) = ([|12, 3, 5|]) = (Q_5)^{-1}$; $(Q_3)(Q_4)^{-1} = \{(Q_3)(Q_4)\}^{-1} = (Q_5)$. Also, ${}^t(T^2)Q_7 T^2 = [|7, 35, 52|]$ and ${}^t(T^4)Q_4 T^4 = [|4, 35, 91|]$ give $(Q_7)(Q_4) = ([|28, 35, 13|]) = ([|13, -35, 28|]) = ([|13, -9, 6|]) = ([|6, 9, 13|]) = (Q_6)^{-1}$. Further, from $(Q_8) = ([|8, -5, 8|])$ and $(Q_4) = ([|4, -5, 16|])$ follows $(Q_8)(Q_4) = ([|32, -5, 2|]) = ([|2, 5, 32|]) = (Q_2)$. We have found that

$$(Q_7)\mathcal{K}^2 = \left\{ (Q_7), (Q_6)^{-1}, (Q_6) \right\} : \ (80.6) \ (4),$$
$$(Q_8)\mathcal{K}^2 = \left\{ (Q_8), (Q_2)^{-1}, (Q_2) \right\} : \ (80.6) \ (2).$$

454　　　　　　　　　　　　　　　*Quadratic Forms*

Therefore, the classification (80.3)–(80.6) is equivalent to the coset decomposition

$$\mathcal{K} = \mathcal{K}^2 \sqcup (Q_3)\mathcal{K}^2 \sqcup (Q_7)\mathcal{K}^2 \sqcup (Q_8)\mathcal{K}^2$$
$$= \mathcal{P} \sqcup (Q_2)\mathcal{P} \sqcup (Q_5)\mathcal{P} \sqcup (Q_{10})\mathcal{P}.$$

The lower line is due to $(Q_8)\mathcal{K}^2 \ni (Q_2)$, $(Q_3)\mathcal{K}^2 \ni (Q_5)$, and $(Q_7)\mathcal{K}^2 \ni (Q_{10})$, with $Q_{10} = [|10, -3, 6|] \equiv Q_6 \bmod \Gamma$.

[91.8]　Also, the set of the congruence relations (80.6) is an instance of (91.34): We have

$$P = \left\{ m \bmod 231 : \kappa_{-3}(m) = 1, \kappa_{-7}(m) = 1, \kappa_{-11}(m) = 1 \right\}$$
$$= \left\{ n^2 \bmod 231 : \langle n, 231 \rangle = 1 \right\}$$
$$= \left\{ 2^{2s} \bmod 231 : s \bmod 15 \right\}.$$

For in the decomposition $(\mathbb{Z}/231\mathbb{Z})^* = (\mathbb{Z}/3\mathbb{Z})^* \times (\mathbb{Z}/7\mathbb{Z})^* \times (\mathbb{Z}/11\mathbb{Z})^*$, the factors on the right side are generated, respectively, by the primitive roots 2 mod 3, 2 mod 7, 2 mod 11; and the square classes are generated by 2^2 mod 3, 2^2 mod 7, 2^2 mod 11 which form a cyclic group of order $1 \cdot 3 \cdot 5 = 15$. Hence,

$$K = P \sqcup \varrho_1 P \sqcup \varrho_2 P \sqcup \varrho_3 P; \quad \mathcal{G} \cong G \cong \{\pm 1\}^2;$$
$$\varrho_1 = 2, \varrho_2 = 5, \varrho_3 = 10, \kappa_D(\varrho_j) = +1,$$
$$\kappa_{-3}(\varrho_1) = -1, \kappa_{-7}(\varrho_1) = +1, \kappa_{-11}(\varrho_1) = -1,$$
$$\kappa_{-3}(\varrho_2) = -1, \kappa_{-7}(\varrho_2) = -1, \kappa_{-11}(\varrho_2) = +1,$$
$$\kappa_{-3}(\varrho_3) = +1, \kappa_{-7}(\varrho_3) = -1, \kappa_{-11}(\varrho_3) = -1.$$

Note that $Q_2(1,0) = \varrho_1$, $Q_5(1,0) = \varrho_2$, $Q_{10}(1,0) = \varrho_3$. Incidentally, since $K^2 = P$, we have here an analogue of the duplication theorem (91.14), which is rather an exceptional situation, however.

[91.9]　Euler (1778a, p.208) contains a table of *numerorum idoneorum*. The last among the integers depicted there is $d = 1848$. The discriminant $D = -4d = -7392$ belongs to the type (x) above. We have

$$D = 4D_0, \; D_0 = 8E_0, \; E_0 = -231, \; \tau^* = 5 \;\; \Rightarrow \;\; |\mathfrak{g}(D)| = 16.$$

The representative forms $\{C = [|a, b, c|]\}$ of $\mathcal{K}(D)$ are as follows:

$$C_0^{(1)} = [|1, 0, 1848|], \; C_0^{(3)} = [|3, 0, 616|], \quad C_0^{(7)} = [|7, 0, 264|],$$
$$C_0^{(8)} = [|8, 0, 231|], \; C_0^{(11)} = [|11, 0, 168|], \; C_0^{(21)} = [|21, 0, 88|],$$
$$C_0^{(24)} = [|24, 0, 77|], \; C_0^{(33)} = [|33, 0, 56|], \quad C_2 = [|43, 2, 43|],$$
$$C_4 = [|4, 4, 463|], \quad C_8 = [|8, 8, 233|], \quad C_{12} = [|12, 12, 157|],$$
$$C_{24} = [|24, 24, 83|], \quad C_{28} = [|28, 28, 73|], \quad C_{38} = [|47, 38, 47|],$$
$$C_{44} = [|44, 44, 53|].$$

That is, $\mathfrak{h}(D) = 16$. Because the class number and the number of genera coincide, we have the relation $\mathcal{K}(D) \cong \mathcal{G}(D) \cong \{\{\pm 1\}^4\}$. In particular, every form in the above table satisfies $(C)^2 = 1$; namely, $\mathcal{G}(D) \cong \mathcal{A}(D)$, which Theorem 95 implies, and of course readily follows from Theorem 73 together with $(C_2)^2 = 1, (C_{38})^2 = 1$ a consequence of (90.43). To confirm this directly, note that the homomorphism Θ takes the values

$$\Theta(\mathcal{C}) = \{\kappa_{-4}(m), \kappa_8(m), \kappa_{-3}(m), \kappa_{-7}(m), \kappa_{-11}(m)\}, \quad \kappa_D(m) = 1,$$

where $m > 0, \langle m, D \rangle = 1$ is represented properly by a form in the class $\mathcal{C}$. Use those m which are equal to $C_0^{(*)}(1, 1)$ and $C_*(0, 1)$. We get 16 different elements of $\{\{\pm 1\}^5\}$; and $\mathcal{P}(D) = \{(C_0^{(1)})\}$, so the present Θ is indeed an isomorphism.

[91.10] All other d's in Euler's table mentioned above satisfy $\mathfrak{h}(-4d) = \mathfrak{g}(-4d)$ (see [DA, art.303]). Each genus contains only one class, and genera are determined solely by residue classes mod $|D|$; hence, which form represents a given prime $p \nmid 2d$ is determined by p mod $4d$ only in the context of (91.34). The $d = -1848$ of the preceding Note is related to $E_0 = -231$ treated in §82 and in Notes [91.7]–[91.8], but their nature are quite different; because of this fact Euler described his 65 integers as *idonei*. For instance, $p \equiv 1$ mod 7392 corresponds to $\Theta(\mathcal{C}) = \{1\}^5$; all such primes should be represented by the principal form $C_0^{(1)}$ in the preceding Note. An instance of three consecutive primes in this arithmetic progression is as follows:

$$739820929 = 19711^2 + 1848 \cdot 436^2, \quad 739865281 = 26993^2 + 1848 \cdot 78^2,$$
$$739909633 = 24641^2 + 1848 \cdot 268^2,$$

which are equal to $1 + (100084 + \nu) \cdot 7392, \nu = 0, 6, 12$, respectively.

[91.11] If D is such that $\mathcal{K}^2(D) = 1$ as in the case $D = -7392$, then we are able to determine whether (71.2) has a solution or not solely via the residue class m mod $|D|$. Here arises an interesting problem about whether there exist infinitely many d's $(D = -4d)$ which have this property. In the three articles succeeding (1778a) Euler made the relevant investigation, but he did not encounter any d greater than 1848. It is conjectured that Euler's table of idonei numbers should be complete. By the way, according to Weinberger (1973), there exists an absolute constant c_0 such that

$$\text{GRH} \Rightarrow \min \{Z > 0 : \mathcal{C}^Z = 1, \forall \mathcal{C} \in \mathcal{K}(-4d)\} > c_0 \log d / \log \log d,$$

where $-4d < 0$ are fundamental discriminants. Therefore, idonei integers appear to be highly exceptional. His proof relies on Theorem 96 in the next section.

456 *Quadratic Forms*

§92

The aim of the present section is to develop an account of the class number formulas of Dirichlet (1839/1840a; 1863/1894, Fünfter Abschnitt). They are at a beautiful confluence point of number theory and analysis, just in the same way as his prime number theorem (88.1).

We repeat the core part of §74 from a somewhat different point of view. Thus, we let $\{Q\}$ be the representative forms of the classes in $\mathcal{K}(D)$; $|\{Q\}|$ is equal to the class number $\mathfrak{h}(D)$, with the understanding of (90.1). For each representative form Q, we consider the decomposition

$$\mathbb{Z}^2 - {}^{t}\{0,0\} = \bigsqcup_{m \neq 0} N_Q(m),$$
$$N_Q(m) = \{{}^{t}\{n_1, n_2\} : Q(n_1, n_2) = m\}; \tag{92.1}$$

that is, we view the whole of representations by Q, disregarding whether proper or improper; the left side of the first line means that (71.3) is observed. If Q is positive definite, then $m > 0$; on the other hand, if Q is indefinite, then m can be either positive or negative. Also, there are cases where $N_Q(m) = \varnothing$. We have

$$N_Q(m) = \bigsqcup_{{}^{t}\{u_1, u_2\} \in \mathcal{Y}_Q(m)} \{V \cdot {}^{t}\{u_1, u_2\} : V \in \mathrm{Aut}_Q\},$$
$$\mathcal{Y}_Q(m) = \mathrm{Aut}_Q \backslash N_Q(m), \tag{92.2}$$

where ${}^{t}\{u_1, u_2\}$ runs over all representative (seed) vectors. Then, taking into account proper representations, we have

$$\mathcal{Y}_Q(m) = \bigsqcup_{g^2 | m} g \cdot \mathcal{Y}_Q^*(m/g^2),$$
$$\mathcal{Y}_Q^*(m) = \{{}^{t}\{u_1, u_2\} \in \mathcal{Y}_Q(m) : \langle u_1, u_2 \rangle = 1\}, \tag{92.3}$$

where $g > 0$. Since $\mathcal{Y}_Q^*(m) = \mathrm{Aut}_Q \backslash \mathcal{S}_Q(m)$ according to (74.2), we have

$$\mathcal{Y}_Q(m) = \bigsqcup_{g^2 | m} g \cdot \mathrm{Aut}_Q \backslash \mathcal{S}_Q(m/g^2). \tag{92.4}$$

By (74.16)–(74.20), we find that

$$(73.1) \Rightarrow \sum_Q |\mathcal{Y}_Q(m)| = \sum_{g^2 | m} \sum_{\mathcal{C} \in \mathcal{K}(D)} |\mathcal{M}_{m/g^2}(D, \mathcal{C})|$$
$$= \sum_{d|m} \kappa_D(d) \sum_{d=g^2 f} \mu^2(f) = \sum_{d|m} \kappa_D(d), \tag{92.5}$$

$$§92 \qquad\qquad 457$$

in which we have applied the same reasoning as in (78.13). It should be noted that the condition (73.1) imposed in (92.5) means that $m > 0$ is supposed. When dealing with $D > 0$, this restriction causes a minor complexity to be cleared at (92.34).

We then introduce, for any $\mathcal{C} \in \mathcal{K}(D)$, $Q \in \mathcal{C}$, and $\operatorname{Re} s > 1$,

$$
\begin{aligned}
Z(s; \mathcal{C}) &= \sum_{m=1}^{\infty} \sum_{\{u_1, u_2\} \in \mathcal{Y}_Q(m)} \frac{J_D(Q(u_1, u_2))}{Q(u_1, u_2)^s}, \\
&= \sum_{m=1}^{\infty} \frac{J_D(m)}{m^s} \big|\mathcal{Y}_Q(m)\big|,
\end{aligned}
\tag{92.6}
$$

where J_D is the trivial character $\bmod\ |D|$, and its presence indicates that (73.1) is taken into account. Absolute convergence is obvious since (92.5) implies that $|\mathcal{Y}_Q(m)|$ is not greater than the number of divisors of m, that is, the sum of absolute values of each term of $Z(s; \mathcal{C})$ is not greater than $\zeta^2(\sigma)$, $\operatorname{Re} s = \sigma$; see $(16.15)_{k=2}$. Also, $Z(s; \mathcal{C})$ is independent of the choice of the representative form Q because of (74.17) and (92.4).

We shall show that via analytic continuation $Z(s; \mathcal{C})$ has a simple pole at $s = 1$; in a neighborhood of $s = 1$ it holds that

$$
Z(s; \mathcal{C}) = \frac{\mathfrak{z}(D)}{s - 1} + O(1),
\tag{92.7}
$$

with the residue $\mathfrak{z}(D) > 0$ that does not depend on $\mathcal{C}$ but only on D. On the other hand, (92.5) implies that

$$
\begin{aligned}
\sum_{\mathcal{C} \in \mathcal{K}(D)} Z(s; \mathcal{C}) &= \sum_{m=1}^{\infty} \frac{J_D(m)}{m^s} \sum_{d \mid m} \kappa_D(d) \\
&= L(s, J_D)\lfloor \kappa_D \rfloor(s) = \zeta(s)\lfloor \kappa_D \rfloor(s) \prod_{p \mid D} \left(1 - \frac{1}{p^s}\right).
\end{aligned}
\tag{92.8}
$$

Here is a subtle point: the relation (73.5) is not used; that is, instead of the Dirichlet L-function $L(s, \kappa_D)$, merely the Dirichlet series $\lfloor \kappa_D \rfloor(s)$ associated with the arithmetic function κ_D is invoked. We have not appealed to the reciprocity law yet; that will be made much later at (92.49). The aim of this procedure of ours will be made apparent with the discussion [C] expanded in the next section, especially in its CONCLUDING REMARK. Anyway, from the combination of (12.2), (92.7), and (92.8) arises

$$
\text{Dirichlet's idea:} \quad \mathfrak{h}(D) = \frac{\varphi(|D|)}{\mathfrak{z}(D)|D|} \lfloor \kappa_D \rfloor(1),
\tag{92.9}
$$

458   *Quadratic Forms*

where φ is the Euler function. Note that we write $\lfloor \kappa_D \rfloor(1)$ to indicate that $\lim_{s \to 1+0} \lfloor \kappa_D \rfloor(s)$ should exist as $\mathfrak{h}(D)$ is finite.

Thus we shall evaluate the residue $\mathfrak{z}(D)$. We need to consider two cases separately according to the sign of D.

(I) The case $D < 0$:

We note first that (76.9) and the definition (92.2) imply

$$\left| N_Q(m) \right| = \left| \mathrm{Aut}_Q \right| \left| \mathcal{Y}_Q(m) \right| = \eta_D \left| \mathcal{Y}_Q(m) \right|, \quad \eta_D = \begin{cases} 6 & D = -3, \\ 4 & D = -4, \\ 2 & D < -4. \end{cases} \quad (92.10)$$

Hence, (92.6) can be written as

$$\eta_D Z(s; \mathcal{C}) = \sum_{\substack{\{n_1, n_2\} \neq \{0, 0\} \\ n_1, n_2 \in \mathbb{Z}}} \frac{J_D(Q(n_1, n_2))}{Q(n_1, n_2)^s}. \quad (92.11)$$

To analyze this right side, we shall employ a reasoning that is an extension of (61.14)–(61.15). Thus, we introduce the function

$$\vartheta_Q(z; v_1, v_2) = \sum_{n_1, n_2 \in \mathbb{Z}} \exp\left(-\pi z Q(n_1 + v_1, n_2 + v_2) \right), \quad (92.12)$$

where $|\arg z| < \pi/2$ and $v_j \in \mathbb{R}$ are arbitrary; see (61.7). We apply Poisson's sum formula (60.1) to each of the sums over n_j. Arguing in much the same way as in what follows (61.7), we have

$$\begin{aligned} \vartheta_Q(z; v_1, v_2) = \sum_{n_1, n_2 \in \mathbb{Z}} & \exp\left(-2\pi i(n_1 v_1 + n_2 v_2) \right) \\ & \times \iint_{\mathbb{R}^2} \exp\left(-\pi z Q(x, y) + 2\pi i(n_1 x + n_2 y) \right) dx \, dy. \end{aligned} \quad (92.13)$$

To this double integral we apply the orthonormal change of variables that transforms Q into a diagonal matrix; after some computation we obtain the theta transformation formula:

$$\begin{aligned} \vartheta_Q(z; v_1, v_2) = \frac{2}{z|D|^{1/2}} \sum_{n_1, n_2 \in \mathbb{Z}} & \exp\left(-2\pi i(n_1 v_1 + n_2 v_2) \right) \\ & \times \exp\left(-(4\pi/(z|D|)) Q^{(-1)}(n_2, n_1) \right). \end{aligned} \quad (92.14)$$

Here

$$Q = [|a, b, c|] \implies Q^{(-1)} = [|a, -b, c|]; \quad (92.15)$$

thus, $(Q^{(-1)}) = (Q)^{-1}$ by (90.39); note the allocation of the integer variables n_1, n_2 in (92.14).

Before moving to the next step, it should be remarked that we may suppose that

$$\begin{cases} D \equiv 1 \bmod 4 \Rightarrow \langle a, D \rangle = 1, \langle b, D \rangle = 1, \\ D \equiv 0 \bmod 4 \Rightarrow \langle a, D \rangle = 1, \langle b, D \rangle = 2. \end{cases} \tag{92.16}$$

For the restriction $\langle a, D \rangle = 1$ can be imposed because of Note [74.4]; and if $2 \nmid D$, then apply to Q the transformation $b \mapsto b + 2ak \equiv 1 \bmod |D|$; also, if $2 | D$, then on noting $2 | b$ apply the transformation $b \mapsto b + 2ak \equiv 2 \bmod |D|$. With this, we consider

$$\vartheta_Q^*(z) = \sum_{n_1, n_2 \in \mathbb{Z}} J_D\big(Q(n_1, n_2)\big) \exp\big(-\pi z Q(n_1, n_2) \big). \tag{92.17}$$

By (18.2),

$$\vartheta_Q^*(z) = \sum_{g | D} \mu(g) \sum_{\substack{n_1, n_2 \in \mathbb{Z} \\ g | Q(n_1, n_2)}} \exp\big(-\pi z Q(n_1, n_2) \big). \tag{92.18}$$

The inner sum is equal to

$$\begin{aligned} \sum_{\substack{l_1, l_2 \bmod g \\ Q(l_1, l_2) \equiv 0 \bmod g}} \sum_{\substack{n_1 \equiv l_1 \bmod g \\ n_2 \equiv l_2 \bmod g}} & \exp\big(-\pi z Q(n_1, n_2) \big) \\ = \sum_{\substack{l_1, l_2 \bmod g \\ Q(l_1, l_2) \equiv 0 \bmod g}} & \vartheta_Q\big(g^2 z; l_1 / g, l_2 / g \big). \end{aligned} \tag{92.19}$$

By the transformation formula (92.14), the right side becomes

$$\frac{2}{z g^2 |D|^{1/2}} \sum_{n_1, n_2 \in \mathbb{Z}} C_g\big(n_1, n_2; Q\big) \exp\left(-(4\pi / (z g^2 |D|)) \, Q^{(-1)}(n_2, n_1) \right), \tag{92.20}$$

where

$$C_g\big(n_1, n_2; Q\big) = \sum_{\substack{l_1, l_2 \bmod g \\ Q(l_1, l_2) \equiv 0 \bmod g}} \exp\big(-2\pi i (n_1 l_1 + n_2 l_2) / g \big). \tag{92.21}$$

In order to compute this exponential sum, we decompose it via Theorem 31:

$$C_g\big(n_1, n_2; Q\big) = \prod_{p | g} C_p\big(n_1, n_2; Q\big), \quad g | D, \ \mu^2(g) = 1. \tag{92.22}$$

We shall evaluate each $C_p(n_1, n_2; Q)$ explicitly:

(1) $p \geq 3$. On noting (92.16), that is, $p \nmid 2ab$, we apply the change of variables $l_1 \mapsto bl_1$, $l_2 \mapsto 2al_2$ in C_p, and get

$$C_p(n_1, n_2; Q) = \sum_{\substack{l_1, l_2 \bmod p \\ l_1 + l_2 \equiv 0 \bmod p}} \exp\left(-2\pi i(bn_1 l_1 + 2an_2 l_2)/p\right), \qquad (92.23)$$

for

$$Q(bl_1, 2al_2) \equiv 0 \bmod p$$
$$\Leftrightarrow 4aQ(bl_1, 2al_2) = 4(ab)^2(l_1 + l_2)^2 - 4a^2 Dl_2^2 \equiv 0 \bmod p \qquad (92.24)$$
$$\Leftrightarrow l_1 + l_2 \equiv 0 \bmod p.$$

Namely, $bn_1 l_1 + 2an_2 l_2 \equiv (2an_2 - bn_1)l_2$; hence, if $2an_2 - bn_1 \equiv 0 \bmod p$, then $C_p(n_1, n_2; Q) = p$; otherwise $= 0$. Therefore, via the identity $4aQ^{(-1)}(n_2, n_1) = (2an_2 - bn_1)^2 - Dn_1^2$, we find that

$$C_p(n_1, n_2; Q) = \begin{cases} p & Q^{(-1)}(n_2, n_1) \equiv 0 \bmod p, \\ 0 & Q^{(-1)}(n_2, n_1) \not\equiv 0 \bmod p. \end{cases} \qquad (92.25)$$

(2) $p = 2$. Let $D = 4D_1$. If $2 \nmid D_1$, then on noting $(b/2)^2 - ac = D_1$ and the lower line of (92.16), we have $ac \equiv 0 \bmod 2$, that is, $2|c$. Hence, in the sum defining (92.21)$_{g=2}$, we have $2|Q(l_1, l_2) \Leftrightarrow 2|l_1$. This means that if $Q^{(-1)}(n_2, n_1) \equiv 0 \bmod 2$, then $2|n_2$, and $C_2(n_1, n_2; Q)$ is equal to the number of $l_2 \bmod 2$, that is, equal to 2; otherwise, $2 \nmid n_2$, and $C_2(n_1, n_2; Q) = 0$. In other words, (92.25)$_{p=2}$ holds. In the remaining case, we have $2|D_1$, and via $(b/2)^2 - ac \equiv 0 \bmod 2$; and again by the lower line of (92.16) we see that $2 \nmid ac$. Hence, $2|Q(l_1, l_2) \Leftrightarrow 2|(l_1 + l_2)$. Also, $Q^{(-1)}(n_2, n_1) \equiv n_1 + n_2 \bmod 2$. These imply that if $Q^{(-1)}(n_2, n_1) \equiv 0 \bmod 2$, then $C_2(n_1, n_2; Q) = 2$, and otherwise $= 0$. In short, (92.25)$_{p=2}$ holds again.

In this way, we find that

$$C_g(n_1, n_2; Q) = \begin{cases} g & Q^{(-1)}(n_2, n_1) \equiv 0 \bmod g, \\ 0 & Q^{(-1)}(n_2, n_1) \not\equiv 0 \bmod g. \end{cases} \qquad (92.26)$$

Summing up the discussion following (92.17), we see that

$$\vartheta_Q^*(z) = \frac{2}{z|D|^{1/2}} \sum_{g|D} \frac{\mu(g)}{g}$$
$$\times \sum_{\substack{n_1, n_2 \in \mathbb{Z} \\ g|Q^{(-1)}(n_2, n_1)}} \exp\left(-(4\pi/(zg^2|D|))Q^{(-1)}(n_2, n_1)\right), \qquad (92.27)$$

In particular, it holds that with a constant $c > 0$

$$\vartheta_Q^*(z) = 2\varphi(|D|)/(|D|^{3/2}z) + O(\exp(-c/z)), \quad \text{as } z \to +0, \tag{92.28}$$

where the main term corresponds to $\{n_1, n_2\} = \{0, 0\}$.

We return to the function $Z(s; \mathcal{C})$. By the definition (94.21) of the Gamma-function, we have

$$\eta_D Z(s; \mathcal{C}) = \frac{\pi^s}{\Gamma(s)} \int_0^\infty z^{s-1} \vartheta_Q^*(z) dz, \quad \text{Re } s > 1. \tag{92.29}$$

The change of the order of integration and summation is legitimate because of the absolute convergence of (92.6). We divide this integral at $z = 1$; to the part with $0 < z < 1$ we apply the change of variable $z \mapsto 1/z$, and get

$$\eta_D Z(s; \mathcal{C}) = \frac{\pi^s}{\Gamma(s)} \int_1^\infty z^{s-1} \vartheta_Q^*(z) dz + \frac{\pi^s}{\Gamma(s)} \int_1^\infty z^{-s-1} \vartheta_Q^*(1/z) dz. \tag{92.30}$$

The integrand of the first term on the right is $O(\exp(-cz))$ with a constant $c > 0$, for in (92.17) we have $j_D(0) = 0$. As for the second integral, we apply the transformation formula (92.27) and the asymptotic formula (92.28). Hence, we find that in (92.7)

$$D < 0 \implies \mathfrak{z}(D) = 2\pi \varphi(|D|)/(\eta_D |D|^{3/2}). \tag{92.31}$$

Therefore, (92.9) yields the following assertion due to Dirichlet (1839, p.361):

Theorem 96 *For any discriminant $D < 0$, we have*

$$\mathfrak{h}(D) = \frac{\eta_D}{2\pi} |D|^{1/2} \lfloor \kappa_D \rfloor (1). \tag{92.32}$$

(II) The case $D > 0$:

Since Aut_Q is an infinite group, we have to alter essentially the argument after (92.10). We take a form $Q = [|a, b, c|]$ representing a class $\mathcal{C} \in \mathcal{K}(D)$. According to Note [74.4], we may suppose that $a > 0$, $\langle a, D \rangle = 1$, which of course does not affect $Z(s, \mathcal{C})$ itself. Despite the difference in the sign of the discriminant, we may suppose also that (92.16) is still fulfilled. With this, by the decomposition (92.1)–(92.2) we have

462   *Quadratic Forms*

$$\left\{ 2an_1 + (b + \sqrt{D})n_2 : {}^t\{n_1, n_2\} \in \mathbb{Z}^2 - {}^t\{0, 0\} \right\}$$

$$= \bigsqcup_{m \neq 0} \left\{ 2an_1 + (b + \sqrt{D})n_2 : {}^t\{n_1, n_2\} \in N_Q(m) \right\}$$

$$= \bigsqcup_{m \neq 0} \bigsqcup_{{}^t\{u_1, u_2\} \in \mathcal{Y}_Q(m)} \left\{ 2an_1 + (b + \sqrt{D})n_2 : \right.$$

$$\left. {}^t\{n_1, n_2\} = V^t\{u_1, u_2\}, V \in \mathrm{Aut}_Q \right\}$$

$$= \bigsqcup_{m \neq 0} \bigsqcup_{{}^t\{u_1, u_2\} \in \mathcal{Y}_Q(m)} \bigsqcup_{\pm} \left\{ \pm \epsilon_D^\ell (2au_1 + (b + \sqrt{D})u_2) : \ell \in \mathbb{Z} \right\}.$$

$$\tag{92.33}$$

In the last line, Note [76.3] and (85.16) are invoked. It should also be remarked that the relation (72.25) is exploited.

We then decompose (92.6) as follows: For Re $s > 1$,

$$Z(s; \mathcal{C}) = \tfrac{1}{2}\Big(Z(s; \mathcal{C}; 0) + Z(s; \mathcal{C}; 1) \Big),$$

$$Z(s; \mathcal{C}; \delta) = \sum_{m \neq 0} \sum_{{}^t\{u_1, u_2\} \in \mathcal{Y}_Q(m)} \frac{J_D(Q(u_1, u_2))}{|Q(u_1, u_2)|^s} \left(\frac{Q(u_1, u_2)}{|Q(u_1, u_2)|} \right)^\delta. \tag{92.34}$$

The factor $(Q(u_1, u_2)/|Q(u_1, u_2)|)^\delta$, $\delta = 0, 1$, is inserted to sift out the terms $Q(u, v) < 0$, which together with the factor $J_D(Q(u_1, u_2))$ makes the condition (73.1) fulfilled.

We are going to construct analogues of ϑ_Q and ϑ_Q^* which are defined by (92.12) and (92.17), respectively. Naturally, it is of no sense to adopt the same in our new situation; a genuinely new procedure is required here. Hecke (1917a) resolved this difficulty by means of a fine idea: We have, for arbitrary real numbers $\omega_1, \omega_2 \neq 0$ and Re $s > 0$,

$$|\omega_1 \omega_2|^{-s} = \frac{\pi^s}{\Gamma^2(s/2)} \int_0^\infty \int_0^\infty (xy)^{s/2-1} \exp\left(-\pi(\omega_1^2 x + \omega_2^2 y) \right) dx dy. \tag{92.35}$$

Applying the change of variables $x \mapsto ze^\tau$, $y \mapsto ze^{-\tau}$,

$$|\omega_1 \omega_2|^{-s} = \frac{2\pi^s}{\Gamma^2(s/2)} \int_{-\infty}^\infty \int_0^\infty z^{s-1} \exp\left(-\pi z(\omega_1^2 e^\tau + \omega_2^2 e^{-\tau}) \right) dz d\tau. \tag{92.36}$$

Dissecting the outer integral along with $\mathbb{R} = \bigsqcup_{\ell \in \mathbb{Z}} \left[\log \epsilon_D^{(2\ell-1)}, \log \epsilon_D^{(2\ell+1)} \right)$,

$$|\omega_1 \omega_2|^{-s} = \frac{\pi^s}{\Gamma^2(s/2)} \sum_{\pm} \int_{-\log \epsilon_D}^{\log \epsilon_D} \int_0^\infty z^{s-1}$$

$$\times \sum_{\ell \in \mathbb{Z}} \exp\left(-\pi z((\pm \epsilon_D^\ell \omega_1)^2 e^\tau + (\pm \epsilon_D^{-\ell} \omega_2)^2 e^{-\tau}) \right) dz d\tau. \tag{92.37}$$

We modify this by the changes $s \mapsto s + \delta$, $\operatorname{Re} s > 1$, and $\omega_1 \mapsto 2au_1 + (b + \sqrt{D})u_2$, $\omega_2 \mapsto 4am/\omega_1$ with $'\{u_1, u_2\} \in \mathcal{Y}_Q(m)$; then multiply the resulting identity by the factor $j_D(m)m^\delta$, and sum over $m \neq 0$. Following, in the opposite way, the decomposition process (92.33), we get the expression

$$Z(s; \mathcal{C}; \delta) = \frac{(4a\pi)^{s+\delta}}{\Gamma^2((s+\delta)/2)} \int_{-\log \epsilon_D}^{\log \epsilon_D} \int_0^\infty z^{s+\delta-1} \vartheta_Q^*(z, \tau; \delta) dz d\tau. \quad (92.38)$$

Here

$$\vartheta_Q^*(z, \tau; \delta) = \sum_{n_1, n_2 \in \mathbb{Z}} \frac{j_D(Q(n_1, n_2))(Q(n_1, n_2))^\delta}{\exp\left(\pi z H_{Q, \tau}(n_1, n_2)\right)}, \quad (92.39)$$

$$H_{Q, \tau} = {}^t W_Q \begin{pmatrix} e^\tau & 0 \\ 0 & e^{-\tau} \end{pmatrix} W_Q, \quad W_Q = \begin{pmatrix} 2a & b + D^{1/2} \\ 2a & b - D^{1/2} \end{pmatrix}. \quad (92.40)$$

The form $H_{Q, \tau}$ (with coefficients in $\mathbb{R}$) is positive definite; for W_Q, see (85.31).

In the same way as in (92.18)–(92.19), we have

$$\vartheta_Q^*(z, \tau; \delta) = \sum_{g \mid D} \mu(g) g^{2\delta} \sum_{\substack{l_1, l_2 \bmod g \\ Q(l_1, l_2) \equiv 0 \bmod g}} \vartheta_Q(g^2 z, \tau; l_1/g, l_2/g; \delta),$$

$$\vartheta_Q(z, \tau; v_1, v_2; \delta) = \sum_{n_1, n_2 \in \mathbb{Z}} \frac{(Q(n_1 + v_1, n_2 + v_2))^\delta}{\exp\left(\pi z H_{Q, \tau}(n_1 + v_1, n_2 + v_2)\right)}. \quad (92.41)$$

By Poisson's sum formula (60.1),

$$\vartheta_Q(z, \tau; v_1, v_2; \delta) = \sum_{n_1, n_2 \in \mathbb{Z}} \exp\left(-2\pi i(n_1 v_1 + n_2 v_2)\right)$$

$$\times \iint_{\mathbb{R}^2} (Q(x, y))^\delta \exp\left(-\pi z H_{Q, \tau}(x, y) + 2\pi i(n_1 x + n_2 y)\right) dx dy. \quad (92.42)$$

Applying the change of variables $W_Q \cdot {}^t\{x, y\} = {}^t\{X, Y\}$ this double integral becomes

$$\frac{(4a)^{-1-\delta}}{D^{1/2}} \iint_{\mathbb{R}^2} (XY)^\delta \frac{\exp(2\pi i(n_{1*} X + n_{2*} Y))}{\exp\left(\pi z(e^\tau X^2 + e^{-\tau} Y^2)\right)} dX dY$$

$$= \frac{(4a)^{-1-\delta}}{z D^{1/2}} \left(-\frac{n_{1*} \cdot n_{2*}}{z^2}\right)^\delta \exp\left(-(\pi/z)(e^{-\tau} n_{1*}^2 + e^\tau n_{2*}^2)\right) \quad (92.43)$$

$$= \frac{(4a)^{-1-\delta}}{z D^{1/2}} \left(\frac{Q^{(-1)}(n_2, n_1)}{4a z^2 D}\right)^\delta \exp\left(-(\pi/(z(4a)^2 D))H_{Q^{(-1)}, -\tau}(n_2, n_1)\right).$$

464 *Quadratic Forms*

This ought to be explained: We have $4aQ(x,y) = XY$ and $\{n_1, n_2\} \cdot W_Q^{-1} = \{n_{1*}, n_{2*}\}$, i.e.,

$$n_{1*} = (2an_2 + (-b + \sqrt{D})n_1)/4a\sqrt{D},$$
$$n_{2*} = -(2an_2 + (-b - \sqrt{D})n_1)/4a\sqrt{D}. \tag{92.44}$$

Thus, $n_{1*} \cdot n_{2*} = -Q^{(-1)}(n_2, n_1)/4aD$, and

$$e^{-\tau} n_{1*}^2 + e^{\tau} n_{2*}^2 = H_{Q^{(-1)}, -\tau}(n_2, n_1)/(4a\sqrt{D})^2, \tag{92.45}$$

adopting the notations (92.15) and (92.40). Following the argument leading to (92.28), we get the transformation formula

$$\vartheta_Q^*(z, \tau; \delta) = \frac{1}{(4azD^{1/2})^{1+2\delta}} \sum_{g|D} \frac{\mu(g)}{g^{1-2\delta}} \sum_{\substack{n_1, n_2 \in \mathbb{Z} \\ g|Q^{(-1)}(n_2, n_1)}} (Q^{(-1)}(n_2, n_1))^{\delta}$$
$$\times \exp\left(- (\pi/(z(4ag)^2 D))H_{Q^{(-1)}, -\tau}(n_2, n_1)\right). \tag{92.46}$$

It remains to discuss the inner integral of (92.38), which is analogous to (92.30); but only the case $\delta = 0$ matters, as $Q^{(-1)}(0,0) = 0$ in (92.46). In this way we find, via (92.34), that in (92.7)

$$D > 0 \implies \mathfrak{z}(D) = \varphi(D)D^{-3/2} \log \epsilon_D. \tag{92.47}$$

Therefore, (92.9) yields the following assertion due to Dirichlet (1839, p.365):

Theorem 97 *For any discriminant $D > 0$, we have*

$$\mathfrak{h}(D) = \frac{1}{\log \epsilon_D} D^{1/2} \lfloor \kappa_D \rfloor (1). \tag{92.48}$$

Next, we shall transform, into a closed form, the factor $\lfloor \kappa_D \rfloor (1)$ which are in (92.32) and (92.48); closed, in the sense that $\lfloor \kappa_D \rfloor (1)$ will be given a form which can be computed in a finite number of steps, at least in principle. Thus, we note first that by (73.5) and Theorem 72 we have

$$\lfloor \kappa_D \rfloor (1) = L(1, \kappa_{D_0}) \prod_{p|R} \left(1 - \frac{\kappa_{D_0}(p)}{p}\right); \tag{92.49}$$

that is, we have just appealed to the reciprocity law. Then, we may assume, without loss of generality, that

$$D \text{ is a fundamental discriminant.} \tag{92.50}$$

Thus κ_D is a primitive character mod $|D|$ (Theorem 71). Hence, by Theorem 51, §55,

$$\kappa_D(n) = \frac{G_D}{|D|} \sum_{d \bmod |D|} \kappa_D(d) e(-dn/|D|),$$

$$G_D = \sum_{l \bmod |D|} \kappa_D(l) e(l/|D|).$$

(92.51)

We have

$$L(1,\kappa_D) = \frac{G_D}{|D|} \sum_{d=1}^{|D|} \kappa_D(d) \sum_{n=1}^{\infty} \frac{e(-dn/|D|)}{n}$$

$$= -\frac{G_D}{|D|} \sum_{d=1}^{|D|} \kappa_D(d) \{ \log(\sin(\pi d/|D|)) - i\pi d/|D| \},$$

(92.52)

since

$$\sum_{n=1}^{\infty} \frac{e(-n\theta)}{n} = -\log\left(1 - e(-\theta)\right)$$

$$= -\log(2 \sin \pi \theta) + i\pi \left(\theta - \tfrac{1}{2}\right), \quad 0 < \theta < 1.$$

(92.53)

We need to compute the Gauss sum G_D; the result is

$$G_D = \sqrt{D} = \begin{cases} \sqrt{D} & D > 0, \\ i\sqrt{|D|} & D < 0. \end{cases}$$

(92.54)

To confirm this, we note that the product D of two coprime fundamental discriminants D_1, D_2 is again a fundamental discriminant because of Theorem 67. In this situation, we have, by Note [55.4],

$$G_{D_1 D_2} = \kappa_{D_1}(|D_2|)\kappa_{D_2}(|D_1|)G_{D_1}G_{D_2}.$$

(92.55)

Hence, if

$$\kappa_{D_1}(|D_2|)\kappa_{D_2}(|D_1|) = \begin{cases} -1 & D_1, D_2 < 0, \\ 1 & \text{otherwise,} \end{cases}$$

(92.56)

then by induction we shall have $G_{D_1 D_2} = \sqrt{D_1 D_2}$ in the sense (92.54); it suffices to confirm (92.55) with $D = p^*$, which for $p > 2$ follows from (59.1), (61.1), (61.3), and for $p = 2$ we have, by a direct computation, $G_{-4} = 2i$, $G_8 = 2\sqrt{2}$, $G_{-8} = 2i\sqrt{2}$. Thus we shall verify (92.56). According to (73.7),

$$2 \nmid D_1 D_2 \implies \kappa_{D_1}(|D_2|)\kappa_{D_2}(|D_1|) = \left(\frac{|D_2|}{|D_1|}\right)\left(\frac{|D_1|}{|D_2|}\right).$$

(92.57)

466   *Quadratic Forms*

The reciprocity law $(62.2)\,(3)$ yields (92.56). On the other hand, by (73.8)–(73.9),

$$2 \nmid D_1,\ D_2 = 2^b f,\ 2 \nmid f \ \Rightarrow$$

$$\kappa_{D_1}(|D_2|)\kappa_{D_2}(|D_1|) = \left(\frac{|f|}{|D_1|}\right)\left(\frac{|D_1|}{|f|}\right)(-1)^{(f-1)(|D_1|-1)/4} \tag{92.58}$$

$$= (-1)^{(f+|f|-2)(|D_1|-1)/4},$$

which gives (92.56) again. We end the confirmation of (92.54).

We now obtain the renowned class number formulas of Dirichlet (1839; 1840a, pp.151–152; 1863/1894, §104):

Theorem 98 *Expressing a discriminant D as $D = D_0 R^2$ with D_0 a fundamental discriminant, we have*

(1) $D < 0$:

$$\mathfrak{h}(D) = \frac{\eta_D \phi(R; \kappa_{D_0})}{2(2 - \kappa_{D_0}(2))} \sum_{r \le |D_0|/2} \kappa_{D_0}(r). \tag{92.59}$$

(2) $D > 0$:

$$\mathfrak{h}(D) = -2\frac{\phi(R; \kappa_{D_0})}{\log \epsilon_D} \sum_{r \le D_0/2} \kappa_{D_0}(r) \log(\sin(\pi r/D_0)). \tag{92.60}$$

where

$$\phi(R; \chi) = R \prod_{p|R}\left(1 - \frac{\chi(p)}{p}\right). \tag{92.61}$$

Proof Summing up (92.32), (92.48), (92.49), (92.52), and (92.54), we find that

$$D < 0 \ \Rightarrow\ \mathfrak{h}(D) = -\frac{\eta_D \phi(R; \kappa_{D_0})}{2|D_0|} \sum_{d=1}^{|D_0|-1} \kappa_{D_0}(d)d, \tag{92.62}$$

and

$$D > 0 \ \Rightarrow\ \mathfrak{h}(D) = -\frac{\phi(R; \kappa_{D_0})}{\log \epsilon_D} \sum_{d=1}^{D_0-1} \kappa_{D_0}(d) \log(\sin(\pi d/D_0)). \tag{92.63}$$

To verify (92.59), we consider first the case where D_0 is odd. In the sum on the right side of (92.62), we separate the terms according to the parity of d. Then the sum is equal to

$$\sum_{k \le |D_0|/2} \{\kappa_{D_0}(2k)(2k) + \kappa_{D_0}(|D_0| - 2k)(|D_0| - 2k)\}. \tag{92.64}$$

Also, the symmetry with respect to $k \mapsto |D_0| - k$ gives that the sum is equal to

$$\sum_{k \leq |D_0|/2} \left\{ \kappa_{D_0}(k)k + \kappa_{D_0}(|D_0| - k)(|D_0| - k) \right\}. \tag{92.65}$$

Subtracting the $2\kappa_{D_0}(2)$-times of the latter from the former, we find that the original sum is equal to

$$-\frac{\kappa_{D_0}(-2)|D_0|}{1 - 2\kappa_{D_0}(2)} \sum_{k \leq |D_0|/2} \kappa_{D_0}(k). \tag{92.66}$$

On noting that $\kappa_{D_0}(-2) = -\kappa_{D_0}(2)$ by (73.7), we end the proof. On the other hand, when D_0 is even, apply (73.20). As for positive D's, note that $\sin(\pi d/D_0) = \sin(\pi(D_0 - d)/D_0)$ and $\kappa_{D_0}(D_0 - d) = \kappa_{D_0}(d)$. This ends the proof.

Notes

[92.1]   The fusion of number theory and analysis. The eulogy for Dirichlet pronounced by Kummer (1861) with deep respect to the deceased and with passion for number theory contains the following passage (p.18): Descartes opened a new discipline by applying analysis to geometry ... The method of Dirichlet *auf alle Probleme der Zahlentheorie gleichmässig erstreckten*. Exactly in this new stream of mathematics the contribution of Riemann (1860) to the theory of the distribution of prime numbers was achieved. Continued in Chapter 5. See also Note [93.4].

[92.2]   Gauss appears to have had essentially the same result as Dirichlet's class number formulas (for even discriminants, of course); see his posthumous article (1863c: written in 1834/1837) and Dedekind's comment attached to it. Some people thus attribute the formulas to Dirichlet and Gauss. However, the view of Smith (1859/1869, p.208) on this matter seems to be quite reasonable.

[92.3]   The argument which Dirichlet (1839; 1863/1894, Fünfter Abschnitt) employed in his own proof of his class number formulas is geometrical; he computed asymptotically the number of ordinary lattice points inside a plane figure expanding along with m appearing in (92.1). For $D < 0$, the figure is ellipse, and for $D > 0$ it is defined by a part of a hyperbola which is indicated in the later part of Note [92.5] below. He computed the same number in a different way using the arithmetical fact (74.20); a comparison of resulting asymptotic formulas yields his class number formulas. His argument is adopted in most number theory literature. We have proceeded differently in order to make explicit the connection of the class number formulas and automorphic forms; the use of the theta-series was, in fact, suggested already by Dirichlet

468   *Quadratic Forms*

(1840a, pp.8–10), but naturally for even $D < 0$ only. In this context, it should be stressed that there exists a fascinating theory stemmed from the constant term of the expansion (92.7); see Note [93.4] below. The discussion related to the principal character j_D usually follows Dirichlet's original treatment (1840a, §5); see also Landau (1927, Vierter Teil, Kapitel 5). Our treatment is more explicit, as the problem is transformed into the evaluation of an exponential sum.

[92.4]   The assertion (92.59)–(92.61), which is nowadays commonly attributed to Dirichlet, is in fact a generalization of the original. He (1840a) discussed in detail only the case $D = 4f$, $\mu^2(f) = 1$; it led him, for instance, to the discovery of the marvelously simple formula:

$$\mathfrak{h}(-4\ell) = \sum_{r<\ell/2} \left(\frac{r}{\ell}\right) \text{ for positive } \ell \equiv -1 \bmod 4, \ \mu^2(\ell) = 1.$$

This is included in (92.59); note that here we need to invoke (73.8) and (76.9).

[92.5]   The idea (92.35)–(92.37) of Hecke that connects indefinite forms with positive definite forms is simple and decisive. It extends readily to indefinite quadratic forms of many variables; especially, together with the unit theorem of Dirichlet (1846; 1871, §177; 1894, §183) which is an extension of Theorem 88, it led Hecke to a proof (1917b) of the functional equation for the Dedekind zeta-function of any algebraic number field, a genuine generalization of (12.8), and is related to the prime ideal theorem for general algebraic number fields (but see Note [95.6] below). However, it should be noted also that Hecke's idea can be dispensed with if one is concerned solely with the proof of the class number formula (92.48), as described in Landau (1932, Vierter Teil), for instance: Obviously we may restrict ourselves to the situation where $Q = [|a,b,c|] \in \mathcal{Q}(D)$, $D > 0$, is reduced so that (82.22) is satisfied. Let $Q(u_0, v_0) = m > 0$ be a representation. We may assume that $u_0, v_0 > 0$; if $u_0 v_0 < 0$, then the form is to be replaced by $[|a, -b, c|]$. With this, let $\lambda^{\pm}(x,y) = 2ax + (b \pm \sqrt{D})y$, and let $\epsilon_D^{\pm v} \lambda^{\pm}(u_0, v_0) = \lambda^{\pm}(u_v, v_v)$ with signs corresponding, so we are following the splitting argument (92.33). Then there exists a v such that

$$\frac{\lambda^+}{\lambda^-}(u_v, v_v) = \epsilon_D^{2v} \frac{\lambda^+}{\lambda^-}(u_0, v_0) \text{ lies in the interval } [1, \epsilon_D^2],$$

Since we have $u_v, v_v > 0$ under the current condition on Q, we are led to the assertion that in order to find seed solutions to $Q(x,y) = m$, it suffices to look into the lattice points $\{x,y\}$ which are on the arc of the hyperbola $\lambda^+(x,y)\lambda^-(x,y) = 4am$ defined by

$$\lambda^-(x,y) \leq \lambda^+(x,y) \leq \epsilon_D^2 \lambda^-(x,y), \quad x,y > 0.$$

Thus $2(am)^{1/2} \le \lambda^+(x,y) \le 2\epsilon_D(am)^{1/2}$. Although finite, the number of the points to be examined via this inequality can be huge since ϵ_D can be so (see Remark 3, §85). Lagrange's method developed in §87 is certainly superior as far as the purpose is concerned to explicitly fix all the seed-solutions to $Q(x,y) = m$.

[92.6] The formula (92.59) implies that

$$D_0 < 0 \quad \Rightarrow \quad \frac{\mathfrak{h}(D)}{\mathfrak{h}(D_0)} = \frac{\eta_D}{\eta_{D_0}} \phi(R; \kappa_{D_0}).$$

Gauss [DA, arts.253–256] proved essentially the same; but he used his analysis of the structure of $\mathcal{K}(D)$ based on his composition theory. At the same place (Werke I, p.283), it was remarked that any similar relation between class numbers of indefinite forms was unknown. However, the formula (92.60) yields

$$D_0 > 0 \quad \Rightarrow \quad \frac{\mathfrak{h}(D)}{\mathfrak{h}(D_0)} = \frac{1}{A} \phi(R; \kappa_{D_0}).$$

Here $\epsilon_D = \epsilon_{D_0}^A$, $A | \varphi(R; \kappa_{D_0})$; the existence of this exponent A can be proved in just the same way as (84.29) above. Thus, it can be said that Dirichlet (1840a; 1863/1894, §100) solved Gauss' problem concerning indefinite forms by an analytic argument. Later Lipschitz (1857) found a genuinely algebraic proof of these relations between class numbers. In his argument, Note [5.3] above played an important rôle. See Smith (1859/1869, art.113). As a numerical example, $\mathfrak{h}(-231) = 12$, $\mathfrak{h}(-23100) = 12 \cdot 10(1 - 1/2)(1 - 1/5) = 48$. Also, by Note [83.5], $\mathfrak{h}(377) = 4$ and $\epsilon_{377} = 233 + 12\sqrt{377}$. Since $\epsilon_{37700} = 50596649 + 2605860\sqrt{377} = \epsilon_{377}^3$, we get $\mathfrak{h}(37700) = (4/3) \cdot 10(1 - 1/2)(1 + 1/5) = 8$.

[92.7] Knowing approximately the structure of $\mathcal{K}(D)$ can yield a decomposition of D. This idea is due to Šimerka (1858); see Crandal and Pomerance (2005, pp.248–251). Having a set of relatively small primes p_j which are represented by forms in $\mathcal{Q}(D)$, we generate the set of classes (90.44). If the set $\{(M_{p_j, \xi_j})\}$ is close to a generator set of $\mathcal{K}(D)$, then we may hit an ambiguous class with a high probability; and we shall be led to a decomposition of D because of (75.2). By the way, as a matter of fact, almost nothing is available about the structure of the group $\mathcal{K}(D)$ if the prime power decomposition of D is unknown. Even if that is known, it is still highly problematic for one to go beyond the genus theory. In other words, the theory of quadratic forms has a lot yet to be discovered.

[92.8] Gauss [DA, art.303] stated that it is plausible that $\mathfrak{h}(D)$ tends to infinity as $D \to -\infty$. This has been solved in an effective way in Goldfeld's

470   *Quadratic Forms*

epoch-making work (1976) together with the later construction by B. Gross and D.B. Zagier of an elliptic curve that Goldfeld had needed to complete his theory: Thus, for every $\varepsilon > 0$ there exists an explicitly computable constant $c_\varepsilon > 0$ such that for every $D < 0$

$$\mathfrak{h}(D) > c_\varepsilon (\log |D|)^{1-\varepsilon}.$$

For the fascinating history of Gauss' class number problem see the readable exposition (1985) due to Goldfeld himself. Gauss [DA, art.304] stated also that there would exist infinitely many $D > 0$ such that each genus of $\mathcal{K}(D)$ contains only one class; thus, probably $\mathfrak{h}(p) = 1$ for infinitely many $p \equiv 1 \bmod 4$. This conjecture is still open. The main difficulty lies in that the formula (92.60) contains $\log \epsilon_D$ in a way which makes hard to separate it from other factors, as Gauss already indicated vaguely. See Takhtajan–Vinogradov (1982).

[92.9]   This is a continuation of Note [35.4]. It is known that for $p \equiv -1 \bmod 4, p \geq 7$,

$$((p-1)/2)! \equiv (-1)^{(1+\mathfrak{h}(-p))/2} \bmod p.$$

That $\mathfrak{h}(-p)$ is odd can be shown by (91.30) or rather by (92.59), as it implies that

$$\mathfrak{h}(-p) = \frac{1}{2 - \left(\frac{2}{p}\right)} \sum_{r \leq (p-1)/2} \left(\frac{r}{p}\right).$$

For instance, $\mathfrak{h}(-23) = 3$ and $11! = 1 + 1735513 \cdot 23$; also $23! \equiv -1 \bmod 47$, for $\mathfrak{h}(-47) = 5$. See Mordell (1961). By the way, that $\mathfrak{h}((-1)^{(p-1)/2}p)$ is odd can be proved by a combination of Note [85.2] and (††) of Note [91.6].

§93

In this final section of the present chapter, we shall extend Dirichlet's analytic method, which has been developed in the preceding section, into three directions: [A] a prime number theorem for primitive forms, [B] an alternative derivation of the genus number, and [C] another proof of the principal genus theorem. It should be stressed that the quadratic reciprocity law is applied in [A] and [B], but [C] is independent of it. The latter fact yields a significant consequence as stated in CONCLUDING REMARK given at the end of [C].

[A]   We shall establish the prime number theorem due to Dirichlet (1840b) and Weber (1882):

Theorem 99 *Any primitive form represents infinitely many prime numbers.*

Proof Let $\lambda \in \widehat{\mathcal{K}}(D)$, and let $\lambda(Q) = \lambda((Q))$ for any form $Q \in \mathcal{Q}(D)$. We then introduce the following analogue of the Dirichlet L-function for $\mathcal{Q}(D)$:

$$\mathcal{L}(s, \lambda) = \sum_{\mathcal{C} \in \mathcal{K}(D)} \lambda(\mathcal{C})Z(s; \mathcal{C}), \tag{93.1}$$

with $Z(s; \mathcal{C})$ as in (92.6). We have, by (92.7),

$$\mathcal{L}(s, \lambda) = \frac{\mathfrak{z}(D)}{s-1}\mathfrak{h}(D)\delta_\lambda + O(1), \quad \delta_\lambda = \begin{cases} 1 & \lambda \equiv 1, \\ 0 & \lambda \not\equiv 1, \end{cases} \tag{93.2}$$

as $s \to 1 + 0$. On the other hand, we have, for $\operatorname{Re} s > 1$,

$$\mathcal{L}(s, \lambda) = \sum_{m=1}^{\infty} \frac{J_D(m)}{m^s} \sum_{g^2 | m} X_\lambda(m/g^2), \quad X_\lambda(m) = \sum_{\xi \bmod 2m} \lambda(M_{m, \xi}), \tag{93.3}$$

with $M_{m, \xi}$ as in (72.5). Although this is obvious, we shall anyway confirm it: On noting (74.16)–(74.20) as well as (92.3), we have, with any $Q \in \mathcal{C}$,

$$\sum_{g^2 | m} X_\lambda(m/g^2) = \sum_{\mathcal{C} \in \mathcal{K}(D)} \lambda(\mathcal{C}) \sum_{g^2 | m} |\mathcal{M}_{m/g^2}(D, \mathcal{C})|$$

$$= \sum_{\mathcal{C} \in \mathcal{K}(D)} \lambda(\mathcal{C}) \sum_{g^2 | m} |\operatorname{Aut}_Q \backslash S_Q(m/g^2)| \tag{93.4}$$

$$= \sum_{\mathcal{C} \in \mathcal{K}(D)} \lambda(\mathcal{C}) \sum_{g^2 | m} |\mathcal{Y}_Q^*(m/g^2)| = \sum_{\mathcal{C} \in \mathcal{K}(D)} \lambda(\mathcal{C})|\mathcal{Y}_Q(m)|.$$

Then by (92.6) we get (93.3). Hence,

$$\mathcal{L}(s, \lambda) = L(2s, J_D)\mathcal{L}^*(s, \lambda), \quad \mathcal{L}^*(s, \lambda) = \sum_{m=1}^{\infty} \frac{J_D(m)}{m^s}X_\lambda(m). \tag{93.5}$$

The arithmetic function X_λ is multiplicative. For if $\langle m_1, m_2 \rangle = 1$, then $M_{m_1 m_2, \xi} = M_{m_1, \xi_1} \circ M_{m_2, \xi_2}$ with $\xi_j \equiv \xi \bmod 2m_j$; and $X_\lambda(m_1 m_2) = X_\lambda(m_1)X_\lambda(m_2)$. Thus, we are led to the Euler product,

$$\mathcal{L}^*(s, \lambda) = \prod_{\kappa_D(p)=1} \left(\sum_{l=0}^{\infty} \frac{1}{p^{ls}}X_\lambda(p^l) \right), \quad X_\lambda(1) = 1. \tag{93.6}$$

Here the restriction $\kappa_D(p) = 1$ is due to Lagrange's principle (72.2). Then, by (90.43) and (90.46),

$$X_\lambda(p^l) = \lambda^l(\mathbf{p}) + \lambda^{-l}(\mathbf{p}), \quad \mathbf{p} = (M_{p, \eta}), \tag{93.7}$$

472   *Quadratic Forms*

where η varies with p of course. A computation gives

$$\mathcal{L}^*(s,\lambda) = \prod_{\kappa_D(p)=1} \left(1 - \frac{\lambda(\mathbf{p})}{p^s}\right)^{-1} \left(1 - \frac{\overline{\lambda}(\mathbf{p})}{p^s}\right)^{-1} \left(1 - \frac{1}{p^{2s}}\right). \qquad (93.8)$$

By means of this Euler product expansion, we shall show that for any given class $\mathcal{C} \in \mathcal{K}(D)$

$$\left|\{p : \mathbf{p} = \mathcal{C}\}\right| = \infty, \qquad (93.9)$$

which amounts to the statement of the present theorem. To this end, we generalize the reasoning (88.2)–(88.6) in an obvious way; so it suffices to confirm that

$$\lambda \neq 1 \;\Rightarrow\; \mathcal{L}^*(1,\lambda) \text{ is bounded and does not vanish.} \qquad (93.10)$$

If λ is complex, then the argument of Note [88.2] extends immediately to the present situation; an analogous observation on the Euler product for $\prod_\lambda \mathcal{L}^*(s,\lambda)$ works well, as $\mathbf{p}^l = 1$ if $\mathfrak{h}(D)|l$ because of a basic fact in the theory of finite groups. On the other hand, if $\lambda \neq 1$ is a real, i.e., genus character $[\kappa_{D_1}]$ on $\mathcal{K}(D)$ (see (91.39) for this notation), then in view of (93.8) we need to show that the value at $s = 1$ of the function

$$\prod_{\kappa_D(p)=1} \left(1 - \frac{\kappa_{D_1}(p)}{p^s}\right)^{-2}, \quad D_1 \neq 1, D_0, \qquad (93.11)$$

is finite and does not vanish; this function is the same as $\mathcal{L}(s, [\kappa_{D_1}])$ save for a factor which is inessential for our present purpose. Further, (93.11) is the same as the following product of Dirichlet L-functions, again save for an inessential factor: Applying (73.5),

$$L(s,\kappa_{D_1})L(s,\kappa_{D_0}\kappa_{D_1}) = \prod_{p} \left(1 - \frac{\kappa_{D_1}(p)}{p^s}\right)^{-1} \left(1 - \frac{\kappa_{D_0}\kappa_{D_1}(p)}{p^s}\right)^{-1}, \qquad (93.12)$$

where p runs over all primes, for those p with $\kappa_{D_0}(p) = -1$ contributes to (93.12) by a factor positive for $s > \frac{1}{2}$; this is of course in accord to (91.37). Here we observe that $\kappa_{D_0}\kappa_{D_1} \neq 1$ as a character. Indeed, if the primitive Dirichlet characters $\kappa_{D_0}, \kappa_{D_1}$ are identical, then we would have $D_0 = D_1$ in view of Theorem 48 (2), §53, violating our assumption made in (93.11). Therefore, $\kappa_{D_0}\kappa_{D_1}$ is a non-trivial Kronecker symbol, and we end the proof by means of (92.9). In the above, (91.37) or the genus character theorem was required in an essential way, i.e., we needed to identify all real elements in $\widehat{\mathcal{K}}(D)$. Continued in Note [93.4].

§93 473

[B] We shall give an alternative proof of the basic theorem (91.30) on the genus numbers of quadratic forms, following Dirichlet's discussion (1839, §§3, 6): We consider the relation

$$\sum_{\kappa \in W} \mathcal{L}(s, [\kappa]) = \sum_{\mathcal{C} \in \mathcal{K}(D)} Z(s; \mathcal{C}) \prod_{\substack{\kappa \in W \\ \text{prime character}}} (1 + [\kappa](\mathcal{C}))$$

$$= 2^{\tau^*} \sum_{\mathcal{C} \in \mathcal{P}(D)} Z(s; \mathcal{C}) = \frac{1}{s-1} 2^{\tau^*} |\mathcal{P}(D)| \mathfrak{z}(D) + O(1), \tag{93.13}$$

as $s \to 1 + 0$. The set W is as in (91.37), and the notation $[\kappa]$ is as before. On noting that the principal genus $\mathcal{P}(D)$ is the collection of $\mathcal{C} \in \mathcal{K}(D)$ such that $\kappa(\mathcal{C}) = 1$ for each prime character $\kappa \in W$, we see that the product in the first line is equal to 2^{τ^*} if $\mathcal{C} \in \mathcal{P}(D)$, and $= 0$ otherwise. The second line depends on (92.7). It should be noted that here we do not need Theorem 95 but only the definition of W. On the other hand, the sum over all $\kappa \in W$ is divided into two parts

$$\sum_{\kappa = 1, \, \kappa_D} \mathcal{L}(s, [\kappa]) + \sum_{\kappa \neq 1, \, \kappa_D} \mathcal{L}(s, [\kappa]). \tag{93.14}$$

The first sum is $2\mathfrak{z}(D)\mathfrak{h}(D)/(s-1) + O(1)$ because $[\kappa_D] = 1$ on $\mathcal{K}(D)$, and the second sum is bounded as has been proved above. Hence, we obtain $2^{\tau^*} |\mathcal{P}(D)| = 2\mathfrak{h}(D) = 2|\mathcal{K}(D)|$, that is, $|\mathcal{K}(D)|/|\mathcal{P}(D)| = 2^{\tau^*-1}$, i.e., (91.30).

[C] Next, we shall give an analytic proof of the principal genus theorem, reworking Kronecker (1865), although he actually considered the estimation of the number of ambiguous classes in a way different from Gauss'. We shall, however, restrict ourselves to the case

$$D \equiv 1 \bmod 4, \tag{93.15}$$

for the sake of simplicity. Note that Kronecker treated even discriminants; he stayed in the Gauss–Dirichlet theory. The discussion will proceed in 16 steps.

1. Following Kronecker (1865, p.297), we begin with an observation which may appear trivial outwardly: Let P be the product of all primes less than a sufficiently large limit so that $D|P^\infty$; and let $\mathfrak{s}$ be the characteristic function of the set of all squares in $\mathbb{N}$. Then, it holds that

$$\sum_{m=1}^{\infty} J_P(m)\mathfrak{s}(m) \frac{r_D(m)}{m^{s/2}} = \sum_{n=1}^{\infty} J_P(n) \frac{r_D(n)}{n^s}, \quad s > 1,$$

$$r_D(m) = \sum_{f|m} \mu^2(f)\kappa_D(f), \tag{93.16}$$

since $r_D(n^2) = r_D(n)$. The sum over n is, as $s \to 1 + 0$,

$$\frac{1}{\zeta(2s)} \prod_{p|P} \left(1 - \frac{1}{p^{2s}}\right)^{-1} L(s, {}_{J}P) \lfloor {}_{J}P\kappa_D \rfloor (s) = \frac{\mathfrak{h}(D)\mathfrak{z}P(D)}{s-1} + O(1),$$

$$\mathfrak{z}P(D) = \frac{6}{\pi^2} \prod_{p|P} \left(1 - \frac{1}{p^2}\right)^{-1} \left(1 - \frac{{}_{J}D(p)}{p}\right) \left(1 - \frac{\kappa_D(p)}{p}\right) \cdot \mathfrak{z}(D) \tag{93.17}$$

by (92.9).

2. However, if the relation $r_D(n^2) = r_D(n)$ is not taken into account, then the analysis of the sum over m in (93.16) becomes nontrivial and interesting: First of all, the presence of the factor $\mathfrak{s}(m)$ means that we are dealing exclusively with the classes in $\mathcal{K}(D)$ which represent squares; see (74.20). Recalling (91.18) then, these classes are in $\mathcal{K}^2(D)$. In other words,

$$\exists m \text{ such that } {}_{J}D(m)\mathfrak{s}(m)|\mathrm{Aut}_Q \backslash S_Q(m)| \neq 0 \implies (Q) \in \mathcal{K}^2(D). \tag{93.18}$$

Conversely, let $\mathcal{C} = ([|a, b, c|])^2$, $\langle a, 2D \rangle = 1$. We may suppose that not merely $b^2 \equiv D \bmod 4|a|$ but $(b + 2ak)^2 \equiv D \bmod 4a^2$ holds with a k since this is equivalent to $bk \equiv -c \bmod |a|$; note that $\langle a, b \rangle = 1$, as $\langle a, D \rangle = 1$. The change b by $b + 2ak$ does not alter the class $([|a, b, c|])$. Thus, with the new b, we have $b^2 \equiv D \bmod 4a^2$, and $\mathcal{C} = ([|a^2, b, (b^2 - D)/4a^2|])$ by (90.41). So, there exists an index set $\{A > 0\}$ such that

$$\mathcal{K}^2(D) = \{([|A^2, B, C|])\},$$

$$|\{A\}| = |\mathcal{K}^2(D)|, \ \langle A, 2D \rangle = 1, \ |A| \neq 1. \tag{93.19}$$

The last condition is due to Notes [74.4]–[74.5]; that is, the form $[|a, b, c|]$ can properly represent a sufficiently large integer which is coprime to a given integer. Hence, to analyze the sum over m of (93.16), we may confine ourselves to the forms listed in (93.19) or to the proper representations of a square with each representative Q_0:

$$Q_0(X, Y) = [|A^2, B, C|](X, Y) = n^2, \ \langle X, Y \rangle = 1, \ {}_{J}P(n) = 1. \tag{93.20}$$

We shall, in step 4 below, apply a classification to ${}^!\{X, Y\} \in S_{Q_0}(n^2)$. To this end, we need to make P more precise: we suppose that P is such that

$$AD|P^{\infty} \tag{93.21}$$

for any A chosen at (93.19). In particular, ${}_{J}P(r) = 1$ implies $\langle 2AD, r \rangle = 1$ for each A. The introduction of the character ${}_{J}P$ is made with an essential aim, which will become apparent in the course of discussion. In addition, it should be noted that n is always assumed to be positive, and such that ${}_{J}P(n) = 1$.

$$\S 93 \qquad\qquad 475$$

3. Then, with each ${}^t\{X, Y\} \in \mathcal{S}_{Q_0}(n^2)$ we analyze the equation

$$DY^2 = (2A^2X + BY + 2An)(2A^2X + BY - 2An)$$
$$= L_+(X, Y)L_-(X, Y), \tag{93.22}$$

say; Kronecker (1865, p.299). Let

$$h = \langle L_+(X, Y), L_-(X, Y) \rangle. \tag{93.23}$$

Thus, $h|(4An)$; so $h = 2^\nu A_0 n_0$ with $\nu \le 2$ and $A_0|A$, $n_0|n$ since $2 \nmid An$ and $\langle A, n \rangle = 1$ because of $_Jp(n) = 1$. Also, we have $h|Y$, as $h^2|DY^2$ and $\langle D, h \rangle = 1$ (D is odd: (93.15)). Then, $n_0|Y$ implies that $n_0|2A^2X \Rightarrow n_0|X$, and $n_0 = 1$ since $\langle X, Y \rangle = 1$; further, $A_0|Y$, and the equation in (93.20) implies $A_0|n^2$, so $A_0 = 1$. Hence, $h = 2^\nu$, and there should exist σ, θ, and $\lambda|D$ such that

$$Y = 2^\nu \sigma \theta, \quad \langle \lambda \sigma^2, D\theta^2/\lambda \rangle = 1, \ \sigma > 0, \tag{93.24}$$

with which the following splitting of (93.22) holds:

$$L_+(X, Y) = 2^\nu \lambda \sigma^2, \ L_-(X, Y) = 2^\nu D\theta^2/\lambda. \tag{93.25}$$

It should be noted here that the specification of sgn σ in (93.24) is to make the pair $\{\sigma, \theta\}$ well-defined. Then, multiply both of the last two equations by λ and add. Using the expression of Y given in (93.24), we get

$$2^{2-\nu}\lambda A^2X = (\lambda\sigma - B\theta)^2 + (D - B^2)\theta^2. \tag{93.26}$$

It follows that

$$\nu \text{ is equal to either 0 or 2.} \tag{93.27}$$

For, if $\nu = 1$, then Y is even; and λA^2X is odd because of (93.15), $2 \nmid A$ (see (93.19)), and $\langle X, Y \rangle = 1$. Besides, $4|(D - B^2)$. Thus, $2\|(\lambda\sigma - B\theta)^2$, which is impossible.

4. We now consider the following classification of ${}^t\{X, Y\} \in \mathcal{S}_{Q_0}(n^2)$:

$$\mathcal{S}_{Q_0}(n^2) = \bigsqcup_{\substack{d|D \\ \langle d, D/d \rangle = 1}} \mathcal{S}_{Q_0}^{(d)}(n^2), \ d > 0, \tag{93.28}$$

where

$$\mathcal{S}_{Q_0}^{(d)}(n^2) = \{{}^t\{X, Y\} \in \mathcal{S}_{Q_0}(n^2) : \langle L_+(X, Y), D \rangle = d\}. \tag{93.29}$$

We have taken into account that (93.24) implies $\langle 2^\nu \lambda \sigma^2, D \rangle = |\lambda|\langle \sigma^2, D/\lambda \rangle = |\lambda|$ as well as $\langle \lambda, D/\lambda \rangle = 1$. In $\mathcal{S}_{Q_0}^{(d)}(n^2)$, we have, by (93.25), the following two (i.e., either $(+)$ or $(-)$) possibilities

476 *Quadratic Forms*

$$(+) \begin{cases} L_+(X,Y) = 2^v d\sigma^2, \\ L_-(X,Y) = 2^v D\theta^2/d, \end{cases} \qquad (-) \begin{cases} L_+(X,Y) = -2^v d\sigma^2, \\ L_-(X,Y) = -2^v D\theta^2/d. \end{cases} \tag{93.30}$$

With each ${}^t\{X,Y\} \in S_{Q_0}^{(d)}(n^2)$, only one of $(+)$ and $(-)$ is possible, for otherwise $L_+(X,Y) = 0$, and $Y = 0$ by (93.22), so $|X| = 1$ as $\langle X,Y \rangle = 1$; then, $n^2 = Q_0(X,0) = A^2$, which contradicts $_{JP}(n) = 1$, as $|A| \neq 1$. Hence,

$$\begin{aligned} S_{Q_0}^{(d)}(n^2) &= S_{Q_0}^{(d,+)}(n^2) \sqcup S_{Q_0}^{(d,-)}(n^2), \\ S_{Q_0}^{(d,\pm)}(n^2) &= \{{}^t\{X,Y\} : (93.30)_\pm\}, \end{aligned} \tag{93.31}$$

where the signs correspond.

5. In $S_{Q_0}^{(d,+)}(n^2)$, if $v = 0$, then we have

$$4dAn = d(L_+(X,Y) - L_-(X,Y)) = (d\sigma)^2 - D\theta^2. \tag{93.32}$$

By (93.26) with $v = 0$ and $\lambda = d$, we have $d\sigma - B\theta \equiv 0 \bmod 2A$; and we apply the replacement $\{\sigma,\theta\} \mapsto \{\xi,\theta\}$:

$$\sigma = 2A\xi + \delta B\theta, \quad \delta d \equiv 1 \bmod 2A, \tag{93.33}$$

with an integer ξ, which is possible, for $\langle d, 2A \rangle = 1$. The relation (93.32) becomes equivalent to

$$\begin{aligned} v = 0 &\Rightarrow n = Q_d(\xi,\theta), \\ Q_d = [|dA, \delta dB, C_d|], \quad C_d &= ((\delta dB)^2 - D)/4dA; \end{aligned} \tag{93.34}$$

Kronecker (1865, p.300). Observe that Q_d comes solely from the pair $\{Q_0, d\}$. Here $Q_d \in \mathcal{Q}(D)$ because C_d is obviously an integer and $p|\langle dA, \delta dB, C_d \rangle \Rightarrow p|\langle dA, n \rangle$ contradicting $_{JP}(n) = 1$. Also, the class (Q_d) depends on the residue class $\delta \bmod 2A$. Moreover, $\langle \xi, \theta \rangle = 1$ since $\langle \sigma, \theta \rangle = 1$ by (93.24). Namely, ${}^t\{\xi,\theta\} \in S_{Q_d}(n)$. On the other hand, for the case $v = 2$, we have, instead of (93.32),

$$dAn = \tfrac{1}{4}d(L_+(X,Y) - L_-(X,Y)) = (d\sigma)^2 - D\theta^2. \tag{93.35}$$

This time (93.26) with $v = 2$ and $\lambda = d$ implies $d\sigma - B\theta \equiv 0 \bmod A$, so

$$\sigma = A\xi + \delta B\theta, \tag{93.36}$$

with the same $\delta \bmod 2A$ as in (93.33). Corresponding to (93.34), we get

$$v = 2 \Rightarrow n = Q_d(\xi, 2\theta), \tag{93.37}$$

with the same form Q_d. Here $\langle \xi, 2\theta \rangle = 1$. For, otherwise $2|\xi$, as $\langle \xi, \theta \rangle = 1$. Then, $2|n$, which contradicts $_{JP}(n) = 1$. Hence ${}^t\{\xi, 2\theta\} \in S_{Q_d}(n)$.

$$\S93 \qquad\qquad 477$$

6. Further, if ${}^{t}\{X,Y\} \in \mathcal{S}_{Q_0}^{(d,-)}(n^2)$, then $d\sigma + B\theta \equiv 0 \bmod 2^{(2-\nu)/2}A$ via $(93.26)_{\lambda=-d}$. We apply the replacement

$$\sigma = -2^{(2-\nu)/2}A\xi - \delta B\theta \qquad (93.38)$$

with the same δ as in (93.33). Then, we get a proper representation of $-n$ by the same form Q_d:

$$-n = Q_d(\xi, 2^{\nu/2}\theta). \qquad (93.39)$$

7. Thus, we have divided $\mathcal{S}_{Q_0}(n^2)$ as in (93.28) and (93.31); and the following assignment has been made:

$$\begin{aligned} \mathcal{S}_{Q_0}^{(d,\pm)}(n^2) \ni {}^{t}\{X,Y\} &\mapsto {}^{t}\{x,y\} \in \mathcal{S}_{Q_d}(\pm n), \\ {}^{t}\{x,y\} &= {}^{t}\{\xi, 2^{\nu/2}\theta\}, \ \nu = 0, 2. \end{aligned} \qquad (93.40)$$

Here the signs correspond, and ${}^{t}\{x,y\}$ is derived from ${}^{t}\{X,Y\}$ via the pair $\{(93.33),(93.36)\}$ and (93.38), respectively. The above procedure makes ${}^{t}\{x,y\}$ a function of ${}^{t}\{X,Y\}$; i.e., the mapping (93.40) is well-defined.

8. We ought to look into (93.40) taking into account the action of Aut_{Q_0} since in (93.16) we deal not with individual elements of $\mathcal{S}_{Q_0}(n^2)$ but with equivalent classes $\bmod \mathrm{Aut}_{Q_0}$; see (74.17) and (74.20). Then we shall consider the following three cases separately:

(i) $D < 0$,

(ii) $D > 0$, $\mathrm{pell}_D(-4)$ has no solution, $\qquad (93.41)$

(iii) $D > 0$, $\mathrm{pell}_D(-4)$ has a solution.

In the present and the next three steps we shall consider the case (i), which is the simplest among the three. Nevertheless, our treatment of it is a prototype to discuss the other two cases. Thus, our discussion on (i) will be elaborate much more than actually needed; in particular, we shall proceed, through (93.50), without taking (76.9) into account: First of all, we note that $\mathcal{S}_{Q_d}(-n) = \varnothing$ since Q_d is positive definite on (i); hence, $\mathcal{S}_{Q_0}^{(d,-)}(n^2) = \varnothing$ by (93.40), or $\mathcal{S}_{Q_0}^{(d)}(n^2) = \mathcal{S}_{Q_0}^{(d,+)}(n^2)$ in (93.31). Then, let ${}^{t}\{X_j, Y_j\} \in \mathcal{S}_{Q_0}^{(d)}(n^2)$, $j = 1, 2$, be equivalent $\bmod \mathrm{Aut}_{Q_0}$, and let ${}^{t}\{x_j, y_j\} \in \mathcal{S}_{Q_d}(n)$, respectively, be as in $(93.40)_+$. We note that $\langle Y_j, n \rangle = 1$; for, if $p | \langle Y_j, n \rangle$, then $p | \langle AX_j, n \rangle$, so $p | \langle X_j, n \rangle$ as $_{JP}(n) = 1$, contradicting $\langle X_j, Y_j \rangle = 1$. Analogously one may show that $\langle y_j, n \rangle = 1$. Hence, by $(93.40)_+$ with $Y_j = 2^{\nu}\sigma_j\theta_j$, i.e., (93.24), we have, either by (93.33) or (93.36),

$$2A^2 X_j Y_j^{-1} + B \equiv d\sigma_j\theta_j^{-1} \equiv 2dA x_j y_j^{-1} + \delta dB \bmod n, \qquad (93.42)$$

478   *Quadratic Forms*

regardless of the value of v. By Note [76.4], $X_1 Y_1^{-1} \equiv X_2 Y_2^{-1} \mod n^2$, so $x_1 y_1^{-1} \equiv x_2 y_2^{-1} \mod n$; hence, $^{\mathrm{t}}\{x_1, y_1\} \equiv {}^{\mathrm{t}}\{x_2, y_2\} \mod \mathrm{Aut}_{Q_d}$. That is, (93.40) with $D < 0$ should be applied via its consequence:

$$\mathrm{Aut}_{Q_0} \backslash \mathcal{S}_{Q_0}^{(d)}(n^2) \to \mathrm{Aut}_{Q_d} \backslash \mathcal{S}_{Q_d}(n). \tag{93.43}$$

9. However, the last diagram is in fact misleading. It is tacitly assumed that any element of Aut_{Q_0} could participate on the left side. Namely, we have not taken into account the possibility that some elements of Aut_{Q_0} may move $\mathcal{S}_{Q_0}^{(d)}(n^2)$ not into itself. In other words, a subgroup of Aut_{Q_0} whose elements do not move $\mathcal{S}_{Q_0}^{(d)}(n^2)$ as a whole has to be identified. To this end, we utilize the matrix representation of quadratic forms (§72): Thus, $(93.40)_+$ implies that

$$\mathcal{S}_{Q_0}^{(d)}(n^2) \ni {}^{\mathrm{t}}\{X, Y\} \Rightarrow$$
$$d\big(A^2 X\mathfrak{e} + \tfrac{1}{2} Y(B\mathfrak{e} + \mathfrak{d})\big) = \big(dAx\mathfrak{e} + \tfrac{1}{2} y(\delta dB\mathfrak{e} + \mathfrak{d})\big)^2, \quad \mathfrak{d}^2 = D\mathfrak{e}; \tag{93.44}$$

see Note [93.3]. Indeed, if $v = 0$, then we have, via $(93.26)_{\lambda=d}$ and (93.33),

$$4d(A^2 X\mathfrak{e} + \tfrac{1}{2} Y(B\mathfrak{e} + \mathfrak{d})) = (d\sigma)^2 \mathfrak{e} + (\theta\mathfrak{d})^2 + 2d\sigma\theta\mathfrak{d}$$
$$= \big(d\sigma\mathfrak{e} + \theta\mathfrak{d}\big)^2 = 4\big(dA\xi\mathfrak{e} + \tfrac{1}{2}\theta(\delta dB\mathfrak{e} + \mathfrak{d})\big)^2. \tag{93.45}$$

The case $v = 2$ is analogous; use (93.36) instead of (93.33). Then, suppose that $^{\mathrm{t}}\{X_j, Y_j\} \in \mathcal{S}_{Q_0}^{(d)}(n^2), j = 1, 2$, are mapped, respectively, to $^{\mathrm{t}}\{x_j, y_j\} \in \mathcal{S}_{Q_d}(n)$, $j = 1, 2$, which are equivalent $\mod \mathrm{Aut}_{Q_d}$. The combination of (93.44) and Note [76.4] implies that there exists $\{t, u\}$ such that $t^2 - Du^2 = 4$, and

$$A^2 X_1 \mathfrak{e} + \tfrac{1}{2} Y_1(B\mathfrak{e} + \mathfrak{d}) = \big(\tfrac{1}{2}(t\mathfrak{e} + u\mathfrak{d})\big)^2 \cdot \big(A^2 X_2 \mathfrak{e} + \tfrac{1}{2} Y_2(B\mathfrak{e} + \mathfrak{d})\big). \tag{93.46}$$

That is, $^{\mathrm{t}}\{X_j, Y_j\}, j = 1, 2$, ought to be equivalent $\mod \mathrm{Aut}_{Q_0}^2$, where $\mathrm{Aut}_{Q_0}^2 = \{V^2 : V \in \mathrm{Aut}_{Q_0}\}$. Conversely, the action of $\mathrm{Aut}_{Q_0}^2$ does not affect the classification (93.28): The equation (93.46) implies, after a rearrangement, that

$$L_+(X_1, Y_1) = L_+(X_2, Y_2) + \tfrac{1}{2}\big(L_+(X_2, Y_2)u + Y_2 t - 2Anu\big)uD; \tag{93.47}$$

note that $L_+(X_2, Y_2)u$ and $Y_2 t$ are of the same parity. Hence, the value of $\langle L_+(X, Y), D \rangle$ is stable against the action of $\mathrm{Aut}_{Q_0}^2$, or

$$V^2 \cdot \mathcal{S}_{Q_0}^{(d)}(n^2) = \mathcal{S}_{Q_0}^{(d)}(n^2), \quad \forall d | D, \ \forall V \in \mathrm{Aut}_{Q_0}. \tag{93.48}$$

Namely, Aut_{Q_0} in (93.43) should be replaced by $\mathrm{Aut}_{Q_0}^2$.

10. We have reached a crucial assertion: If $D < 0$, then the map (93.40) induces

$$\mathrm{Aut}_{Q_0}^2 \backslash \mathcal{S}_{Q_0}^{(d)}(n^2) \to \mathrm{Aut}_{Q_d} \backslash \mathcal{S}_{Q_d}(n) \quad \text{(injective)}. \tag{93.49}$$

The desired amendment of (93.43) has been done. Here, whether (93.49) is surjective or not is immaterial; what is essential for our purpose is that (93.46) implies that the map (93.49) is injective: Namely, via (93.28), we are led to the inequality

$$\Lambda_D |\mathrm{Aut}_{Q_0}\backslash\mathcal{S}_{Q_0}(n^2)| \leq \sum_{\substack{d|D \\ \langle d,D/d\rangle=1}} |\mathrm{Aut}_{Q_d}\backslash\mathcal{S}_{Q_d}(n)|, \tag{93.50}$$

where $\Lambda_D = |\mathrm{Aut}_{Q_0}/\mathrm{Aut}_{Q_0}^2|$. We now take (76.9) into account and observe that

$$\Lambda_D = 2, \tag{93.51}$$

and, for $s > 1$,

$$2\sum_{m=1}^{\infty} \frac{J_P(m)\mathfrak{s}(m)}{m^{s/2}} |\mathrm{Aut}_{Q_0}\backslash\mathcal{S}_{Q_0}(m)|$$

$$\leq \sum_{\substack{d|D \\ \langle d,D/d\rangle=1}} \sum_{n=1}^{\infty} \frac{J_P(n)}{n^s} |\mathrm{Aut}_{Q_d}\backslash\mathcal{S}_{Q_d}(n)|. \tag{93.52}$$

This sum over n is equal to

$$\frac{1}{\eta_D} \sum_{\substack{n_1,n_2 \\ \langle n_1,n_2\rangle=1}} \frac{J_P(Q_d(n_1,n_2))}{(Q_d(n_1,n_2))^s}$$

$$= \frac{1}{\eta_D \zeta(2s)} \prod_{p|P} \left(1 - \frac{1}{p^{2s}}\right)^{-1} \sum_{\{n_1,n_2\}\neq\{0,0\}} \frac{J_P(Q_d(n_1,n_2))}{(Q_d(n_1,n_2))^s}, \tag{93.53}$$

where we have applied (18.2), and η_D is as in (92.10).

11. We then adopt the computation starting at (92.11); it should be sufficient to explain only salient points: First the form Q_d needs to be replaced by an equivalent form $Q = [|a,b,c|]$ such that $\langle ab,P\rangle = 1$; if necessary, replace b by $b + 2ak \equiv 1 \bmod P/2$, and note that b is odd because of (93.15). The replacement of J_D by J_P requires an additional discussion. To get (92.25), we computed the trigonometric sum $C_p(n_1,n_2;Q)$. However, it was actually enough for us to have (92.25) with $\{n_1,n_2\} = \{0,0\}$, as the derivation of (92.28) indicates well. Thus, we shall show the following partial extension of (92.25): dropping temporarily the present assumption $D < 0$, we have, for any $p|P$,

$$C_p(0,0;Q) = \begin{cases} p, \\ 2p-1, \quad \text{if } \kappa_D(p) = \begin{cases} 0, \\ 1, \\ -1, \end{cases} \tag{93.54} \\ 1, \end{cases}$$

480   *Quadratic Forms*

respectively. The case $\kappa_D(p) = 0$ is obviously the same as the first line of (92.25) with $\{n_1, n_2\} = \{0,0\}$. We consider the remaining two cases:

(1) $p \geq 3$. As in (92.24), we have $Q(bl_1, 2al_2) \equiv 0 \Leftrightarrow b^2(l_1 + l_2)^2 \equiv Dl_2^2$ mod p. If $\kappa_D(p) = 1$, then we take an ω such that $\omega^2 \equiv D$ mod p, and find that

$$C_p(0,0;Q) + 1 = \left|\{\{l_1^+, l_2^+\} \text{ mod } p : b(l_1^+ + l_2^+) \equiv \omega l_2^+ \text{ mod } p\}\right|$$
$$+ \left|\{\{l_1^-, l_2^-\} \text{ mod } p : b(l_1^- + l_2^-) \equiv -\omega l_2^- \text{ mod } p\}\right|, \tag{93.55}$$

where the $+1$ on the left side stands for the correction of double counting of the pair $\{0,0\}$ on the right side; the joint of these sets contains only this pair. Since both of the sets on the right side contain p pairs, we get the second assertion of (93.54). On the other hand, if $\kappa_D(p) = -1$, then there exists no pair $\{l_1, l_2\}$ other than $\{0,0\}$ to count; hence, we get the third assertion of (93.54).

(2) $p = 2$. If $\kappa_D(2) = 1$, then $D = b^2 - 4ac \equiv 1$ mod 8, and $2 \nmid ab$, so $2|c$, i.e., $Q(l_1, l_2) \equiv l_1^2 + l_1 l_2$ mod 2. We have

$$C_2(0,0;Q) = \left|\{\{l_1, l_2\} \text{ mod } 2 : l_1(l_1 + l_2) \equiv 0 \text{ mod } 2\}\right|, \tag{93.56}$$

which implies the second assertion of $(93.54)_{p=2}$. On the other hand, if $\kappa_D(2) = -1$, then $D = b^2 - 4ac \equiv 5$ mod 8, so $2 \nmid abc$. We have

$$C_2(0,0;Q) = \left|\{\{l_1, l_2\} \text{ mod } 2 : l_1^2 + l_1 l_2 + l_2^2 \equiv 0 \text{ mod } 2\}\right|, \tag{93.57}$$

which implies the third assertion of $(93.54)_{p=2}$.

Hence, we get, regardless of the sign of D,

$$\sum_{g|P} \frac{\mu(g)}{g^2} C_g(0,0;Q) = \frac{\varphi(|D|)}{|D|} \prod_{p|P} \left(1 - \frac{J_D(p)}{p}\right)\left(1 - \frac{\kappa_D(p)}{p}\right). \tag{93.58}$$

12. Returning to the situation $D < 0$, we find that

$$\lim_{s \to 1+0} (s-1) \times (93.53) = \mathfrak{z}_P(D), \tag{93.59}$$

in which $\mathfrak{z}_P(D)$ is the same as (93.17) but with $\mathfrak{z}(D)$ defined by (92.31). We sum both sides of (93.52) over all representative forms listed in (93.19). The sum on the left side becomes the same as the right side of (93.16) because of (74.20); indeed, taking the sum over Q_0 inside, we get the same as the sum over all forms representing the classes in $\mathcal{K}(D)$, precisely due to the presence of the factor $\mathfrak{s}(m)$ as asserted by (93.18). Hence, from (93.17) and (93.59), we find that

$$2\mathfrak{h}(D) \leq \sum_{Q_0} \sum_{\substack{d|D \\ \langle d, D/d \rangle = 1}} 1 = |\mathcal{K}^2(D)| \cdot 2^\tau, \tag{93.60}$$

where τ is as in (91.4). That is, $|\mathcal{K}(D)/\mathcal{K}^2(D)| \leq 2^{\tau-1}$. However, obviously $\mathcal{K}^2(D) \subseteq \mathcal{P}(D)$, and we have already shown at (91.13) or rather in Part [B] above that $|\mathcal{K}(D)/\mathcal{P}(D)| = 2^{\tau-1}$, under the present premise (93.15). Therefore, we conclude that $|\mathcal{K}(D)/\mathcal{K}^2(D)| = 2^{\tau-1}$, namely $\mathcal{P}(D) = \mathcal{K}^2(D)$. This ends an alternative proof of the principal genus theorem (91.14) in the case of negative odd discriminants.

13. Next, we shall deal with the case (93.41) (ii) in the present and the next two steps. Thus, it holds that

$$\mathrm{Aut}^*_{Q_d} = \varnothing, \quad \forall\, d|D, \tag{93.61}$$

with the definition (76.13). We shall follow the argument of the last four steps while considering how to adjust the details to (93.41) (ii). This time, we shall appeal to Notes [76.4]–[76.5]: Let ${}^{\mathfrak{t}}\{X_1, Y_1\} \in \mathcal{S}^{(d,+)}_{Q_0}(n^2)$ and ${}^{\mathfrak{t}}\{X_2, Y_2\} \in \mathcal{S}^{(d,-)}_{Q_0}(n^2)$ be equivalent $\mathrm{mod}\,\mathrm{Aut}_{Q_0}$, and let ${}^{\mathfrak{t}}\{x_j, y_j\} \in \mathcal{S}_{Q_d}(\pm n)$ be their respective images via the map $(93.40)_\pm$, with the signs corresponding. We have $\langle Y_j y_j, n \rangle = 1$ as before. In place of (93.42), it holds that

$$2A^2 X_j Y_j^{-1} + B \equiv (-1)^{j-1} d\sigma_j \theta_j^{-1} \equiv 2dAx_j y_j^{-1} + \delta dB \bmod n, \tag{93.62}$$

regardless of the value of v; the minus-sign case, i.e., $j = 2$, is due to (93.38). We have presently $X_1 Y_1^{-1} \equiv X_2 Y_2^{-1} \bmod n^2$; so $x_1 y_1^{-1} \equiv x_2 y_2^{-1} \bmod n$, and ${}^{\mathfrak{t}}\{x_j, y_j\}$, $j = 1, 2$, have to be equivalent $\bmod \mathrm{Aut}^*_{Q_d}$ by Note [76.5]. However, such a possibility is negated by (93.61). Then, with an obvious reasoning, we conclude that (93.40) should be applied in the following way: provided (93.41) (ii),

$$\mathrm{Aut}_{Q_0}\backslash\mathcal{S}^{(d,\pm)}_{Q_0}(n^2) \to \mathrm{Aut}_{Q_d}\backslash\mathcal{S}_{Q_d}(\pm n) \tag{93.63}$$
$$\text{with signs corresponding.}$$

This is an analog of (93.43) and has to be amended as before.

14. We need to find a subgroup of Aut_{Q_0} whose elements do not move each of $\mathcal{S}^{(d,\pm)}_{Q_0}(n^2)$ as a whole. Thus, we note that as an analogue of (93.44) it holds that

$$\mathcal{S}^{(d,\pm)}_{Q_0}(n^2) \ni {}^{\mathfrak{t}}\{X, Y\} \Rightarrow$$
$$\pm d\big(A^2 X\mathfrak{e} + \tfrac{1}{2}Y(B\mathfrak{e} + \mathfrak{d})\big) = \big(dAx\mathfrak{e} + \tfrac{1}{2}y(\delta dB\mathfrak{e} + \mathfrak{d})\big)^2, \tag{93.64}$$

where (93.39) is taken into account, and the signs correspond; see Note [93.3]. With this, assume that ${}^{\mathfrak{t}}\{X_j^\pm, Y_j^\pm\} \in \mathcal{S}^{(d,\pm)}_{Q_0}(n^2)$, $j = 1, 2$, are respectively mapped to ${}^{\mathfrak{t}}\{x_j^\pm, y_j^\pm\} \in \mathcal{S}_{Q_d}(\pm n)$, $j = 1, 2$, which are equivalent $\bmod \mathrm{Aut}_{Q_d}$; here the signs correspond of course. Then, (93.64) together with Note [76.4] implies that ${}^{\mathfrak{t}}\{X_j^\pm, Y_j^\pm\}$, $j = 1, 2$, ought to be equivalent $\bmod \mathrm{Aut}^2_{Q_0}$. Again, the

482 *Quadratic Forms*

action of $\mathrm{Aut}^2_{Q_0}$ does not affect the classification (93.28): The equation (93.47) extends to the present situation without any change. It is implied, in view of (93.63), that the classification (93.31) is stable under the action of $\mathrm{Aut}^2_{Q_0}$. Hence, in (93.63) the group Aut_{Q_0} needs to be replaced by $\mathrm{Aut}^2_{Q_0}$. Summing up, we are led to the following assertion: Provided (93.41) (ii), the map (93.40) induces

$$\mathrm{Aut}^2_{Q_0} \backslash \mathcal{S}^{(d,\pm)}_{Q_0}(n^2) \to \mathrm{Aut}_{Q_d} \backslash \mathcal{S}_{Q_d}(\pm n) \quad \text{(injective)}$$
$$\text{with signs corresponding.} \tag{93.65}$$

This is an analogue of (93.49).

15. On noting (93.31), we get, by (93.65),

$$\Lambda_D |\mathrm{Aut}_{Q_0} \backslash \mathcal{S}_{Q_0}(n^2)| \le \sum_{\substack{d|D \\ \langle d, D/d \rangle = 1}} \sum_{\pm} |\mathrm{Aut}_{Q_d} \backslash \mathcal{S}_{Q_d}(\pm n)|, \tag{93.66}$$

where

$$\Lambda_D = |\mathrm{Aut}_{Q_0}/\mathrm{Aut}^2_{Q_0}| = 4. \tag{93.67}$$

Thus, in place of (93.52), we have, for $s > 1$,

$$4 \sum_{m=1}^{\infty} \frac{J_P(m)\mathfrak{s}(m)}{m^{s/2}} |\mathrm{Aut}_{Q_0} \backslash \mathcal{S}_{Q_0}(m)|$$
$$\le \sum_{\substack{d|D \\ \langle d, D/d \rangle = 1}} \sum_{\pm} \sum_{n=1}^{\infty} \frac{J_P(n)}{n^s} |\mathrm{Aut}_{Q_d} \backslash \mathcal{S}_{Q_d}(\pm n)|. \tag{93.68}$$

We sum both sides of this inequality over all representative forms listed in (93.19). The sum on the left side becomes the same as the right side of (93.16) due to just the same mechanism, i.e., (93.18), as indicated in step 12. As for the sum over n, we adopt the computation §92 (II). A minor difference is caused by the terms with $-n$; however, they can be handled by considering instead the expression $\frac{1}{2}(Z(s; \mathcal{C}; 0) - Z(s; \mathcal{C}; 1))$, in place of (92.34). Hence, by (93.17), (93.58), and (93.68), we find that

$$4\mathfrak{h}(D) \le \sum_{Q_0} \sum_{\substack{d|D \\ \langle d, D/d \rangle = 1}} \sum_{\pm 1} 1 = |\mathcal{K}^2(D)| \cdot 2^\tau \cdot 2. \tag{93.69}$$

This ends an alternative proof of the principal genus theorem (91.14) in the case of any positive odd discriminant D with which $\mathrm{pell}_D(-4)$ is not solvable.

16. Finally, we shall deal with (93.41) (iii). After the above discussion, the matter has become relatively easy to deal with. Anyway, this time, we have

$$\mathrm{Aut}^*_{Q_d} \neq \varnothing, \quad \forall d|D, \tag{93.70}$$

instead of (93.61). We first introduce

$$\mathcal{S}_{|Q_d|}(n) = \mathcal{S}_{Q_d}(n) \sqcup \mathcal{S}_{Q_d}(-n); \tag{93.71}$$

namely we consider the set of all proper solutions to $|Q_d(x,y)| = n$. Both Aut_{Q_d} and $\mathrm{Aut}^*_{Q_d}$ act on $\mathcal{S}_{|Q_d|}(n)$, so we introduce also

$$\mathrm{Aut}_{|Q_d|} = \{\pm V^k_* : k \in \mathbb{Z}\}, \tag{93.72}$$

with an arbitrary $V_* \in \mathrm{Aut}^*_{Q_d}$. Note that $\mathrm{Aut}_{Q_d} = \pm\mathrm{Aut}^2_{|Q_d|}$, which is (85.22) with $Q = Q_d$. This time the discussion in step 13 implies readily that we have, in place of (93.63),

$$\mathrm{Aut}_{Q_0}\backslash\mathcal{S}^{(d)}_{Q_0}(n^2) \to \mathrm{Aut}_{|Q_d|}\backslash\mathcal{S}_{|Q_d|}(n). \tag{93.73}$$

However, an injective mapping is desired. We thus observe that the identity (93.64) holds without change. Let $^t\{X_j, Y_j\} \in \mathcal{S}^{(d)}_{Q_0}(n^2)$, $j = 1,2$, be mapped to two elements of $\mathcal{S}_{|Q_d|}(n)$ which are equivalent either (1) mod Aut_{Q_d} or (2) mod $\mathrm{Aut}^*_{Q_d}$. Then

$$A^2 X_1 \mathfrak{e} + \tfrac{1}{2}Y_1(B\mathfrak{e} + \mathfrak{d}) = \left\{ \begin{matrix} (1) \\ (2) \end{matrix} \right\} \cdot \left(A^2 X_2 \mathfrak{e} + \tfrac{1}{2}Y_2(B\mathfrak{e} + \mathfrak{d})\right). \tag{93.74}$$

Here (1) : $\left(\tfrac{1}{2}(t_*\mathfrak{e} + u_*\mathfrak{d})\right)^4$ and (2) : $-\left(\tfrac{1}{2}(t_*\mathfrak{e} + u_*\mathfrak{d})\right)^2$ with $t^2_* - Du^2_* = -4$; (1) is of course the same as the action of $\mathrm{Aut}^2_{Q_0}$. With (2) we have

$$L_+(X_1, Y_1) = L_+(X_2, Y_2) - \tfrac{1}{2}\left(L_+(X_2, Y_2)u_* + Y_2 t_* - 2Anu_*\right)u_* D, \tag{93.75}$$

which is to be compared with (93.47). That is, any element of $-(\mathrm{Aut}^*_{Q_0})^2$, maps $\mathcal{S}^{(d)}_{Q_0}(n^2)$ into itself. We are led to the group

$$\mathrm{Aut}^\sharp_{Q_0} = \{(-V^2_*)^k : k \in \mathbb{Z}\} \subset \mathrm{Aut}_{Q_0}, \tag{93.76}$$

where V_* is as in (93.71). In this way, we obtain

$$\mathrm{Aut}^\sharp_{Q_0}\backslash\mathcal{S}^{(d)}_{Q_0}(n^2) \to \mathrm{Aut}_{|Q_d|}\backslash\mathcal{S}_{|Q_d|}(n) \quad \text{(injective)}. \tag{93.77}$$

Then we note that

$$\left|\mathrm{Aut}_{|Q_d|}\backslash\mathcal{S}_{|Q_d|}(n)\right| = |\mathrm{Aut}_{Q_d}\backslash\mathcal{S}_{Q_d}(n)|. \tag{93.78}$$

484　　　　　　　　　　　　　　*Quadratic Forms*

Hence, in place of (93.66), we get

$$\Lambda_D|\mathrm{Aut}_{Q_0}\backslash\mathcal{S}_{Q_0}(n^2)| \le \sum_{\substack{d|D \\ \langle d, D/d \rangle = 1}} |\mathrm{Aut}_{Q_d}\backslash\mathcal{S}_{Q_d}(n)|, \qquad (93.79)$$

where

$$\Lambda_D = |\mathrm{Aut}_{Q_0}/\mathrm{Aut}_{Q_0}^{\sharp}| = 2, \qquad (93.80)$$

since $\mathrm{Aut}_{Q_0}/\mathrm{Aut}_{Q_0}^{\sharp} = \{\pm 1\}$ by (85.22). Therefore, under (93.41) (iii), we are led to the same expression as (93.52), although their actual contents are different. The rest of the argument can be omitted.

Thus we have obtained $|\mathcal{K}(D)/\mathcal{K}^2(D)| \le 2^{\tau-1}$. Combining this with the result in [B], we end an alternative proof of the principal genus theorem for arbitrary odd discriminants.

CONCLUDING REMARK: We have proved the assertion (††) of Note [91.6]; hence, our version of Gauss' 2nd proof of the reciprocity law has been completed.

Notes

[93.1]　Theorem 99 is a crystal emerged from Gauss' arithmetic theory of quadratic forms and Dirichlet's analytic number theory. As a matter of fact, Dirichlet (1840) dealt with only the case where $D = -4p, p \equiv -1 \bmod 4$ (thus §91 (ii)), and moreover $\mathcal{K}(D) = \mathcal{P}(D)$ is supposed to be cyclic. Its extension to general situations was achieved by Weber (1882); for that purpose he created the theory of Abelian groups and their characters, as we have mentioned already a couple of times. In Dirichlet's original assertion, which is at the beginning of a letter (1840) of his to J. Liouville, it is claimed that the prime number theorem (88.1) should still hold with respect to the values taken by a given primitive quadratic form at integral points. This extension of Theorem 99 to any arithmetic progression can be proved with the replacement of j_D in (93.3) by appropriate Dirichlet characters; see Bachmann (1894, Zehnter Abschnitt) for details. A quantitative assertion, which is uniform with respect to D and the difference of the arithmetic progression, can be attained by means of an appropriate modification of the argument of our §106.

[93.2]　Dirichlet's discussion of the genus numbers (1839, §§3, 6) gives (91.13) almost immediately, as we have seen in the above, although the argument via his prime number theorem, given in our §91, is shorter, as Dedekind remarked (Dirichlet (1871/1894, the footnote at the end of §125)).

[93.3]　The notable aspect of Kronecker's article (1865) is in that the theory of ternary quadratic forms, which played an essential rôle in both Gauss' and

$\S93$ 485

Arndt's proofs, is dispensed with. The argument in [C] above is in fact of very basic nature. By the way, the identities (93.44) and (93.64)$_+$ are typical instances of Note [90.5] (*); one may confirm, after some rearrangements, that Theorem 92 yields indeed $(Q_d(x,y))^2 = Q_0(X,Y)$ with the assignment of $\{x,y\}$ given in (93.40)$_+$. On the other hand, with (93.40)$_-$, one gets $(Q_d(x,y))^2 = Q_0(-X,-Y)$, and thus (93.64)$_-$ follows.

[93.4] We are moving to the theory of the distribution of prime numbers. It is one of the most fundamental issues in number theory. However, the methods to be developed in the next chapter are considerably different from those we have employed so far. This difference is certainly an indication of the variety and wealth of investigations in number theory. Then, before closing the present chapter, it should be appropriate to make a short digression on the very beginning of the number theory properly in the automorphic context; today it is a great field one may proceed to after finishing the present volume. Thus, we refer to the report by Smith (1859/1869, $\S\S 24$–38) concerning the cubic and biquadratic reciprocity laws, i.e., beyond the quadratic reciprocity; he gave a clear overview of a series of works by Gauss, Dirichlet, Jacobi, and Eisenstein (incidentally, in p.72, the death of Dirichlet on 5 May 1859 was mentioned with deep laments). In $\S\S 33$–38 there, Smith accounted, with adequate details, how Jacobi and Eisenstein captured these higher reciprocity laws by means of elliptic functions in a way analogous to the use of the exponential function in establishing the quadratic reciprocity law. Amid such developments, Kronecker started exploring Dirichlet's analytic theory of quadratic forms. He discovered among other things that a transcendental function related to elliptic functions appears also in Dirichlet's theory of class numbers of quadratic forms: Following Kronecker (1863), we let $D_1 < 0$, $D_2 > 0$ be coprime fundamental discriminants and combine (92.32) and (92.48) to get

$$(\star) \qquad \lfloor \kappa_{D_1} \rfloor (1) \lfloor \kappa_{D_2} \rfloor (1) = \frac{2\pi \log \epsilon_{D_2}}{\eta_{D_1} \sqrt{|D_1 D_2|}} \mathfrak{h}(D_1) \mathfrak{h}(D_2).$$

A sheer triviality. However, since $D = D_1 D_2$ is a (negative) fundamental discriminant, the left side is equal to $\lfloor \kappa_{D_1} \rfloor (1) \lfloor \kappa_{D_1} \kappa_D \rfloor (1)$, save for an inessential factor. This is but the same as (93.12)$_{s=1}$; thus, (*) is $\mathcal{L}(1, \kappa_{D_1})$, with $\mathcal{L}$ as in (93.1), again apart from an inessential factor. Hence, (*) is a sum of $\mathfrak{h}(D)$ terms, each of which is a multiple of the constant term of (92.7); and it is expressible by means of logarithms of special values of a modular form, i.e., the Dedekind η-function (Kronecker's limit formula (1863, p.46): the details were supplied later in (1883/1889); see also Siegel (1961, Chapter 1)). That a modular form appears here is anyway natural since (92.7) itself is Γ-invariant. Then, solving $(\star)$ in $\log \epsilon_{D_2}$ and exponentiating, one finds that the least solution to $\mathrm{pell}_{D_2}(4)$ can be expressed in terms of the η-function; a mysterious arithmetic–analytic phenomenon which has been lifted to a means to investigate algebraic number fields in an amazingly constructive way as is masterly demonstrated by Siegel (1861, Chapter 2). This is indeed an impressive episode of automorphic context. In the next chapter as well, we shall encounter similar impacts, although somewhat remotely. These trends in number theory originated when Lagrange grasped, as an entity, the set of all integral binary quadratic forms.

5

Distribution of Prime Numbers

§94

Our aim in the present chapter is to develop, in a succinct and accessible way, a modern theory of the distribution of prime numbers in intervals and in arithmetic progressions. This field has a great hoard of old and new conjectures. However, we shall be occupied with very basic facts and methods, the knowledge of which should be fundamental for anyone to embark on a deeper study.

Our discussion consists of four parts: §§94–101, §§102–105, §106, and §107. The first part is preparatory with an insertion in §94 on Selberg's zeta-function viewed in the light of RH. In the second part, we shall discuss the duality of the two sieve ideas due to Linnik and Seberg; an outgrowth is Bombieri's prime number theorem, a true marvel in number theory. In the third part a method will be developed to tightly unite each L-function with these sieves; as a consequence we shall lay out a structured proof for Linnik's prime number theorem, another marvel in number theory. Both of these great theorems would follow immediately from GRH. We shall, however, show how to dispense with the extraordinary hypotheses; indeed, the present chapter is a narrative of the essence of manifold efforts by many experts into this aim. The fourth part is an account of expository nature on a recent development concerning bounded gaps between primes, which one may regard perchance as a prelude that has come finally towards the resolution of the twin prime conjecture. The argument there will be found to be firmly in the framework created by Linnik and Selberg, thus essentially elementary.

Remark The plan of §§94–106 is, in fact, close to that of the second part in *Tata Lecture Notes in Math. Phy., 72*. We shall augment it with thorough proofs on all of its supporting devices which had been largely omitted in the original; in doing this, we shall naturally take into account recent trends as well. Thus the present chapter, together with the earlier chapters, can be regarded as a self-contained treatise on the core part of the modern theory of the distribution of prime numbers. The convention introduced in Notes [12.7]–[12.8] will be in effect throughout the rest. Attaining numerical precision will not be our principal concern as such bounds often turn out to be transitory, particularly in this field of analytic number theory. We shall be content with showing arguments that should yield procedures for effective estimations of basic constants and parameters, save for the τ_ε contained in (101.35) below and those depending on it. Historical comments will be kept to a minimum, unlike in the foregoing chapters.

We shall begin our discussion with a description of the paradigm that is established by the monumental article Riemann (1860). There we shall largely stay throughout the present chapter; on this point Note [95.8] below is to give an additional explanation.

Thus, the function $\zeta(s)$ defined as (11.6) continues meromorphically to the whole of the complex plane; hereafter, we shall denote our principal complex variable by $s = \sigma + it$ with $\mathrm{Re}\, s = \sigma$ and $\mathrm{Im}\, s = t$. The resulting function, denoted still by $\zeta(s)$, is singular only at $s = 1$ which is a simple pole with residue equal to 1. The logarithmic derivatives of both sides of the Euler product representation (11.7) give, for $\sigma > 1$,

$$
\begin{aligned}
-\frac{\zeta'}{\zeta}(s) &= \sum_p \frac{\log p}{p^s} + \sum_p \frac{\log p}{p^{2s} - p^s} \\
&= \sum_{n=1}^{\infty} \frac{\Lambda(n)}{n^s},
\end{aligned}
\tag{94.1}
$$

which is the same as (17.15); the second sum in the first line is absolutely convergent and regular for $\sigma > \frac{1}{2}$. Employing Chebyshev's function ψ defined by (17.18), we have, for $\pi(x)$ the function counting prime numbers,

$$
\begin{aligned}
\pi(x) &= \int_2^x \frac{d\psi(u)}{\log u} - \sum_{\nu=2}^{\infty} \frac{1}{\nu} \pi(x^{1/\nu}) \\
&= \int_2^x \frac{d\psi(u)}{\log u} + O(x^{1/2}/\log x),
\end{aligned}
\tag{94.2}
$$

where Chebyshev's bound (12.14) is applied.

We note that (12.2) gives

$$
\lim_{s \to 1+0} \sum_{n=1}^{\infty} \frac{\Lambda(n) - 1}{n^s} = - \lim_{s \to 1+0} \left(\frac{\zeta'}{\zeta}(s) + \zeta(s) \right) = -2c_E,
\tag{94.3}
$$

§94   489

which suggests that $\Lambda(n)$ is equal to 1 on average, and $\psi(x) = (1 + o(1))x$. Then, (94.2) together with integration by parts will imply the prime number theorem (12.9). In order to make this heuristics rigorous, we need to have some details of the analytic behavior of the meromorphic function $(\zeta'/\zeta)(s)$ in the half plane $\sigma \leq 1$; naturally it ought to be more involved than that of $\zeta(s)$, $\sigma > 1$.

To this end, we exploit the functional equation of $\zeta(s)$. Thus, let

$$\xi(s) = \tfrac{1}{2}s(s - 1)\pi^{-s/2}\Gamma\left(\tfrac{1}{2}s\right)\zeta(s). \tag{94.4}$$

Then (12.8) or rather (61.15) implies that

$$\xi(s) \text{ is an entire function satisfying } \xi(s) = \xi(1 - s). \tag{94.5}$$

Because of (11.8) and the fact that $\Gamma(s)$ has no zero as we shall show later in the present section, the combination (94.4)–(94.5) implies that $\xi(s)$ does not vanish either for $\sigma > 1$ or for $\sigma < 0$. Hence,

$$\xi(\rho) = 0, \rho = \beta + i\gamma \Leftrightarrow \zeta(\rho) = 0, 0 \leq \beta \leq 1, \gamma \neq 0. \tag{94.6}$$

Here, that $\gamma \neq 0$ follows from (12.21) and the fact that $\xi(0) \neq 0$, as (12.2), (94.4), and (94.29) below imply $\xi(1) = \tfrac{1}{2}$. These ρ's are called either complex or non-trivial zeros of the zeta-function; then, $\overline{\rho}$ (complex conjugate), $1 - \rho$, $1 - \overline{\rho}$ are also complex zeros because of the Schwartz reflection principle. All zeros of $\zeta(s)$ outside the critical strip $0 \leq \sigma \leq 1$ coincide with the poles of $\Gamma(\tfrac{1}{2}s)$ in the half-plane $\sigma < 0$, so

$$\zeta(-2m) = 0, m \in \mathbb{N}, \tag{94.7}$$

in view of (94.30) below. We call $s = -2m, m \in \mathbb{N}$, trivial zeros of the zeta-function; they are all simple since poles of $\Gamma(s)$ are so.

With this, Riemann stated the following without proof:

(R$_1$)   Denoting by $N(T)$ the number of complex zeros $\rho = \beta + i\gamma$ such that $0 < \gamma < T$, we have, for $T \geq 2$,

$$N(T) = \frac{T}{2\pi} \log \frac{T}{2\pi} - \frac{T}{2\pi} + O(\log T). \tag{94.8}$$

(R$_2$)   It is *sehr wahrscheinlich* that for every complex zero

$$\beta = \tfrac{1}{2}. \tag{94.9}$$

(R$_3$)   There exist two absolute constants a, b such that for all complex s

$$\xi(s) = e^{a+bs} \prod_{\rho}\left(1 - \frac{s}{\rho}\right)e^{s/\rho}. \tag{94.10}$$

490 Distribution of Prime Numbers

(R$_4$) The function $\psi(x)$ admits the expansion, for $x > 1$,

$$\psi(x) = x - \sum_{\rho} \frac{x^{\rho}}{\rho} - \frac{\zeta'}{\zeta}(0) - \tfrac{1}{2}\log\left(1 - x^{-2}\right). \tag{94.11}$$

These four statements are modern adaptations of Riemann's originals, arranged in the same order as he did. Riemann proved actually nothing explicit concerning the asymptotic nature of $\pi(x)$; nevertheless, he established an everlasting framework by describing in an amazing clarity the relation between the theory of the distribution of primes and the theory of the zeta-function. We shall prove (R$_1$) and (R$_3$) in the present section and (R$_4$) in the next section; in fact, we shall make (94.11) more precise with respect to the mode of summation over all complex zeros ρ. As for (R$_2$), it is the Riemann Hypothesis $=$ RH, i.e., (12.18). It has been defying all attempts to conquer and has become only more prominent.

We begin with some preparations. They are quite basic implements from complex analysis: first some fine consequences of the maximum modulus principle and then an account of the Gamma function will be given.

Theorem 100 *Let $f(z)$ be regular for $|z| \leq R$, and $|f(z)| \leq M$ on $|z| = R$. Then, provided $f(0) = 0$, we have*

$$|f(z)| \leq M|z|/R, \quad |z| \leq R. \tag{94.12}$$

Proof This is called Schwartz's lemma. It suffices to apply the maximum modulus principle to $f(z)/z$. We end the proof.

Theorem 101 *Let $f(z)$ be regular for $|z| \leq R$, and $|f(z)| \leq M$ on $|z| = R$. Then, provided $f(0) \neq 0$, the number of zeros of $f(z)$, including multiplicities, on the disk $|z| \leq r < R$ is not greater than*

$$\frac{1}{\log(R/r)} \log\left(M/|f(0)|\right). \tag{94.13}$$

Proof This is called the Jensen inequality concerning the number of zeros of an analytic function. Let z_j, $1 \leq j \leq J$, be the zeros of $f(z)$ on the disk $|z| \leq r$, and put

$$g(z) = f(z) \prod_{j=1}^{J} \left(\frac{R^2 - \bar{z}_j z}{R(z_j - z)} \right). \tag{94.14}$$

Then $g(z)$ is regular for $|z| \leq R$, and $|g(z)| = |f(z)| \leq M$ on $|z| = R$. In particular, $|g(0)| \leq M$ by the maximum modulus principle. On noting that $g(0) = f(0) \prod(R/z_j)$ and $R/|z_j| \geq R/r$, we end the proof.

Theorem 102 *Let $f(z)$ be regular for $|z| \leq R$, and $f(0) = 0$. Let $\mathcal{R}$ be the maximum of $\mathrm{Re}\,\{f(z)\}$ on $|z| = R$. Then, $\mathcal{R} \geq 0$, and*

$$|f(z)| \leq \frac{2|z|}{R - |z|}\mathcal{R}, \quad |z| < R. \tag{94.15}$$

Proof This is called the Borel–Carathéodory theorem. The fact that $\mathrm{Re}\,\{f(z)\} \leq \mathcal{R}$ for $|z| \leq R$, thus in particular $\mathcal{R} \geq 0$, is verified by an application of the maximum modulus principle to $\exp(f(z))$. We may suppose that $f(z)$ is not a constant, so $\mathcal{R} > 0$. With this, we put $g(z) = f(z)/(2\mathcal{R} - f(z))$. Then, $g(z)$ is regular for $|z| \leq R$, and $|g(z)| \leq 1$ on $|z| = R$. Hence, Theorem 100 implies that $|g(z)| \leq |z|/R$. On noting that $|f(z)| = |2\mathcal{R}g(z)/(1 + g(z))|$, we end the proof.

Theorem 103 *Let $f(z)$ be regular, $f(0) \neq 0$, and $|f(z)/f(0)| \leq \exp(S)$ for $|z| \leq K$. Let $\{z_j\}$ be the set of all zeros of $f(z)$ on the disk $|z| \leq K/2$, including multiplicities. Then there exists an absolute constant $a > 0$ with which the following three assertions (i)–(iii) hold:*

(i)

$$\left| \frac{f'}{f}(z) - \sum_j \frac{1}{z - z_j} \right| < a\frac{S}{K}, \quad |z| \leq \tfrac{1}{4}K. \tag{94.16}$$

(ii) *If $f(z) \neq 0$, $\mathrm{Re}\,z \geq 0$, then*

$$-\,\mathrm{Re}\,\frac{f'}{f}(0) < a\frac{S}{K}. \tag{94.17}$$

Also, if $f(z_0) = 0$ with $-K/2 < z_0 < 0$ under the same assumption, then

$$-\,\mathrm{Re}\,\frac{f'}{f}(0) < a\frac{S}{K} + \frac{1}{z_0}. \tag{94.18}$$

(iii) *If $f(z) \neq 0$ for $\mathrm{Re}\,z \geq -2r$, with $0 < r < K/8$, then*

$$\left| \frac{f'}{f}(z) \right| < 2a\frac{S}{K} + 3\left| \frac{f'}{f}(0) \right|, \quad |z| \leq r. \tag{94.19}$$

Proof This is due to Landau (1924). The function $g(z) = f(z)\prod_j(z - z_j)^{-1}$ is regular for $|z| \leq K$ and $g(z) \neq 0$ for $|z| \leq K/2$. Also, $|g(z)/g(0)| \leq \exp(S)$ on $|z| = K$, and the same holds for $|z| \leq K$. The function $h(z) = \log(g(z)/g(0))$, $h(0) = 0$, is regular and $\mathrm{Re}\,h(z) \leq S$ for $|z| \leq K/2$. Hence, by Theorem 102, we have $|h(z)| < 6S$ for $|z| \leq 3K/8$. Further, via Cauchy's integral representation of derivatives, $|h'(z)| < 48S/K$ for $|z| \leq K/4$. This gives (i) with $a = 48$. Next, under the assumption in (ii) we have $\mathrm{Re}\,z_j < 0$, and

490 *Distribution of Prime Numbers*

(94.17)–(94.18) follows from (94.16). Similarly, under the assumption in (iii), we have, from (94.16),

$$- \operatorname{Re} \frac{f'}{f}(z) < a \frac{S}{K}, \quad |z| = 2r. \tag{94.20}$$

We apply Theorem 102 to the function $-(f'/f)(z) + (f'/f)(0)$ with $R = 2r$. We end the proof.

Theorem 104 *If $f(z)$, $z = x + iy$, is regular and $O(\exp(\eta y))$, for every $\eta > 0$, uniformly in the region $x_1 \le \operatorname{Re} z \le x_2$, $2 \le y_1 \le y$, and if $f(x_j + iy) \ll y^{\kappa_j}(\log y)^a$ for $y_1 \le y$ with constants $a, \kappa_j \in \mathbb{R}$, then it holds that $f(z) \ll y^{\kappa(x)}(\log y)^a$ uniformly in the vertical strip $x_1 \le \operatorname{Re} z \le x_2$; here $\kappa(x)$ is the linear function whose value at x_j is equal to κ_j, $j = 1, 2$.*

Proof This is a version of the Phragmen–Lindelöf theorem; $y^{\kappa(x)}(\log y)^a$ is called the convex bound for $f(z)$. Let $g(z) = (\log(-iz))^a \exp\left(\kappa(z)\log(-iz)\right)$, in which the branch $\log(1) = 0$ is chosen, and let $h(z) = f(z)/g(z)$. Since $|g(z)| = (\log y)^a \exp(\kappa(x)\log y + O(1))$ as $y \to +\infty$, we see that $|h(z)|$ is bounded by a constant H, of course independent of η, on the three lines: $\operatorname{Re} z = x_j$, $j = 1, 2$, and $\operatorname{Im} z = y_1$. Then the function $\exp(2i\eta z)h(z)$ is bounded by H on the edges of the rectangle defined by the straight lines $\operatorname{Re} z = x_j$, $\operatorname{Im} z = y_j$, $j = 1, 2$, provided y_2 is sufficiently large. In particular, $|\exp(2i\eta z)h(z)| \le H$ throughout this rectangle by the maximum modulus principle. While fixing z there, we let η tend to 0, which ends the proof, as y_2 is arbitrary.

We turn to a theory of the Gamma function $\Gamma(s)$. It was introduced by Euler (1738, written in 1729), and in the modern notation

$$\Gamma(s) = \int_0^\infty e^{-u} u^{s-1} du, \quad \operatorname{Re} s > 0. \tag{94.21}$$

Integrating by parts, we have $\Gamma(s) = \Gamma(s+1)/s$, so $\Gamma(n) = (n-1)!$ for $n \in \mathbb{N}$; and for any sufficiently large $m \in \mathbb{N}$

$$\Gamma(s) = \frac{m^{s+m} e^{-m}}{s(s+1)\cdots(s+m-1)} \int_{-1}^\infty \left(e^{-w}(w+1)\right)^m (w+1)^{s-1} dw, \tag{94.22}$$

which implies that $\Gamma(s)$ continues meromorphically to the whole of $\mathbb{C}$. Considering the Taylor expansion of $e^{-w}(w+1)$, we find that as m tends to infinity the main part of the integral comes from the interval $[-m^{-2/5}, m^{-2/5}]$. Replacing the integrand by $\exp(-mw^2/2)\left(1 + O(m^{-1/5})\right)$ in this subinterval, we get

$$\Gamma(s) = \frac{\sqrt{2\pi}\, m^{s+m-1/2} e^{-m}}{s(s+1)\cdots(s+m-1)}\left(1 + O(m^{-1/5})\right); \tag{94.23}$$

the implied constant may depend on s, which is irrelevant to our current purpose. In particular,

$$
\begin{aligned}
&\Gamma(s)\Gamma\left(s+\tfrac{1}{2}\right) \\
&= \pi^{1/2}2^{1-2s}\frac{\sqrt{2\pi}\,(2m)^{2s+2m-1/2}e^{-2m}}{(2s)(2s+1)\cdots(2s+2m-1)}\left(1+O\!\left(m^{-1/5}\right)\right),
\end{aligned}
\tag{94.24}
$$

which yields the duplication formula: for all complex s

$$
\Gamma(s)\Gamma\left(s+\tfrac{1}{2}\right) = \pi^{1/2}2^{1-2s}\Gamma(2s).
\tag{94.25}
$$

Assuming $s > 0$ for a while, we have, by (94.23),

$$
\log\Gamma(s) = \tfrac{1}{2}\log(2\pi) + \lim_{m\to\infty}\left\{\left(s+m-\tfrac{1}{2}\right)\log m - m - \sum_{k=0}^{m-1}\log(s+k)\right\}.
\tag{94.26}
$$

Differentiating this,

$$
\begin{aligned}
\frac{\Gamma'}{\Gamma}(s) &= \lim_{m\to\infty}\left\{\log m - \sum_{k=0}^{m-1}\frac{1}{s+k}\right\} \\
&= -\frac{1}{s} + \lim_{m\to\infty}\left\{\log m - \sum_{k=1}^{m-1}\frac{1}{k} - \sum_{k=1}^{m-1}\left(\frac{1}{s+k} - \frac{1}{k}\right)\right\} \\
&= -\frac{1}{s} - c_E - \sum_{k=1}^{\infty}\left(\frac{1}{s+k} - \frac{1}{k}\right),
\end{aligned}
\tag{94.27}
$$

where c_E is the Euler constant; the procedure is legitimate, as the convergence is uniform. On integrating and exponentiating, we get the product representation

$$
\frac{1}{\Gamma(s)} = se^{c_E s}\prod_{n=1}^{\infty}\left(1+\frac{s}{n}\right)e^{-s/n}, \quad \forall s \in \mathbb{C},
\tag{94.28}
$$

which converges absolutely for all bounded s and represents an entire function. Also, we have

$$
\Gamma(\tfrac{1}{2}) = \sqrt{\pi}, \quad \Gamma'(1) = -c_E, \quad \Gamma'\!\left(\tfrac{1}{2}\right) = -(c_E + 2\log 2)\sqrt{\pi},
\tag{94.29}
$$

$$
\Gamma(s)\Gamma(1-s) = \frac{\pi}{\sin(\pi s)}.
\tag{94.30}
$$

The first formula in (94.29) follows from (94.25)$_{s=\frac{1}{2}}$; the second from (94.27); and the third from (94.25). As for (94.30), it is a consequence of (94.28) and the well-known product representation for $\sin(\pi s)$. In particular, $\Gamma(s)$ never

Distribution of Prime Numbers

vanishes; and its singularities are only poles of order one at non-positive integers $-j$ with the residue $(-1)^j/j!$. Also, (94.30) implies the integral representation due to Hankel

$$\frac{1}{\Gamma(s)} = \frac{i}{2\pi} \int_C e^{-u}(-u)^{-s} du, \quad s \in \mathbb{C}, \tag{94.31}$$

where the contour C is the same as the one employed in (12.5). Hence, $\{\Gamma(s)\}^{-1}$ is an entire function, as has been shown already at (94.28).

In passing, we add the Beta-integral formula of Euler: for $\operatorname{Re}\alpha, \operatorname{Re}\beta > 0$

$$\frac{\Gamma(\alpha)\Gamma(\beta)}{\Gamma(\alpha + \beta)} = \int_0^1 x^{\alpha-1}(1-x)^{\beta-1} dx = \int_0^\infty \frac{y^{\alpha-1}}{(y+1)^{\alpha+\beta}} dy. \tag{94.32}$$

The change of variable $x \mapsto y/(y+1)$ gives the equivalence of these integrals. Then, we represent $\Gamma(\alpha + \beta)/(y+1)^{\alpha+\beta}$ by (94.21) with e^{-u} and s replaced by $e^{-(y+1)u}$ and $\alpha + \beta$, respectively; the rest of argument can be omitted. Sometimes this integral is compared with the Jacobi sum introduced in Note [61.3].

The use of Poisson's sum formula (60.1) shows that the sum over k in (94.26) is equal to

$$\frac{1}{2}\log s + \frac{1}{2}\log(s+m-1) + \int_0^{m-1} \log(s+x) dx$$
$$+ 2\sum_{j=1}^\infty \int_0^{m-1} \log(s+x)\cos(2\pi jx) dx. \tag{94.33}$$

We apply integration by parts three times to each summand in the infinite sum, and insert the result into (94.26). Via analytic continuation, we obtain the following two useful expansions: In the region $|\arg(s)| < \pi$

$$\log\Gamma(s) = \frac{1}{2}\log(2\pi) + \left(s - \frac{1}{2}\right)\log s - s + \frac{1}{12s} \tag{V.1}$$

$$- \frac{1}{2\pi^3}\sum_{j=1}^\infty \frac{1}{j^3}\int_0^\infty \frac{\sin(2\pi jx)}{(s+x)^3} dx, \tag{94.34}$$

$$\frac{\Gamma'}{\Gamma}(s) = \log s - \frac{1}{2s} - \frac{1}{12s^2} + \frac{3}{2\pi^3}\sum_{j=1}^\infty \frac{1}{j^3}\int_0^\infty \frac{\sin(2\pi jx)}{(s+x)^4} dx, \tag{94.35}$$

where $\log s$ is on its principal branch, i.e., $|\operatorname{Im} \log s| < \pi$. Repeating integration by parts, one may obtain asymptotic expansions.

The Stirling formulas for the complex variable follow from (94.34):

$$\Gamma(s) = \sqrt{2\pi}\, s^{s-1/2} e^{-s}\left(1 + \frac{1}{12s} + O(|s|^{-2})\right), \tag{94.36}$$

$$|\Gamma(s)| = \sqrt{2\pi}\,|t|^{\sigma-1/2}e^{-\pi|t|/2}\left(1 + O\!\left(|t|^{-2}\right)\right);\qquad(94.37)$$

the error term of (94.36) is uniform when $|s| \to \infty$ in the angular region $|\arg s| < \pi - \varepsilon$, while (94.37) is uniform for $|\sigma| < 1/\varepsilon$, $|t| > \varepsilon$. A consequence of (94.37) is Mellin's inversion of (94.21):

$$\exp(-x) = \frac{1}{2\pi i}\int_{(a)}\Gamma(s)x^{-s}ds,\quad a > 0,\ |\arg x| < \tfrac{1}{2}\pi.\qquad(94.38)$$

The contour (a) is the vertical line $\operatorname{Re} s = a$, a basic notation to be used frequently in what follows without mention. To verify this formula, note that (94.37) implies that the contour integral on the right converges absolutely and uniformly in the region $|\arg x| < \tfrac{1}{2}\pi$, where it represents a regular function $f(x)$. On shifting the contour to $\left(-\tfrac{1}{2}\right)$ and computing the residue at $s = 0$, we find that $\lim_{x\to 0}f(x) = 1$. On the other hand, $f'(x) = -f(x)$. Hence we have (94.38).

Also, we have

$$\frac{\Gamma'}{\Gamma}(s) \ll \log|s|.\qquad(94.39)$$

This holds save for the neighborhoods $|s + j| < \varepsilon$, $j = 0, 1, 2, \ldots$; the implied constant depends effectively on ε. The proof follows from (94.35) when $\operatorname{Re} s \geq \tfrac{1}{2}$, and via (94.30) and (94.35)$_{s\mapsto 1-s}$ when $\operatorname{Re} s \leq \tfrac{1}{2}$. In particular, from (94.5) and (94.39) it follows that the bound

$$\frac{\zeta'}{\zeta}(s) \ll \log|s|,\quad \sigma \leq -1,\qquad(94.40)$$

holds save for the neighborhoods $|s + 2j| < \varepsilon$, $j \in \mathbb{N}$; the implied constant depends only on ε.

We are now ready to start the proof of Riemann's statements (R$_1$) and (R$_3$). We note first that

$$N(T + 1) - N(T) \ll \log T,\quad T \geq 2,\qquad(94.41)$$

with the implied constant being absolute. This is an immediate consequence of Theorem 101; the necessary bound $\zeta(s) = O(t^{2A})$ for $\sigma \geq -A, t \geq 1$, with any fixed $A \geq 1$, follows from (12.1) if $\sigma \geq \tfrac{1}{2}$, and from the functional equation (94.5) together with (94.37) if $\sigma \leq \tfrac{1}{2}$, or rather one may invoke Note [60.2]. Also, by Theorem 103 (i) with $f(z) = \zeta(z + 2 + it_0)$, $t_0 \geq 20$, and $K = 16$, we get

$$\frac{\zeta'}{\zeta}(s) = \sum_{\rho}\frac{1}{s - \rho} + O(\log t_0),\qquad \begin{array}{l}|s - (2 + it_0)| \leq 4,\\ |\rho - (2 + it_0)| \leq 8,\end{array}\qquad(94.42)$$

496   *Distribution of Prime Numbers*

Note that the pole $s = 1$ is outside the disk $|s - (2 + it_0)| \leq 16$ and that the implied constant is absolute. We next remark that

$$\zeta(\sigma + iT) \neq 0, \ \forall \sigma \in \mathbb{R} \ \Rightarrow \ N(T) = \frac{1}{2\pi} \Delta \arg \xi(s). \tag{94.43}$$

Here, Δ indicates the variation while s moves continuously in the positive direction on the edges of the rectangle defined by the four vertices $\frac{3}{2}$, $\frac{3}{2} + iT$, $-\frac{1}{2} + iT$, $-\frac{1}{2}$. On the bottom edge, $\xi(s)$ is real and does not vanish; hence the corresponding variation is equal to zero. Also, the variation from $\frac{1}{2} + iT$ to $-\frac{1}{2}$ is the same as that from $\frac{3}{2}$ to $\frac{1}{2} + iT$ since (94.5) implies that $\xi(\sigma + it) = \xi(1 - \sigma + it)$. Hence, denoting the variation from $\frac{3}{2}$ to $\frac{1}{2} + iT$ by Δ^*, we have

$$N(T) = \frac{1}{\pi} \Delta^* \arg \xi(s). \tag{94.44}$$

Invoking Striling's formula (94.36) or rather (94.34), we find that

$$N(T) = \frac{T}{2\pi} \log \frac{T}{2\pi} - \frac{T}{2\pi} + \frac{7}{8} + S(T) + O(T^{-1}), \tag{94.45}$$

where

$$S(T) = \frac{1}{\pi} \Delta^* \arg \zeta(s). \tag{94.46}$$

On the other hand, when s moves on the edges of the rectangle defined by the vertices $\frac{3}{2}$, $\frac{3}{2} + iT$, $a + iT$, $a (> \frac{3}{2})$, the variation of $\arg \zeta(s)$ is obviously equal to zero. Hence,

$$S(T) = \frac{1}{\pi} \arg \zeta(\tfrac{1}{2} + iT) \text{ on the condition given in (94.43)}, \tag{94.47}$$

where the value of the argument is the result of the continuous variation along the horizontal line from $+\infty + iT$ down to $\frac{1}{2} + iT$; the initial value is equal to zero.

Before proceeding further, we shall fix the meaning of the logarithm of $\zeta(s)$; naturally, we ought to discuss with care. Thus, we first note that we have $\zeta(s) = U(s) \exp(iV(s))$ at least for $\text{Re}(s) > 1$, where $U(s) = |\zeta(s)|$ and $V(s) = \arg \zeta(s)$; this $\arg \zeta(s)$ is the same as in (94.47) but with s in place of $\frac{1}{2} + iT$. Then, we consider

$$\int_{+\infty}^{s} \frac{\zeta'}{\zeta}(w)dw, \tag{94.48}$$

where the contour is the horizontal half-line connecting $+\infty$ and s; the integration can be performed as far as no singularity of the integrand is encountered. More precisely, let S be the region that is the result of subtracting, from $\mathbb{C}$, every horizontal half-line starting either at $s = 1$ or at a complex zero of $\zeta(s)$ and tending to $-\infty$; by Cauchy's integral theorem, (94.48) is readily seen to be a single-valued regular function over S. Moreover, the above expression $\zeta(s) = U(s)\exp(iV(s))$ holds throughout S by analytic continuation; and there

$$\frac{\zeta'}{\zeta}(\alpha + it) = \left(U^{-1}\frac{\partial U}{\partial \alpha} + i\frac{\partial V}{\partial \alpha}\right)(\alpha + it), \tag{94.49}$$

since $\zeta'(s) = (\partial/\partial\sigma)\zeta(s)$. Inserting (94.49) into (94.48) with $w = \alpha + it$ and $dw = d\alpha$, we find that the integral is identical to

$$\log U(s) + iV(s), \quad s \in S, \tag{94.50}$$

where $\log U(s)$ is the ordinary logarithm of real values. On the other hand, if s is on the half-plane $\operatorname{Re} s > 1$, then (94.48) becomes

$$\sum_{n=2}^{\infty}\frac{\Lambda(n)}{n^s \log n} = -\sum_{p}\log\left(1 - \frac{1}{p^s}\right), \tag{94.51}$$

where each logarithm on the right side is of the principal value, i.e., real for $s > 1$. This is, of course, the same as an analytic continuation of the ordinary logarithm of $\zeta(s)$, $s \in (1, +\infty)$; a further continuation to S is precisely identical to (94.48). Therefore, we now understand that

$$\begin{aligned}\log\zeta(s), s \in S, \text{ is defined by (94.48)};\\ \log\zeta(s) = \log|\zeta(s)| + i\arg\zeta(s),\end{aligned} \tag{94.52}$$

with the above definition of $\arg\zeta(s)$. One may, of course, consider the values of $\log\zeta(s)$ on those deleted half-lines as well, but that should be done via the analytic continuation thus discussed.

With this, we may return to the estimation of $S(T)$. On the assumption made in (94.43), we have, by (94.52),

$$S(T) = \frac{1}{\pi}\operatorname{Im}\,\log\zeta\left(\tfrac{1}{2} + iT\right). \tag{94.53}$$

We take $s = \frac{1}{2} + iT$ in (94.48), and use $|(\zeta'/\zeta)(\alpha + iT)| \leq -(\zeta'/\zeta)(\alpha)$ for $2 < \alpha < \infty$, and $(94.42)_{t_0=T}$ with $s = \alpha + iT$ for $\frac{1}{2} \leq \alpha \leq 2$. Since each term $1/(\alpha + iT - \rho)$ in the latter contributes less than π in absolute value, we find, on noting (94.41), that

$$S(T) = O(\log T). \tag{94.54}$$

This ends the proof of (R_1).

We turn to (R_3): Let $\eta(s)$ be the product over ρ, which converges absolutely and uniformly for all finite s because of (94.41). We take a sequence $\{G \geq 2\}$ tending to infinity such that

Distribution of Prime Numbers

$$\max_{\rho} \frac{1}{|G - \gamma|} \ll \log G. \tag{94.55}$$

The existence of $\{G\}$ is geometrically obvious, again because of (94.41). Let C_G be the edge of the square defined with the vertices $\frac{1}{2} + (\pm 1 \pm i)G$. Then

$$\frac{\xi'}{\xi}(w) \ll (\log G)^2, \quad \frac{\eta'}{\eta}(w) \ll (\log G)^2, \quad w \in C_G. \tag{94.56}$$

Here the bound for $(\xi'/\xi)(w)$, $\operatorname{Re} w \geq \frac{1}{2}$, is a consequence of (94.35) and (94.41)–(94.42); the functional equation implies the same bound for $\operatorname{Re} w \leq \frac{1}{2}$. As for $(\eta'/\eta)(w)$, we deal with the three cases $|\rho| \leq |w|/2$, $|w|/2 < |\rho| < 2|w|$, and $|\rho| \geq 2|w|$, separately; we omit the details as the reasoning is routine. Then, we find that $(\xi'/\xi)(s) - (\eta'/\eta)(s)$ is a constant since it is entire by definition, and

$$\lim_{G \to \infty} \int_{C_G} \frac{(\xi'/\xi)(w) - (\eta'/\eta)(w)}{w(w - s)} dw = 0. \tag{94.57}$$

We end the proof of (R_3).

The explicit values of a, b in (94.10) have no importance for our purpose in the present chapter. Nevertheless, it should be expedient to see how to fix them: We note first that (12.5) gives

$$\zeta(0) = \frac{1}{2\pi i} \int_C \frac{dx}{x(e^x - 1)} = -\frac{1}{2}; \tag{94.58}$$

hence,

$$a = -\log 2. \tag{94.59}$$

Also, (94.10) itself implies that throughout $\mathbb{C}$

$$\frac{\zeta'}{\zeta}(s) = \frac{1}{2}\log \pi + b - \frac{1}{s-1} - \frac{1}{2}\frac{\Gamma'}{\Gamma}\left(\tfrac{1}{2}s + 1\right) + \sum_{\rho}\left(\frac{1}{s-\rho} + \frac{1}{\rho}\right), \tag{94.60}$$

so

$$b = -1 - \frac{1}{2}\log \pi + \frac{\zeta'}{\zeta}(0) + \frac{1}{2}\frac{\Gamma'}{\Gamma}(1). \tag{94.61}$$

On the other hand, the functional equation (12.7) gives

$$-\frac{\zeta'}{\zeta}(1-s) = -\log(2\pi) - \frac{1}{2}\pi \tan\left(\tfrac{1}{2}\pi s\right) + \frac{\Gamma'}{\Gamma}(s) + \frac{\zeta'}{\zeta}(s), \tag{94.62}$$

so

$$\frac{\zeta'}{\zeta}(0) = \log(2\pi) - \frac{\Gamma'}{\Gamma}(1) - \lim_{s \to 1}\left(\frac{\zeta'}{\zeta}(s) + \frac{1}{s-1}\right). \tag{94.63}$$

This and (12.2) as well as (94.29) yield

$$\frac{\zeta'}{\zeta}(0) = \log(2\pi), \quad b = \tfrac{1}{2}\log(4\pi) - 1 - \tfrac{1}{2}c_E. \tag{94.64}$$

With Riemann's (94.10), or rather the expansion (94.60), we are now ready to begin

Selberg's zeta-function for Γ (part 2)

We suppose RH, i.e., (12.18), or that complex zeros of $\zeta(s)$ are $\{\tfrac{1}{2}\pm i\gamma\}, \gamma > 0$. Then (94.60) implies that for all $s_1, s_2 \in \mathbb{C}$

$$\frac{1}{2s_1 - 1}\frac{\zeta'}{\zeta}(s_1) - \frac{1}{2s_2 - 1}\frac{\zeta'}{\zeta}(s_2) = \sum_{\gamma>0} g(s_1, s_2; \gamma) + *, \tag{94.65}$$

where

$$g(s_1, s_2; \tau) = \frac{1}{(s_1 - \tfrac{1}{2})^2 + \tau^2} - \frac{1}{(s_2 - \tfrac{1}{2})^2 + \tau^2}. \tag{94.66}$$

Here and below the symbol $*$ indicates that the omitted parts are not essential for our present purpose. The γ-sum is a meromorphic function over $\mathbb{C}^2$ because of (94.8) or rather (94.41).

Amazingly, a closely resembling expansion holds with the logarithmic derivative of the Selberg zeta-function $\zeta_\Gamma(s)$ introduced at (85.23): for all $s_1, s_2 \in \mathbb{C}$

$$\frac{1}{2s_1 - 1}\frac{\zeta'_\Gamma}{\zeta_\Gamma}(s_1) - \frac{1}{2s_2 - 1}\frac{\zeta'_\Gamma}{\zeta_\Gamma}(s_2) = \sum_{\omega>0} g(s_1, s_2; \omega) + *, \tag{94.67}$$

where $\{\tfrac{1}{4} + \omega^2\}$ is the discrete spectrum of the Laplace–Beltrami operator Δ on the Riemann surface $\mathcal{F}^*$, which has been defined adjacent to (85.25). The ω-sum is meromorphic over $\mathbb{C}^2$ since it is known that $|\{0 < \omega < H\}| \ll H^2$. The assertion (85.25) is a consequence of (94.67). The precise form of the last $*$ is indicated in Note [94.4] below.

Actually, (94.67) is a specialization of Selberg's trace formula for Γ, as we shall briefly explain presently. All the basic assertions required below, such as (94.70), are stated with complete proofs in the first chapter of Motohashi (1997), although ζ_Γ itself is not treated there.

(i) We return to the Smith reduction (82.34) of indefinite quadratic forms. The circular curve $\mathfrak{l}_Q$ on $\mathcal{H}$ passes through the Γ-images of $\mathcal{F}$ which are adjacent to each other; accordingly it is divided into infinitely many arcs. The set of their image arcs on $\mathcal{F}^*$, all of which correspond to Smith reduced forms, is of course

500   *Distribution of Prime Numbers*

finite and thus periodic. Hence, the image of $\mathfrak{l}_Q$ on $\mathcal{F}^*$ is a closed curve, called as a closed geodesic on $\mathcal{F}^*$. It winds around $\mathcal{F}^*$ infinitely many times. Any winding corresponds to the mapping of a form equivalent to Q into itself with an element of Γ, so it is a mapping arising from an element in a Γ-conjugate of Aut_Q, i.e., (74.21); and any single winding corresponds to the least solution ϵ_D to $\mathrm{pell}_D(4)$. Hence, we get the following correspondences: with $Q \in \mathcal{Q}_\pm(D)$

$$\text{the image geodesic on } \mathcal{F}^* \text{ of } \mathfrak{l}_Q$$
$$\leftrightarrow \text{ the } \Gamma\text{-conjugacy class of } \mathrm{Aut}_Q \tag{94.68}$$
$$\leftrightarrow \text{ the class in } \mathcal{K}_\pm(D) \text{ containing } Q.$$

Considering the same with respect to all forms in $\mathcal{Q}_\pm(D)$, we find, of course, that the number of closed geodesics thus arising is equal to the class number $\mathfrak{h}_\pm(D)$. However, the second line (94.68) or rather (85.25) has turned out to be imprecise; and (85.30) is the right formulation. By the way, the length of any of these closed geodesics with respect to the hyperbolic metric $|dz|/y$ of Beltrami is invariably equal to $2 \log \epsilon_D$, as is implied by (85.31). By the way, the change of notation made at (90.1) is ignored because of an obvious reason.

(ii) Let $L^2(\mathcal{F}, d\mu) = L^2(\mathcal{F}^*, d\mu)$, $d\mu(z) = dx\,dy/y^2$, stand for the Hilbert space composed of all Γ-automorphic f's such that $\|f\|^2 = ((f,f)) < +\infty$, where

$$((u,v)) = \int_{\mathcal{F}} u(z)\overline{v(z)}d\mu(z). \tag{94.69}$$

Then there exists an orthonormal system $\{\psi_j : j \geq 0\}$ such that $\psi_0 \equiv (3/\pi)^{\frac{1}{2}}$ and $\Delta\psi_j = (\frac{1}{4}+\omega_j^2)\psi_j, j \geq 1$, with which it holds that for any $u, v \in L^2(\mathcal{F}, d\mu)$

$$((u,v)) = \sum_{j=0}^{\infty} ((u,\psi_j))\overline{((v,\psi_j))} + \frac{1}{2\pi}\int_0^\infty \mathcal{E}(t,u)\overline{\mathcal{E}(t,v)}dt, \tag{94.70}$$

where

$$\mathcal{E}(t,f) = \underset{Y\to\infty}{\mathrm{l.i.m.}} \int_{\mathcal{F}_Y} f(z)E(z,\tfrac{1}{2}-it)\,d\mu(z), \tag{94.71}$$

with $E(z,s)$ as in Note [61.8], $\mathcal{F}_Y = \mathcal{F} \cap \{z : \mathrm{Im}\, z < Y\}$, and the limit taken in $L^2(0,\infty)$.

(iii) To prove the spectral expansion (94.70), we consider first the free-space (i.e., over $\mathcal{H}$) resolvent kernel $r_\alpha(z,w)$ of $\Delta + \alpha(\alpha-1)$. As suggested by the ordinary potential theory over the Euclidean plane, $r_\alpha(z,w)$ should depend only on the hyperbolic distance between z and w which is a function of

$$\varrho(z,w) = \frac{|z-w|^2}{4(\mathrm{Im}\, z)(\mathrm{Im}\, w)} \qquad (z,w \in \mathcal{H}); \tag{94.72}$$

$$\S 94 \qquad\qquad 501$$

so $r_\alpha(z,w) = k_\alpha(\varrho(z,w))$ with a certain k_α, and it is a point-pair invariant, i.e., $r_\alpha(\gamma(z),\gamma(w)) = r_\alpha(z,w)$ for any $\gamma \in \mathrm{PSL}(2,\mathbb{R})$. Since $(\Delta+\alpha(\alpha-1))_w r_\alpha = 0$ for $z \neq w$, as r_α being a resolvent, we are led to the differential equation

$$\left[(\varrho^2 + \varrho)(d/d\varrho)^2 + (1 + 2\varrho)(d/d\varrho) + \alpha(1 - \alpha)\right]k_\alpha(\varrho) = 0. \qquad (94.73)$$

Hence, taking it into account that by definition k_α should have a prescribed singularity at the origin, we have, with the hypergeometric function and its Mellin–Barnes integral representation,

$$\begin{aligned}
k_\alpha(\varrho) &= \frac{\Gamma^2(\alpha)}{4\pi\,\Gamma(2\alpha)}\varrho^{-\alpha}{}_2F_1\left(\alpha,\alpha;2\alpha;-1/\varrho\right)\\
&= \frac{1}{8\pi^2 i}\int_{(-\frac{1}{4})}\frac{\Gamma^2(\xi+\alpha)\Gamma(-\xi)}{\Gamma(\xi+2\alpha)}\varrho^{-\xi-\alpha}d\xi,
\end{aligned} \qquad (94.74)$$

in which the path is the vertical line $\mathrm{Re}\,\xi = -\frac{1}{4}$; the integral converges quite rapidly because of (94.37).

(iv) Then we consider folding r_α along the action of Γ, or the following transformation of $f \in L^2(\mathcal{F},d\mu)$:

$$\begin{aligned}
\mathcal{R}_\alpha f(z) &= \int_\mathcal{F} R_\alpha(z,w)f(w)d\mu(w),\\
R_\alpha(z,w) &= \sum_{U\in\Gamma} r_\alpha(z,U(w)),
\end{aligned} \qquad (94.75)$$

where the sum is absolutely convergent for $\alpha > 1$, $z \not\equiv w \bmod \Gamma$. An essential point to be noted here is in that we need only the fact that $\mathcal{R}_\alpha$ is the left-inverse of $\Delta + \alpha(\alpha - 1)$, i.e., $\mathcal{R}_\alpha(\Delta + \alpha(\alpha - 1)) = 1$ on the subspace spanned by test functions on $\mathcal{F}$, or rather on $\mathcal{F}^*$. Thus, we have anyway

$$\Delta f = \lambda f \;\Rightarrow\; \mathcal{R}_\alpha f = (\lambda + \alpha(\alpha - 1))^{-1}f, \qquad (94.76)$$

provided α is sufficiently large. It follows that non-trivial eigen-functions of Δ in $L^2(\mathcal{F},d\mu)$, if exist, should be smooth and of rapid decay; and λ is to be of the shape $\frac{1}{4} + \omega^2$ with $\omega > 3\pi^2/2$.

(v) One may then aim to diagonalize the kernel $R_\alpha(z,w)$ by means of the Hilbert–Schmidt theory on integral transformations with bounded continuous symmetric kernels. However, although symmetric, $R_\alpha(z,w)$ is neither bounded nor continuous. We need to apply a sort of surgery to $R_\alpha(z,w)$, and for this purpose the above Mellin–Barnes integral representation turns out to be quite handy. Thus, to eliminate the discontinuity, we consider

$$\mathcal{R}_{\alpha,1} = (2\alpha)^{-1}\left\{\mathcal{R}_\alpha - \mathcal{R}_{\alpha+1}\right\}. \qquad (94.77)$$

502 *Distribution of Prime Numbers*

For the corresponding kernel

$$R_{\alpha,1}(z,w) = (2\alpha)^{-1}\{R_\alpha(z,w) - R_{\alpha+1}(z,w)\} \tag{94.78}$$

is continuous, while $(2\alpha)^{-1}(k_\alpha(\varrho) - k_{\alpha+1}(\varrho))$ admits a Mellin–Barnes integral representation similar to the above one. At the same time $\mathcal{R}_{\alpha,1}$ is obviously diagonal over the space spanned by eigen-functions of Δ. However, $R_{\alpha,1}$ is still unbounded. So, to get a compactification of $\mathcal{R}_{\alpha,1}$, we should subtract from $R_{\alpha,1}$, a certain symmetric Γ-automorphic kernel $C_{\alpha,1}$ of simple nature. The asymptotic character of $C_{\alpha,1}$ should naturally be the same as that of $R_{\alpha,1}$. Moreover, this modification of $R_{\alpha,1}$ should not alter the action of $\mathcal{R}_{\alpha,1}$ too much. It is optimal to have a $C_{\alpha,1}$ which annihilates $\{\psi_j : j \geq 0\}$, and the rest $R_{\alpha,1} - C_{\alpha,1}$ is bounded over $\mathcal{F}^* \times \mathcal{F}^*$. It turns out that $C_{\alpha,1}$ comes from the continuous spectrum which is spanned by the Eisenstein series. Then the discreteness of the eigenvalues of Δ becomes a simple consequence of Bessel's inequality. The rest of the proof of the spectral decomposition is a typical instance of the application of the Hilbert–Schmidt theory on the existence of eigenvalues of integral transformations. As a matter of fact, we need to iterate sufficiently many times the above modification of the resolvent to attain a smoother kernel.

(vi) We now apply the spectral decomposition (94.70) to the function

$$R(z,w) = R_{s_1}(z,w) - R_{s_2}(z,w), \quad s_1 \neq s_2, \ \mathrm{Re}\, s_1 > 1, \ \mathrm{Re}\, s_2 > 1. \tag{94.79}$$

We have, rather obviously,

$$R(z,w) = \tilde{R}(z,w) + \mathcal{E}, \quad \tilde{R}(z,w) = \sum_{j=1}^{\infty} g(s_1, s_2; \omega_j)\psi_j(z)\overline{\psi_j(w)}, \tag{94.80}$$

where $\mathcal{E}$ corresponds to the contribution of the Eisenstein series, which we shall ignore for the sake of simplicity; it should not cause any loss of precision as far as our present aim is concerned. The function $\tilde{R}(z,w)$ is continuous and bounded, and its contribution to the trace, i.e., the integral of $\tilde{R}(z,z)$ over $\mathcal{F}$ is exactly equal to the sum on the right side of (94.67).

(vii) With this, we consider the trace of the part $R^{(h)}(z,w)$ of $R(z,w)$ pertaining to the set $\Gamma^{(h)}$ which is introduced after the classification (85.32): Namely, by virtue of the decomposition (85.36), we have

$$\int_{\mathcal{F}} R^{(h)}(z,z)d\mu(z) = \sum_{D} \sum_{\mathcal{C}\in\mathcal{K}(D)} \sum_{n=1}^{\infty} \int_{[V]\backslash\mathcal{H}} k\big(z, V^n z\big)d\mu(z), \tag{94.81}$$

where $k = k_{s_1} - k_{s_2}$, and $V = V_Q, Q \in \mathcal{C}$; we have used the relations

$$k\big(z, U^{-1}V^n U(z)\big) = k\big(U(z), V^n U(z)\big), \quad \bigsqcup_{U\in[V]\backslash\Gamma} U(\mathcal{F}) = [V]\backslash\mathcal{H}. \tag{94.82}$$

Then, with $\phi = \phi_Q$ as in (85.24),

$$\int_{[V]\backslash\mathcal{H}} k\big(z, V^n(z)\big)d\mu(z) = \int_{\phi([V]\backslash\mathcal{H})} k\big(\phi^{-1}(z), V^n\phi^{-1}(z)\big)d\mu(z)$$

$$= \int_{[\phi V\phi^{-1}]\backslash\mathcal{H}} k\big(z, \phi V^n\phi^{-1}(z)\big)d\mu(z) = \iint_{\substack{1\leq|z|\leq\epsilon_D^2 \\ 0<\mathrm{Im}\,z}} k\big(z, \epsilon_D^{2n}z\big)\frac{dxdy}{y^2}, \quad (94.83)$$

where we have used the fact that

$$\mathcal{H} = \bigsqcup_{n=-\infty}^{\infty} (\phi V^n\phi^{-1})\phi([V]\backslash\mathcal{H}) \;\Rightarrow\; \phi([V]\backslash\mathcal{H}) = [\phi V\phi^{-1}]\backslash\mathcal{H}. \quad (94.84)$$

Further, with $\mathrm{Re}\,\alpha > 1$, $a,b > 1$, and $B = \big(\tfrac{1}{2}(b^{1/2} - b^{-1/2})\big)^2$,

$$\iint_{\substack{1\leq|z|\leq a \\ 0<\mathrm{Im}\,z}} k_\alpha\big(\varrho(z,bz)\big)\frac{dxdy}{y^2} = 2\log a \int_0^{\frac{1}{2}\pi} k_\alpha\left(B\sin^{-2}\theta\right)\frac{d\theta}{\sin^2\theta}$$

$$= \frac{\log a}{4\pi^2 i}\int_{\left(-\frac{1}{4}\right)} \frac{\Gamma^2(\xi+\alpha)\Gamma(-\xi)}{\Gamma(\xi+2\alpha)}B^{-(\xi+\alpha)}\int_0^{\frac{1}{2}\pi}(\sin\theta)^{2(\xi+\alpha-1)}d\theta\,d\xi$$

$$= \frac{\log a}{8\pi^{3/2}i}\int_{\left(-\frac{1}{4}\right)} \frac{\Gamma(\xi+\alpha)\Gamma\big(\xi+\alpha-\frac{1}{2}\big)\Gamma(-\xi)}{\Gamma(\xi+2\alpha)}B^{-(\xi+\alpha)}d\xi \qquad (94.85)$$

$$= \frac{\log a}{4\pi^{1/2}}\frac{\Gamma\big(\alpha-\frac{1}{2}\big)\Gamma(\alpha)}{\Gamma(2\alpha)}B^{-\alpha}\,{}_2F_1\left(\alpha-\tfrac{1}{2},\alpha;2\alpha;-1/B\right)$$

$$= \frac{b^{1/2-\alpha}\log a}{(2\alpha-1)(b^{1/2}-b^{-1/2})},$$

where we have applied the duplication formula (94.25) and the famous quadratic transformation of ${}_2F_1$, i.e.,

$${}_2F_1(a,b;2b;z) = w^{-2a}\,{}_2F_1\left(a,a-b+\tfrac{1}{2},b+\tfrac{1}{2};(1-1/w)^2\right) \qquad (94.86)$$

with $w = \tfrac{1}{2}(1 + (1-z)^{1/2})$.

(viii) Collecting these assertions we reach

$$\int_{\mathcal{F}} R^{(h)}(z,z)d\mu(z) = \frac{1}{2s_1-1}\sum_{D>0}\mathfrak{h}_\pm(D)\sum_{n=0}^{\infty}\frac{\log\epsilon_D^2}{\epsilon_D^{2s_1+2n}-1}$$

$$-\frac{1}{2s_2-1}\sum_{D>0}\mathfrak{h}_\pm(D)\sum_{n=0}^{\infty}\frac{\log\epsilon_D^2}{\epsilon_D^{2s_2+2n}-1}. \qquad (94.87)$$

This ends our brief derivation of (94.67). Continued to Note [94.4].

Notes

[94.1]   A detailed description of the background of the earlier part of the present chapter can be found in Montgomery–Vaughan (2006), which contains highly reliable historical accounts as well. Iwaniec–Kowalski (2004) is a further reading.

[94.2]   Landau's Theorem 103 is an excellent means in the investigation of local behavior of functions similar to the zeta-function. Riemann's statement (R_3) is proved usually by means of Hadamard's theory of entire functions of finite order, as in Titchmarsh (1951, pp.29–31). In the above, we have exploited instead particularities of the zeta-function; this approach is made possible by Landau's Theorem 103.

[94.3]   That $\zeta_\Gamma(s)$ satisfies an analogue of RH is indeed awe-inspiring. However, there exist fundamental differences between $\zeta(s)$ and $\zeta_\Gamma(s)$. For instance, $\zeta(1) = \infty$, whereas $\zeta_\Gamma(1) = 0$ as is induced by the next Note. The most significant difference is closely related to the discussion made in the paragraph following (12.9). In the case of $\zeta_\Gamma(s)$, the definition is made in terms of a quasi-Euler product (85.23) that resembles (11.7) or rather (17.3), but its expansion does not appear to lead to any expression comparable to (11.6). Moreover, the situation with (94.67) is quite opposite to that of (94.65). Namely, in (94.67) the set of zeros $\{\frac{1}{2}\pm i\omega\}$ comes first, and the analogue of the explicit formula (94.60) follows in the form of a complete spectral expansion of a specific function on $\mathcal{F}^*$; then the expansion is transformed into (85.23) by means of the hyperbolic geometry. Looking into this mechanism, one may perceive that the zeros of $\zeta_\Gamma(s)$ constitute of genuinely intrinsic existences, as they are related to the physical nature of $\mathcal{F}^*$. We are allured to search for a way to view the zeros of $\zeta(s)$ equally intrinsic; prime numbers might be captured as geometric objects, but so far no clue has been detected.

[94.4]   The identity (94.84) seems due to Lebedev (1972, pp.253–254). The discussion in §§2.4–2.5 of *Analytic number theory*, II, which is mentioned in the preamble of the present essays, contains the full details of the part following (94.67). Thus, let

$$Y_\Gamma(s) = \frac{\zeta_\Gamma'}{\zeta_\Gamma}(s) - 2\frac{\zeta'}{\zeta}(2s)$$

$$- \frac{1}{s-1} - \frac{1}{s} - \frac{2}{2s-1} + \log(2\pi) - 2\frac{\Gamma'}{\Gamma}(2s) + Z(s)$$

with

$$Z(s) = -\tfrac{1}{6}(2s-1)\frac{\Gamma'}{\Gamma}(s) + \tfrac{1}{4}\frac{\Gamma'}{\Gamma}\left(\tfrac{1}{2}(s+1)\right) - \tfrac{1}{4}\frac{\Gamma'}{\Gamma}\left(\tfrac{1}{2}s\right)$$

$$+ \tfrac{2}{9}\frac{\Gamma'}{\Gamma}\left(\tfrac{1}{3}(s+2)\right) - \tfrac{2}{9}\frac{\Gamma'}{\Gamma}\left(\tfrac{1}{3}s\right).$$

Then we have, more precisely than (94.67),

$$\frac{Y_\Gamma(s_1)}{2s_1 - 1} - \frac{Y_\Gamma(s_2)}{2s_2 - 1} = \sum_\omega g(s_1, s_2; \omega).$$

From this, it follows that the function

$$\xi_\Gamma(s) = \frac{(2\pi)^s \zeta_\Gamma(s)}{\zeta(2s)\Gamma(2s)s(s-1)(2s-1)} \exp\left(\int_0^{s-\frac{1}{2}} Z(\eta + \tfrac{1}{2})d\eta\right)$$

is entire, and satisfies the functional equation

$$\xi_\Gamma(s) = \xi_\Gamma(1 - s).$$

Equivalently, with

$$z_p(s) = -\frac{(2\pi)^{1-2s}\zeta(2s)\Gamma(2s)}{\zeta(2(1-s))\Gamma(2(1-s))},$$

$$z_e(s) = \left(\frac{\sin\frac{1}{2}\pi(1-s)}{\sin\frac{1}{2}\pi s}\right)^{1/2}\left(\frac{\sin\frac{1}{3}\pi(1-s)}{\sin\frac{1}{3}\pi s}\right)^{2/3},$$

$$z_1(s) = \exp\left(\tfrac{1}{3}\pi \int_0^{s-\frac{1}{2}} \eta \tan(\pi\eta)d\eta\right),$$

we have

$$\zeta_\Gamma(s) = z_p(s)z_e(s)z_1(s)\zeta_\Gamma(1 - s).$$

Here z's correspond, respectively, to the contributions of the parabolic, elliptic, and unit elements of Γ; see (85.32). By the way, Selberg's discovery of $\zeta_\Gamma(s)$ is certainly related to his deep experience with $\zeta(s)$ such as the article (1946a).

§95

We shall next give a proof of the statement (R$_4$) by means of the expansions (94.27) and (94.60) coupled with the inversion formula of Perron (1908): It holds uniformly for $a > 0$, $T \geq 2$ that

$$\frac{1}{2\pi i}\int_{a-iT}^{a+iT} y^s \frac{ds}{s} = \begin{cases} 1 + O\left(y^a/(T\log y)\right), & 1 < y, \\ O\left(y^a/(T|\log y|)\right), & 0 < y < 1, \end{cases} \tag{95.1}$$

with the implied constants being absolute. This is actually a truncated form of Perron's original assertion. In the case $y > 1$, we shift the contour towards $-\infty$. The arising integral along the two horizontal lines $(-\infty \pm iT, a \pm iT]$ is estimated by means of integration by parts. In the case $y < 1$, we shift the contour towards $+\infty$.

506 *Distribution of Prime Numbers*

We take, in (95.1), $y = x/n$, $x = [x] + \frac{1}{2} > 1$ with an arbitrary $a > 1$; multiplying both sides by $\Lambda(n)$, we sum over n. Then, via (17.18) and (94.1), we get

$$\psi(x) = -\frac{1}{2\pi i} \int_{a-iT}^{a+iT} \frac{\zeta'}{\zeta}(s) x^s \frac{ds}{s} + O\left(\frac{x^a}{T} \sum_{n=1}^{\infty} \frac{\Lambda(n)}{n^a |\log x/n|}\right). \tag{95.2}$$

In the last sum over n, the parts with $n \leq \frac{1}{2}x$ and $n \geq 2x$ are obviously $O(|(\zeta'/\zeta)(a)|)$. To the remaining part we apply the bound $|\log(x/n)|^{-1} \ll n/|x-n|$, and obtain

$$\psi(x) = -\frac{1}{2\pi i} \int_{a-iT}^{a+iT} \frac{\zeta'}{\zeta}(s) x^s \frac{ds}{s} + O\left(\frac{x^a}{T}\left|\frac{\zeta'}{\zeta}(a)\right| + \frac{x}{T}(\log x)^2\right), \tag{95.3}$$

which is uniform in $a > 1, T \geq 2, x = [x] + \frac{1}{2}$. We note here that the bound (94.40) holds for $\sigma \leq -1$ save for neighborhoods of trivial zeros and that for any given $T \geq 2$ there exists a T_0, $T \leq T_0 \leq T + 1$, such that $(\zeta'/\zeta)(\sigma + iT_0) \ll \log^2 T$ for $-1 \leq \sigma \leq 2$ as is implied by (94.41)–(94.42). Thus, we take $T = T_0$, $a = 1 + (\log x)^{-1}$ in (95.3), and shift the contour to $\sigma = -V$ with V a positive odd integer. Applying (94.27) and (94.60), we see that

$$\psi(x) = x - \sum_{\substack{\rho \\ |\gamma| < T_0}} \frac{x^\rho}{\rho} - \frac{\zeta'}{\zeta}(0) + \sum_{m < V/2} \frac{x^{-2m}}{2m}$$

$$- \frac{1}{2\pi i} \int_{-V-iT_0}^{-V+iT_0} \frac{\zeta'}{\zeta}(s) x^s \frac{ds}{s} + O\left(\frac{x}{T}(\log xT)^2\right), \tag{95.4}$$

where the implied constant is independent of V; we have omitted the estimation of the contribution of the horizontal lines $[-V \pm iT_0, a \pm iT_0]$, for it is immediate. Because of (94.40), the last integral converges to 0 as $V \to +\infty$. Invoking (94.41) and (94.64), we obtain the following truncated version of (R_4):

Theorem 105 *Uniformly for any $x = [x] + \frac{1}{2} > 1$ and $T \geq 2$, it holds that*

$$\psi(x) = x - \sum_{\substack{\rho \\ |\gamma| < T}} \frac{x^\rho}{\rho} - \log(2\pi) - \frac{1}{2}\log\left(1 - x^{-2}\right) + O\left(\frac{x}{T}(\log xT)^2\right). \tag{95.5}$$

The condition that $x > 1$ be half an odd integer makes no essential difference. Neither the third term nor the fourth on the right side of (95.5) has any particular significance. However, when T tends to $+\infty$, they become significant:

$$\psi(x) = x - \lim_{T \to \infty} \sum_{\substack{\rho \\ |\gamma| < T}} \frac{x^\rho}{\rho} - \log(2\pi) - \frac{1}{2}\log\left(1 - x^{-2}\right). \tag{95.6}$$

$\S95$ 507

This expansion, which is due to von Mangoldt (1895, p.294), holds, in fact, unless x is a power of a prime. Since the function $\psi(x)$ is discontinuous at such a value of x, this sort of restriction of x in (95.6) is certainly necessary. More precisely, (95.6) holds for any $x > 1$ if $\psi(x)$ is understood to be $\frac{1}{2}(\psi(x-0) + \psi(x+0))$, which amounts the inclusion of the case $y = 1$ into (95.1). This reminds us of a well-known fact in the theory of Fourier expansions; see (60.1). Indeed, describing his own version of (R4), Riemann (1860, p.679) used the appellation *periodischen Glieder* for the terms involving complex zeros. With the understanding of these details, (R4) is accepted as a correct assertion.

One more point should be mentioned on (95.5)–(95.6): In our investigation on the asymptotic nature of the functions $\pi(x)$ and $\psi(x)$ that is to be developed below, we shall not require the full force of these assertions. It suffices to have

$$\psi(x) = x - \sum_{\substack{\rho \\ |\gamma| < T}} \frac{x^\rho}{\rho} + O\left(\frac{x}{T}(\log x)^2\right), \quad 2 \leq T \leq x. \tag{95.7}$$

This is readily accessible via the combination of (94.6), (94.39), (94.41), (94.42), and (95.1). In other words, the functional equation (94.5), Theorem 103 and basic properties of the Gamma function suffice to derive the fundamental expansion (95.7).

We now turn to the horizontal distribution of complex zeros, which is more essential than the vertical one (that is, $N(T)$) when dealing with the distribution of primes, as will be revealed in due course. Note that the use of the term horizontal distribution implies per se that we shall proceed without supposing RH.

To this end, we quote a famous inequality that seems to have originated with F. Mertens: For $\sigma > 1$, $t \in \mathbb{R}$,

$$-3\frac{\zeta'}{\zeta}(\sigma) - 4\operatorname{Re}\frac{\zeta'}{\zeta}(\sigma + it) - \operatorname{Re}\frac{\zeta'}{\zeta}(\sigma + 2it) \geq 0; \tag{95.8}$$

the left side is equal to $2\sum_{n=1}^{\infty} \Lambda(n)(1 + \cos(t \log n))^2 n^{-\sigma}$, which is non-negative. Let $\rho = \beta + i\gamma$, $\gamma > 0$, be a complex zero; of course $\gamma \geq \alpha$ with a certain constant $0 < \alpha \leq 1$. We apply Theorem 103 to $f(z) = \zeta(\sigma_0 + i\gamma + z)$ with $K = \frac{1}{2}\alpha \leq \frac{1}{2}$ and

$$\sigma_0 = 1 + \frac{\tau}{\log(\gamma + 2)}, \quad t = \gamma, \tag{95.9}$$

with an independent variable $\tau > 0$ such that $\tau(\log(\gamma + 2))^{-1} \leq \frac{1}{24}\alpha$; note that the pole $s = 1$ is well outside the disk $|s - (\sigma_0 + i\gamma)| \leq K$. It should be stressed that the easy bound

$$\zeta(s) \ll |s|, \quad \tfrac{1}{2} \le \sigma, \ \alpha \le t, \tag{95.10}$$

which follows from (12.1), suffices for our present purpose. By (94.17)–(94.18) coupled with (95.10) we find that there exists an absolute constant $c > 0$ such that, provided $\beta > 1 - \tfrac{5}{24}\alpha$,

$$-\operatorname{Re}\frac{\zeta'}{\zeta}(\sigma_0 + i\gamma) < c\log(\gamma + 2) - \frac{1}{\sigma_0 - \beta},$$
$$-\operatorname{Re}\frac{\zeta'}{\zeta}(\sigma_0 + 2i\gamma) < c\log(\gamma + 2). \tag{95.11}$$

Inserting these into (95.8), we find that $3/(\sigma_0-1)-4/(\sigma_0-\beta)+c\log(\gamma+2) > 0$ with another absolute constant $c > 0$. Hence, by fixing τ appropriately, we get

$$\beta < 1 - (\sigma_0 - 1)\frac{1 - c(\sigma_0 - 1)\log(\gamma + 2)}{3 + c(\sigma_0 - 1)\log(\gamma + 2)} < 1 - \frac{\tau}{4\log(\gamma + 2)}. \tag{95.12}$$

We are thus led to the following zero-free region of de la Vallée Poussin (1900, p.37):

Theorem 106 *There exists an absolute constant $c > 0$ such that*

$$\left|\frac{\zeta'}{\zeta}(s) + \frac{1}{s-1}\right| \ll \log(|t| + 2), \quad \sigma \ge 1 - \frac{c}{\log(|t| + 2)}, \tag{95.13}$$

with the implied constant being absolute as well.

Proof Let α, τ, and K be as above. If $0 \le t \le \alpha$, then this bound for $(\zeta'/\zeta)(s)$ is immediate. If $t \ge \alpha$, then on noting (95.12) we apply Theorem 103 (iii) to $f(z) = \zeta(\sigma_1+it+z)$ with $\sigma_1 = 1+\tau/(8\log(t+2))$ and $r = 3\tau/(16\log(t+2))$. Thus one may take $c = \tau/16$. We end the proof. This proof is different from the original.

Remark In particular, $\zeta(s)$ does not vanish on the vertical line $\sigma = 1$, and thus on $\sigma = 0$ either, via the functional equation. Hence, we may make the notion of complex/non-trivial zeros of $\zeta(s)$ that has been defined by (94.6) more precise by restricting β as $0 < \beta < 1$.

We are now able to prove the prime number theorem of de la Vallée Poussin (1900, p.63):

Theorem 107 *There exists an absolute constant $c > 0$, such that for all $x \ge 2$*

$$\psi(x) = x + O\big(x\exp\big(-c(\log x)^{1/2}\big)\big). \tag{95.14}$$

Hence,

$$\pi(x) = \operatorname{li}(x) + O\big(x\exp(-c(\log x)^{1/2})\big), \tag{95.15}$$

where $\operatorname{li}(x)$ *is defined as (12.12).*

$$\S 95 \qquad\qquad 509$$

Proof In either (95.5) or (95.7) we take $T = \exp\left((\log x)^{1/2}\right)$, and observe that the zero-free region (95.13) implies that $x^\rho \ll x \exp\left(-c(\log x)^{1/2}\right)$. Since $\sum_{|\gamma|<T} 1/|\rho| \ll (\log T)^2$ because of (94.41), we end the proof of (95.14), and consequently that of (95.15).

Alternative Proof: We shift the contour of (95.3) with $a = 1 + (\log x)^{-1}$ to $[a-iT,b-iT]\cup[b-iT,b+iT]\cup[b+iT,a+iT]$, where $b = 1 - c(2\log T)^{-1}$ and $T = \exp\left((\log x)^{1/2}\right)$; the residue at the pole $s = 1$ is equal to $-2\pi i x$. The bound $(\zeta'/\zeta)(s) \ll \log T$ on the new contour is implied by (95.13). Hence, the resulting integral is $\ll x^b \log^2 T + x/T$. This ends the proof.

In passing, we remark that

$$\text{RH} \iff \psi(x) = x + O\left(x^{1/2}(\log x)^2\right); \qquad (95.16)$$

see (12.19). The sufficiency part is immediate, as one needs only to take $T = x$ in either (95.5) or (95.7). The necessity part follows from

$$-\frac{\zeta'}{\zeta}(s) = \frac{s}{s-1} + s\int_1^\infty \left(\psi(x) - x\right)\frac{dx}{x^{s+1}}; \qquad (95.17)$$

the right side of (95.16) implies that this integral is regular for $\sigma > \frac{1}{2}$. Actually, (95.16) is equivalent to that $\psi(x) = x + O\left(x^{1/2+\varepsilon}\right)$ holds for every $\varepsilon > 0$, as it implies RH via (95.17). This fact might appear somewhat bizarre.

There exists a variety of results which are known to be equivalent to or rather in the same depth as (95.14)–(95.15). Among them we shall show only the following two facts, though we omit the discussion of the equivalence assertion. These are chosen because their proofs are typical instances of arguments which can be applied to a wide family of problems involving prime numbers.

Theorem 108 *For any $x > 2$, we have*

$$\prod_{p\leq x}\left(1 - \frac{1}{p}\right) = \frac{e^{-c_E}}{\log x}\left(1 + O\left(\exp\left(-c(\log x)^{1/2}\right)\right)\right), \qquad (95.18)$$

$$\sum_{n\leq x}\mu(n) \ll x\exp\left(-c(\log x)^{1/2}\right), \qquad (95.19)$$

where c_E is the Euler constant, μ the Möbius function, and both c's are absolute constants.

Proof Note that (95.18) proves Euler's claim (11.11). The ordinary logarithm of the left side of (95.18) is equal to

$$-\sum_{p\leq x}\sum_{m=1}^\infty \frac{1}{mp^m} = -\sum_{n\leq x}\frac{\Lambda(n)}{n\log n} + \sum_{p\leq x}\sum_{p^m>x}\frac{1}{mp^m}. \qquad (95.20)$$

The second sum on the right side is $\sum_{m=2}^{[\log x / \log 2]} m^{-1} \sum_{x^{1/m} < p \le x} p^{-m}$, which is $O(x^{-1/2})$. Thus, it suffices to show that

$$\sum_{n \le x} \frac{\Lambda(n)}{n \log n} = \log \log x + c_E + O\left(\exp\left(-c(\log x)^{1/2}\right)\right). \tag{95.21}$$

We apply (95.1) to (94.51)–(94.52) but with $x = [x] + \frac{1}{2}$, and get

$$\frac{1}{2\pi i} \int_{a-iT}^{a+iT} \log \zeta(s+1) \frac{x^s}{s} ds = \sum_{n \le x} \frac{\Lambda(n)}{n \log n} + O\big((\log x)/T\big), \tag{95.22}$$

where $a = (\log x)^{-1}$ and $T = \exp\big((\log x)^{1/2}\big)$. The treatment of this integral is somewhat delicate. First we write it as

$$\frac{1}{2\pi i} \int_{a-iT}^{a+iT} (h(s) - \log s) \frac{x^s}{s} ds, \tag{95.23}$$

where $\log s$ is on the principal branch and $h(s) = \log \zeta(s+1) + \log s$. Obviously, $h(s)$ continues analytically to the domain which is the shift of $\mathcal{S}$, introduced after (94.48), to the left by 1. Also, it continues to a neighborhood of the origin. To see this, it suffices to observe that for $s > 0$ we have $h(s) = \log(s\zeta(s+1))$ with the ordinary logarithm of positive reals and that with small $s > 0$

$$h(s) = \sum_{j=1}^{\infty} \frac{(-1)^{j-1}}{j} (s\zeta(s+1) - 1)^j. \tag{95.24}$$

Consequently we have $h(0) = 0$ as well. With this, we shift the contour in (95.23) to the oriented path $L = L_{-2} + L_{-1} + C + L_1 + L_2$, where $L_{-2} = [a - iT, -b - iT]$, $L_{-1} = [-b - iT, -b]$, $C = \{|s| = b\}$, $L_1 = [-b, -b + iT]$, $L_2 = [-b + iT, a + iT]$, where $b = c(\log T)^{-1}$ with c as above. The expression (95.23) becomes

$$\frac{1}{2\pi i} \int_{L_0} h(s) \frac{x^s}{s} ds - \frac{1}{2\pi i} \int_L (\log s) \frac{x^s}{s} ds, \tag{95.25}$$

where $L_0 = L \setminus C$. In order to estimate the first integral, we note that there exists an absolute constant $c_1 > 0$ such that

$$|\log \zeta(s)| \le \log \zeta\left(1 + (\log |t|)^{-1}\right) + c_1,$$
$$\sigma \ge 1 - \frac{c}{\log |t|}, \quad |t| \ge 2, \tag{95.26}$$

as can be confirmed by dividing the integral (94.48) at $w = 1 + (\log |t|)^{-1} + it$ and applying (95.13). Thus, $h(s) \ll \log T$ for $s \in L_0$, $|t| \ge 2$. Since $h(s)$ is of course bounded for $s \in L_0$, $|t| \le 2$, we find that

$$\frac{1}{2\pi i} \int_{L_0} h(s) \frac{x^s}{s} ds \ll x^{-b/2}. \tag{95.27}$$

As for the second term in (95.25), its part other than that corresponding to C is readily estimated to be $O(x^{-b/2})$. Then, to the integral over C, we apply the change of variable $s = b \exp(i\theta)$, $-\pi < \theta < \pi$; we see that it is equal to

$$\begin{aligned} -\log b - \sum_{k=1}^{\infty} \frac{(-b \log x)^k}{k! \cdot k} &= -\log b - \int_0^{b \log x} (e^{-u} - 1) \frac{du}{u} \\ &= \log\log x - \int_0^1 (e^{-u} - 1) \frac{du}{u} - \int_1^{b \log x} e^{-u} \frac{du}{u} \\ &= \log\log x + c_E + O\left(\exp\left(-c(\log x)^{1/2}\right)\right), \end{aligned} \tag{95.28}$$

in which we have used the relation

$$\begin{aligned} -\int_0^1 (e^{-u} - 1) \frac{du}{u} - \int_1^{\infty} e^{-u} \frac{du}{u} \\ = -\int_0^{\infty} e^{-u} \log u \, du = -\Gamma'(1) = c_E; \end{aligned} \tag{95.29}$$

see (94.29). We finish the discussion on (95.18). On the other hand, to verify (95.19), it suffices to note that (95.26) gives, for $t \in \mathbb{R}$,

$$\zeta(s)^{-1} \ll \log(|t| + 2), \quad \sigma \geq 1 - \frac{c}{\log(|t| + 2)}. \tag{95.30}$$

This is trivial for $|t| \leq 2$; if $|t| \geq 2$, then note that

$$|\zeta(s)|^{-1} = \exp(-\log|\zeta(s)|) \leq \exp(|\log|\zeta(s)||) \leq \exp(|\log\zeta(s)|). \tag{95.31}$$

We end the proof.

For the sake of a later purpose, we add here the following on the tapered Möbius function:

Theorem 109 *With a variable $w > 1$ and a constant $\vartheta > 0$, we let, for $k \geq 1$,*

$$\mu^{(k)}(r) = \frac{1}{k!} (\vartheta \log w)^{-k} \sum_{j=0}^{k} (-1)^{k-j} \binom{k}{j} \mu^{(j,k)}(r), \tag{95.32}$$

where

$$\mu^{(j,k)}(r) = \begin{cases} \mu(r) \left(\log \dfrac{w^{1+j\vartheta}}{r} \right)^k & r \leq w^{1+j\vartheta}, \\ 0 & r > w^{1+j\vartheta}. \end{cases} \tag{95.33}$$

512   *Distribution of Prime Numbers*

Then we have

$$\mu^{(k)}(r) = \mu(r), \quad r \le w. \tag{95.34}$$

Also, we have, for any constant $\alpha > 0$,

$$\sum_{n=1}^{\infty} d_k(n) \left(\sum_{r|n} \mu^{(k)}(r) \right)^2 n^{-\theta} \ll 1, \quad \theta \ge 1 + \alpha/\log w, \tag{95.35}$$

where d_k is the divisor function defined as (16.11), and the implied constant depends on k, α, ϑ effectively.

Proof   If $r \le w$, then the sum in (95.32) is

$$\mu(r) \sum_{l=0}^{k} (-1)^{k-l} \binom{k}{l} (\log w)^l (\log r)^{k-l} \sum_{j=0}^{k} (-1)^{k-j} \binom{k}{j} (1 + j\vartheta)^l. \tag{95.36}$$

The inner sum equals $[(d/dw)^l e^w (e^{\vartheta w} - 1)^k]_{w=0}$, so it is $k! \, \vartheta^k$ if $l = k$, and 0 if $l < k$, which confirms (95.34). Then, let $\xi(r) = \mu(r)(\log z/r)^k$, $r \le z$, and $\xi(r) = 0, r > z$, with a sufficiently large variable $z > 0$. We consider the sum

$$P = \sum_{n=1}^{\infty} d_k(n) \left(\sum_{r|n} \xi(r) \right)^2 n^{-\eta}, \quad \eta = 1 + \frac{c}{\log z}, \tag{95.37}$$

with a constant $c > 0$. To show (95.35), it suffices to prove $P \ll (\log z)^{2k}$. Expanding the squares, we see that

$$P = \zeta(\eta)^k \sum_{r_1, r_2 \le z} \xi(r_1)\xi(r_2) \prod_{p|[r_1, r_2]} \left(1 - \left(1 - \frac{1}{p^\eta} \right)^k \right), \tag{95.38}$$

where we have used the fact that for a square-free a and for $\mathrm{Re}\, s > 1$

$$\sum_{n=1}^{\infty} \frac{d_k(an)}{n^s} = \zeta^k(s) \prod_{p|a} \left\{ \left(1 - \frac{1}{p^s} \right)^k \sum_{v=0}^{\infty} \frac{d_k(p^{v+1})}{p^{vs}} \right\}$$

$$= \zeta^k(s) a^s \prod_{p|a} \left(1 - \left(1 - \frac{1}{p^s} \right)^k \right); \tag{95.39}$$

see (15.18). Then by the same separation-device as the one employed in Note [18.7], i.e., (18.3), we get, with $f(r) = \prod_{p|r} \left(1 - (1 - p^{-\eta})^k \right)^{-1}$,

$$P = \zeta^k(\eta) \sum_{r_1, r_2 \leq z} \frac{\xi(r_1)\xi(r_2)}{f(r_1)f(r_2)} f(\langle r_1, r_2 \rangle)$$

$$= \zeta^k(\eta) \sum_{r_1, r_2 \leq z} \frac{\xi(r_1)\xi(r_2)}{f(r_1)f(r_2)} \sum_{r|r_1, r|r_2} (\mu * f)(r) \tag{95.40}$$

$$= \zeta^k(\eta) \sum_{r \leq z} \mu^2(r) \frac{(\mu * f)(r)}{f^2(r)} \left\{ R_r(z/r) \right\}^2,$$

where

$$R_r(y) = \sum_{\substack{u \leq y \\ \langle u, r \rangle = 1}} \frac{\mu(u)}{f(u)} \left(\log \frac{y}{u} \right)^k. \tag{95.41}$$

On noting that for $k \geq 1$

$$\frac{k!}{2\pi i} \int_{2-i\infty}^{2+i\infty} \frac{y^s}{s^{k+1}} ds = \begin{cases} (\log y)^k & y \geq 1, \\ 0 & y < 1, \end{cases} \tag{95.42}$$

as can be shown analogously to (95.1), we have

$$R_r(y) = \frac{k!}{2\pi i} \int_{2-i\infty}^{2+i\infty} \frac{U_r(s, \eta)}{\zeta^k(s + \eta)} \frac{y^s}{s^{k+1}} ds, \tag{95.43}$$

where

$$\frac{U_r(s, \eta)}{\zeta^k(s + \eta)} = \sum_{\substack{u=1 \\ \langle u, r \rangle = 1}}^{\infty} \frac{\mu(u)}{f(u)u^s}, \tag{95.44}$$

with

$$U_r(s, \eta) = \prod_{p|r} \left(1 - \frac{1}{p^{s+\eta}} \right)^{-k}$$

$$\times \prod_{p \nmid r} \left(1 - \frac{1}{p^{s+\eta}} \right)^{-k} \left(1 - \frac{1}{p^s} \left(1 - \left(1 - \frac{1}{p^\eta} \right)^k \right) \right), \tag{95.45}$$

which converges absolutely for $\sigma > -\frac{1}{2}$, for the last p-factor is $1 + O(p^{-2\sigma - 2\eta})$. Observing (95.30), we move the contour in (95.43) to $s = -c_2/\log(|t|+2)+it$ and find that

$$R_r(y) = k! \operatorname*{Res}_{s=0} \frac{U_r(s, \eta)}{\zeta(s + \eta)^k} \frac{y^s}{s^{k+1}} + O\left(\prod_{p|r} \left(1 + p^{-1/2} \right)^k \right). \tag{95.46}$$

To estimate this residue we take the contour integral on $|s| = (\eta - 1)/2$ and see that it is

514 Distribution of Prime Numbers

$$\ll y^{(\eta-1)/2} \prod_{p|r} \left(1 + p^{-1/2}\right)^k, \tag{95.47}$$

since $\zeta^k(s+\eta)s^{k+1} \gg (\eta-1)$; one may, of course, get a bound better than (95.47) by considering the Taylor expansion, but this suffices for our present purpose. Further we note that

$$\mu^2(r)\frac{(\mu * f)(r)}{f^2(r)} = \mu^2(r) \prod_{p|r} \left(1 - \frac{1}{p^\eta}\right)^k \left(1 - \left(1 - \frac{1}{p^\eta}\right)^k\right). \tag{95.48}$$

Inserting (95.46)–(95.48) into (95.40), we obtain $P \ll z^{\eta-1}\zeta^k(\eta)\zeta^k(2\eta-1)$, which is $O\left((\log z)^{2k}\right)$. We end the proof.

The assertion (95.34)–(95.35) will have an important application in §106.

Notes

[95.1]   Here we make a trivial but an essential observation: If there were only a finite number of complex zeros, then the right side of (95.6) would be continuous for any $x > 1$, which is absurd. That is to say, there ought to exist infinitely many complex zeros: in order to capture a single prime number, say 2, all of infinitely many complex zeros of the zeta-function have to be mobilized as far as one relies on (95.6). The same is pronounced more explicitly by the following result due to Landau (1911, p.553):

$$\sum_{0<\gamma<T} x^\rho = \begin{cases} -\dfrac{T}{2\pi} \log p + O(\log T) & \text{for } x = p^m, \\ O(\log T) & \text{for } x \neq p^m \end{cases}.$$

Note that this is not uniform in $x > 1$: unlike (95.5), the implied constants are dependent on x. It is suggested anyway that the beautiful formula (95.6) should be regarded as a statistical fact concerning the set of all complex zeros. We shall see, in §98 and subsequent sections, that the application of the statistical viewpoint towards complex zeros is indeed fundamental.

[95.2]   The alternative proof given above of Theorem 102 does not require any global nature of the zeta-function; indeed we do not need even the functional equation, while in his own proof of the theorem de la Vallée Poussin relied on it. What we have really required are Landau's Theorem 103, the inequality (95.8), and the almost trivial bound (95.10); see also Note [95.6].

[95.3]   The assertion (95.8) is equivalent to

$$\zeta(\sigma)^3|\zeta(\sigma+it)|^4|\zeta(\sigma+2it)| \geq 1, \quad \sigma > 1,$$

and this has an arithmetical interpretation which can be compared with (88.8): for any real number ω and $\operatorname{Re} s > 1$,

$$\sum_{n=1}^{\infty} \frac{|\sigma_{i\omega}(n)|^4}{n^s} = \zeta(s)^6 \zeta(s+i\omega)^4 \zeta(s-i\omega)^4 \zeta(s+2i\omega)\zeta(s-2i\omega)Y(s,\omega),$$

where $Y(s,\omega)$ is a Dirichlet series which converges absolutely for $\operatorname{Re} s > \frac{1}{2}$. Both imply immediately $\zeta(s) \neq 0$ on the line $\operatorname{Re} s = 1$, a fact that is known to be equivalent to the prime number theorem (12.9) according to the Tauberian theorem of S. Ikehara; that is, without entering into the critical strip we are able to establish the prime number theorem (12.9). See Montgomery–Vaughan (2006, Section 8.3 and p.277).

[95.4]   Riemann (1860, pp.679–680) mentioned a numerical observation due to Gauss and Goldschmidt that $\pi(x)$ is smaller than Li(x) up to $x = 3 \cdot 10^6$ with Li(x) as in Note [12.3]; and he continued that in any further study on this matter one should take account the influence of *periodischen Glieder* (see the explanation after (95.6)). This plan of Riemann was executed later by Littlewood (1914) and resulted in the following conclusion: As x tends to $+\infty$,

$$\pi(x) - \operatorname{Li}(x) = \Omega_{\pm} \left(x^{1/2}(\log x)^{-1} \log\log\log x \right),$$

without any hypothesis. Here, $f(x) = \Omega_{\pm}(g(x))$, $g(x) > 0$, means that both

$$\limsup f(x)/g(x) > 0 \quad \text{and} \quad \liminf f(x)/g(x) < 0$$

hold as x tends to a limit. See Montgomery–Vaughan (2006, pp.475–479) for a detailed proof of this remarkable discovery of Littlewood. The smallest x at which $\pi(x) - \operatorname{Li}(x)$ changes its sign from negative to positive is called the Skews number, as Skewes (1933) gave the first estimation of such an x. His bound was literally gigantic, and the latest improved bound is still far beyond the power of any modern computer to actually fix the Skews number. By the way, in order to establish Ω-results as this in analytic number theory one requires most always the existence of explicit expansions like (95.6) of the error-terms in question. Typical instances of such formulas other than that for $\psi(x)$ are in Voronoï (1904) and in Motohashi (1993, 1994). See Kaczorowski–Szydło (1997).

[95.5]   Let $N_0(T)$ denote the number of those complex zeros $\rho = \frac{1}{2} + i\gamma$, $0 < \gamma \leq T$. As a supporting evidence for RH, Selberg (1942) established $N_0(T) \gg T \log T$, so there exists an absolute constant $c_0 > 0$ such that $N_0(T)/N(T) > c_0$. His argument is closely related to his sieve method. By a method different from Selberg's, Levinson (1974) obtained $c_0 > \frac{1}{3}$, and Conrey (1989) improved it to $c_0 > \frac{2}{5}$. However, it is not known yet whether

516   *Distribution of Prime Numbers*

or not such assertions concerning the zeta-zeros on the critical line have any implication on the distribution of primes.

[95.6] We should add a simple proof due to Landau (1903a) of the prime number theorem

$$\pi(x) = \text{li}(x) + O\left(x\exp\left(-c(\log x)^{1/10}\right)\right),$$

which depends only on (95.8) and trivial bounds for $\zeta(s)$, $\zeta'(s)$ in a narrow neighborhood of the vertical line $\text{Re}\, s = 1$; so it is independent even of Theorem 103. It runs as follows: Suppose that $t > 0$ is sufficiently large. We take $N = [t]$ in (96.12) and its differentiated version, getting

$$\zeta(\sigma + it) \ll \log t, \quad \zeta'(\sigma + it) \ll (\log t)^2, \quad \sigma > 1 - (\log t)^{-1}.$$

On the other hand, by the inequality stated in Note [95.3], we get

$$|\zeta(\sigma + it)| \gg (\sigma - 1)^{3/4}(\log t)^{-1/4}, \quad \sigma > 1.$$

Also, for $\sigma > 1 - (\log t)^{-1}$,

$$\zeta(\sigma + it) - \zeta(1 + it) = \int_1^\sigma \zeta'(\alpha + it)d\alpha \ll |\sigma - 1|(\log t)^2.$$

Thus, there exist two absolute constants $a_1, a_2 > 0$ such that

$$\zeta(1 + it) > a_1(\sigma - 1)^{3/4}(\log t)^{-1/4} - a_2(\sigma - 1)(\log t)^2, \quad \sigma > 1.$$

Taking $\sigma = 1 + b(\log t)^{-9}$ with a small absolute constant $b > 0$, we obtain $\zeta(1 + it) \gg (\log t)^{-7}$. Hence, by the last integration formula we have, with absolute constants $a_3, a_4 > 0$,

$$\zeta(\sigma + it) > a_3(\log t)^{-7} - a_4|\sigma - 1|(\log t)^2, \quad \sigma > 1 - (\log t)^{-1}.$$

It is implied that with $a = a_3/2a_4$

$$\zeta(s)^{-1} \ll (\log t)^7, \quad \frac{\zeta'}{\zeta}(s) \ll (\log t)^9, \quad \sigma > 1 - a(\log t)^{-9},$$

which can be used in place of (95.30) and (95.13), respectively. The rest of the argument is omitted. The second half of this article of Landau marked a fundamental development in the history of the theory of algebraic numbers: As an immediate extension of the first half, which is given above, Landau established the prime ideal theorem for general algebraic number fields, much prior to Hecke's proof (1917b) of the functional equation for Dedekind zeta-functions. By the way, the use of the function $\sum_{n=1}^{\infty} |\sigma_{i\omega}(n)|^{2k}n^{-s}$, $\text{Re}\,(s) > 1$, with an integer $k \geq 3$ yields a better result than Landau's.

[95.7] The bound (95.35) for $k = 1$ is due to Selberg (1946b, §4)) and the general case to Motohashi (1978, Lemma 5). It indicates a remarkable cancellation in the sum $\sum_{r|n} \mu^{(k)}(r)$. The same does not take place for $k = 0$, a fact that can be inferred from the asymptotical nature of

$$\lim_{x \to \infty} \frac{1}{x} \sum_{n \leq x} \left(\sum_{r|n, r \leq z} \mu(r) \right)^{\ell} = \sum_{r_j \leq z, \, 1 \leq j \leq \ell} \frac{\mu(r_1)\mu(r_2) \cdots \mu(r_\ell)}{[r_1, r_2, \ldots, r_\ell]},$$

as $z \to \infty$, where the denominator is as in (14.1). See Balazard et al (2008).

[95.8] In the later part of §12 we faintly indicated that the approach to the distribution of prime numbers via the analytic continuation of $\zeta(s)$ to the half-plane $\mathrm{Re}\, s \leq 1$ is shrouded by a sort of indefinite uneasiness: it sounds worthwhile to ponder why one should know the zeros of $\zeta(s)$ while they do not have any immediately evident relevance to the concept of prime numbers. As a matter of fact, the use of contour integration like that in (95.1) can be replaced by real analytic means. Thus, it is inferred that we should be able to develop our discussion strictly in the region of absolute convergence of the Dirichlet series in question. The basis of this research framework was created by Halász (1968). His theory is capable to yield, among other things, the asymptotic formula

$$\sum_{n \leq x} f(n) = Cx + o(x)$$

for any multiplicative function f such that $|f(n)| \leq 1$ for all $n \in \mathbb{N}$ and its Dirichlet series $\lfloor f \rfloor(s)$ does not need to be presupposed to have analytic continuation to $\mathrm{Re}\, s < 1$ but only to satisfy the simple condition

$$\lfloor f \rfloor(s) = C(s-1)^{-1} + o((\sigma - 1)^{-1}), \quad \sigma \to 1 + 0,$$

uniformly for any bounded $|t|$. For instance, because of Note [95.3], the Möbius function is in this category, and thus, via (17.13), various sums of the von Mangoldt function can be regarded to be in the framework. It is, of course, quite interesting to compose the whole or rather the basic theory of the distribution of prime numbers solely by real analytic means, for it will not only remove, to a certain extent, that indefinite uneasiness but may also open a new perspective that gives us a hope to be able to enter into the realm that traditional approaches have so far failed to penetrate. Nevertheless, at least presently, we ought to point it out that the approach via analytic continuation has a definitive advantage since it is capable of producing both asymptotic formulas with explicit error terms as well as those exquisitely delicate Ω-results indicated in Note [95.4]. Because of this, in the present essays we shall stay in the paradigm established by Riemann.

518 *Distribution of Prime Numbers*

§96

The error term in the prime number theorem (95.14)–(95.15), which indicates the depth of the approximation to $\pi(x)$ by $\mathrm{li}(x)$, is a direct consequence of the zero-free region (95.13); and the latter depends on the estimation of $\zeta(s)$ in the critical strip that follows readily from (12.1). Although being a modest bound, without it the basic inequality (95.8) coupled with the function theoretical assertions in §94 is unable to yield (95.13). With this fact in mind, we shall set forth the methods due to Weyl and van der Corput, and Vinogradov, I.M., in the present and the next sections, respectively. The former gives a bound for $\zeta(\frac{1}{2} + it)$ that is classified as sub-convexity in the sense to be explained shortly; it is not visibly related to the issue concerning the zero-free region but has a definite relation to the distribution of the complex zeros, as will be made precise in due course. On the other hand, Vinogradov's method is effective when applied to the situation with s close to the right edge of the critical strip and produces a zero-free region that yields the hitherto best global bound for $|\pi(x) - \mathrm{li}(x)|$. Further, entering into §§98–100 on the asymptotic distribution of primes in short intervals, one will see that a combination of these two results and a few others yields remarkable consequences.

We state first the convexity bound:

$$\zeta\left(\tfrac{1}{2} + it\right) \ll t^{1/4} \log t, \quad t \geq 2. \tag{96.1}$$

This is implied immediately by Theorem 104 since $\zeta(s) \ll \log t$ on $\sigma = 1$, and $\zeta(s) \ll t^{1/2} \log t$ on $\sigma = 0$, both for $t \geq 2$; the former has been indicated in Note [95.6], and the latter is a consequence of (94.5) and (94.37). Because of this fact, any assertion $\zeta\left(\frac{1}{2} + it\right) \ll t^{\theta}$, $t \geq 2$, with $\theta < \frac{1}{4}$ is called a sub-convexity bound. We are going to establish such a bound.

To this end, we shall consider the estimation of a general exponential sum

$$F(N) = \sum_{N < n \leq 2N} e(f(n)), \quad e(x) = \exp(2\pi i x), \tag{96.2}$$

where $N \in \mathbb{N}$, and $f(x)$ is a real-valued function that is sufficiently smooth. If the points $\{e(f(n)) : N < n \leq 2N\}$ are distributed uniformly on the unit circle, then a significant cancellation among the summands is expected, and $|F(N)|$ will turn out to be markedly less than the trivial bound N. It was Weyl (1916) who devised the first effective method which enables one to detect a non-trivial cancellation even if $\{e(f(n))\}$ are not uniformly distributed. His idea is quite simple: Divide the interval $(N, 2N]$ into short intervals, and consider, instead of $F(N)$, the sub-sums

$$\sum_{0<u\leq U} e\big(f(u+M)\big), \quad M \approx N, \tag{96.3}$$

where $U = N^\alpha$, $0 < \alpha < 1$, is to be chosen appropriately. If f is of C^∞ class, as is usual in practice, then $f(u+M)$ can be well approximated by a polynomial $g(u) \in \mathbb{R}[u]$; hence, instead of (96.3), the sum

$$G(U) = \sum_{0<u\leq U} e\big(g(u)\big) \tag{96.4}$$

is to be estimated; this is called a Weyl sum. Then, via the identity (the Weyl shift)

$$|G(U)|^2 = \sum_r \sum_u e\big(g(u+r) - g(u)\big), \tag{96.5}$$

with an obvious abbreviation, the problem is reduced to that concerning the polynomial $g(u+r) - g(u)$ of u, the order of which is lower than that of g. Hence, repeating the same procedure, we eventually arrive at

$$S = \sum_{a<u\leq b} e\,(\alpha u + \beta), \quad \alpha, \beta \in \mathbb{R},$$

$$|S| \leq \min\left\{b - a, \frac{1}{|\sin\pi\alpha|}\right\} \leq \min\left\{b - a, \frac{1}{2\{\alpha\}}\right\}, \tag{96.6}$$

with the notation $\{\alpha\}$ as in (60.6). From this we are led to a bound for $|F(N)|$; its application to the function $(t/2\pi) \log x$ should yield a non-trivial bound for $\zeta(s)$.

Alternatively, one may apply the sum formula (60.1) of Poisson to $F(N)$, which is the argument of van der Corput (1921):

$$F(N) = \sum_{n=-\infty}^{\infty} \int_N^{2N} e\big(f(x) + nx\big)dx + \tfrac{1}{2}\big(e(f(2N) - e(f(N)))\big). \tag{96.7}$$

Or rather,

$$F(N) = \int_N^{2N} e\big(f(x)\big)dx$$
$$- \frac{1}{2\pi i} \sum_{\substack{n=-\infty \\ n\neq 0}}^{\infty} \frac{1}{n} \int_N^{2N} \frac{f'(x)}{f'(x) + n} d\big(e(f(x) + nx)\big) + O(1). \tag{96.8}$$

For instance, if

$$f'(x) \text{ is monotonic, and } |f'(x)| \leq \eta < 1, N \leq x \leq 2N, \tag{96.9}$$

520   *Distribution of Prime Numbers*

then $f'(x)/(f'(x)+n)$ is also monotonic, and each integral in this infinite sum is estimated to be $O(1/|n|)$ by the second mean value theorem for integrals. Hence, on (96.9)

$$F(N) = \int_N^{2N} e(f(x))dx + O(1), \tag{96.10}$$

where the implied constant may depend on η. On the other hand, if $f'(x)+\nu=0$ with an integer $\nu \neq 0$ has a solution in the interval $[N,2N]$, then the method of steepest descent comes into play. In any event, the estimation of $F(N)$ can be transformed into that of certain integrals. We shall describe the details of this procedure in the proof of the following sub-convexity bound:

Theorem 110

$$\zeta\left(\tfrac{1}{2}+it\right) \ll t^{1/6}(\log t)^{3/2}, \quad t \geq 2. \tag{96.11}$$

Proof   We first modify (12.1) as

$$\zeta(s) = \sum_{n=1}^N \frac{1}{n^s} + \frac{N^{1-s}}{s-1} - \frac{1}{2}N^{-s} + s\int_N^\infty \frac{[x]-x+\tfrac{1}{2}}{x^{s+1}}dx, \tag{96.12}$$

with $N>1$, $\sigma>0$ and $t \geq 1$. By the expansion (60.2) together with its bounded convergence (60.6), we have

$$\int_{N_1}^{N_2} \frac{[x]-x+\tfrac{1}{2}}{x^{s+1}}dx = \frac{1}{\pi}\sum_{m=1}^\infty \frac{1}{m}\int_{N_1}^{N_2} \frac{\sin(2m\pi x)}{x^{s+1}}dx. \tag{96.13}$$

If $t \leq N_1 < N_2$, then by integration by parts

$$\int_{N_1}^{N_2} x^{-\sigma-1}\exp\left(it\log x \pm 2m\pi ix\right)dx \ll N_1^{-\sigma-1}m^{-1}. \tag{96.14}$$

This implies that (96.12) yields analytic continuation of $\zeta(s)$ to the domain $\sigma > -1$; and

$$\zeta(s) = \sum_{n<t} \frac{1}{n^s} + O(t^{-\sigma}), \quad -1 < \sigma; 1 \leq t. \tag{96.15}$$

If $\sigma = \tfrac{1}{2}$, then the part $n < t^{1/3}$ can be ignored obviously; also, we may assume that t is sufficiently large.

Thus, we are concerned with the sum

$$Z(N,t) = \sum_{N<n\leq 2N} n^{it} = \sum_{N<n\leq 2N} e\left(\frac{t}{2\pi}\log n\right), \quad t^{1/3} \leq N < t. \tag{96.16}$$

In order to apply the Weyl shift (96.5), we modify (96.16) as

$$Z(N,t) = \frac{1}{U} \sum_{N < n \leq 2N} \sum_{0 < u \leq U} (n+u)^{it} + O(U), \qquad (96.17)$$

with

$$U = Nt^{-1/3} < N^{1/2}t^{1/6}. \qquad (96.18)$$

We have, by Cauchy's inequality,

$$|Z(N,t)| \ll \frac{N^{1/2}}{U} \left(\sum_{0 < u < v \leq U} |Z(N,t;u,v)| \right)^{1/2} + N^{1/2}t^{1/6}. \qquad (96.19)$$

Here

$$Z(N,t;u,v) = \sum_{N < n \leq 2N} e\big(h(n)\big),$$
$$h(n) = (t/2\pi) \log \big((n+u)/(n+v)\big). \qquad (96.20)$$

We have, by (96.8),

$$Z(N,t;u,v) = \int_N^{2N} e\big(h(x)\big)dx$$
$$- \frac{1}{2\pi i} \sum_{n \neq 0} \frac{1}{n} \int_N^{2N} \frac{h'(x)}{h'(x)+n} de\big(h(x)+nx\big) + O(1). \qquad (96.21)$$

The saddle point x_n of the integrand corresponding to n satisfies

$$\frac{t}{2\pi} \cdot \frac{(v-u)}{(x_n+u)(x_n+v)} + n = 0. \qquad (96.22)$$

Thus, if x_n is contained in the interval of integration, then n should be in the interval

$$J = \left[-(1/\pi)(v-u)tN^{-2}, \ -(1/4\pi)(v-u)tN^{-2} \right]. \qquad (96.23)$$

Because of (96.18) we need to consider the application of the method of steepest descent only when $N \leq t^{2/3}$, for otherwise (96.23) will not contain any integer.

We shall consider the estimation of the saddle terms in (96.21), that is, those with $n \in J$. We have obviously

$$\frac{1}{n} \int_N^{2N} h'(x)e\big(h(x)+nx\big)dx = -\int_N^{2N} e\big(h(x)+nx\big)dx + O\big(1/|n|\big); \qquad (96.24)$$

Distribution of Prime Numbers

and to the integral on the right side we apply the division

$$\left(\int_N^{x_n-\delta} + \int_{x_n-\delta}^{x_n+\delta} + \int_{x_n+\delta}^{2N} \right) e\big(h(x) + nx\big)dx, \tag{96.25}$$

where $\delta > 0$ is to be determined later. We write the first integral as

$$\frac{1}{2\pi i} \int_N^{x_n-\delta} \frac{1}{h'(x) + n} de\big(h(x) + nx\big) \tag{96.26}$$

and note that

$$\left| h'(x) + n \right| = \left| \int_{x_n}^{x} h''(\xi)d\xi \right| \gg \delta(v - u)tN^{-3}. \tag{96.27}$$

Hence, (96.26) is $O\left(N^3/(\delta t(v - u))\right)$ by the second mean value theorem for integrals, as $h'(x) + n$ is obviously monotonic in the relevant interval. The same bound holds with the third integral in (96.25). The second integral is $O(\delta)$. We optimize δ by $\delta = N^{3/2}(t(v - u))^{-1/2}$, which is well less than N the length of the integration interval.

We now have

$$Z(N, t; u, v) = \int_N^{2N} e\big(h(x)\big)dx$$

$$- \frac{1}{2\pi i} \sum_{\substack{n \notin J \\ n \neq 0}} \frac{1}{n} \int_N^{2N} \frac{h'(x)}{h'(x) + n} de\big(h(x) + nx\big) + O\left((t(v - u))^{1/2}N^{-1/2}\right). \tag{96.28}$$

Applying integration by parts to the first integral, we see that

$$Z(N, t; u, v)$$

$$\ll N^2(t(v - u))^{-1} + \sum_{\substack{n \notin J \\ n \neq 0}} \frac{1}{|n|} \sup_{N \leq x \leq 2N} \frac{|h'(x)|}{|h'(x) + n|} \tag{96.29}$$

$$+ (t(v - u))^{1/2}N^{-1/2}$$

$$\ll N^2(t(v - u))^{-1} + \log(2 + t(v - u)N^{-2}) + (t(v - u))^{1/2}N^{-1/2},$$

in which the second term is the result of estimating the sum over n by splitting it into two parts according as $|n|$ is greater or less than $2 + t(v - u)/N^2$. Hence, if $N \leq t^{2/3}$, then

$$\sum_{0 < u < v \leq U} |Z(N, t; u, v)| \ll N^2 t^{-1} U \log U + U^2 \log t + N^{-1/2} t^{1/2} U^{5/2} \tag{96.30}$$

$$\ll U^2 t^{1/3} \log t.$$

On the other hand, if $t^{2/3} < N < t$, then there exists no saddle term, that is, $J \cap \mathbb{Z}$ is empty and the last term in (96.28) does not appear. We get the same bound as (96.30).

Therefore, returning to (96.19), we find that

$$\sum_{N<n\le 2N} n^{it} \ll N^{1/2} t^{1/6} (\log t)^{1/2}, \quad 2 \le N < t; \tag{96.31}$$

hence

$$\sum_{n\le M} n^{it} \ll M^{1/2} t^{1/6} (\log t)^{1/2}, \quad 2 \le M < t, \tag{96.32}$$

as can be seen by dissecting the interval $[1, M]$ into subintervals $(N, 2N]$ and summing over N. Inserting this into (96.15), we end the proof of (96.11).

Notes

[96.1] The reasoning (96.14)–(96.15) can be completed by an explicit application of the method of steepest descent. Such a discussion is developed in Titchmarsh (1951, Chapter 4). It yields the approximate functional equation for the zeta-function: In the typical situation $s = \frac{1}{2} + it$, $t \ge 2$, it asserts that

$$\zeta\left(\tfrac{1}{2} + it\right) = \sum_{n\le(t/2\pi)^{1/2}} \frac{1}{n^{1/2+it}} + X(\tfrac{1}{2} + it) \sum_{n\le(t/2\pi)^{1/2}} \frac{1}{n^{1/2-it}} + O(t^{-1/4}),$$

where $X(s) = 2^s \pi^{s-1} \sin(\pi s/2) \Gamma(1-s)$; note that this implies $\zeta\left(\tfrac{1}{2}+it\right) \ll t^{1/4}$, which is slightly better than (96.1). A *Nachlass* of Riemann contains an asymptotic expansion of this O-term, as has been revealed in its strenuous rework by Siegel (1932). It is indeed a breathtaking display of Riemann's prowess in theoretical and numerical computation of the zeta-function based on the integral representation (12.5) and the method of steepest decent. It must have served as an empirical support for his statement (R_2) or RH, for it allowed him to fix a few initial complex zeros and led him to the fateful discovery that they are all on the critical line $\operatorname{Re} s = \frac{1}{2}$, as we have indicated already in §12. One should know that Riemann's formula is still the basis in today's gigantic computation of values, especially zeros, of his zeta-function. A readable account can be found in Edwards (1974); see also the readable account by Borwein et al. (2000) on the various aspects of the computational theory of the zeta-function. By the way, for an extension of Riemann's asymptotic formula see Motohashi (1987): it is shown that

$$\zeta^2\left(\tfrac{1}{2} + it\right) = \sum_{n\le t/2\pi} \frac{d(n)}{n^{1/2+it}} + X^2(\tfrac{1}{2} + it) \sum_{n\le t/2\pi} \frac{d(n)}{n^{1/2-it}} + O(t^{-1/6}),$$

524 *Distribution of Prime Numbers*

where $d(n)$ is the divisor function. This O-term as well admits an asymptotic expansion: It begins with the term $-2X(\frac{1}{2} + it)\Delta(t/2\pi)(t/2\pi)^{-1/2}$ with Δ as in Note [16.10], so the above exponent $-\frac{1}{6}$ could be replaced by a smaller one. Any result of similar nature is not known yet for powers higher than 2 of the zeta-function.

[96.2]   The sub-convexity exponent $\frac{1}{6}$ in (96.11) was first obtained by Weyl (1921), rather implicitly though; and later by van der Corput (1921) by the argument given above (see Titchmarsh (1951, Chapter 5)); a comprehensive account is developed in Graham–Kolesnik (1991). In this context, the Lindelöf exponent refers to

$$\mu(\sigma) = \inf\left\{\xi : \zeta(\sigma + it) \ll t^{\xi \mid \varepsilon}, \forall t \ge 2\right\};$$

here we adopt the notation commonly used, despite conflict with that for the Möbius function. We have $\mu(0) = \frac{1}{2}$, $\mu(1) = 0$; thus Theorem 104 implies that $\mu(\sigma) \le \frac{1}{2}(1 - \sigma)$ for $0 \le \sigma \le 1$; and in particular $\mu(\frac{1}{2}) \le \frac{1}{4}$, which is essentially equivalent to (96.1). It is conjectured that $\mu(\sigma) = \frac{1}{2} - \sigma$ for $0 \le \sigma \le \frac{1}{2}$ and $= 0$ for $\sigma \ge \frac{1}{2}$, which is attributed to Lindelöf (1908). Its truth is known to follow from that of RH (Titchmarsh (1951, Chapters XIII and XIV)); that is, the bound $\zeta\left(\frac{1}{2} + it\right) \ll (|t| + 1)^{\varepsilon}, \forall t \in \mathbb{R}$, is predicted to hold. Because of this, the efforts to achieve anything lower than the sub-convexity exponent $\frac{1}{6}$ have resulted in a collection of methods which can be applied to a variety of exponential sums. A modern innovation in this field is due to Bombieri et al (1986); for a readable summary of the salient points of their idea, see Ivić (1991, pp.25–27) as well as the monograph of Graham and Kolesnik cited above. This method was pursued, more recently, infusing the new idea called decoupling curves; thus, Bourgain (2017) achieved hitherto the best exponent $\mu(\frac{1}{2}) \le \frac{13}{84} = \frac{1}{6} - \frac{1}{84}$.

[96.3]   The exponent $\frac{1}{6}$ in (96.11) may be regarded as a universal constant because of the following evidence: The vast family of automorphic L-functions, all of which satisfy functional equations and Euler product expansions similar to those of the square of the zeta-function (Euler products of degree 2), have the common sub-convexity exponent not greater than $\frac{1}{3} = 2 \times \frac{1}{6}$ in a quite uniform way. See Ivić (2001) and Jutila–Motohashi (2005). See also Note [101.1].

§97

The method of van der Corput described in the preceding section may turn out to be ineffective if $|f'(x)|$ is large, for then there can appear uncontrollably many saddle terms. For instance, if t is large and N is relatively small, then the point $e(h(n))$ in (96.20) revolves too rapidly for the method to detect cancellation, as the interval (96.23) contains too many integers which offset the saving arising from (96.25). This fact is reflected in that (96.29) and thus (96.31) is trivial if $N \ll t^{1/3}$.

Vinogradov's method (1937) is effective in such cases as $N \ll t^{1/K}$ with K large. In the heart of his idea is a reduction of the problem to counting the number of solutions to a system of indefinite equations which can be treated with simple arithmetical means.

Thus, if $g(x) = \sum_{l=0}^{k} a_l x^l \in \mathbb{R}[x]$ in the Weyl sum (96.4), then for any integer $q \geq 1$

$$|G(U)|^{2q} = \sum_{\mathbf{h}} J_{q,k}(U, \mathbf{h}) \prod_{j=1}^{k} e(a_j h_j), \tag{97.1}$$

where $\mathbf{h} = \{h_1, h_2, \ldots, h_k\} \in \mathbb{Z}^k$, and $J_{q,k}(U, \mathbf{h})$ is the number of integer solutions to the following system of indefinite equations:

$$\left(x_1^l + x_2^l + \cdots + x_q^l\right) - \left(x_{q+1}^l + x_{q+2}^l + \cdots + x_{2q}^l\right) = h_l, \tag{97.2}$$
$$1 \leq x_j \leq U, \quad 1 \leq j \leq 2q, \quad 1 \leq l \leq k.$$

The reduction to (96.6) means that Weyl used (97.1) with $q = 2^{k-2}$. The number of the participating variables increases exponentially along with k an indicator of the revolving speed of the points $e(g(n))$, and the use of (96.6) becomes wasteful if k is large; observe that if we apply Weyl's method to the sum (96.20) under the situation where $\log t / \log N$ is large, then the order of the polynomial, which approximates $h(n)$ well, ought to be large. Vinogradov discovered, via the analysis of $J_{q,k}(U, \mathbf{h})$, that it is not necessary to take q so large as Weyl did. Vinogradov's choice of q is of polynomial order in k; that is a great, or more correctly a tremendous improvement over Weyl's. In what follows, we shall appreciate the essentials of Vinogradov's method in the light of its application to the sum $Z(N, t)$.

Assuming that N is sufficiently large and $N \leq t$, we begin with the expression

$$|Z(N,t)| \leq \frac{1}{U^2} \sum_{N < n \leq 2N} |Y(n,t)| + O(U^2), \tag{97.3}$$

where

$$Y(n,t) = \sum_{u=1}^{U} \sum_{v=1}^{U} e\left(\frac{t}{2\pi} \log\left(1 + \frac{uv}{n}\right)\right), \tag{97.4}$$

526   *Distribution of Prime Numbers*

with any $U \in \mathbb{N}$. Also, we put, for any $k \in \mathbb{N}$,

$$W(n,t) = \sum_{u=1}^{U} \sum_{v=1}^{U} e\left(\alpha_1 uv + \alpha_2 (uv)^2 + \cdots + \alpha_k (uv)^k\right),$$

$$\alpha_v = (-1)^{v-1} \frac{t}{2\pi v n^v}. \tag{97.5}$$

Then

$$Y(n,t) = W(n,t) + O\left(tU^{2(k+2)} n^{-k-1}\right). \tag{97.6}$$

Thus,

$$|Z(N,t)| \le \frac{1}{U^2} \sum_{N < n \le 2N} |W(n,t)| + O\left(U^2 + tN(U^2/N)^{k+1}\right). \tag{97.7}$$

We let

$$U = [N^{\delta/2}], \quad k = \left[\frac{\log t}{(1-\delta)\log N}\right] + 1, \quad 0 < \delta < 1, \tag{97.8}$$

so that

$$|Z(N,t)| \le \frac{1}{U^2} \sum_{N < n \le 2N} |W(n,t)| + O\left(N^\delta\right). \tag{97.9}$$

With any $q \in \mathbb{N}$ we have, by the Hölder inequality,

$$|W(n,t)|^{4q^2} U^{2q-4q^2}$$

$$\le \left(\sum_{u=1}^{U} \left|\sum_{v=1}^{U} e\left(\alpha_1 uv + \alpha_2 (uv)^2 + \cdots + \alpha_k (uv)^k\right)\right|^{2q}\right)^{2q}$$

$$\le \left(\sum_{\mathbf{h}} J_{q,k}(U,\mathbf{h}) \left|\sum_{u=1}^{U} e\left(\alpha_1 h_1 u + \alpha_2 h_2 u^2 + \cdots + \alpha_k h_k u^k\right)\right|\right)^{2q}$$

$$\le U^{4q^2 - 2q} \sum_{\mathbf{h}} J_{q,k}(U,\mathbf{h}) \left|\sum_{u=1}^{U} e\left(\alpha_1 h_1 u + \alpha_2 h_2 u^2 + \cdots + \alpha_k h_k u^k\right)\right|^{2q}. \tag{97.10}$$

We note that

$$J_{q,k}(U,\mathbf{h}) \le J_{q,k}(U,\mathbf{0}) = J_{q,k}(U), \tag{97.11}$$

with the zero-vector $\mathbf{0}$, as is obvious from

$$J_{q,k}(U,\mathbf{h}) = \int_{[0,1]^k} \left|\sum_{u=1}^{U} e\left(\theta_1 u + \theta_2 u^2 + \cdots + \theta_k u^k\right)\right|^{2q}$$

$$\times e\left(-\theta_1 h_1 - \theta_2 h_2 - \cdots - \theta_k h_k\right) d\theta_1 d\theta_2 \cdots d\theta_k. \tag{97.12}$$

Hence,

$$|W(n,t)|^{4q^2}U^{4q-8q^2}$$

$$\le J_{q,k}(U)\left|\sum_{\mathbf{h},\,\mathbf{h}'}J_{q,k}(U,\mathbf{h}')e(\alpha_1 h_1 h_1' + \alpha_2 h_2 h_2' + \cdots + \alpha_k h_k h_k')\right| \tag{97.13}$$

$$\le \{J_{q,k}(U)\}^2\prod_{\nu=1}^k S_\nu,$$

where

$$S_\nu = \sum_{|\mu|<U_\nu}\min\left(2U_\nu,\frac{1}{2\{\alpha_\nu\mu\}}\right), \quad U_\nu = qU^\nu; \tag{97.14}$$

see (96.6). To estimate S_ν, we write

$$\alpha_\nu = (-1)^{\nu-1}\frac{t}{2\pi\nu n^\nu} = \frac{(-1)^{\nu-1}}{l} + \frac{\theta}{l^2}, \tag{97.15}$$

$$l = [2\pi\nu n^\nu/t], \quad |\theta| \le 1.$$

Assuming that l is sufficiently large,

$$S_\nu \ll (U_\nu/l+1)\max_Q\sum_{\mu=1}^l\min\left(U_\nu,\frac{1}{\{\alpha_\nu(\mu+lQ)\}}\right)$$

$$= (U_\nu/l+1)\max_Q\sum_{\mu=1}^l\min\left(U_\nu,\frac{1}{\{a(\mu)/l+b(\mu)/l^2\}}\right), \tag{97.16}$$

where $a(\mu) = (-1)^{\nu-1}\mu + [\theta Q]$ and $b(\mu) = \theta\mu + (\theta Q - [\theta Q])l$. We classify $a(\mu)$ with respect to the modulus l, and note that $|b(\mu)|/l^2 \le 2/l$, and if $a(\nu) \equiv j \bmod l$, $3 \le j \le l-3$, then $\{a(\mu)/l+b(\mu)/l^2\} \gg j/l$; thus,

$$S_\nu \ll (U_\nu/l+1)(U_\nu + l\log l) \ll U_\nu^2\left(1/l + 1/U_\nu + l/U_\nu^2\right)\log l. \tag{97.17}$$

We observe here that if ν is such that with a small $\delta > 0$

$$N^{(1-\delta/3)\nu} \le t \le N^{(1-\delta/4)\nu}, \tag{97.18}$$

then l is large enough, and we get

$$S_\nu \le U_\nu^2 t^{-\delta/5}. \tag{97.19}$$

Returning to (97.13), we use this bound for those ν satisfying (97.18), and the trivial bound $S_\nu \le 4U_\nu^2$ otherwise; note that there exist $\nu < k$ that satisfy (97.18) because of (97.8). We obtain

$$|W(n,t)|^{4q^2} \le (2q)^{2k}U^{8q^2-4q+k(k+1)}\{J_{q,k}(U)\}^2 t^{-c\delta^2\log t/\log N} \tag{97.20}$$

with an absolute constant $c > 0$.

528 Distribution of Prime Numbers

Our next task is to estimate $J_{q,k}(U)$; repeating the definition (97.2), it is to bound the number of integer solutions to the system of indefinite equations

$$x_1^l + x_2^l + \cdots + x_q^l = x_{q+1}^l + x_{q+2}^l + \cdots + x_{2q}^l,$$
$$1 \le x_j \le U, \ 1 \le j \le 2q, \ 1 \le l \le k. \tag{97.21}$$

To this end, we shall develop a reduction procedure: The estimation will be transformed into that of $J_{q-k,k}([U/p]+1)$ with an appropriately chosen prime p; and it will be seen that an essential ingredient of our argument is the following assertion:

Theorem 111 *Let p be a prime number greater than k. Let X be the number of solutions of the system of congruence equations*

$$x_1^l + x_2^l + \cdots + x_k^l \equiv \lambda_l \bmod p^l, \quad 1 \le l \le k, \tag{97.22}$$

under the condition

$$D \le x_j < D + Mp^k, \quad 1 \le j \le k; \quad x_j \not\equiv x_{j'} \bmod p, \quad j \ne j'. \tag{97.23}$$

Then, we have

$$X \le k! M^k p^{k(k-1)/2}. \tag{97.24}$$

Proof We may suppose, without loss of generality, that $D = 0$; to confirm this, replace x_j by $x_j' + D$ and rewrite (97.22) in terms of $\{x_j'\}$. Let

$$x_j = x_{j0} + x_{j1}p + \cdots + x_{jk}p^k,$$
$$x_{j0} \ne x_{j'0}, \quad j \ne j', \tag{97.25}$$
$$0 \le x_{jr} < p, \ 1 \le r < k; \ 0 \le x_{jk} < M.$$

Then by (97.22)

$$x_{10}^l + \cdots + x_{k0}^l \equiv \lambda_l \bmod p, \quad 1 \le l \le k. \tag{97.26}$$

Thus $\{x_{10}, x_{20}, \dots, x_{k0}\}$ is a set of all different solutions to a congruence equation of order k with respect to the modulus p. Since the coefficients of this equation are uniquely determined by $\{\lambda_j\} \bmod p$ (see Note [97.4]), the set of its solutions is unique, save for permutations; hence the number of solutions to (97.26) is at most $k!$. Next, suppose that l is such that $1 \le l \le k - 1$ and the column vectors ${}^t\{x_{jh} : 1 \le j \le k\}, 0 \le h \le l - 1$, are all fixed; and consider the system of congruence equations

$$x_1^m + \cdots + x_k^m \equiv \lambda_m \bmod p^{l+1}, \quad l+1 \le m \le k. \tag{97.27}$$

This is equivalent to the system of linear congruence equations

$$x_{10}^{m-1}x_{1l} + \cdots + x_{k0}^{m-1}x_{kl} \equiv \lambda_m' \bmod p, \quad l+1 \le m \le k, \tag{97.28}$$

where

$$\lambda'_m p^l = \lambda_m - \sum_{j=1}^{k} \left(x_{j0} + x_{j1}p + \cdots + x_{jl-1}p^{l-1} \right)^m. \tag{97.29}$$

The rank of the coefficient matrix of the system (97.28) is equal to $k - l$ (see Note [97.4] again). Hence, the number of solutions ${}^t\{x_{jl} : 1 \leq j \leq k\}$ is equal to p^l. Further, the number of possible ${}^t\{x_{jk} : 1 \leq j \leq k\}$ is less than or equal to M^k. This ends the proof.

We return to (97.21) and show the following crucial fact:

$$\begin{array}{c} \text{let } k \text{ be sufficiently large and satisfy} \\ (2k)^{3k} \leq U, \quad k(k+1) \leq q. \end{array} \tag{97.30}$$

Then there exists a U_1 such that $U^{1-1/k} \leq U_1 \leq 4U^{1-1/k}$, and

$$J_{q,k}(U) < 2^{4q} U^{2q/k+(3k-5)/2} J_{q-k,k}(U_1). \tag{97.31}$$

To begin with, we take a prime p such that $\frac{1}{2}U^{1/k} < p < U^{1/k}$; see (12.15) or rather (95.15). We put $U_1 = [U/p] + 1$. Obviously we have $J_{q,k}(U) \leq J_{q,k}(pU_1)$. This is the same as to deal with, instead of (97.21), the system

$$\begin{array}{c} \left(x_1 + py_1 \right)^l + \cdots + \left(x_q + py_q \right)^l \\ = \left(x_{q+1} + py_{q+1} \right)^l + \cdots + \left(x_{2q} + py_{2q} \right)^l, \\ 1 \leq x_j \leq p, \ 0 \leq y_j < U_1, 1 \leq j \leq 2q, \ 1 \leq l \leq k. \end{array} \tag{97.32}$$

Two cases are to be considered separately: (I) both the sets $\{x_j\}$ and $\{x_{q+j}\}$, $1 \leq j \leq q$, contain at least k distinct elements, and (II) otherwise.

We shall estimate the number J_1 of the solutions to (97.32) in the case (I). Let J'_1 be the number of solutions to (97.32) which satisfy

$$x_j \neq x_{j'}, \quad x_{q+j} \neq x_{q+j'}, \ 1 \leq j < j' \leq k. \tag{97.33}$$

Obviously $J_1 \leq \binom{q}{k}^2 J'_1 \leq q^{2k} J'_1$. We then introduce

$$S(x) = \sum_{0 \leq y < U_1} e\left(\theta_1 (x + py) + \theta_2 (x + py)^2 + \cdots + \theta_k (x + py)^k \right). \tag{97.34}$$

We have

$$\begin{aligned} J'_1 &= \int_{[0,\,1]^k} \left| \sum_{x_1,\dots,x_k} S(x_1) \cdots S(x_k) \right|^2 \left| \sum_{x=1}^{p} S(x) \right|^{2q-2k} d\theta_1 \cdots d\theta_k \\ &\leq p^{2q-2k} J''_1, \end{aligned} \tag{97.35}$$

with $\{x_j\}$ as in (97.33) and

$$J_1'' = \max_{1 \le x \le p} \int_{[0,1]^k} \left| \sum_{x_1,\ldots,x_k} S(x_1)\cdots S(x_k) \right|^2 |S(x)|^{2q-2k} d\theta_1 \cdots d\theta_k, \quad (97.36)$$

where Hölder's inequality has been applied. The last integral equals the number of solutions $\{x_j, y_j : 1 \le j \le 2q\}$ to the system

$$\begin{aligned}
\left(x_1 - x + py_1\right)^l &+ \cdots + \left(x_k - x + py_k\right)^l \\
&- \left(x_{q+1} - x + py_{q+1}\right)^l - \cdots - \left(x_{q+k} - x + py_{q+k}\right)^l \quad (97.37) \\
&= p^l\left(y_{k+1}^l + \cdots + y_q^l - y_{q+k+1}^l - \cdots - y_{2q}^l\right), \quad 1 \le l \le k,
\end{aligned}$$

under the condition (97.33), while x is fixed, as can be seen by expanding the expression on the left side. Let $J_1''(\mathbf{h})$ with $\mathbf{h} = \{h_1, \ldots, h_k\}$ be the number of solutions to the system

$$\begin{aligned}
\left(x_1 - x + py_1\right)^l &+ \cdots + \left(x_k - x + py_k\right)^l \quad (97.38) \\
&- \left(x_{q+1} - x + py_{q+1}\right)^l - \cdots - \left(x_{q+k} - x + py_{q+k}\right)^l = p^l h_l, \ 1 \le l \le k.
\end{aligned}$$

Then

$$J_1'' \le \sum_{\mathbf{h}} J_1''(\mathbf{h}) J_{q-k,k}(U_1, \mathbf{h}) \le J_{q-k,k}(U_1) \sum_{\mathbf{h}} J_1''(\mathbf{h}). \quad (97.39)$$

The last sum equals the number of solutions to the system

$$\begin{aligned}
\left(x_1 - x + py_1\right)^l &+ \cdots + \left(x_k - x + py_k\right)^l \equiv \left(x_{q+1} - x + py_{q+1}\right)^l + \cdots \\
&+ \left(x_{q+k} - x + py_{q+k}\right)^l \bmod p^l, \quad 1 \le l \le k,
\end{aligned}$$

$$(97.40)$$

under (97.33). When any one of the sequences $\{x_{q+1}, \ldots, x_{q+k}, y_{q+1}, \ldots, y_{q+k}\}$, the number of which is not greater than $(pU_1)^k$, is fixed, the number of solutions to (97.40) is bounded by (97.24). That is, we have

$$J_1'' \le k! \left([U/p^k] + 1\right)^k (pU_1)^k p^{k(k-1)/2} J_{q-k,k}(U_1). \quad (97.41)$$

Collecting these, we get

$$\begin{aligned}
J_1 &\le q^{2k} p^{2q-2k} k! \left([U/p^k] + 1\right)^k (pU_1)^k p^{k(k-1)/2} J_{q-k,k}(U_1) \\
&< 2^{4q-1} U^{2q/k+(3k-5)/2} J_{q-k,k}(U_1),
\end{aligned} \quad (97.42)$$

where we have used the easy bounds $q^{2k} \le 2^q$, $k! \le 2^q$, $([U/p^k] + 1)^k \le (2^k + 1)^k \le 2^q$, $(pU_1)^k \le (U + p)^k < 2U^k$ following from (97.30) and the choice of p. This ends the treatment of the case (I).

The estimation of the number J_2 of the solutions in the case (II) is of no difficulty: Let either $\{x_j\}$ or $\{x_{q+j}\}$, $1 \leq j \leq q$, contains at most $k - 1$ different elements. Then

$$
\begin{aligned}
J_2 &= \sum_{\substack{x_1,\ldots,x_q \\ x_{q+1},\ldots,x_{2q}}} \int_{[0,1]^k} S(x_1)\cdots S(x_q)\overline{S(x_{q+1})}\cdots\overline{S(x_{2q})}d\theta_1\cdots d\theta_k \\
&\leq \frac{1}{2q} \sum_{\substack{x_1,\ldots,x_q \\ x_{q+1},\ldots,x_{2q}}} \int_{[0,1]^k} \left\{|S(x_1)|^{2q} + \cdots + |S(x_{2q})|^{2q}\right\} d\theta_1\cdots d\theta_k \quad (97.43) \\
&= J_{q,k}(U_1) \sum_{\substack{x_1,\ldots,x_q \\ x_{q+1},\ldots,x_{2q}}} 1 \leq J_{q,k}(U_1)\cdot 2p^q \binom{p}{k-1}(k-1)^q,
\end{aligned}
$$

as is readily seen that the integral of $|S(x_\nu)|^{2q}$ is equal to $J_{q,k}(U_1)$ and that the total number of ways of choosing q times one of pre-selected $k - 1$ elements from a set of p elements is equal to $\binom{p}{k-1}(k-1)^q$. Hence, on noting that $J_{q,k}(U_1) \leq U_1^{2k}J_{q-k,k}(U_1)$, we are led to

$$
\begin{aligned}
J_2 &< 2(k-1)^q p^{q+k-1} U_1^{2k} J_{q-k,k}(U_1) \\
&< 2^{4q-1} U^{2q/k+(3k-5)/2} J_{q-k,k}(U_1),
\end{aligned} \quad (97.44)
$$

where we have used (97.30) again; in particular $k^q < U^{q/3k}$. We end the proof of (97.31).

As a consequence of (97.31) we obtain Vinogradov's mean value theorem; this naming comes from the integral representation $(97.12)_{\mathbf{h=0}}$:

Theorem 112 *Let the real number U and the integer k be sufficiently large, and let the integer q satisfy*

$$
k(k + \tau) \leq q \quad (97.45)
$$

with an integer $\tau \geq 0$. Then we have

$$
J_{q,k}(U) \leq (4k)^{4q\tau} U^{2q-k(k+1)/2+\eta(k,\tau)}, \quad (97.46)
$$

with

$$
\eta(k,\tau) = \tfrac{1}{2}k(k+1)\left(1 - \frac{1}{k}\right)^\tau. \quad (97.47)
$$

Remark 1 Our argument provides U and k with explicit lower bounds, although we skip the details. The coefficient $(4k)^{4q\tau}$ is not best possible but suffices for our purpose. Also, it should be mentioned that heuristically the optimal bound is $J_{q,k}(U) \ll U^{2q-k(k+1)/2}$, as can be seen by counting the

532   *Distribution of Prime Numbers*

constraints imposed on the $2q$ variables by each of the lth equation of (97.2); hence, the bound (97.46) is quite effective, provided k, τ are both large.

Proof Since (97.46) is trivial if $\tau = 0$, we shall proceed by induction with respect to τ. Thus, we assume that the above assertion is valid when $\tau = m$; and we consider the situation with $k(k + m + 1) \leq q$. We shall deal first with the case

$$(2k)^{3k(k/(k-1))^m} \leq U. \tag{97.48}$$

We apply (97.31). Since $k(k + m) \leq q - k$, we may use (97.46) to bound $J_{q-k,q}(U_1)$. Hence

$$J_{q,k}(U) \leq 2^{4q} U^{2q/k+(3k-5)/2} (4k)^{4(q-k)m} U_1^{2(q-k)-k(k+1)/2+\eta(k,m)}$$

$$\leq 2^{4q} U^{2q/k+(3k-5)/2} (4k)^{4(q-k)m} 2^{4(q-k)} U^{(1-1/k)(2q-2k-k(k+1)/2+\eta(k,m))}$$

$$\leq (4k)^{4q(m+1)} U^{2q-k(k+1)/2+\eta(k,m+1)}. \tag{97.49}$$

Therefore, if (97.48) holds, then we get $(97.46)_{\tau=m+1}$. If (97.48) is not satisfied, then the trivial bound $J_{q,k}(U) \leq U^{2k} J_{q-k,k}(U)$ suffices:

$$J_{q,k}(U) \leq (4k)^{4(q-k)m} U^{2q-k(k+1)/2+\eta(k,m)}$$

$$= (4k)^{4q(m+1)} U^{2q-k(k+1)/2+\eta(k,m+1)} (4k)^{-4q-4km} U^{\eta(k,m)/k}, \tag{97.50}$$

which is less than the right side of (97.46) with $\tau = m+1$. This ends the proof.

We may now return to the estimation of the zeta-sum.

Theorem 113

$$\sum_{n \leq M} n^{it} \ll M^{1-c(\log M/\log t)^2}, \quad 1 \leq M < t. \tag{97.51}$$

where $c > 0$ and the implied constant are both absolute.

Remark 2 This is far stronger than (96.32), in the sense that it is non-trivial even when M is so small as $\exp\left((\log t)^{2/3}\right)$.

Proof It suffices to deal with $Z(N,t)$ defined by (96.16), as we have seen at (96.32). A combination of (97.20) and Theorem 107 yields

$$W(n,t) \ll U^{2+\frac{1}{2}\eta(k,\tau)q^{-2}} t^{-c\delta^2 q^{-2}(\log t/\log N)}, \quad q = k(k+\tau), \tag{97.52}$$

in which U, δ, k are as in (97.8). Since

$$U^{\frac{1}{2}\eta(k,\tau)} \leq t^{(\delta/(7(1-\delta)^2))(1-1/k)^\tau (\log t/\log N)}, \tag{97.53}$$

$$\S 97 \qquad\qquad 533$$

if we let $\tau = Ck$ with a sufficiently large $C > 0$ and take δ appropriately, then

$$W(n,t) \ll U^2 t^{-c(\log t/\log N)^{-3}}, \tag{97.54}$$

with both $c > 0$ and the implied constant being absolute. We insert this into (97.9) and end the proof.

Theorem 114 *There exists an absolute constant $c > 0$ such that uniformly for* $\frac{1}{2} \leq \sigma \leq 1, t \geq 2$

$$\zeta(s) \ll t^{c(1-\sigma)^{3/2}} (\log t)^{2/3}. \tag{97.55}$$

Proof This is essentially due to Vinogradov (1985). We may assume that t is sufficiently large. By Note [13.8], (96.15), and (97.51),

$$\begin{aligned}
\zeta(\sigma + it) &= \sigma \int_{1-0}^{t} \left(\sum_{n \leq M} n^{-it} \right) M^{-\sigma-1} dM + O(t^{1-\sigma-c}) \\
&\ll \int_{0}^{\log t} \exp\left((1-\sigma)\xi - c\xi^3/(\log t)^2 \right) d\xi + t^{1-\sigma-c},
\end{aligned} \tag{97.56}$$

where c is the same as in (97.51). Then, with $\xi_0 = (2(1-\sigma)/c)^{1/2} \log t$, we get readily

$$\begin{aligned}
\int_{0}^{\xi_0} \cdots d\xi &\ll (\exp((1-\sigma)\xi_0) - 1)/(1-\sigma), \\
\int_{\xi_0}^{\log t} \cdots d\xi &\ll (\log t)^{2/3},
\end{aligned} \tag{97.57}$$

which ends the proof, as can be seen by considering separately the cases where σ is either not greater than or not less than $1 - (\log t)^{-2/3}$.

In this way, we are led to Vinogradov's zero-free region:

Theorem 115 *There exists an absolute constant $c > 0$ such that $\zeta(s)$ does not vanish in the region*

$$\sigma > 1 - \frac{c}{(\log t)^{2/3}(\log\log t)^{1/3}}, \quad t \geq 3, \tag{97.58}$$

and there

$$\frac{1}{\zeta(s)}, \quad \frac{\zeta'}{\zeta}(s) \ll (\log t)^{2/3}(\log\log t)^{1/3}. \tag{97.59}$$

Proof The reasoning is similar to that on Theorem 101. This time we take

$$\sigma_0 = 1 + \frac{\tau}{(\log \gamma)^{2/3}(\log\log \gamma)^{1/3}}, \tag{97.60}$$

534 *Distribution of Prime Numbers*

where $\zeta(\beta + i\gamma) = 0$ with a sufficiently large $\gamma > 0$, and $\tau > 0$ is a small variable. We apply Theorem 103 with $K = (\log \log \gamma / \log \gamma)^{2/3}$ and $S = c \log \log \gamma$. We get, by virtue of (97.55),

$$
\begin{aligned}
-\operatorname{Re}\frac{\zeta'}{\zeta}(\sigma_0 + i\gamma) &< \frac{c}{K}\log \log \gamma - \frac{1}{\sigma_0 - \beta}, \\
-\operatorname{Re}\frac{\zeta'}{\zeta}(\sigma_0 + 2i\gamma) &< \frac{c}{K}\log \log \gamma,
\end{aligned}
\tag{97.61}
$$

provided $\sigma_0 - \beta \leq K/2$. Then, by (95.8),

$$
\beta < 1 - (\sigma_0 - 1)\frac{1 - c(\sigma_0 - 1)K^{-1}\log \log \gamma}{3 + c(\sigma_0 - 1)K^{-1}\log \log \gamma}.
\tag{97.62}
$$

We end the proof.

In just the same way as either of the two proofs for Theorem 107, we obtain the prime number theorem of Vinogradov, the hitherto best approximation to $\pi(x)$ by $\mathrm{li}(x)$:

Theorem 116 *There exists an absolute constant $c > 0$ such that uniformly for $x > 3$*

$$
\pi(x) = \mathrm{li}(x) + O\Big(x\exp\Big(-c(\log x)^{3/5}(\log \log x)^{-1/5}\Big)\Big).
\tag{97.63}
$$

Notes

[97.1]   We emphasize what is stated in the paragraph following (97.2). Vinogradov's improvement upon Weyl's method is one of the rarest events in the history of number theory: reducing exponential growth to polynomial growth. Shor's Theorem 43 is another instance. See also Note [105.1].

[97.2]   The method of Vinogradov (1935, 1936a, 1936b, 1937) has gone through numerous improvements and simplifications by himself and by his followers. However, the framework of the theory has remained largely the same as that of the original. Thus, his main ideas are well visible in the above application to the estimation of zeta-sums as well. They consist of the double-sum strategy taken at (97.3)–(97.4) and the reduction (97.31) based on the classification of variables (97.32) with respect to $\mathrm{mod}\,p$. Concerning the latter, Linnik's article (1943) is a notable contribution as it brought forth a considerable simplification by the use of his own bound (97.24). In this section we adopted Karatsuba's lucid presentation (1975, Chapter VI) of these essentials.

$\S 98$ 535

[97.3] The double-sum strategy, a typical instance of which is Vinogradov's, is a fundamental device in modern analytic number theory. The heart of the method is, if rendered very roughly, in the deletion of arithmetic constraints imposed on outer summands by means of either Cauchy–Schwartz or Hölder inequalities which, with the succeeding exchange of the order of summation, makes it possible to reduce the original problem to the treatment of smoother expressions so that the detection of non-trivial cancellations can be achieved as was witnessed at (97.13). Naturally, how to bring a particular issue to a form suitable to the application of the method can be tricky; Vinogradov's idea (97.3) is the key in the case of zeta-sums (96.16) and in his original treatment of Weyl sums. See Notes 19.8, [104.3], and [105.5] for typical applications of the double sum strategy in a somewhat different context. See Linnik (1963, Introduction) for a general view.

[97.4] What is asserted immediately after (97.26) may need a verification. It is a consequence of the Girard–Newton formula; namely, elementary symmetric polynomials can be expressed in terms of polynomials over $\mathbb{Q}$ of the power sums of the variables. This is but apparent from the identity

$$\exp\left(\sum_{l=1}^{\infty} \frac{(-1)^{l-1}}{l} T_l x^l\right) = \prod_{j=1}^{k}\left(1 + \lambda_j x\right), \quad T_l = \sum_{j=1}^{k} \lambda_j^l.$$

The denominators of coefficients of such polynomials are not divisible by p if $k < p$ (a basic assumption in Theorem 111). Hence, elementary polynomials of $\{\lambda_j\}$ are uniquely determined in terms of $\{T_j\}$ with respect to modulo p. As for the rank of the system (97.28), it is considered in the finite field $\mathbb{F}_p$. Because of the second line of (97.25) we may suppose, without loss of generality, that $x_{10}x_{20}\cdots x_{k-l0} \neq 0$ in $\mathbb{F}_p$; hence the coefficient matrix of order $k - l$, that is,

$$\left(x_{b0}^a\right), \quad l \leq a \leq k - 1, \ 1 \leq b \leq k - l$$

is non-singular in $\mathbb{F}_p$, as it is essentially a Vandermonde matrix composed of x_{b0}, $1 \leq b \leq k - l$, which are assumed to be mutually different in $\mathbb{F}_p$. By the way, an application of the Girard–Newton formula is made in Gauss [DA, art.338].

$\S 98$

We turn to the asymptotic distribution of primes in short intervals; the discussion will be developed in the present and the next two sections.

536   *Distribution of Prime Numbers*

Our aim is to establish

$$\pi(x+y) - \pi(y) = \left(1 + o(1)\right)\frac{y}{\log x},$$

or equivalently $\psi(x+y) - \psi(x) = \left(1 + o(1)\right)y,$ (98.1)

with y/x being small.

Here RH becomes relevant; if it turns out to be correct, then the assertion (95.16) will imply that $y = x^{1/2}(\log x)^3$ is admissible in (98.1). While the proof of RH is still in total darkness, there exists a set of arguments, collectively called the zero-density method, with which one can prove results of the same nature, that is, (98.1) with y/x being of the order of a negative power of x, without recourse to any hypothesis. Such a possibility is not indicated by the investigation on $\psi(x)$ which we have developed in the preceding sections; either Theorem 107 or even Theorem 116, which is the deepest result so far attained for the full interval $[1, x]$, asserts only that one may take $y/x \approx \exp(-(\log x)^{\alpha})$ with an $\alpha > 0$ less than 1.

Hoheisel (1930) broke this impasse by the following discovery:

Theorem 117 *Let $N(\alpha, T)$ be the number of complex zeros of $\zeta(s)$ which are in the region $\alpha \leq \sigma$, $|t| \leq T$. If there exists absolute constants $\theta \geq 2$ and $A \geq 0$ such that*

$$N(\alpha, T) \ll T^{\theta(1-\alpha)} \log^A T, \tag{98.2}$$

uniformly for $\frac{1}{2} \leq \alpha \leq 1$, $2 \leq T$, then it holds for any fixed ϖ, $1 - 1/\theta < \varpi < 1$, that

$$\pi(x+y) - \pi(x) = \left(1 + o(1)\right)\frac{y}{\log x}, \quad y = x^{\varpi}, \tag{98.3}$$

as x tends to infinity.

Proof In the formula (95.5) or rather (95.7) we take $T = x^{1-\omega} \log^3 x$, $\omega = 1 - 1/\theta + \varepsilon$. Then we have, for any $0 < y < x$,

$$\psi(x+y) - \psi(x) = y - \int_x^{x+y} \left(\sum_{\rho, \, |\gamma| < T} u^{\rho-1} \right) du + o(x^{\omega}). \tag{98.4}$$

We have, by summation by parts (Note [13.8]),

$$\left| \sum_{\rho, \, |\gamma| < T} u^{\rho-1} \right| \leq -\int_0^1 u^{\alpha-1} dN(\alpha, T)$$

$$= (\log u) \int_{\frac{1}{2}}^1 N(\alpha, T) u^{\alpha-1} d\alpha + O\left(N(T)u^{-1/2}\right), \tag{98.5}$$

where $N(T)$ is as in (94.8). By virtue of Theorem 110, we have, with $\eta = (\log T)^{-3/4}$,

$$\int_{\frac{1}{2}}^{1} N(\alpha, T) u^{\alpha-1} d\alpha \ll (\log x)^c \int_{\frac{1}{2}}^{1-\eta} \left(x^{-1} T^\theta\right)^{1-\alpha} d\alpha \tag{98.6}$$

$$\ll (\log x)^c x^{(\theta(1-\omega)-1)\eta} .$$

This ends the proof.

Hoheisel's own assertion was weaker than (98.3) simply because the zero-free region of Vinogradov's type, that is, $\zeta(s) \neq 0$ for $\sigma > 1 - c(\log t)^{-\kappa}$, $t \geq 2$, with $0 < \kappa < 1$ was not available yet.

Before Hoheisel's discovery, asymptotic formulas like (98.3) had been believed to be accessible only under

the quasi-Riemann hypothesis:
there exists a σ_*, $\frac{1}{2} \leq \sigma_* < 1$, such that
$$\zeta(s) \neq 0, \ \sigma_* < \sigma, \tag{98.7}$$

since this implies obviously $\psi(x) = x + O(x^{\sigma_* + \varepsilon})$ and $\pi(x) = \text{li}(x) + O(x^{\sigma_* + \varepsilon})$; these are in fact equivalent to (98.7), as can be shown by (95.17). Moreover, the truth of

the zero density hypothesis:
$$N(\alpha, T) \ll T^{2(1-\alpha)} \log^c T, \ \frac{1}{2} \leq \alpha \leq 1, \tag{98.8}$$

will imply that RH can be dispensed with as far as the asymptotic distribution of primes in short intervals is concerned, for then $y = x^{1/2+\varepsilon}$ will become admissible in (98.1).

Hoheisel thus originated a reasoning which enables one to avoid such a strong assumption as (98.7) in order to investigate $\pi(x + y) - \pi(x)$. It has been naturally placed at the heart of the prime number theory since then and has evolved into a method effective in a quite wide context as will become apparent in the course of our discussion.

The origin of Hoheisel's method can actually be traced back to the capture of a statistical evidence of RH by Bohr and Landau (1914b): for any fixed $\delta > 0$

$$N\left(\tfrac{1}{2} + \delta, T\right) = o(T), \tag{98.9}$$

as $T \to \infty$. Namely, compared with (94.9), almost all complex zeros are confined in an arbitrarily narrow neighborhood of the critical line $\text{Re } s = \frac{1}{2}$. The most notable aspect of their argument is in that it is related to sieve. We shall make this feature precise, as its recognition is very basic to understand one of the principal motivations of the present chapter: Bohr and Landau

538   *Distribution of Prime Numbers*

(1914a) had proved that $N(\frac{1}{2} + \delta, T) = O(T)$, i.e., not $o(T)$, prior to (98.9). Although it is also a result supporting RH statistically, the same holds, in fact, for various Dirichlet series which lack Euler product representations. Hence, it may be asserted that the genuinely arithmetical peculiarity of the zeta-function was not well exploited in the article (1914a). Because of this, in the immediately adjacent article (1914b), they considered instead the product $\zeta(s)P_X(s)$ with the mollifier $P_X(s) = \prod_{p \leq X}(1 - p^{-s})$ which reminds us of the sifting mechanism in the sense of §§17–19. The function $\zeta(s)P_X(s)$, $\sigma > 1$, is close to 1 provided X is taken sufficiently large, that is, $\zeta(s)$ is well tamed by $P_X(s)$ because of the Euler product of $\zeta(s)$; and if this tendency continues to hold for all s such that $\frac{1}{2} < \alpha \leq \sigma < 1$, then it should follow that $N(\alpha, T) = 0$ for any T, which is, of course, identical to a proof of the quasi-Riemann hypothesis (98.7). Therefore, the problem is now to see whether $|\zeta(s)P_X(s) - 1|$ with $\sigma < 1$ can be small or not; if it is sufficiently small, then the zeta-function does not vanish at such an s. In other words, our task is to see how far the Euler product influences the analytic behavior of the zeta-function in the half plane $\sigma < 1$. The novelty of Bohr–Landau (1914b) is in that the general smallness of $|\zeta(s)P_X(s) - 1|$ is detected statistically while it is virtually hopeless to achieve the same for individual s, still today: They proved that for every α with $\frac{1}{2} < \alpha < 1$ and for an appropriately chosen X

$$\int_{-T}^{T} \left|\zeta(\alpha + it)P_X(\alpha + it) - 1\right|^2 dt = o(T). \tag{98.10}$$

Therefore, $|\zeta(\alpha + it)P_X(\alpha + it) - 1|$ is small on average, and hence, it is heuristically evident that $\zeta(s)$ vanishes seldom in the vicinity of the line $\sigma = \alpha$; see Note [102.3]. Actually, the assertion (98.9) was derived from (98.10) by means of a general function theoretical device which is a consequence of Jensen's inequality (94.13).

After these initial developments, a notable change was applied to the mollifier by Carleson (1921); he replaced $P_X(s)$ by the Dirichlet polynomial

$$M_X(s) = \sum_{n \leq X} \frac{\mu(n)}{n^s}, \quad \mu\text{: the Möbius function.} \tag{98.11}$$

This turned out to be more efficient and has become a standard choice since then. Also, a new function theoretical means was devised by Littlewood (1924) to count the number of zeros of a regular function in a rectangle, replacing the device of Bohr–Landau mentioned above; see also Titchmarsh (1951, pp.187–188). In the framework thus established, Ingham (1937) proved, among other things,

$$N(\alpha, T) \ll T^{2(1+2\eta+\varepsilon)(1-\alpha)} \log^c T; \tag{98.12}$$

here η is the Lindelöf exponent $\mu\left(\frac{1}{2}\right)$ defined in Note [96.2]; for a proof see Note [100.2]. Hence, only if Vinogradov's zero-free region had already been available, then by means of (96.12) Ingham could have accomplished (98.1) with $y = x^{5/8+\varepsilon}$; see Note [100.1]. Moreover, the proof of the Lindeöf hypothesis will virtually imply (98.8), that is, $y = x^{1/2+\varepsilon}$ in (98.1), which will most probably close the investigation of the asymptotic distribution, i.e., (98.3), of primes in short intervals.

Then, in the late 1960s, a drastic change took place in the art of counting zeros of the zeta and L-functions, which is nowadays called the large-moduli method following its inventor Montgomery (1971, Chapter 8). We shall relate salient points of this fundamental method in the subsequent sections. It should be noted in advance that the large-moduli method can be regarded as belonging to the scheme of sieve; indeed an origin of it might be traced back to the large sieve of Linnik (1941), a modern account of which will be a main subject of the later part of the present chapter.

Thus, we begin with relevant basic devices from analysis:

Theorem 118 *There exists a non-decreasing function $\phi(x)$ of C^∞ class such that*

$$\phi(x) = \begin{cases} 1 & x \geq 1, \\ 0 & x \leq 0. \end{cases} \tag{98.13}$$

Proof We take

$$u(x) = \begin{cases} 0 & x \leq 0, \\ \exp(-1/x) & x \geq 0. \end{cases} \tag{98.14}$$

Then the function

$$\phi(x) = C \int_{-\infty}^{x} u(\xi)u(1 - \xi)d\xi, \tag{98.15}$$

with C such that $\phi(1) = 1$, has the desired property. We end the proof.

Theorem 119 *For any function $g(x)$ which is smooth in the interval $[0, \tau]$, $\tau > 0$, we have*

$$\left| g\left(\tfrac{1}{2}\tau\right) \right| \leq \frac{1}{\tau} \int_0^\tau \left(|g(x)| + \tfrac{1}{2}\tau|g'(x)| \right) dx. \tag{98.16}$$

Proof This is often called Sobolev's inequality, and is a consequence of the identity

$$\tau g(x) = \int_0^\tau g(y)dy + \int_0^x yg'(y)dy + \int_x^\tau (y - \tau)g'(y)dy. \tag{98.17}$$

We end the proof.

540   *Distribution of Prime Numbers*

Theorem 120 *Let $f(z)$ be regular and $|f(z)| \leq M$, $M > 1$, on the disk $|z| \leq r$. Then we have, for any $A \geq 1$,*

$$f(0) \ll (A/r) \log M \int_{-r}^{r} |f(iy)| dy + M^{-A}, \tag{98.18}$$

with the implied constant being absolute.

Proof This is due to Balasubramanian and Ramachandra (1989). We have, for any $U > 0$,

$$2\pi i f(0) = \int_{-ri}^{ri} f(z)\big((U^z - U^{-z})/z\big) dz$$
$$+ \int_{C^-} f(z) \frac{U^z}{z} dz + \int_{C^+} f(z) \frac{U^{-z}}{z} dz, \tag{98.19}$$

where $C^- = \{|z| = r, \operatorname{Re} z \leq 0\}$ and $C^+ = \{|z| = r, \operatorname{Re} z \geq 0\}$ both in positive direction; to show this formula, it suffices to replace the path in the first integral by C^+. Then, take $U = \exp(v_1 + v_2 + \cdots + v_k)$ and integrate over $0 < v_j < V$ for all j. On noting that $|(U^z - U^{-z})/z| \leq 2kV$ in the first integral on the right side of (98.19), we have

$$2\pi V^k |f(0)| \leq 2kV^{k+1} \int_{-r}^{r} |f(iy)| dy$$
$$+ \int_{C^-} |f(z)| \big|(e^{Vz} - 1)/z\big|^k \frac{|dz|}{|z|} + \int_{C^+} |f(z)| \big|(e^{-Vz} - 1)/z\big|^k \frac{|dz|}{|z|}. \tag{98.20}$$

Since $|e^{\pm Vz} - 1| \leq 2$ on $C^{\mp}$, respectively,

$$|f(0)| \leq \pi^{-1} kV \int_{-r}^{r} |f(iy)| dy + M\big(2/rV\big)^k. \tag{98.21}$$

Taking $V = 16/r$, $k = [A \log M]$, we end the proof.

Theorem 121 *Let $f(x)$ be a C^{∞} function compactly supported on $\mathbb{R}$. Then we have*

$$f(x) = \int_{-\infty}^{\infty} \widehat{f}(u) e(xu) du,$$
$$\int_{-\infty}^{\infty} |f(x)|^2 dx = \int_{-\infty}^{\infty} |\widehat{f}(u)|^2 du, \tag{98.22}$$

where

$$\widehat{f}(u) = \int_{-\infty}^{\infty} f(y) e(-uy) dy, \quad e(y) = \exp(2\pi i y). \tag{98.23}$$

Proof The first assertion is the inverse Fourier transform and the second is the Plancherel theorem. The domain of applicable functions can be enlarged

$\S 98$ 541

greatly if one appeals to the L^2-theory with respect to Lebesque integration; but the above suffices for our purpose. To confirm the first, we note that the right side is equal to

$$\lim_{K \to \infty} \int_{-K}^{K} \widehat{f}(u)e(xu)du, \tag{98.24}$$

since $\widehat{f}(u) \ll (|u| + 1)^{-2}$ as can be seen applying integration by parts twice in (98.23). This integral is equal to

$$\frac{1}{\pi} \int_{-\infty}^{\infty} f(y) \frac{\sin(2\pi K(x - y))}{x - y} dy \tag{98.25}$$

which, with K sufficiently large, we divide into two parts according as $|x-y| < K^{-1/2}$ and $|x - y| \geq K^{-1/2}$. The first part equals

$$\int_{-K^{-1/2}}^{K^{-1/2}} (f(x) + O(|y|)) \frac{\sin(2\pi Ky)}{y} dy$$
$$= f(x) \int_{-2\pi K^{1/2}}^{2\pi K^{1/2}} \frac{\sin(y)}{y} dy + O(K^{-1/2}) = \pi f(x) + O(K^{-1/2}) \tag{98.26}$$

(see (60.5)). As for the second part, we apply integration by parts again and find that it is $O(K^{-1/2})$. This ends the confirmation. The discussion of the second assertion is analogous. We end the proof.

Theorem 122 *Let f be a C^∞ function compactly supported on $(0, \infty)$. Let $\widetilde{f}(s)$ be its Mellin transform*

$$\widetilde{f}(s) = \int_0^\infty f(u)u^{s-1}du. \tag{98.27}$$

Then we have, for any $a > 0$,

$$f(x) = \frac{1}{2\pi i} \int_{(a)} \widetilde{f}(s)x^{-s}ds, \tag{98.28}$$

where the contour is the vertical line $\operatorname{Re} s = a$.

Proof This is called the inversion formula of Mellin, but actually equivalent to the inverse Fourier transform as an appropriate change of variables shows. Alternatively, integrating twice by parts, we have

$$\widetilde{f}(s) = \frac{1}{s(s + 1)} \int_0^\infty f^{(2)}(u)u^{s+1}dx, \tag{98.29}$$

which implies that the right side of (98.28) equals

$$\frac{1}{2\pi i} \int_0^\infty uf^{(2)}(u) \int_{(a)} \frac{(u/x)^s}{s(s + 1)} dsdu, \tag{98.30}$$

542 *Distribution of Prime Numbers*

because of absolute convergence. The inner integral is equal to $2\pi i(1 - x/u)$ if $x \leq u$, and to 0 otherwise. This ends the proof.

Theorem 123 *Let $a^{(j)}$, $1 \leq j \leq J$, be non-zero elements of a Hilbert space equipped with the inner product $((\cdot, \cdot))$. Then, for any element b in the space, it holds that*

$$\sum_{j \leq J} \frac{|((\mathbf{b}, \mathbf{a}^{(j)}))|^2}{\sum_{j' \leq J} |((\mathbf{a}^{(j)}, \mathbf{a}^{(j')}))|} \leq \|\mathbf{b}\|^2, \tag{98.31}$$

where $\|\mathbf{b}\| = ((\mathbf{b}, \mathbf{b}))^{1/2}$. Consequently we have

$$\sum_{j \leq J} |((\mathbf{b}, \mathbf{a}^{(j)}))| \leq \|\mathbf{b}\| \left(\sum_{j, j' \leq J} |((\mathbf{a}^{(j)}, \mathbf{a}^{(j')}))| \right)^{1/2}, \tag{98.32}$$

$$\sum_{j \leq J} |((\mathbf{b}, \mathbf{a}^{(j)}))|^2 \leq \|\mathbf{b}\|^2 \max_{j \leq J} \sum_{j' \leq J} |((\mathbf{a}^{(j)}, \mathbf{a}^{(j')}))|. \tag{98.33}$$

Remark The collection of these inequalities is regarded as the basis in the modern theory of Linnik's large sieve. The main assertion (98.31) is called the Selberg inequality (Bombieri (1971, Proposition 1); Montgomery (1971, Lemma 1.8)); as is well indicated by the case with $\{\mathbf{a}^{(j)}\}$ being an orthonormal system, it is a generalization of Bessel's inequality. The corollaries (98.32) and (98.33) are originally due to Halász (1968, see the bottom line of p.389) and Bombieri (Montgomery (1971, (1.12))), respectively; (98.31) became available later than their works.

Proof We use the trivial fact that for any $\xi_j \in \mathbb{C}$

$$\left\| \mathbf{b} - \sum_{j \leq J} \xi_j \mathbf{a}^{(j)} \right\|^2 \geq 0, \tag{98.34}$$

as in the proof of Bessel's inequality. Expanding out the square, we have

$$\|\mathbf{b}\|^2 - 2\mathrm{Re} \sum_{j \leq J} \overline{\xi}_j((\mathbf{b}, \mathbf{a}^{(j)})) + \sum_{j, j' \leq J} \xi_j \overline{\xi}_{j'}((\mathbf{a}^{(j)}, \mathbf{a}^{(j')})) \geq 0. \tag{98.35}$$

To this double sum, we apply $|\xi_j \xi_{j'}| \leq \frac{1}{2}(|\xi_j|^2 + |\xi_{j'}|^2)$, and have

$$\|\mathbf{b}\|^2 - 2\mathrm{Re} \sum_{j \leq J} \overline{\xi}_j((\mathbf{b}, \mathbf{a}^{(j)})) + \sum_{j \leq J} |\xi_j|^2 \sum_{j' \leq J} |((\mathbf{a}^{(j)}, \mathbf{a}^{(j')}))| \geq 0. \tag{98.36}$$

We take

$$\xi_j = \frac{((\mathbf{b}, \mathbf{a}^{(j)}))}{\sum_{j' \leq J} |((\mathbf{a}^{(j)}, \mathbf{a}^{(j')}))|}, \tag{98.37}$$

and obtain (98.31). This ends the proof.

Theorem 124 *For any $T \geq 1$ and any complex sequence $\{a_n\}$, we have*

$$\int_{-T}^{T} \left| \sum_{n=1}^{\infty} a_n n^{it} \right|^2 dt \ll \sum_{n=1}^{\infty} (n+T)|a_n|^2, \tag{98.38}$$

provided the right side converges. The implied constant is absolute.

Proof This is due to Gallagher (1970). We consider the mean square of the function

$$F(u) = \sum_{\omega} c_{\omega} e(\omega u), \tag{98.39}$$

where $\{c_{\omega}\}$ is a finite sequence in $\mathbb{C}$. With any $\tau > 0$, we put

$$v(x) = \begin{cases} \tau^{-1} & |x| \leq \tau/2, \\ 0 & |x| > \tau/2. \end{cases} \tag{98.40}$$

We have

$$\int_{-\infty}^{\infty} |F(u)\widehat{v}(u)|^2 du = \int_{-\infty}^{\infty} \left| \sum_{\omega} c_{\omega} v(x+\omega) \right|^2 dx, \tag{98.41}$$

where $\widehat{v}(u) = (1/\pi\tau u) \sin(\pi\tau u)$. To show this we replace $v(x)$ by

$$v_{\delta}(x) = \tau^{-1} \phi((x + \tau/2 + \delta)/\delta) \phi((\tau/2 + \delta - x)/\delta), \tag{98.42}$$

where ϕ is defined by (98.13) and $\delta > 0$ is sufficiently small. Then $\sum_{\omega} c_{\omega} v_{\delta}(x+\omega)$ is of C^{∞} class and compactly supported. Its Fourier transform is $F(u)\widehat{v_{\delta}}(u)$. Hence by Theorem 121 the identity (98.41) holds but with v_{δ} in place of v. We have

$$\begin{aligned} \widehat{v_{\delta}}(u) &= \widehat{v}(u) + 2\frac{\delta}{\tau} \int_0^1 \phi(x) \cos\left(2\pi\delta ux - \pi u(\tau + 2\delta)\right) dx \\ &= \widehat{v}(u) + O\left(\delta/(1 + \delta|u|)\right), \end{aligned} \tag{98.43}$$

where integration by parts has been applied. Thus both sides of (98.41) with $v = v_{\delta}$ converge uniformly in δ, and we obtain (98.41). With this, since $\widehat{v}(u) \geq 2/\pi$ for $-1/2\tau \leq u \leq 1/2\tau$, we have

$$\int_{-1/2\tau}^{1/2\tau} |F(u)|^2 du \ll \tau^{-2} \int_{-\infty}^{\infty} \left| \sum_{|\omega - x| \leq \tau/2} c_{\omega} \right|^2 dx; \tag{98.44}$$

the right side comes from that of (98.41) but with x replaced by $-x$. We then take $\tau = 1/2T$, $\omega = (\log n)/2\pi$, $c_{\omega} = a_n$, and find, after a rearrangement, that

$$\int_{-T}^{T} \left| \sum_{n=1}^{\infty} a_n n^{it} \right|^2 dt \ll T^2 \int_1^{\infty} \left| \sum_{x \leq n \leq x \exp(\pi/T)} a_n \right|^2 \frac{dx}{x}, \tag{98.45}$$

544   *Distribution of Prime Numbers*

provided that $\{a_n\}$ is a finite sequence. To the last integrand we apply the Cauchy–Schwartz inequality and obtain (98.38). We end the proof.

Theorem 125 *Let*

$$\text{Re } s_j \geq 0, \quad |\text{Im } s_j| \leq T, \quad T \geq 2,$$
$$|\text{Im } (s_j - s_k)| \geq 1, \quad j \neq k \leq J. \tag{98.46}$$

Then it holds, for any complex sequence $\{a_n\}$ and any integer $N \geq 2$, that

$$\sum_{j \leq J} \left| \sum_{N < n \leq 2N} \frac{a_n}{n^{s_j}} \right|^2 \ll (T + N)(\log N) \sum_{N < n \leq 2N} |a_n|^2, \tag{98.47}$$

$$\sum_{j \leq J} \left| \sum_{N < n \leq 2N} \frac{a_n}{n^{s_j}} \right|^2 \ll \left(N + JT^{1/2} \log T \right) \sum_{N < n \leq 2N} |a_n|^2, \tag{98.48}$$

with implied constants being absolute.

Proof This is due to Montgomery (1971, Chapter 8). By summation by parts (Note [13.8]), we have, for $s = \sigma + it$,

$$\sum_{N < n \leq 2N} \frac{a_n}{n^s} = (2N)^{-\sigma} \sum_{N < n \leq 2N} a_n n^{-it}$$
$$+ \sigma \int_N^{2N} x^{-\sigma - 1} \sum_{N < n \leq x} a_n n^{-it} dx. \tag{98.49}$$

Thus, we may suppose that $\text{Re } s_j = 0, j \leq J$, i.e., $s_j = it_j, j \leq J$. We set, in (98.16), with $\tau = 1$,

$$g(x) = F(x - \tfrac{1}{2} + t_j), \quad F(x) = \left(\sum_{N < n \leq M} a_n n^{-ix} \right)^2, \tag{98.50}$$

with $N < M \leq 2N$, so that

$$|F(t_j)| \ll \int_{t_j - \frac{1}{2}}^{t_j + \frac{1}{2}} \left(|F(t)| + |F'(t)| \right) dt. \tag{98.51}$$

Hence,

$$\sum_{j \leq J} |F(t_j)| \ll \int_{-T - \frac{1}{2}}^{T + \frac{1}{2}} \left(|F(t)| + |F'(t)| \right) dt. \tag{98.52}$$

To this integral we apply (98.38) and get (98.47). On the other hand, to show (98.48), let $k(x)$ be of C^∞ class, supported in the interval $[\tfrac{1}{2}N, \tfrac{5}{2}N]$, and equal to 1 on the interval $[N, 2N]$ as well as $k^{(\nu)}(x) \ll N^{-\nu}$ for bounded $\nu \geq 0$; one may readily construct such a k by means of Theorem 118; see (98.42). Further, let $a_n = 0$ for $n \notin (N, 2N]$. With this, we take, in (98.33),

$$\mathbf{b} = \left(k(n)^{1/2} a_n\right), \quad \mathbf{a}^{(j)} = \left(k(n)^{1/2} n^{it_j}\right). \tag{98.53}$$

Then, with respect to the ordinary inner product, we are led to

$$\sum_{j \leq J} |F(t_j)| \leq \max_{j \leq J} \sum_{j' \leq J} \left| \sum_n k(n) n^{i(t_j - t_{j'})} \right| \sum_{N < n \leq 2N} |a_n|^2. \tag{98.54}$$

Our problem becomes now to estimate the sum

$$K(u) = \sum_n k(n) n^{iu}, \quad u \geq 0. \tag{98.55}$$

Let $\widetilde{k}$ be the Mellin transform of k. Then, for any fixed integer $v \geq 0$,

$$\widetilde{k}(s) \ll N^{\sigma} (|s| + 1)^{-v}. \tag{98.56}$$

By means of Theorem 122, we have

$$\begin{aligned} K(u) &= \frac{1}{2\pi i} \int_{(2)} \zeta(s - iu) \widetilde{k}(s) ds \\ &= \frac{1}{2\pi} \int_{(0)} \zeta(i(t - u)) \widetilde{k}(it) dt + O\left(N(u + 1)^{-v}\right). \end{aligned} \tag{98.57}$$

As we have indicated after (96.1), $\zeta(it) \ll t^{1/2} \log t$ for $t \geq 2$. Hence

$$K(u) \ll (u + 1)^{1/2} \log(u + 2) + N(u + 1)^{-v}. \tag{98.58}$$

Inserting this into (98.54), we obtain (98.48) and end the proof. It should be remarked that the shift of path in (98.57) to the imaginary axis is by no means a sole way to take; see Note [100.3] below.

Theorem 126 *There exists an absolute constant $c_0 > 0$ such that under the assumption* (98.46) *and*

$$\left| \sum_{N < n \leq 2N} \frac{a_n}{n^{s_j}} \right| \geq V \geq c_0 A^{1/2}, \quad A = \sum_{N < n \leq 2N} |a_n|^2, \tag{98.59}$$

we have

$$J \ll NAV^{-2} + NT(\log T)^2 A^3 V^{-6}. \tag{98.60}$$

Proof This is due to Huxley (1972). We may take $c_0^2 = c_1 \cdot 2^{3/2} \log 2$, where c_1 is the implied constant in (98.48). For then there exists $T_0 \geq 2$ such that $V^2 = 2c_1 T_0^{1/2} (\log T_0) A \geq c_0^2 A$; and the number of s_j, whose imaginary parts

546   *Distribution of Prime Numbers*

are contained in an arbitrary interval of length T_0 is $O(NAV^{-2})$ because of (98.48). Hence,

$$J \ll NAV^{-2}(T/T_0 + 1), \tag{98.61}$$

which ends the proof.

Notes

[98.1] The zero-density method is an art based on counting objects which are, in fact, expected not to exist in the light of RH. It can appear to be artificial. Partly because of this, a few ways have been devised with the aim to analyze the distribution of primes without using any estimation of $N(\alpha, T)$ but still relying on the devices supporting the zero-density method like the last theorems. Hence, with this wider perspective in mind, we adopt the term

the Hoheisel scheme

in order to describe the paradigm in which we shall hereafter develop our study of the distribution of prime numbers in short intervals and arithmetic progressions, regardless of whether or not any zero-density estimate is actually exploited. The Hoheisel scheme, with appropriate modification, is known to be applicable to wider situations which are controlled by Euler products or like; for a discussion in the context of automorphic L-functions see Motohashi (2015) for instance.

[98.2] It is not known yet if (98.3) implies (98.7) or anything similar. The quasi-RH appears to be far deeper than the issue of the existence of primes in short intervals.

[98.3] Although somewhat out of context, we shall here confirm the assertion made immediately after (61.15) that the functional equation of the zeta-function and the sum formula of Poisson are equivalent to each other. We assume thus that the function $f(x)$ is the same as in Theorem 122 and show first that

$$\int_0^\infty f(x) \cos(2\pi nx)dx = \frac{1}{2\pi i} \int_{(\alpha)} \widetilde{f}(1 - s) \cos\left(\tfrac{1}{2}\pi s\right) \Gamma(s)(2\pi n)^{-s}ds,$$

where $\widetilde{f}$ is as in (98.26), and (α) the vertical line $\mathrm{Re}\, s = \alpha > 0$. On shifting the contour to (β) with $-1 < \beta < -\tfrac{1}{2}$ and computing the residue at the pole $s = 0$, we see that the right side equals

$$\widetilde{f}(1) + \frac{1}{2\pi i} \int_0^\infty f(x) \int_{(\beta)} \cos\left(\tfrac{1}{2}\pi s\right) \Gamma(s)(2\pi nx)^{-s}dsdx,$$

in which we have applied an exchange of the order of integration; the absolute convergence of the double integral is guaranteed by (94.39). The inner integral

$$§98 \qquad\qquad 547$$

can be computed by shifting the contour to $-\infty$ and summing the resulting residues while invoking (94.33) and (94.38); we find that it is equal to $2\pi i(\cos(2\pi nx) - 1)$. This obviously ends the required confirmation. Hence, we have

$$\sum_{n=1}^{\infty} \int_0^{\infty} f(x) \cos(2\pi nx) dx = \frac{1}{2\pi i} \int_{(\alpha)} \widetilde{f}(1-s) \cos\left(\tfrac{1}{2}\pi s\right) \Gamma(s)(2\pi)^{-s} \zeta(s) ds,$$

provided $\alpha > 1$. We now use the functional equation of the zeta-function and see that the right side equals

$$\frac{1}{4\pi i} \int_{(\alpha)} \widetilde{f}(1-s)\zeta(1-s) ds = -\tfrac{1}{2}\widetilde{f}(1) + \frac{1}{4\pi i} \int_{(\beta)} \widetilde{f}(1-s)\zeta(1-s) ds$$

$$= -\tfrac{1}{2}\widetilde{f}(1) + \tfrac{1}{2} \sum_{n=1}^{\infty} f(n),$$

where we have applied the inversion formula (98.27). Therefore it holds that

$$\sum_{n=1}^{\infty} f(n) = \sum_{n=-\infty}^{\infty} \int_0^{\infty} f(x) e(nx) dx.$$

We omit the deduction of (60.1) from this. In general, functional equations of functions similar to the zeta-function are equivalent to sum formulas which can be regarded as extensions of Poisson's. For instance, the sum formula associated with $\zeta^2(s)$ is attributed to Voronoï (1904), as mentioned in Note [16.10]. Thus, we have the Voronoï sum formula: with f as above,

$$\sum_{n=1}^{\infty} d(n) f(n) = \int_0^{\infty} (\log x + 2c_E) f(x) dx$$

$$+ 4 \sum_{n=1}^{\infty} d(n) \int_0^{\infty} \left\{ K_0\left(4\pi\sqrt{nx}\right) - \tfrac{1}{2}\pi Y_0\left(4\pi\sqrt{nx}\right) \right\} f(x) dx,$$

which converges absolutely. Also, we have the Voronoï explicit formula: Let $\Delta(x)$ be as in Note [16.10]. Then

$$\Delta(x) = x(\log x + 2c_E - 1) + \tfrac{1}{4}$$

$$- \frac{2}{\pi} x^{1/2} \sum_{n=1}^{\infty} \frac{d(n)}{n^{1/2}} \left\{ K_1\left(4\pi\sqrt{nx}\right) + \tfrac{1}{2}\pi Y_1\left(4\pi\sqrt{nx}\right) \right\};$$

the convergence is the same as that of (60.2). Here K_ν, Y_ν are the Bessel functions in the common notation, which are, in fact, Mellin inversions of products of two values of the Gamma function appearing in the functional equations for $\zeta^2(s)$ and the like; see §§1.1–1.3 of *Analytic number theory*, II,

548　　　　　　　　　　*Distribution of Prime Numbers*

referred to in Note [94.4] above. A greatly simplified proof of the last explicit formula for $\Delta(x)$ has been achieved by Meurman (1992); it should be noted that he dispensed with the functional equation for $\zeta^2(s)$.

§99

The inequalities stated in the preceding section are all highly instrumental to work in the Hoheisel scheme. However, the real core of the theory is in the mean values of the zeta-function, as is apparent already in (98.10) of Bohr and Landau, to whom the concept of zero density is naturally attributed. The aim of the present section is to develop the basics of the theory on the mean values of the zeta-function. More precisely, we shall study

$$I_k(T) = \int_{-T}^{T} \left|\zeta\left(\tfrac{1}{2} + it\right)\right|^{2k} dt \tag{99.1}$$

for the three cases $k = 1, 2, 6$.

We begin with the fourth moment $I_2(T)$: Generalizing it slightly, we shall show

Theorem 127 *For any sufficiently large $T > 0$ we have*

$$\int_{-T}^{T} \left|\zeta\left(\tfrac{1}{2} + it\right)\right|^{4} dt \ll T(\log T)^6. \tag{99.2}$$

Also, under the assumption

$$\left|\zeta\left(\tfrac{1}{2} + it_j\right)\right| \geq V > 0,$$
$$|t_j| \leq T, \quad 1 \leq j \leq J; \quad |t_j - t_k| \geq 1, \quad j \neq k, \tag{99.3}$$

we have

$$J \ll TV^{-4}(\log T)^7. \tag{99.4}$$

Proof　Let $d(n)$ be the divisor function. Combining $(16.15)_{k=2}$ and (94.42), we have

$$\frac{1}{2\pi i} \int_{(2)} \zeta^2(w + \xi)\Gamma(w)T^w dw = \sum_{n=1}^{\infty} \frac{d(n)}{n^{\xi}} e^{-n/T}, \quad \xi = \tfrac{1}{2} + it. \tag{99.5}$$

Shifting the contour to $\mathrm{Re}\, w = -\tfrac{3}{4}$,

$$\zeta^2(\xi) = \sum_{n=1}^{\infty} \frac{d(n)}{n^{\xi}} e^{-n/T} - R(t, T) + \frac{1}{2\pi i} \int_{(-\frac{3}{4})} \zeta^2(w + \xi)\Gamma(w)T^w dw, \tag{99.6}$$

§99   549

where $R(t, T)$ is the residue at the double pole $w = 1 - \xi$. We write the functional equation (12.5) as $\zeta(s) = X(s)\zeta(1 - s)$ with $X(s) = 2^s \pi^{s-1} \sin(\pi s/2)\Gamma(1 - s)$. Then the last integrated term equals

$$\frac{1}{2\pi i} \int_{(-\frac{3}{4})} X^2(w + \xi)\Gamma(w)T^w \left(\sum_{n=1}^{\infty} \frac{d(n)}{n^{1-w-\xi}} \right) dw. \tag{99.7}$$

In the part with $n < T$ we shift the contour to $\operatorname{Re} w = a = -(\log T)^{-1}$ and in the remaining part to $\operatorname{Re} w = b = -\frac{1}{2} - (\log T)^{-1}$, so the third term on the right side of (99.6) is equal to

$$\frac{1}{2\pi i} \int_{(a)} X^2(w + \xi)\Gamma(w)T^w \left(\sum_{n<T} \frac{d(n)}{n^{1-w-\xi}} \right) dw$$
$$+ \frac{1}{2\pi i} \int_{(b)} X^2(w + \xi)\Gamma(w)T^w \left(\sum_{n\geq T} \frac{d(n)}{n^{1-w-\xi}} \right) dw. \tag{99.8}$$

Here $X^2(w + \xi)\Gamma(w)T^w$ is $O(e^{-|w|} \log T)$ on (a) and $O(e^{-|w|}T^{1/2})$ on (b) because of (94.37). Hence, (99.6) implies that

$$|\zeta(\xi)|^2 \ll e^{-|\xi|}T^{1/2} \log T + \left| \sum_{n=1}^{\infty} \frac{d(n)}{n^{\xi}} e^{-n/T} \right|$$
$$+ \log T \int_{(a)} \left| \sum_{n<T} \frac{d(n)}{n^{1-w-\xi}} \right| e^{-|w|} |dw| \tag{99.9}$$
$$+ T^{1/2} \int_{(b)} \left| \sum_{n\geq T} \frac{d(n)}{n^{1-w-\xi}} \right| e^{-|w|} |dw|.$$

The integral in (99.2) is

$$\ll T \log^2 T + \int_{-T}^{T} \left| \sum_{n=1}^{\infty} \frac{d(n)}{n^{\xi}} e^{-n/T} \right|^2 dt$$
$$+ \log^2 T \int_{(a)} e^{-|w|} \int_{-T}^{T} \left| \sum_{n<T} \frac{d(n)}{n^{1-w-\xi}} \right|^2 dt|dw| \tag{99.10}$$
$$+ T \int_{(b)} e^{-|w|} \int_{-T}^{T} \left| \sum_{n\geq T} \frac{d(n)}{n^{1-w-\xi}} \right|^2 dt|dw|.$$

To these three integrals with respect to t we apply Theorem 124. Invoking the bounds (16.17)–(16.18) with $k, l = 2$, we obtain the assertion (99.2).

As for (99.4), we apply Theorem 120, so

$$\left|\zeta\left(\tfrac{1}{2}+it_j\right)\right|^4 \ll \log T \int_{t_j-\frac{1}{2}}^{t_j+\frac{1}{2}} \left|\zeta\left(\tfrac{1}{2}+it\right)\right|^4 dt + T^{-A}. \tag{99.11}$$

Hence, we have

$$V^4 J \ll \log T \int_{-T-1}^{T+1} \left|\zeta\left(\tfrac{1}{2}+it\right)\right|^4 dt + T^{1-A} \ll T(\log T)^7. \tag{99.12}$$

This ends the proof.

We shall next investigate the mean square or, more precisely, the weighted mean square of the zeta-function on the critical line:

$$\int_{-\infty}^{\infty} \left|\zeta\left(\tfrac{1}{2}+it\right)\right|^2 g(t)dt. \tag{99.13}$$

It is supposed that

$$\begin{array}{c} \text{the function } g(z) \text{ is real on } \mathbb{R}, \text{ and} \\ \text{regular as well as } O\big((|z|+1)^{-E}\big) \text{ for } |\text{Im } z| \le E, \\ \text{with a sufficiently large constant } E > 0. \end{array} \tag{99.14}$$

Our analysis will become quite involved compared with the above discussion on the fourth moment, for we need to develop a detailed asymptotics of (99.13), not merely bounding it.

To begin with, we put

$$\mathcal{I}(u,v;g) = \int_{-\infty}^{\infty} \zeta(u+iz)\zeta(v-iz)g(z)dz, \quad \text{Re } u, \text{ Re } v > 1. \tag{99.15}$$

Shifting the contour to Im $z = \tfrac{1}{2}E$, we see that $\mathcal{I}(u,v;g)$ continues meromorphically to the symmetric domain

$$|u|, |v| < E_0 = cE, \tag{99.16}$$

where $c > 0$ is supposed to be sufficiently small, yet E_0 is sufficiently large. With this, if Re u, Re $v < 1$, then shifting the contour back to the original we have

$$\begin{aligned} \mathcal{I}(u,v;g) = &\int_{-\infty}^{\infty} \zeta(u+iz)\zeta(v-iz)g(z)dz \\ &+ 2\pi\zeta(u+v-1)\left\{g((u-1)i) + g((1-v)i)\right\}. \end{aligned} \tag{99.17}$$

Hence, for any $0 < \alpha < 1$

$$\int_{-\infty}^{\infty} |\zeta(\alpha+it)|^2 g(t)dt = \mathcal{I}(\alpha,\alpha;g) - 4\pi\zeta(2\alpha-1)\text{Re}\left\{g((\alpha-1)i)\right\}; \tag{99.18}$$

we stress that this $\mathcal{I}(\alpha,\alpha;g)$ is the result of the analytic continuation of (99.15). In particular,

$$\int_{-\infty}^{\infty} \left|\zeta\left(\tfrac{1}{2}+it\right)\right|^2 g(t)dt = \mathcal{I}\left(\tfrac{1}{2},\tfrac{1}{2};g\right) + 2\pi \operatorname{Re}\left\{g\left(\tfrac{1}{2}i\right)\right\}, \tag{99.19}$$

where (94.58) has been used.

We shall show an alternative way to analytically continue $\mathcal{I}(u,v;g)$ to the domain (99.16); the result will give a practical expression for (99.13) via (99.19). Thus, we decompose (99.15) as follows: For $\operatorname{Re} u, \operatorname{Re} v > 1$

$$\mathcal{I}(u,v;g) = \left\{\sum_{m=n}+\sum_{m<n}+\sum_{m>n}\right\}m^{-u}n^{-v}\widehat{g}\big((\log(m/n))/2\pi\big) \tag{99.20}$$
$$= \zeta(u+v)\widehat{g}(0) + \mathcal{J}(u,v;g) + \overline{\mathcal{J}(\overline{v},\overline{u};g)},$$

say, where $\widehat{g}$ is the Fourier transform of g. We shall prove that the function $\mathcal{J}(u,v;g)$ continues analytically to (99.16) so that (99.20) holds there with an addition of a correction term of the same nature as the second term on the right side of (99.17). To this end, we note that for $\operatorname{Re} u, \operatorname{Re} v > 1$

$$\mathcal{J}(u,v;g) = \sum_{m,n>0} m^{-u-v}(1+n/m)^{-v}\widehat{g}\big(-(\log(1+n/m))/2\pi\big). \tag{99.21}$$

Then, in order to separate these m,n, we introduce the Mellin transform

$$g^*(s,\lambda) = \int_0^{\infty} y^{s-1}(1+y)^{-\lambda}\widehat{g}\big(-(\log(1+y))/2\pi\big)dy; \tag{99.22}$$

note that

$$\widehat{g}\big(-(\log(1+y))/2\pi\big) = \int_{-\infty+Ei}^{\infty+Ei} g(z)(1+y)^{iz}dz \ll (1+y)^{-E}. \tag{99.23}$$

We assert that $g^*(s,\lambda)/\Gamma(s)$ continues analytically to the region

$$|\operatorname{Re} s| \leq \tfrac{1}{3}E, \quad |\operatorname{Re}\lambda| \leq \tfrac{1}{3}E; \tag{99.24}$$

moreover, if λ remains bounded and s tends to infinity in this region, then

$$g^*(s,\lambda) \ll |s|^{-E/2}. \tag{99.25}$$

To confirm the former claim, we note that the definition (99.22) implies that provided $\operatorname{Re} s > 0$

$$g^*(s,\lambda) = \int_0^{\infty} y^{s-1}(1+y)^{-\lambda}\int_{-\infty+Ei}^{\infty+Ei} g(z)(1+y)^{iz}dzdy$$
$$= \Gamma(s)\int_{-\infty+Ei}^{\infty+Ei} \frac{\Gamma(\lambda-iz-s)}{\Gamma(\lambda-iz)}g(z)dz, \tag{99.26}$$

552 Distribution of Prime Numbers

where the Beta integral formula (94.32) is applied. The regularity of the function $g^*(s,\lambda)/\Gamma(s)$ in (99.24) is now obvious in view of (94.37) and (99.14). As for (99.25), we note that

$$
g^*(s,\lambda) = \frac{1}{s(s+1)\cdots(s+v-1)} \int_0^\infty \frac{y^{s+v-1}}{(1+y)^{\lambda+v}}
$$
$$
\times \int_{-\infty+Ei}^{\infty+Ei} (\lambda-iz)(\lambda-iz+1)\cdots(\lambda-iz+v-1)g(z)(1+y)^{iz}dzdy;
$$

(99.27)

and it suffices to take $v = [E/2] + 1$.

We return to the identity (99.21). We have, by the Mellin inversion,

$$
(1+h)^{-v}\widehat{g}(-(\log(1+h))/2\pi) = \frac{1}{2\pi i}\int_{(2)} g^*(s,v)h^{-s}ds, \quad h > 0. \quad (99.28)
$$

This does not directly follow from Theorem 122; nevertheless, the reasoning (98.28)–(98.29) works well with $\widetilde{f}(s)$ being replaced by (99.27)$_{v=2}$. Thus, we may take $h = n/m$, and sum over m, n; we get, for $\operatorname{Re} u, \operatorname{Re} v > \frac{3}{2}$,

$$
\mathcal{J}(u,v;g) = \frac{1}{2\pi i}\int_{(2)} \zeta(s)\zeta(u+v-s)g^*(s,v)ds, \quad (99.29)
$$

in which (99.25) is applied. Shifting the contour to $\operatorname{Re} s = 3E_0$,

$$
\mathcal{J}(u,v;g) = \zeta(u+v-1)g^*(u+v-1,v)
$$
$$
+ \frac{1}{2\pi i}\int_{(3E_0)} \zeta(s)\zeta(u+v-s)g^*(s,v)ds. \quad (99.30)
$$

This shows that $\mathcal{J}(u,v;g)$ continues to the symmetric domain (99.16); hence, the decomposition (99.20) holds in (99.16), as has been claimed.

We then apply the functional equation (12.5) to the factor $\zeta(u+v-s)$ of (99.30) and use (16.7); we have, for u, v in (99.16),

$$
\mathcal{J}(u,v;g) = \zeta(u+v-1)g^*(u+v-1,v)
$$
$$
- 2i(2\pi)^{u+v-2}\sum_{n=1}^\infty \sigma_{u+v-1}(n)\int_{(3E_0)} (2\pi n)^{-s}\sin\left(\tfrac{1}{2}(u+v-s)\pi\right)
$$
$$
\times \Gamma(s+1-u-v)g^*(s,v)ds,
$$

(99.31)

where the necessary absolute convergence is secured by (99.25).

§99   553

We combine the identities (99.18), (99.20), and (99.31), and obtain, for $0 < \alpha < 1$,

$$\int_{-\infty}^{\infty} |\zeta(\alpha + it)|^2 g(t)dt = \zeta(2\alpha)\widehat{g}(0)$$

$$+ 2\zeta(2\alpha - 1)\text{Re}\{g^*(2\alpha - 1, \alpha)\} - 4\pi\zeta(2\alpha - 1)\text{Re}\{g((\alpha - 1)i)\}$$

$$+ 4(2\pi)^{2\alpha - 2}\text{Im}\left[\sum_{n=1}^{\infty} \sigma_{2\alpha - 1}(n) \int_{(2)} (2\pi n)^{-s} \sin\left(\tfrac{1}{2}(2\alpha - s)\pi\right)\right.$$

$$\left. \times \Gamma(s + 1 - 2\alpha)g^*(s, \alpha)ds\right].$$

$$(99.32)$$

To transform this infinite sum into a practical form, we consider

$$P_{\pm}(\delta) = \frac{1}{2i}\int_{(2)} (2\pi n)^{-s}e^{\pm\pi i(2\alpha - s)/2}\Gamma(s + 1 - 2\alpha)g^*_\delta(s, \alpha)ds, \qquad (99.33)$$

where $g_\delta(z) = e^{-\delta z^2}g(z)$ with $\delta \geq 0$. We have

$$\lim_{\delta \to 0^+} \left(P_+(\delta) - P_-(\delta)\right)$$

$$= \int_{(2)} (2\pi n)^{-s} \sin\left(\tfrac{1}{2}\pi(2\alpha - s)\right)\Gamma(s + 1 - 2\alpha)g^*(s, \alpha)ds,$$

$$(99.34)$$

since the bound (99.25) is valid as far as (99.14) is fulfilled. In the definition of $g^*_\delta(s, \alpha)$, that is, in (99.22) with g being replaced by g_δ, we turn the y-contour so that provided $\delta > 0$ and $0 < \theta < \tfrac{1}{2}\pi$ we have

$$P_+(\delta) = \frac{e^{\pi i\alpha}}{2i}\int_{(2)} (2\pi n)^{-s}e^{-\pi is/2}\Gamma(s + 1 - 2\alpha)$$

$$\times \int_0^{e^{i\theta}\infty} y^{s-1}(1 + y)^{-\alpha}\widehat{g}_\delta(-(\log(1 + y))/2\pi)dyds;$$

$$(99.35)$$

note that the condition $\delta > 0$ will be in force as far as (99.37). This double integral converges absolutely, for the integral representation (99.23) with g_δ in place of g holds, we have $\widehat{g}_\delta(-(\log(1 + y))/2\pi) \ll (1 + |y|)^{-E}$; and (94.39) implies that the integrand is $\ll (1 + |y|)^{-E}\exp(-\theta|s|/2)$ uniformly in y and s. Thus, on exchanging the order of integration, we have, by (94.42),

$$P_+(\delta) = i\pi(2\pi n)^{1-2\alpha}\int_0^{e^{i\theta}\infty} y^{2\alpha - 2}(1 + y)^{-\alpha}e(-n/y)$$

$$\times \widehat{g}_\delta\left(-(\log(1 + y))/2\pi\right)dy.$$

$$(99.36)$$

We apply integration by parts with respect to the factor $e(-n/y)$ and see that this integral converges uniformly for $0 \leq \theta < \tfrac{1}{2}\pi$; note that $\alpha > 0$ is supposed.

554   *Distribution of Prime Numbers*

Hence, we may turn the y-contour back to the original. Further, applying the change of variable $y \mapsto 1/y$, we get

$$P_+(\delta) = i\pi(2\pi n)^{1-2\alpha} \int_0^\infty (y(1+y))^{-\alpha} e(-ny)$$
$$\times \widehat{g_\delta}\big(-(\log(1+1/y))/2\pi\big)dy. \tag{99.37}$$

This converges uniformly for $\delta \geq 0$, as can be confirmed by an application of integration by parts with respect to the factor $e(-ny)$. We transform $P_-(\delta)$ in just the same way.

Therefore, via (99.32) and (99.34), we have, for $0 < \alpha < 1$,

$$\int_{-\infty}^\infty |\zeta(\alpha+it)|^2 g(t)dt = \zeta(2\alpha)\widehat{g}(0)$$
$$+ 2\zeta(2\alpha-1)\mathrm{Re}\{g^*(2\alpha-1,\alpha)\} - 4\pi\zeta(2\alpha-1)\mathrm{Re}\{g((\alpha-1)i)\}$$
$$+ 4\sum_{n=1}^\infty \sigma_{1-2\alpha}(n) \int_0^\infty (y(y+1))^{-\alpha}$$
$$\times \mathrm{Re}\{\widehat{g}(-(\log(1+1/y))/2\pi)\} \cos(2\pi ny)dy.$$
$$\tag{99.38}$$

Invoking (94.59), (94.64), and (99.26), we obtain the following explicit formula for the weighted mean square of the zeta-function:

Theorem 128 *If the weight function g satisfies (99.14), then we have*

$$\int_{-\infty}^\infty \left|\zeta\left(\tfrac{1}{2}+it\right)\right|^2 g(t)dt$$
$$= \int_{-\infty}^\infty \left[\mathrm{Re}\left\{\frac{\Gamma'}{\Gamma}\left(\tfrac{1}{2}+it\right)\right\} + 2c_E - \log(2\pi)\right]g(t)dt + 2\pi\mathrm{Re}\{g(\tfrac{1}{2}i)\}$$
$$+ 4\sum_{n=1}^\infty d(n) \int_0^\infty (y(y+1))^{-1/2} g_c\big(\log(1+1/y)\big)\cos(2\pi ny)dy,$$
$$\tag{99.39}$$

where

$$g_c(Y) = \int_{-\infty}^\infty g(z)\cos(Yz)dz. \tag{99.40}$$

We now specialize (99.39) with

$$g(z) = \left(\pi^{1/2}G\right)^{-1}\exp\left(-((z-T)/G)^2\right), \tag{99.41}$$

where the variables $T, G > 0$ are supposed to be sufficiently large. Thus, we shall consider

$$I_1(T,G) = \left(\pi^{1/2}G\right)^{-1} \int_{-\infty}^{\infty} \left|\zeta\left(\tfrac{1}{2} + i(T+t)\right)\right|^2 \exp\left(-(t/G)^2\right)dt. \quad (99.42)$$

Since

$$g_c(Y) = \exp\left(-\tfrac{1}{4}(GY)^2\right)\cos(TY), \quad (99.43)$$

we have

$$I_1(T,G) = (\pi^{1/2}G)^{-1} \int_{-\infty}^{\infty} \left[\operatorname{Re}\frac{\Gamma'}{\Gamma}\left(\tfrac{1}{2} + i(T+t)\right)\right]\exp\left(-(t/G)^2\right)dt$$

$$+ 4\sum_{n=1}^{\infty} d(n)Q(n;T,G) + 2c_E - \log(2\pi)$$

$$+ 2\pi^{1/2}G^{-1}\cos\left(TG^{-2}\right)\exp\left(-G^{-2}(T^2 - \tfrac{1}{4})\right), \quad (99.44)$$

where

$$Q(n;T,G) = \int_0^{\infty} \left(y(y+1)\right)^{-1/2}\cos\left(T\log(1+1/y)\right)\cos(2\pi ny)$$
$$\times \exp\left(-\tfrac{1}{4}(G\log(1+1/y))^2\right)dy. \quad (99.45)$$

Hereafter we shall consider $Q(n;T,G)$ on the assumption

$$T^\varepsilon < G < T/\log T. \quad (99.46)$$

We put

$$Q_\pm(n;T,G) = \tfrac{1}{2}\operatorname{Re}\int_0^{\infty} \left(y(y+1)\right)^{-1/2}\exp\left(2\pi iny \pm Ti\log(1+1/y)\right)$$
$$\times \exp\left(-\tfrac{1}{4}(G\log(1+1/y))^2\right)dy, \quad (99.47)$$

so that

$$Q(n;T,G) = Q_+(n;T,G) + Q_-(n;T,G). \quad (99.48)$$

To estimate $Q_-(n;T,G)$, we turn the y-contour to $y = xe^{i\theta}, 0 < x < \infty$, with a small constant $\theta > 0$. On the new contour the integrand is

$$\ll \exp\left(-cR(x)\right)(x(x+1))^{-1/2},$$
$$R(x) = nx + T/(x+1) + (G\log(1+1/x))^2, \quad (99.49)$$

with a constant $c > 0$. Since $nx + T/(x+1) \geq (nT)^{1/2}(x/(x+1))^{1/2}$, we have $R(x) \gg (nT)^{1/4}$ for $x \geq (nT)^{-1/2}$; and $R(x) \gg (G\log(nT))^2$ for $x \leq (nT)^{-1/2}$. Hence, we have, for any fixed $A > 0$,

$$Q_-(n; T, G) \ll (nT)^{-A}. \tag{99.50}$$

As for $Q_+(n; T, G)$, we shall apply the method of steepest descent. The saddle point $y = y_0$ satisfies $y_0(y_0 + 1) = T/(2\pi n)$; thus,

$$y_0 = -\tfrac{1}{2} + \left(T/(2\pi n) + \tfrac{1}{4}\right)^{1/2} = \frac{T/(\pi n)}{1 + (2T/(\pi n) + 1)^{1/2}}. \tag{99.51}$$

We move the y-contour to the sum of three straight lines $L_1 \cup L_2 \cup L_3$:

$$\begin{aligned} L_1 &= [0, y_0(1 - e^{\pi i/4} r_+)], \\ L_2 &= [y_0(1 - e^{\pi i/4} r_+), y_0(1 + e^{\pi i/4} r_-)], \\ L_3 &= [y_0(1 + e^{\pi i/4} r_-), e^{\theta i} \infty), \quad r_\pm = 2^{1/2} \tan\theta/(1 \pm \tan\theta), \end{aligned} \tag{99.52}$$

where $\theta > 0$ is again a small constant. On L_1, we have $y = x(1 - e^{\pi i/4} r_+)$, $0 \le x \le y_0$; note that $\arg(y) = -\theta$; so

$$\begin{aligned} \operatorname{Re} &\left\{2\pi i n y + T i \log(1 + 1/y) - \tfrac{1}{4}(G\log(1 + 1/y))^2\right\} \\ &= 2^{-1/2} r_+ \left(2\pi n x - T/(1 + x)\right) - \frac{x + \tfrac{1}{2}}{(1 + x)^2} r_+^2 T(1 + O(\theta)) \\ &\quad - \tfrac{1}{4}(G\log(1 + 1/x))^2(1 + O(\theta)) \\ &< -\frac{y_0 + \tfrac{1}{2}}{2(1 + y_0)^2} r_+^2 T - \tfrac{1}{8}(G\log(1 + 1/y_0))^2. \end{aligned} \tag{99.53}$$

If $n \ll T$, then this is $< -c_1(nT)^{1/2}$, and otherwise $< -c_2 T - c_3(G\log(n/T))^2$ with constants $c_j > 0$; hence,

$$\int_{L_1} \ll (nT)^{-A}. \tag{99.54}$$

On L_3, we have $y = x(1 + e^{\pi i/4} r_-)$, $y_0 \le x$; note that $\arg(y) = \theta$; so the same expression as the leftmost side of (99.53) is equal to

$$\begin{aligned} -&2^{-1/2} r_- \left(2\pi n x - T/(1 + x)\right) - \frac{x + \tfrac{1}{2}}{(1 + x)^2} r_-^2 T(1 + O(\theta)) \\ &- \tfrac{1}{4}(G\log(1 + 1/x))^2(1 + O(\theta)). \end{aligned} \tag{99.55}$$

If $2y_0 \le x$, then the situation becomes analogous to that of $Q_-(n; T, G)$; and if $y_0 \le x < 2y_0$, then it is similar to the treatment of the contribution of L_1. Hence,

$$\int_{L_3} \ll (nT)^{-A}. \tag{99.56}$$

As for L_2, we put $y = y_0(1 + e^{\pi i/4} r)$, $-r_+ \leq r \leq r_-$. Then the same expression as the leftmost side of (99.53) is $< -\frac{1}{4}(G\log(1 + 1/y_0))^2(1 + O(|r|))$. If $T^2 \ll n$, then this is $< -c(G\log n)^2$ with a constant $c > 0$. Thus, the corresponding $Q_+(n; T, G)$ is negligible. Among $n \ll T^2$, those satisfying $\log(1 + 1/y_0) \gg G^{-1}\log T$ are also negligible. Hence, we can restrict our discussion to those n such that

$$n \ll TG^{-2}(\log T)^2, \quad y_0 \approx (T/n)^{1/2} \gg G/\log T. \tag{99.57}$$

In particular, we may suppose, instead of (99.46),

$$T^\varepsilon < G < T^{1/2+\varepsilon}. \tag{99.58}$$

Under (99.57)–(99.58), we have, on L_2,

$$
\begin{aligned}
2\pi i n y &+ Ti\log(1 + 1/y) - \tfrac{1}{4}(G\log(1 + 1/y))^2 \\
&= 2\pi i n y_0 + Ti\log(1 + 1/y_0) - \tfrac{1}{4}(G\log(1 + 1/y_0))^2 \\
&\quad - \frac{y_0 + \frac{1}{2}}{(y_0 + 1)^2} Tr^2 + h_1 r + h_2 r^2 + h_3 r^3 + O\big((nT)^{1/2} r^4\big),
\end{aligned} \tag{99.59}
$$

where h_j are functions of n, T; and h_1, $h_2 = O(\log^2 T)$, $h_3 = O((nT)^{1/2})$. In the fourth term on the right side of (99.59), we have $(y_0 + \frac{1}{2})(1 + y_0)^{-2}T \approx (nT)^{1/2}$; hence, the contribution of the part of L_2 corresponding to $|r| \geq (nT)^{-1/5}$ is $O(\exp(-(nT)^{1/10}))$, and negligible. Thus, we may further restrict our consideration to $|r| < (nT)^{-1/5}$. Under this assumption, we have

$$
\begin{aligned}
(y(y+1))^{-1/2} &\exp\big(2\pi i n y + Ti\log(1 + 1/y) - \tfrac{1}{4}(G\log(1 + 1/y))^2\big) \\
&= (y_0(y_0 + 1))^{-1/2}\exp\big(2\pi i n y_0 + Ti\log(1 + 1/y_0) - \tfrac{1}{4}(G\log(1 + 1/y_0))^2\big) \\
&\quad \times \exp\big(-(y_0 + \tfrac{1}{2})(1 + y_0)^{-2}Tr^2\big) \\
&\quad \times \big\{1 + k_1 r + k_2 r^2 + k_3 r^3 + O((nT)^{1/2} r^4)\big\},
\end{aligned} \tag{99.60}
$$

where $k_1 = O(\log^2 T)$, $k_2 = O(\log^4 T)$, $k_3 = O((nT)^{1/2}\log^2 T)$. Therefore,

$$
\begin{aligned}
\int_{L_2} &= e^{\frac{1}{4}\pi i}\left(\frac{\pi y_0(1 + y_0)}{T(y_0 + \frac{1}{2})}\right)^{1/2} \exp\big(2\pi i n y_0 + Ti\log(1 + 1/y_0)\big) \\
&\quad \times \exp\big(-\tfrac{1}{4}(G\log(1 + 1/y_0))^2\big)\big(1 + O((nT)^{-1/2}\log^4 T)\big).
\end{aligned} \tag{99.61}
$$

Obviously we may now discard the assumption (99.57)–(99.58); namely, (99.61) holds on the original assumption (99.46).

558 Distribution of Prime Numbers

Theorem 129 *Let $I_1(T,G)$ be defined by (99.42). Then we have, on the condition $T^\varepsilon \leq G \leq T(\log T)^{-1}$,*

$$I_1(T,G) = \log(T/2\pi) + 2c_E +$$
$$+ 2\pi^{1/2} \sum_{n=1}^{\infty} \frac{(-1)^{n-1}d(n)}{(2\pi nT + \pi^2 n^2)^{1/4}} \sin(F(T,n)) \tag{99.62}$$
$$\times \exp(-(\text{Garcsinh}((\pi n/(2T))^{1/2}))^2) + O(T^{-1}\log^6 T),$$

where

$$F(T,n) = 2T\text{arcsinh}((\pi n/(2T))^{1/2}) + (2\pi nT + \pi^2 n^2)^{1/2} - \tfrac{1}{4}\pi. \tag{99.63}$$

Proof Collecting all the assertions obtained after (99.44), we have

$$I_1(T,G) = \log(T/2\pi) + 2c_E$$
$$+ 2\sum_{n=1}^{\infty} d(n)\left(\frac{\pi y_0(1+y_0)}{T(y_0 + \tfrac{1}{2})}\right)^{1/2} \cos\left(2\pi n y_0 + T\log(1 + 1/y_0) + \tfrac{1}{4}\pi\right)$$
$$\times \exp\left(-\tfrac{1}{4}(G\log(1 + 1/y_0))^2\right) + O(T^{-1}\log^6 T). \tag{99.64}$$

In view of (99.51), we have

$$\frac{y_0(1+y_0)}{T(y_0 + \tfrac{1}{2})} = \frac{1}{(2\pi nT + \pi^2 n^2)^{1/2}}, \tag{99.65}$$
$$1 + 1/y_0 = \left((\pi n/2T)^{1/2} + (\pi n/2T + 1)^{1/2}\right)^2.$$

On noting $\text{arcsinh}(\xi) = \log(\xi + (\xi^2 + 1)^{1/2})$, we end the proof.

The explicit formula (99.62) for the weighted mean square of the zeta-function yields the following fact on the twelfth moment of the zeta-function:

Theorem 130 *We have, for $T \geq 2$,*

$$I_6(T) = \int_{-T}^{T} \left|\zeta\left(\tfrac{1}{2} + it\right)\right|^{12} dt \ll T^2 \log^{21} T. \tag{99.66}$$

Proof More precisely, we shall prove the following assertion: Let the sequence $\mathcal{Q} \subset [T, 2T]$ be such that

$$\xi \in \mathcal{Q} \Rightarrow \left|\zeta\left(\tfrac{1}{2} + i\xi\right)\right| \geq V > T^\varepsilon; \tag{99.67}$$
$$\xi,\xi' \in \mathcal{Q}, \xi \neq \xi' \Rightarrow |\xi - \xi'| \geq 1.$$

Then we have

$$|\mathcal{Q}| \ll T^2 V^{-12} \log^{20} T. \tag{99.68}$$

$$\S99 \qquad\qquad 559$$

From this, (99.66) follows immediately. Thus, let $T^\varepsilon \le H < T$, and let $\mathcal{R}$ be the subsequence of $\mathcal{Q}$ that is contained in an arbitrary interval $[U, U + H]$, $T \le U < 2T$. Obviously

$$|\mathcal{Q}| \ll TH^{-1} \max |\mathcal{R}|. \qquad (99.69)$$

Further, we divide the interval $[U, U + H]$ into short intervals of length G, $T^\varepsilon \le G < \frac{1}{2}H$. The parameters G, H are to be fixed later. Among these short intervals, we select the ones that actually intersect with $\mathcal{R}$; and suppose that their middle points make up the set $\mathcal{S}$. Let $\tau \in \mathcal{S}$, and let $\mathcal{R}(\tau)$ be the subsequence of $\mathcal{R}$ that is contained in the short interval whose middle point is τ. We have

$$|\mathcal{R}(\tau)| V^2 \ll \sum_{\xi \in \mathcal{R}(\tau)} \left|\zeta\left(\tfrac{1}{2} + i\xi\right)\right|^2. \qquad (99.70)$$

To this we apply Theorem 120 with an obvious specialization and find that

$$\begin{aligned}|\mathcal{R}(\tau)| V^2 &\ll \log T \int_{\tau - G}^{\tau + G} \left|\zeta\left(\tfrac{1}{2} + it\right)\right|^2 dt \\ &\ll (G \log T) I_1(\tau, G).\end{aligned} \qquad (99.71)$$

On the assumption

$$G \log^2 T \ll V^2, \qquad (99.72)$$

we have, by Theorem 129 and summation by parts,

$$\begin{aligned}|\mathcal{R}(\tau)| &V^2 (G \log T)^{-1} \\ &\ll \sum_N \int_0^N \frac{\exp(-(\mathrm{Garcsinh}((\pi(N + x)/(2\tau))^{1/2}))^2)}{(2\pi(N + x)\tau + \pi^2(N + x)^2)^{1/4}} dS(x, \tau) \quad (99.73) \\ &\ll \sum_N (NT)^{-1/4} \left(|S(N, \tau)| + N^{-1} \int_0^N |S(x, \tau)| dx\right) \exp(-G^2 N/T);\end{aligned}$$

here

$$S(x, \tau) = \sum_{N < n \le N + x} (-1)^n d(n) \exp(iF(\tau, n)), \qquad (99.74)$$

$$T^\varepsilon \le N = 2^\nu \le TG^{-2} \log T, \quad \nu \in \mathbb{N}$$

because of the last factor in (99.73). Then, by the large sieve inequality (98.32) and by (16.17) with $k, l = 2$,

$$\sum_{\tau \in \mathcal{S}} |S(x,\tau)| \ll (N \log^3 N)^{1/2}$$

$$\times \left(N|\mathcal{S}| + \sum_{\substack{\tau \neq \tau' \\ \tau, \tau' \in \mathcal{S}}} \left| \sum_{N < n \leq N+x} \exp(i(F(\tau,n) - F(\tau',n))) \right| \right)^{1/2}. \tag{99.75}$$

To this sum over n we apply the transformation (96.8). Hence, by (99.63), $F(\tau,n) + \frac{1}{4}\pi$ is equal to $2(2\pi n\tau)^{1/2}$ times a power series of $\pi n/2\tau$, and thus

$$\left| \left(\frac{\partial}{\partial \xi} \right)^{\nu} (F(\tau,\xi) - F(\tau',\xi)) \right| \approx |\tau - \tau'| T^{-1/2} N^{1/2-\nu},$$

$$\tau \neq \tau'; \, \xi \approx N. \tag{99.76}$$

We proceed in much the same way as (96.21)–(96.30), and find that

$$\sum_{N < n \leq N+x} \exp\left(i(F(\tau,n) - F(\tau',n)) \right)$$

$$\ll \frac{(TN)^{1/2}}{|\tau - \tau'|} + |\tau - \tau'|^{1/2} T^{-1/4} N^{1/4}, \quad \tau \neq \tau'. \tag{99.77}$$

We insert this into (99.75) and get, via (99.73),

$$\frac{V^2}{G \log T} \sum_{\tau \in \mathcal{S}} |\mathcal{R}(\tau)| \ll |\mathcal{S}|^{1/2} T^{1/2} G^{-3/2} \log^2 T$$

$$+ |\mathcal{S}| H^{1/4} G^{-3/4} \log^{3/2} T, \tag{99.78}$$

as the contribution of the term $N|\mathcal{S}|$ in (99.75) is negligible. Provided

$$H \ll G^3 (\log T)^{-6}, \tag{99.79}$$

the left side of (99.78) is greater than the second term on its right side. Namely, if (99.79) is valid, then

$$|\mathcal{R}| = \sum_{\tau \in \mathcal{S}} |\mathcal{R}(\tau)| \ll |\mathcal{S}|^{1/2} T^{1/2} V^{-2} G^{-1/2} \log^3 T. \tag{99.80}$$

Hence,

$$|\mathcal{R}| \ll TV^{-4} G^{-1} \log^6 T. \tag{99.81}$$

We now make the specialization $G \approx V^2 (\log T)^{-2}$; in view of (99.79), this means that under the assumption $H \ll V^6 (\log T)^{-12}$ we have, by (99.69) and (99.81),

$$|\mathcal{Q}| \ll T^2 V^{-6} H^{-1} \log^8 T. \tag{99.82}$$

$$\S 99 \qquad 561$$

Since the bound (96.11) implies that the specialization $H \approx V^6(\log T)^{-12}$ satisfies the initial condition $T^\varepsilon \le H < T$, we have obtained (99.68). We end the proof.

Notes

[99.1] The proof of Theorem 127 is due to Ramachandra (1974); the exponent of the log-factor is not optimal, which is irrelevant to our later applications, however. His method greatly simplified the bounding procedure of the fourth power moment of the zeta-function. As for earlier works on this subject see Titchmarsh (1951, Chapter 7).

[99.2] Theorem 128 is adopted from Motohashi (1997, pp.145–150). The expansion (99.62) is, in fact, a localized version of the fundamental work on $I_1(T)$ by Atkinson (1949). For our purpose, (99.62) suffices. By the way, in applying the method of steepest descent, one may make use of a general treatment of the contribution of integrals around saddle points. See for instance Titchmarsh (1951, Lemma 4.6). Also, in dealing with off-saddle integrals, an effective general argument is available; see Jutila–Motohashi (2005, Lemma 6). However, we are in the opinion that each application of the method of steepest descent should better be performed from scratch, especially when approximations uniform with respect to parameters involved are targeted. Such situations are plenty in analytic number theory, and the procedure leading to (99.62) can be an instructive example to deal with them.

[99.3] Atkinson's article (1949) contains an important idea; it is employed at (99.20). It may be regarded as an analytic completion of the Weyl shift (96.5) and is called the Atkinson dissection. Repeating what has been noted in Note [5.4]: A natural extension of this device to the situation of four integer variables was stated in Motohashi (1993, p.193; 1997, pp.150–153), which led him to an analogue of Theorem 128 for

$$\int_{-\infty}^{\infty} \left| \zeta \left(\tfrac{1}{2} + it \right) \right|^4 g(t) dt.$$

There the ordinary Fourier analysis, which is visible in (99.39), is replaced by the spectral analysis with respect to an orthonormal system of automorphic forms that is briefly indicated in the later part of §94.

[99.4] It should be remarked that (99.62) implies (96.11) with an ignorable difference in the log-factor: If $G < T^{1/2}$, then

$$I_1(T,G) \ll \log T + \sum_{n < TG^{-2} \log^2 T} d(n)(nT)^{-1/4} \exp(-G^2 n/T)$$
$$\ll \log T + T^{1/2} G^{-3/2} \log T,$$

562 *Distribution of Prime Numbers*

where we have used (16.17). That is, provided $G \gg T^{1/3}$, we have

$$\int_T^{T+G} \left|\zeta\left(\tfrac{1}{2} + it\right)\right|^2 dt \ll G \log T.$$

Then, via Theorem 120 we get $\zeta\left(\tfrac{1}{2} + it\right) \ll t^{1/6} \log t$, $t \geq 2$. In particular, the short argument following (99.82) is in fact redundant.

[99.5] Theorem 130 is due to Heath-Brown (1978). He exploited the difficult explicit formula for $I_1(T)$ of Atkinson (1949), which itself relies on another difficult result, i.e., Voronoï's sum formula (Note [98.3]). In contrast, the above argument is straightforward as it starts from (99.15) and dispenses with the works of Voronoï and Atkinson. Yet another way, via the explicit formula for the fourth moment stated in the preceding Note, to prove (99.66) can be found in Ivić–Motohashi (1994).

[99.6] The theory of mean values of L-functions, i.e., those with Euler product expansions in the automorphic context, has become, in recent years, a major field of analytic number theory. Its origin can be traced back to Bohr–Landau (1914b). Ivić (1991) is a readable account of the classical phase of the mean-values activity. See Michel (2022) for a more recent perspective.

§100

We now return to the estimation of $N(\alpha, T)$. Thus, let $M_X(s)$ be as in (98.11). Then, we have, by (17.2),

$$M_X(s)\zeta(s) = 1 + \sum_{X<n} \frac{a(n)}{n^s}, \quad \sigma > 1; \quad a(n) = \sum_{\substack{d|n \\ d \leq X}} \mu(d). \tag{100.1}$$

By (94.38) we get the identity

$$e^{-1/Y} = -\sum_{X<n} \frac{a(n)}{n^\rho} e^{-n/Y} + \frac{1}{2\pi i} \int_{(2)} M_X(s+\rho)\zeta(s+\rho)\Gamma(s)Y^s ds, \tag{100.2}$$

for an arbitrary complex zero $\rho = \beta + i\gamma$, $\beta \geq \tfrac{1}{2}$, and arbitrary X, $Y \geq 2$. On noting that this integrand is regular at $s = 0$, the shift of the path of integration to $\sigma = \tfrac{1}{2} - \beta$ gives

$$e^{-1/Y} = W_1(\rho) + W_2(\rho) + M_X(1)\Gamma(1 - \rho)Y^{1-\rho}, \tag{100.3}$$

where

$$W_1(\rho) = -\sum_{X<n} \frac{a(n)}{n^\rho} e^{-n/Y},$$

$$W_2(\rho) = \frac{1}{2\pi} \int_{-\infty}^{\infty} M\left(\tfrac{1}{2} + i(\gamma + t)\right) \zeta\left(\tfrac{1}{2} + i(\gamma + t)\right)$$
$$\times \Gamma\left(\tfrac{1}{2} - \beta + it\right) Y^{1/2-\beta+it} dt. \tag{100.4}$$

The parameters X, Y are to be fixed later; initially we assume that $T^\varepsilon \leq X \leq Y \leq T^c$ with a sufficiently large c, while T tends to infinity. Also, we suppose that $|\gamma| \geq (\log T)^2$, which causes a negligible error $O(\log^3 T)$. In particular, the third term on the right of (100.3) is negligible because of (94.37). Hence, (100.3) implies that either $|W_1(\rho)| \geq \tfrac{1}{4}$ or $|W_2(\rho)| \geq \tfrac{1}{4}$. In what follows we shall show that both of $W_j(\rho)$ are, in fact, small statistically, and consequently the following zero-density estimate will be derived:

Theorem 131 *Uniformly for* $\tfrac{1}{2} \leq \alpha \leq 1$ *and* $T \geq 2$,

$$N(\alpha, T) \ll T^{(D(\alpha)+\varepsilon)(1-\alpha)}, \quad D(\alpha) = \begin{cases} \dfrac{3}{2-\alpha} & \tfrac{1}{2} \leq \alpha \leq \tfrac{3}{4}, \\[2mm] \dfrac{3}{3\alpha-1} & \tfrac{3}{4} \leq \alpha \leq 1, \end{cases} \tag{100.5}$$

where the implied constant depends on ε *at most. Hence,*

$$N(\alpha, T) \ll T^{(12/5+\varepsilon)(1-\alpha)}. \tag{100.6}$$

Remark By virtue of Theorem 115, the bound (98.2) is essentially equivalent to $N(\alpha, T) \ll T^{(\theta+\varepsilon)(1-\alpha)}$, as far as the consequence (98.3) is concerned; one may take $c(\log\log T)^{4/3}/(\log T)^{1/3}$ instead of ε. This observation is incorporated into the statement (100.5)–(100.6).

Proof We note first that we may assume that $\tfrac{1}{2} + (\log T)^{-1} \leq \alpha \leq \beta$ because of (94.8). With this, (100.3), $W_1(\rho), W_2(\rho)$ are replaced, respectively, by

$$\tfrac{1}{2} \leq |U_1(\rho)| + |U_2(\rho)|, \tag{100.7}$$

$$U_1(\rho) = -\sum_{X<n\leq Y\log^2 T} \frac{a(n)}{n^\rho} e^{-n/Y}, \tag{100.8}$$

$$U_2(\rho) = \frac{1}{2\pi} \int_{-\log^2 T}^{\log^2 T} M_X\left(\tfrac{1}{2} + i(\gamma + t)\right) \zeta\left(\tfrac{1}{2} + i(\gamma + t)\right) \tag{100.9}$$
$$\times \Gamma\left(\tfrac{1}{2} - \beta + it\right) Y^{1/2-\beta+it} dt,$$

in which the last line depends on (94.37). Let $\mathcal{R}$ be the set of complex zeros in the rectangle $\{\alpha \leq \sigma \leq 1, T \leq t \leq 2T\}$ such that if ρ and ρ' are different elements in $\mathcal{R}$, then $|\rho - \rho'| \geq \log^3 T$. By (94.41), we have

$$N(\alpha, 2T) - N(\alpha, T) \ll |\mathcal{R}| \log^4 T. \tag{100.10}$$

Let $\mathcal{R}_\nu$, $\nu = 1, 2$, be the subset of those $\rho \in \mathcal{R}$ such that $|U_\nu(\rho)| \geq \frac{1}{4}$ so that $\mathcal{R} \subseteq \mathcal{R}_1 \cup \mathcal{R}_2$. By (98.47), we have

$$|\mathcal{R}_1| \ll \sum_{\rho \in \mathcal{R}_1} |U_1(\rho)|^2$$

$$\ll \log^3 T \max_{X < N \leq Y \log^2 T} (N + T)e^{-N/Y} \sum_{N < n \leq 2N} \frac{d^2(n)}{n^{2\alpha}} \tag{100.11}$$

$$\ll (Y^{2(1-\alpha)} + TX^{1-2\alpha}) \log^7 T,$$

where (16.17) has been used. Also, on noting $\left|\Gamma\left(\frac{1}{2} - \beta + it\right)\right| \ll \log T$ in (100.9), we have

$$|\mathcal{R}_2| \ll (Y^{\frac{1}{2}-\alpha} \log T)^{4/3}$$

$$\times \sum_{\rho \in \mathcal{R}_2} \left(\int_{-\log^2 T}^{\log^2 T} \left|\zeta\left(\tfrac{1}{2} + i(\gamma + t)\right)M_X\left(\tfrac{1}{2} + i(\gamma + t)\right)\right| dt \right)^{4/3}$$

$$\ll Y^{2(1-2\alpha)/3} \log^2 T \int_T^{2T} \left|\zeta\left(\tfrac{1}{2} + it\right)M_X\left(\tfrac{1}{2} + it\right)\right|^{4/3} dt, \tag{100.12}$$

where we have used Hölder's inequality and invoked the definition of $\mathcal{R}$. Thus,

$$|\mathcal{R}_2| \ll Y^{2(1-2\alpha)/3} \log^2 T$$

$$\times \left(\int_T^{2T} \left|\zeta\left(\tfrac{1}{2} + it\right)\right|^4 dt \right)^{1/3} \left(\int_T^{2T} \left|M_X\left(\tfrac{1}{2} + it\right)\right|^2 dt \right)^{2/3}, \tag{100.13}$$

again by Hölder's inequality. Hence, the combination of Theorems 124 and 127 gives

$$|\mathcal{R}_2| \ll T^{1/3}(X + T)^{2/3} Y^{2(1-2\alpha)/3} \log^5 T. \tag{100.14}$$

Collecting (100.10)–(100.14) and taking $X = T$, $Y = T^{3/(4-2\alpha)}$, which fulfill the initial condition on X, Y, we obtain

$$N(\alpha, T) \ll T^{3(1-\alpha)/(2-\alpha)} \log^c T. \tag{100.15}$$

Next, we shall employ Theorems 126 and 130 to estimate $|\mathcal{R}_\nu|$. This time we take $X = T^\varepsilon$. We have first

$$|\mathcal{R}_2| \ll Y^{6-12\alpha} T^\varepsilon \int_T^{2T} \left|\zeta\left(\tfrac{1}{2} + it\right)\right|^{12} dt \ll Y^{6-12\alpha} T^{2+\varepsilon}. \tag{100.16}$$

As for $\mathcal{R}_1$, we set, for any $N \in [X, Y \log^2 T)$,

$$\mathcal{R}_1(N) = \left\{ \rho \in \mathcal{R}_1 : \sum_{N < n \leq 2N} \frac{a(n)}{n^\rho} e^{-n/Y} \gg \frac{1}{\log T} \right\}, \tag{100.17}$$

so that

$$|\mathcal{R}_1| \ll (\log T) \max_N |\mathcal{R}_1(N)|. \tag{100.18}$$

Then we take the integer ℓ which satisfies

$$N^\ell < Y^2 (\log T)^4 \leq N^{\ell+1}, \quad 2 \leq \ell \ll 1; \tag{100.19}$$

note that

$$\left(Y^2 (\log T)^4\right)^{2/3} \leq \left(Y^2 (\log T)^4\right)^{\ell/(\ell+1)} \leq N^\ell < Y^2 (\log T)^4. \tag{100.20}$$

If $\mathcal{R}_1(N) \ni \rho$, then

$$(\log T)^{-\ell} \ll \sum_{N^\ell < n \leq (2N)^\ell} \frac{b(n)}{n^\rho}, \quad b(n) \ll d_{2\ell}(n), \tag{100.21}$$

where $d_{2\ell}(n)$ is as in (16.11). By (98.60) with $a(n) = b(n)n^{-\alpha}$,

$$\begin{aligned} |\mathcal{R}_1(N)| &\ll N^{2\ell(1-\alpha)} T^\varepsilon + N^{2\ell(2-3\alpha)} T^{1+\varepsilon} \\ &\ll Y^{4(1-\alpha)} T^\varepsilon + Y^{(8/3)(2-3\alpha)} T^{1+\varepsilon}, \end{aligned} \tag{100.22}$$

provided $\alpha > \frac{2}{3}$; note that both the lower and the upper bounds for N^ℓ given in (100.20) have been used. We take finally $Y = T^{3/(12\alpha-4)}$ and end the proof.

The combination of Theorems 117 and 131 yields Huxley's prime number theorem (1972):

Theorem 132 *We have*

$$\pi(x+y) - \pi(x) = (1 + o(1)) \frac{y}{\log x}, \quad x^\varpi < y < x, \tag{100.23}$$

whenever $\varpi > \frac{7}{12}$.

Notes

[100.1] As has been indicated already in §98, the zero-detecting method employed in this section is due to Montgomery (1969b; 1971, Chapter 12); he actually dealt more generally with the zeros of Dirichlet L-functions. It should be recorded that Montgomery accomplished the exponent $\frac{3}{5}+\varepsilon$ in (98.3) improving upon Ingham's $\frac{5}{8} + \varepsilon$ that had resisted attempts to lower for more than three decades. The assertion (100.5) is due to Huxley (1972); actually his result does not have the factor $T^{\varepsilon(1-\alpha)}$, which makes but no difference as

566   *Distribution of Prime Numbers*

far as the consequence (100.23) is concerned. He utilized the difficult bound
$\zeta\left(\frac{1}{2} + it\right) \ll t^{6/37}(\log t)^c$, $t \geq 2$, due to Haneke (1963) which was then the
deepest available; see Note [96.2] for the latest bound. The above proof of
the second part of (100.5) is adopted from Ivić (1985, Chapter 11); the use
of Haneke's bound is replaced by the more accessible result (99.66). Also,
he used the clever trick (100.19) which originated with Jutila (1977a); it had
played a rôle in Huxley's work as well. It should be observed that the bound
(100.16) becomes irrelevant when $\alpha > \frac{5}{6}$. The bound (100.15) is originally
due to Ingham (1940; Titchmarsh (1951, Theorem 9.19 (B))) but proved here
in a different way. The history of the theory of the zero-density estimates for
the zeta-function is well documented in Titchmarsh (1951), Pracher (1957),
Montgomery (1971), and Ivić (1985, 1991).

[100.2] This and the next Notes are additions to the above discussion: To
prove the bound (98.12) of Ingham (1937), we keep (100.11) but replace
(100.12) by

$$|\mathcal{R}_2| \ll \left(Y^{\frac{1}{2}-\alpha} \log T\right)^2 \sum_{\rho \in \mathcal{R}_2} \left(\int_{-\log^2 T}^{\log^2 T} \left|\zeta\left(\tfrac{1}{2} + i(\gamma + t)\right)M_X\left(\tfrac{1}{2} + i(\gamma + t)\right)\right| dt\right)^2$$

$$\ll T^{2\eta+\varepsilon} Y^{1-2\alpha} \log^4 T \int_T^{2T} \left|M_X\left(\tfrac{1}{2} + it\right)\right|^2 dt$$

$$\ll T^{2\eta+\varepsilon} Y^{1-2\alpha} (X + T) \log^5 T.$$

We choose $X = T$, $Y = T^{1+2\eta+\varepsilon}$, and end the proof. See also Titchmarsh
(1951, Theorem 9.18).

[100.3] Halász and Turán (1969) proved that

$$\text{Lindelöf hypothesis: } \mu\left(\tfrac{1}{2}\right) = 0 \;\Rightarrow\; N\left(\tfrac{3}{4} + \varepsilon\right) \ll T^\varepsilon.$$

It was a pivotal event in analytic number theory since their way of exploiting
Halász's inequality (98.32) inspired Montgomery (1969a, 1971) to devise a
drastically new approach to the Hoheisel scheme. Among a variety of results
thus he obtained, we state, in particular, the assertion (1969b, (9); 1971,
$(12.16)_{\alpha=1/2}$) with a minor alteration: for $\alpha > \frac{3}{4}$

$$N(\alpha, T) \ll T^{4\kappa \frac{2(3\alpha-2)(1-\alpha)}{(4\alpha-3)(2\alpha-1)}} \log^c T, \quad \kappa = \mu\left(\tfrac{1}{2}\right) + \varepsilon^2,$$

from which the result of Halász and Turán follows immediately. The three ε's
in the above are identical. To prove this, we first replace (98.48) by

$$\sum_{j\leq J}\left|\sum_{N\leq n<2N}\frac{a_n}{n^{s_j}}\right|^2 \ll \left(N+JN^{1/2}T^\kappa\right)\sum_{N\leq n<2N}|a_n|^2,$$

which can be obtained by choosing, in (98.57), the path $\sigma = \frac{1}{2}$ instead of $\sigma = 0$. Then, in place of (100.11), we have provided $\alpha \geq \frac{3}{4}$,

$$|\mathcal{R}_1| \ll \left(Y^{2(1-\alpha)} + |\mathcal{R}_1|X^{3/2-2\alpha}T^\kappa\right)\log^7 T.$$

That is, if $X^{3/2-2\alpha}T^\kappa \leq (\log T)^{-8}$, then

$$|\mathcal{R}_1| \ll Y^{2(1-\alpha)}\log^7 T.$$

As for $\rho \in \mathcal{R}_2$, we have, by (100.9),

$$Y^{\alpha-1/2}T^{-\kappa} \ll \int_{-\log^2 T}^{\log^2 T}\left|M_X(\tfrac{1}{2}+i(\gamma+t))\right|dt,$$

since $\Gamma(\frac{1}{2} - \beta + it) \ll 1$ by (94.37). Hence,

$$|\mathcal{R}_2|\left(Y^{\alpha-1/2}T^{-\kappa}\right)^2 \ll \left(X + |\mathcal{R}_2|X^{1/2}T^\kappa\right)\log^7 T.$$

That is, if $Y^{2\alpha-1}T^{-2\kappa} \geq X^{1/2}T^\kappa\log^8 T$, then

$$|\mathcal{R}_2| \ll XY^{1-2\alpha}T^{2\kappa}\log^7 T.$$

We set $X^{3/2-2\alpha}T^\kappa = (\log T)^{-8}$ and $Y^{2\alpha-1}T^{-2\kappa} = X^{1/2}T^\kappa\log^8 T$, which ends the proof.

[100.4]   The theory of zero-density is constrained by the great difficulty in estimating the mean values of powers of the zeta-function on the critical line. That is, the most essential issue is to establish bounds like $I_k(T) \ll T^{1+\varepsilon}$ for a $k \geq 3$; there exist a few proposals of possible methods to attack this problem but they have remained yet speculative.

[100.5]   In the above we are exclusively concerned with asymptotic distribution of primes in short intervals and have shown the hitherto essentially best result (100.23). Instead, if the existence of primes in short intervals is considered, then a result much deeper than (100.23) has been known: Baker et al (2001) established that

$$\pi(x+y) - \pi(x) > 0, \quad y > x^{21/40},$$

as x tends to infinity. Their argument is a highly involved elaboration of that of Iwaniec–Jutila (1979), a fusion of the Hoheisel scheme and the combinatorial linear sieve of Iwaniec (1980). Also, a theorem of Watt (1995) on a mean value containing the fourth power of the zeta function is a vital ingredient. We have to skip the details of this impressive advancement though, for it requires a

568   *Distribution of Prime Numbers*

full development of the combinatorial sieve method and the spectral theory of automorphic forms over congruence subgroups, which is beyond the scope of the present essays.

[100.6] The bound (100.6) would correspond to $\mu(\frac{1}{2}) \leq \frac{1}{10}$, if viewed via (98.12); and the assertion of the last Note to $\mu(\frac{1}{2}) \leq \frac{1}{38}$ if naively argued. These estimates of the Lindelöf exponent are truly further-reaching than today's state of the art in analytic number theory is capable. One must be impressed by the power of the sieve method combined with the theory of mean values of Dirichlet series and, most importantly, of the zeta-function.

§101

In the present section, we shall give a basic account of the distribution of prime numbers in arithmetic progressions, extending the core part of what has been developed in §§94–95 on $\pi(x)$ and $\psi(x)$, that is, the semi-explicit formula (95.7) and Theorem 107.

Our interest is in the asymptotic nature of the functions

$$
\begin{aligned}
\pi(x; q, \ell) &= \sum_{\substack{p \leq x \\ p \equiv \ell \bmod q}} 1, \\
\psi(x; q, \ell) &= \sum_{\substack{p \leq x \\ p \equiv \ell \bmod q}} \Lambda(n),
\end{aligned}
\qquad \langle q, \ell \rangle = 1,\ q \geq 3,
\qquad (101.1)
$$

as x tends to infinity; naturally, the requirement that $\ell \bmod q$ be a reduced residue class causes no loss of generality. We shall be concerned mostly with the function $\psi(x; q, \ell)$; the translation of assertions thus acquired into those on $\pi(x; q, \ell)$, via the formula

$$
\pi(x; q, \ell) = \int_2^x \frac{d\psi(u; k, \ell)}{\log u} - \sum_{\nu=2}^{\infty} \frac{1}{\nu} \sum_{\substack{h=1 \\ h^\nu \equiv \ell \bmod q}}^{q} \pi(x^{1/\nu}; q, h),
\qquad (101.2)
$$

will be considered appropriately. The most salient point of our discussion in this section is in that for the sake of applications to various problems in number theory we ought to make every assertion as uniform as possible with respect to moduli q varying along with x. Hence, our discussion becomes subtler and more interesting than that on the primary function $\psi(x)$.

$$\S101 \qquad\qquad 569$$

We shall proceed to the analogue of (95.7) for $\psi(x; q, \ell)$ via the relation

$$\psi(x; q, \ell) = \frac{1}{\varphi(q)} \sum_{\chi \bmod q} \overline{\chi}(\ell)\psi(x, \chi),$$

$$\psi(x, \chi) = \sum_{n \leq x} \chi(n)\Lambda(n), \tag{101.3}$$

and the expansion (52.8); that is, we shall show a representation of $\psi(x, \chi)$ in terms of the zeros of $L(s, \chi)$ defined by (52.6). Thus we shall first make the notion of these zeros precise. On noting the decomposition (53.8), we may restrict our attention to the primitive character χ^* which induces χ, for the zeros of the factor $\prod_{p|q}(1 - \chi^*(p)/p^s)$ are easy to handle and contributes negligibly: this is the same as to take into account the simple fact that $\psi(x, \chi) = \psi(x, \chi^*) + O(\nu(q)\log x)$ where $\nu(q)$ is as in Note [18.6]. Thus, we have, on noting Theorem 48(1),

$$\psi(x; q, \ell) = \frac{\psi(x)}{\varphi(q)} + \frac{1}{\varphi(q)} \sum_{\substack{k|q \\ 3 \leq k}} \sideset{}{^*}\sum_{\chi \bmod k} \overline{\chi}(\ell)\psi(x, \chi) + O\big((\log q)(\log x)\big),$$

$$\tag{101.4}$$

uniformly in $q, x \geq 1$, where the asterisk indicates the restriction that characters be primitive; this error term will make a negligible contribution in our later discussion. Hence, we shall suppose, hereafter up to (101.19), that χ's be as in (101.4) unless otherwise stated; in particular, they are non-principal.

Then we write the functional equation (55.11) as

$$\xi(1 - s, \overline{\chi}) = \frac{i^{\delta_\chi} k^{1/2}}{G_\chi} \xi(s, \chi),$$

$$\xi(s, \chi) = (k/\pi)^{(s+\delta_\chi)/2} \Gamma\big(\tfrac{1}{2}(s + \delta_\chi)\big) L(s, \chi). \tag{101.5}$$

Analogously to what is stated on $\xi(s)$ defined by (94.4), it can be asserted that $\xi(s, \chi)$ is an entire function whose zeros are all in the critical strip $0 \leq \sigma \leq 1$. They are zeros of $L(s, \chi)$ as well since the Gamma-factor never vanishes. Conversely, $\xi(s, \chi)$ vanishes at zeros of $L(s, \chi)$ in the critical strip minus $\{s = 0\}$. They are called non-trivial zeros, and each of them is denoted as $\rho = \beta + i\gamma$ generically; note that we do not use the term complex-zeros because the possibility of the existence of zeros in the interval $(\frac{1}{2}, 1)$ has not been eliminated yet; details on this matter will be given later. However, the use of the same notation for non-trivial zeros, as in (94.6), should not cause any confusion. All zeros of $L(s, \chi)$ other than ρ's are called trivial; they are $\{-2m - \delta_\chi : m = 0, 1, 2, \ldots\}$ which are all simple. Note that the origin is

570   *Distribution of Prime Numbers*

counted as a trivial zero of $L(s, \chi)$ if and only if $\chi(-1) = 1$; its simplicity is due to the fact $L(1, \chi) \neq 0$ that was proved in §88[A].

Remark 1 As for $L(s, \chi)$ with χ being possibly non-primitive, non-trivial zeros are defined to be those of $L(s, \chi^*)$; all other zeros of $L(s, \chi)$ are called trivial.

With this, we begin discussing the zero-free region for $L(s, \chi)$ (with χ as in (101.4)) by means of Theorem 103. So we note the bound

$$L(s, \chi) \ll (q(|t| + 1))^{2A}, \quad -A \leq \sigma, \ t \in \mathbb{R}, \tag{101.6}$$

for any $A \geq 1$ with the implied constant depending effectively on A, as can be readily shown by means of either (101.5) or Note [60.3]; but see Note [101.1]. It should be stressed that stating (101.6) we have used the fact $k|q$, so $k \leq q$; the same practice will be employed hereafter.

The following counterpart (101.7)–(101.8) of (94.41)–(94.42) is the basis of our argument: Let $N(T, \chi)$ be the number of non-trivial zeros ρ of $L(s, \chi)$ in the domain $|t| \leq T$. Then

$$N(T + 1, \chi) - N(T, \chi) \ll \log(q(T + 1)) \tag{101.7}$$

and for any $t_0 \in \mathbb{R}$

$$\frac{L'}{L}(s, \chi) = \sum_{\rho} \frac{1}{s - \rho} + O(\log q(|t_0| + 1)), \quad \begin{aligned} |s - (2 + it_0)| &\leq 4, \\ |\rho - (2 + it_0)| &\leq 8; \end{aligned} \tag{101.8}$$

the implied constants are absolute. With (101.6), these are immediate consequences of Theorems 101 and 103(i), respectively.

Then, we twist the inequality (95.8) by χ so that for $\sigma > 1$ and $t \in \mathbb{R}$

$$-3\frac{L'}{L}(\sigma, J_k) - 4\mathrm{Re}\,\frac{L'}{L}(\sigma + it, \chi) - \mathrm{Re}\,\frac{L'}{L}(\sigma + 2it, \chi^2) \geq 0; \tag{101.9}$$

the left side equals $2\sum_n J_k(n)\Lambda(n)(1 + \cos(\theta_n - t\log n))^2 n^{-\sigma}$, where J_k is the principal character mod k and $\chi(n) = \exp(i\theta_n)$.

Case (1): χ is complex.

We note that $\chi^2 \neq J_k$ and $(L'/L)(\sigma, J_k) = (\zeta'/\zeta)(\sigma) + O(\log\log 3q)$, for $\sigma > 1$, where the error-term is actually equal to $\sum_{p|k}(\log p)/p^{\sigma} + O(1)$ which can be estimated by the reasoning employed in Note [18.6]. We do not need to essentially alter the argument for (95.11); in fact, the present case is easier since $L(s, \chi)$ is entire unlike $\zeta(s)$. Thus, let $\beta + i\gamma$ be a non-trivial zero of $L(s, \chi)$. We apply Theorem 103 to $f(z) = L(\sigma_0 + i\gamma + z, \chi)$ with $K = \frac{1}{2}$ and

$$\sigma_0 = 1 + \frac{\tau}{\log(q(|\gamma| + 1))}. \tag{101.10}$$

§101   571

Hereafter up to (101.19), $\tau > 0$ is an independent variable such that $\tau/\log q \le \frac{1}{24}$. By (94.17)–(94.18) we find that there exists an absolute constant $c > 0$ such that, provided $\beta > \frac{19}{24}$ or $|\sigma_0 + i\gamma - \rho| \le \frac{1}{2}K$,

$$-\mathrm{Re}\,\frac{L'}{L}(\sigma_0 + i\gamma, \chi) < c\log(q(|\gamma| + 1)) - \frac{1}{\sigma_0 - \beta},$$
$$-\mathrm{Re}\,\frac{L'}{L}(\sigma_0 + 2i\gamma, \chi^2) < c\log(q(|\gamma| + 1)); \tag{101.11}$$

note that the former depends on (101.6) and the latter on the fact that the same holds with $L(s, \chi^2)$, as can be seen by Note [60.3]. Taking $\sigma = \sigma_0$, $t = \gamma$ in (101.9), we get $3/(\sigma_0 - 1) - 4/(\sigma_0 - \beta) + c\log(q(|\gamma| + 1)) > 0$ with another absolute constant $c > 0$. So, provided τ is taken sufficiently small,

$$\beta < 1 - (\sigma_0 - 1)\frac{1 - c(\sigma_0 - 1)\log(q(|\gamma| + 1))}{3 + c(\sigma_0 - 1)\log(q(|\gamma| + 1))}$$
$$< 1 - \frac{\tau}{4\log(q(|\gamma| + 1))}. \tag{101.12}$$

Hence, via Theorem 103(iii) we are led to the assertion that if χ is complex, then

$$\frac{L'}{L}(s, \chi) \ll \log(q(|t| + 1)), \quad \sigma > 1 - \frac{\tau}{16\log(q(|t| + 1))}, \ t \in \mathbb{R}, \tag{101.13}$$

where the implied constant depends on τ effectively, i.e., explicitly computable as a function of τ only.

Case (2): χ is real.

(a) Let $\beta + i\gamma$ be a zero of $L(s, \chi)$ such that $|\gamma| \ge \tau/(2\log q)$. We have, with σ_0 as in (101.10),

$$-\mathrm{Re}\,\frac{L'}{L}(\sigma_0 + 2i\gamma, \chi^2) = -\mathrm{Re}\,\frac{\zeta'}{\zeta}(\sigma_0 + 2i\gamma) + O(\log\log 3q)$$
$$< \frac{2}{3(\sigma_0 - 1)}, \tag{101.14}$$

provided τ is sufficiently small; we have used (95.13). Then we find that essentially the same as (101.12) holds; and $\beta < 1 - \tau/(12\log(q(|\gamma| + 1)))$. Hence

$$\frac{L'}{L}(s, \chi) \ll \log(q(|t| + 1)),$$
$$\sigma > 1 - \frac{\tau}{48\log(q(|t| + 1))}, \ |t| \ge \frac{\tau}{2\log q}, \tag{101.15}$$

where the implied constant depends on τ effectively as before.

572   *Distribution of Prime Numbers*

(b) Let $\beta + i\gamma$ be a zero of $L(s, \chi)$ such that

$$0 \le 1 - \beta < \frac{\tau}{12 \log q}, \quad |\gamma| < \frac{\tau}{2 \log q}. \tag{101.16}$$

If $\gamma \ne 0$, then $\beta - i\gamma$ is also a zero of $L(s, \chi)$ as χ is real. By Theorem 103(i), we have, with the same reasoning as that for (94.20),

$$-\frac{L'}{L}(\sigma_0, \chi) < -\mathrm{Re}\left(\frac{1}{\sigma_0 - \beta - i\gamma} + \frac{1}{\sigma_0 - \beta + i\gamma}\right) + c \log q, \tag{101.17}$$

with $\sigma_0 = 1 + \tau/\log q$. By $(101.9)_{t=0}$, we have

$$\frac{4}{\sigma_0 - 1} - \frac{8(\sigma_0 - \beta)}{(\sigma_0 - \beta)^2 + \gamma^2} + c \log q > 0. \tag{101.18}$$

On the assumption (101.16), the left side is lesser than $(c - 2/\tau) \log q$. It is implied that provided $\tau > 0$ is sufficiently small, (101.16) yields that $\gamma = 0$; then $\beta < 1$, of course. With this, suppose that there exist two real zeros ρ_1, ρ_2 greater than $1 - \tau/(12 \log q)$. We have, by Theorem 103(i) and $(101.9)_{t=0}$,

$$\frac{4}{\sigma_0 - 1} - \frac{4}{\sigma_0 - \rho_1} - \frac{4}{\sigma_0 - \rho_2} + c \log q > 0, \tag{101.19}$$

with σ_0 as in (101.17). This yields a contradiction.

Therefore, we are led to the basic assertion:

Theorem 133 *There exists an effective absolute constant $c_F > 0$ such that it holds, for any non-principal χ mod q, $q \ge 3$, which is not necessarily supposed to be primitive, that*

$$L(s, \chi) \ne 0 \text{ in the region } \sigma > 1 - \frac{c_F}{\log(q(|t| + 1))}, \quad t \in \mathbb{R},$$
$$\text{possibly except for a zero.} \tag{101.20}$$

If such an exceptional zero $\rho(q)$ exists, then χ is real and unique among all the characters mod *q; moreover, $\rho(q)$ is real, simple. Also, in the same region, we have*

$$\frac{L'}{L}(s, \chi) - \frac{\eta_\chi}{s - \rho(q)} \ll \log(q(|t| + 1)), \tag{101.21}$$

where $\eta_\chi = 1$ if $\rho(q)$ exists, and $= 0$ otherwise; and the implied constant is effective.

Remark 2 We use the term effective according to the sense that the numerical value of c_F can be explicitly evaluated; it will turn out to be relatively small, as indicated in the proof below. Also, $\rho(q)$ and the corresponding χ are called the q-exceptional zero and character.

Remark 3 By the way, the assertion (101.20) implies, in particular, that non-trivial zeros, in the sense of Remark 1, of $L(s, \chi)$ with any χ are in the open vertical strip $0 < \sigma < 1$.

Proof We apply the foregoing discussion to $L(s, \chi^*)$, so the conductor of χ is k dividing q. Then the existence c_F has been proved already; one may take $c_F = \tau/12$. Also, (101.21) is immediate since its confirmation amounts to an application of Theorem 103(iii) to either $L(s, \chi^*)/(s - \rho(q))$ if $\eta_\chi = 1$ or to $L(s, \chi^*)$ otherwise in much the same way as the second part of the proof of Theorem 101. It remains to confirm the uniqueness assertion. Let χ, χ' be different real characters mod q such that their associated L-functions have real zeros ρ, $\rho' \geq 1 - c_F/\log q$, respectively. Then for $\sigma > 1$

$$-\frac{\zeta'}{\zeta}(\sigma) - \frac{L'}{L}(\sigma, \chi) - \frac{L'}{L}(\sigma, \chi') - \frac{L'}{L}(\sigma, \chi\chi') > 0, \tag{101.22}$$

since the left side equals $\sum_n \Lambda(n)(1 + \chi(n))(1 + \chi'(n))n^{-\sigma}$. On noting that $\chi\chi'$ is non-principal, we see by Theorem 103 (ii) that for $\sigma_1 = 1 + 2c_F/\log q$

$$\frac{1}{\sigma_1 - 1} - \frac{1}{\sigma_1 - \rho} - \frac{1}{\sigma_1 - \rho'} + c\log q > 0. \tag{101.23}$$

This is a contradiction, as c_F can be assumed to be sufficiently small; we end the proof.

We are now ready to extend (95.7) to $\psi(x, \chi)$ appearing in (101.4), so here χ mod k, $k|q$, is primitive. We have, with $2 \leq T \leq x$,

$$\psi(x, \chi) = -\eta_\chi \frac{x^{\rho(q)}}{\rho(q)} - \sum_{\substack{\rho \neq \rho(q), 1-\rho(q) \\ |\gamma| < T}} \frac{x^\rho}{\rho}$$
$$+ O\left(\eta_\chi x^{1-\rho(q)} \log x + \frac{x}{T}(\log qx)^2\right), \tag{101.24}$$

where the implied constant is effective. Here ρ are non-trivial zeros of $L(s, \chi)$, and η_χ is as in (101.21). We shall verify this. Thus, let $x = [x] + \frac{1}{2}$. By (95.1)

$$\psi(x, \chi) = -\frac{1}{2\pi i} \int_{a-iT}^{a+iT} \frac{L'}{L}(s, \chi)x^s \frac{ds}{s} + O\left(\frac{x}{T}(\log x)^2\right), \tag{101.25}$$

with $a = 1 + (\log x)^{-1}$ and T satisfying

$$\frac{L'}{L}(\sigma \pm iT, \chi) \ll (\log qT)^2, \quad -\frac{1}{2} \leq \sigma \leq a, \tag{101.26}$$

which is a consequence of (101.7)–(101.8). Also

$$\frac{L'}{L}(s, \chi) \ll \log q|s|, \quad \sigma = -\frac{1}{2}, \tag{101.27}$$

574   *Distribution of Prime Numbers*

which is a restricted analogue of (94.40) and the proof is essentially the same. Shifting the path in (101.25) to $\operatorname{Re} s = -\frac{1}{2}$, we have

$$\psi(x, \chi) = - \sum_{\substack{\rho \\ |\gamma| < T}} \frac{x^\rho}{\rho} - (1 - \delta_\chi) \log x - \delta'_\chi + O\left(\frac{x}{T} (\log qx)^2\right), \quad (101.28)$$

where

$$\frac{L'}{L}(s, \chi) = \frac{1 - \delta_\chi}{s} + \delta'_\chi + O(|s|), \quad s \to 0. \quad (101.29)$$

We note that

$$\frac{1}{2\pi i} \int_C \frac{L'}{L}(s, \chi) \frac{ds}{s} = \delta'_\chi + \sum_{|\gamma| < 1/2} \frac{1}{\rho}. \quad (101.30)$$

The contour C is the positively oriented boundary of the rectangle $\{-\frac{1}{2} \leq \sigma \leq \frac{3}{2}, |t| \leq \frac{1}{2}\}$ on which it can be assumed that $L(s, \chi) \neq 0$ without loss of generality. This integral is $O\left((\log q)^2\right)$ because of (101.27) and a bound similar to (101.26). Hence, dropping the condition on x, we have

$$\psi(x, \chi) = - \sum_{\substack{\rho \\ |\gamma| < T}} \frac{x^\rho}{\rho} + \sum_{|\gamma| < 1/2} \frac{1}{\rho} + O\left(\frac{x}{T} (\log qx)^2\right). \quad (101.31)$$

If χ is exceptional, then $L(1 - \rho(q), \chi) = 0$, and

$$-\frac{x^{1-\rho(q)}}{1 - \rho(q)} + \frac{1}{1 - \rho(q)} \ll x^{1-\rho(q)} \log x. \quad (101.32)$$

On noting that the second sum on the right of (101.31) is equal to $1/(1 - \rho(q)) + O\left((\log q)^2\right)$ because of (101.7) and (101.20), we end the proof of (101.24).

Inserting (95.7), (101.24) into (101.4), we obtain, after minor rearrangements,

Theorem 134 *Let* $2 \leq T \leq x$ *and* $\langle q, \ell \rangle = 1$. *Then,*

$$\begin{aligned}
\psi(x; q, \ell) = {} & \frac{x}{\varphi(q)} - \chi_1(\ell) \frac{x^{\rho(q)}}{\varphi(q)\rho(q)} \\
& - \frac{1}{\varphi(q)} \sum_{\chi \bmod q} \overline{\chi}(\ell) \sum_{\substack{\rho \neq \rho(q), 1-\rho(q) \\ |\gamma| < T}} \frac{x^\rho}{\rho} \\
& + O\left(\frac{x^{1-\rho(q)}}{\varphi(q)} \log x + \frac{x}{T} (\log qx)^2\right).
\end{aligned} \qquad (101.33)$$

Here $\rho(q)$ *and* χ_1 *are the* q-*exceptional zero and character; if* χ_1 *does not exist, then the terms involving* $\rho(q)$ *do not appear; in the double sum* $\rho = \beta + i\gamma$ *runs*

$\S101$ 575

*over non-trivial zeros of $L(s, \chi)$ in the sense of Remark 1 above. The implied
constant is effectively computable.*

We now take $T = \exp((\log x)^{1/2})$ in (101.33). Then, via Theorem 133,
the Landau–Page prime number theorem follows immediately: Uniformly for
$q \leq \exp((\log x)^{1/2})$, $\langle q, \ell \rangle = 1$,

$$
\pi(x; q, \ell) = \frac{1}{\varphi(q)} \int_2^x \frac{du}{\log u} - \frac{\chi_1(\ell)}{\varphi(q)} \int_2^x \frac{u^{\rho(q)-1}}{\log u} du
$$
$$
+ O\left(x \exp\left(-c(\log x)^{1/2}\right)\right),
\tag{101.34}
$$

where $c > 0$ and the implied constant are effectively computable.

This is obviously an extension of (95.15) to arithmetic progressions. However, it does not have much practical value because of the dubious situation
relevant to the possible existence of the exceptional zero $\rho(q)$. We need to
make its location more precise, if ever exists. The following assertion due to
Siegel (1935) is a partial answer to this very basic issue:

Theorem 135 *For each $\varepsilon > 0$, there exists a constant $\tau_\varepsilon > 0$, depending only
on ε, such that for all $q > 0$*

$$
\rho(q) < 1 - \frac{\tau_\varepsilon}{q^\varepsilon}.
\tag{101.35}
$$

*In particular, we have, for each $A \geq 1$ and uniformly for $1 \leq q \leq \log^A x$,
$\langle q, \ell \rangle = 1$,*

$$
\psi(x; q, \ell) = \frac{x}{\varphi(q)} + O\left(x \exp\left(-c_A(\log x)^{1/2}\right)\right),
\tag{101.36}
$$

$$
\pi(x; q, \ell) = \frac{1}{\varphi(q)} \mathrm{li}(x) + O\left(x \exp\left(-c_A(\log x)^{1/2}\right)\right),
\tag{101.37}
$$

where the implied constant and $c_A > 0$ may depend on A at most.

Proof We shall first prove (101.36); then (101.37) follows immediately. If
$q = 1, 2$, then (101.36) is the same as (95.14). Thus we may assume that
$3 \leq q \leq \log^A x$. By (95.1), We have, in (101.4),

$$
\psi(x, \chi) = -\frac{1}{2\pi i} \int_{a-iT}^{a+iT} \frac{L'}{L}(s, \chi) \frac{x^s}{s} ds + O\left(x \exp\left(-\tfrac{1}{2}(\log x)^{1/2}\right)\right),
\tag{101.38}
$$

with $x = [x] + \tfrac{1}{2}$, $a = 1 + 1/\log x$, and $T = \exp\left((\log x)^{1/2}\right)$. We take $\varepsilon = 1/(2A)$ in (101.35) so that $L(s, \chi) \neq 0$ for $\sigma \geq 1 - c(\log x)^{-1/2}$, $|t| \leq T$,
with a $c > 0$ depending on A. We shift the contour to $\sigma = 1 - c(\log x)^{-1/2}$,
$|t| \leq T$. We obtain, by (101.21), $\psi(x, \chi) \ll x \exp\left(-c_A(\log x)^{1/2}\right)$, from which
(101.36) follows. Alternatively one may apply (101.33) of course.

576   *Distribution of Prime Numbers*

We turn to a proof of (101.35). To begin with, we assume that there exists a real character χ_0 mod q_0 such that $L(\rho_0, \chi_0) = 0$, $1 - \frac{1}{10}\varepsilon < \rho_0 < 1$; if there exists no real character satisfying this, then we do not need to discuss any further. We shall regard q_0, ρ_0 as constants and suppose that $q > q_0$ is sufficiently large. Obviously it suffices to discuss the $L(s, \chi)$ associated with a real and primitive χ mod q. With this, we consider the function, with $a(n, \chi)$ as in (88.7),

$$H(s) = \sum_{n=1}^{\infty} \frac{a(n, \chi)a(n, \chi_0)}{n^s} = \frac{\zeta(s)L(s, \chi)L(s, \chi_0)L(s, \chi\chi_0)}{L(2s, \chi\chi_0)}, \qquad (101.39)$$

which should be compared with (88.8). In a small neighborhood of $s = 2$, we have the Taylor expansion

$$H(s) = \sum_{j=0}^{\infty} b_j(2 - s)^j, \quad b_j = \frac{1}{j!} \sum_{n=1}^{\infty} \frac{a(n, \chi)a(n, \chi_0)}{n^2}(\log n)^j. \qquad (101.40)$$

We have also

$$H(s) - \frac{\omega}{s - 1} = \sum_{j=0}^{\infty} (b_j - \omega)(2 - s)^j, \quad |s - 2| < \tfrac{3}{2};$$

$$\omega = \frac{L(1, \chi)L(1, \chi_0)L(1, \chi\chi_0)}{L(2, \chi\chi_0)}, \qquad (101.41)$$

for the function $H(s) - \omega/(s - 1)$ is regular inside the indicated disk. We apply Cauchy's integral formula to this function, with the contour $|s - 2| = \frac{4}{3}$; we get $|b_j - \omega| \ll (\frac{3}{4})^j q$ uniformly for j, q, where the bound (101.6) and the remark adjacent to (101.11) have been taken into account; note that the current assumption on χ implies that $\chi\chi_0$ is non-principal. In particular, on the disk $|s - 2| \le \frac{10}{9}$, it holds that

$$H(s) - \frac{\omega}{s - 1} = \sum_{j=0}^{J-1} (b_j - \omega)(2 - s)^j + O\big((5/6)^J q\big), \qquad (101.42)$$

uniformly for $J \ge 1$ and q. On noting that $b_0 = H(2) \ge 1$ and $b_j \ge 0$, as $a(1, \chi)$, $a(1, \chi') = 1$ and $a(n, \chi)$, $a(n, \chi') \ge 0$, we have, in the interval $\frac{8}{9} < s < 1$,

$$H(s) \ge 1 - \frac{\omega}{1 - s}(2 - s)^J + O\big((5/6)^J q\big) \ge \tfrac{1}{2} - \frac{\omega}{1 - s}q^{6(1-s)}, \qquad (101.43)$$

in which the second line is the result of the specialization $J = [6 \log q]$. Since $H(\rho_0) = 0$, we get $\omega > \frac{1}{2}(1 - \rho_0)q^{-6(1-\rho_0)}$; thus

$$L(1, \chi) \gg q^{-4\varepsilon/5}, \qquad (101.44)$$

$$§101 \qquad 577$$

since $L(1, \chi_0) \ll 1$ and $L(1, \chi \chi_0) \ll \log q$. On the other hand, if $L(\rho, \chi) = 0$ with $\rho > 1 - (\log q)^{-1}$, then again by Cauchy's integral formula

$$L(1, \chi) = \frac{1}{2\pi i} \int_{\rho}^{1} \int_{|w-1|=2/\log q} \frac{L(w, \chi)}{(w - \eta)^2} dw d\eta \ll (1 - \rho)(\log q)^2.$$

$$(101.45)$$

This ends the proof. For an alternative proof of (101.35) see Note [106.2] (1).

Although some uniformity, better than that in (101.33), with respect to moduli has been attained, Theorem 135 contains a serious drawback: The constant τ_ε is not effective. Namely, any of the arguments that have so far been devised to prove (101.35) is unable to yield an algorithm with which one may fix the actual value of τ_ε for each ε. We are certain only of the existence of τ_ε. Consequentially, the same applies to c_A in (101.36)–(101.37). Therefore, this prime number theorem, which is commonly attributed to Siegel (1935) and Walfisz (1936), is unable to yield any realistic/practical answer even to such a basic question as when $\pi(x; q, \ell)$ becomes positive individually, that is, the whereabout of the least prime congruent to ℓ mod q. We shall be able to give, only in §106, a method which solves this problem without any hypothesis.

Stating (101.36), the main assertion in the present section, we have to satisfy ourselves with the uniformity that is assured only for those moduli $1 \leq q \leq (\log x)^A$ with A being an arbitrary large constant. We face thence a major obstacle that is yet to be cleared away from analytic number theory, that is, the problem of how to go substantially beyond such a modest range of moduli for individual $\psi(x; q, \ell)$. In the light of this situation of stalemate for so many years, it is a truly fortunate outcome of an impressive series of investigations originating in Linnik (1941) and Rényi (1948) that if one looks into the statistical behavior of $\psi(x; q, \ell)$ along with varying moduli, then a totally new vista opens up. The details will be described in §105, and we shall witness that the Hoheisel scheme framed in Note [98.1] works in a wider perspective. Moreover, as is indicated above, in §106 we shall show that the scheme works impressively even for those individual moduli q which are of the size of certain positive powers of x, without being affected by the possible presence of q-exceptional zeros.

Notes

[101.1] The bound (101.6) can of course be refined considerably. For instance, with a non-principal χ mod q and s in the critical strip, the combination of the representation (53.8), the functional equation (55.11), the convexity principle (Theorem 104), and Stirling's formula (94.37) yields, for $0 \leq \sigma \leq 1$,

$$L(s,\chi) \ll \Big(\prod_{p|q/q^*} (1 + p^{-\sigma}) \Big) (q^*(|t| + 1))^{(1-\sigma)/2} \log(q^*(|t| + 1)),$$

where q^* is the conductor of χ. Thus, $L(\frac{1}{2} + it, \chi) \ll (q(|t| + 1))^{1/4+\varepsilon}$. This exponent $\frac{1}{4}$ can be replaced by a subconvexity exponent. Michel (2022), already referred to in Note [99.6], is a readable report on this subject as well.

[101.2] If χ is q-exceptional, then χ^* is q^*-exceptional; and $\rho(q) = \rho(q^*)$. However, the converse does not hold always. Here arises an annoying complication: Suppose that $\chi \bmod q$ is primitive as well as q-exceptional with $\rho(q)$ the exceptional zero in the sense defined by Remark 2 above. With this, take an arbitrary $\tilde{\chi} \bmod \tilde{q}$ which is induced by χ. Then $L(\rho(q), \tilde{\chi}) = 0$ of course. So if $\rho(q) > 1 - c_F/\log \tilde{q}$, then $\rho(q) = \rho(\tilde{q})$. However, for instance if $\rho(q) < 1 - \mu c_F/\log q$ with a small $\mu > 0$, then $\rho(q)$ may not be the $\tilde{q}$-exceptional zero: since $\tilde{q}$ can be arbitrarily large as far as $\tilde{q} \equiv 0 \bmod q$, we may have $\rho(q) < 1 - c_F/\log \tilde{q}$; i.e., if $\tilde{q} > q^{1/\mu}$, then $\tilde{\chi}$ is not $\tilde{q}$-exceptional in the sense of Remark 2. One way to make the situation clearer is to restrict our consideration to primitive characters. Namely, if the set of all zeros of all L-functions associated with primitive characters is taken into consideration, then one may assert the following: With the same absolute constant c_F as in Theorem 133,

$$\prod_{3 \le q \le Q} \prod_{\chi \bmod q}^* L(s,\chi) \ne 0, \quad \sigma > 1 - \frac{c_F}{\log(Q(|t| + 1))}, \quad t \in \mathbb{R},$$

possibly except for a real zero; here $\prod^*$ indicates that the product is restricted to primitive characters. If such an exceptional zero $\tilde{\rho}(Q)$ exists, then it is simple and $L(\tilde{\rho}(Q), \chi_1) = 0$ with a real χ_1 which is unique among primitive characters whose moduli are not greater than Q. The proof is omitted, as it is an easy extension of that for Theorem 133; note that this uniqueness assertion depends on Theorem 48 (2).

[101.3] It is, in fact, expected that there should not exist any exceptional character:

$$\text{Conjecture}: \quad \begin{array}{c} \text{there exists an effective constant } c_0 > 0 \text{ such that} \\ L(s,\chi) \ne 0, 1 - c_0/\log q < s < 1, \\ \text{for any real } \chi \bmod q, q \ge 3. \end{array}$$

This is one of the most fundamental problems in analytic numbers.

[101.4] The above proof of (101.35) is due to Estermann (1948). Elaborating his argument, Tatuzawa (1951) showed, among other things, that for each $\varepsilon > 0$

$$L(s,\chi) \ne 0, \quad 1 - \frac{\varepsilon}{10q^\varepsilon} < s < 1,$$

for all primitive real χ mod q, with at most one exception. However, it is not known yet how to bound the exceptional modulus in terms of ε. Throughout his research life, Tatuzawa was deeply occupied by the problem of making Siegel's theorem effective.

[101.5] It is annoying that in the effectivity context our knowledge is presently confined essentially to the following assertion: There exists an effective constant $c > 0$ such that for any real primitive character χ mod q, $q \geq 3$,

$$L(1, \chi) \geq \frac{c}{q^{1/2}},$$

and thus $\rho(q) < 1 - c/(q^{1/2}(\log q)^2)$ with another effective $c > 0$ as (101.45) implies. Of course this follows also from Theorems 71, 97, and 98, as is customarily mentioned in literature. However, here is an independent confirmation: We consider the integral

$$\frac{1}{2\pi i} \int_{(2)} \zeta(s) L(s, \chi) ((2X)^s - X^s) \Gamma(s) ds.$$

Shifting the contour to $(-\frac{1}{2})$, we get the identity

$$\sum_{n=1}^{\infty} a(n, \chi) (e^{-n/(2X)} - e^{-n/X}) = L(1, \chi) X$$

$$+ \frac{1}{2\pi i} \int_{(-\frac{1}{2})} \zeta(s) L(s, \chi) ((2X)^s - X^s) \Gamma(s) ds,$$

where $a(n, \chi)$ is as in (88.7). We note that $a(n, \chi) \geq 0$, $a(n^2, \chi) \geq 1$, and $L\left(-\frac{1}{2} + it, \chi\right) \ll q(|t| + 1)$ by $(101.5)_{k=q}$. Hence,

$$X^{1/2} \ll L(1, \chi) X + qX^{-1/2}.$$

Taking $X = cq$ with a sufficiently large $c > 0$, we end the confirmation. See also Goldfeld (1976, 1985).

[101.6] On the generalized Riemann hypothesis (GRH), i.e., (53.9), we have, uniformly for any coprime pair $\{q, \ell\} \in \mathbb{N}^2$,

$$\psi(x; q, \ell) = \frac{x}{\varphi(q)} + O(x^{1/2}(\log qx)^2), \quad x > 1.$$

This is an immediate consequence of the combination of Theorems 102, 103 (i), and (95.1). In particular,

on GRH the least prime congruent to ℓ mod q is $O(q^2(\log q)^3)$.

We shall see in §106 what can be done without GRH.

580 · Distribution of Prime Numbers

[101.7]   Returning to Note [32.6], here we give a proof of Ankeny's theorem:
Let H be an arbitrarily given proper subgroup of $G = (\mathbb{Z}/q\mathbb{Z})^*$. The assertion
Note [51.5] (1) implies that $G/H \cong \{\chi \bmod q : \chi \equiv 1 \text{ on } H\}$. Hence, there
should exist a non-principal character $\chi \bmod q$ such that $\chi(h) = 1$ for all
$h \in H$. Assume now that GRH is valid with $L(s, \chi)$. Let $k(x)$ be as in (98.53),
with N being sufficiently large. We consider the sum

$$U = \sum_{n=1}^{\infty} \chi(n) \Lambda(n) k(n).$$

We have

$$U = -\frac{1}{2\pi i} \int_{(2)} \frac{L'}{L}(s, \chi) \tilde{k}(s) ds \ll N^{1/2} \log q,$$

with the Mellin transform $\tilde{k}$ of k. This bound is the result of shifting the contour
to $\mathrm{Re}\, s = \frac{1}{4}$; the necessary facts on $(L'/L)(s, \chi)$ can readily be obtained by
means of Theorems 101 and 103 (i). Let $\{n \bmod q : k(n) \neq 0, \langle q, n \rangle = 1\}$
be contained in H. Then we have $U \gg N - O((\log N)(\log q))$; this O-term is
actually equal to $\sum_n \Lambda(n) k(n)$, $\langle n, q \rangle \neq 1$. Namely, if $N > A(\log q)^2$ with a
constant $A > 0$, then we would be led to a contradiction. All implied constants
in the above are effectively computable, so is A. This ends the proof. See
Montgomery (1971, pp.123–125).

[101.8]   Continuing the discussion of §37, we assert that it holds on GRH that

$$(1) \qquad q \in \bigcap_{\substack{a \leq A(\log q)^2 \\ \langle a, q \rangle = 1}} \mathrm{spp}(a) \;\Rightarrow\; q \text{ is a prime power,}$$

where A is the same constant as the one in the preceding Note; thus, $A = 2$ is
admitted, if Bach's estimation is employed (Note [46.1]). First a preparation:
Let $q - 1 = 2^\mu t$, $2 \nmid t$. Then, under the assumption $q \in \mathrm{spp}(a)$ (in particular,
$\langle a, q \rangle = 1$),

(2) $\left(\frac{a}{p}\right) = -1$ for a certain $p | q$, $p - 1 = 2^\beta u$, $2 \nmid u \;\Rightarrow\; a^{2^{\beta-1} t} \equiv -1 \bmod q$;

(3) $a^{2^{\beta-1} t} \equiv -1 \bmod q \;\Rightarrow\; \left(\frac{a}{p}\right) = -1$ for all $p | q$ such that $p - 1 = 2^\beta u$, $2 \nmid u$.

We consider (2). If $a^t \equiv 1 \bmod q$, then $\left(\frac{a}{p}\right) = 1$, $p | q$, since t is odd, so this
case is discarded. Thus, there exists a γ such that $a^{2^\gamma t} \equiv -1 \bmod q$ with
$0 \leq \gamma \leq \mu - 1$. Suppose that $\gamma < \beta - 1$. Then $a^{2^{\beta-1} t} \equiv 1 \bmod q$. With
$b = a^{2^{\beta-1}}$, we have $b^t \equiv 1 \bmod p$ as well as $b^u \equiv -1 \bmod p$, for the condition
in (2) means that $a^{2^{\beta-1} u} \equiv -1 \bmod p$; see (57.3). Let the order of $b \bmod p$
be v, so $v | t$. Also we have $v | 2u$. That is, $v | u$. This is a contradiction. Hence,

$\beta - 1 \leq \gamma$. On the other hand, we have $\gamma \leq \beta - 1$, for $a^{2^{\gamma}t} \equiv -1 \bmod p$, so $2^{\beta}u/\langle 2^{\gamma}t, 2^{\beta}u\rangle$ ought to be even by (43.2). This finishes the confirmation of (2). To verify (3), suppose that $\left(\frac{a}{p}\right) = 1$, i.e., $a^{2^{\beta-1}u} \equiv 1 \bmod p$. Then, with $b = a^{2^{\beta-1}}$ again, the order of $b \bmod p$ is a divisor w of u. However, since $b^{t} \equiv -1 \bmod p$, we have $w|2t$, so $w|t$, and we get again a contradiction. This ends the proof of (2)–(3). Now, we start the proof of (1). For this sake, suppose that $p_1p_2|q, p_1 \neq p_2, p_j - 1 = 2^{\beta_j}u_j, 2 \nmid u_j$. We consider the character $\chi_1(n) = \left(\frac{n}{p_1}\right)$. By the preceding Note, under GRH there exists an m such that $\chi_1(m) = -1, \langle m,q \rangle = 1, m \leq A(\log q)^2$. By the assumption, we have $q \in \mathrm{spp}(m)$. Then, by (2), $m^{2^{\beta_1-1}t} \equiv -1 \bmod p_2$. This implies, by (43.2) as before, that $\beta_1 - 1 < \beta_2$, i.e., $\beta_1 \leq \beta_2$. Hence, by symmetry we should have $\beta_1 = \beta_2$. We then consider the character $\chi_2(n) = \left(\frac{n}{p_1p_2}\right)$. Under GRH, there exists an m_1 such that $\chi_2(m_1) = -1, \langle m_1,q \rangle = 1, m_1 \leq A(\log q)^2$. By the assumption, $q \in \mathrm{spp}(m_1)$. We have, for instance, $\left(\frac{m_1}{p_1}\right) = -1$ and $\left(\frac{m_1}{p_2}\right) = 1$. Then, by (2), $m_1^{2^{\beta_1-1}t} \equiv -1 \bmod q$. Since $\beta_1 = \beta_2$, we have, by (3), $\left(\frac{m_1}{p_2}\right) = -1$, a contradiction. That is, we have shown that if q satisfies the left side of (1), then it has only one prime factor. This ends the proof of (1). In the above we have followed Crandall–Pomerance (2005, p.141).

[101.9] To negate GRH, it suffices to find a single pair $\{p_1,p_2\}, p_1 \neq p_2$, such that $q = p_1p_2$ satisfies the inclusion condition of (1) in the preceding Note. This fact may lead one to the conclusion that the reliability of the primality test (1) should be quite high.

§102

We now enter into an investigation on the behavior of the function $\pi(x; q, \ell)$ with the varying modulus q that can be as large as a positive power of x. Our approach to this fundamental theme in analytic number theory is, in essence, a direct infusion of sieves of Linnik (1941) and Selberg (1947) into the structure of the family of Dirichlet L-functions.

It is thus expedient to begin on an exposition of their fundamental sieve ideas. To this end, we return to §19 and consider the interval situation

$$\mathcal{A} = \mathcal{N} = (M, M+N] \cap \mathbb{Z}, \quad M \in \mathbb{Z}, N \in \mathbb{N}. \tag{102.1}$$

We alter also the notion of the map Ω into a more practical one as follows. We fix first a set $\mathcal{P}$ of primes as its domain and let $\Omega(p), p \in \mathcal{P}$, be a preassigned set of residue classes mod p; so $\Omega(p)$ can be taken for a set of $\mathbb{Z}$ as before but

582   *Distribution of Prime Numbers*

hereafter it will also be regarded as a set of residue classes. We let $n \in \Omega(p)$ denote that $n \bmod p$ is in $\Omega(p)$, while $n \notin \Omega(p)$ is its negation. We shall then consider, in place of (19.1),

$$\mathcal{S}(\mathcal{N}, \Omega, z) = \{n \in \mathcal{N} : n \notin \Omega(p), \forall p \leq z, p \in \mathcal{P}\} :$$
$$\text{abbreviated to } \mathcal{S}(\mathcal{N}, z). \tag{102.2}$$

Here the parameter $z > 0$ is supposed to tend to infinity taking non-integral values in all expressions within the present section. Our sieve problem is then defined as the estimation of $|\mathcal{S}(\mathcal{N}, z)|$.

We suppose additionally that

$$0 < |\Omega(p)| < p, \quad \forall p \in \mathcal{P}. \tag{102.3}$$

This does not cause any loss of generality. For, if $|\Omega(p_0)| = 0$, i.e., $\Omega(p_0) = \varnothing$, then p_0 does not actually participate in the current sifting procedure and ought to be excluded from $\mathcal{P}$. On the other hand, if $|\Omega(p_1)| = p_1$, then as soon as z becomes greater than p_1 the sifting results in $\mathcal{S}(\mathcal{N}, z) = \varnothing$, and terminates.

[CONVENTION]   In what follows we shall not always mention explicitly the dependency on $\mathcal{P}$ of our argument, for the nature of $\mathcal{P}$ can be read in the behavior of Ω.

For instance, in the case of the twin prime conjecture mentioned as an example after (19.1), $|\Omega(p)|$ is small, i.e., 2 for all odd p, so the combinatorial sieve is able to produce non-trivial results as has been indicated in Notes of §19. One may ask, on the other hand, whether it is possible to construct any sieve which is effective for the situation where $|\Omega(p)|$ grows rapidly with p. For instance, if one has to sift out all non-quadratic residues $\bmod p$ for each odd p, then the combinatorial sieve does not seem to be capable to yield any meaningful conclusion since the requirement imposed in Note [19.6] is not fulfilled at all.

It was Linnik who invented a fundamental device to deal with such extreme situations. His idea, which is called the large sieve because $|\Omega(p)|$ can be large in his method, was to employ Fourier analysis in the sense of §50. With a later sophistication, it runs as follows: Let ϖ_Ω be the characteristic function of the set $\{n \in \mathbb{Z} : n \notin \Omega(p), \forall p \leq z\}$. Then we introduce the Fourier series

$$U(\theta) = \sum_{M < n \leq M+N} \varpi_\Omega(n) e(n\theta),$$

$$U(\theta; q, \ell) = \sum_{\substack{M < n \leq M+N \\ n \equiv \ell \bmod q}} \varpi_\Omega(n) e(n\theta), \tag{102.4}$$

$$\S102 \qquad 583$$

where $e(x) = \exp(2\pi i n x)$ as before; note that ϖ_Ω depends on z and the above convention is being observed. Since

$$U\left(\theta + \frac{t}{p}\right) = \sum_{a=1}^{p} U(\theta; p, a) e\left(\frac{at}{p}\right), \qquad (102.5)$$

we have, by the orthogonality (50.8),

$$\sum_{t=1}^{p} \left| U\left(\theta + \frac{t}{p}\right) \right|^2 = p \sum_{a=1}^{p} |U(\theta; p, a)|^2. \qquad (102.6)$$

On the other hand, $U(\theta; p, a) = 0$ for $a \in \Omega(p)$ and $p \le z$, so

$$|U(\theta)|^2 = \left| \sum_{a=1}^{p} U(\theta; p, a) \right|^2 \le (p - |\Omega(p)|) \sum_{a=1}^{p} |U(\theta; p, a)|^2, \qquad p \le z. \qquad (102.7)$$

From these two, we get

$$|U(\theta)|^2 \frac{|\Omega(p)|}{p - |\Omega(p)|} \le \sum_{t=1}^{p-1} \left| U\left(\theta + \frac{t}{p}\right) \right|^2, \qquad p \le z, \qquad (102.8)$$

to which (102.3) is relevant. Taking a different prime $p' \le z$,

$$\left| U\left(\theta + \frac{t'}{p'}\right) \right|^2 \frac{|\Omega(p)|}{p - |\Omega(p)|} \le \sum_{t=1}^{p-1} \left| U\left(\theta + \frac{t'}{p'} + \frac{t}{p}\right) \right|^2. \qquad (102.9)$$

Summing over $t' = 1, 2, \ldots, p' - 1$,

$$|U(\theta)|^2 \frac{|\Omega(p')|}{p' - |\Omega(p')|} \frac{|\Omega(p)|}{p - |\Omega(p)|} \le \sum_{t'=1}^{p'-1} \sum_{t=1}^{p-1} \left| U\left(\theta + \frac{t'}{p'} + \frac{t}{p}\right) \right|^2. \qquad (102.10)$$

Repeating the same procedure, we are led to

$$|U(\theta)|^2 \prod_{p|r} \frac{|\Omega(p)|}{p - |\Omega(p)|} \le \sum_{\substack{t \bmod r \\ \langle t, r \rangle = 1}} \left| U\left(\theta + \frac{t}{r}\right) \right|^2, \qquad r|P(z), \qquad (102.11)$$

where $P(z) = \prod_{p \le z} p$, $p \in \mathcal{P}$, and (32.3) has been applied. Further, letting $\theta = 0$ and summing the result over square-free $r \le z$, we find that

$$G(z)|\mathcal{S}(\mathcal{N}, z)|^2 \le \sum_{r \le z} \sum_{\substack{t \bmod r \\ \langle t, r \rangle = 1}} \left| \sum_{M < n \le M + N} \varpi_\Omega(n) e\left(\frac{t}{r} n\right) \right|^2, \qquad (102.12)$$

584   *Distribution of Prime Numbers*

with

$$G(z) = \sum_{r \leq z} \mu(r)^2 H(r), \qquad (102.13)$$

$$H(r) = \prod_{p|r} \frac{|\Omega(p)|}{p - |\Omega(p)|}. \qquad (102.14)$$

It should be noted that r's of (102.12) are not restricted to divisors of $P(z)$.

Thus our sieve problem has been transformed into the estimation of the nested sum in (102.12) over the Farey sequence $\{t/r\}$ of order z; see (20.1). We shall show in the next section how to deal with this new problem. It suffices, for the moment, to know that our sieve problem has been brought into Fourier analysis in the sense of §50, specifically, that of Note [50.4], so a situation like the Parseval formula, which will later be termed L^2-inequalities, comes into view. Thus, here we state only the end result (a consequence of (103.20) below):

Theorem 136

$$|S(\mathcal{N},z)| \leq \frac{1}{G(z)}(N + 2z^2). \qquad (102.15)$$

Observe that for the above example related to non-quadratic residues, this gives a satisfactory bound: Since $G(z) \approx z$, we have $|S(\mathcal{N},z)| \ll N/z + z$. Hence $|S(\mathcal{N}, \sqrt{N})| \ll \sqrt{N}$, which should be the correct order, provided $M \ll N$, because all squares in such an $\mathcal{N}$ survive the sifting.

Next, we shall show the idea of Selberg that is called the Λ^2-sieve: We apply first a multiplicative extension to the map Ω as can be inferred from (102.11); that is, we let $n \in \Omega(u)$, for square-free u, stand for that $n \in \Omega(p), \forall p|u$; naturally, $\Omega(1) = \mathbb{Z}$. With this, take an arbitrary real-valued arithmetic function λ such that $\lambda(1) = 1$, and then observe that

$$\varpi_\Omega(n) \leq \left(\sum_{n \in \Omega(u)} \lambda(u) \right)^2 = \sum_{n \in \Omega([u_1,u_2])} \lambda(u_1)\lambda(u_2), \qquad (102.16)$$

where u, u_1, u_2 are divisors of $P(z)$ which is defined above. Indeed, if $\varpi_\Omega(n) = 0$, then this inequality is obvious, and if $\varpi_\Omega(n) = 1$, then it becomes an identity. Trivial! Nevertheless, the idea (102.16) leads us to an amazing collection of conclusions, a part of which will be presented in the remaining five sections.

To proceed along Selberg's reasoning, we assume for the sake of simplicity that

$$\lambda(u) = 0 \text{ either if } u > z \text{ or if } \mu(u) = 0. \qquad (102.17)$$

We sum (102.16) over $n \in \mathcal{N}$, and have

$$|\mathcal{S}(\mathcal{N}, z)| \leq N \cdot S + R, \tag{102.18}$$

where

$$\begin{aligned}
S &= \sum_{u_1, u_2} \frac{|\Omega([u_1, u_2])|}{[u_1, u_2]} \lambda(u_1) \lambda(u_2), \\
|R| &\leq \sum_{u_1, u_2} |\Omega([u_1, u_2])| |\lambda(u_1) \lambda(u_2)|.
\end{aligned} \tag{102.19}$$

Selberg determined the minimum of the quadratic form S of λ's on the side conditions $\lambda(1) = 1$ and (102.17). For this purpose, he diagonalized S with a very basic yet highly ingenious argument, which we have, in fact, adopted already in Note [18.7] and in the proof of Theorem 109, i.e., at (95.40): On noting (10.6) and (102.3),

$$\frac{|\Omega([u_1, u_2])|}{[u_1, u_2]} = \frac{|\Omega(u_1)|}{u_1} \cdot \frac{|\Omega(u_2)|}{u_2} \cdot \frac{\langle u_1, u_2 \rangle}{|\Omega(\langle u_1, u_2 \rangle)|}. \tag{102.20}$$

Then, the Möbius inversion (17.5) or rather a simple computation gives, with H as in (102.14),

$$\frac{\langle u_1, u_2 \rangle}{|\Omega(\langle u_1, u_2 \rangle)|} = \sum_{r|u_1, \, r|u_2} \frac{\mu^2(r)}{H(r)}, \tag{102.21}$$

because $\mu(u_1)\mu(u_2) \neq 0$ by (102.17). We get

$$S = \sum_r H(r) \tau^2(r), \quad \tau(r) = \frac{\mu(r)}{H(r)} \sum_{r|g} \frac{|\Omega(g)|}{g} \lambda(g), \tag{102.22}$$

where (102.17) is taken into account; note that $\tau(r) = 0$ if $\mu(r) = 0$. The linear transformation $\lambda \mapsto \tau$ is invertible:

$$\lambda(u) = \frac{u\mu(u)}{|\Omega(u)|} \sum_{u|r} H(r)\tau(r), \tag{102.23}$$

for this sum is equal to

$$\begin{aligned}
\sum_{u|r} \mu(r) &\sum_{r|g} \frac{|\Omega(g)|}{g} \lambda(g) \\
&= \sum_{u|g} \frac{|\Omega(g)|}{g} \lambda(g) \sum_{u|r, \, r|g} \mu(r) = \frac{|\Omega(u)|}{u} \mu(u)\lambda(u);
\end{aligned} \tag{102.24}$$

see Note [102.4].

586   *Distribution of Prime Numbers*

The expression (102.23) transforms the side condition $\lambda(1) = 1$ into

$$1 = \sum_{r \leq z} H(r)\tau(r), \tag{102.25}$$

which is used in turn to transform (102.22) into

$$S = \frac{1}{G(z)} + \sum_{r \leq z} \mu^2(r)H(r)\left(\tau(r) - \frac{1}{G(z)}\right)^2, \tag{102.26}$$

where $G(z)$ is as in (102.13). Therefore the optimal τ is given by

$$\tau(r) = \frac{\mu^2(r)}{G(z)}, \quad r \leq z. \tag{102.27}$$

Namely, via (102.23) we find that the optimal λ is

$$\lambda(u) = \frac{\mu(u)}{G(z)} \prod_{p|u} \frac{p}{p - |\Omega(p)|} \cdot \sum_{\substack{r \leq z/u \\ \langle r,u \rangle = 1}} \mu^2(r)H(r), \tag{102.28}$$

which yields

$$S = \frac{1}{G(z)}. \tag{102.29}$$

The specification (102.28) is in accord with the side conditions $\lambda(1) = 1$ and (102.17). Also, taking an arbitrary $u \leq z$, $\mu(u) \neq 0$, and classifying the terms in (102.13) with respect to the value of $\langle u, r \rangle$, we have

$$G(z) = \sum_{g|u} H(g) \sum_{\substack{r \leq z/g \\ \langle r,u \rangle = 1}} \mu^2(r)H(r). \tag{102.30}$$

Restricting the inner sum to $r \leq z/u$, we get, via (102.14),

$$G(z) \geq \prod_{p|u}(1 + H(p)) \cdot \sum_{\substack{r \leq z/u \\ \langle r,u \rangle = 1}} \mu^2(r)H(r)$$

$$= \prod_{p|u} \frac{p}{p - |\Omega(p)|} \cdot \sum_{\substack{r \leq z/u \\ \langle r,u \rangle = 1}} \mu^2(r)H(r). \tag{102.31}$$

Hence (102.28) implies further

$$|\lambda(u)| \leq \mu^2(u). \tag{102.32}$$

Therefore the device (102.16) leads us to

$$|\mathcal{S}(\mathcal{N},z)| \leq \frac{N}{G(z)} + R, \quad |R| \leq \sum_{u \leq z^2} d_3(u)|\Omega(u)|, \tag{102.33}$$

as the number of solutions $u = [u_1, u_2]$ is not greater than $d_3(u)$; see (16.11).

The assertion (102.33) is but weaker than (102.15), especially when $|\Omega(p)|$ are allowed to become large. For instance, the last example concerning non-quadratic residues now gets the bound $\ll N/z + z^4 (\log z)^2$, as $|\Omega(u)| \leq u$, which is far weaker than what is provided by (102.15). However, this impression is just superficial: In the next section we shall prove that (102.16) as well is capable to yield (102.15), only if a Fourier analysis and an associated quasi-Parseval assertion are applied to (102.16) instead of expanding the squares as is done in (102.18); it is technically an application of (103.21) below, an inequality dual to (103.20) mentioned prior to the last theorem.

One of the primary purposes in the present chapter is to appreciate the power of Selberg's fundamental sieve idea (102.16) by employing it as a basic tool in the investigation of Dirichlet L-functions. Our actual application of the Λ^2-sieve will, however, be made not in its sheer form but in a hybridized version of it with the large sieve of Linnik. Indeed, as we shall show in the next section there is a duality relation between these two fundamental sieve ideas, because of which they admit of a fruitful unification. Consequentially, we shall attain a set of arithmetic means or a variety of L^2-inequalities embracing sieve effects, further in later sections. With them, we shall construct an analytic and arithmetic theory of L-functions, which is fairly accessible as it is well-structured.

However, before entering into the discussion of this main project of ours, it should be appropriate to present the following Tauberian theorem concerning Dirichlet series of non-negative coefficients since it has a wide application in asymptotically evaluating sums which occur in the Λ^2-sieve:

Theorem 137 *Let* $C(s) = \sum_{n=1}^{\infty} c_n n^{-s}$ *with* $c_n \geq 0$ *be such that*

(1) $C(s)$ *converges for* $\mathrm{Re}\, s > 0$;

(2) $C(s) = As^{-\alpha} + O(B|s|^{1-\alpha})$ *for* $\mathrm{Re}\, s > 0$, $|\mathrm{Im}\, s| \leq 1$; $\qquad$ (102.34)

(3) $C(s) = O(B|s|^{\nu})$ *for* $\mathrm{Re}\, s > 0$, $|\mathrm{Im}\, s| \geq 1$,

where $\alpha > 0, A > 0, B > 1, \nu > 0, s^{-\alpha} = \exp(-\alpha \log s), |\mathrm{Im} \log s| < \frac{1}{2}\pi$. *Then, it holds that*

$$\sum_{n \leq N} c_n = \frac{A}{\Gamma(\alpha + 1)} (\log N)^{\alpha} + O\big((A + B)D_{\alpha}(N)\big), \qquad (102.35)$$

as N tends to infinity; here

$$D_{\alpha}(N) = \begin{cases} 1 & \alpha < 1, \\ \log \log N & \alpha = 1, \\ (\log N)^{\alpha - 1} & \alpha > 1. \end{cases} \qquad (102.36)$$

588 *Distribution of Prime Numbers*

Remark In the proof below, A, B are treated as constants. In applications they are, in fact, parameters. The implied constant in (102.35) depends on α, v, and the two implicit constants in the condition (102.34) but neither on A nor on B.

Proof We use Cesàro's k^{th} mean of $\{c_n\}$, which is defined as

$$\begin{aligned} S_k(x) &= \sum_{n \le x} c_n (1 - n/x)^k \\ &= \frac{k!}{2\pi i} \int_{(a)} \frac{C(s)}{s(s+1)\cdots(s+k)} x^s ds, \end{aligned} \tag{102.37}$$

with x sufficiently large, while $a > 0$ is arbitrary, and $k > v$; so the integral converges absolutely. We let $a = (\log x)^{-1}$ and divide the contour into three parts $(a - i\infty, a - i]$, $[a - i, a + i]$, $[a + i, a + i\infty)$. The lower and the upper parts contribute $O(B)$ by (3) of (102.34). In the middle part, we have

$$\frac{C(s)}{s(s+1)\cdots(s+k)} = \frac{A}{s^{\alpha+1}k!} + O((A+B)|s|^{-\alpha}). \tag{102.38}$$

The corresponding contribution to (102.37) is

$$\frac{A}{2\pi i} \int_{a-i}^{a+i} \frac{x^s}{s^{\alpha+1}} ds + O\left((A+B) \int_{-1}^{1} \frac{dt}{(a^2+t^2)^{\alpha/2}}\right). \tag{102.39}$$

The last integral is $O(D_\alpha(x))$, and the other is equal to

$$\int_{(a)} \frac{x^s}{s^{\alpha+1}} ds + O(1); \tag{102.40}$$

so

$$S_k(x) = A \frac{(\log x)^\alpha}{\Gamma(\alpha+1)} + O((A+B)D_\alpha(x)); \tag{102.41}$$

see (94.31). Also, for $h \ge 1$,

$$\int_1^x S_{h-1}(y) y^{h-1} dy = \frac{x^h}{h} S_h(x), \tag{102.42}$$

and thus

$$S_h(x) \le S_{h-1}(x) \le \frac{2^h S_h(2x) - S_h(x)}{2^h - 1}, \tag{102.43}$$

because of the non-negativity of c_n's. Starting with $h = k$ and (102.41), we repeatedly apply (102.43) and end the proof.

$\S102$ 589

Notes

[102.1]   The term large sieve is used nowadays in a quite wide sense. It indicates the whole variety of L^2-inequalities that have stemmed from Linnik's use of Fourier analysis in his pioneering article (1941). Perhaps because of this, the sieve effect of the large sieve is often regarded like an incidental consequence, although it is indeed at the heart of the method. See Note [103.7].

[102.2]   The procedure leading to (102.12) is adopted from Bombieri (1987, §3). However, the bound (102.15) itself is originally due to Montgomery (1968).

[102.3]   The Λ^2- sieve device cannot be made simpler. Yet, by no means appears its power to have been exhausted, especially in view of the recent developments described in §107 below; see Note [102.4] as well. Its origin can be found in Selberg's investigation (1942, Section 7) on the distribution of zeros of the Riemann zeta-function in the vicinity of the critical line. Refining the earlier theory of the zero-density of the zeta-function, which has been briefly described in §98, Selberg was, in retrospect, led to the problem of making the following quadratic form

$$\int_{-T}^{T} \left| \zeta\left(\tfrac{1}{2} + it\right) \sum_{u \leq z} \frac{\lambda(u)}{u^{1/2+it}} - 1 \right|^2 dt$$

of $\{\lambda(u) \in \mathbb{R}\}$ as small as possible on the natural side condition $\lambda(1) = 1$, where z is to be taken suitably in connection with T which tends to infinity. He could reduce the problem essentially to the determination of the minimum value of the form

$$\sum_{u_1, u_2 \leq z} \frac{\lambda(u_1)\lambda(u_2)}{[u_1, u_2]},$$

which obviously corresponds to the first line of (102.19) with $\Omega(p) = \{0 \bmod p\}$ for every p. The sieve-effect of the argument with which Selberg solved this extremal problem was explicitly formulated on various general settings in his later papers (1947, 1949, 1950b, 1972, 1977, 1991). It should be especially kept in mind that the Λ^2-sieve was thus created in the course of deeper studies of the analytic behavior of the Riemann zeta-function in the critical strip. Note that as stated after (98.9) Bohr–Landau's idea of exploiting (98.10) is already related to sieve, that is, the Eratosthenes–Legendre sieve. Our aim is to try to extend this traditional scheme to all the L-functions, although in the present essays we shall restrict ourselves to Dirichlet's L-functions.

[102.4]   It should be worth stressing that the identity (102.23) yields a way to define $\lambda(u)$ by means of an arbitrary real-valued function τ supported on the

590   *Distribution of Prime Numbers*

interval $[0, z]$; the λ thus defined satisfies (102.17), although the result may not satisfy side condition $\lambda(1) = 1$. This observation readily extends to a more general situation to be discussed in §107 at the end of the present volume and yields a deep assertion on gaps between prime numbers.

[102.5] One may want to see how well the right side of (102.16) approximates to the left side. For this, we refer to an identity due to H. Halberstam (Motohashi (1983, p.66)) which is a specialization of (19.9) and gives an explicit representation of the difference between the two sides, revealing the mechanism behind Selberg's device which at first might appear somewhat ad hoc. In other words, (102.16) can be regarded as a way to construct upper-bound sieve weights in the sense of (19.10). Thus one may say that (102.16) as well is included in the sieve framework initiated by Brun, although the sieve weights thus obtained are not characteristic functions any more.

[102.6] Naturally, the Λ^2-sieve is applicable to any small sieve situation, i.e., the one in which $\omega(p) = |\Omega(p)|$ fulfills the requirement Note [19.6] with a constant κ. However, one should be aware that its sole use is capable to yield only upper bounds, whereas the combinatorial sieve usually yields both upper and lower bounds on its own. To get lower-bound results via the Λ^2-sieve, one needs to try to combine it with other devices like the Buchstab identity (19.5); see Jurkat and Richert (1965) for details.

[102.7] The above proof of Theorem 137 is a rework of the one by Onishi (1973). We stress that any analytic continuation of $C(s)$ to the half plane $\operatorname{Re} s < 0$ is not supposed. In sieve literature more general sum formulas are commonly utilized; see for instance Halberstam–Richert (1974, Chapter 5). However, for most applications, (102.35) appears to be sufficient.

[102.8] It is customary to reveal how effective the Selberg sieve is by applying it to the estimation of $\pi_2(x)$ the number of twin primes $p \leq x$: Thus, let $M = 0, N = [x], \Omega(2) = \{0 \bmod 2\}, \Omega(p) = \{0, -2 \bmod p\}$ for all odd p. Then (102.33) gives

$$\pi_2(x) \leq |\mathcal{S}(\mathcal{N}, z)| + z \leq \frac{x}{G(z)} + \sum_{u \leq z^2} d_3(u)d(u) + z,$$

since $|\Omega(u)| \leq d(u)$ presently. This sum over u is $O(z^2(\log z)^5)$, for the Dirichlet series $\lfloor d_3 \cdot d \rfloor(s)$ is essentially identical to $\zeta^6(s)$. Also, to estimate

$$G(z) = \sum_{g \leq z} \mu^2(g) \prod_{\substack{p|g \\ p>2}} \frac{2}{p-2},$$

we consider the function

$$\sum_{g=1}^{\infty} \frac{\mu^2(g)}{g^s} \prod_{\substack{p|g \\ p>2}} \frac{2}{p-2} = \zeta^2(s+1)\left(1 + \frac{1}{2^s}\right)\left(1 - \frac{1}{2^{s+1}}\right)^2$$

$$\times \prod_{p>2}\left(1 - \frac{1}{p^{s+1}}\right)^2\left(1 + \frac{2}{p^s(p-2)}\right), \quad \mathrm{Re}\, s > 0.$$

By (102.35), we find readily that

$$G(z) = \tfrac{1}{4}(\log z)^2 \prod_{p>2}\left(1 - \frac{1}{(p-1)^2}\right)^{-1} + O(\log z).$$

Thus we obtain

$$\pi_2(x) \le \left(16 + o(1)\right)\frac{x}{(\log x)^2}\prod_{p>2}\left(1 - \frac{1}{(p-1)^2}\right);$$

see Selberg (1949, p.16). Brun's assertion $\sum(\text{twin prime})^{-1} < \infty$ (Note [19.5]) is now immediate. In the light of the hypothetical asymptotic formula

$$\pi_2(x) = \left(2 + o(1)\right)\frac{x}{(\log x)^2}\prod_{p>2}\left(1 - \frac{1}{(p-1)^2}\right)$$

of Hardy–Littlewood (1923, (5.311)), one will be amazed by the power of Selberg's truly simple idea (102.16).

§103

In the present section we shall first prove L^2-inequality concerning additive characters of $\mathbb{Z}$ that provides the right side of (102.12) with a bound, i.e., (103.20), which immediately gives the fundamental sieve assertion (102.15). We shall next supply details to a duality phenomenon that was mentioned in the ending paragraph of the text of the preceding section. Further, we shall discuss L^2-inequalities which can be viewed as hybrids of the two approaches (102.12) and (102.16), i.e., via the Linnik and the Selberg sieves, to (102.15).

We begin with

Theorem 138 *Let $\mathcal{D}$ be a linear operator mapping a finite-dimensional Hilbert space into another. Let $[[\mathcal{D}]]$ be its norm. Then we have*

$$[[\mathcal{D}]] = [[\mathcal{D}^*]], \tag{103.1}$$

where $\mathcal{D}^*$ is the adjoint operator of $\mathcal{D}$.

592 — Distribution of Prime Numbers

Proof This is a well-known duality principle and an immediate consequence of the definitions of the norm $[[\cdot]]$ and adjoint operators; the finite dimensionality is of no essential restriction: Let the norm of a Hilbert space be denoted by $\|\cdot\|$ generically. Then $[[\mathcal{D}]]$ is defined as $\sup \|\mathcal{D}\mathbf{a}\|/\|\mathbf{a}\|$ where $\mathbf{a}$ runs over all non-zero vectors of the domain space. Let $\mathbf{b}$ run over all non-zero vectors of the codomain. We have, by the Cauchy–Schwartz inequality,

$$\|\mathcal{D}^*\mathbf{b}\|^2 \le \|\mathcal{D}\mathcal{D}^*\mathbf{b}\|\|\mathbf{b}\| \le [[\mathcal{D}]]\|\mathcal{D}^*\mathbf{b}\|\|\mathbf{b}\|. \tag{103.2}$$

Hence $\|\mathcal{D}^*\mathbf{b}\|/\|\mathbf{b}\| \le [[\mathcal{D}]]$, that is, $[[\mathcal{D}^*]] \le [[\mathcal{D}]]$. By symmetry the opposite inequality holds as well. This ends the proof.

Theorem 139 *Let* $\{e(\theta_j) : 1 \le j \le J\}$, $e(x) = \exp(2\pi i x)$ *be a set of distinct points on the unit circle on the complex plane such that the least difference between their angles is* $2\pi\delta$. *Then we have, for any complex sequences* $\{a_n\}$ *and* $\{b_j\}$,

$$\sum_{j=1}^{J}\left|\sum_{M<n\le M+N} a_n e(n\theta_j)\right|^2 \le \left(N + 2\delta^{-1}\right)\sum_{M<n\le M+N}|a_n|^2, \tag{103.3}$$

$$\sum_{M<n\le M+N}\left|\sum_{j=1}^{J} b_j e(n\theta_j)\right|^2 \le \left(N + 2\delta^{-1}\right)\sum_{j=1}^{J}|b_j|^2, \tag{103.4}$$

where M, N *are as in* (102.1).

Remark These inequalities hold for $J = 1$ trivially. We note that $\delta = \min\{\min_{n\in\mathbb{Z}}|\theta_j - \theta_k - n| : j \ne k\}$, which is called the discrepancy constant of the points $\{\theta_j\}$. If these points are contained in the Farey sequence of order z, then $\delta > z^{-2}$ according to the discussion of §20.

Proof If the operator $\mathcal{D}$ of the last lemma is specialized as to be

$$\mathcal{D}\left({}^{\mathrm{t}}(b_j)_{1\le j\le J}\right) = {}^{\mathrm{t}}\left(\sum_{j=1}^{J} b_j e(n\theta_j)\right)_{M<n\le M+N}, \tag{103.5}$$

in which vectors are denoted with an obvious abbreviation, that is, $\mathcal{D} : \mathbb{C}^J \to \mathbb{C}^N$. Then (103.4) means that $[[\mathcal{D}]]^2 \le N + 2\delta^{-1}$ under the ordinary Euclid metric. On the other hand, the left side of (103.3) is

$$\left\|\mathcal{D}^*\left({}^{\mathrm{t}}(\overline{a}_n)_{M<n\le M+N}\right)\right\|^2 \le [[\mathcal{D}^*]]^2 \sum_{M<n\le M+N}|a_n|^2, \tag{103.6}$$

and via (103.1) we are led to the inequality (103.3).

Hence, it suffices to prove (103.4). For $J = 2$ the left side is

$$N(|b_1|^2 + |b_2|^2) + 2\text{Re}\left\{b_1\overline{b_2}\sum_{n=M+1}^{M+N} e((\theta_1 - \theta_2)n)\right\}$$
$$\leq N(|b_1|^2 + |b_2|^2) + \frac{2|b_1 b_2|}{|\sin(\pi(\theta_1 - \theta_2))|}. \tag{103.7}$$

The lower bound $|\sin(\pi(\theta_1 - \theta_2))| \geq 2\delta$ implies $(103.4)_{J=2}$. Thus we shall proceed on the assumption

$$3 \leq J \leq \delta^{-1}. \tag{103.8}$$

With this, we shall show that for each integer $H > 0$

$$\sum_{-H \leq n \leq H} \left|\sum_{j=1}^{J} b_j e(n\theta_j)\right|^2 \leq (2H + 1 + 2(2/3)^{1/2}\delta^{-1})\sum_{j=1}^{J}|b_j|^2. \tag{103.9}$$

If $N = 2H + 1$, then (103.4) follows immediately. If $N = 2H$, then we enlarge the range of summation on the left side of (103.4) to $M \leq n \leq N + M$ and apply (103.9). On noting that $\delta \leq \frac{1}{3}$, we see that

$$2H+1 + 2(2/3)^{1/2}\delta^{-1} = N + 1 + 2(2/3)^{1/2}\delta^{-1}$$
$$\leq N + (1/3 + 2(2/3)^{1/2})\delta^{-1} < N + 2\delta^{-1}. \tag{103.10}$$

We shall then prove, instead of (103.9),

$$\sum_{n=-\infty}^{\infty} w(n)\left|\sum_{j=1}^{J} b_j e(n\theta_j)\right|^2 \leq (2H + 1 + 2(2/3)^{1/2}\delta^{-1})\sum_{j=1}^{J}|b_j|^2, \tag{103.11}$$

where w is the Fejér tapering:

$$w(x) = \begin{cases} 1, & |x| \leq H, \\ 0, & H + K \leq |x|, \\ (H + K - |x|)/K, & H \leq |x| \leq H + K, \end{cases} \tag{103.12}$$

with the integer $K > 0$ to be fixed later. The left side of (103.11) is

$$\sum_{j,k} b_j\overline{b_k}W(\theta_j - \theta_k) \leq \left(W(0) + \max_j \sum_{k \neq j}|W(\theta_j - \theta_k)|\right)\sum_{j=1}^{J}|b_j|^2, \tag{103.13}$$

where

$$W(\theta) = \frac{1}{K}\left\{\left(\frac{\sin(\pi(H + K)\theta)}{\sin(\pi\theta)}\right)^2 - \left(\frac{\sin(\pi H\theta)}{\sin(\pi\theta)}\right)^2\right\}. \tag{103.14}$$

Distribution of Prime Numbers

We note that

$$|W(0)| = 2H + K,$$

$$|W(\theta)| \le \frac{1}{K(\sin \pi \theta)^2}, \quad \theta \notin \mathbb{Z}. \tag{103.15}$$

Then, we observe, for $0 < \theta \le \frac{1}{2}$, that

$$\frac{1}{(\sin \pi \theta)^2} = \frac{1}{(\pi \theta)^2} + \frac{1}{\pi^2} \sum_{m=1}^{\infty} \left(\frac{1}{(m+\theta)^2} + \frac{1}{(m-\theta)^2} \right)$$

$$\le \frac{1}{(\pi \theta)^2} + \frac{1}{\pi^2} \sum_{m=1}^{\infty} \left(\frac{1}{(m+\frac{1}{2})^2} + \frac{1}{(m-\frac{1}{2})^2} \right) \tag{103.16}$$

$$= \frac{1}{(\pi \theta)^2} + \frac{8}{\pi^2} \left(1 - \frac{1}{2^2} \right) \zeta(2) - \frac{4}{\pi^2} < \frac{1}{(\pi \theta)^2} + 1$$

On noting (103.8), we see that

$$\sum_{k \ne j} |W(\theta_j - \theta_k)| \le \frac{2}{\pi^2 K} \sum_{\ell=1}^{\infty} \frac{1}{(\ell \delta)^2} + \frac{J}{K} \tag{103.17}$$

$$\le \frac{1}{3K\delta^2} + \frac{1}{K\delta} \le \frac{2}{3K\delta^2}.$$

Therefore,

$$\sum_{-H \le n \le H} \left| \sum_j b_j e(n\theta_j) \right|^2 \le \left(2H + K + \frac{2}{3K\delta^2} \right) \sum_{j=1}^{J} |b_j|^2. \tag{103.18}$$

Taking $K = [(2/3)^{\frac{1}{2}} \delta^{-1}] + 1$, we end the proof of (103.3)–(103.4).

Alternatively, one may prove (103.3) directly, that is, without using the duality (103.1): Take, in (98.31), the vectors

$$\mathbf{a}^{(j)} = (\ldots, w(n)^{1/2} e(n\theta_j), \ldots), \quad \mathbf{b} = (\ldots, a_n w(n)^{-1/2}, \ldots), \tag{103.19}$$

under the assumption that (103.8) holds and that $a_n = 0$ for $n \notin [-H, H]$. The rest of argument can be skipped.

As an immediate consequence of the last theorem, we state that with any $Q \ge 1$

$$\sum_{q \le Q} \sum_{\substack{h=1 \\ \langle h, q \rangle = 1}}^{q} \left| \sum_{M < n \le M+N} a_n e\left(\frac{h}{q} n \right) \right|^2 \le (N + 2Q^2) \sum_{M < n \le M+N} |a_n|^2, \tag{103.20}$$

for any complex sequence $\{a_n\}$, and its dual

$$\sum_{M<n\leq M+N} \left| \sum_{\substack{q\leq Q}} \sum_{\substack{h=1\\ \langle h,q\rangle=1}}^{q} b_{h/q} e\left(\frac{h}{q}n\right) \right|^2$$

$$\leq (N + 2Q^2) \sum_{\substack{q\leq Q}} \sum_{\substack{h=1\\ \langle h,q\rangle=1}}^{q} |b_{h/q}|^2,$$

$$(103.21)$$

for any complex sequence $\{b_{h/q}\}$. In particular, applying (103.20) to the right side of (102.12) we obtain Theorem 136 immediately.

Now, by means of (103.21) we are able to make precise the duality phenomenon that we asserted at the end of the text of the preceding section: The Fourier expansion of the characteristic function of the set $\{n \in \mathbb{Z} : n \in \Omega(u)\}$ in the sense of §50 is

$$\frac{1}{u} \sum_{v \bmod u} \sum_{a\in\Omega(u)} e\left(\frac{v}{u}(n-a)\right)$$

$$= \frac{1}{u} \sum_{r|u} \sum_{\substack{t \bmod r\\ \langle t,r\rangle=1}} \left(\sum_{a\in\Omega(u)} e\left(-\frac{t}{r}a\right) \right) \cdot e\left(\frac{t}{r}n\right).$$

$$(103.22)$$

Inserting this into (102.16) we have

$$|S(\mathcal{N},z)| \leq \sum_{M<n\leq M+N} \left| \sum_{r\leq z} \sum_{\substack{t \bmod r\\ \langle t,r\rangle=1}} b_{t/r} e\left(\frac{t}{r}n\right) \right|^2,$$

$$(103.23)$$

$$b_{t/r} = \sum_{\substack{u\leq z\\ u\equiv 0 \bmod r}} \frac{\lambda(u)}{u} \sum_{a\in\Omega(u)} e\left(-\frac{t}{r}a\right).$$

$$(103.24)$$

The inequality (103.21) implies that

$$|S(\mathcal{N},z)| \leq (N + 2z^2) \sum_{r\leq z} \sum_{\substack{t \bmod r\\ \langle t,r\rangle=1}} \left| b_{t/r} \right|^2.$$

$$(103.25)$$

This double sum is equal to

$$\sum_{u_1,u_2\leq z} \frac{\lambda(u_1)\lambda(u_2)}{u_1 u_2} \sum_{a_1\in\Omega(u_1)} \sum_{a_2\in\Omega(u_2)} \sum_{r|\langle u_1,u_2\rangle} \sum_{\substack{t \bmod r\\ \langle t,r\rangle=1}} e\left(\frac{t}{r}(a_1-a_2)\right). \quad (103.26)$$

The inner quadruple sum is decomposed as

$$\prod_{p_1|u_1, p_2|u_2} \left[\sum_{a_1\in\Omega(p_1)} \sum_{a_2\in\Omega(p_2)} \sum_{r|\langle p_1,p_2\rangle} \sum_{\substack{t \bmod r\\ \langle t,r\rangle=1}} e\left(\frac{t}{r}(a_1-a_2)\right) \right]. \quad (103.27)$$

Distribution of Prime Numbers

If $p_1 \neq p_2$, then this new quadruple sum equals $|\Omega(p_1)||\Omega(p_2)|$, and if $p_1 = p_2$, then it is

$$\sum_{a_1 \in \Omega(p_1)} \sum_{a_2 \in \Omega(p_1)} \sum_{r=1,\, p_1} \sum_{\substack{t \bmod r \\ \langle t, r \rangle = 1}} e\left(\frac{t}{r}(a_1 - a_2)\right) \tag{103.28}$$

$$= |\Omega(p_1)|^2 + |\Omega(p_1)|(p_1 - |\Omega(p_1)|).$$

Hence, (103.26) becomes

$$\sum_{u_1, u_2 \leq z} \frac{\lambda(u_1)\lambda(u_2)}{u_1 u_2} \prod_{p_1,\, p_2 \,\big|\, \frac{[u_1, u_2]}{\langle u_1, u_2 \rangle}} |\Omega(p_1)||\Omega(p_2)| \prod_{p \,|\, \langle u_1, u_2 \rangle} p|\Omega(p)|. \tag{103.29}$$

This is exactly the same as S defined in (102.19). Hence, (103.25) can be written as

$$|\mathcal{S}(\mathcal{N}, z)| \leq (N + 2z^2)S. \tag{103.30}$$

Therefore we are led to (102.15) again.

We have taken another look at Selberg's sieve via Linnik's, and have obtained an alternative proof of (102.15). Comparing the two arguments, we immediately notice the duality between these two fundamental sieve ideas. This is indeed remarkable, as they were invented independently of each other.

We shall exploit the relation between the two sieve methods further and show a refinement of (103.20)–(103.21). Thus, we return to (102.28). The optimum $\lambda(u)$'s imply that

$$\sum_{n \in \Omega(u)} \lambda(u) = \frac{1}{G(z)} \sum_{r \leq z} \mu(r)^2 H(r) \Psi_r(n),$$

$$\Psi_r(n) = \prod_{\substack{p | r \\ n \in \Omega(p)}} \left(\frac{-1}{H(p)}\right), \tag{103.31}$$

where $H(p)$ is as in (102.14); observe that the function Ψ_r is finite, provided (102.3); here is a notational conflict with the one in §50, which should, however, cause no harm. One may say thus that the argument leading to (103.30) is essentially the same as to a proof of the inequality

$$\sum_{M < n \leq M+N} \left(\sum_{r \leq z} \mu(r)^2 H(r) \Psi_r(n)\right)^2 \leq (N + 2z^2) \sum_{r \leq z} \mu^2(r) H(r). \tag{103.32}$$

This indicates that the linear operator

$$\left(\mu^2(r) H(r)^{1/2} \Psi_r(n)\right), \quad M < n \leq M + N, \ r \leq z, \tag{103.33}$$

$$\S 103 \qquad\qquad 597$$

is working behind (103.32). Namely, the system $\{\mu^2(r)H(r)^{1/2}\,\Psi_r(n)\}$ should involve a kind of orthogonality like characters; note that Ψ_r is actually a function over $\mathbb{Z}/r\mathbb{Z}$. Hence we surmise that the following pair of L^2-inequalities should hold: For any sequences $\{a_n\}, \{b(r)\} \subset \mathbb{C}$,

$$\sum_{r \leq z} \mu^2(r)H(r) \left| \sum_{M < n \leq M+N} a_n \Psi_r(n) \right|^2 \tag{103.34}$$
$$\leq (N + 2z^2) \sum_{M < n \leq M+N} |a_n|^2,$$

$$\sum_{M < n \leq M+N} \left| \sum_{r \leq z} b(r)\mu^2(r)H(r)^{1/2}\Psi_r(n) \right|^2 \tag{103.35}$$
$$\leq (N + 2z^2) \sum_{r \leq z} \mu^2(r)|b(r)|^2.$$

Indeed we are able to show the following hybrid of the Selberg sieve and the L^2-inequalities (103.20)–(103.21):

Theorem 140 *Suppose* (102.3). *Let* $H(r)$ *and* $\Psi_r(n)$ *be defined by* (102.14) *and* (103.31), *respectively. Then, for arbitrary sequences* $\{a_n\}, \{b_{h/q}(r)\}$ *in* $\mathbb{C}$ *and for any* $Q \geq 1$, *it holds that*

$$\sum_{\substack{qr \leq Q \\ \langle q,r \rangle = 1}} \mu^2(r)H(r) \sum_{\substack{h=1 \\ \langle h,q \rangle = 1}}^{q} \left| \sum_{M < n \leq M+N} a_n \Psi_r(n) e\left(\frac{h}{q}n\right) \right|^2 \tag{103.36}$$
$$\leq (N + 2Q^2) \sum_{M < n \leq M+N} |a_n|^2,$$

and

$$\sum_{M < n \leq M+N} \left| \sum_{\substack{qr \leq Q \\ \langle q,r \rangle = 1}} \sum_{\substack{h=1 \\ \langle h,q \rangle = 1}}^{q} \mu^2(r)b_{h/q}(r)H(r)^{1/2}\Psi_r(n)e\left(\frac{h}{q}n\right) \right|^2 \tag{103.37}$$
$$\leq (N + 2Q^2) \sum_{\substack{qr \leq Q \\ \langle q,r \rangle = 1}} \sum_{\substack{h=1 \\ \langle h,q \rangle = 1}}^{q} \mu^2(r)|b_{h/q}(r)|^2.$$

Proof By virtue of Theorem 138, it suffices to show the second assertion. We observe first that the Fourier expansion

$$\Psi_r(n) = \frac{\mu(r)}{|\Omega(r)|} \sum_{\substack{t=1 \\ \langle t,r \rangle = 1}}^{r} \left(\sum_{a \in \Omega(r)} e\left(-\frac{t}{r}a\right) \right) e\left(\frac{t}{r}n\right) \tag{103.38}$$

598 Distribution of Prime Numbers

holds. Indeed, due to the multiplicativity, we see that the right side is equal to

$$\frac{\mu(r)}{|\Omega(r)|} \prod_{p|r} \sum_{a\in\Omega(p)} \sum_{t=1}^{p-1} e\left(\frac{t}{p}(n-a)\right)$$

$$= \frac{\mu(r)}{|\Omega(r)|} \prod_{\substack{p|r \\ n\in\Omega(p)}} (p-|\Omega(p)|) \prod_{\substack{p|r \\ n\notin\Omega(p)}} (-|\Omega(p)|) = \prod_{\substack{p|r \\ n\in\Omega(p)}} \frac{|\Omega(p)|-p}{|\Omega(p)|},$$

$$(103.39)$$

which is $\Psi_r(n)$. Thus,

$$\sum_{\substack{qr\leq Q \\ \langle q,r\rangle=1}} \sum_{\substack{h=1 \\ \langle h,q\rangle=1}}^{q} \mu^2(r)b_{h/q}(r)H(r)^{1/2}\Psi_r(n)e\left(\frac{h}{q}n\right)$$

$$= \sum_{\substack{qr\leq Q \\ \langle q,r\rangle=1}} \sum_{\substack{h=1 \\ \langle h,q\rangle=1}}^{q} \sum_{\substack{t=1 \\ \langle t,r\rangle=1}}^{r} e\left(\left(\frac{h}{q}+\frac{t}{r}\right)n\right) \qquad (103.40)$$

$$\times \left[\mu(r)b_{h/q}(r)\frac{H(r)^{1/2}}{|\Omega(r)|} \sum_{a\in\Omega(r)} e\left(-\frac{t}{r}a\right)\right].$$

Applying (103.21), we find that the left side of (103.39) is

$$\leq (N+2Q^2) \sum_{\substack{qr\leq Q \\ \langle q,r\rangle=1}} \sum_{\substack{h=1 \\ \langle h,q\rangle=1}}^{q} \mu^2(r)|b_{h/q}(r)|^2 \frac{H(r)}{|\Omega(r)|^2}$$

$$\times \sum_{\substack{t=1 \\ \langle t,r\rangle=1}}^{r} \left|\sum_{a\in\Omega(r)} e\left(-\frac{t}{r}a\right)\right|^2. \qquad (103.41)$$

This lower line is equal to $|\Omega(r)|^2/H(r)$. We end the proof.

The assertion (103.36) implies a refinement of (102.15): Let $a_n = 0$ if there exists a p such that $n \in \Omega(p), p \leq z$. Then

$$\sum_{q\leq z} G_q(z/q) \sum_{\substack{h=1 \\ \langle h,q\rangle=1}}^{q} \left|\sum_{M<n\leq M+N} a_n e\left(\frac{h}{q}n\right)\right|^2$$

$$\leq (N+2z^2) \sum_{M<n\leq M+N} |a_n|^2, \qquad (103.42)$$

where

$$G_q(x) = \sum_{\substack{r\leq x \\ \langle q,r\rangle=1}} \mu^2(r)H(r). \qquad (103.43)$$

The part with $q = 1$ of (103.42) implies (102.15). In other words, (103.36) contains the sieve bound (102.15).

We remark that for any $k \geq 1$

$$G_k(x) \geq \prod_{p \mid k} \left(1 - \frac{|\Omega(p)|}{p} \right) \cdot G(x), \tag{103.44}$$

which is to be compared with (102.31). For

$$
\begin{aligned}
G(x) = G_1(x) &= \sum_{u \mid k} \sum_{\substack{n \leq x \\ \langle n, k \rangle = u}} \mu^2(n) H(n) \\
&= \sum_{u \mid k} \sum_{\substack{m \leq x/u \\ \langle m, k \rangle = 1}} \mu^2(mu) H(mu) \\
&\leq \sum_{u \mid k} \mu^2(u) H(u) G_k(x) = \prod_{p \mid k} \frac{p}{p - |\Omega(p)|} \cdot G_k(x).
\end{aligned}
\tag{103.45}
$$

The purpose of our discussion made after (103.30) was to throw light on the arithmetic function $\Psi_r(n)$. Its origin is in the optimization procedure (102.20)–(102.29). What one should notice here is the fact that the same could be applied not only to the linear set $\mathcal{N}$ but also to a far wider variety of curved or arithmetically twisted situations. We shall see in later sections that this observation results in a structural approach to the basic theory of L-functions.

Notes

[103.1]   The principal aim in the above was to render, from our view point, the duality relation between Linnik's large sieve and Selberg's Λ^2-sieve. This duality was perceived simultaneously by not a few people in either published or unpublished forms.

[103.2]   The argument (103.9)–(103.18) is due to Bombieri (1971). The bound $[[\mathcal{D}]] \leq (N + 2\delta^{-1})^{1/2}$ for $\mathcal{D}$ as in (103.5) suffices for most applications to number theoretical problems; indeed, even the bound $[[\mathcal{D}]] \ll (N + \delta^{-1})^{1/2}$ with an implicit multiplier often works well. Nevertheless, it is naturally preferable to have a bound as sharp and explicit as possible. Thus, Montgomery and Vaughan (1973) proved $[[\mathcal{D}]] \leq (N + \delta^{-1})^{1/2}$. Then Selberg (see Montgomery (1978, §7)) obtained, with a highly sophisticated argument,

$$[[\mathcal{D}]] \leq (N + \delta^{-1} - 1)^{1/2},$$

which is sharp (see Montgomery (1971, p.14)). However, it is known that the bound of Mongomery and Vaughan readily yields Selberg's; see Montgomery

(1978, p.559). The argument of Montgomery and Vaughan (1973) depends on their fundamental inequality (1974): Let $\{\theta_j\}$ be as before and $\{c_j\}$ be arbitrary sequence in $\mathbb{C}$. Then

$$(1) \qquad \left| \sum_{j \neq k} c_j c_k / (\theta_j - \theta_k) \right| \leq \frac{\pi}{\delta} \sum_j |c_j|^2,$$

which is an extension of the classical Hilbert's inequality. This implies

$$(2) \qquad \left| \sum_{j \neq k} c_j c_k / \sin(\pi(\theta_j - \theta_k)) \right| \leq \frac{1}{\delta} \sum_j |c_j|^2,$$

which yields $[[\mathcal{D}]] \leq (N + \delta^{-1})^{1/2}$. Montgomery and Vaughan actually proved more than (1):

$$(3) \qquad \left| \sum_{j \neq k} c_j c_k / (\theta_j - \theta_k) \right| \leq \tfrac{3}{2}\pi \sum_j \delta_j^{-1} |c_j|^2,$$

where $\delta_j = \min_{k \neq j} \min_{n \in \mathbb{Z}} |\theta_j - \theta_k - n|$; namely, they took into account the irregularity of the distribution of the points $\{e(\theta_j)\}$ on the unit circle. From this it follows, via the duality principle, that

$$\left[\left[\left((N + \tfrac{3}{2}\delta_j^{-1})^{-1/2} e(n\theta_j) \right) \right] \right] \leq 1 \quad \text{(a matrix of dimension } N \times J \text{)}.$$

Hence, the following refinement of (103.23) holds: for any $\{b_{h/q}\}$ in $\mathbb{C}$

$$(4) \qquad \sum_{M < n \leq M+N} \left| \sum_{q \leq Q} \sum_{\substack{h=1 \\ \langle h, q \rangle = 1}}^{q} b_{h/q} e\left(\frac{h}{q} n \right) \right|^2 \leq \sum_{q \leq Q} \sum_{\substack{h=1 \\ \langle h, q \rangle = 1}}^{q} (N + \tfrac{3}{2} qQ) |b_{h/q}|^2.$$

[103.3] The arguments (103.22)–(103.30) is adopted from Motohashi (1977b, II). Its most salient point is in that it does not need any explicit representation of optimum $\lambda(u)$'s at the outset. Namely, the inequality (103.30) holds for any $\lambda(u)$'s satisfying only the conditions $\lambda(1) = 1$ and (102.17). The combination of the inequality (103.21) and the Fourier expansion (103.22) almost immediately hits the target (103.30); and the rest is the same as Selberg's optimization procedure. This feature of the article yields an interesting consequence as is explained in the next Note. A Fourier analysis of the Λ^2-sieve is given by Selberg (1991, §§19–20) from a different viewpoint.

[103.4] The inequality in Note [103.2](4) has a noteworthy implication to the Λ^2-sieve: It gives rise to a new optimal set of $\lambda(u)$'s. Thus, we apply it to (103.25). Then, proceeding analogously as (103.24)–(103.32), we find that $|\mathcal{S}(\mathcal{N}, z)| \leq \widetilde{S}$ with

$$\widetilde{S} = \sum_{u_1, u_2 \leq z} \left(N + \tfrac{3}{2} z \langle u_1, u_2 \rangle \right) \frac{|\Omega([u_1, u_2])|}{[u_1, u_2]} \lambda(u_1) \lambda(u_2).$$

The diagonalization of $\widetilde{S}$ can be performed as we did for S in §102; so

$$\widetilde{S} = \sum_{r \leq z} \left(N + \tfrac{3}{2} z r \right) \mu^2(r) H(r) \tau^2(r),$$

where $H(r)$ and $\tau(r)$ are as before. On the side conditions $\lambda(1) = 1$ or (102.25) and (102.17) we find that

$$|\mathcal{S}(\mathcal{N}, z)| \leq 1/\widetilde{G}(z), \quad \widetilde{G}(z) = \sum_{r \leq z} \frac{\mu^2(r) H(r)}{N + \tfrac{3}{2} z r}.$$

The new optimal $\lambda(u)$'s are

$$\widetilde{\lambda}(u) = \frac{\mu(u)}{\widetilde{G}(z)} \prod_{p \mid u} \frac{p}{p - |\Omega(p)|} \cdot \sum_{\substack{r \leq z/u \\ \langle r, u \rangle = 1}} \frac{\mu(r)^2 H(r)}{N + \tfrac{3}{2} z u r},$$

which corresponds to (102.28). Compare the above with Montgomery–Vaughan (1973, p.120). Continued in Note [104.2].

[103.5] The prototype $\psi_r(n) = \mu(\langle r, n \rangle) \varphi(\langle r, n \rangle)$ of $\Psi_r(n)$ defined by (103.33) was introduced by Selberg (1972, (3.4)) without indicating how he came to it (here is a notational conflict, though harmless, with the one for the additive characters introduced in §50). Indeed, if $\Omega(p) = \{0 \bmod p\}$ for all p, then $\Psi_r(n) = \psi_r(n)$. He called it a pseudocharacter because it behaves like a Dirichlet character: being multiplicative and depending on the residue class $n \bmod r$. The identity (103.31) reveals that $\Psi_r(n)$ came from the optimized weights of the Λ^2-sieve; this observation is due to Kobayashi (1973, pp.4–5). However, the above argument leading to the identity should be compared with his. The recognition of the operator (103.33) was made by Motohashi (1977b, II); it yielded the last theorem as well as the later developments accounted in §106.

[103.6] It is possible to extend the domain of the map Ω to the set of all prime powers $\{p^a\}$ so that $\Omega(p^a)$ is a pre-assigned set of residues classes $\bmod p^a$. See Selberg (1977). With this extension, we have still an analogue of (103.33) and the rest; see Motohashi (1983, §§1.1–1.2).

[103.7] One might sense that the large sieve of Linnik is nothing else than an analytic principle to estimate L^2-expressions of trigonometrical sums or like in a near optimal manner. Indeed, its sieve effect is the consequence of the fact that a certain expression is biased to be large when an element of a set is to survive a particular sifter while the statistical size, i.e., the L^2-mean of the

602   *Distribution of Prime Numbers*

expression remains stable; so a fine upper bound is attained for the number of
the surviving elements. One may notice here an amplification effect similar to
that of the Fourier transform F_L mentioned at the end of §47. The large moduli
method from the title of Montgomery (1971, Chapter 8) would characterize
this modern paradigm in analytic number theory better than the large sieve
method. However, the latter conveys a spirit of the good old days in analytic
number theory.

§104

In the present section, which is of semi-expository nature, we shall show, in
part [A], that sieve method yields a genuinely non-trivial bound (104.9) for
individual $\pi(x; k, a)$ with k being as large as a positive power of x, which
is thought to be out of reach for purely analytic means as has been tacitly
indicated in §101; the use of the coprime pair $\{k, a\}$ in place of the former
$\{q, \ell\}$ is for local convenience. We shall also show, in part [B], a useful result
(104.13) of a similar trait concerning multiplicative functions.

[A] We apply (103.34) to the interval $\mathcal{N} = (M, M + N]$ $(M, N \in \mathbb{N})$,
with the sifting specification $\mathcal{P} = \{p : p \nmid k\}$, $1 \leq a < k$, $\langle k, a \rangle = 1$,
$\Omega(p) = \{-ak^{-1} \bmod p\}$. We have, with the notation (102.2) and CONVENTION
following (102.3),

$$S(\mathcal{N}, z) = \{M < n \leq M + N : p \nmid (kn + a), \forall p \leq z\}, \tag{104.1}$$

since the condition $p \nmid k$ in $\mathcal{P}$ is immaterial here, as $\langle k, a \rangle = 1$. Note that

$$\{kM + a < \text{prime} \equiv a \bmod k \leq k(M + N) + a\}$$
$$\subset \{kn + a : n \in S(\mathcal{N}, z)\} \cup \{u \leq z : u \equiv a \bmod k\}. \tag{104.2}$$

Provided $\langle k, r \rangle = 1$, we have $H(r) = 1/\varphi(r)$ by (102.14), and

$$\Psi_r(n) = \mu(\langle kn + a, r \rangle)\varphi(\langle kn + a, r \rangle) \tag{104.3}$$

by (103.31), that is, $\psi_r(kn + a)$ with ψ_r as in Note [103.5]. Hence, with
$a_n = \varpi_\Omega$ (see (102.4)) we get by (103.34)

$$|S(\mathcal{N}, z)| \leq \frac{N + 2z^2}{G_k(z)}, \quad G_k(z) = \sum_{\substack{r \leq z \\ \langle r, k \rangle = 1}} \frac{\mu^2(r)}{\varphi(r)}, \tag{104.4}$$

which is, of course, the same as a specialization of the large sieve bound
(102.15). The assertion (103.44) provides an effective lower bound for $G_k(z)$,
and we have $G_k(z) \geq (\varphi(k)/k)G_1(z)$; further,

$$G_1(z) = \sum_{n \leq z} \frac{\mu^2(n)}{n} \prod_{p|n} \left(1 - \frac{1}{p}\right)^{-1} = \sum_{n \leq z} \frac{\mu^2(n)}{n} \sum_{v|n^\infty} \frac{1}{v}$$

$$\geq \sum_{n \leq z} \frac{1}{n} \geq \int_1^z \frac{d\xi}{\xi}. \tag{104.5}$$

Thus, we have, in (104.4),

$$G_k(z) \geq \frac{\varphi(k)}{k} \log z. \tag{104.6}$$

By the way, (102.35) gives

$$G_k(z) = \frac{\varphi(k)}{k} \log z + O\big((\log \log z)^2\big), \tag{104.7}$$

provided $\log k \ll \log z$. The necessary bound for $\prod_{p|k}(1 - 1/p^{s+1})$ with $\mathrm{Re}\, s = 1/\log z$ can be obtained by following Note [18.6].

For instance, let $M = 0$ and $N = [(x - a)/k]$. Then, on noting (104.2), we get

$$\pi(x; k, a) \leq \min_z \left(|\mathcal{S}(\mathcal{N}, z)| + \frac{z}{k} + 1\right)$$

$$\leq \min_z \left(\frac{x/k + 2z^2}{(\varphi(k)/k)\log z} + \frac{z}{k} + 1\right). \tag{104.8}$$

Hence we obtain

Theorem 141 *Uniformly for any coprime pair $k, a \in \mathbb{N}$*

$$\pi(x; k, a) \leq (2 + o(1))\frac{x}{\varphi(k)\log(x/k)}, \tag{104.9}$$

provided x/k tends to infinity.

[B] Since the prototype of the bound (104.9), without any explicit multiplier constant, is said to have been obtained first by E.C. Titchmarsh with Brun's sieve (see §19), the bounds of this type are collectively called the Brun–Titchmarsh theorem nowadays. Among them, in the case of sums of arithmetic functions, the most basic assertion is due to Shiu (1980):

Theorem 142 *Let f be a non-negative multiplicative function on $\mathbb{N}$ satisfying*

$$f(p^v) \leq A^v \tag{104.10}$$

and

$$f(n) \leq A_\varepsilon n^\varepsilon \tag{104.11}$$

604   *Distribution of Prime Numbers*

for all p, v, n and for every $\varepsilon > 0$ with constants $A, A_\varepsilon > 0$. Let

$$1 \le k \le y^\alpha, \ x^\beta \le y \le x, \ \langle k, a \rangle = 1, \ 0 < \alpha, \ \beta < 1. \tag{104.12}$$

Then it holds that

$$\sum_{\substack{x-y<n\le x \\ n\equiv a \bmod k}} f(n) \ll \frac{y}{\varphi(k)\log x} \exp\left(\sum_{p\le x, \, p\nmid k} \frac{f(p)}{p}\right), \tag{104.13}$$

as x tends to infinity; the implied constant depends effectively on A, A_ε, α, β, and ε.

Proof   All constants involved in our discussion below are effective in the same way as in (104.13). Let the prime power decomposition of n be $p_1^{e_1} p_2^{e_2} \cdots p_J^{e_J}$, $p_1 < p_2 < \cdots < p_J$. We take H such that $1 \le H \le J+1$ and

$$n = uv : \quad u = p_1^{e_1} p_2^{e_2} \cdots p_{H-1}^{e_{H-1}} \le w < u p_H^{e_H}, \ w = y^{(1-\alpha)/2}, \tag{104.14}$$

where $p_0 = 1$, $p_{J+1} = +\infty$, and either u or v can be equal to 1. Let

$$S = \left\{ n : x - y < n \le x, \ n \equiv a \bmod k, \ p_H > w^{1/2} \right\}. \tag{104.15}$$

Observe that if $H = J + 1$, then $v = 1$, and if $n \in S$, then $\sum_{j=H}^{J} e_j \le 2\log x / \log w \ll 1$, so $f(v) \ll 1$ for all $n \in S$ because of (104.10). Thus, the contribution of S is

$$\sum_{n\in S} f(u)f(v) \ll \sum_{\substack{u\le w \\ \langle u,k\rangle=1}} f(u)\left(1 + \sum_m 1\right), \tag{104.16}$$

where $(x-y)/u < m \le x/u$, $m \equiv au^{-1} \bmod k$, $p|m \Rightarrow p > w^{1/2}$ with $uu^{-1} \equiv 1 \bmod k$. By (104.4) and (104.6) with $M = [x/2ku]$, $N = [2y/ku] \ge [2w]$, and $z = w^{1/2}$, the last expression is

$$\ll \sum_{\substack{u\le w \\ \langle u,k\rangle=1}} f(u)\left(1 + \frac{y + kuw}{\varphi(k)u\log w}\right) \ll \frac{y}{\varphi(k)\log x} \sum_{\substack{u\le w \\ \langle u,k\rangle=1}} \frac{f(u)}{u}, \tag{104.17}$$

where we have used $kuw \le y^\alpha w^2 = y$. Thus, we see that

$$\sum_{n\in S} f(n) \ll \frac{y}{\varphi(k)\log x} \exp\left(\sum_{p\le x, \, p\nmid k} \frac{f(p)}{p}\right), \tag{104.18}$$

since the sum over u in (104.17) is, by (104.11),

$$\le \prod_{p\le x, \, p\nmid k}\left(1 + \sum_{l=1}^{\infty} \frac{f(p^l)}{p^l}\right) \le \exp\left(\sum_{p\le x, \, p\nmid k}\sum_{l=1}^{\infty} \frac{f(p^l)}{p^l}\right), \tag{104.19}$$

and the part with $l \ge 2$ in the last double sum is $O(1)$.

Next, let

$$T = \{n : x - y < n \le x, \ n \equiv a \bmod k, \ p_H \le w^{1/2}\} = T_1 + T_2, \quad (104.20)$$

where

$$T_1 = \{n \in T : u \le w^{1/2}\}, \quad T_2 = \{n \in T : w^{1/2} < u \le w\}; \quad (104.21)$$

we have now $H \le J$. Further, let

$$T_2 = T_{21} + T_{22}, \quad (104.22)$$

where

$$\begin{aligned} T_{21} &= \{n \in T_2 : p_H \le E\}, \\ T_{22} &= \{n \in T_2 : p_H > E\}, \end{aligned} \quad (104.23)$$

with a sufficiently large constant E to be fixed later.

We shall show that the contributions of T_1 and T_{21} are both negligible. Thus, if $n \in T_1$, then $p_H \le w^{1/2}$ and $p_H^{eH} > w/u \ge w^{1/2}$; that is, n has a prime factor p such that $p \le w^{1/2}$ and $p^l \| n$ with $p^l > w^{1/2}$. Hence, on noting (104.11), we see that

$$\sum_{n \in T_1} f(n) \ll x^\varepsilon \sum_{p \le w^{1/2}} \sum_l \left(\frac{y}{kp^l} + 1 \right), \quad (104.24)$$

where $w^{1/2} < p^l \le x$. This double sum is less than

$$\sum_{p \le w^{1/2}} \frac{\log x}{\log p} + \frac{y}{k} \sum_{2 \le l \le \log x} \sum_{w^{1/2l} < p} \frac{1}{p^l} \ll w^{1/2} + \frac{y}{k} w^{-1/4}. \quad (104.25)$$

Hence, (104.24) is negligible. As for the contribution of T_{21}, we have, again by (104.11), $\sum_{n \in T_{21}} \le x^\varepsilon |T_{21}|$, and

$$|T_{21}| \le \sum_u \left(\frac{y}{ku} + 1 \right) \le w + \frac{y}{k} w^{-1/6} \sum_u \frac{1}{u^{2/3}}, \quad (104.26)$$

where $p|u \Rightarrow p \le E$. We have

$$\sum_u \frac{1}{u^{2/3}} \ll \exp \left(\sum_{p \le E} \frac{1}{p^{2/3}} \right) \ll \exp \left(E^{1/3} \right). \quad (104.27)$$

Hence, $\sum_{n \in T_{21}}$ is also negligible.

It remains for us to consider the contribution of T_{22}. Let v be such that $\max\{E, w^{1/(t+1)}\} < p_H \le w^{1/t}$, $2 \le t \le t_0$, with $t \in \mathbb{N}$ and $t_0 = [\log w / \log E]$.

We note that $\sum_{j=H}^{J} e_j \ll t$, so by (104.10) we have $f(v) \le B^t$ with a constant $B \ge 2$ which depends on A, α, β only. Hence,

$$\sum_{n \in T_{22}} f(u)f(v) \ll \sum_t B^t \sum_u f(u) \sum_m 1, \tag{104.28}$$

where $w^{1/2} < u \le w$, $\langle k, u \rangle = 1$, $p|u \Rightarrow p < w^{1/t}$, and $(x - y)/u < m \le x/u$, $m \equiv au^{-1} \bmod k$, $p|m \Rightarrow w^{1/(t+1)} < p$. In the same way as (104.17), we see that

$$\sum_{n \in T_{22}} f(u)f(v) \ll \frac{y}{\varphi(k) \log x} \sum_{t-2}^{t_0} tB^t \sum_u \frac{f(u)}{u}, \tag{104.29}$$

as $\log w^{1/(t+1)} \gg (\log x)/t$. This sum over u is less than

$$w^{(\eta-1)/2} \sum_u \frac{f(u)}{u^\eta} \ll w^{(\eta-1)/2} \exp\left(\sum_p \frac{f(p)}{p^\eta}\right), \tag{104.30}$$

where $p < w^{1/t}$, $p \nmid k$, and $\frac{2}{3} < \eta \le 1$. We take $\eta = 1 - t(\log E)^{1/2}/\log w$; note that $\eta > 1 - (\log E)^{-1/2} > \frac{2}{3}$ as $E \, (< w^{1/t})$ is supposed to be sufficiently large. Then we observe that

$$\begin{aligned}
\sum_p \frac{f(p)}{p^\eta} &= \sum_p \frac{f(p)}{p} + \sum_p \frac{f(p)}{p}(p^{1-\eta} - 1) \\
&= \sum_p \frac{f(p)}{p} + O\big(\exp\big(2(\log E)^{1/2}\big)\big),
\end{aligned} \tag{104.31}$$

since

$$p^{1-\eta} - 1 \le ((1 - \eta)\log p)p^{1-\eta} \le \frac{(\log E)^{1/2}}{\log w^{1/t}} \exp((\log E)^{1/2}) \log p \tag{104.32}$$

and $\sum_p (f(p)/p) \log p \ll \log(w^{1/t})$ because of (13.7) and (104.10). Collecting these, we find that

$$\begin{aligned}
\sum_{n \in T_{22}} f(u)f(v) &\ll \frac{y}{\varphi(k) \log x} \exp\left(\sum_{p \le x, \, p \nmid k} \frac{f(p)}{p}\right) \\
&\quad \times \sum_{t=1}^\infty t \exp\left(-\tfrac{1}{2}t((\log E)^{1/2} - 2\log B)\right).
\end{aligned} \tag{104.33}$$

Taking $E = \exp\big(C(\log B)^2\big)$ with a sufficiently large $C > 0$, we end the proof of (104.13).

In particular, we have, for any real $\lambda \geq 0$ and integer $r \geq 0$,

$$\sum_{\substack{x-y<n\leq x \\ n\equiv a \bmod k}} d_r^\lambda(n) \ll \frac{y}{k}\left(\frac{\varphi(k)}{k}\log x\right)^{r^\lambda-1}, \tag{104.34}$$

provided (104.12); this is sharp apart from the implied constant. For we have, in (104.13) with $f = d_r^\lambda$,

$$\sum_{p\leq x,\, p\nmid k} \frac{d_r^\lambda(p)}{p} = r^\lambda\left(\sum_{p\leq x}\frac{1}{p} - \sum_{p\mid k}\frac{1}{p}\right) \tag{104.35}$$

$$= r^\lambda\big(\log\log x + \log(\varphi(k)/k) + O(1)\big).$$

Notes

[104.1] It should be stressed that no known method other than sieve has ever been able to produce bounds for $\pi(x;k,a)$ which remain non-trivial for so large k as in (104.9). Because of this feature, it occupies a special position in the theory of the distribution of primes. It has been used as a principal means in such situations that complex analysis becomes powerless in dealing with primes in arithmetic progressions. One of the earliest and most notable applications of this sort took place in Linnik (1944) as we shall explain in Note [106.1], which is a striking instance of fruitful combinations of sieve and analysis. With analysis one may tailor various problems in number theory so that sieve method becomes able to play a decisive rôle.

[104.2] Hence, the Brun–Titchmarsh theorem has been drawing constant interest. In this context Montgomery–Vaughan (1973) proved the beautiful bound

$$\pi(x+y;k,a) - \pi(x;k,a) \leq \frac{2y}{\varphi(k)\log(y/k)},$$

provided $x > 0$ and $y > k$ only, which contains (104.8) of course. They used their own result (3) of Note [103.2] or, more precisely, the dual of (4); it is based also on a careful numerical study.

[104.3] However, the above bounds for $\pi(x;k,a)$ are not the best that one can produce by means of sieve method: The first substantial improvements upon (104.9) have been obtained by Motohashi (1974); it is shown among other things that uniformly for any $\langle k,a\rangle = 1$, $k \leq x^{2/5}$,

$$\pi(x;k,a) \leq (2+o(1))\frac{x}{\varphi(k)\log(x/\sqrt{k})}.$$

The core of his argument is in the fact that the error term in the Λ^2-sieve, as he noticed, can be transformed into a bilinear form so that the cancellation

608   *Distribution of Prime Numbers*

is effectively detected, in accord with the double-sum strategy (Note [97.3]).
Also, in his later article (1999) a further improvement is obtained:

$$\pi(x;k,a) \leq (2+o(1))\frac{x}{\varphi(k)\log(x/k^{3/8})}, \quad k \leq x^{9/20}.$$

Actually, Iwaniec (1982) had achieved the same earlier by means of his own
combinatorial linear sieve (1980c) with an error term of a form of highly
flexible bilinear expression. On the other hand, Motohashi relied on the Λ^2-
sieve with an error term of a bilinear structure which is more effective than
his former version (1974); it is an infusion of the dissection argument given
in Note [19.7] into (102.16) so that the weights $\lambda(u)$ are replaced by those
on the set of products of short intervals, in other words, the Λ^2-sieve for
sifting products of short intervals. The interest of these improvements is in
the following fact: It has, thus, turned out that the dual pair of the large
sieve inequalities (103.36)–(103.37) does not remain to be the best possible
if the interval is replaced by an arithmetic progression with a relatively large
difference and Ψ_r is specified as (104.3): We have, for any a with $\langle k,a \rangle = 1$
and any complex sequence $\{a_n\}$

$$\sum_{\substack{r \leq Q \\ \langle k,r \rangle=1}} \mu^2(r)H(r)\left|\sum_{n \leq N} a_n \Psi_r(kn+a)\right|^2 \leq Y\sum_{n \leq N}|a_n|^2,$$

$$Y = N + O\left(\frac{Q^{1+\varepsilon}}{\sqrt{k}}(Q+k)(\log N)^4\right).$$

If $\sqrt{k} \leq Q$, then this is a clear improvement upon (103.34) under the present
specification. When $k \leq Q$ with k large, there exists a cancellation far more
than the plain large sieve inequality (103.34) asserts, that is, a decrement by
the factor $k^{-1/2}$ takes place; see Motohashi (1977b). At any event, it transpires
that the large sieve inequalities, in general, should probably be improved by
taking into account arithmetical peculiarities of either the sequence to be sifted
or the points $\{\theta_j\}$ or both. What we shall develop in §106 should be viewed
from such a perspective. In this context, it is perhaps interesting to know that
the logarithmic factor in the denominator of the above bound for $\pi(x;k,a)$ can
be replaced by $\log x$, uniformly for $k \leq x^\theta$ with an effectively computable
constant $\theta > 0$. It is a consequence of the assertion (106.13).

[104.4] Comparing the assertion of Theorem 123 with (104.9), one may
surmise that the latter should have any implication about the exceptional zeros
of L-functions, as it holds for moduli far greater than that in the former. This is
true, albeit the constant multiplier 2 needs to be replaced by any one explicitly
smaller. More precisely, if (104.9) is ever improved to

$$\pi(x;k,a) \leq (2-\delta)\frac{x}{\varphi(k)\log x/k}, \quad k \leq x^{\tau},$$

with absolute constants $\delta, \tau > 0$, then it can be concluded that any $L(s,\chi)$, χ mod k, does not admit exceptional zero, as (101.35) implies readily. It should be added that this hypothetical bound is needed for not all reduced residue classes mod k but only more than half of them; thus, statistical improvements, e.g., for almost all a mod k, of (104.9) are of considerable interest.

[104.5] The above proof of (104.13) is a minor simplification of Shiu's original. See also Srinivasan (1997). Nair and Tenenbaum (1998) significantly extended (104.13). In view of (104.34), it might be a surprise to know that the determination of any asymptotic formula for the sum

$$\sum_{\substack{n \leq x \\ n \equiv a \bmod k}} d(n), \quad \langle a,k \rangle = 1,$$

is an extremely hard problem, if uniformity for $k > x^{2/3}$ is stipulated.

§105

The two L^2-inequalities shown in Theorem 139 concern additive characters. In the present section we shall turn to the multiplicative analogue of the theorem, namely, the corresponding L^2-inequalities involving primitive Dirichlet characters. However, this transfer causes no extra difficulty as will be seen in a moment.

It should be remembered how decisive Dirichlet's idea of multiplicative characters was to investigate the distribution of primes in arithmetic progressions and in values of primitive quadratic forms. We shall see that the multiplicative analogue of the last theorem and their immediate corollaries result in a statistical assertion concerning $\psi(x;q,\ell)$ with q moving freely in a surprisingly wide range. On the other hand, the asymptotical study of $\psi(x;q,\ell)$ with individual q so large as a positive power of x poses a deeper problem and the next section §106 will be devoted to the issue.

Theorem 143 *Let $H(r)$ and $\Psi_r(n)$ be defined by (102.14) and (103.31), respectively; and let $\sum^*$ indicate the restriction of the sum to primitive characters. Then, we have, for any $Q \geq 1$ and any $\{a_n\} \subset \mathbb{C}$,*

$$\sum_{\substack{qr \le Q \\ \langle q,r \rangle = 1}} \mu^2(r) H(r) \frac{q}{\varphi(q)} \sum_{\chi \bmod q}^{*} \left| \sum_{M < n \le M+N} a_n \Psi_r(n) \chi(n) \right|^2 \tag{105.1}$$

$$\le \left(N + 2Q^2 \right) \sum_{M < n \le M+N} |a_n|^2.$$

As a corollary, we have, for any $T \ge 1$,

$$\sum_{\substack{qr \le Q \\ \langle q,r \rangle = 1}} \mu^2(r) H(r) \frac{q}{\varphi(q)} \sum_{\chi \bmod q}^{*} \int_{-T}^{T} \left| \sum_{n=1}^{\infty} a_n \Psi_r(n) \chi(n) n^{it} \right|^2 dt \tag{105.2}$$

$$\ll \sum_{n=1}^{\infty} \left(n + Q^2 T \right) |a_n|^2,$$

where the right side is assumed to be finite; the implied constant is absolute.

Remark 1 The dual inequality of (105.1) is not displayed since the principle (103.1) gives it immediately.

Proof The assertion (105.2) is a consequence of the combination of (98.45) and (105.1). As for (105.1) itself, we have for a primitive $\chi \bmod q$

$$\sum_{M < n \le M+N} a_n \Psi_r(n) \chi(n)$$

$$= \frac{1}{G(\overline{\chi})} \sum_{h \bmod q} \overline{\chi}(h) \sum_{M < n \le M+N} a_n \Psi_r(n) e\left(\frac{h}{q} n \right), \tag{105.3}$$

because of (55.4)–(55.5). Then we note that

$$\sum_{\chi \bmod q}^{*} \left| \sum_{M < n \le M+N} a_n \Psi_r(n) \chi(n) \right|^2$$

$$\le \frac{1}{q} \sum_{\chi \bmod q} \left| \sum_{h \bmod q} \overline{\chi}(h) \sum_{M < n \le M+N} a_n \Psi_r(n) e\left(\frac{h}{q} n \right) \right|^2 \tag{105.4}$$

$$= \frac{\varphi(q)}{q} \sum_{\substack{h=1 \\ \langle h,q \rangle = 1}}^{q} \left| \sum_{M < n \le M+N} a_n \Psi_r(n) e\left(\frac{h}{q} n \right) \right|^2.$$

Applying (103.36) to this, we end the proof.

The assertion (105.1) implies the multiplicative analogue of (103.42): Let $a_n = 0$ if there exists a p such that $n \in \Omega(p), p \le z$. Then

$$\sum_{q \leq z} G_q(z/q) \frac{q}{\varphi(q)} \sum_{\chi \bmod q}^{*} \left| \sum_{M < n \leq M+N} a_n \chi(n) \right|^2 \tag{105.5}$$
$$\leq (N + 2z^2) \sum_{M < n \leq M+N} |a_n|^2.$$

For the main purpose of the present section, that is, the proof of the great prime number theorem Theorem 147, we actually need the following simpler version of (105.1)–(105.2):

$$\sum_{q \leq Q} \sum_{\chi \bmod q}^{*} \left| \sum_{M < n \leq M+N} a_n \chi(n) \right|^2 \ll \left(N + Q^2 \right) \sum_{M < n \leq M+N} |a_n|^2, \tag{105.6}$$

$$\sum_{q \leq Q} \sum_{\chi \bmod q}^{*} \int_{-T}^{T} \left| \sum_{n=1}^{\infty} a_n \chi(n) n^{it} \right|^2 dt \ll \sum_{n=1}^{\infty} \left(n + Q^2 T \right) |a_n|^2. \tag{105.7}$$

We shall apply these L^2-inequalities to the statistical study of $\psi(x; q, \ell)$. For this sake, we first extend Theorems 125 and 127:

Theorem 144 *Let $N, Q, T \geq 2$ be arbitrary. Let the finite sequences $S_\chi = \{s_j\} \subset \mathbb{C}$ be such that*

$$\operatorname{Re} s_j \geq 0, \quad |\operatorname{Im} s_j| \leq T, \quad |\operatorname{Im}(s_j - s_k)| \geq 1, \, j \neq k. \tag{105.8}$$

Then we have, for any $\{c(n)\} \subset \mathbb{C}$,

$$\sum_{Q < q \leq 2Q} \sum_{\chi \bmod q}^{*} \sum_{s \in S_\chi} \left| \sum_{N < n \leq 2N} c(n) \chi(n) n^{-s} \right|^2 \tag{105.9}$$
$$\ll \left(N + Q^2 T \right) \log N \sum_{N < n \leq 2N} |c(n)|^2.$$

Proof The argument for (98.47) works here as well, except for an application of (105.7) instead of (98.38). This ends the proof.

Theorem 145 *Let $Q, T \geq 2$ be arbitrary. Then, we have*

$$\sum_{Q < q \leq 2Q} \sum_{\chi \bmod q}^{*} \int_{-T}^{T} |L(\tfrac{1}{2} + it, \chi)|^4 dt \ll Q^2 T (\log QT)^6. \tag{105.10}$$

Also, under the assumption

$$|L(\tfrac{1}{2} + it_{j,\chi}, \chi)| \geq V > 0,$$
$$|t_{j,\chi}| \leq T, \quad 1 \leq j \leq J_\chi; \quad |t_{j,\chi} - t_{k,\chi}| \geq 1, \quad j \neq k, \tag{105.11}$$

612   *Distribution of Prime Numbers*

we have

$$\sum_{Q<q\leq 2Q}\sum_{\chi \bmod q}{}^{*} J_\chi \ll Q^2 TV^{-4}(\log QT)^7. \tag{105.12}$$

Proof The second assertion can be derived from the first in much the same way as (99.4) from (99.2). The first assertion is proved by a slight extension of the argument for Theorem 127. Let $\chi \bmod q$ be as in (105.10). We note

$$\frac{1}{2\pi i}\int_{(2)} L^2(w+\xi,\chi)\Gamma(w)(QT)^w dw$$
$$= \sum_{n=1}^{\infty}\frac{d(n)\chi(n)}{n^\xi}e^{-n/(QT)}, \quad \xi = \tfrac{1}{2}+it. \tag{105.13}$$

We shift the contour to $\operatorname{Re} w = -\tfrac{3}{4}$, while writing the functional equation (55.10) as $L(s,\chi) = \lambda(s,\chi)L(1-s,\overline{\chi})$.

$$L^2(\xi,\chi) = \sum_{n=1}^{\infty}\frac{d(n)\chi(n)}{n^\xi}e^{-n/(QT)}$$
$$+ \frac{1}{2\pi i}\int_{(-\frac{3}{4})}\lambda^2(w+\xi,\chi)\Gamma(w)(QT)^w\left(\sum_{n=1}^{\infty}\frac{d(n)\overline{\chi}(n)}{n^{1-w-\xi}}\right)dw. \tag{105.14}$$

We divide the integrated term into two:

$$\frac{1}{2\pi i}\int_{(a)}\lambda^2(w+\xi,\chi)\Gamma(w)(QT)^w\left(\sum_{n<QT}\frac{d(n)\overline{\chi}(n)}{n^{1-w-\xi}}\right)dw$$
$$+ \frac{1}{2\pi i}\int_{(b)}\lambda^2(w+\xi,\chi)\Gamma(w)(QT)^w\left(\sum_{n\geq QT}\frac{d(n)\overline{\chi}(n)}{n^{1-w-\xi}}\right)dw, \tag{105.15}$$

where $a = -(\log QT)^{-1}$ and $b = -\tfrac{1}{2}-(\log QT)^{-1}$. On noting that the Stirling formula (94.37) gives

$$\lambda^2(w+\xi,\chi)\Gamma(w)(QT)^w \ll \begin{cases} e^{-|w|}\log QT & \operatorname{Re} w = a, \\ e^{-|w|}(QT)^{1/2} & \operatorname{Re} w = b, \end{cases} \tag{105.16}$$

with implied constants being absolute, we see that

$$|L(\xi,\chi)|^2 \ll \left|\sum_{n=1}^{\infty}\frac{d(n)\chi(n)}{n^\xi}e^{-n/(QT)}\right|$$
$$+ (\log QT)\int_{(a)}\left|\sum_{n<QT}\frac{d(n)\overline{\chi}(n)}{n^{1-w-\xi}}\right|e^{-|w|}|dw| \tag{105.17}$$
$$+ (QT)^{1/2}\int_{(b)}\left|\sum_{n\geq QT}\frac{d(n)\overline{\chi}(n)}{n^{1-w-\xi}}\right|e^{-|w|}|dw|.$$

$$\S105 \qquad\qquad 613$$

Squaring both sides and integrating over the interval $-T \le t \le T$ while using (105.7), we end the proof.

Theorem 146 *Let $Q, T \ge 2$ and $\frac{1}{2} \le \alpha \le 1$. Let $N(\alpha, T; \chi)$ be the number of zeros of $L(s, \chi)$ in the rectangular domain $\alpha \le \sigma \le 1$, $|t| \le T$. Then we have*

$$\sum_{Q < q \le 2Q} \sum_{\chi \bmod q}^{*} N(\alpha, T; \chi) \ll (Q^2 T)^{3(1-\alpha)/(2-\alpha)} (\log QT)^{11}. \qquad (105.18)$$

Proof This is an easy extension of the argument for (100.15). Thus, since (101.10) implies that $N(T, \chi) \ll T \log qT$ uniformly for T and $\chi \bmod q$, we may suppose $\alpha \ge \frac{1}{2} + (\log QT)^{-1}$. Then we twist the mollifier (98.11) as

$$M_X(s, \chi) = \sum_{n \le X} \frac{\mu(n) \chi(n)}{n^s}, \qquad X \ge 2. \qquad (105.19)$$

It holds that for any non-trivial zero $\rho = \beta + i\gamma$, $\beta \ge \alpha$, of $L(s, \chi)$ and for any $Y \ge 2$

$$e^{-1/Y} = - \sum_{n \ge X} \frac{a(n) \chi(n)}{n^{\rho}} e^{-n/Y}$$

$$+ \frac{1}{2\pi i} \int_{(2)} M_X(s + \rho, \chi) L(s + \rho, \chi) \Gamma(s) Y^s ds, \qquad (105.20)$$

where $a(n)$ as in (100.1) and $(QT)^{\varepsilon} \le X \le Y \le (QT)^A$ with a constant $A > 0$; naturally T is supposed to be sufficiently large. Following (100.7)–(100.9), we get

$$|U_1(\rho, \chi)| + |U_2(\rho, \chi)| \ge \tfrac{1}{2}, \qquad (105.21)$$

$$U_1(\rho, \chi) = - \sum_{X \le n < Y \log^2 QT} \frac{a(n) \chi(n)}{n^{\rho}} e^{-n/Y}, \qquad (105.22)$$

$$U_2(\rho, \chi) = \frac{1}{2\pi} \int_{-\log^2 QT}^{\log^2 QT} M_X(\tfrac{1}{2} + i(\gamma + t), \chi) L(\tfrac{1}{2} + i(\gamma + t), \chi)$$

$$\times \Gamma(\tfrac{1}{2} - \beta + it) Y^{1/2 - \beta + it} dt. \qquad (105.23)$$

Let $\mathcal{R}(\chi)$ be the set of those zeros of $L(s, \chi)$ in the region $\{\alpha \le \sigma \le 1, T \le |t| \le 2T\}$ such that if $\mathcal{R}(\chi) \ni \rho, \rho', \rho \ne \rho'$, then $|\rho - \rho'| \ge (\log QT)^3$. Again using (101.10), we have

$$N(\alpha, 2T; \chi) - N(\alpha, T; \chi) \ll |\mathcal{R}(\chi)| (\log QT)^4. \qquad (105.24)$$

Further, let $\mathcal{R}_\nu(\chi) = \{\rho \in \mathcal{R}(\chi) : |U_\nu(\rho, \chi)| \geq \frac{1}{4}\}$. Then by (105.9)

$$\sum_{Q<q\leq 2Q} \sideset{}{^*}\sum_{\chi \bmod q} |\mathcal{R}_1(\chi)|$$

$$\ll \sum_{Q<q\leq 2Q} \sideset{}{^*}\sum_{\chi \bmod q} \sum_{\rho\in\mathcal{R}_1(\chi)} |U_1(\rho, \chi)|^2 \tag{105.25}$$

$$\ll (\log QT)^3 \max_{X<N\leq Y\log^2 T} (N + Q^2 T)e^{-N/Y} \sum_{N<n\leq 2N} \frac{d^2(n)}{n^{2\alpha}}$$

$$\ll (Y^{2(1-\alpha)} + Q^2 TX^{1-2\alpha})(\log QT)^7,$$

where we have used (16.18). On the other hand, following (100.12),

$$|\mathcal{R}_2(\chi)| \ll Y^{2(1-2\alpha)/3}(\log QT)^2 \int_T^{2T} |L(\tfrac{1}{2} + it, \chi)M_X(\tfrac{1}{2} + it, \chi)|^{4/3} dt; \tag{105.26}$$

and via (105.7) and (105.10)

$$\sum_{Q<q\leq 2Q} \sideset{}{^*}\sum_{\chi \bmod q} |\mathcal{R}_2(\chi)| \tag{105.27}$$
$$\ll (Q^2 T)^{1/3}(X + Q^2 T)^{2/3} Y^{2(1-2\alpha)/3}(\log QT)^5.$$

Taking $X = Q^2 T$ and $Y = (Q^2 T)^{3/(4-2\alpha)}$, we end the proof.

The following implication of this zero-density theorem is called the prime number theorem of Bombieri:

Theorem 147 *Let*

$$E(y; q, \ell) = \psi(y; q, \ell) - \frac{y}{\varphi(q)}, \tag{105.28}$$

and let $A \geq 1$ be an arbitrary constant. Then, with

$$Q = x^{1/2}(\log x)^{-A-16}, \tag{105.29}$$

we have

$$\sum_{q\leq Q} \max_{(q,\ell)=1} \max_{y\leq x} |E(y; q, \ell)| \ll \frac{x}{(\log x)^A}, \tag{105.30}$$

and hence

$$\sum_{q\leq Q} \max_{(q,\ell)=1} \max_{y\leq x} \left|\pi(y; q, \ell) - \frac{\mathrm{li}\, y}{\varphi(k)}\right| \ll \frac{x}{(\log x)^A}, \tag{105.31}$$

where the implied constants depend only on A.

Remark 2 This celebrated assertion is also called the Bombieri–Vinogradov prime number theorem because of the historical fact stated in Notes [105.1] and [105.3].

Remark 3 The parameter Q in (105.30)–(105.31) is often called the level of the equidistribution of prime numbers in arithmetic progressions; and the same practice is applied to the exponent $\frac{1}{2}$ appearing in (105.29). Naturally, these terms are used in analogous situations as well, for instance, in the statistical study of sums of arithmetic functions over arithmetic progressions.

Remark 4 The implied constants in (105.30)–(105.31) are, in fact, not effectively computable in terms of A since the proof below incorporates Theorem 135 which is ineffective.

Proof On noting (101.4) with $q \le Q$, we have

$$E(y; q, \ell) = \frac{1}{\varphi(q)} \sum_{\substack{\chi \bmod q \\ \chi \ne J_q}} \overline{\chi}(\ell) \psi(y, \chi^*) + O\big((\log Qy)^2\big); \qquad (105.32)$$

thus, by Theorem 48,

$$\max_{\langle q, \ell \rangle = 1} \max_{y \le x} \big|E(y; q, \ell)\big| \ll \frac{\log x}{q} \sum_{\substack{k \mid q \\ k \ge 3}} \sideset{}{^*}\sum_{\chi \bmod k} \max_{y \le x} |\psi(y, \chi)| + (\log x)^2,$$

$$(105.33)$$

where we have used the easy bound $\varphi(q)^{-1} \ll (\log 2q)/q$. Hence,

$$\sum_{q \le Q} \max_{\langle q, \ell \rangle = 1} \max_{y \le x} \big|E(y; q, \ell)\big|$$

$$\ll (\log x)^3 \max_{3 \le K \le Q} \frac{1}{K} \sum_{K < k \le 2K} \sideset{}{^*}\sum_{\chi \bmod k} \max_{y \le x} |\psi(y, \chi)| + Q(\log x)^2.$$

$$(105.34)$$

We shall estimate the last double sum. Let $K_0 = (\log x)^{A+16}$, and let us consider first the part $K \le K_0$. According to (101.38),

$$\sum_{K < k \le 2K} \sideset{}{^*}\sum_{\chi \bmod k} \max_{y \le x} |\psi(y, \chi)| \ll x \exp(-c(\log x)^{1/2}), \quad K \le K_0. \quad (105.35)$$

where the constant $c > 0$ depends only on A but ineffectively. Anyway the part for $K \le K_0$ can be ignored, and hereafter we shall assume $K_0 \le K$. Also, we may assume, without loss of generality, that there exists no exceptional character. This is because as has been remarked in Note [101.2] there exists at most one exceptional character mod k such that $K < k \le 2K$ and its contribution to (105.34) is $O(\psi(x)(\log x)^3 K_0^{-1})$, which is negligible.

Then by (101.28) with $T = [y]$,

$$\psi(y, \chi) = -\sum_{|\gamma| < y} \frac{y^\rho}{\rho} + O((\log ky)^2), \quad \chi \bmod k, \tag{105.36}$$

where $\rho = \beta + i\gamma$ as usual. On noting (101.10), we have

$$\max_{y \leq x} |\psi(y, \chi)| \ll \sum_{|\gamma| < x} \frac{x^\beta}{|\rho|} + (\log x)^2$$

$$\ll \sum_{\substack{|\gamma| < x \\ \frac{1}{2} \leq \beta}} \frac{x^\beta}{|\rho|} + x^{1/2}(\log x)^2, \tag{105.37}$$

since the part with $\beta < \frac{1}{2}$ is $\ll x^{1/2} \sum_{|\gamma| \leq 1} |\rho|^{-1} + x^{1/2}(\log x)^2$ because of (101.7), and this sum is $\ll (\log x)^2$ as the current assumption on exceptional characters implies $(\log K)^{-1} \ll |\rho|$. The sum in the lower line of (105.37) is

$$\ll \left\{ \sum_{\substack{|\gamma| \leq 1 \\ \frac{1}{2} \leq \beta}} + (\log x) \max_{1 \leq U \leq x} \sum_{\substack{U < |\gamma| \leq 2U \\ \frac{1}{2} \leq \beta}} \right\} \frac{x^\beta}{|\rho|}$$

$$\ll (\log x) \max_{1 \leq U \leq x} \frac{1}{U} \int_1^{\frac{1}{2}} x^\alpha \, dN(\alpha, 2U; \chi) \tag{105.38}$$

$$\ll (\log x)^2 \max_{1 \leq U \leq x} \frac{1}{U} \int_{\frac{1}{2}}^1 N(\alpha, 2U; \chi) x^\alpha \, d\alpha + x^{1/2}(\log x)^2,$$

where the last term depends on (101.7) and (101.20). Applying the zero-density estimate (105.18), we have, for $K_0 \leq K \leq Q$,

$$\frac{1}{K} \sum_{K < k \leq 2K} \sideset{}{^*}\sum_{\chi \bmod k} \max_{y \leq x} |\psi(y, \chi)|$$

$$\ll (\log x)^{13} \max_{1 \leq U \leq x} \frac{1}{KU} \int_{\frac{1}{2}}^1 (K^2 U)^{3(1-\alpha)/(2-\alpha)} x^\alpha \, d\alpha + Qx^{1/2}(\log x)^2$$

$$\ll (\log x)^{13} \int_{\frac{1}{2}}^1 K^{(4-5\alpha)/(2-\alpha)} x^\alpha \, d\alpha + Qx^{1/2}(\log x)^2, \tag{105.39}$$

as $3(1 - \alpha)/(2 - \alpha) \leq 1$ for $\frac{1}{2} \leq \alpha \leq 1$. This integral is

$$\ll \int_{\frac{4}{5}}^1 K_0^{(4-5\alpha)/(2-\alpha)} x^\alpha \, d\alpha + \int_{\frac{1}{2}}^{\frac{4}{5}} Q^{(4-5\alpha)/(2-\alpha)} x^\alpha \, d\alpha$$

$$\ll xK_0^{-1} + x^{1/2}Q + x^{7/8}, \tag{105.40}$$

$$\S 105 \qquad\qquad 617$$

since the second integrand is $\leq x^{1/2}(Q + x^{3/8})$, as can be seen by considering the cases $Q \geq x^{3/8}$ and $\leq x^{3/8}$ separately. We end the proof.

The prime number theorem (105.31) is amazing, even though it does not assert anything about individual $\pi(x; q, \ell)$. If one employs the format (105.31), then GRH implies only that Q can be taken as large as $x^{1/2}(\log x)^{-A-2}$, which does not appear to be impressive in the light of (105.29); however, the implied constants corresponding to those in (105.30)–(105.31) will become effective. For a variety of problems which can be decomposed into sums of $\pi(x; q, \ell)$ with varying moduli, the theorem is employed as a replacement of GRH; the replacement is powerful to such an extent that it is still unknown whether or not the use of GRH is able to yield anything definitely better than what can be obtained via Theorem 147.

The statistical approach to the Riemann hypothesis started by Bohr and Landau brought Hoheisel's method, and being combined with Linnik's large sieve method has reached Bombieri's prime number theorem. This is certainly the most pleasant avenue to a great discovery in analytic number theory.

The equidistribution in arithmetic progressions displayed above is not unique to the sequence of primes or rather the arithmetic function $\Lambda(n)$, as indicated in Remark 3. It holds actually with a wide variety of arithmetic functions, and among them the following induction principle holds:

Let f be an arithmetic function defined over $\mathbb{N}$, which need not be multiplicative; and let

$$E_f(x, \chi) = \sum_{n \leq x} f(n)\chi(n),$$

$$E_f(x; q, \ell) = \sum_{\substack{n \leq x \\ n \equiv \ell \bmod q}} f(n) - \frac{1}{\varphi(q)} \sum_{\substack{n \leq x \\ \langle n, q \rangle = 1}} f(n), \quad \langle q, \ell \rangle = 1. \tag{105.41}$$

Then we say that f is in U-class if it satisfies the following three conditions:

(1) There exists an integer $\tau \geq 0$ and a constant $r \geq 0$ such that

$$f(n) \ll (d(n))^{\tau} (\log 2n)^r. \tag{105.42}$$

(2) For each constant $A, B > 0$

$$E_f(x, \chi) \ll \frac{x}{(\log x)^B}, \tag{105.43}$$

provided the conductor of the non-principal character χ is not greater than $(\log x)^A$.

618   *Distribution of Prime Numbers*

(3) For each constant $A > 0$ there exists a constant $C > 0$ such that

$$\sum_{q \leq x^{1/2}/(\log x)^C} \max_{\langle q, \ell \rangle = 1} \max_{y \leq x} |E_f(y; q, \ell)| \ll \frac{x}{(\log x)^A}. \tag{105.44}$$

Theorem 148 *If arithmetic functions f, g are both in U-class, then their convolution $f * g$ is also in U-class.*

Proof It is obvious that $f * g$ satisfies the condition (1). As for the condition (2), it suffices to note the decomposition

$$E_{f*g}(x, \chi) = \sum_{m \leq x^{1/2}} f(m) \chi(m) E_g(x/m, \chi)$$
$$+ \sum_{n \leq x^{1/2}} g(n) \chi(n) E_f(x/n, \chi) - E_f(x^{1/2}, \chi) E_g(x^{1/2}, \chi). \tag{105.45}$$

In order to confirm that $f * g$ satisfies (3), we let δ_R be the characteristic function of the interval $(R, 2R]$ and observe that

$$E_{f*g}(y; q, \ell) = \sum_{M,N} \Delta_{M,N}(y; q, \ell), \tag{105.46}$$

$$\Delta_{M,N}(y; q, \ell) = \sum_{\substack{mn \leq y \\ mn \equiv \ell \bmod q}} \delta_M(m) \delta_N(n) f(m) g(n)$$
$$- \frac{1}{\varphi(q)} \sum_{\substack{mn \leq y \\ \langle mn, q \rangle = 1}} \delta_M(m) \delta_N(n) f(m) g(n). \tag{105.47}$$

Here M, N belong to the sequence $\{2^\nu : \nu = 0, 1, 2, \ldots\}$ and $MN \leq x$; and also we may obviously suppose that $x^{1/3} < y \leq x$. Namely, we have

$$\max_{\langle q, \ell \rangle = 1} \max_{y \leq x} |E_{f*g}(y; q, \ell)|$$
$$\ll \sum_{\substack{M,N \\ x^{1/3} < MN \leq x}} \max_{\langle q, \ell \rangle = 1} \max_{y \leq x} |\Delta_{M,N}(y; q, \ell)| + O(x^{1/3+\varepsilon}). \tag{105.48}$$

Moreover, we may restrict ourselves to the situation where

$$(\log x)^R < M, N \leq \frac{x}{(\log x)^R} \tag{105.49}$$

with an arbitrary $R > 0$ by means of the identity

$$\Delta_{M,N}(y; q, \ell) = \sum_{\substack{n \leq y \\ \langle n, q \rangle = 1}} \delta_N(n) g(n)$$
$$\times \{E_f(\min(2M, y/n); q, \ell\bar{n}) - E_f(\min(M, y/n); q, \ell\bar{n})\}, \tag{105.50}$$

where $n\bar{n} \equiv 1 \bmod q$, the part corresponding to $N \leq (\log x)^R$ can be discarded by the assumption (3) for f and the same can be said for the part with $M \leq (\log x)^R$.

Next

$$\sum_{Q<q\leq 2Q} \max_{(q,\ell)=1} \max_{y\leq x} |\Delta_{M,N}(y;q,\ell)|$$

$$\leq \sum_{Q<q\leq 2Q} \frac{1}{\varphi(q)} \sum_{\substack{\chi \bmod q \\ \chi \neq J_q}} \max_{y\leq x} |\Delta_{M,N}(y,\chi)|, \tag{105.51}$$

where J_q is the principal character $\bmod\, q$, and

$$\Delta_{M,N}(y,\chi) = \sum_{mn\leq y} \delta_M(m)\delta_N(n)f(m)g(n)\chi(mn), \tag{105.52}$$

in which the conductor of χ can be assumed to be not less than $(\log x)^A$, for both the functions $\delta_M f$ and $\delta_N g$ satisfy the condition (2) one may argue just in the same way as in (105.45). Namely, the right side of (105.51) is

$$\leq \sum_{r\leq 2Q/(\log x)^A} \frac{1}{\varphi(r)} \sum_{Q/r<q\leq 2Q/r} \frac{1}{\varphi(q)} \sum_{\chi \bmod q}^* \max_{y\leq x} |\Delta_{M,N}(y, J_r\chi)|, \quad (105.53)$$

for $\varphi(qr) \geq \varphi(q)\varphi(r)$. Hence, our problem has been reduced to the estimation of

$$\frac{1}{Q} \sum_{Q<q\leq 2Q} \sum_{\chi \bmod q}^* \max_{y\leq x} |\Delta_{M,N}(y, J_r\chi)|. \tag{105.54}$$

provided (105.49) and $Q > \frac{1}{2}(\log x)^A$.

Then, by Perron's inversion formula (95.1), we see that (105.54) is

$$\ll 1 + \frac{1}{Q} \sum_{Q<q\leq 2Q} \sum_{\chi \bmod q}^* \int_{a-iT}^{a+iT} |F(s, J_r\chi)G(s, J_r\chi)| \frac{|ds|}{|s|}. \tag{105.55}$$

Here $T = x^2$ and $a = (\log x)^{-1}$; $F(s,\chi) = \sum_n \delta_M(m)f(m)\chi(m)m^{-s}$, and $G(s,\chi)$ is analogous. Thus, invoking (105.7), we find that (105.55) is

$$\ll 1 + \frac{(\log x)^b}{Q}\left((M + Q^2)(N + Q^2)MN\right)^{1/2}, \tag{105.56}$$

where the constant b depends only on the condition (1); see (16.17). Collecting these estimations, we end the proof.

For instance, since the divisor function $d_k(n)$ is the k-fold convolution of $f(n) \equiv 1$, it is in U-class; then via (16.22)–(16.23), one may readily conclude

620   *Distribution of Prime Numbers*

that $d_k^l(n)$ as well is in U-class for any $k, l \in \mathbb{N}$. Similar examples are obtained from the explicit formula (74.20); thus, in particular, the function $r(n)$ defined by (78.7) is in U-class. Continued in Note [105.6].

Notes

[105.1]   Theorem 147 is due to Bombieri (1965), although the above argument incorporates later developments. The corresponding assertion by Vingradov, A. I. (1965) is, in fact, weaker than Bombieri's as it requires $Q \leq x^{1/2-\varepsilon}$; nevertheless, in applications to traditional sieve problems this causes no essential differences from Bombieri's. They relied on the zero-density argument, i.e., Hoheisel's scheme; to this end, Bombieri used the L^2-large sieve inequalities, while Vinogradov appealed to the dispersion method. More details are in Notes [105.3]–[105.4]. Thus, the two fundamental ideas, the large sieve and the dispersion method, both due to Linnik, resulted simultaneously in the far-reaching prime number theorem. Amazing!

[105.2]   The origin of the prime number theorem of the type of (105.31) can be traced back to Rényi (1948); his result holds for $Q \leq x^{\vartheta}$ with a relatively small absolute constant $\vartheta > 0$. For this sake he employed a zero-density estimate of his own which can be regarded as a distant prototype of (105.18). Before Rényi, Linnik (1944) had discovered a way to avoid the quasi-GRH, the analogue of (98.7) for every L-function, while dealing with a single modulus in his effort to prove his celebrated theorem on the least prime number in an arithmetic progression, which we shall deal with in the next section. Hence, the task of Rényi was to extend Linnik's argument to many moduli; and he discovered that the large sieve was the key for his aim. The relevant history between Rényi (1948) and Bombieri (1965)–Vinogradov (1965) is detailed in Barban (1966, written in 1964). The history afterward can be found in Montgomery (1971, Chapters 15 and 17) and in the added part of Bombieri (1987).

[105.3]   The critical exponent $\frac{1}{2}$ in (105.29) stems from the fact that the zero-density estimate (105.18) implies that in the region

$$\tfrac{1}{2} + H \frac{\log\log Q}{\log Q} \leq \sigma \leq 1, \ |t| \leq (\log Q)^H,$$

with a sufficiently large constant H,

almost all $L(s, \chi)$, $\chi \bmod q, q \leq Q$, do not have any zero. That is, GRH nearly holds in a restricted sense. This also means that the T-aspect in (105.18) is, in fact, not essential but the Q-aspect matters, as it can be confirmed by using the Riesz typical mean or rather the Mellin transform of a suitable weight function in place of Perron's inversion formula. In Vinogradov (1965) the zero-density estimate

$$\sum_{Q<q\leq2Q} \sideset{}{^*}\sum_{\chi \bmod q} N(\alpha,T;\chi) \ll Q^{3-2\alpha+\delta}(T\log Q)^{c/\delta^4}$$

is shown; here $\delta > 0$ is an arbitrary fixed constant. This is definitely weaker than (105.29), specifically in the T-aspect. Nevertheless, because of the reasons indicated above, it is still capable of yielding the exponent $\frac{1}{2} - \varepsilon$ mentioned in Note [105.1]. Yet, it should be remarked that to have the T-aspect like that in (105.18) becomes essential if one considers the analogue of Theorem 144 for short intervals, i.e., $\pi(x + y; q, \ell) - \pi(x; q, \ell)$, $y \approx x^\theta$, with an absolute constant $\theta < 1$.

[105.4]   The dispersion method of Linnik (1963) is described, very roughly, as a way to deal with asymptotic problems concerning arithmetic sums via the analysis of the dispersions of those sums which are arithmetically perturbed ones of the original sums. Relevantly, we add a remark to Theorem 148 which is due to Motohashi (1976): The assumption (3) was formulated, just bearing in mind the application of the large sieve inequalities (105.6)–(105.7). In other words, if some methods other than the large sieve inequalities are taken into account, as actually the author envisaged implicitly, then there should arise the possibility to go beyond the level $x^{1/2}(\log x)^{-B}$ of the equidistribution. He contemplated that one of such methods to be tried should be the dispersion method, as the appearance of the double sum structure in the above discussion suggests naturally. Indeed, with this scheme, Bombieri et al (1986) achieved the extremely impressive level $Q = x^{4/7}$ for $\pi(x; q, \ell)$, albeit ℓ is fixed; see also Bombieri (1987, §12).

[105.5]   As for the proof of Theorem 147, Gallagher (1968) demonstrated that the use of the zero-density can in fact be discarded. He utilized instead the trivial identity

$$\frac{L'}{L} = \frac{L'}{L}(1 - LM)^2 + 2L'M - LL'M, \quad M = M_X(s, \chi),$$

with an obvious abbreviation and applied the multiplicative large sieve inequality as well as an elementary Tauber argument on the Riesz typical mean. This identity is apparently related to the famed decomposition of the von Mangoldt function Λ by Vaughan (1977) into a sum of convolutions of arithmetic functions which are more manageable than Λ itself, and thus gave rise to a number of remarkable simplifications in discussions related to prime numbers; it is another instance of the double sum strategy. We shall employ in the next section an elaborated version of Gallagher's identity, so the details of his alternative approach to Theorem 147 are skipped; but see Bombieri (1987).

622   *Distribution of Prime Numbers*

[105.6]   Wolke (1973) generalized Theorem 147 to sums of fairly general multiplicative functions over arithmetic progressions. His result includes, for instance, $d_k^l(n)$ with any $l \in \mathbb{C}$, which is of course not manageable with our Theorem 148.

§106

Theorem 147 is definitely a marvel in mathematics: it is capable of replacing GRH at decisive stages in the investigation on classical problems involving primes in arithmetic progressions with varying moduli; a typical example will be shown in the next section. However, it is also true that the theorem is unable to yield any non-trivial consequence concerning the distribution of primes in individual arithmetic progressions, especially if the differences are not small enough compared with the length of progressions. In order to fill this lacuna, we shall present Theorem 150 that is of a structure similar to Theorem 147 yet contains the celebrated Linnik's prime number theorem for short arithmetic progressions that holds without any hypothesis in the same sense as Hoheisel's Theorem 117 does in the case of short intervals:

Theorem 149 *There exists an effectively computable absolute constant $\mathcal{L}$ such that uniformly for any coprime $\{q, \ell\}$*

$$\pi\left(q^{\mathcal{L}}; q, \ell\right) > 0. \tag{106.1}$$

As a matter of fact, this assertion itself had been shown much prior to Theorem 147: Linnik (1944) established it, and the exponent $\mathcal{L}$ is called the Linnik constant. In his proof is an extremely delicate argument concerning the zeros of all Dirichlet L-functions $\bmod q$, and at its core is a crucial application of the Brun–Titchmarsh theorem (a version of (104.9) without a precise multiplier constant); therefore, a sieve mechanism is working there in a decisive but rather fortuitous way. We shall enhance this salient aspect of Linnik's proof so that the sieve mechanism is to play the principal rôle from the very outset of our discussion. It should be stressed that Linnik's original proof of (106.1) and ours are both independent of Siegel's Theorem 135, whence they produce $\mathcal{L}$ an explicitly computable constant.

If the argument of the preceding section is to be pursued further in the present context, then we should first discuss the specific zero-density assertion that Linnik devised. However, we shall take a route avoiding the zero-density. It is an extension of the argument mentioned in Note [105.5]; we assert that it is more appropriate for our aim of attaining a structural approach to Theorem

$$\S 106 \qquad\qquad 623$$

149. Nevertheless, we shall state the relevant zero-density assertion at the end of the present section; it reveals, in particular, a highly bizarre nature of exceptional zeros.

To begin with, we make a few of our earlier notions precise, disregarding repetitions. Thus, we let $\mathbf{x}(q)$ and $\mathbf{x}^*(q)$ be the sets of all characters and all primitive characters mod q, respectively. Let Q be a sufficiently large positive parameter, about which we shall be more precise later. Let

$$\mathbf{X}(Q) = \bigcup_{q \leq Q} \mathbf{x}(q), \quad \mathbf{X}^*(Q) = \bigcup_{q \leq Q} \mathbf{x}^*(q). \tag{106.2}$$

We consider the following sets of non-trivial zeros ρ of L-functions:

$$\mathbf{z}(q,Q) = \bigcup_{\chi \in \mathbf{x}(q)} \{\rho = \beta + i\gamma : L(\rho,\chi) = 0, |\gamma| \leq Q^7\},$$

$$\mathbf{z}^*(q,Q) = \bigcup_{\chi \in \mathbf{x}^*(q)} \{\cdots\}, \quad \mathbf{Z}^*(Q) = \bigcup_{\chi \in \mathbf{X}^*(Q)} \{\cdots\}, \tag{106.3}$$

with an obvious abbreviation. Note that $0 < \beta < 1$ and that the height Q^7 is taken just for the sake of convenience. Obviously,

$$\mathbf{x}(q) = \bigcup_{k|q} J_q \cdot \mathbf{x}^*(k), \quad \mathbf{z}(q,Q) = \bigcup_{k|q} \mathbf{z}^*(k,Q), \tag{106.4}$$

where J_q is the principal character mod q as usual.

In view of Note [101.2], there exists an effectively computable absolute constant $\kappa \in (0,1]$ such that at most one element of $\mathbf{Z}^*(Q)$ does not satisfy

$$\beta \leq 1 - \frac{\kappa}{\log Q}. \tag{106.5}$$

According to Theorem 133 and Note [101.2], this exceptional element, if ever exists, is a real zero β_1 of $L(s,\chi_1)$ with χ_1 mod q_1 which is unique in $\mathbf{X}^*(Q)$ and also real. However, we refine the situation of being exceptional by the stricter criterion:

$$\chi_1 \colon Q\text{-exceptional} \quad \Leftrightarrow \quad 1 - \frac{\kappa}{4 \log Q} \leq \beta_1 \leq 1 - \frac{1}{Q}, \tag{106.6}$$

where the upper bound follows from the assertion shown in Note [101.5]; it can be replaced by a better estimate, though. Thus, if the exceptional element in the sense of Theorem 133 exists in $\mathbf{Z}^*(Q)$ but it does not satisfy this inequality, then it should not be regarded as genuinely exceptional. Hereafter, χ_1 mod q_1 stands for the Q-exceptional character, and β_1 for the associated exceptional zero. We see via (101.21) that for any $\chi \in \mathbf{X}(Q)$

624   *Distribution of Prime Numbers*

$$\frac{L'}{L}(\sigma_0 + it, \chi) \ll \log Q, \quad |t| \le Q^7, \tag{106.7}$$

with

$$\sigma_0 = \begin{cases} 1 - \kappa/(8 \log Q) & \text{if } \chi^* \ne \chi_1, \\ 1 - \kappa/(2 \log Q) & \text{if } \chi^* = \chi_1, \end{cases} \tag{106.8}$$

where $\chi^* \in \mathbf{X}^*(Q)$ induces χ, and the implied constant in (106.7) is effective.
  Accordingly, we let

$$\widetilde{\psi}(y, \chi) = \begin{cases} \psi(y, \chi) & \chi^* \ne J_1, \chi_1, \\ \psi(y, \chi) - y & \chi^* = J_1, \\ \psi(y, \chi) + y^{\beta_1}/\beta_1 & \chi^* = \chi_1, \end{cases} \tag{106.9}$$

where $\chi^* = J_1$ means that $\chi = J_q$ with a $q \le Q$. Then we shall show

Theorem 150 *There exist effectively computable constants* $a_0, a_1, a_2, a_3 > 0$
such that provided

$$a_0 < Q^{a_1} \le x \le \exp\left((\log Q)^2\right) \tag{106.10}$$

it holds that

$$\sum_{q \le Q} \sum_{\chi \bmod q}^* \max_{y \le x} \left|\widetilde{\psi}(y, \chi)\right| \le a_2 \Delta_Q \exp\left(-a_3 \frac{\log x}{\log Q}\right) x, \tag{106.11}$$

with

$$\Delta_Q = \begin{cases} 1 & \text{if } \chi_1 \text{ does not exist,} \\ (1 - \beta_1) \log Q & \text{if } \chi_1 \text{ exists.} \end{cases} \tag{106.12}$$

Let $q \ge 3$ *be arbitrary. Then, provided* $a_0 < q^{a_1} \le x \le \exp((\log q)^2)$, *we
have, for each reduced residue* $\ell \bmod q$,

$$\left|\psi(x; q, \ell) - \frac{x}{\varphi(q)} + \chi_1(\ell)\frac{x^{\beta_1}}{\varphi(q)\beta_1}\right| \le 2a_2 \Delta_q \exp\left(-a_3 \frac{\log x}{\log q}\right)\frac{x}{\varphi(q)}, \tag{106.13}$$

where the term involving β_1 *occurs only when there exists a* $\chi \bmod q$ *which is
induced by* χ_1 *the q-exceptional character.*

Remark 1 Note that in (106.13) we have employed the definition (106.6) with
$Q = q$. If $\exp((\log q)^2) \le x$, then the Page–Landau prime number theorem
(101.34) or rather (101.33) holds.

Proof of Theorem 149: In (106.13), which follows immediately from (101.4)
and (106.11), we take $x = q^C$ with a sufficiently large C. If χ_1 does not exist,
then we end the proof. If χ_1 exists, then we have

$$\psi(x; q, \ell) \geq \Big(1 - 2a_2((1 - \beta_1) \log q) \exp(-a_3 C) - \beta_1^{-1} \exp(-(1 - \beta_1)C \log q)\Big) \frac{x}{\varphi(q)}. \tag{106.14}$$

If $(1 - \beta_1)C \log q \leq 1$, then

$$\psi(x; q, \ell) > ((1 - \beta_1) \log q) \times \Big(C/(e\beta_1) - 2a_2 \exp(-a_3 C) - 1/(\beta_1 \log q)\Big) \frac{x}{\varphi(q)} > 0, \tag{106.15}$$

where we have used $\exp(-\xi) < 1 - \xi/e$ for $0 < \xi \leq 1$. On the other hand, if $(1 - \beta_1)C \log q > 1$, then

$$\psi(x; q, \ell) > \Big(1 - 2a_2\kappa \exp(-a_3 C) - 1/(e\beta_1)\Big) \frac{x}{\varphi(q)} > 0, \tag{106.16}$$

with κ as in $(106.5)_{Q=q}$. Any upper bound of the current C works as a value of $\mathcal{L}$ in (106.1). We end the proof.

Remark 2 The combination of Note [101.5] and (106.13) yields that uniformly for $x \geq q^C$

$$\pi(x; q, \ell) \gg \frac{x}{q^{3/2}(\log x)^2}, \tag{106.17}$$

with the implied constant depending on C effectively. We have not paid any particular attention to the choices of parameters that should yield smaller values of C; we are content with attaining a well-structured proof of the existence of an explicitly computable C. It should be noted that Xylouris (2011) claims $\mathcal{L} \leq 5$ in (106.1). His argument, which relies on the Λ^2-sieve as ours does, is extremely involved because of his efforts at numerical precision throughout.

Proof of Theorem 150:   Our proof of the main assertion (106.11) comprises several steps. In doing this, we suppose that χ_1 exists. The discussion of the contrary case is omitted, for it is just a far easier version of what we develop below. Our main parameter is Q, which is supposed to be sufficiently large while satisfying the first inequality of (106.10); naturally a_0 ought to be large but can be determined effectively anyway.

(i) We introduce first the multiplicative function

$$f(n) = \sum_{d|n} \chi_1(d)d^{-\theta}, \quad \theta = 1 - \beta_1, \tag{106.18}$$

626   *Distribution of Prime Numbers*

with which we shall materialize the idea that was set out in the last paragraph of the text of §103: We consider the Selberg sieve situation

$$Y(\lambda) = \sum_{n \leq N} f(n) \left(\sum_{u|n} \lambda(u) \right)^2, \tag{106.19}$$

where N tends to infinity. We suppose customarily that $\lambda(u) \in \mathbb{R}$ and

$$\lambda(1) = 1; \quad \lambda(u) = 0, u > z; \quad |\lambda(u)| \leq |\mu(u)|. \tag{106.20}$$

Our later specialization of $\{\lambda(u)\}$ will satisfy these; meanwhile, z is to be fixed appropriately in terms of Q. We observe the following:

$$f(n) \text{ is positive for all } n \in \mathbb{N}; \tag{106.21}$$

$$F_p = \sum_{l=0}^{\infty} f(p^l) p^{-l} > 1 + p^{-7/3} \text{ for each } p; \tag{106.22}$$

$$\sum_{n \leq N} \chi(n) f(n) = \mathcal{F} E_\chi F_q^{-1} N + O\big((q^2 Q)^{1/3} N^{2/3}\big), \tag{106.23}$$

where $\mathcal{F} = L(1 + \theta, \chi_1)$; $\chi \bmod q$, $q \leq Q$; $F_q = \prod_{p|q} F_p$; $E_\chi = 1$ if $\chi = J_q$ and $= 0$ otherwise. The assertion (106.21) is equivalent to $f(p^a) > 0$ for any integer $a \geq 0$, which is but trivial; on the other hand, (106.22) follows from

$$F_p = \left(1 - \frac{1}{p}\right)^{-1} \left(1 - \frac{\chi_1(p)}{p^{1+\theta}}\right)^{-1} > \left(1 - \frac{1}{p^{2(1+\theta)}}\right)^{-1}. \tag{106.24}$$

As for (106.23), we apply (95.1) to the function $L(s, \chi) L(s + \theta, \chi \chi_1)$ with $T = N^{1/2} / (q^2 Q)^{1/4} + 1$:

$$\frac{1}{2\pi i} \int_{a-iT}^{a+iT} L(s, \chi) L(s + \theta, \chi \chi_1) \frac{N^s}{s} ds + O\left(N^{1+\varepsilon}/T\right), \tag{106.25}$$

where $a = 1 + \varepsilon$, and N can be assumed to be half an odd integer without any loss of generality. Shifting the contour to $\operatorname{Re} s = \varepsilon$ and taking into account the convexity bound for L-functions (Note [101.1]), we obtain (106.23) but with the error term $O(((q^2 Q)^{1/4} N^{1/2})^{1+\varepsilon})$; note that if χ is induced by χ_1, then the simple pole at $s = 1 - \theta$ of $L(s + \theta, \chi \chi_1)$ is canceled out by the zero β_1 of $L(s, \chi)$. The assertions (106.22) and (106.23) are stated in the forms convenient for our succeeding discussion; they are not the best attainable.

(ii) We shall next compute $Y(\lambda)$, beginning with

$$Y(\lambda) = \sum_{u_1, u_2 \leq z} \lambda(u_1) \lambda(u_2) \sum_{n \leq N/u} f(un), \quad u = [u_1, u_2]; \tag{106.26}$$

note that $\mu(u) \neq 0$. Let $f_1 = \mu_f$ be the convolution inverse of the present f (see (17.8)) so that for each p and $\operatorname{Re} s > 0$

$$\sum_{l=0}^{\infty} \frac{f_1(p^l)}{p^{ls}} = \left(\sum_{l=0}^{\infty} \frac{f(p^l)}{p^{ls}} \right)^{-1} = \left(1 - \frac{1}{p^s} \right) \left(1 - \frac{\chi_1(p)}{p^{s+\theta}} \right). \tag{106.27}$$

Then

$$f(un) = \mu(u) \sum_{v \mid \langle n, u^{\infty} \rangle} f\left(\frac{n}{v} \right) f_1(uv), \tag{106.28}$$

as is readily verified via the Dirichlet series $\sum_{n=1}^{\infty} f(un) n^{-s}$. Hence, by (106.23) with $\chi \equiv j_1 \equiv 1$,

$$\sum_{n \leq N/u} f(un) = \mu(u) \sum_{\substack{v \leq N/u \\ v \mid u^{\infty}}} f_1(uv) \left\{ \mathcal{F} \frac{N}{uv} + O\left(Q^{1/3} \left(\frac{N}{uv} \right)^{2/3} \right) \right\}. \tag{106.29}$$

We observe that

$$\mathcal{F} N \frac{\mu(u)}{u} \left(\sum_{v \mid u^{\infty}} - \sum_{\substack{v > N/u \\ v \mid u^{\infty}}} \right) \frac{f_1(uv)}{v}$$

$$= \mathcal{F} N \prod_{p \mid u} \left(1 - F_p^{-1} \right) + O\left(\mathcal{F} N^{2/3} \sum_{v \mid u^{\infty}} \frac{|f_1(uv)|}{(uv)^{2/3}} \right), \tag{106.30}$$

since if $v > N/u$, then $N/uv < (N/uv)^{2/3}$. We get

$$\sum_{n \leq N/u} f(un) = \mathcal{F} N \prod_{p \mid u} (1 - F_p^{-1}) + O\left(Q^{1/3} N^{2/3} \sum_{v \mid u^{\infty}} \frac{|f_1(uv)|}{(uv)^{2/3}} \right), \tag{106.31}$$

for $\mathcal{F} \ll \log Q$ trivially. Inserting this into (106.26) while noting (106.20), we find that

$$Y(\lambda) = \mathcal{F} N \sum_{u_1, u_2 \leq z} \lambda(u_1) \lambda(u_2) \prod_{p \mid [u_1, u_2]} \left(1 - F_p^{-1} \right)$$

$$+ O\left(Q^{1/3} (Nz)^{2/3} (\log z)^8 \right). \tag{106.32}$$

This error term needs to be explained: In view of (106.27) and $\mu(u) \neq 0$, the sum over v in (106.31) is

$$\prod_{p \mid u} \left(\frac{|f_1(p)|}{p^{2/3}} + \frac{|f_1(p^2)|}{p^{4/3}} \right) < \prod_{p \mid u} \left(\frac{2}{p^{2/3}} + \frac{1}{p^{4/3}} \right) < \frac{d_3(u)}{u^{2/3}}, \tag{106.33}$$

with the divisor function d_3. Because of (106.20), the relevant sum is not greater than $\sum_{n \leq z^2} d_3^2(n) n^{-2/3}$ (see (102.33)); and this is $\ll z^{2/3} (\log z)^8$ by (16.17).

628   *Distribution of Prime Numbers*

(iii) We diagonalize the quadratic form $Y^*(\lambda)$ over u_1, u_2 on the right side of (106.32). On noting

$$\prod_{p|[u_1,u_2]} \left(1 - F_p^{-1}\right) = \prod_{p|u_1} \left(1 - F_p^{-1}\right) \prod_{p|u_2} \left(1 - F_p^{-1}\right) \sum_{\substack{v|u_1 \\ v|u_2}} \prod_{p|v} \left(F_p - 1\right)^{-1},$$

(106.34)

we have

$$Y^*(\lambda) = \sum_{v \leq z} \mu^2(v) K(v) \xi(v)^2, \quad \xi(v) = \frac{\mu(v)}{K(v)} \sum_{\substack{g \leq z \\ g \equiv 0 \bmod v}} \frac{K(g)}{F_g} \lambda(g), \quad (106.35)$$

where

$$K(g) = \prod_{p|g} \left(F_p - 1\right). \tag{106.36}$$

Since

$$\lambda(g) = \mu(g) \frac{F_g}{K(g)} \sum_{g|v} \mu^2(v) K(v) \xi(v), \tag{106.37}$$

the side condition $\lambda(1) = 1$ is expressed in terms of ξ; and we have

$$Y^*(\lambda) = \frac{1}{G(z)} + \sum_{v \leq z} \mu^2(v) K(v) \left(\xi(v) - \frac{1}{G(z)}\right)^2, \tag{106.38}$$

where

$$G(z) = \sum_{v \leq z} \mu^2(v) K(v). \tag{106.39}$$

Hence, the optimal λ is given by

$$\lambda(g) = \frac{\mu(g)}{G(z)} \frac{F_g}{K(g)} \sum_{g|v} \mu^2(v) K(v), \tag{106.40}$$

and then

$$Y^*(\lambda) = \frac{1}{G(z)}. \tag{106.41}$$

In much the same way as in (102.30)–(102.31), one may confirm that (106.40) fulfills (106.20). Also, analogously to (103.45), we have, for $g, x \geq 1$,

$$\sum_{\substack{v \leq x \\ \langle g, v \rangle = 1}} \mu^2(v) K(v) \geq G(x) F_g^{-1}. \tag{106.42}$$

$$\S106 \qquad\qquad 629$$

(iv) We are now led to the following counterpart of (103.31):

$$\sum_{g\mid n} \lambda(g) = \frac{1}{G(z)} \sum_{r\le z} \mu^2(r)K(r)\Phi_r(n), \quad \Phi_r(n) = \frac{\mu(\langle r,n\rangle)}{K(\langle r,n\rangle)}. \tag{106.43}$$

The Φ_r is multiplicative on $\mathbb{Z}/r\mathbb{Z}$; here is a harmless notational conflict with (67.4). The argument so far developed has been aimed at the identification of this quasi-character Φ_r. Then, corresponding to (105.1), we assert that it holds, for any complex sequence $\{a_n\}$ and $N \ll M$, that

$$\sum_{\substack{q\le Q;\, r\le z \\ \langle q,r\rangle=1}} \mu^2(r)K(r)F_q \sum_{\chi \bmod q}^{*} \left| \sum_{M<n\le M+N} a_n\chi(n)\Phi_r(n)(f(n))^{1/2} \right|^2$$

$$\le \big(\mathfrak{F}N + Z(M;Q,z)\big) \sum_{M<n\le M+N} |a_n|^2, \tag{106.44}$$

where

$$Z(M;Q,z) \ll Q^4 z^{13/6} M^{2/3}. \tag{106.45}$$

Also, corresponding to (105.2), we have, for any $T \ge 1$,

$$\sum_{\substack{q\le Q;\, r\le z \\ \langle q,r\rangle=1}} \mu^2(r)K(r)F_q \sum_{\chi \bmod q}^{*} \int_{-T}^{T} \left| \sum_{n=1}^{\infty} a_n\chi(n)\Phi_r(n)(f(n))^{1/2}n^{it} \right|^2 dt$$

$$\ll \sum_{n=1}^{\infty} \big(\mathfrak{F}n + TZ(n;Q,z)\big)|a_n|^2, \tag{106.46}$$

provided the right side converges. The implied constants in (106.45) and (106.46) are effective. The bound (106.45) for Z can be replaced by a smaller one as is obvious from the next step.

(v) We shall verify (106.44)–(106.46). The assertion (106.46) can be readily shown by the combination of (98.45) and (106.44). Thus, it suffices to deal with the dual of (106.44): With an arbitrary complex sequence $\{b(r,\chi)\}$, we consider

$$\sum_{M<n\le M+N} f(n) \left| \sum_{\substack{q\le Q;\, r\le z \\ \langle q,r\rangle=1}} \sum_{\chi \bmod q}^{*} \mu^2(r)(K(r)F_q)^{1/2}\Phi_r(n)\chi(n)b(r,\chi) \right|^2$$

$$= \sum_{\substack{q,q'\le Q;\, r,r'\le z \\ \langle q,r\rangle=1,\, \langle q',r'\rangle=1}} \sum_{\substack{\chi \bmod q \\ \chi' \bmod q'}}^{*} \mu^2(r)\mu^2(r')\{K(r)K(r')F_qF_{q'}\}^{1/2}b(r,\chi)\overline{b(r',\chi')}$$

$$\times \Big(S(M+N, \chi\overline{\chi'};r,r') - S(M, \chi\overline{\chi'};r,r') \Big). \tag{106.47}$$

Here

$$S(x, \chi; r, r') = \sum_{n \leq x} \chi(n)(\Phi_r \Phi_{r'})(n) f(n), \qquad (106.48)$$

with $(\Phi_r \Phi_{r'})(n) = \Phi_r(n) \Phi_{r'}(n)$. We then consider the function

$$\sum_{n=1}^{\infty} \chi(n)(\Phi_r \Phi_{r'})(n) f(n) n^{-s}, \qquad (106.49)$$

which converges absolutely for $\operatorname{Re} s > 1$ since $|\Phi_r(n)| \leq \prod_{p|r} F_p/(F_p - 1)$. As it can be supposed that $\mu(r)\mu(r') \neq 0$, we have the decomposition $F_1 F_2 F_3$ of (106.49): With

$$F(s, \chi) = \prod_p F_p(s, \chi), \quad F_p(s, \chi) = \sum_{l=0}^{\infty} \chi(p^l) f(p^l) p^{-ls}, \qquad (106.50)$$

we have, for $\operatorname{Re} s > 1$,

$$F_1 = F(s, \chi) \prod_{p|rr'} F_p(s, \chi)^{-1},$$

$$F_2 = \prod_{p \mid \frac{[r,r']}{\langle r, r' \rangle}} \left(1 - (F_p - 1)^{-1}(F_p(s, \chi) - 1)\right), \qquad (106.51)$$

$$F_3 = \prod_{p \mid \langle r, r' \rangle} \left(1 + (F_p - 1)^{-2}(F_p(s, \chi) - 1)\right).$$

Then, defining f_2 via the relation

$$F_1 F_2 F_3 = F(s, \chi) A_{r, r'}(s, \chi), \quad A_{r, r'}(s, \chi) = \sum_{n \mid (rr')^{\infty}} \chi(n) f_2(n) n^{-s}, \qquad (106.52)$$

we have

$$(\Phi_r \Phi_{r'})(n) f(n) = \sum_{d \mid \langle n, (rr')^{\infty} \rangle} f\left(\frac{n}{d}\right) f_2(d). \qquad (106.53)$$

By this representation and (106.23), we have, for $\chi \bmod q$,

$$S(x, \chi; r, r') = \mathcal{F} E_\chi F_q^{-1} A_{r, r'}(1, \chi) x$$
$$+ O\left((q^2 Q)^{1/3} x^{2/3} \sum_{d \mid (rr')^{\infty}} \frac{|f_2(d)|}{d^{2/3}}\right), \qquad (106.54)$$

where the same argument as that of deducing (106.31) has been applied. From (106.27), (106.51), and (106.52), we have

$$\sum_{d|(rr')^\infty} \frac{|f_2(d)|}{d^{2/3}} < \prod_{p|rr'} \left(1 + \frac{1}{p^{2/3}}\right)^2 \prod_{p|\frac{[r,r']}{\langle r,r'\rangle}} \left(1 + \frac{(F_p(\frac{2}{3}, J_1) - 1)}{F_p - 1}\right)$$

$$\times \prod_{p|\langle r,r'\rangle} \left(1 + \frac{(F_p(\frac{2}{3}, J_1) - 1)}{(F_p - 1)^2}\right) \ll \frac{d_v(rr')}{K(r)K(r')[r,r']^{2/3}}, \tag{106.55}$$

with an integer $v > 0$; note that $|F_p(s, \chi) - 1| \le F_p(\sigma, J_1) - 1$ for $\mathrm{Re}\, s = \sigma > 0$. Also,

$$\langle q, rr'\rangle = 1 \implies A_{r,r'}(1, J_q) = \frac{\delta_{r,r'}}{K(r)}, \tag{106.56}$$

with the Kronecker delta; this is the quasi-orthogonality of the quasi-characters $\{\Phi_r\}$. Inserting these into (106.47), we find that the main term is equal to

$$\mathfrak{J}N \sum_{\substack{q \le Q;\ r \le z \\ \langle q, r\rangle = 1}} \sideset{}{^*}\sum_{\chi \bmod q} \mu^2(r)|b(r, \chi)|^2. \tag{106.57}$$

On the other hand, the error term is, after some rearrangement,

$$\ll Q^{1/3} M^{2/3} \sum_{\substack{q, q' \le Q;\ r, r' \le z \\ \langle q, r\rangle, \langle q', r'\rangle = 1}} \frac{(qq')^{2/3}\{F_q F_{q'}\}^{1/2}\mu^2(r)\mu^2(r')d_v(rr')}{(K(r)K(r'))^{1/2}[r, r']^{2/3}}$$

$$\times \sideset{}{^*}\sum_{\substack{\chi \bmod q \\ \chi' \bmod q'}} |b(r, \chi)|^2 \tag{106.58}$$

$$\ll Q^4 M^{2/3} z^{13/6} \sum_{\substack{q \le Q;\ r \le z \\ \langle q, r\rangle = 1}} \sideset{}{^*}\sum_{\chi \bmod q} \mu^2(r)|b(r, \chi)|^2,$$

in which we have used the lower bound $K(r) > r^{-7/3}$ implied by (106.22), and

$$\sum_{r' \le z} \mu^2(r)\mu^2(r')d_v(rr')(rr')^{1/2}\langle r, r'\rangle^{2/3} \ll z^{13/6}, \tag{106.59}$$

via the inversion formula (18.3). We end the verification of (106.44)–(106.46).

(vi) Next, we shall show two preparatory assertions: It holds that

$$\mu^2(r) = 1,\ \omega_d \ll \mu^2(d) \implies$$

$$\sum_{n=1}^{\infty} \chi(n)\Phi_r(n)f(n)\left(\sum_{d|n} \omega_d\right)n^{-s} = F(s, \chi)M_r(s, \chi; \omega), \tag{106.60}$$

632   *Distribution of Prime Numbers*

with Re $s > 1$ and

$$M_r(s, \chi; \omega) = \frac{F_r}{K(r)} \sum_{d=1}^{\infty} \omega_d \frac{\mu(\langle r, d \rangle)}{F_{\langle r, d \rangle}}$$
$$\times \prod_{p|d}(1 - F_p(s, \chi)^{-1}) \prod_{p|r/\langle r,d \rangle} (F_p(s, \chi)^{-1} - F_p^{-1}).$$

(106.61)

Also, with $\mu^{(k)}$ as in Theorem 109, we assert that

provided $Q < w^c$ with a constant $c > 0$,

$$\sum_{r \leq w^{1+2\vartheta}} \mu^2(r) K(r) \big(M_r(1, J_1; \mu^{(2)})\big)^2 \ll \frac{1}{\mathscr{F} \log w},$$

(106.62)

where w, ϑ are the same as those in (95.32); the implied constant is effective. To confirm (106.60), we note that $\Phi_r(dn) = \Phi_r(d)\Phi_u(n)$, $u = r/\langle r, d \rangle$, so the sum of (106.60) becomes

$$\sum_{d=1}^{\infty} \chi(d)\omega_d \Phi_r(d)d^{-s} \sum_{n=1}^{\infty} \chi(n)\Phi_u(n)f(dn)n^{-s}.$$

(106.63)

The inner sum equals

$$\left(\sum_{\langle m, d \rangle = 1} \chi(m)\Phi_u(m)f(m)m^{-s} \right)\left(\sum_{n|d^{\infty}} \chi(n)f(dn)n^{-s} \right),$$

(106.64)

for $\Phi_u(n) = 1$ if $\mu^2(r) = 1$ and $n|d^{\infty}$. This is

$$\overline{\chi}(d)d^s \prod_{p \nmid d}\left(1 + \Phi_u(p)(F_p(s, \chi) - 1)\right) \prod_{p|d}(F_p(s, \chi) - 1)$$
$$= \overline{\chi}(d)d^s F(s, \chi) \prod_{\substack{p \nmid d \\ p|r}}(F_p - 1)^{-1}(F_p(s, \chi)^{-1}F_p - 1)$$
$$\times \prod_{p|d}(1 - F_p(s, \chi)^{-1}),$$

(106.65)

which implies (106.61); it should be noted that there the factor $J_q(d) = 1$ has been eliminated because of an obvious reason. As for the sum in (106.62), which is denoted as R, we note that with $\{s, \chi\} = \{1, J_1\}$ the last product factor of (106.61) induces that $r|d$. Hence

$$R = \sum_{r \leq w^{1+2\vartheta}} \frac{\mu^2(r)}{K(r)}\left(\sum_{d \leq w^{1+2\vartheta}, \, d|r} \mu^{(2)}(d)K(d)/F_d \right)^2$$

(106.66)

§106   633

Comparing this with (106.19), (106.32), and (106.35), we find that

$$\sum_{n \le N} f(n) \left(\sum_{d|n} \mu^{(2)}(d) \right)^2 = \mathcal{F}RN + O\left(Q^{1/3}N^{2/3}w^{2(1+2\vartheta)/3}(\log w)^8\right).$$
(106.67)

Hence, summing by parts, we get, for any $b > 0$ and $a > 1$,

$$\sum_{n \ge w^b} f(n) \left(\sum_{d|n} \mu^{(2)}(d) \right)^2 n^{-a}$$
$$= (a-1)^{-1}\mathcal{F}Rw^{b(1-a)} + O\left(Q^{1/3}w^{b(2/3-a)+2(1+2\vartheta)/3}(\log w)^8\right).$$
(106.68)

We take $a = 1 + (\log w)^{-1}$ and $b = 2(1+2\vartheta)+2c$ with c as in (106.62). Then the right side of (106.68) becomes

$$e^{-b}\left(\mathcal{F}R \log w + O(w^{-c/4})\right).$$
(106.69)

On the other hand, this is $O(1)$ because of Theorem 109. We end the verification of (106.62).

(vii) Returning to (106.39), we shall show that with Δ_Q as in (106.12)

$$z = Q^{2/3} \implies G(z) \gg \Delta_Q^{-1}\mathcal{F} \log Q,$$
(106.70)

which represents the sieve effect that is implicit in (106.44). Thus, we note the trivial inequalities

$$G(z) > z^{-2\theta} \sum_{g \le z} \mu^2(g)K(g)g^{2\theta} > z^{-2\theta}G(z),$$
(106.71)

and consider the function

$$\sum_{g=1}^{\infty} \mu^2(g)K(g)g^{2\theta-s} = \zeta(s+1-2\theta)L(s+1-\theta,\chi_1)B(s),$$
(106.72)

for $\sigma > 2\theta$; here $B(s)$ is absolutely convergent and bounded for $\sigma \ge -\frac{1}{2} + \varepsilon$, and $B(2\theta) = 1$, as can be seen via the Euler product expansion of the left side. By (95.1), the sum in (106.71) is equal to

$$\frac{1}{2\pi i}\int_{a-iT}^{a+iT} \zeta(s+1-2\theta)L(s+1-\theta,\chi_1)B(s)\frac{z^s}{s}ds + O(z^\varepsilon/T),$$
(106.73)

where $a = \varepsilon$, $T = (z/Q^{1/2})^{1/3}$, and z can be assumed to be half an odd integer without loss of generality. Shifting the contour to $\sigma = -\frac{1}{2} + \varepsilon$, while noting

634   *Distribution of Prime Numbers*

the convexity bound for the zeta- and the L-functions, we find that (106.73) equals

$$\frac{z^{2\theta}}{2\theta}\mathcal{F} + O\big((Q^{1/2}/z)^{2/5}\big). \tag{106.74}$$

We replace the sum of (106.71) by this and note that $G(z) > 1$. We see that if $z = Q^{2/3}$, then $G(z) \approx \mathcal{F}/\theta$ via (106.71), so (106.70) has been verified.

(viii) We are now ready to deal with the sum in (106.11). The combination of (101.21), (101.25)$_{x=y}$, and (106.7)–(106.9) gives, with $U = Q^7$,

$$\widetilde{\psi}(y, \chi) = -\frac{1}{2\pi i}\int_{\sigma_0 - iU}^{\sigma_0 + iU} \frac{L'}{L}(s, \chi)\frac{y^s}{s}ds + O\big(xQ^{-7}(\log Q)^2\big), \tag{106.75}$$

where σ_0 is as in (106.8); hereafter we shall have $y \leq x$ always. Then we apply, with an elaboration, the modification of the last integrand that has been suggested in Note [105.5] so that

$$\begin{aligned}\widetilde{\psi}(y, \chi) = {}&- \frac{1}{2\pi i}\int_{\sigma_0 - iU}^{\sigma_0 + iU} \frac{L'}{L}(s, \chi)\big(Y_r(s, \chi)\big)^2\frac{y^s}{s}ds \\ &+ \frac{1}{2\pi i}\int_{\sigma_0 - iU}^{\sigma_0 + iU} W_r(s, \chi)\frac{y^s}{s}ds + O\big(xQ^{-7}(\log Q)^2\big),\end{aligned} \tag{106.76}$$

with

$$Y_r(s, \chi) = F(s, \chi)M_r(s, \chi; \mu^{(2)}) - 1, \tag{106.77}$$

$$W_r(s, \chi) = (Y_r(s, \chi) - 1)M_r(s, \chi; \mu^{(2)})L'(s, \chi)L(s + \theta, \chi \chi_1). \tag{106.78}$$

Here M_r is as in (106.61) with $r \leq z = Q^{2/3}$ according to (106.70), and $\mu^{(2)}$ is the specialization of (95.32)$_{k=2}$ with

$$w = Q^{20}, \ \vartheta = \varepsilon; \tag{106.79}$$

the reason for this choice of w will become obvious at (106.93). Moving the contour of the second integral in (106.76) to $\sigma = \varepsilon$, $|t| \leq U$, we do not encounter any singularity. On the new contour we have $W_r(s, \chi) \ll w^2 z^5 Q^3 U^2$; this is a rough bound but suffices for our purpose. Indeed, it follows readily via a combination of the convexity bound for L-functions and the fact $F_p(s, \chi)^{-1} = 1 + O(p^{-\sigma})$ for $\sigma > 0$ as well as $K(r) > r^{-7/3}$ because of (106.22). We shall omit the details of similar estimation argument in what follows. Thus, the second integral in (106.76) is $O(x^\varepsilon Q^{61})$; and provided

$$Q^{70} \leq x \leq \exp((\log Q)^2), \tag{106.80}$$

we have

$$\widetilde{\psi}(y,\chi) = -\frac{1}{2\pi i}\int_{\sigma_0-iU}^{\sigma_0+iU}\frac{L'}{L}(s,\chi)\big(Y_r(s,\chi)\big)^2\frac{y^s}{s}\,ds$$
$$+ O\big(xQ^{-7}(\log Q)^2\big).$$

(106.81)

Invoking (106.7)–(106.8), we have, with parameters so far chosen,

$$\max_{y\le x}|\widetilde{\psi}(y,\chi)| \ll (\log Q)\exp\Big(-\frac{\kappa}{2}\frac{\log x}{\log Q}\Big)x$$
$$\times \int_{-U}^{U}|Y_r(\sigma_0+it,\chi)|^2\frac{dt}{|t|+1} + xQ^{-7}(\log Q)^2.$$

(106.82)

We multiply both sides by the factor $\mu^2(r)K(r)F_q$ and sum over $r \le z = Q^{2/3}$, $\langle r,q\rangle = 1$, and further over primitive characters $\chi \bmod q$, $q \le Q$. Since the left side is independent of the variable r, we find that

$$\sum_{q\le Q}F_q G_q \sum_{\chi\bmod q}^{*}\max_{y\le x}|\widetilde{\psi}(y,\chi)|$$
$$\ll (\Psi\log Q)\exp\Big(-\frac{\kappa}{2}\frac{\log x}{\log Q}\Big)x + xQ^{-4},$$

(106.83)

where

$$G_q = \sum_{\substack{r\le z\\ \langle q,r\rangle=1}}\mu^2(r)K(r),$$

(106.84)

$$\Psi = \sum_{\substack{q\le Q;\ r\le z\\ \langle q,r\rangle=1}}\mu^2(r)F_q K(r)\sum_{\chi\bmod q}^{*}\int_{-U}^{U}|Y_r(\sigma_0+it,\chi)|^2\frac{dt}{|t|+1}.$$

(106.85)

Hence, by (106.42) and (106.70) as well as $G(z) > 1$, we get

$$\sum_{q\le Q}\sum_{\chi\bmod q}^{*}\max_{y\le x}|\widetilde{\psi}(y,\chi)|$$
$$\ll \Psi\mathcal{F}^{-1}\Delta_Q\exp\Big(-\frac{\kappa}{2}\frac{\log x}{\log Q}\Big)x + xQ^{-4}.$$

(106.86)

(ix) Our problem has now been reduced to the proof of the crucial

$$\Psi \ll \mathcal{F}.$$

(106.87)

This will close our discussion, as $\Delta_Q\exp(-\kappa(\log x)/(2\log Q)) \ge Q^{-1-\kappa/2}$, provided (106.10). Thus, to confirm (106.87), we apply (94.38) and consider

$$X_r^{(1)}(s,\chi) = \frac{1}{2\pi i}\int_{2-i\infty}^{2+i\infty}Y_r(s+\xi,\chi)\Gamma(\xi)V^\xi\,d\xi, \quad V = Q^{35},$$

(106.88)

636   *Distribution of Prime Numbers*

where $s = \sigma_0 + it$, $|t| \leq U$. By the identity (106.60) with $\omega = \mu^{(2)}$ and by (95.34),

$$X_r^{(1)}(s, \chi) = \sum_{w \leq n} \chi(n) f(n) \Phi_r(n) \left(\sum_{u|n} \mu^{(2)}(u) \right) n^{-s} e^{-n/V}. \qquad (106.89)$$

Moving the contour in (106.88) to $\operatorname{Re} \xi = -\sigma_0 + \varepsilon$ and noting that $Y_r(\varepsilon + it, \chi) \ll wz^{8/3} Q^{3/2}(|t| + 1)$, we have

$$Y_r(s, \chi) = X_r^{(1)}(s, \chi) - X_r^{(2)}(s, \chi) + O(Q^{-4}), \qquad (106.90)$$

with

$$X_r^{(2)}(s, \chi) = E_\chi \mathcal{F} F_q^{-1} M_r(1, \chi; \mu^{(2)}) \Gamma(1 - s) V^{1-s}, \qquad (106.91)$$

so that denoting by $\Psi^{(j)}$ the result of replacing Y_r by $X_r^{(j)}$ in (106.85)

$$\Psi \ll \Psi^{(1)} + \Psi^{(2)} + Q^{-5}. \qquad (106.92)$$

We apply (106.46) and get

$$\begin{aligned} \Psi^{(1)} &\ll \sum_{w \leq n} (\mathcal{F} n + Q^{49/9} n^{2/3}) f(n) \left(\sum_{u|n} \mu^{(2)}(u) \right)^2 n^{-2\sigma_0} e^{-2n/V} \\ &\ll \mathcal{F} \sum_{n=1}^{\infty} d(n) \left(\sum_{u|n} \mu^{(2)}(u) \right)^2 n^{1-2\sigma_0} e^{-2n/V}, \end{aligned} \qquad (106.93)$$

since we now have $Z(n; Q, z) = Q^{49/9} n^{2/3}$ as $z = Q^{2/3}$. The lower line depends on the fact that if $n \geq w = Q^{20}$, then $\mathcal{F} n \gg Q^{49/9} n^{2/3}$, for (106.71) and (106.74) imply that $\mathcal{F}^{-1} \ll \theta^{-1}$, and moreover $\theta^{-1} \ll Q$ by (106.6). In (106.93) we may decrease the exponent $1 - 2\sigma_0$ to $-1 - (\log Q)^{-1}$ but with an increase of the implied constant. Then, by (95.35) we find that $\Psi^{(1)} \ll \mathcal{F}$. As for $\Psi^{(2)}$, we observe that

$$\int_{\sigma_0 - iU}^{\sigma_0 + iU} |\Gamma(1 - s)|^2 \frac{|ds|}{|s|} \ll \log Q. \qquad (106.94)$$

So, on noting that $E_\chi = 1$ with a primitive χ means that $\chi = j_1$,

$$\Psi^{(2)} \ll \mathcal{F}^2 (\log Q) \sum_{r \leq z} \mu^2(r) K(r) |M_r(1, j_1; \mu^{(2)})|^2. \qquad (106.95)$$

Since $z < w$, we get $\Psi^{(2)} \ll \mathcal{F}$ by (106.62). Thus, $\Psi \ll \mathcal{F} + Q^{-5} \ll \mathcal{F}$, as $\mathcal{F} \gg \theta > Q^{-1}$. Therefore, (106.87) has been verified. This completes the proof of Theorem 150.

We have dispensed with the zero-density argument in order to prove Linnik's prime number theorem (106.1). It should, however, be appropriate to indicate here what the combination of devices applied in the present section and in the proof of Theorem 150 is able to yield on the distribution of non-trivial and non-Q-exceptional zeros $\rho = \beta + i\gamma$ of L-functions associated with primitive characters mod q, $q \leq Q$: Thus, we assert that

$$\left|\{\rho \in \mathbf{Z}^*(Q) : \alpha \leq \beta, \rho \neq \beta_1\}\right| \ll \Delta_Q Q^{c_0(1-\alpha)}, \tag{106.96}$$

where c_0 and the implied constant are effectively computable. This implies, in particular, that every $L(s, \chi)$, $\chi \in \mathbf{X}^*(Q)$, has no zero other than β_1 in the rectangle

$$1 - c_1 \frac{\log(2/\Delta_Q)}{\log Q} \leq \sigma \leq 1, \; |t| \leq Q^7, \tag{106.97}$$

with c_1 an effective constant. If one supposes that it ever happens, for instance, that $1 - Q^{-1/4} \leq \beta_1$, then the bizarreness of the occurrence of exceptional zeros could be well discerned. Continued in Note [106.2].

Notes

[106.1]   This section is an adaptation of the core part of the discussions developed in Motohashi (1975; 1977a; 1977b; 1978; 1983, Part II).

[106.2]   A brief history related to Linnik's prime number theorem: We confine ourselves strictly to the Hoheisel scheme; thus other known approaches are not taken into account. Although the single modulus situation was discussed originally, we shall employ the expression in the large sieve context as we did in the above.

(1) Linnik's own proof of Theorem 149 depends on his two fundamental discoveries (1944, (I) and (II)). However, it is hard to describe them in his own formulation; thus we shall rely on the later understanding: They are (I) the zero-density estimate of the type (105.18) but lacking the log-factor, i.e., essentially the same as (106.96) without the factor Δ_Q, and (II) the assertion (106.97), which Linnik proved separately from (I), calling it the Deuring–Heilbronn phenomenon because of a historical reason; however, we propose instead the Linnik phenomenon as it is certainly more appropriate. In passing, we note that Siegel's theorem is contained in (106.96); it suffices to consider the situation that $\beta + i\gamma$, $\beta \leq 1 - \varepsilon$, $|\gamma| \leq Q^7$, is an element other than β_1 of $\mathbf{Z}^*(Q)$.

(2) Rodosskii (1954) is said to have made Linnik's formidable argument accessible. Pracher (1957, Kapitell X) adopted it, because of which the present author, in early 1960s, became aware of Linnik's theorem (106.1).

(3) Turán (1961) obtained a truly alternative proof of Linnik's zero-density estimate. He applied his own invention the power-sum method (1953) in order

638　　　　　　　　　　*Distribution of Prime Numbers*

to detect zeros: Let $\{z_j : 1 \le j \le J\} \subset \mathbb{C}$ be arbitrary, and let $J \le K$. Then there exists a $k \in [K, 2K]$ such that

$$|z_1^k + z_2^k + \cdots + z_J^k| \ge (|z_1|/50)^k,$$

which indicates that the high derivatives $(L'/L)^{(k)}(s, \chi)$ become the principal object in his argument. Knapowski (1962) followed Turán and obtained a new proof of (106.97), whence Linnik's theorem (106.1) became indeed accessible.

(4) Fogels (1965) improved Turán's zero-density assertion, still using the power sum method, so that he (p.84) could prove a short interval version of (106.1); by the way, the footnote on the first page of this article is worth noticing in the context of the present note. Then, Gallagher (1970) brought the matter into the truly large sieve context together with his own device Theorem 124 from Fourier analysis. The power sum method was again a main device to detect zeros. Our Theorem 150 is essentially the same as Gallagher's theorem 7; actually he treated the short interval situation, i.e., $\widetilde{\psi}(x + y, \chi) - \widetilde{\psi}(x, \chi)$, but this is not an essential matter as we have mentioned already. It should be noted that Gallagher needed (106.97) (à la Knapowski) to complete his work. We have dispensed with both the power sum method and (106.97).

(5) As has been remarked in Note [103.5], Selberg (1972, the ending paragraph) introduced the pseud-character into the large sieve and asserted that it yields the zero-density bound of the Linnik type, that is, lacking logarithmic factors. A detailed proof of this was given in Selberg's lectures at Princeton in 1973/1974, a simplified version of which is contained in Montgomery's unpublished manuscript of 1974. Later Montgomery gave its copy to D. Wolke who in turn shared it with Jutila and the present author. Jutila (1977b) and the present author (the above articles) independently worked out an extension of Selberg's idea so that the Linnik phenomenon is also included in the new theory. As a closer comparison should reveal, the difference between Jutila's and our arguments is in that ours starts from the Λ^2-sieve itself and thus makes apparent the reason why it works. Because of this feature, ours generalizes readily to more delicate situations; for instance, it works well in the automorphic context; see Motohashi (2015). The Linnik phenomenon extends to a vast family of functions with Euler products, which, of course, makes the existence of exceptional zeros further less feasible.

§107

The aim of the present section, meant in fact to be an appendix, is to indicate the core part of the historic discovery, made independently by Maynard (2013)

§107   639

and by Tao (unpublished, but see Polymath (2014b)), concerning bounded gaps between prime numbers. Based on previous contributions by Goldston, Pintz, and Yıldırım (2005), both Maynard and Tao considered the asymptotic evaluation of the expression

$$\sum_{\substack{N < n \leq 2N \\ n \equiv c_0 \bmod Y}} \left(\varpi(n+h_1) + \varpi(n+h_2) + \cdots + \varpi(n+h_k) - \varrho \right)$$

$$\times \left(\sum_{\substack{u_1 | (n+h_1),\, u_2 | (n+h_2),\, \ldots,\, u_k | (n+h_k) \\ u_1 u_2 \cdots u_k \leq z}} \lambda(u_1, u_2, \ldots, u_k) \right)^2, \qquad (107.1)$$

as N tends to infinity. Here $\varrho \geq 1$; ϖ is the characteristic function of the set of all prime numbers; $\mathcal{H} = \{h_1, h_2, \ldots, h_k\} \subset \mathbb{N}$ is a sequence of k distinct integers such that $|\{h_j \bmod p : h_j \in \mathcal{H}\}| < p$ for every prime p; as for z, c_0, and Y, we let

$$\log z \approx \log N, \quad Y = \prod_{p \leq X} p, \quad X = \log \log N, \qquad (107.2)$$

and c_0 satisfy $\langle Y, (c_0 + h_1)(c_0 + h_2) \cdots (c_0 + h_k) \rangle = 1$, which is obviously possible. The rôle of Y or rather that of X is to make easy to attain the co-primality requirement among summation variables in various sums to be handled. The prime number theorem implies $Y \ll (\log N)^2$, which can be regarded as negligibly small in our discussion. Further, the rôle of ϱ is to describe the lower bound of the number of primes in $\{n + h_j : h_j \in \mathcal{H}\}$; namely, if (107.1) is positive then there is an $n \in (N, 2N]$ such that at least $\lfloor \varrho \rfloor + 1 (\geq 2)$ prime numbers appear among $n + h_j, j \leq k$. Thus, in particular, if (107.1) diverges to $+\infty$, then

$$\liminf_{v \to +\infty} (p_{v+1} - p_v) < +\infty, \qquad (107.3)$$

which is a capturing of infinitely many bounded gaps among primes, a realization of a desideratum in mathematics.

Although the Maynard–Tao argument is not the first to achieve (107.3), a notably pleasant feature of theirs is in that it is essentially elementary, that is, save for an appeal to the mean prime number theorem, it relies solely on manipulations of sums of very basic arithmetic functions. Nevertheless, the procedure becomes inevitably involved if k is arbitrary. Hence, we begin our discussion with the situation $k = 2$, as it is easy to deal with, and still capable to readily extend to the general case; that is, after studying this section up to (107.44), there should never be any essential difficulty in dealing with the general case with sufficiently large $k \geq 3$.

640   *Distribution of Prime Numbers*

Remark 1 We shall ignore the numerical issue about the lower bound of k, which is discussed by Maynard and Polymath. The tuple $\{p_{\pi(k)+1}, \ldots, p_{\pi(k)+k}\}$ serves anyway for the purpose of $\mathcal{H}$.

1. Thus, let $k = 2$ in (107.1). We shall try to see to what extent the argument (102.17)–(102.33) should be altered. To this end, we shall first compute asymptotically, as N tends to infinity, the ϱ-part of $(107.1)_{k=2}$:

$$\sum_{\substack{N < n \leq 2N \\ n \equiv c_0 \bmod Y}} \sum_{\substack{u_1 | (n+h_1),\, u_2 | (n+h_2), \\ v_1 | (n+h_1),\, v_2 | (n+h_2), \\ u_1 u_2 \leq z,\, v_1 v_2 \leq z}} \lambda(u_1, u_2) \lambda(v_1, v_2), \qquad (107.4)$$

with a fixed tuple $\{h_1, h_2\}$ of different positive integers such that $h_1 = h_2 \bmod 2$, i.e., Note [102.8]. We shall work on the assumption:

$$\begin{aligned} &\lambda(u_1, u_2) = 0 \text{ if any of the following holds} \\ &u_1 u_2 > z,\ \mu(u_1 u_2) = 0,\ \langle u_1 u_2, Y \rangle \neq 1, \end{aligned} \qquad (107.5)$$

which should be compared with (102.17).

Remark 2 Other side conditions than this are not imposed on λ; for instance, we do not suppose $\lambda(1, 1) = 1$ which corresponds to $\lambda(1) = 1$ of (102.16). Thus, (107.4) is regarded just as a quadratic form of the variables $\{\lambda(a, b)\}$, and our interest is in its asymptotic evaluation.

Because of the choice of c_0 and since N is large, we have, in (107.4),

$$\langle n + h_1, n + h_2 \rangle = 1, \quad \langle u_1 v_1, u_2 v_2 \rangle = 1. \qquad (107.6)$$

For, if $p | \langle n + h_1, n + h_2 \rangle$, then $p | (h_2 - h_1)$, so $p < X$, as h_j are supposed to be constants, which contradicts to the restriction $n \equiv c_0 \bmod Y$. Exchanging the order of summation in (107.4), we see that the sum equals

$$(N/Y) S_0 + O\big(\lambda_{\max}^2 (z \log z)^2\big), \qquad (107.7)$$

where $\lambda_{\max} = \sup |\lambda(u_1, u_2)|$ and, by (107.6),

$$S_0 = \sum_{\substack{u_1, u_2, v_1, v_2 \\ \langle u_1 v_1, u_2 v_2 \rangle = 1}} \frac{\lambda(u_1, u_2) \lambda(v_1, v_2)}{[u_1, v_1][u_2, v_2]}; \qquad (107.8)$$

the error term in (107.7) is actually $\leq \lambda_{\max}^2 \big(\sum_{u_1 u_2 \leq z} 1\big)^2$ in absolute value. Here we observe that $\langle u_1, u_2 \rangle = 1$, $\langle v_1, v_2 \rangle = 1$ are implicit in (107.4); so the condition $\langle u_1 v_1, u_2 v_2 \rangle = 1$ in (107.8) is simplified to $\langle u_1, v_2 \rangle \langle u_2, v_1 \rangle = 1$. Thus,

$$S_0 = \sum_{\substack{u_1, u_2, v_1, v_2 \\ \langle u_1, v_2 \rangle \langle u_2, v_1 \rangle = 1}} \frac{\lambda(u_1, u_2) \lambda(v_1, v_2)}{u_1 u_2 v_1 v_2} \sum_{\substack{w_1 | \langle u_1, v_1 \rangle,\, w_2 | \langle u_2, v_2 \rangle}} \varphi(w_1) \varphi(w_2)$$

$$= \sum_{w_1, w_2} \varphi(w_1)\varphi(w_2) \sum_{\substack{u_1, u_2, v_1, v_2 \\ \langle u_1, v_2 \rangle \langle u_2, v_1 \rangle = 1 \\ w_1 | \langle u_1, v_1 \rangle, \ w_2 | \langle u_2, v_2 \rangle}} \frac{\lambda(u_1, u_2)\lambda(v_1, v_2)}{u_1 u_2 v_1 v_2}$$

$$= \sum_{w_1, w_2} \varphi(w_1)\varphi(w_2) \sum_{\substack{u_1, u_2, v_1, v_2 \\ w_1 | \langle u_1, v_1 \rangle, \ w_2 | \langle u_2, v_2 \rangle}} \frac{\lambda(u_1, u_2)\lambda(v_1, v_2)}{u_1 u_2 v_1 v_2}$$

$$\times \sum_{t_1 | \langle u_1, v_2 \rangle, \ t_2 | \langle u_2, v_1 \rangle} \mu(t_1)\mu(t_2) \tag{107.9}$$

$$= \sum_{w_1, w_2, t_1, t_2} \varphi(w_1)\varphi(w_2)\mu(t_1)\mu(t_2) \sum_{\substack{u_1, u_2, v_1, v_2 \\ w_1 | \langle u_1, v_1 \rangle, \ w_2 | \langle u_2, v_2 \rangle \\ t_1 | \langle u_1, v_2 \rangle, t_2 | \langle u_2, v_1 \rangle}} \frac{\lambda(u_1, u_2)\lambda(v_1, v_2)}{u_1 u_2 v_1 v_2}$$

$$= \sum_{w_1, w_2, t_1, t_2} \varphi(w_1)\varphi(w_2)\mu(t_1)\mu(t_2) \sum_{\substack{u_1, u_2 \\ w_1 t_1 | u_1 \\ w_2 t_2 | u_2}} \frac{\lambda(u_1, u_2)}{u_1 u_2} \sum_{\substack{v_1, v_2 \\ w_1 t_2 | v_1 \\ w_2 t_1 | v_2}} \frac{\lambda(v_1, v_2)}{v_1 v_2}.$$

The last line depends on (107.6); for instance, if $w_1 | v_1, t_1 | v_2$, then $\langle w_1, t_1 \rangle = 1$. Hence, we introduce, analogously to (102.22), the change of variables

$$\tau(r_1, r_2) = \varphi(r_1)\varphi(r_2)\mu(r_1)\mu(r_2) \sum_{\substack{g_1, g_2 \\ r_1 | g_1, \ r_2 | g_2}} \frac{\lambda(g_1, g_2)}{g_1 g_2}, \tag{107.10}$$

and rewrite (107.9) as

$$S_0 = \sum_{\substack{w_1, w_2, t_1, t_2 \\ \langle w_1 w_2 t_1 t_2, Y \rangle = 1}} \mu^2(w_1 w_2 t_1 t_2)\mu(t_1)\mu(t_2)$$

$$\times \frac{\tau(w_1 t_1, w_2 t_2)\tau(w_1 t_2, w_2 t_1)}{\varphi(w_1)\varphi(w_2)(\varphi(t_1)\varphi(t_2))^2} \tag{107.11}$$

Applying the Möbius inversion to (107.10), we have

$$\lambda(u_1, u_2) = \mu(u_1)\mu(u_2)u_1 u_2 \sum_{\substack{u_1 | r_1, u_2 | r_2 \\ \langle r_1 r_2, Y \rangle = 1}} \mu^2(r_1 r_2) \frac{\tau(r_1, r_2)}{\varphi(r_1)\varphi(r_2)}, \tag{107.12}$$

which should be compared with (102.23). In particular

$$\lambda_{\max} \ll (\log z)^3 \tau_{\max}. \tag{107.13}$$

We make here a crucial observation on the nature of (107.12): the condition (107.5) is well satisfied by

$$\text{any } \tau, \text{ as far as } \tau(r_1, r_2) = 0 \text{ for } r_1 r_2 > z; \tag{107.14}$$

642 *Distribution of Prime Numbers*

see Note [102.4]. In other words, the arithmetic restrictions, i.e., those other than the geometric $u_1 u_2 \leq z$, which are imposed on the variables $\{\lambda(u_1, u_2)\}$ in (107.5) are now translated into those implicit in the expression (107.12) via (107.10) an invertible change of variables so that the new variables $\{\tau(r_1, r_2)\}$ are to satisfy solely the primitive restriction (107.14). Anyway, one may regard (107.12) with (107.14) as the definition of λ's; this shall we do in the sequel.

Then, (107.9) implies that

$$S_0 = \sum_{\substack{w_1, w_2 \\ \langle w_1 w_2, Y \rangle = 1}} \mu^2(w_1 w_2) \frac{\tau^2(w_1, w_2)}{\varphi(w_1)\varphi(w_2)} + O\left(\tau_{\max}^2 (\log z)^2 / X\right), \qquad (107.15)$$

for we have $\sum_{w \leq z} 1/\varphi(w) \ll \log z$ and $\sum_{t > 1, \langle t, Y \rangle = 1} 1/\varphi(t)^2 \ll X^{-1}$. Further, the factor $\mu^2(w_1 w_2)$ can be replaced by $\mu^2(w_1)\mu^2(w_2)$ since $\mu(w_1 w_2) = 0$ implies that w_1 and w_2 have a common factor $> X$ and such terms can be discarded in much the same way.

We now specify τ by

$$\tau(r_1, r_2) = F\left(\frac{\log r_1}{\log z}, \frac{\log r_2}{\log z}\right),$$
$$F(\xi_1, \xi_2) \text{ being supported on } \{\xi_1 \geq 0, \xi_2 \geq 0, \xi_1 + \xi_2 \leq 1\}. \qquad (107.16)$$

Remark 3 We suppose that F is bounded including its partial derivatives. This does not cause any loss of generality.

We are to evaluate asymptotically the sum

$$\sum_{\substack{w_1, w_2 \\ \langle w_1 w_2, Y \rangle = 1}} \frac{\mu^2(w_1)\mu^2(w_2)}{\varphi(w_1)\varphi(w_2)} F^2\left(\frac{\log w_1}{\log z}, \frac{\log w_2}{\log z}\right). \qquad (107.17)$$

Thus, we need to compute

$$\sum_{\substack{w \\ \langle w, Y \rangle = 1}} \frac{\mu^2(w)}{\varphi(w)} G\left(\frac{\log w}{\log z}\right), \qquad (107.18)$$

where $G(\xi)$ is supported in the interval $[0, 1]$. For this sake, we consider the function

$$\sum_{\substack{n=1 \\ \langle n, Y \rangle = 1}}^{\infty} \frac{\mu^2(n)}{\varphi(n)n^s} = \zeta(s+1)P(s)R(s), \qquad (107.19)$$

where

$$P(s) = \prod_{p|Y}\left(1 - \frac{1}{p^{s+1}}\right),$$

$$R(s) = \prod_{p\nmid Y}\left(1 - \frac{1}{p^{s+1}}\right)\left(1 + \frac{1}{p^{s+1}(1 - 1/p)}\right). \tag{107.20}$$

The function (107.19) satisfies (102.34) with $\alpha = 1$, $A = \varphi(Y)/Y$, $B = \log\log N$, $v = 1$. Hence, via summation by parts, (107.18) is readily seen to be equal to

$$(\log z)\frac{\varphi(Y)}{Y}\int_0^1 G(\xi)d\xi + O\big((\log\log N)^2\big), \tag{107.21}$$

where the error term contains actually the factor $\sup|G'(\xi)|$, but we have omitted it because of an obvious extension of the last remark. We need perhaps to explain our choice of B: it stems from the fact that

$$P(s) - P(0) = \int_0^s P'(\omega)d\omega \ll |s|(\log X)^2, \tag{107.22}$$

since $P'(\omega) = P(\omega)(P'/P)(\omega)$ and $P(\omega) \ll \log X$, $(P'/P)(\omega) \ll \log X$ because of the discussion in Note [13.8]; here the integral is of course taken along the straight line connecting the origin and s. Obviously we may omit the treatment of $R(s)$.

Therefore, applying (107.21) doubly, we conclude that if F is non-trivial, while satisfying the second line of (107.16), then we have, with λ defined by (107.12) and (107.16),

$$(107.4) = \big(1 + o(1)\big)N(\log z)^2\frac{\varphi^2(Y)}{Y^3}\int_0^1\int_0^1 F^2(\xi_1,\xi_2)d\xi_1 d\xi_2, \tag{107.23}$$

provided

$$z \le N^{1/2}(\log N)^{-C} \tag{107.24}$$

with a large constant $C > 0$. We have taken into account (107.2), (107.7), and (107.13); and non-trivial means that this double integral does not vanish.

2. Continuing discussion, we consider

$$\sum_{\substack{N < n \le 2N \\ n \equiv c_0 \bmod Y}} \varpi(n + h_1)\left(\sum_{\substack{u_1|(n+h_1),\, u_2|(n+h_2) \\ u_1 u_2 \le z}} \lambda(u_1, u_2)\right)^2. \tag{107.25}$$

644   *Distribution of Prime Numbers*

Obviously, it makes no difference if the condition $u_1 | (n + h_1)$ is replaced by $u_1 = 1$, since (107.24) is now imposed; so (107.25) equals

$$\frac{1}{\varphi(Y)} (\mathrm{li}(2N) - \mathrm{li}(N))S_1 + O(\lambda_{\max}^2 \widetilde{E}_3(2N, z^2 Y)), \tag{107.26}$$

where

$$S_1 = \sum_{u,v} \frac{\lambda(1,u)\lambda(1,v)}{\varphi([u,v])} \tag{107.27}$$

and

$$\widetilde{E}_l(x,Q) = \sum_{q \leq Q} d_l(q) \max_{\langle a,q \rangle = 1} \left| \pi(x; q, a) - \frac{\mathrm{li}(x)}{\varphi(q)} \right|. \tag{107.28}$$

Here $d_l(q)$ is the l-fold divisor function (16.11).

We then introduce the fundamental hypothesis:

$$\widetilde{E}_l(x, x^\vartheta) \ll x(\log x)^{-L}. \tag{107.29}$$

It is understood that while l is fixed, there should exist a positive $\vartheta < 1$ for each $L > 0$; the implied constant depends only on L and l. However, this is a consequence of another hypothesis:

$$\sum_{q \leq x^\vartheta} \max_{\langle q,a \rangle = 1} \max_{y \leq x} \left| \pi(y; q, a) - \frac{\mathrm{li}\, y}{\varphi(q)} \right| \ll \frac{x}{(\log x)^L}. \tag{107.30}$$

For the part of $\widetilde{E}_l(x, x^\vartheta)$ with $d_l(q) \leq (\log x)^K$ is $\ll x(\log x)^{K-L}$ by (107.30), and the remaining part is $\ll x(\log x)^{-K} \sum_{q \leq x^\vartheta} d_l^2(q)/\varphi(q) \ll x(\log x)^{l^2+1-K}$ by (16.18); the rest of the argument can be omitted.

Using the relation

$$\frac{\varphi(u)\varphi(v)}{\varphi([u,v])} = \sum_{w | \langle u,v \rangle} \psi(w), \quad \psi(w) = \prod_{p | w} (p - 2) \neq 0, \tag{107.31}$$

we have

$$S_1 = \sum_{\substack{w \\ \langle w, Y \rangle = 1}} \mu^2(w) \frac{\tau_1^2(w)}{\psi(w)}, \quad \tau_1(w) = \mu(w)\psi(w) \sum_{w | u} \frac{\lambda(1,u)}{\varphi(u)}. \tag{107.32}$$

Inserting (107.10), with $u_1 = 1, u_2 = u$, into the last sum, we get, after an arrangement,

$$\tau_1(w) = w\mu(w)\psi(w) \sum_{\substack{\langle r_1 r_2, Y\rangle=1 \\ w|r_2}} \mu^2(r_1 r_2)\frac{\tau(r_1, r_2)\mu(r_2)}{\varphi(r_1)\varphi^2(r_2)}$$

$$= \frac{w\psi(w)}{\varphi^2(w)} \sum_{\langle r_1, Y\rangle=1} \mu^2(r_1 w)\frac{\tau(r_1, w)}{\varphi(r_1)} + O(\tau_{\max}(\log z)/X),$$

(107.33)

where we have $\langle w, Y\rangle = 1$ because of (107.32). This error term is due to the fact that if $r_2 \neq w$, then $r_2/w > X$. Further, on noting that, since $\langle w, Y\rangle = 1$ as just has been mentioned,

$$\frac{w\psi(w)}{\varphi^2(w)} = \prod_{p|w}\left(1 - \frac{1}{(p-1)^2}\right) = 1 + O(1/X),$$

(107.34)

we have, in (107.32),

$$\tau_1(w) = \sum_{\langle r_1, wY\rangle=1} \mu^2(r_1)\frac{\tau(r_1, w)}{\varphi(r_1)} + O(\tau_{\max}(\log z)/X),$$

(107.35)

to which (107.16) is to be inserted.

Thus, we need to asymptotically evaluate the sum

$$\sum_{\langle r_1, wY\rangle=1} \frac{\mu^2(r_1)}{\varphi(r_1)}F\left(\frac{\log r_1}{\log z}, \frac{\log w}{\log z}\right).$$

(107.36)

To this end, we consider the function

$$\sum_{\langle r_1, wY\rangle=1} \frac{\mu^2(r_1)}{\varphi(r_1)r_1^s} = \zeta(s+1)P_w(s)R_w(s),$$

(107.37)

where

$$P_w(s) = P(s)W(s), \quad W(s) = \prod_{p|w}\left(1 - \frac{1}{p^{s+1}}\right)$$

(107.38)

with $P(s)$ as in (107.20), and $R_w(s)$ is the same as $R(s)$ there but with Y being replaced by wY. The function (107.37) satisfies (102.34) with $\alpha = 1$, $A = \varphi(wY)/wY$, $B = (\log\log N)^3$, $\nu = 1$. As for B, we imitate (107.22):

$$P_w(s) - P_w(0) = \int_0^s \left(P'(\eta)W(\eta) + P(\eta)W'(\eta)\right)d\eta.$$

(107.39)

The argument of Note [18.6] yields that

$$\sum_{p|w}\frac{1}{p} \leq \log\log\log w + O(1), \quad \sum_{p|w}\frac{\log p}{p} \leq \log\log w + O(1);$$

(107.40)

646　　　　　　　　　　*Distribution of Prime Numbers*

that is, $W(\eta) \ll \log\log w$, $W'(\eta) \ll (\log\log w)^2$. Hence,

$$\tau_1(w) = (\log z)\frac{\varphi(wY)}{wY}\int_0^1 F(\xi_1, \log w/\log z)d\xi_1 + O((\log\log N)^4).$$

$$(107.41)$$

Inserting this into the expression (107.32) for S_1, we see that

$$S_1 = (1 + o(1))(\log z)^3 \left(\frac{\varphi(Y)}{Y}\right)^3 \int_0^1 \left(\int_0^1 F(\xi_1,\xi_2)d\xi_1\right)^2 d\xi_2. \quad (107.42)$$

We have skipped the necessary discussion on the function

$$\sum_{\substack{\langle w,Y\rangle=1}} \frac{\mu^2(w)}{\psi(w)w^s}\left(\frac{\varphi(w)}{w}\right)^2, \qquad (107.43)$$

as it is obviously analogous to (107.19). Hence, we are led to

$$(107.25) = (1 + o(1))\frac{N(\log z)^3}{\log N}\frac{\varphi^2(Y)}{Y^3}\int_0^1\left(\int_0^1 F(\xi_1,\xi_2)d\xi_1\right)^2 d\xi_2$$
$$+ O\left((\log z)^6\widetilde{E}_3(2N, z^2Y)\right),$$

$$(107.44)$$

where (107.13) has been taken into account.

3. Combining (107.23) and (107.44), we find that under (107.30) it holds that

$$(107.1)_{k=2} = (1 + o(1))N(\log z)^2\frac{\varphi^2(Y)}{Y^3}J_0^{(2)}\left(\tfrac{1}{2}\vartheta\mathcal{F}_2 - \varrho\right),$$
$$z = N^{\vartheta/2}/(\log N)^2, \quad \mathcal{F}_2 = (J_1^{(2)} + J_2^{(2)})/J_0^{(2)}.$$

$$(107.45)$$

Here $J_0^{(2)}$ is the integral in (107.23); $J_1^{(2)}$ is the one in (107.42), to which $J_2^{(2)}$ is analogous. The function F is as in (107.16), while Remark 3 is observed, and further the condition $J_0^{(2)} \neq 0$ is imposed.

However, the assertion (107.45) is of no avail for our purpose towards the establishment of (107.3): We have $\vartheta < 1$, and $\mathcal{F}_2 \leq 2$ as the Cauchy–Schwartz inequality implies; thus $(107.1)_{k=2}$ tends to $-\infty$.

4. Hence, in view of the experience of Goldston et al (2005), we consider (107.1) with a sufficiently large k: The relevant arguments developed so far are generalized without any new difficulty at all. We are led to a generalization of (107.10)–(107.16):

$$\lambda(u_1,\dots,u_k) = \mu(u_1)\cdots\mu(u_k)u_1\cdots u_k$$
$$\times \sum_{\substack{u_1|w_1,\dots,u_k|w_k \\ \langle w_1\cdots w_k,Y\rangle=1}} \frac{\mu^2(w_1\cdots w_2)}{\varphi(w_1)\cdots\varphi(w_k)}F\left(\frac{\log w_1}{\log z},\dots,\frac{\log w_k}{\log z}\right), \qquad (107.46)$$

where

$$F(\xi_1, \ldots, \xi_k) \text{ is supported}$$
$$\text{on } \{\xi_1, \ldots, \xi_k \geq 0, \xi_1 + \cdots + \xi_k \leq 1\}, \tag{107.47}$$

to which Remark 3 above is extended. With this, we evaluate

$$\sum_{\substack{w_1, \ldots, w_k \\ \langle w_1 \cdots w_k, Y \rangle = 1}} \frac{\mu^2(w_1 \cdots w_k)}{\varphi(w_1) \cdots \varphi(w_k)} F^2\left(\frac{\log w_1}{\log z}, \ldots, \frac{\log w_k}{\log z}\right). \tag{107.48}$$

Thus, via a multiple application of (107.18)–(107.22), we see that the ϱ-part of (107.1) with (107.46)–(107.47) equals

$$(1 + o(1))N(\log z)^k \frac{\varphi^k(Y)}{Y^{k+1}} J_0^{(k)},$$
$$J_0^{(k)} = \int_0^1 \cdots \int_0^1 F^2(\xi_1, \ldots, \xi_k) d\xi_1 \cdots d\xi_k \tag{107.49}$$

provided (107.24). Also, the ϖ $(n + h_v)$-part of (107.1) equals

$$(1 + o(1)) \frac{N(\log z)^{k+1}}{\log N} \frac{\varphi^k(Y)}{Y^{k+1}} J_v^{(k)}$$
$$+ O\big((\log z)^{2k+2} \widetilde{E}_{3(k-1)}(2N, z^2 Y)\big), \tag{107.50}$$

where

$$J_v^{(k)} = \int_0^1 \cdots \int_0^1 \left(\int_0^1 F(\xi_1, \ldots, \xi_k) d\xi_v\right)^2 d\xi_1 \cdots d\xi_{v-1} d\xi_{v+1} \cdots d\xi_k. \tag{107.51}$$

We omit the details of the derivation of (107.49)–(107.51), save for the appearance $\widetilde{E}_{3(k-1)}(2N, z^2 Y)$: This is due to the counting of the moduli

$$[u_1, v_1] \cdots [u_{v-1}, v_{v-1}][u_{v+1}, v_{v+1}] \cdots [u_k, v_k], \tag{107.52}$$

which induces the $(k - 1)$-fold convolution of d_3.

Thus, we find, under the specification (107.46)–(107.47) as well as the hypothesis (107.29) or rather (107.30), that

$$(107.1) = (1 + o(1))N(\log z)^k \frac{\varphi^k(Y)}{Y^{k+1}} J_0^{(k)}\left(\tfrac{1}{2}\vartheta \mathcal{F}_k - \varrho\right),$$
$$z = N^{\vartheta/2}/(\log N)^2, \quad \mathcal{F}_k = \sum_{v=1}^{k} J_v^{(k)}/J_0^{(k)}, \tag{107.53}$$

with $J_0^{(k)} \neq 0$. In other words, if there exists a non-trivial F such that

$$\tfrac{1}{2}\vartheta \mathcal{F}_k > \varrho, \tag{107.54}$$

then the assertion (107.3) should follow, provided (107.30).

5. We shall demonstrate that for each sufficiently large k

there exists an F such that
$$\mathcal{F}_k > \log k - 4 \log \log k - 1.$$
(107.55)

To this end, we first make an intuitive observation on the nature of $\lambda(u_1, \ldots, u_k)$ in (107.1). Thus, the variables $u_1, \ldots, u_k$ are arithmetically independent of each other but only geometrically constrained as $u_1 \cdots u_k \le z$. In the case $k = 2$, this independence is transferred, via (107.10), to the variables of the function F introduced at (107.16). Of course the same can be said about the F introduced at (107.46). Hence, it should be reasonable to specify F as follows:

$$F(\xi_1, \ldots, \xi_k) = \begin{cases} k^{k/2} f(k\xi_1) \cdots f(k\xi_k) & \text{if } \xi_1 + \cdots + \xi_k \le 1, \\ 0 & \text{otherwise,} \end{cases}$$
(107.56)

where

$$f(\eta) \text{ is supported on } [0, \ell] \text{ with an } \ell < k,$$
(107.57)

$$\text{and normalized as } \int_0^\ell f^2(\eta) d\eta = 1.$$
(107.58)

The actual value of ℓ is to be fixed later in terms of k.

The condition (107.58) implies that

$$J_0^{(k)} \le 1.$$
(107.59)

On the other hand, it is readily seen that $J_\nu^{(k)} = J_k^{(k)}$, and

$$J_k^{(k)} \ge k^{-1} \left(\int \cdots \int_{\eta_1 + \cdots + \eta_{k-1} \le k - \ell} (f(\eta_1) \cdots f(\eta_{k-1}))^2 d\eta_1 \cdots d\eta_{k-1} \right)$$
$$\times \left(\int_0^\ell f(\eta) d\eta \right)^2$$
(107.60)
$$= k^{-1} H_k \phi^2,$$

say. We have, with (107.57)–(107.59),

$$\mathcal{F}_k \ge H_k \phi^2.$$
(107.61)

We need to have a good lower bound for H_k and ϕ. To this end, we shall appeal to the probability theory and to the calculus of variations, respectively.

We regard H_k as the probability of the event $\eta_1 + \cdots + \eta_{k-1} \le k - \ell$ in the space $[0, \ell]^{k-1}$ equipped with the density $(f(\eta_1) \cdots f(\eta_{k-1}))^2$. The mean value of the random variable $\eta_1 + \cdots + \eta_{k-1}$ is equal to $(k-1)\omega_1$, where

$$\omega_1 = \int_0^\ell \eta f^2(\eta) d\eta, \tag{107.62}$$

and the corresponding variance is $(k-1)(\omega_2 - \omega_1^2)$, where

$$\omega_2 = \int_0^\ell \eta^2 f^2(\eta) d\eta \leq \ell\omega_1. \tag{107.63}$$

Since we naturally expect that the F defined by (107.56)–(107.58) is such that H_k is close to 1, the event $\eta_1 + \cdots + \eta_{k-1} \leq k - \ell$ should well cover the event $\eta_1 + \cdots + \eta_{k-1} \leq (k-1)\omega_1$; that is, we ought to suppose that

$$(k-1)\omega_1 < k - \ell. \tag{107.64}$$

With this, the inequality of Chebyshev gives

$$\begin{aligned} H_k &\geq 1 - \frac{(k-1)(\omega_2 - \omega_1^2)}{(k - \ell - (k-1)\omega_1)^2} \\ &\geq 1 - \frac{(k-1)\ell\omega_1}{(k - \ell - (k-1)\omega_1)^2}, \end{aligned} \tag{107.65}$$

where the second line depends on (107.63).

We next apply the calculus of variations to optimize ϕ under the constraint that (107.58) holds while ω_1 is fixed in (107.62). The relevant Euler–Lagrange equation with two Lagrange multipliers yields immediately that the extremal f is given by

$$f(\eta) = \frac{V}{1 + U\eta} \tag{107.66}$$

with constants U, V. Then, we have

$$\begin{aligned} \phi &= \frac{V}{U} \log(1 + U\ell), \quad 1 = \frac{V^2\ell}{1 + U\ell}, \\ \omega_1 &= \left(\frac{V}{U}\right)^2 \left(\log(1 + U\ell) - \frac{U\ell}{1 + U\ell}\right). \end{aligned} \tag{107.67}$$

Here we take, with a sufficiently large k,

$$\ell = \frac{k}{(\log k)^3}, \quad U = \log k, \quad V^2 = (\log k)\left(1 + \frac{(\log k)^2}{k}\right) \tag{107.68}$$

650 *Distribution of Prime Numbers*

so that

$$\phi^2 = \log k - 4\log\log k + 4\frac{(\log\log k)^2}{\log k} + O\left(\frac{(\log k)^3}{k}\right),$$

$$\omega_1 = 1 - 2\frac{\log\log k}{\log k} - \frac{1}{\log k} + O\left(\frac{(\log k)^2}{k}\right) < \frac{k}{k-1}\left(1 - \frac{\ell}{k}\right);$$

(107.69)

the lower line implies that (107.64) is fulfilled. Hence, we have

$$H_k > 1 - \frac{1}{3(\log k)(\log\log k)^2},$$

(107.70)

and

$$H_k\phi^2 > \log k - 4\log\log k + O\left(\frac{1}{(\log\log k)^2}\right),$$

(107.71)

which ends the proof of (107.55).

6. We have found, far more than (107.3), that for each $m \geq 1$

$$\liminf_{\nu\to+\infty}(p_{\nu+m} - p_\nu) < +\infty,$$

(107.72)

since we are now able to find a k such that $\frac{1}{2}\vartheta\mathcal{F}_k > m$ with any combination of the constants $\{\vartheta, m\}$. Therefore, the mean prime number theorem of Rényi (1948) mentioned in Note [105.2] already suffices to capture infinitely often bounded gaps between primes to such a wide extent as (107.72).

Notes

[107.1] While dealing with the twin prime conjecture, Selberg (1991, p.245; in fact, apparently in early 1950s) pondered on the expression

$$\sum_{N<n\leq 2N}\left(d(n) + d(n+2) - \varrho\right)\left(\sum_{\substack{u_1|n,\,u_2|(n+2)\\u_1u_2\leq z}}\lambda(u_1, u_2)\right)^2,$$

where d is the divisor function. This is analogous to (107.1)$_{k=2}$, though the divergence to $-\infty$ is aimed; in fact, the ϱ-term is not included in his formulation, but the formula following (23.38) there indicates well that he had the above in his mind. The implied modification of the λ-weight had remained just as an important suggestion left by the creator of the Λ^2-sieve until Maynard and Tao ventured independently upon the investigation of (107.1).

[107.2] The use of the probability theory in the last part of the fifth step in the above discussion is due to Tao (Polymath (2014b)). We adopted it in order to add some theoretical flavor. The corresponding argument of

$$\S107 \qquad\qquad 651$$

Maynard (2013, §8) is essentially the same; he applied the procedure which is used to prove Chebyshev's inequality. Our end result (107.71) is not the best that could be derived by our argument; but it suffices for our purpose. As for the case of small k, the analysis is, of course, focused on attaining numerical precision as good as possible and inevitably becomes extremely involved. In such a way Maynard discovered $\liminf (p_{\nu+1} - p_\nu) \le 600$, and Polymath (2014b) squeezed it to achieve the upper bound 246.

[107.3] A brief historical account should be in order.

(a) The pivotal event in this field was achieved by GPY (Goldston–Pinz–Yıldırım (2005))): they established

$$(*) \qquad \liminf_{\nu \to +\infty} \frac{p_{\nu+1} - p_\nu}{\log p_\nu} = 0$$

by investigating, in effect, the expression

$$\sum_{N < n \le 2N} \left(\sum_{h_j \in \mathcal{H}} \varpi(n + h_j) - 1 \right) \left(\sum_{\substack{u \mid (n+h_1)\cdots(n+h_k) \\ u \le z}} \lambda(u; k + l) \right)^2.$$

Here $\mathcal{H}$ is the same as in (107.1), and $\lambda(u; \nu) = \mu(u)((\log z/u)/\log z)^\nu$. This sieve weight λ is essentially equal to the optimal one in the Λ^2-sieve (102.16) applied to the situation $|\Omega(p)| = \nu$ for each sufficiently large p, a ν-dimensional sieve problem, as can be seen via Theorem 137 and (103.31). A decisive innovation of GPY is in the use of $\lambda(u; k + l)$ instead of $\lambda(u; k)$ while the sum itself belongs to a k-dimensional sieve situation. The attachment of the taper $((\log z/u)/\log z)^l$ induces a surprising effect; an optimal choice of l in the relation with k yielded $(*)$.

(b) However, the most basic fact on which $(*)$ stands is the Bombieri–Vinogradov mean prime number theorem, i.e., $\vartheta = \frac{1}{2} - \varepsilon$ in (107.30). Remarkable is that if (107.30) holds with a $\vartheta > \frac{1}{2}$, then the argument of GPY should yield (107.3). Hence, such an improvement of the Bombieri–Vinogradov theorem became an issue. In this circumstance, MP (Motohashi–Pintz (2006)) made the following observation: According to a sophisticated formulation due to Polymath (2014a), what MP stated is essentially the same as that the mean prime number theorem

$$(**) \qquad \sum_{\substack{q \le x^{1/2+\theta} \\ p \mid q \Rightarrow p \le x^\omega}} \left| \pi(x; a_q, q) - \frac{\mathrm{li}(x)}{\varphi(q)} \right| \ll x(\log x)^{-L}$$

should yield (107.3). Here, $a_p \bmod p$ is fixed at each p, and $a_q \bmod q$ is its multiplicative extension; and it is conjectured that $(**)$ should hold uniformly with respect to the set $\{a_p \bmod p\}$, while $L > 0$ is arbitrary, and $\theta, \omega > 0$ are absolute constants. This means that MP found that the GPY sieve can be smoothed; that is, even if the sieve weight $\lambda(u; k + l)$ is restricted to the set of ω-smooth integers u, the GPY sieve still works fine, and thus $(**)$ should yield (107.3).

(c) Zhang (2013) published a proof of a variant of $(**)$ and became the first to detect infinitely often bounded gaps between primes. His argument depends on the RH concerning algebraic varieties over finite fields, and is extremely involved. A complete proof of $(**)$ itself has been achieved by Polymath (2014a).

(d) However, before the present author had finished thoroughly studying either article, an unexpected development took place: Maynard and Tao independently discovered that $(**)$ could be totally discarded, as indicated above.

A detailed account of (a)–(d), together with rich historical comments, can be found in Broughan (2021).

Bibliography

Abel N. H. (1828): Aufgabe. *J. reine angew Math.*, **3**, 212.

(1829): Mémoire sur une classe particulière d'équations résolubles algébriquement. *J. reine angew Math.*, **4**, 131–156.

Agărgün A. G. and Özkan E. M. (2001): A historical survey of the fundamental theorem of arithmetic. *Historia Math.*, **28**, 207–214.

Agrawal M., Kayal N., and Saxena N. (2004): PRIMES is in P. *Annals of Math.*, **160**, 781–793.

Ahmes (ca 1650 BCE): The Rhind mathematical papyrus (British Museum 10057 and 10058). Free translation and commentary by A.B. Chace (1927). *Math. Ass. America*, **1**.

Alford W. R., Granville A., and Pomerance C. (1994): There are infinitely many Carmichael numbers. *Annals of Math.*, **140**, 703–722.

al-Khwarizmi (ca 820a): *The algebra*. Translation by F. Rosen from an Arabic edition. The Oriental Translation Fund, London, 1831.

(ca 820b): Thus spake al-Khwārismī. A translation of the text of Cambridge University Library ms. Ii.vi.5. Translation by J. N. Clossrey and A. S. Henry. *Historia Math.*, **17** (1990), 103–131.

Ankeny N. C. (1952): The least quadratic non residue. *Annals of Math.*, **55**, 65–72.

Anonymous (1864): On primes and proper primes. *The Oxford, Cambridge, and Dublin Messenger of Math.*, **2**, 1–6.

Arcozzi N. (2012): Beltrami's models of non-Euclidean geometry. In: *Mathematics in Bolonia 1861–1960*, Birkhäuser, Basel, pp.1–30.

Arndt A. F. (1846): Disquisitiones de residuis cujusvis ordinis. *J. reine angew. Math.*, **31**, 333–342.

(1859a): Auflösung einer Aufgabe in der Composition der quadratischen Formen. *J. reine angew Math.*, **56**, 64–71.

(1859b): Ueber die Anzahl der Genera der quadratischen Formen. *J. reine angew Math.*, **56**, 72–78.

(1859c): Einfacher Beweis für die Irreductibilität einer Gleichung in der Kreistheilung. *J. reine angew Math.*, **56**, 178–181.

Artin, E. (1971): *Galois theory*. University of Notre Dame Press, London.

Bibliography

Aryabhata I (499 CE): *The āryabhaṭīya. An ancient Indian work on mathematics and astronomy.* Translated with notes by W. E. Clark, The University of Chicago Press, Chicago, 1930.

Atkinson F. V. (1949). The mean-value of the Riemann zeta function. *Acta Math.*, **81**, 353–376.

Ayyangar A. A. K. (1941): Theory of the nearest square continued fraction. *Jo. of the Mysore Univ.*, **1**, 21–32, 97–117.

Bach E. (1990): Explicit bounds for primality testing and related problems. *Math. Comp.*, **55**, 355–380.

Bachmann P. (1872): *Die Lehre von der Kreistheilung und ihre Beziehungen zur Zahlentheorie.* B.G. Teubner, Leipzig.

(1894): *Die analytische Zahlentheorie.* B.G. Teubner, Leipzig.

(1898): *Die Arithmetik der quadratischen Formen.* Erste Abt. B.G. Teubner, Leipzig; Zweite Abt. 1923.

(1902): *Niedere Zahlentheorie.* Erster Teil. B.G. Teubner, Leipzig; Zweiter Teil. 1910.

(1911): *Über Gauß' zahlentheoretische Arbeiten.* Reprinted in: *Gauss Werke* X-2, pp.1–69.

Baker R. C., Harman G., and Pintz J. (2001): The difference between consecutive primes. II. *Proc. London Math. Soc.*, **83**, 532–562.

Balasubramanian R. and Ramachandra K. (1989): A lemma in complex function theory I. *Hardy–Ramanujan J.*, **12**, 1–5.

Balazard E., Naimi M., and Pétermann Y.-F. S. (2008): Étude d'une somme arithmétique multiple liée à la fonction de Möbius. *Acta Arith.*, **132**, 245–298.

Barban M. B. (1966): The large sieve method and its application to number theory. *Uspehi Mat. Nauk*, **21**, 51–102. (Russian)

Barenco A. C., Bennett H., Cleve R., Di Vincenzo D. P., Margolus N., Shor P., Sleator T., Smolin J. A., and Weinfurter H. (1995): Elementary gates for quantum computation. *Phys. Rev. A*, **31** (5), 3457–3467.

Bateman P. T. (1951): On the representations of a number as the sum of three squares. *Trans. Amer. Math. Soc.*, **71**, 70–101.

Baumgart O. (1885): *Über das quadratische Reciprocitätsgesetz.* B.G. Teubner, Leipzig.

Beeger N. G. W. H. (1922): On a new case of the congruence $2^{p-1} \equiv 1 \bmod p^2$. *Messenger of mathematics*, **51**, 149–150.

Beltrami E. (1868): Teoria fondamentale degli spazii di curvatura costante. *Ann. Mat. Pura App.*, **2**, 232–255. (Translated: *Ann. Sci. de l'É.N.S.*, **6** (1869), 347–375)

Bennett C. H. (1973): Logical reversibility of computation. *IBM J. Res. Develop.*, **17**, 525–532.

Berndt B. C. and Evans R. J. (1981): The determination of Gauss Sums. *Bull. Amer. Math. Soc.*, **5**, 107–129. Corrigendum. **7** (1982), 441.

Bhargava M. (2001): Higher Composition Laws. Ph.D. Thesis, Princeton University.

Bhaskara II (1150): Bija Ganita. In: *Algebra of the Hindus from a Persian manuscript of 1634*, translated by E. Strachey (1813). The Honourable East Indian Company, London; Lilavati, Vijaganita. In: *Algebra with arithmetic and mensuration from the Sanscrit.* Translated by H. T. Colebrooke (1817). J. Murray, London, pp.1–276.

Bibliography

Binet J. P. M. (1831): Sur la résolution des équations indéterminées du premier degrée en nombres entirers. *J. École Polyt.*, **13**, 289–296.

Bohr H. und Landau E. (1914a): Ein Satz über Dirichletsche Reihen mit Anwendung auf die ζ-Funktion und die L-Funktionen. *Rend. di Palermo*, **37**, 269–272.

(1914b): Sur les zèros de la fonction $\zeta(s)$ de Riemann. *Comptes rendus*, **158**, 106–110.

Bombelli R. (1579): *L'Algebra*. G. Rossi, Bologna.

Bombieri E. (1965). On the large sieve. *Mathematika*, **12**, 201–225.

(1971): A note on the large sieve. *Acta arith.*, **18**, 401–404.

(1987): *Le grand crible dan la théorie analytique des nombres*. Seconde édition. Astérisque, **18**, Paris.

Bombieri E., Friedlander J. B., and Iwaniec H. (1986): Primes in arithmetic progressions to large moduli. *Acta Math.*, **156**, 203–251; Part II. *Math. Ann.*, **277** (1987), 361–393; Part III. *J. Amer. Math. Soc.*, **2** (1989), 215–224.

Bombieri E. and Iwaniec H. (1986): On the order of $\zeta\left(\frac{1}{2} + it\right)$. *Ann. Scuola Norm. Sup. Pisa Cl. Sci.*, **13**, 449–486.

Borwein J., Bradley D., and Crandall R. (2000): Computational strategies for the Riemann zeta function. *J. Comp. App. Math.*, **121**, 247–296.

Botts T. (1967): A chain reaction process in number theory. *Math. Magazine*, **40**, 55–65.

Bouniakowsky V. (1857): Sur les diviseurs numériqes invariables des fonctions rationnelles entières. *Mém. Acad. Impér. Sci. Saint-Pétersbourg, sixème série, Sci. Math. Phys.*, **VI**, 306–329.

Bourgain J. (2017): Decoupling, exponential sums and the Riemann zeta function. *J. Amer. Math. Soc.*, **30**, 205–224.

Brahmagupta (628): Ganita, Cuttaca. In: *Algebra with arithmetic and mensuration from the Sanscrit*, translated by H. T. Colebrooke (1817). J. Murray, London, pp.277–378.

Brillhart J. (1972): Note on representing a prime as a sum of two squares. *Math. Computation*, **26**, 1011–1013.

Broughan K. (2021): *Bounded gaps between primes. The epic breakthroughs of the early twenty-first century*. Cambridge Univ. Press, Cambridge.

Bruggeman R. W. (1978): Fourier coefficients of cusp forms. *Invent. math.*, **45**, 1–18.

Brun V. (1915): Über das Goldbasche Gesetz und die Anzahl der Primzahlpaare. *Arkiv for Math. og Natur.*, **34**, 8–19.

(1919): La série $\frac{1}{5} + \frac{1}{7} + \frac{1}{11} + \frac{1}{13} + \frac{1}{17} + \frac{1}{19} + \frac{1}{29} + \frac{1}{31} + \frac{1}{41} + \frac{1}{43} + \frac{1}{59} + \frac{1}{61} + \cdots$ oú les dénominateurs sont "nombres premiers jumeaux" est convergente ou finie. *Bull. Sci. Math.*, (2) **43**, 100–104, 124–128.

(1920). Le crible d'Eratosthène et le théorème de Goldbach. *Videnskaps. Skr., Mat. Natur. Kl. Kristiana*, No. 3.

(1925): Untersuchungen über das Siebverfahren des Eratosthenes. *Jahresbericht der Deutschen Math. Verein.*, **33**, 81–96.

Buchstab A. A. (1937): Asymptotische Abschätzung einer allgemeinen zahlentheoretischen Funktion. *Mat. Sbornik*, (2) **44**, 1239–1246. (Russian with German résumé)

Caldwell C. K., Reddick A., Xiong Y., and Keller W. (2012): The history of the primality of one: a selection of sources. *J. Integer Sequences*, **15**, article 12.9.8.

Cardano G. (1545): *Artis magnæ, sive de regulis algebraicis, liber unus*. Petreius, Nürnberg. (*Ars Magna*. Dover Publ., New York, 1993.)

Bibliography

Carlson F. (1921): Über die Nullstellen der Dirichletschen Reihen und der Riemannschen ζ-Funktion. *Arkiv for Mat. Ast. och Fysik.*, **15**, No. 20.

Carmichael R. D. (1907): On Euler's ϕ-function. *Bull. Amer. Math. Soc.*, **13**, 241–243; Note on Euler's ϕ-function. **28** (1922), 109–110.

(1910): Note on a new number theory function. *Bull. Amer. Math. Soc.*, **16**, 232–238.

Cataldi P. A. (1613): *Trattato del modo brevissimo di trouare la radice quadra delli numeri.* Bartolomeo Cochi, Bologna.

Cauchy A.-L. (1816): Démonstration d'un théorème curieux sur les nombres. *Bull. Sci. Soc. Philom. Paris*, année 1816, 133–135.

(1825): *Mémoire sur les intégrales définies, prises entre des limites imaginaires.* Chez de Bure Frères, Libraires du doi et de la bibliothèque du roi, Paris.

(1829): *Exercices de mathématiques. Quartiéme année.* Chez de Bure Frère, Libraires du doi et de la bibliothèque du roi, Paris.

(1840): Méthode simple et nouvelle pour la détermination complète des sommes alternees formées avec les racines primitives des équations binomes. *J. math. pures et appliq.*, **5**, 154–168.

Charves (L. Charve) (1877): Démonstration de la périodicité des fractions continues, engendrées par les racines d'une équation du deuxème degré. *Bull. Sci. Math. Astron.*, (2) **1**, 41–43.

Chebyshev P. L. (1848): *Teoria sravneny.* St. Petersburg Univ. (Russian): German translation. *Theorie der Congruenzen.* Mayer & Müller, Berlin 1889.

(1851): Sur la fonction qui détemine la totalité des nombres premiers inférieur à une limite donnée. *Mémoire présentés à la Acad. Impériale de St. Pétersbourg par divers savants*, **VI**, 141–157. Also: *J. math. pures et appliq.*, **XVII**, 1852, 341–365. (*Œuvres* I, pp.29–48)

(1854): Sur nombres premiers. *Mémoire présentés à la Acad. Impériale de St. Pétersbourg par divers savants*, **VII**, 17–33. Also: *J. math. pures et appliq.*, **XVII**, 1852, 366–390. (*Œuvres* I, pp.51–70)

Cipolla M. (1903): Un metodo per la risolutione della congruenza di secondo grado. *Rendiconto dell' Accademia delle Scienze Fisiche e Matematiche, Napoli*, Ser. 3, **IX**, 154–163.

(1907): Sulla risoluzione apiristica delle congruenze binomie secondo un modulo primo. *Math. Ann.*, **63**, 54–61.

Chen J. R. (1973): On the representation of a larger even integer as the sum of a prime and the product of at most two primes. *Sci. Sinica*, **16**, 157–176.

Conrey B. (1989): More than two fifths of the zeros of the Riemann zeta function are on the critical line. *J. reine angew. Math.*, **399**, 1–26.

Coppersmith D. (1994): An approximate Fourier transform useful in quantum factoring. *IBM Research Report*, RC 19642 (07/12/94), Mathematics. IBM Research Division, T. J. Watson Research Center, Yorktown Heights, New York.

Cornacchia G. (1908): Su di un metodo per la risoluzione in numeri interi dell'equazione $\sum_{h=0}^{n} C_h x^{n-h} y^h = P$. *Giornale di matematiche di Battaglini*, **46**, 33–90.

Crandall R. and Pomerance C. (2005): *Prime numbers. A computational perspective.* Second edition. Springer-Verlag, New York.

Bibliography 657

Datta B. and Singh A. N. (1935): *History of Hindu mathematics. A source book.* Part I. Motilal Banarasidas, Lahore, and Part II. 1938; Single volume edition. Asia Publishing House, Bombay, 1962.

Dedekind R. (1857a): Abriss einer Theorie der höhern Congruenzen in Bezug auf einen reellen Primzahl-Modulus. *J. reine angew. Math.*, **54**, 1–26.

(1857b): Beweis für die Irreductibilität der Kreistheilungs-Gleichungen. *J. reine angew. Math.*, **54**, 27–30.

(1873): Die Lehre von der Kreistheilung und ihre Beziehungen zur Zahlentheorie. Akad. Vorlesungen von Dr. Paul Bachmann. *Literaturzeitung. Zeits. für Math. Physik*, **18**, 14–24.

(1876): Sur la théorie des nombres entiers algébrique. *Bull. Sci. Math. Astron.*, (1) **11**, 278–288; (2) **1** (1877a), 17–41, 69–92, 144–164, 207–248.

(1877b): Schreiben an Herrn Borchardt über die Theorie der elliptischen Modul-Functionen. *J. reine angew. Math.*, **83**, 265–292.

(1877c): Über die Anzahl der Ideal-Klassen in den verschiedenen Ordnungen eines endlichen Zahlkörpers. In: *Festschrift der Tech. Hochsch. Braunschweig zur Säkularfeier des Gebrustages von C. F. Gauss*, Braunschweig, pp.1–51.

(1892): *Stetigkeit und irrationale Zahlen*. Friedrich Biewerg, Braunschweig. (Continuity and irrational numbers. In: *Essays on the theory of numbers*, pp.1–13, The Open Court Publ., Chicago 1901)

de la Vallée Poussin C.-J. (1896): Recherches analytiques sur la théorie des nombres premiers. *Ann. Soc. Sci. Bruxelles*, **20**, 183–256; Reprinted by *Acad. Royal de Belgique, Bruxelles*, 1897.

(1900): Sur la fonction $\zeta(s)$ de Riemann et le nombres des nombres premiers inférieurs à une limite donnée. *Mém. Courronnés et Autres Mém. Publ. Acad. Roy. Sci., des Lettres Beaux-Arts de Belgique*, **59**, Nr.1.

Desmarest E. (1852): *Théorie des nombres. Traité de l'analyse indéterminée du second degré a deux inconnues*. Librairie de L. Hachette, Paris.

Dickson L. E. (1911): Notes on the theory of numbers. *Amer. Math. Monthly*, **18**, 109–111.

(1919): *History of number theory*. I. Carnegie Inst., Washington; II. 1920; III. 1923.

(1929): *Introduction to the theory of numbers*. The Univ. Chicago Press, Chicago.

Diffie W. and Hellman M. E. (1976): New directions in cryptography. *IEEE Trans. Inf. Theory*, **22**, 644–654.

Diophantus (ca 250): *Arithmetica*.

(1) C.G. Bachet de Méziriac (1621): *Diophanti Alexandrini arithmeticorum libri sex et de numeris multangulis liber unus*. Lutetia Parisiorum.

(2) T. L. Heath (1910): *Diophantus of Alexandria. A study in the history of Greek algebra*. Second edition. Cambridge Univ. Press, Cambridge.

Dirichlet P. G. L. (1828): Démonstrations nouvelles de quelques théorèmes relatifs aux nombres. *J. reine angew Math.*, **3**, 390–393. (*Werke* I, pp.99–104)

(1834): Einige neue Sätze über unbestimmte Gleichungen. *Abh. Königl. Preuss Akad. Wissens.*, 649–664. (*Werke* I, pp.219–236)

(1835): Über eine neue Anwendung bestimmter Integrale auf die Summation endlicher order unendlicher Reihen. *Abh. Königl. Preuss Akad. Wissens.*, 391–407. (*Werke* I, pp.237–256)

658 Bibliography

(1837a): Beweis des Satzes, dass jede unbegrenzte arithmetische Progression, deren erstes Glied und Differenz ganze Zahlen ohne gemeinschaftlichen Factor sind, unendlich viele Primzahlen enthält. *Abh. Königl. Preuss Akad. Wissens.*, 45–81. (*Werke* I, pp.313–342)

(1837b): Sur la manière de résoudre l'équation $t^2 - pu^2 = 1$ au moyen des fonctions circulaires. *J. reine angew. Math.*, **17**, 286–290. (*Werke* I, pp.343–350)

(1838): Sur l'usage des séries infinies dans la théorie des nombres. *J. reine angew. Math.*, **18**, 259–274. (*Werke* I, pp.357–374)

(1839): Recherches sur diverses applications de l'analyse infinitésimale à la théorie des nombres. Premièr part. *J. reine angew. Math.*,**19**, 324–369; (1840a) Seconde partier. **21**, 1–12 and 134–155. (*Werke* I, pp.411–496)

(1840b): Auszug aus einer der Akademie der Wissenschaften zu Berlin am 5ten März 1840 vorgelesenen Abhandlung. *J. reine angew. Math.*, **21**, 98–100; Extrait d'une lettere de M. Lejeune-Dirichlet à M. Liouville. *C.R. Acad. Sci.*, **10** (1840), 285–288.

(1842): Recherches sur les formes quadratiques à coëfficients et à indéterminées complexes. *J. reine angew. Math.*, **24**, 291–371. (*Werke* I, pp.533–618)

(1846): Zur Theorie der complexen Einheiten. *Abh. König. Preuss. Akad. Wiss.*, 103–107. (*Werke* I, pp.640–644)

(1849): Über die Bestimmung der mittleren Werthe in der Zhalentheorie. *Abh. König. Preuss. Akad. Wiss.*, 69–83. (*Werke* II, pp.49–66)

(1851): De formarum binariarum secundi gradus compositione. Commentatio qua ad audiendam orationem pro loco in facultate philosophica. *Berolini Typis Academicis.* (*Werke* II, pp.105–114; Traduit: *J. math. pures et appliq.*, 4 (1859), 389–398)

(1854a): Über den ersten der von Gauss gegebenen Beweise des Reciprocitäts-gesetzes in der Theorie der quadratischen Reste. *J. reine angew. Math.*, **47**, 139–150. (*Werke* II, pp.121–138)

(1854b): Vereinfachung der Theorie der binären quadratischen Formen von positiver Determinante. *Abh. König. Preuss. Akad. Wiss.*, 99–115. (*Werke* II, pp.139–158)

(1857): Démonstration nouvelle d'une proposition relative à la théorie des formes quadratiques. *J. math. pures et appliq.*, sér. II, **1**, 273–276. (*Werke* II, pp.209–214)

(1863): *Vorlesungen über Zahlentheorie.* Herausgegeben von R. Dedekind. Friedrich Vieweg und Sohn, Braunschweig; Zweite Auflage. 1871; Dritte Auflage. 1879; Vierte Auflage. 1894.

Dress F. and Olivier M. (1999): Polynômes prenant des valeurs premières. *Experiment. Math.*, **8**, 319–338.

du Bois-Reymond P. (1883): Ueber den Gültigkeitsbereich der Taylor'schen Reihenen-twickelung. *Math. Ann.*, **21**, 109–117.

Dunnington G. W. (2004): *Carl Friedrich Gauss. Titan of Science.* Second edition. The Math. Assoc. America, Washington, DC.

Edwards H. M. (1974): *Riemann's zeta-function.* Academic Press, New York; Reprint: Dover Publ., Inc., Mineola, New York, 2001.

Ellis J. H. (1987): The story of non-secret encryption. (private document)

Elsholtz C. and Tao T. (2013): Counting the number of solutions to the Erdős–Straus equation on unit fractions. *J. Austral. Math. Soc.*, **94**, 50–105.

Elstrodt J. (2007): The life and work of Gustav Lejeune Dirichlet (1805–1859). *Clay Math. Proc.*, **7**, 1–37.

Bibliography

659

Erdős P. (1950): Az $1/x_1 + 1/x_2 + \cdots + 1/x_n = a/b$ egyenlet egész számú megoldásairól. *Mat. Lapok*, **1**, 192–210.

Estermann T. (1931): Über die Darstellung einer Zahl als Differenz von zwei Produkten. *J. reine angew. Math.*, **164**, 173–182.

— (1948): On Dirichlet's *L*-functions. *J. London Math. Soc.*, **23**, 275–279.

Euclid (ca 300 BCE): *ΣTOIXEIA* (the *Elements*).

(1) Campano da Novara (1482): *Opus elementorum Euclidis megarensis in geometriam artem in id quoque Campani perspicacissimi commentationes finiunt.* E. Randolt, Venetiis.

(2) H. Billingsley (1570): *The elements of geometrie of the most auncient philosopher Euclide of Megara.* J. Daye, London.

(3) T. L. Heath (1956): *The thirteen books of Euclid's Elements translated from the text of Heiberg.* Second edition. Vols. I–III. Dover, New York.

Euler L. (1729): De progressionibus transcendentibus seu quarum termini generales algebraice dari nequeunt. *Comm. Acad. Sci. Petropolitanae*, **5** (1738), 36–57.

— (1732a): Observationes de theoremate quodam Fermatiano, aliisque ad numeros primos spectantibus. (*Comm. Arith. Collect.*, I, pp.1–3)

— (1732b): Methodus generalis summandi progressiones. *Comm. Acad. Sci. Petropolitanae*, **6** (1738), pp.68–97.

— (1733a): De solutione problematum Diophanteorum per numeros integros. (*Comm. Arith. Collect.*, I, pp.4–10)

— (1733b): Solutio problematis arithmetici de inveniendo numero, qui per datos numeros divisus, relinquat data residua. (*Comm. Arith. Collect.*, I, pp.11–20)

— (1735): De summis serierum reciprocarum. *Comm. Acad. Sci. Petropolitanae*, **7** (1740), 123–134.

— (1736): Theorematum quorundam ad numeros primos spectantium demonstratio. (*Comm. Arith. Collect.*, I, pp.21–23)

— (1737a): De fractionibus continuis dissertatio. *Comm. Acad. Sci. Petropolitanae*, **9** (1744), 98–137.

— (1737b): Variae observationes circa series infinitas. *Comm. Acad. Sci. Petropolitanae*, **9** (1744), 160–188.

— (1740): De extractione radicum ex quantitatibus irrationalibus. *Comm. Acad. Sci. Petropolitanae*, **13** (1751), 16–60.

— (1747): Theoremata circa divisores numerorum. (*Comm. Arith. Collect.*, I, pp.50–61)

— (1748a): Theoremata circa divisores numerorum in hac forma $paa \pm qbb$ contentorum. (*Comm. Arith. Collect.*, I, pp.35–49)

— (1748b): *Introductio in analysin infinitorum.* M.M. Bousquet & Socios, Lausannæ.

— (1749a): De numeris, qui sunt aggregata duorum quadratorum. (*Comm. Arith. Collect.*, I, pp.155–173)

— (1749b): Remarques sur un beau rapport entre les séries des puissances tant directes que réciproques. *Mem. Acad. Sci. Berlin*, **17** (1768), 83–106.

— (1751): Demonstratio theorematis Fermatiani, omnem numerum primum formae $4n + 1$ esse summam duorum quadratorum. (*Comm. Arith. Collect.*, I, pp.210–233)

— (1752): Observatio de summis divisorum. (*Comm. Arith. Collect.*, I, pp.146–154)

— (1755): Theoremata circa residua ex divisione potestatum relicta. (*Comm. Arith. Collect.*, I, pp.260–273)

660   *Bibliography*

(1758a): Theoremata arithmetica nova methodo demonstrata. (*Comm. Arith. Collect.*, I, pp.274–286)

(1758b): De resolutione formularum quadraticarum indeterminatarum per numeros integros. (*Comm. Arith. Collect.*, I, pp.297–315)

(1759): De usu novi algorithmi in problemate Pelliano solvendo. (*Comm. Arith. Collect.*, I, pp.316–336)

(1760): De numeris primis valde magnis. (*Comm. Arith. Collect.*, I, pp.356–378)

(1765): Quomodo numeri praemagni sint explorandi, utrum sint primi nec ne? (*Comm. Arith. Collect.*, I, pp.379–390)

(1771): *Vollständige Anleitung zur Algebra.* Kaiser. Akad. Wiss., St. Petersburg. (English translation: *Elements of algebra.* Third edition. Longman, Rees, Omre, and Co., London, 1822)

(1772a): Observationes circa divisionem quadratorum per numeros primos. (*Comm. Arith. Collect.*, I, pp.477–486; *Opusc. Analy.*, I, pp.64–84)

(1772b): Disquisitio accuratior circa residua ex divisione quadratorum altiorumque potestatum per numeros primos relicta. (*Comm. Arith. Collect.*, I, pp.487–506; *Opusc. Analy.*, I, pp.121–156)

(1772c): Demonstrationes circa residua ex divisione potestatum per numeros primos resultantia. (*Comm. Arith. Collect.*, I, pp.516–537)

(1772d): Novae demonstrationes circa resolutionem numerorum in quadrata. (*Comm. Arith. Collect.*, I, pp.538–548)

(1772e): De criteriis aequationis $fxx + gyy = hzz$, utrum ea resolutionem admittat, nec ne? (*Comm. Arith. Collect.*, I, pp.556–569; *Opusc. Analy.*, I, pp.211–241)

(1772f): Extrait d'une letter à M. Bernoulli. (*Comm. Arith. Collect.*, I, p.584)

(1773a): De quibusdam eximiis proprietatibus circa divisores potestatum occurrentibus. (*Comm. Arith. Collect.*, II, pp.1–26; *Opusc. Analy.*, I, pp.242–267)

(1773b): Nova subsidia pro resolutione formulae $axx + 1 = yy$. (*Comm. Arith. Collect.*, II, pp.35–43; *Opusc. Analy.*, I, pp.310–328)

(1773c): Miscellanea analytica. (*Comm. Arith. Collect.*, II, pp.44–52; *Opusc. Analy.*, I, pp.329–344)

(1774): De tabula numerorum primorum, usque ad millionem et ultra continuanda; in qua simul omnium numerorum non primorum minimi divisores exprimantur. (*Comm. Arith. Collect.*, II, pp.64–91)

(1775): Speculationes circa quasdam insignes proprietates numerorum. (*Comm. Arith. Collect.*, II, pp.127–133)

(1778a): De variis modis numeros praegrandes examinandi, utrum sint primi nec ne? (*Comm. Arith. Collect.*, II, pp.198–214)

(1778b): Utrum hic numerus: 1000009 sit primus, nec ne, inquiritur. (*Comm. Arith. Collect.*, II, pp.243–248)

(1783): *Opuscula analytica.* Tomus primus. Acad. Imper. Sci., Petropoli; Tomus secundus. 1785.

(1849a): Tractatus de numerorum doctrina capita XVI, quae supersunt. (*Comm. Arith. Collect.*, II, pp.503–575)

(1849b): De numeris amicabilibus. (*Comm. Arith. Collect.*, II, pp.627–636)

(1849c): *Commentationes arithmeticae collectae.* Tomus prior. Acad. Imper. Sci., Petropoli; Tomus posterior. (Ed. P. H. Fuss and N. Fuss; V. Bouniakowsky and P. Tchébychew)

Farey J. (1816): On a curious property of vulgar fractions. *Philos. Mag. J.*, **47**, 385–386.

Fermat P. (1679): *Varia opera mathematca.* Joannem Pech, Tolosæ. (Edited by author's son S. Fermat)

(1891): *Œuvres.* Tome premier. Gauthier–Villars, Paris; Tome deuxième. 1894; Tome troisième. 1896; Tome quatrième. 1912.

Fibonacci (1202): Liber abbaci. In: *Scritti di Leonardo Pisano publ. da B. Boncompagni.* Vol. I. Tipografia delle scienze matematiche e fisiche, Roma 1857.

Fogels E. (1965): On the zeros of L-functions. *Acta Arith.*, **11**, 67–96.

Fowler D. H. (1979): Ratio in early Greek mathematics. *Bull. Amer. Math. Soc.*, **1**, 807–846.

Franel J. (1925): Les suites de Farey et le problème des nombres premiers. *Nachr. Ges. Wiss. Göttingen Math.-Phys. Kl., J.* 1924, 198–201.

Friedlander J. B. and Iwaniec H. (2010): *Opera de cribro.* Amer. Math. Society, Providence, RI.

Frobenius G. (1879): Theorie der linearen Formen mit ganzen Coefficienten. *J. reine angew. Math.*, **86**, 146–208.

Fuss P. H. (1843): *Correspondance mathématique et physique de quelques célèbres géomètres du XVIIIéme siècle.* Tomes I et II. Académie Impériale des Sciences, St.-Pétersbourg.

Gallagher P. X. (1968): Bombieri's mean value theorem. *Mathematika*, **15**, 1–6.

(1970): A large sieve density estimate near $\sigma = 1$. *Invent. math.*, **11**, 329–339.

Galois É. (1828): Démonstration d'un théorème sur les fractions continues périodiques. (*Œuvres*, pp.385–392)

(1830): Sur la théorie des nombres. (*Œuvres*, pp.398–407)

(1831): Sur les conditions de résolubilité des équations par radicaux. (*Œuvres*, pp.417–433)

Gandz S. (1937): The origin and development of the quadratic equations in Babylonian, Greek, and early Arabic Algebra. *Osiris*, **3**, 405–557.

Gantmacher F. R. (1959): *The theory of matrices.* Chelsea, New York.

Gauss C. F. (1801): *Disquisitiones arithmeticae.* Fleischer, Lipsiae. (*Werke* I, pp.1–478);

(a) French edition. Courcier, Paris 1807.

(b) German edition. Julius Springer, Berlin 1889.

(c) English edition. Springer Verlag, New York 1986.

(d) *The shaping of arithmetic after C. F. Gauss's Disquisitiones Arithmeticae.* Edited by C. Goldstein, N. Schappacher, and J. Schwermer. Springer Verlag, Berlin 2007.

(1808): Theorematis arithmetici demonstratio nova. (*Werke* II-1, pp.1–8)

(1811): Summatio quarumdam serierum singularium. (*Werke* II-1, pp.9–45)

(1818): Theorematis fundamentalis in doctrina de residuis quadraticis demonstrationes et ampliationes novae.

(1) Demonstratio quinta. (*Werke* II-1, pp.49–54);

(2) Demonstratio sexta. (*Werke* II-1, pp.55–59)

(1828): Theoria residuorum biquadraticorum. Commentatio prima. (*Werke* II-1, pp.65–92); Commentatio secunda. (1832) (*Werke* II-1, pp.93–148).

662 *Bibliography*

(1863a): Analysis residuorum.

   (1) Caput sextum. Pars prior. Solutio congruentiae $x^m - 1 \equiv 0$. (*Werke* II-1, pp.199–211)

   (2) Caput octavum. Disquisitiones generales de congruentiis. (*Werke* II-1, pp.212–242)

(1863b): Disquisitionum circa aequationes puras ulterior evolutio. (*Werke* II-1, pp.243–265)

(1863c): De nexu inter multitudinem classium, in quas formae binariae secundi gradus distribuuntur, earumque determinantem. (*Werke* II-1, pp.269–303)

Goldberg K. (1953): A table of Wilson quotients and the third Wilson prime. *J. London Math. Soc.*, (1) **28**, 252–256.

Goldfeld D. (1976): The class number of quadratic fields and the conjecture of Birch and Swinnerton-Dyer. *Ann. Scuola Norm. Sup. Pisa Cl. Sci.*, (4) **3**, 624–663.

   (1985): Gauss' class number problem for imaginary quadratic number fields. *Bull. Amer. Math. Soc.*, **13**, 23–37.

Goldston D. A., Pintz J., and Yıldırım C. Y. (2005): Primes in tuples. I. arXiv: 0508185 v1. (*Annals of Math.*, (2) **170** (2009), 819–862)

Gradshtein I. S. and Ryzhik I. M. (2007): *Tables of integrals, series and products.* Seventh edition. Academic Press, Elsevier, London.

Graham S. W. and Kolesnik G. (1991): *van der Corput's method of exponential sums.* London Mathematical Society Lecture Note Series, Vol. **126**, Cambridge University Press, Cambridge.

Grandi A. (1883): Dimostrazione di un teorema della teoria dei numeri. *Atti del Reale Istituto Veneto di Scienze, Lettere ed Arti*, ser. 6, **1**, 809–812.

Granville A. (2008): Smooth numbers: computational number theory and beyond. Algorithmic Number Theory. *MSRI Publications*, **44**, 267–323.

Granville A. and Mollin R. A. (2000): Rabinowitsch revisited. *Acta Arith.*, **96**, 139–153.

Greaves G. (2001). *Sieves in number theory.* Springer-Verlag, Berlin.

Hadamard J. (1896): Sur la distribution des zéros de la fonction $\zeta(s)$ et ses conséquences arithmétiques. *Bull. Soc. Math. France*, **24**, 199–220.

   (1954): *An essay on the psychology of invention in the mathematical field.* Dover, New York.

Halász G. (1968): Über die Mittelwerte multiplikativer zahlentheoretischer Funktionen. *Acta Math. Acad. Sci. Hungar.*, **19**, 365–403.

Halász G. and Turán P. (1969): On the distribution of roots of Riemann zeta and allied functions. *J. Number Theory*, **1**, 121–137.

Halberstam H. and Richert H.-E. (1974): *Sieve methods.* Academic Press, London.

Haneke W. (1963): Verschärfung der Abschätzung von $\zeta\left(\frac{1}{2} + it\right)$. *Acta Arith.*, **8**, 357–430.

Hardy G. H. (1914): Sur les zéros de la fonction $\zeta(s)$. *Comptes rendus de l'académie des sciences Paris*, **158**, 1012–1014.

Hardy G. H. and Littlewood J. E. (1923): Some problems of 'Partitio Numerorum'; III: On the expression of a number as a sum of primes. *Acta Math.*, **44**, 1–70.

Hardy K., Muskat J. B., and Williams K. S. (1990): A deterministic algorithm for solving $n = fu^2 + gv^2$ in coprime integers u and v. *Math. Computation*, **55**, 327–343.

Bibliography

Haros C. (1802): Tables pour évaluer une fraction ordinaire avec autant de decimales qu'on voudra; et pour trouver la fraction ordinaire la plus simple, et qui approche sensiblement d'une fraction décimale. *J. de l'École Polytech.*, **4**, 364–368.

Hasse H. (1964): *Vorlesungen über Zahlentheorie.* Grundl. Math. Wiss., **59**, Springer–Verlag, Berlin.

Heath-Brown D. R. (1978): The twelfth power moment of the Riemann zeta-function. *Quart. J. Math. Oxford*, **29**, 443–462.

Hecke E. (1917a): Über die Kroneckersche Grenzformel für reelle quadratische Körper und die Klassenzahl relativ-Abelscher Körper. *Verhandl. der Natur. Gesell. Basel*, **28**, 363–372.

(1917b): Über die Zetafunktion beliebiger algebraischer Zahlkörper. *Nachr. Gesell. Wiss. Göttingen, Math.–Phy. Klasse, J.*, 1917, 77–89.

(1923): *Vorlesung über die Theorie der algebraischen Zahlen.* Akademische Verlagsgesellschaft, Leipzig; English edition. Springer, New York, 1981.

(1937): Über Modulfunktionen und die Dirichletschen Reihen mit Eulerscher Produktentwicklung. I. *Math. Ann.*, **114**, 1–28; II. 316–351.

Heilbronn H. (1934): On the class number in imaginary quadratic fields. *Quart. J. Math.*, **5**, 150–160.

Hejahl D. A. (1983): *The Selberg trace formula for* $\mathrm{PSL}(2, \mathbb{R})$. Vol. 2. Lecture Notes in Math., **1001**, Springer Berlin, Heidelberg.

Hensel K. (1901): Ueber die Entwickelung der algebraischen Zahlen in Potenzreihen. *Math. Ann.*, **55**, 301–336.

Hertzer H. (1908): Über die Zahlen der Form $a^{p-1} - 1$, wenn p eine Primzahl. *Archiv Math. Phys.*, (3) **13**, 107.

Hilbert D. (1897): Die Theorie der algebraischen Zahlkörper. *Jahresbericht der Deutschen Math.*, **4**, 175–546.

Hoheisel G. (1930): Primzahlprobleme in der Analysis. *Sitz. Preuss. Akad. Wiss.*, **33**, 3–11.

Holst E., Strømer C., and Sylow L. (1902): *Niels Henrik Abel. Memorial publié à l'occsion du centenaire de sa naissance.* J. Dybwad, Christiania.

Hooley C. (1967): Artin's conjecture for primitive roots. *J. reine angew. Math.*, **225**, 209–220.

Hölder O. (1936): Zur Theorie der Kreisteilungsgleichung $K_m(x) = 0$. *Prace Matematyczno–Fizyczyne*, **43**, 13–23.

Hua L.-K. (1942): On the least solution of Pell's equation. *Bull. Amer. Math. Soc.*, **42**, 731–735.

Humbert G. (1915): Sur les formes quadratiques binaires positives. *C. R. Acad. Sci.*, **160**, 647–650.

Huxley M. N. (1972): On the difference between consecutive primes. *Invent. math.*, **15**, 164–170.

Huygens C. (1728): Automati planetarii. In: *Opuscula posthuma.* Tomus secundus. Janssonio–Waesbergios, Amstelodami, pp.155–184.

Hyde A. M., Lee P. D., and Spearman B. K. (2014): Polynomials $(x^3 - n)(x^3 + 3)$ solvable modulo any integer. *Amer. Math. Monthly*, **121**, 355–358.

Iamblichus (ca 300 CE): *Life of Pythagoras.* Translated from the Greek by T. Taylor. A.J. Valpy, London, 1818.

Ingham A. E. (1930): Note on Riemann's ζ-function and Dirichlet's L-functions. *J. London Math. Soc.*, **5**, 107–112.

(1937): On the difference between consecutive primes. *Q. J. Math. Oxford*, **8**, 255–266.

(1940): On the estimation of $N(\sigma, T)$. *Q. J. Math. Oxford*, **11**, 291–292.

Ivić A. (1985): *The Riemann zeta-function. Theory and applications.* John Wiley & Sons, New York; Reprint: Dover Publ., Inc., Mineola, New York, 2003.

(1991): *Mean values of the Riemann zeta-function.* Tata Inst. Fund. Res. Lect. Math. Phy., **82**, TIFR, Bombay.

(2001): On sums of Hecke series in short intervals. *J. Théor. Nombres Bordeaux*, **13**, 453–468.

Ivić A. and Motohashi Y. (1994): The mean square of the error term for the fourth power moment of the zeta-function. *Proc. London Math. Soc.*, **69**, 309–329.

Ivory J. (1806): Demonstration of a theorem respecting prime numbers. *New Series of Math. Repository*, **I**, Part II, 6–8.

Iwaniec H. (1971). On the error term in the linear sieve. *Acta Arith.*, **19**, 1–30.

(1980a): Sieve methods. *Intern. Congress of Math. Proc., Helsinki 1978*, Acad. Sci. Fennica, Helsinki, pp.357–364.

(1980b): Rosser's sieve. *Acta Arith.*, **36**, 171–202.

(1980c): A new form of the error term in the linear sieve. *Acta Arith.*, **37**, 307–320.

(1981): Rosser's sieve – bilinear forms of the remainder terms – some applications. In: *Recent progress in analytic number theory.* Vol. 1. Acad. Press, London, pp.203–230.

(1982): On the Brun–Titchmarsh theorem. *J. Math. Soc., Japan*, **34**, 95–123.

Iwaniec H. and Jutila M. (1979): Primes in short intervals. *Arkiv för mathematik*, **17**, 167–176.

Iwaniec H. and Kowalski E. (2004): *Analytic number theory.* Amer. Math. Society, Providence, RI.

Jacobi C. G. J. (1828): Beantwortung der Aufgabe S.212. *J. reine angew. Math.*, **3**, 301–302.

(1837): Über die Kreistheilung und ihre Anwendung auf die Zahlentheorie. *Monatsbericht Akad. Wiss. Berlin*, 127–136. (*J. reine angew. Math.*, **30**, 166–182)

Jacobson M. J., Jr. and Williams H. C. (2000): The Size of the fundamental solutions of consecutive Pell equations. *Experimental Math.*, **9**, 631–640.

Jones J. P., Sato D., Wada H., and Wiens D. (1976): Diophantine representation of the set of prime numbers. *Amer. Math. Monthly*, **83**, 449–464.

Jurkat W. B. and Richert H.-E. (1965): An improvement of Selberg's sieve method. I. *Acta Arith.*, **11**, 217–240.

Jutila M. (1977a): Zero-density estimates for L-functions. *Acta Arith.*, **32**, 56–62.

(1977b): Linnik constant. *Math. Scand.*, **41**, 45–62.

Jutila M. and Motohashi Y. (2005): Uniform bounds for Hecke L-functions. *Acta Math.*, **195**, 61–115.

Kaczorowski J. and Szydło B. (1997): Some Ω-results related to the fourth power moment of the Riemann zeta-function and to the additive divisor problem. *J. Théor. Nombres Bordeaux*, **9**, 41–50.

Karatsuba A. A. (1975): *Elements of analytic number theory.* Nauka, Moscow; Second edition. Fizmatlit, Moscow 1983. (Russian)

Bibliography

Kinkelin H. (1862): *Allgemeine Theorie der harmonischen Reihen mit Angwendung auf die Zahlentheorie.* Schweighauserische Buchdruckerei, Basel.

Klein F. (1890): *Vorlesungen über die Theorie der elliptischen Modulfunctionen.* Ausgearbeitet und vervollständigt von R. Fricke. Erster Band. B.G. Teubner, Leipzig; Zweiter Band. 1892.

Kloosterman H. D. (1926): On the representation of numbers in the form $ax^2 + by^2 + cz^2 + dt^2$. *Acta Math.*, **49**, 407–464.

Knapowski S. (1962): On Linnik's theorem concerning exceptional L-zeros. *Publ. Math. Debrecen*, **9**, 168–178.

Kobayashi I. (1973): A note on the Selberg sieve and the large sieve. *Proc. Japan Acad.*, **49**, 1–5.

Korselt A. R. (1899): Probléme chinois. *L'Intermédiaire des Mathématiciens*, **6**, 142–143.

Kraïtchik M. (1922): *Théorie des nombres.* Tome I. Gauthier-Villars, Paris; Tome II (1926).

Kronecker L. (1845) : Beweis dass für jede Primzahl p die Gleichung $1 + x + x^2 + \ldots + x^{p-1} = 0$ irreductibel ist. *J. reine angew. Math.*, **29**, 280.

(1854): Mémoire sur les facteurs irréductibles de l'expression $x^n - 1$. *J. math. pures et appliq.*, **19**, 177–192.

(1863): Über die Auflösung der Pell'schen Gleichung mittels elliptischer Functionen. *Monats. Königl. Preuss. Akad. Wiss. Berlin.*, a.d.J. 1863, 44–50.

(1865): Über den Gebrauch der Dirichletischen Methoden in der Theorie der quadatischen Formen. *Monats. Königl. Preuss. Akad. Wiss. Berlin.*, a.d.J. 1864, 285–303.

(1876): Zur Geschichte des Reciprocitätsgesetzes. *Monats. Königl. Preuss. Akad. Wiss. Berlin.*, a.d.J. 1875, 267–274.

(1883): Zur Theorie der elliptischen Functionen. *Sitz. König. Preuss. Akad. Wiss. Berlin*, 497–506, 525–530; (1885), 761–784; (1886), 701–780; (1889), 53–63, 123–135.

(1889): Summirung der Gaussschen Reihen $\sum_{h=0}^{h=n-1} e^{2h^2\pi i/n}$. *J. reine angew. Math.*, **105**, 267–268.

Krumbiegel B. and Amthor A. (1880): Das Problema bovinum des Archimedes. *Zeitschrift Math. Phys.*, **25**, Historische-liter. Abt., 121–136 and 153–171.

Kummer E. E. (1860): Über die allgemeinen Reciprocitätsgesetze unter den Resten und Nichtresten der Potenzen, deren Grad eine Primzahl ist. *Abh. Königl. Akad. Wiss. Berlin, J.*, 1859, *Math. Abh.*, 19–159.

(1861): Gedächtnissrede auf Gustav Peter Lejeune Dirichlet. *Abh. Königl. Akad. Wiss. Berlin, J.*, 1860, Historische Einleitung, 1–36. (*Dirichlet Werke* II, pp.309–344)

Kuznetsov N. V. (1977): Petersson hypothesis for forms of weight zero and Linnik hypothesis. *Khabarovsk Complex Res. Inst. Acad. Sci. USSR*, Preprint. (Russian); also in *Math. USSR–Sb.*, **39** (1981), 299–342.

Lagrange J. L. (1768): Solution d'un problème d'arithmétique. (*Œuvres* 1, pp.671–731) (1769): Sur la solution des problèmes indéterminés du second degré. (*Œuvres* 2, pp.377–535)

(1770a): Additions au mémoire sur la résolution des équations numériques. (*Œuvres* 2, pp.581–652)

Bibliography

(1770b): Nouvelle méthode pour résoudre les problèmes indéterminée en nombres entiers. (*Œuvres* 2, pp.655–726)

(1770c): Démonstration d'un théorème d'arithmétique. (*Œuvres* 3, pp.189–201)

(1771a): Réflexions sur la résolution algébrique des équations. (*Œuvres* 3, pp.205–421)

(1771b): Démonstration d'un théorème nouveau concernant les nombres premiers. (*Œuvres* 3, pp.425–438)

(1773): Recherches d'arithmétique. Première partie. (*Œuvres* 3, pp.695–758); Seconde partie. (1775: *Œuvres* 3, pp.759–795)

(1798): Additions aux élémants d'algèbre d'Euler. Analyse indéterminée. (*Œuvres* 7, pp.5–180; English translation in Euler (1771: 1822))

(1808): Traité de la résolution des équations numériques de tous les degrés, avec des notes sur plusieurs points de la théorie des équations algébriques. Quatrième édition. (*Œuvres* 8, pp.9–369)

Lambert J. H. (1770): *Zusätze zu den logaritmischen und trigonometrischen Tabellen zur Erleichterung und Abkürzung der bey Anwendung der Mathematik vorfallenden Berechnungen.* Haude und Spener, Berlin.

Lamé G. (1844): Note sur la limite du nombre des divisions dans la recherche du plus grand commun diviseur entre deux nombres entiers. *C.R. Acad. Sci.*, **19**, 867–870.

Landau E. (1903a): Neuer Beweis des Primzahlsatzes und Beweis des Primidealsatzes. *Math. Ann.*, **56**, 645–670.

(1903b): Über die Klassenzahl der binären quadratischen Fromen von negativer Discriminante. *Math. Ann.*, **56**, 671–676.

(1908a): Nouvelle démonstation pour la formule de Riemann sur le nombre des nombres premiers inférieurs à une limite donnée et démonstation d'une formule plus générale pour le cas des nombres premiers d'une progression arithmétique. *Ann. Sci. l'École Norm. Supér.*, Sér. 3, **25**, 399–442.

(1908b): Über die Einteilung der positiven ganzen Zahlen in vier Klassen nach der Mindestzahl der zu ihrer additiven Zusammensetzung erfolderichen Quadrate. *Archiv der Math. Phy.*, (3) **13**, 305–312.

(1909): *Handbuch der Lehre von der Verteilung der Primzahlen.* Erster Band und Zweiter Band. B.G. Teubner, Leipzig and Berlin.

(1911): Über die Nullstellen der Zetafunktion. *Math. Ann.*, **71**, 548–564.

(1918a): Über imaginär-quadratische Zahlkörper mit gleicher Klassenzahl. *Nachr. Akad. Wiss. Göttingen*, 277–284.

(1918b): Über die Klassenzahl imaginär-quadratischer Zahlkörper. *Nachr. Akad. Wiss. Göttingen*, 285–295.

(1921): Über die Nullstellen der Dirichletschen Reihen und der Riemannschen ζ-Funktion. *Arkiv för Mat. Astr. och Fys.*, **16**, 1–18.

(1924): Über die Wurzeln der Zetafunktion. *Math. Zeit.*, **20**, 98–104.

(1925): Bemerkungen zu der vorstehenden Abhandlung von Herrn Franel. *Nachr. Ges. Wiss. Göttingen Math.-Phys. Kl., J.*, 1924, 202–206.

(1927): *Vorlesungen über Zahlentheorie.* Erster Band. Erster Teil. Aus der elementaren Zahlentheorie. S. Hirzel, Leipzig. (Reprinted by Chelsea, New York, 1950)

(1929): Über die Irreduzibilität der Kreisteilungsgleichung. *Math. Z.*, **29**, 462.

Bibliography

Landauer R. (1961): Irreversibility and heat generation in the computing process. *IBM Journal of Res. Develop.*, **5**, 183–191.

Lau Y.-K. and Tsang K.-M. (2009): On the mean square formula of the error term in the Dirichlet divisor problem. *Math. Proc. Cambridge Phil. Soc.*, **146**, 277–287.

Lebedev N. N. (1972): *Special Functions and Their Applications.* Dover Publ. Inc., Mineola, New York.

Lecerf Y. (1963): Machines de Turing réversibles. Récursive insolubilité en $n \in \mathbb{N}$ de l'équation $u = \theta^n u$, où θ est un «isomorphisme de codes». *C.R. Acad. Sci.*, **257**, 2597–2600.

Lee M. A. (1969): Some irreducible polynomials which are reducible $\bmod p$ for all p. *Amer. Math. Monthly*, **76**, 1125.

Legendre A.-M. (1785): Recherches d'analyse indéterminée. *Histoire de l'Académie Royale des Sciences*, 465–559.

(1798 (An VI)): *Essai sur la théorie des nombres.* Duprat, Paris; Seconde édition. Courcier, Paris, 1808.

(1830): *Théorie des nombres.* Tome I et Tome II. Firmin Didot Frères, Paris.

Legendre A.-M. and Jacobi C. G. J. (1875): Correspondance mathématique entre Legendre et Jacobi. *J. reine angew. Math.*, **80**, 205–279.

Lehmer D. H. (1930): An extended theory of Lucas' functions. *Ann. Math.*, **31**, 419–443.

(1932): Euler's totient function. *Bull. Amer. Math. Soc.*, **38**, 745–751.

Lehmer D. N. (1914): *List of prime numbers from 1 to 10, 006, 721.* Carnegie Institution of Washington, Washington, D.C.

Levinson N. (1974): More than one third of the zeros of Riemann's zeta-function are on $\sigma = 1/2$. *Adv. Math.*, **13**, 383–436.

Lindelöf E. (1908): Quelques remarques sur la croissance de la fonction $\zeta(s)$. *Bull. Sci. Math.*, **32**, 341–356.

Linnik U. V. (1941): The large sieve. *C.R. Acad. Sci. URSS* (N.S.), **30**, 292–294.

(1943): On Weyl's sums. *Rec. Math. (Mat. Sbornik)*, **12**, 28–39.

(1944): On the least prime in an arithmetic progression. I. The basic theorem. *Rec. Math. (Mat. Sbornik)*, **5**, 139–178; II. The Deuring–Heilbronn phenomenon. 347–368.

(1962): Additive problems and eigenvalues of the modular operators. *Proc. Internat. Congress Math., Stockholm*, pp.270–284.

(1963): *Dispersion method in binary additive problems.* Translations Math. Monographs. Vol. **4**. Amer. Math. Soc., Providence.

Liouville J. (1844): Remarques sur des classes très étendues de quantités dont la valeur n'est ni rationnelle ni même réductible à des irrationnelles algébriques. *Comptes Rendus*, **18**, 883–885.

Lipschitz R. (1857): Einige Sätze aus der Theorie der quadratischen Fromen. *J. reine angew. Math.*, **53**, 238–259.

Littlewood J.-E. (1914): Sur la distribution des nombres premiers. *C.R. Acad. Sci.*, **158**, 1869–1872.

(1924): On the zeros of Riemann's zeta function. *Proc. Camb. Phil. Soc.*, **22**, 295–318.

Lucas E. (1867): *Application de l'arithmétique a la construction de l'armure des satins réguliers.* G. Retaux, Paris.

(1878a): Théorèmes d'arithmétique. *Atti della Reale Accademia delle Scienze di Torino*, **13**, 271–284.

(1878b): Théorie des fonctions numériques simplement périodiques. *Amer. J. Math.*, **1**, 184–321.

(1891): *Théorie des nombres*. Tome premier. Gauthier–Villars et Fils, Paris.

von Mangoldt H. (1895): Zu Riemanns Abhandlung "Ueber die Anzahl der Primzahlen unter einer gegebenen Grösse". *J. reine angew. Math.*, **114**, 255–305.

Mathews G. B. (1892): *Theory of numbers*, Part I. Deighton, Bell and Co., Cambridge.

Matthews C. R. (1979): Gauss sums and elliptic functions. I. *Invent. math.*, **52**, 163–185; II. **54**, 23–52.

Matthews K. R. (2002): The Diophantine equation $ax^2 + bxy + cy^2 = N$, $D = b^2 - 4ac > 0$. *J. Théorie des Nombres de Bordeaux*, **14**, 257–270.

Märcker G. (1840): Ueber Primzahlen. *J. reine angew. Math.*, **20**, 350–359.

Maynard J. (2013): Small gaps between primes. arXiv:1311.4600v2. (*Annals of Math.*, **181** (2015), 383–413)

McCurley K. S. (1984): Prime values of polynomials and irreducibility testing. *Bull. Amer. Math. Soc.*, **11**, 155–158.

Meissner W. (1913): Über die Teilbarkeit von $2^p - 2$ durch das Quadrat der Primzahl $p = 1093$. *Sitz. König. Preuss. Akad. Wiss. Akad. Berlin, J. 1913*, 663–667.

Mersenne F. M. (1636): *Harmonicorum libri*. G. Baudy, Lutetiæ Parisiorum.

(1644): *Cogitata physico mathematica*. A. Bertier, Parisiis.

Mertens F. (1874): Ein Beitrag zur analytischen Zahlentheorie. *J. reine angew. Math.*, **78**, 46–62.

(1896): Über die Gaussischen Summen. *Sitz. König. Preuss. Akad. Wiss. Akad. Berlin*, 217–219.

(1897a): Über eine zahltheoretische Aufgabe. *Sitz. Kaiser. Akad. Wiss. Wien*, math.-natur. Classe, **106**-2a, 132–133.

(1897b): Über eine zahlentheoretische Funktion. *Sitz. Kaiser. Akad. Wiss. Wien*, math.-natur. Classe, **106**-2a, 761–830.

Meurman T. (1992): A simple proof of Voronoi's identity. *Astérisque*, **209**, 265–274.

Michel P. (2022): Recent progresses on the subconvexity problem. *Sémi. Bourbaki*, 2021–2022, n°1190.

Milnor J. (1982): Hyperbolic geometry: The first 150 years. *Bull. Amer. Math. Soc.*, **6**, 9–24.

Monier L. (1980): Evaluation and comparison of two efficient probabilistic primality testing algorithms. *Theoret. Comput. Sci.*, **12**, 97–108.

Montgomery H. L. (1968): A note on the large sieve. *J. London Math. Soc.*, **43**, 93–98.

(1969a): Mean and large values of Dirichlet polynomials. *Invent. math.*, **8**, 334–345.

(1969b): Zeros of *L*-functions. *Invent. math.*, **8**, 346–354.

(1971): *Topics in multiplicative number theory*. Lect. Notes in Math., **227**, Springer-Verlag, Berlin.

(1978): The analytic principle of the large sieve. *Bull. Amer. Math. Soc.*, **84**, 547–567.

Montgomery H. L. and Vaughan R. C. (1973): The large sieve. *Mathematika*, **20**, 119–134.

(1974): Hilbert's inequality. *J. London Math. Soc.*, **8**, 73–82.

(2006): *Multiplicative number theory*. I. Cambridge Univ. Press, Cambridge.

Mordell L. J. (1961): The congruence $(p - 1/2)! \equiv \pm 1 \pmod{p}$ [sic]. *Amer. Math. Monthly*, **68**, 145–146.

Morrison M. A. and Brillhart J. (1975): A method of factoring and the factorization of F_7. *Math. Comp.*, **29**, 183–205.

Motohashi Y. (1974): On some improvements of the Brun–Titchmarsh theorem. *J. Math. Soc. Japan*, **26**, 306–323.

(1975): On a density theorem of Linnik. *Proc. Japan Acad.*, **51**, 815–817.

(1976): An induction principle for the generalization of Bombieri's prime number theorem. *Proc. Japan Acad.*, **52**, 273–275.

(1977a): On the Deuring–Heilbronn phenomenon. Part I. *Proc. Japan Acad.*, **53**, 1–2; Part II. 25–27.

(1977b). A note on the large sieve. *Proc. Japan Acad.*, **53**, 17–19; Part II. 122–124.

(1978): Primes in arithmetic progressions. *Invent. math.*, **44**, 163–178.

(1983): *Sieve methods and prime number theory*. Tata Inst. Fund. Res. Lect. Math. Phy., **72**, Tata IFR, Bombay.

(1987): *Riemann–Siegel Formula*. Lect. Notes, Ulam Chair Seminar, Colorado Univ., Boulder.

(1993): An explicit formula for the fourth power mean of the Riemann zeta-function. *Acta Math.*, **170**, 181–220.

(1994): The binary additive divisor problem. *Ann. Sci. École Norm. Sup.*, 4e ser., **27**, 529–572.

(1997): *Spectral theory of the Riemann zeta-function*. Cambridge Tracts in Math., **127**, Cambridge Univ. Press, Cambridge.

(1999): On the error term in the Selberg sieve. In: *Number Theory in Progress. A. Schinzel Festschrift*. Walter de Gruyter, Berlin, pp.1053–1064.

(2007). Sums of Kloosterman sums revisited. In: *The Conference on L-Functions, Fukuoka 2006*, World Scientific, Singapore, pp.141–163.

(2015): On sums of Hecke–Maass eigenvalues squared over primes in short intervals. *J. London Math. Soc.*, **91**, 367–382.

Motohashi Y. and Pintz J. (2006): A smoothed GPY sieve. arXiv: 0602599v2. (*Bull. London Math. Soc.*, **40** (2008), 298–310)

Möbius A. F. (1832): Über eine besondere Art von Umkehrung der Reihen. *J. reine angew. Math.*, **9**, 105–123.

Murty M. R. (1988): Primes in certain arithmetic progressions. *J. Madras Univ.*, Section B, **51**, 161–169. Included in: M. R. Murty and N. Thain (2006): Prime numbers in certain arithmetic progressions. *Functiones et Approximatio*, **35** (2006), 249–259.

Nair M. and Tenenbaum G. (1998): Short sums of certain arithmetic functions. *Acta Math.*, **180**, 119–144.

Nemet-Nejat K. R. (2002): *Daily life in ancient Mesopotamia*. Hendrickson Publishers, Inc., Peabody.

Nesselmann G. H. F. (1842): *Die Algebra der Griechen*. Verlag von G. Reimer, Berlin.

Nicomachus (ca 100 CE): *Arithmetike eisagoge*. (A.M.T.S. Boetii (Boethius) (ca 500): *De institutione arithmetica libri duo, De institutione musica libri quinque*. Teubner B.G., Lipsiæ 1867; D'Ooge M. L. (1926): *Introduction to arithmetic*. Macmillan, New York and London)

670 Bibliography

Niven I., Zuckerman H. S., and Montgomery H. L. (1991): *An introduction to the theory of numbers*. Fifth edition. John Wiley & Sons, New York.

Odlyzko A. M. and te Riele H. J. J. (1984): Disproof of the Mertens conjecture. *J. reine angew. Math.*, **357**, 138–160.

Onishi H. (1973): A Tauberian theorem on Dirichlet series. *J. Number Theory*, **5**, 55–57.

Page A. (1935): On the number of primes in an arithmetic progression. *Proc. London Math. Soc.*, **39**, 116–141.

Pengelley D. and Richman F. (2006): Did Euclid need the Euclidean algorithm to prove unique factorization? *Amer. Math. Monthly*, **113**, 196–205.

Penrose R. (2007): *The road to reality*. Vintage Books, New York.

Pépin T. (1877): Sur la formule $2^{2^n} + 1$. *Comptes Rendus Acad. Sci. Paris*, **85**, 329–333.

Perron O. (1908): Zur Theorie der Dirichletschen Reihen. *J. reine angew. Math.*, **134**, 95–143.

Pieper H. (1997): Über Legendres Versuche, das der quadratische Reziprozitätsgesetz zu beweisen. *Acta hist. Leopoldina*, **27**, 223–237.

Platt D. J. and Trudgian T. S. (2016): Zeros of partial sums of the zeta-function. *London Math. Soc., J. Comput. Math.*, **19**, 37–41.

Plinius G. Secundus (ca 77 CE): *Libros Naturalis Historiae*. Johannes de Spira, Venezia, 1469.

Poincaré H. (1880): Sur un mode nouveau de représentation géométrique des forms quadratiques définies ou indéfinies. *J. l'École Polyt.*, **28**, 177–245.

Poinsot L. (1824): Mémoire sur l'applications de l'algèbre à la théorie des nombres. *Mémoire de l'Acad. Roy. Sci. l'Inst. France* (année 1819 et 1820), **4**, 99–183.

(1845): Réflexions sur les principes fondamentaux de la théorie des nombres. *J. math. pures et appliq.*, **10**, 1–93.

Poisson S. D. (1827): *Mémoire sur le calcul numérique des intégrales définies*. Mémoire de l'Acad. Sci. l'Inst. France (année 1823), **6**, 571–602.

Pollard J. M. (1974): Theorems on factorization and primality testing. *Math. Proc. Cambridge Phil. Soc.*, **76**, 521–528.

(1975): A Monte Carlo method for factorization. *BIT*, **15**, 331–334.

(1978): Monte Carlo methods for index computation (mod$p$). *Math. Comp.*, **32**, 918–924.

Polymath D. H. J. (2014a): New equidistribution estimates of Zhang type. arXiv:1402.0811v3.

(2014b): Variants of the Selberg sieve, and bounded intervals containing many primes. arXiv:1407.4897v4.

Pomerance C. (1985): The quadratic sieve factoring algorithm. *Lect. Notes in Comput. Sci.*, **209**, 169–182.

Pomerance C., Selfridge J. L., and Wagstaff S. S., Jr. (1980): The pseudoprimes to $25 \cdot 10^9$. *Math. Comp.*, **35**, 1003–1026.

Pracher K. (1957): *Primzahlverteilung*. Springer Verlag, Berlin.

Prestet J. (1689): *Nouvaux élémens des mathématiques*. Premier et Second vol. A. Pralard, Paris.

Rabin M. O. (1980): Probabilistic algorithm for testing primality. *J. Number Theory*, **12**, 128–138.

Rabinowicz G. (1913): Eindeutigkeit der Zerlegung in Primzahlfaktoren in quadratischen Zahlkörpern. *J. reine angew. Math.*, **142**, 153–164.

Bibliography

Ramachandra K. (1974): A simple proof of the mean fourth power estimate for $\zeta(\frac{1}{2}+it)$ and $L(\frac{1}{2} + it, \chi)$. *Ann. Scuola Norm. Sup. Pisa*, (4) **1**, 81–97.

Ramanujan S. (1916): Some formulae in the analytic theory of numbers. *Mess. Math.*, **45**, 81–84.

— (1918): On certain trigonometrical sums and their applications in the theory of numbers. *Trans. Cambridge Phil. Soc.*, **22**, 259–276.

Rashed R. (1980): Ibn al-Haytham et le théroème de Wilson. *Arch. History of Exact Sci.*, **22**, 305–315.

Rényi A. (1948): On the representation of an even number as the sum of a prime and an almost prime. *Izv. Akad. Nauk SSSR Ser. Mat.*, **12**, 57–78. (Russian)

Riemann B. (1860): Über die Anzahl der Primzahlen under einer gegebenen Grösse. *Monatsber. Königl. Preuss. Akad. Wiss. Berlin, J.*, 1859, 671–680.

Rivest R. L., Shamir A., and Adleman L. (1978): A method of obtaining digital signatures and public-key cryptosystems. *Comm. Assoc. Comput. Mach.*, **21**, 120–126.

Rodosskii K. A. (1954): On the least prime number in an arithmetic progression. *Mat. Sb.* (N.S.), **34**, 331–356. (Russian)

Rogers K. (1974): Legendre's theorem and quadratic reciprocity. *J. Number Theory*, **6**, 339–344.

Roth K. F. (1955): Rational approximations to algebraic numbers. *Mathematika*, **2**, 1–20.

Sardi C. (1869): Teoremi di aritmetica. *Giornale di matematiche di Battaglini*, **7**, 24–27.

Sarnak P. (1982): Class numbers of indefinite quadratic forms. *J. Number Theory*, **15**, 229–247.

Schaar M. (1850): Recherches sur la théorie des résidus quadratiques. *Mém. Acad. Roy. Sci. Lettres et Beaux Arts Belgique*, **25**, 20 pp.

Schlesinger L. (1912): Über Gauss' Arbeiten zur Funktionentheorie. (*Gauss Werke* X-2, Abhandlung 2)

Schmidt W. M. (2004): *Equations over finite fields. An elementary approach*. Second edition. Kendrick Press, Herber City, UT.

Schoof R. (1985): Elliptic curves over finite fields and the computation of square roots mod p. *Math. Computation*, **44**, 483–494.

Schönemann T. (1846a): Grundzüge einer allgemeinen Theorie der höhern Cogruenzen, deren Modul eine reelle Primzahl ist. *J. reine angew. Math.*, **31**, 269–325; Von denjenigen Moduln, welche Potenzen von Primzahlen sind. **32** (1846b), 93–105.

Schur I. (1921): Über die Gaussschen Summen. *Nachr. Gesell. Wiss. Göttingen, Math.-Phys. Kl.*, 147–153.

— (1929): Zur Irreduzibilität der Kreisteilungsgleichung. *Math. Z.*, **29**, 463.

Selberg A. (1942): On the zeros of Riemann's zeta-function. *Skr. Norske Vid. Akad. Oslo*, No. 10, 1–59. (*Collected papers* I, pp.85–141)

— (1943): On the normal density of primes in small intervals, and the difference between consecutive primes. *Arkiv for Math. og Naturv.*, **47**, No.6, 87–105. (*Collected papers* I, pp.160–178)

— (1946a): Contribution to the theory of the Riemann zeta-function. *Arch. för Math. og Naturv.*, **48**, 89–155. (*Collected papers* I, pp.214–280)

672   Bibliography

(1946b): Contributions to the theory of Dirichlet's L-functions. *Skr. Norske Vid. Akad. Oslo*, No. 3, 1–62. (*Collected papers* I, pp.281–340)

(1946c): The zeta-function and the Riemann hypothesis. *10th. Skand. Math. Kongr.*, 187–200. (*Collected papers* I, pp.341–355)

(1947): On an elementary method in the theory of primes. *Det Kong. Norske Vid. Selsk. Forh., Trondhjem*, **19** , 64–67. (*Collected papers* I, pp.363–366)

(1949): On elementary methods in prime number theory and their limitations. *11th. Skand. Math. Kongr., Trondhjem*, 13–22. (*Collected papers* I, pp.388–397)

(1950a): An elementary proof of the prime-number theorem for arithmetic progressions. *Canadian J. Math.*, **2**, 66–78. (*Collected papers* I, pp.398–410)

(1950b): The general sieve-method and its place in prime number theory. *Proc. Intern. Cong. Math., Cambridge, Mass.*, **1**, 286–292. (*Collected papers* I, pp.411–417)

(1956): Harmonic analysis and discontinuous groups in weakly symmetric Riemannian spaces with applications to Dirichlet series. *J. Indian Math. Soc.*, **20**, 47–87. (*Collected papers* I, pp.423–463)

(1965): On the estimation of Fourier coefficients of modular forms. *Proc. Symp. Pure Math.*, **8**, 1–15. (*Collected papers* I, pp.506–520)

(1972): Remarks on sieves. *Proc. 1972 Number Theory Conf., Boulder*, pp. 205–216. (*Collected papers* I, pp.609–615)

(1977): Remarks on multiplicative functions. *Springer Lect. Notes in Math.*, **626**, 232–241. (*Collected papers* I, pp.616–625)

(1991): Lectures on sieves. In: *Collected Papers*, II, pp.65–247.

Serret J.-A. (1849a): *Cours d'algèbre supérieure*. Bachelier, Paris; Deuxième édition. 1854.

(1849b): Sur un théorème relatif aux nombres entiers. *J. math. pures et appliq.*, **13**, 12–14; Note de C. Hermite. p.15.

Shiu P. (1980): A Brun–Titchmarsh theorem for multiplicative functions. *J. reine angew. Math.*, **313**, 161–170.

Shor P. W. (1994): Algorithms for quantum computation: Discrete logarithms and factoring. *Proc. 35th Ann. Symp. Found. Comp. Sci.*, IEEE Comp. Soc. Press, pp.124–134.

(1997): Polynomial-time algorithms for prime factorization and discrete logarithms on a quantum computer. *SIAM J. Comput.*, **26**, 1484–1509.

Siegel C. L. (1932): Über Riemanns Nachlass zur analytischen Zahlentheorie. *Quellen und Studien zur Geschichte der Math. Astr. und Physik*, Abt. B: Studien, **2**, 45–80.

(1935): Über die Klassenzahl quadratischer Zahlkörper. *Acta Arith.*, **1**, 83–86.

(1957): *Lectures on quadratic forms*. Tata Inst. Fund. Res. Lect. Math. Phy., **7**, TIFR, Bombay.

(1961): *Lectures on advanced analytic number theory*. Tata Inst. Fund. Res. Lect. Math. Phy., **23**, TIFR, Bombay.

Šimerka W. (1858): Die Perioden der quadratischen Zahlformen bei negativen Determinanten. *Sitzungsber. Kaiserl. Akad. Wiss., Math.-Nat. Wiss.*, **31**, 33–67.

(1885): Zbytky z arithmetické posloupnosti. *Časopis pro pěstování mathematiky a fysiky*, **14**, 221–225.

Singh P. (1985): The so-called Fibonacci numbers in ancient and medieval India. *Historia Math.*, **12**, 229–244.

Bibliography

Skewes S. (1933): On the difference $\pi(x) - \mathrm{li}(x)$. *J. London Math. Soc.*, **8**, 277–283; II. **5** (1955), 48–70.

Smith H. J. S. (1855): De compositione numerorum primorum formae $4\lambda + 1$ ex duobus quadratis. *J. reine angew. Math.*, **50**, 91–92. (*Collected papers* I, pp.33–34) (1859/1869): Report on the theory of numbers. (*Collected papers* I, pp.38–364) (1861): On systems of linear indeterminate equations and congruences. *Phil. Trans. Royal Soc. London*, **151**, 293–326. (*Collected papers* I, pp.367–409) (1876): On the value of a certain arithmetical determinant. *Proc. London Math. Soc.*, **7**, 208–212. (*Collected papers* II, pp.161–165) (1877): Mémoire sur les équations modulaires. Abstract presented by Cremona: *Atti della R. Accad. Lincei, Transunti*, Ser. III, **1**, 68–69. (*Collected papers* II, p.240)

Sorenson J. and Webster J. (2017): Strong pseudoprimes to twelve prime bases. *Math. Computation*, **86**, 985–1003.

Srinivasan S.(1997). A weak Brun–Titchmarsh theorem for multiplicative functions. *Proc. Indian Acad. Sci.*, (Math. Sci.), **107**, 387–389.

Stepanov S. A. (1969): On the number of points of a hyperelliptic curve over a finite prime field. *Izv. Akad. Nauk SSSR, ser. Mat.*, **33**, 1171–1181. (Russian) (1994): *Arithmetic of algebraic curves*. Consultants Bureau, New York and London.

Stevin S. (1625): *L'arithmetique de Simon Stevin de Bruges. Annotations par A. Girard*. Elzeviers, Leide.

Stieltjes T.-J. (1894): Recherches sur les fractions continues. *Ann. Facult. Sci. Toulouse*, **8**, no. 4, 1–122; **9**, no. 1 (1895), 5–47.

Sylvester J. J. (1879): On certain ternary cubic-form equations. *American J. Math.*, **2**, 357–393.

Takhtajan L. A. and Vinogradov A. I. (1982): The Gauss–Hasse hypothesis on real quadratic fields with class number one. *J. reine angew. Math.*, **335**, 40–86.

Tatuzawa T. (1951): On a theorem of Siegel. *Japanese J. Math.*, **21**, 163–178.

Thue A. (1902): Et par antydninger til en taltheoretisk metode. *Kra. Vidensk. Selsk. Skrifter. I. Mat. Nat. Kl.*, **7**, 57–75. (*Selected mathematical papers of Axel Thue*. Universitetsforlaget, Oslo, 1977, pp.57–75).

Titchmarsh E. C. (1951): *The theory of the Riemann zeta-function*. Oxford Univ. Press, Oxford.

Toffoli T. (1980): Reversible computing. *Technical Report, MIT Laboratory for computing science*, TM-51.

Tonelli A. (1891): Bemerkung über die Auflösung quadratischer Congruenzen. *Nachrichten Königl. Gesell. Wiss. Georg-Augusts-Univ. Göttingen*, 344–346.

Tsirelson B. (1997): *Quantum information processing*. Lecture notes. Tel Aviv Univ., Tel Aviv.

Turán P. (1953): *Eine Neue Method in der Analysis und deren Anwendungen*. Akademiai Kiado, Budapest. (1961): On a density theorem of Yu.V. Linnik. *Magyár Tud. Akad. Mat. Kutató Intz.* (= *Publ. Math. Inst. Hung. Acad. Sci.*), (2) **6**, 165–179.

van der Corput J. G. (1921): Zhalentheoretische Abschätzungen. *Math. Ann.*, **84**, 53–79.

Vandermonde A.-T. (1774): Mémoire sur la résolution des équations. *Histoire de l'Académie Royale des Sciences*, anné 1771, 365–416.

674 *Bibliography*

Vaughan R. C. (1977): Sommes trigonometriques sur les nombres premiers. *C. R. Acad. Sci. Paris*, Sér. A, **258**, 981–983.

Vinogradov A. I. (1965). The density hypothesis for Dirichlet *L*-series. *Izv. Akad. Nauk SSSR Ser. Mat.*, **29**, 903–934; Corrigendum. **30** (1966), 719–720. (Russian)

Vinogradov I. M. (1935): On Weyl's sums. *Rec. Math.*, (*Mat. Sbornik*), **42**, 521–530.

(1936a): A new method of resolving of certain general questions of the theory of numbers. *Rec. Math.*, (*Mat. Sbornik*), **43**, 9–20.

(1936b): A new method of estimation of trigonometrical sums. *Rec. Math.*, (*Mat. Sbornik*), **43**, 175–188.

(1937): *A new method in analytic number theory*. Akad. Nauk SSSR, Leningrad and Moscow. (Russian)

(1958): A new estimate of the function $\zeta(1 + it)$. *Izv. Akad. Nauk SSSR, Ser. Mat.*, **22**, 161–164. (Russian)

Voronin S. M. (1976): The zeros of zeta-functions of quadratic forms. *Trudy Mat. Inst. Steklov.*, **142**, 135–147. (Russian)

Voronoï G. F. (1904): Sur une fonction transcendante et ses applications à la sommation de quelques séries. *Ann. Écloe Norm.*, **21**, 207–267, 459–533.

Walfisz A. (1936): Zur additiven Zahlentheorie. II. *Math. Z.*, **40**, 592–607.

Wallis J. (1656): *Arithmetica infinitorum*. L. Lichfield, Oxonii.

(1685): *A treatise of algebra both historical and practical. Shewing, the original, progress, and advancement thereof, from time to time; and by what steps it hath attained to the height at which now it is.* Printed by J. Playford for R. Davis, London.

Waring E. (1782): *Meditationes algebricæ*. J. Archdeacon, Cantabrigiæ.

Watt N. (1995): Kloosterman sums and a mean value for Dirichlet polynomials. *J. Number Theory*, **53**, 179–210.

Weber H. (1882): Beweis des Satzes, dass jede eigentlich primitive quadratische Form unendlich viele Primzahlen darzustellen fähig ist. *Math. Ann.*, **20**, 301–329.

(1895): *Lehrbuch der Algebra*. Erster Bd. Friedrich Vieweg und Sohn, Braunschweig; Zweite Bd. 1896; Dritter Bd. 1908.

Weinberger P. J. (1973): Exponents of the class groups of complex quadratic fields. *Acta Arith.*, **22**, 117–124.

Weyl H. (1916): Über die Gleichverteilung von Zahlen mod. Eins. *Math. Ann.*, **77**, 313–352.

(1921): Zur Abschätzung von $\zeta(1 + it)$. *Math. Zeitt.*, **10**, 88–101.

Whittaker E. T. and Watson G. N. (1969): *A course of modern analysis*. Cambridge Univ. Press, Cambridge.

Wieferich A. (1909): Zum letzten Fermatschen Theorem. *J. reine angew. Math.*, **136**, 293–302.

Wiener M. J. (1990): Cryptanalysis of short RSA secret exponents. *IEEE Trans. Information Theory*, **36**, 553–558.

Wolke D. (1973): Über die mittlere Verteilung der Werte zahlertheoretischer Funktionen auf Restklassen. I. *Math. Annalen*, **202**, 1–25.

Wolstenholme J. (1862): On certain properties of prime numbers. *Q. J. Pure Appl. Math.*, **5**, 35–39.

Xylouris T. (2011): *Über die Nullstellen der Dirichletschen L-Funktionen und die kleinste Primzahl in einer arithmetischen Progression.* Dissertation. Rheinischen Friedrich–Wilhelms–Universität, Bonn.

Zhang Y. (2013): Bounded gaps between primes. Draft. (*Annals of Math.*, **179** (2014), 1121–1174)

Index

addition and multiplication
  additive divisor problem, 55
  entanglement, 55
  the *abc* conjecture, 56
algebra
  al-Khwarizmi, 3, 108
  phases of developments, 2
algebraic equations
  Cardano's *Artis Magnæ*, 287
  Euler's impasse, 288
  solubility with nested radicals, 292
    Abel's view, 292
  Vandermonde's breakthrough, 288
    configuration of zeros, 289
    Lagrange's praise, 292
algebraic extensions
  finite fields
    brief history, 276
    cyclic extension, 274
    existence for any degree, 273
    Frobenius map, 274
    irreducibility criterion, 274
algorithm to solve quadratic form $Q(x, y) = m$
  extension to $Q(x, y) + dx + ey + f = 0$, 407
  with reduction
    Lagrange and Legendre, 331, 337, 370
  without reduction
    Cornacchia, 356
    Lagrange with full seed set, 400
ambiguous forms and classes, 327
  cardinality of ambiguous classes, Gauss, 451
  module formulation, 329
analytic continuation
  merit, 517

analytic devices
  Γ-function, 492
    duplication formula, 493
    Mellin inversion, 495
    Stirling's formula, 494
  Balasubramanian–Ramachandra's lemma, 540
  Borel–Carathéodory's theorem, 491
  calculus of variations, 649
  duality principle, 591
  Fourier transform, 540
    Gallagher's lemma, 543
  Hilbert's inequality extended
    Montgomery–Vaughan, 600
  Huxley's inequality, 545
  Jensen's inequality, 490
  Landau's lemma on $(f'/f)(z)$, 491
  Mellin transform, 541
  Montgomery's inequality, 544
  Perron's formula, 505
  Phragmen–Lindelöf's theorem, 492
  Poisson sum formula, 235
  Schwartz's lemma, 490
  Selberg's inequality, 542
    Bombieri's inequality, 542
    Halász's inequality, 542
  Sobolev's inequality, 539
  steepest descent, or saddle point method, 40, 43, 520, 523, 556, 561
  Tauberian theorem, 515, 587
  test function, 539
  Turán's power sum method, 638
Antikythera mechanism, 110
Archimedes' *Problema bovinum*, 384
arithmetic function, 51
  Dirichlet series, 53

676

Index

677

analytic uniqueness, 53
Fourier, or Ramanujan-sum expansion, 202
inversion, 65
multiplication, ordinary, 64
 Rankin convolution, 64
multiplicative, 54
 completely multiplicative, 54
 Euler product, 55
multiplicative convolution, 57
 multiplication of Dirichlet series, 59
Atkinson dissection, 551, 561
Atkinson formula, weighted, 554, 557
automorphic context and function, 11
automorphism group Aut_Q, 320
 Γ-conjugacy classes, 323
 skew automorphisms Aut_Q^*, $\text{pell}_D(-4)$, 332, 334
 structure, $\text{pell}_D(4)$, 330, 333, 390

Babylonian mathematics, 32, 107, 110, 112, 285, 287
best rational approximation, 96, 99
binomial congruences, 140
 algorithm of Cipolla, 278
 algorithm of Tonelli, 174, 252
 group theoretic analysis, 177
 power non-residues, 174
 reciprocity issue, 165
 discovery of Euler, 168
 solubility criterion, 164
bit, 4
Bombieri's prime number theorem, 614
 A.I. Vinogradov's, prime number theorem, 620
 Gallagher's proof, 621
 generalization principle, 618
 Wolke's extension, 621
bounded gaps between primes
 GPY theorem, 651
 MP smoothing, 651
 history, 651
 Maynard–Tao's theorem, 639, 650
 Zhang's theorem, 652
Brahmagupta identity, sign-ambiguity, 308
 Lagrange's identity for $Q(\alpha,\beta)Q(\kappa,\lambda)$, 305, 309
 matrix module, 308
Brun–Titchmarsh theorems, 602, 603
 Montgomery–Vaughan's bound, 607
 improvements, Y.M., 607
 Shiu's bound, 603

cakravâla algorithm, 397
characters (mod q), 200
 additive, 200
 duality, 200
 orthogonality, 200
 genuinely different characters, 224
 multiplicative, 204
 duality, 204
 orthogonality, 204
 quasi-orthogonality, 202
 idea of Linnik, 202
 idea of Rényi, 224
 twists of Dirichlet series by characters, 210
characters of finite Abelian groups, 205
 duality relations, 207
 p-rank, 207
Chebyshev's bound for $\pi(x)$, 39, 46
Chebyshev's function $\psi(x)$, 67, 488
 explicit expansion, 506
 merit, 515
Chen's theorem, 81
class number formula, Dirichlet's, 466
 Gauss' possession of the same, 467
 indefinite forms, Hecke's idea, 462
class number of quadratic forms
 Gauss' class number problems
 discriminant $p \equiv 1 \bmod 4$, 470
 Goldfeld's discovery, 469
 indefinite, $\mathfrak{h}_\pm(D)$, 362, 370
 Lagrange's discovery, 329
 positive definite, $\mathfrak{h}_+(D)$, 335, 337
 relations with $((p-1)/2)! \bmod p$, 470
composition of quadratic forms
 concordant forms, Dirichlet, 432
 Dirichlet's composition identity, 352, 424
 Fermat–Lagrange's observation, 310, 351
 Gauss' class group, 432
 Gauss' formulation, 424
 Gauss' fundamental theorem, Arndt's version, 426
 Legendre's observation, 437
 matrix module approach, 435
 what Legendre missed, 437
computational complexity, 4
 polynomial time, 5, 183, 188, 192
congruence equation, 138
 counting number of solutions, 139
 degree, 138
 fundamental premise, 139
 reason: reduction to binomial congruences, 139

678   *Index*

Hensel's lifting of roots, 143
intersective polynomial, 144
Lagrange's fundamental theorem, 141
   Legendre's recognition, 143
multiple roots, 142
congruence of integers, 116
before Gauss, 118
modular arithmetic, 119
modulus, moduli, 116
reduced residue system, 119
   multiplicative Abelian group, 135
   multiplicative decomposition, 136
   order of a reduced residue, 153
   size of generator set, on GRH, 138
   structure, 170
residue system, classes, 117
   generators, 124
ring decomposition, 129
congruence of polynomials
double moduli, 270
integral moduli, 118
polynomial moduli, 272
   Gauss' 6th proof of reciprocity law, 268
continued fraction, 88
convergents, 89
   Legendre's criterion, 101
   semiconvergents, 100
Euclid's algorithm, 90
   structure, 99
expansion of irrationals, 94
   infinite continued fractions, 95
generic continued fraction, 91
   regular $\leftrightarrow$ half-regular, 92
history, 108
matrix representation, 89
modular deform, 104
negative continued fraction, 114
rational approximation, 96
   Huygens' planetarium, 110
   Lagrange's best approximation theorem, 97
   strata of approximations, 99
   worst approximation, 104
solving $ax + by = c$, 90
uniqueness of expansion, 90, 95
continued fraction expansion of $\sqrt{d}$, 376, 396
factorization of d, 383
Legendre's integer factorization, 385
   Kraïtchik's, 386
palindromic, 103, 382
coprime, coprimality, 6, 15

Dedekind's definition, 29
Euclid's fundamental theorem, 13
   the proof in *Elements*, 14
formulation with the Möbius function, 69
independence of divisibility, 25
two senses, 15, 17
cubic, biquadratic, and higher reciprocity
   laws, 485
cubic congruence, 168, 294
curves over finite fields, 246
counting points, 246
   Stepanov's elementary method, 246
cyclotomic polynomial, 258
cubic, quartic decompositions, 267
irreducibility, 258
linear independence of zeros, 259
origin of Galois theory, 262
quadratic decomposition, 262
regular polygon of 17 edges, 265
regular polygons, Fermat primes, 266
Vandermonde's discovery, 266

decimal expansion of $1/p$, 158
definite, indefinite quadratic forms, 297
Deuring–Heilbronn, or Linnik phenomenon, 637
diagonal ternary quadratic forms
$$ax^2 + by^2 + cz^2 = 0$$
Dedekind's extension
   coprimality property of solutions, 421
Legendre's theorem, 284
Dirichlet character χ mod q often called
   simply a character, 207
conductor, 212
Fourier expansion, 217, 221
periods, 211
primitive character, 212
   attribution, 213
   cardinality, 213
   structure, 214
primitive real character, 215, 250
Dirichlet L-function $L(s, \chi)$, 209
functional equation, 219, 242
GRH, the generalized RH, 213
lower bound for $L(1, \chi)$, 579
primes in an arithmetic progression, 209
Dirichlet's prime number theorem, 409
Ingham's proof of $L(1, \chi) \neq 0$, 410
Legendre's conjecture, 34, 283
Dirichlet–Weber prime number theorem, 470
discrete logarithm Ind_r, 169
   ρ-algorithm, 171

Index

discrete logarithm problem, 171
Gauss' own view on Ind_r, 171
discriminant D, 297, 302
   decomposition, 298
   fundamental discriminant, 298
      Stammdiskriminante, Weber, 304
      Kronecker's original definition, 304
      origin, 304
      quadratic number field, 304
dispersion method of Linnik, 621
   A.I. Vinogradov's prime number theorem, 620
divisibility fundamental, 1
division algorithm, 3
   deviation from being divisible, 3
   Euclid's expression, 4
   extension to polynomials, 271
   matrix formulation, 3
   with the least absolute remainder, 9
divisor function d, 51
   additive divisor problem, 55
   divisor problem, 63
      Dirichlet's splitting method, 60, 63
      statistical assertion, Luo–Tsang, 63
   size of values, 59
   Voronoï formula, 64, 547
divisor sum function σ_α, 57
   Ramanujan series for $\sigma_\alpha \cdot \sigma_\beta$, 74
   Ramanujan's expansion, 72, 202, 247
   resemblance to Hecke operator, 60
double sum method, 83, 534
duality principle
   linear operators of Hilbert spaces, 591

Egyptian mathematics, 4
   digital computer, 4
   Egyptian fraction, 87
      Erdős–Straus conjecture, 87
Eisenstein series, 247
   Ramanujan's expansion of σ_α, 247
equations over finite fields, 270
equivalence, or classification of quadratic
      forms by Γ, 320
   Γ-orbits, 321
   module formulation, 325
equivalence of representations by a quadratic
      form, 320
   in terms of congruence, 333
   module formulation, 333
   seeds, 320
Euclid's algorithm, 6
   identity of Aryabhata I, 7

matrix formulation, 6
   modification, 9
   number of division steps, 8
      Lamé's bound, 8, 104
   original formulation, 7
      anthyphairesis, antenaresis, 7
   structure, 99
   uniqueness, 7
Euclid's *Elements*, 2
   brief history, 107
      Arabic and Greek traditions, 107
      Heiberg edition, 2
      Vatican codex *vat.gr.190*, 107
      Theon edition, 107
Euler phi-function $\varphi(n)$, 70, 120
   Carmichael conjecture, 73
   Lehmer conjecture, 73
   lower bound, 74
exceptional characters, zeros, 572, 637
   Siegel's theorem, 575
   Tatuzawa's theorem, 578
exponential sums, 518
   Bombieri–Iwaniec's method, 524
   Bourgain's decoupling method, 524
   I.M. Vinogradov's method, 525
   Linnik's lemma, 528
   mean value theorem, 531
   transformation into double sums, 525
   van der Corput's method, 519
   Weyl shifts, 519

Farey, or Haros sequence, 84, 86
   rational approximation, 85
   relations with sieves and RH, 86
Fermat numbers F_r, 63
   Euler's factorization of F_5, 160
   Lucas' observation, 232
   Pépin's test, 232
Fermat–Euler theorem, 120
   contraposition of Fermat's theorem, 146
   equivalence, 120
   Euler's first and second proofs, 121
   extensions
      Grandi, 123
      Legendre, 379
      Schönemann, 123
   Fermat quotient, 122
      Wieferich primes, 123
   Fermat's indication of proof, 121
Fibonacci sequence f_n, 103
   $\langle f_a, f_b \rangle = f_{\langle a, b \rangle}$, 103
   Broccolo Romanesco, 103

680 *Index*

Pisano period, 124
finite Abelian group, 132
  invariant factor, elementary divisor, 133
  p-rank, 133, 207
  structure theorem ver.1, 132
  structure theorem ver.2, 132
finite fields, 33, 137
  basic theory of extensions, 273
  $\mathbb{F}_p$, 137
Fourier analysis of reduced residue classes, 202
Fourier expansion of Dirichlet characters, 217
  arbitrary character, explicit formula, 221
  primitive characters, 218
    Gauss sum $G(\chi)$, 218
    vanishing theorem of Kinkelin, 218
fractional linear transformations, 11
free Abelian group of finite rank, 24
  subgroups, 24
fundamental theorem of arithmetic, 29
  unique prime power decomposition theorem, 30
    contributions of Legendre and Gauss, 33

Gauss sum $G(\chi)$, 218
  origin, 285
    resolvents of cyclotomic equations, 291
  over algebraic extensions of a finite field, 274
  relation with Jacobi sums, 244
Gauss sum $\mathfrak{g}_\ell(p)$, 243
  quadratic, 239
  quartic, 244
Gauss' *Caput octavum*,
  varia congruentiarum genera, 33
genus and classes, 354
  Euler's *numerorum idoneorum*, 454
genus theory
  definition of genus, genera, 441
  Gauss' argument, 450
    cardinality of ambiguous classes, 451
    2nd proof of the reciprocity law, 452
  genus characters, 439
    generators, 447, 448
  genus map, 446
  genus numbers, 446
    analytic approach, 473
  glimpse, 354
  Lagrange's difficulty, 449
    Legendre's resolution, 450
  principal genus theorem, 442
    analytic approach, 473

Goldbach conjecture, 28, 81
greatest common divisor $\gcd(a,b) = \langle a,b \rangle$, 5, 14
  basic properties, 13
  sum of modules, 9, 15
  totally coprime representation, 50
GRH, the generalized RH, 213
  implications, 138, 175, 182, 348, 455, 487, 579, 580
group structure over the integers, 11

Halász's method, 517
Hardy–Littlewood conjecture, 591
Hoheisel scheme, 546

ideals, 309
indefinite quadratic forms
  Gauss' reduced forms, 365
  Lagrange reduction à la Gauss, 363
  Lagrange's direct algorithm, 400
    full set of seeds, 402
  Legendre's cf-orbit, 366
    class number $\mathfrak{h}_\pm(D)$, 370
    ending in circular orbits, 369
    parity of circular orbit periods, 370
  reduced quadratic irrationals, 364
    purely periodic continued fractions, 365
  Smith's geometric reduction, 368
Indian mathematics, classic, 7
Indo–Arabic numeral system, 3
  al-Khwarizmi and Fibonacci, 3
  popularization of number theory, 3, 108
integer factorization, 5, 32, 116
  basic ideas, 193
    Euler, 347, 362
    Fermat, 193, 194
    Kraïtchik, 194, 386
    Legendre, 385
    Pollard, 151, 195
    Euler with 1000009, 152, 347
    Märcker's algorithm, 383
  probabilistic, 182
    number of relevant residues, 181
    Shor's quantum algorithm, 183

Jacobi sum, 243
Jacobi symbol
  merit, 248

Kloosterman sum, 246
  Bruggeman–Kuznetsov theory, 55
  Kuznetsov detection of cancellation, 247

Index 681

Kronecker symbol κ_D, 313
  number of solutions to Lagrange
    congruence, 314
  prime discriminants and characters, 317
  real Dirichlet character mod $|D|$, 314
    conductor, 317
    genus character theorem, 449
    L-function $L(s, \kappa_D)$, 314
    relevance to the reciprocity law, 315
    skew symmetry, 318
Kronecker's limit formula, 486

Lagrange's principle to solve quadratic form
  $Q(x, y) = m$
  canonical form $M_{m, \xi}$, 305
  Lagrange congruence, 304
  use of Γ, 306
Landau's notations
  Bachmann's O, Landau's o, 44
  Vinogradov's notation $\ll$, 44
large sieve for intervals
  Bombieri's inequality, 592
  duality between Linnik's and Selberg's
    ideas, 595, 596
  hybrid sieve, 597
  Linnik's sieve, 582, 584
  multiplicative extension, 609
  pseudocharacters
    derived from Λ^2-sieve, 596, 601
  Selberg's Λ^2-sieve, 584, 586
    flexibility, 589
    Montgomery–Vaughan's improvement,
      600
    origin, 589
least common multiple $\mathrm{lcm}(a, b) = [a, b]$, 25,
  49
  $\langle a, b \rangle [a, b] = |ab|$, 25, 124
  totally coprime representation, 50
Lindelöf hypothesis, 524
  Lindelöf constant, 524
linear congruence equations, 125
  lattice points, 127
  system of equations, 129
  Thue's bound, 128
linear continuum
  Euclid, Lagrange, Dedekind, 100
linear indefinite equations, 17
  arithmetic progression in the plane, 18
  chaotic effect of coefficient changes, 18
  system of equations, 21
    Smith's canonical form, 21
Linnik's prime number theorem, 622

history, 637
large sieve version, 624
proof by hybrid large sieve, Y.M., 625
logarithmic integral $\mathrm{li}(x)$, $\mathrm{Li}(x)$, 38
Lucas' primality test for Mersenne numbers,
  385

matrix usage, 11
mean values of $\zeta(\frac{1}{2} + it)$
  fourth power mean, 548
    Watt's bound, 567
  spectral expansion of the weighted fourth
    power mean, Y.M., 561
  twelfth power mean, Heath-Brown, 558
    proofs by Y.M. and Ivić–Y.M., 562
mean values of $L(\frac{1}{2} + it, \chi)$
  fourth power mean, Ramachandra, 611
  recent trends, 562
Mersenne numbers M_p, 62, 232
  Fermat's factorization of M_{37}, 160
  Lucas' primality test, 385
Mertens' idea on $\zeta(s)$, 507
  explicit analytic expression, 515
Mertens' prime number theorem, 509
Möbius function $\mu(n)$, 64
  formulation of coprimality, 69
  Mertens, or Stieltjes conjecture, 69
    disproved by Odlyzko–te Riele, 69
  Möbius inversion, 64
    generalization $\mu_f(n)$, 65
    original statement, 68
  sums, tapered, 511
modular binary exponentiation, 147
modular group Γ, 10
  decomposition of quadruple sums by Γ, 13
  hyperbolic elements, 393
    Sarnak–Siegel decomposition, 394
  parametrization, 127
  the two generators, 10, 105
    extension to higher dimension, 23
modular group's rôle
  Gauss' prescription, 306
modules of matrices, 308
  Gauss and modules, 329, 436
  Poincaré's idea, 313
  principal matrix-module, 309
    quadratic forms as an ideal, 309
  product, conjugation, divisibility, 312
modules over integers, 8
  divisibility of integers, Dedekind's ideals,
    10
  greatest common divisors, 9

682   *Index*

mollifier Dirichlet polynomial, 538
motion of quadratic forms by Γ, 320
  equivalence of forms, 320
    module formulation, 325
  orbits of forms, 321
multiplicative structure of the integer set, 52

order of a reduced residue, 153
  primitive roots, or residues, 154

palindrome, Legendre, 382
Pell equation $\text{pell}_d(n)$, 303
  eponym, 386
$\text{pell}_d(-1)$
  Lagrange's unsolved problem, 382
$\text{pell}_D(-4)$
  $D > 0$ a discriminant, 332, 391
    relation with class numbers, 391
$\text{pell}_d(1)$
  Brahmagupta's composition, 398
  cakravâla algorithm, 395
  chaotic nature of solutions, 380
$\text{pell}_D(4)$
  $D > 0$ a discriminant, 330, 387
    structure of Aut_Q, 330, 390
  the least solution, ϵ_D, 390
  use of Serret's theorem, 388
$\text{pell}_d(\pm 1)$, 376
  Lagrange–Legendre algorithm, 376
    Amthor's solution to *Problema bovinum*, 384
    Legendre extension of Fermat–Euler theorem, 379
    Lucas' primality test of M_p, 385
    structure of solutions, Lagrange, 379
    use of Legendre's criterion for convergents, 376
$\text{pell}_D(\pm 4)$
  Kronecker's solution, 486
$\text{pell}_p(-1)$
  Legendre, 381
$\text{pell}_p(1)$
  relation with cyclotomy, 394
perfect numbers, 61
  Euclid's discussion, 61
    Dickson's proof, 61
  Euler's theorem, 61
  Mersenne primes, 62
pigeon box principle, 87, 128
Poincaré series, 246
Poisson sum formula, 64, 235, 546
positive definite quadratic forms, 334

class number $\mathfrak{h}_+(D)$, 337
Cornacchia's direct algorithm, 356
  equivalence of forms, geometric formulation, 337
  fundamental domain of Γ, 336
  Lagrange's reduction à la Gauss, 334
  ordering representations, 338, 348
power non-residues, 174
  size on GRH, 175
primality test, 29
  deterministic test
    AKS theorem, 182
    Legendre's test, 159
    practical test, 150
  probabilistic test, 180
    implication of GRH, 580
prime counting function $\pi(x)$, Landau, 27
prime number, or prime p, 24
  erroneous definition, 28
  independence, 26
  peculiarity of the prime 2, 28
  polynomial representations, 34
    all the primes, 35
    Bouniakowsky's conjecture, 34
    quadratic forms, 35, 470
  prime numbers and complex variables, 43
  Sumerian encounter, 27
prime number theorem for short intervals, asymptotics
  Hoheisel's discovery, 536
  Huxley's theorem, 565
prime number theorem for short intervals, existence
  Iwaniec–Jutila, 567
  Baker et al, 567
prime number theorem in mean
  A. I. Vinogradov's theorem, 620
  Bombieri–Friedlander–Iwaniec theorem, 621
  Bombieri's theorem, 614
  Polymath's theorem, 652
  Rény's theorem, 620
prime number theorem with an error term
  de la Vallée Poussin, 508
  I. M. Vinogradov, 534
  Littlewood's $\Omega_{\pm}$ theorem, 515
    Skews number, 515
  simple proof, Landau, 516
prime number theorem: $\pi(x) \sim x/\log x$, 38
  Chebyshev's hypothetical proof, 38
  Dirichlet's conjecture, 38

Index

Gauss' conjecture, 38
Hadamard–de la Vallée Poussin's proof, 37
Legendre's conjecture, 38
logarithmic integral
  Dirichlet's conjecture, 38
  Gauss' data, 38
prime number theorem: $\pi(x) \to \infty$, 27
  Chebyshev's bound for $\pi(x)$, 39
  Dirichlet's theorem, 409
  Dirichlet–Weber theorem, 470
  Euclid's proof, 27
    Murty's extension, 29
  Euler's proof, 30
prime numbers in an arithmetic progression
  Dirichlet's theorem, 409
  Linnik's theorem, 622
    Xylouris' exponent, 625
  Siegel–Walfisz theorem, 575
prime power decomposition, uniqueness, 30
  contributions of Legendre and Gauss, 33
primitive, non-primitive quadratic forms, 298
primitive real characters, 215
  Jacobi symbol, 248
  Kronecker symbol, 313
  Legendre symbol, 225
primitive roots, residues, 153
  Artin's conjecture, 159
    Hooley's hypothetical proof, 159
  Chebyshev's observation, 232
  decimal expansion of $1/p$, 158
  Euler *radices primitivas*, 154
  existence proof by Legendre, 155
  extension to prime power moduli, 161
  restriction of moduli, 162
primorial primes, 29
principal quadratic forms, 297, 433
probable pp, pseudo psp, 148
  Carmichael, or Šimerka numbers, 148
    infinitely many, Alford et al, 150
    Korselt's criterion, 169
  strong probable spp, pseudo spsp, 149
    bounds for the number of basis residues, 179
    practical primality test, 150
proper representations by a quadratic form, 296
  automorphism group Aut_Q, 320
  canonical system of representations, 319
  embedding into Γ, 319
  Gauss' term 'belongs to', 323
    Dirichlet's interpretation, 324

injection into Γ/Γ_∞, 319
seeds, 320, 331, 402
  in terms of congruence, 333
  sum of all the set of seeds, 322
    cardinality, 322
Pythagoras' number theory, 106
  monochord, 114
Pythagorean triples
  clay tablet *Plimpton 322*, 48

quadratic congruences, 225
  Schoof's algorithm, 281
  Tonelli's algorithm, 252
quadratic number fields, 300, 304
quadratic order
  conductor, 312
quadratic reciprocity law, 227
  essential contents, 249
  Gauss' eight proofs, 228
    1st proof briefly touched on, 416
    2nd proof, with a twist, 315, 452, 484
    3rd proof, 232
    4th proof, Dirichlet's and Cauchy's versions, 233, 240
    5th proof, not discussed in this volume, 228
    6th proof, Jacobi's version and impact, 267
    7th and 8th proofs, via field extension, 275
    timeline, 277
  Legendre symbol, 225
    significance, 226
  Legendre's incomplete proof, 229, 282
    completed, 283, 418
    merit, 283
  proof via ideal theory, 436
  who discovered, 228, 281
    Euler, Legendre, and Gauss, 282
quantum computation, 5
  finite Fourier transform, 187, 190
  finite dimensional Hilbert space, 184
  measurement procedure, 192
    detecting convergents à la Legendre, 192
    finding the order of a reduced residue, 192
  polynomial time, 192
  quantum mechanism axioms, 185
    Hawking's philosophy, 188
  qubit, 184
  reversible computing, 188
    Landauer's principle, 188

684   *Index*

Shor's factorization algorithm, 183
unitary embedding of functions, 186
universal logic gates, 189
quasi Riemann hypothesis, 537

Ramanujan formulas, 74
Ramanujan sum, 71
   representing divisibility, 202
Rankin convolution, 64
rational approximation, 85
   Roth's theorem, 87
reduced quadratic irrationals, 364
   purely periodic continued fractions, 365
resolvents of algebraic equations, 291
RH, the Riemann hypothesis, 40
   failure with a zeta-like function, 353
   implication on $\pi(x)$, 40
   relation with Farey sequence, 86
   zeros of partial zeta-sums, 43
rho (ρ)-algorithms of Pollard, 151, 171
Riemann's article, 36, 39, 488
   the four main statements, 489
Riemann's *Nachlass*, 39, 523
   Riemann–Siegel formula, 523
      extension to $\zeta^2(s)$, Y.M., 524
RSA public key cryptosystem, 196
   earlier invention story, 196

Saros period, 110
seeds of solutions, 320, 390, 402
Selberg's assertion on Legendre symbol, 411
Selberg's zeta-function ζ_Γ, 392
   analogue of RH holds, 392
   complex zeros of ζ_Γ and eigenvalues of
      Laplace–Beltrami operator, 392
   hyperbolic elements in Γ, 393
   hyperbolic non-Euclidean geometry, history,
      395
   trace formula, 499
sieve, elements, 76
   basic sieve identities, 77, 78, 83
      Buchstab identity, 77
   combinatorial sieve, Brun and Iwaniec, 78
   non-combinatorial sieve, Linnik and
      Selberg, 78
   sieve dimension, 82
   structure of the error terms, 82
      Iwaniec's flexible bilinear form, 83
Smith determinant, 75
Smith reduction of indefinite forms, 368
   hyperbolic geodesics, 368
Smith's canonical form, 21

invariant factors, 133
smooth integers, 84
square-free, sqf, 45
steepest descent, or saddle point method, 40,
   43, 520, 523, 556, 561
subconvexity bounds
   automorphic L-functions, Jutila–Y.M., 524
   recent progress, 578
   Weyl–van der Corput, 520
Sumerian arithmetic, 27
summation by parts, 49
   Stieltjes integration by parts, 49
symmetric polynomials, 123, 258
   Girard–Newton formula, 535

temperaments, 112
   Equal, keyboard, convergent, 113
   Pythagorean, string, coprimality, 112
Turán's power sum method, 638
twin prime conjecture, 76, 81, 487, 590

Vinogradov, A. I., prime number theorem, 620
Vinogradov, I. M., prime number theorem, 534
von Mangoldt function $\Lambda(n)$, 66
   ψ-function, Chebyshev's legacy, 68
   logarithmic derivative of $\zeta(s)$, 67
   Vaughan's decomposition, 621
Voronoï sum formula, 547
   Voronoï's explicit formula
      Meurman's simple proof, 548

weaves and Euclid's algorithm, 128
Wilson's theorem, 144
   $((p-1)/2)!$ mod p, 146, 470
   Chebyshev's proof, 144
   Euler's proof, 173
   Gauss's proof and extensions, 144, 163, 173
   Lagrange's proof, 145

zeta-function $\zeta(s)$, 31
   arg $\zeta(s)$, 496
   asymptotic number of complex zeros, 496
   bounds for log $\zeta(s)$, $\zeta^{-1}(s)$, 510
   critical line, critical strip, 40
   definition of log $\zeta(s)$, 497
   Euler product, 31
   Euler's functional equation, 42
   individual prime number and zeros of $\zeta(s)$,
      Landau, 514
   integral representations, 36, 39
   logarithmic derivative, 68
      explicit expansion, 498
   mean values, 548, 558, 561, 562, 567
   Riemann hypothesis, RH, 40

Index 685

Riemann's first complex-zeros, 40
Riemann's functional equation, 37
Riemann's paradigm, 36, 41, 488, 517
the two expressions, their relation, 40
zero-density theory
  Halász–Turán's discovery, 566
  Huxley's bound, 563
  Ingham's bound, 538

Montgomery's zero-detecting method, 539, 562
origin, Bohr–Landau's statistical proof of RH, 537
zero-free region
  de la Vallée Poussin's region, 508
  I.M. Vinogradov's region, 533
zeros on the critical line
  Conrey's ratio $> \frac{2}{5}$, 515

Printed in the United States
by Baker & Taylor Publisher Services